IET ENERGY ENGINEERING 139

Variability, Scalability and Stability of Microgrids

Other volumes in this series:

Volume 1	**Power Circuit Breaker Theory and Design** C.H. Flurscheim (Editor)
Volume 4	**Industrial Microwave Heating** A.C. Metaxas and R.J. Meredith
Volume 7	**Insulators for High Voltages** J.S.T. Looms
Volume 8	**Variable Frequency AC Motor Drive Systems** D. Finney
Volume 10	**SF_6 Switchgear** H.M. Ryan and G.R. Jones
Volume 11	**Conduction and Induction Heating** E.J. Davies
Volume 13	**Statistical Techniques for High Voltage Engineering** W. Hauschild and W. Mosch
Volume 14	**Uninterruptible Power Supplies** J. Platts and J.D. St Aubyn (Editors)
Volume 15	**Digital Protection for Power Systems** A.T. Johns and S.K. Salman
Volume 16	**Electricity Economics and Planning** T.W. Berrie
Volume 18	**Vacuum Switchgear** A. Greenwood
Volume 19	**Electrical Safety: A guide to causes and prevention of hazards** J. Maxwell Adams
Volume 21	**Electricity Distribution Network Design, 2nd Edition** E. Lakervi and E.J. Holmes
Volume 22	**Artificial Intelligence Techniques in Power Systems** K. Warwick, A.O. Ekwue and R. Aggarwal (Editors)
Volume 24	**Power System Commissioning and Maintenance Practice** K. Harker
Volume 25	**Engineers' Handbook of Industrial Microwave Heating** R.J. Meredith
Volume 26	**Small Electric Motors** H. Moczala *et al.*
Volume 27	**AC–DC Power System Analysis** J. Arrillaga and B.C. Smith
Volume 29	**High Voltage Direct Current Transmission, 2nd Edition** J. Arrillaga
Volume 30	**Flexible AC Transmission Systems (FACTS)** Y.-H. Song (Editor)
Volume 31	**Embedded Generation** N. Jenkins *et al.*
Volume 32	**High Voltage Engineering and Testing, 2nd Edition** H.M. Ryan (Editor)
Volume 33	**Overvoltage Protection of Low-Voltage Systems, Revised Edition** P. Hasse
Volume 36	**Voltage Quality in Electrical Power Systems** J. Schlabbach *et al.*
Volume 37	**Electrical Steels for Rotating Machines** P. Beckley
Volume 38	**The Electric Car: Development and future of battery, hybrid and fuel-cell cars** M. Westbrook
Volume 39	**Power Systems Electromagnetic Transients Simulation** J. Arrillaga and N. Watson
Volume 40	**Advances in High Voltage Engineering** M. Haddad and D. Warne
Volume 41	**Electrical Operation of Electrostatic Precipitators** K. Parker
Volume 43	**Thermal Power Plant Simulation and Control** D. Flynn
Volume 44	**Economic Evaluation of Projects in the Electricity Supply Industry** H. Khatib
Volume 45	**Propulsion Systems for Hybrid Vehicles** J. Miller
Volume 46	**Distribution Switchgear** S. Stewart
Volume 47	**Protection of Electricity Distribution Networks, 2nd Edition** J. Gers and E. Holmes
Volume 48	**Wood Pole Overhead Lines** B. Wareing
Volume 49	**Electric Fuses, 3rd Edition** A. Wright and G. Newbery
Volume 50	**Wind Power Integration: Connection and system operational aspects** B. Fox *et al.*
Volume 51	**Short Circuit Currents** J. Schlabbach
Volume 52	**Nuclear Power** J. Wood
Volume 53	**Condition Assessment of High Voltage Insulation in Power System Equipment** R.E. James and Q. Su
Volume 55	**Local Energy: Distributed generation of heat and power** J. Wood
Volume 56	**Condition Monitoring of Rotating Electrical Machines** P. Tavner, L. Ran, J. Penman and H. Sedding
Volume 57	**The Control Techniques Drives and Controls Handbook, 2nd Edition** B. Drury
Volume 58	**Lightning Protection** V. Cooray (Editor)
Volume 59	**Ultracapacitor Applications** J.M. Miller

Volume 62 **Lightning Electromagnetics** V. Cooray
Volume 63 **Energy Storage for Power Systems, 2nd Edition** A. Ter-Gazarian
Volume 65 **Protection of Electricity Distribution Networks, 3rd Edition** J. Gers
Volume 66 **High Voltage Engineering Testing, 3rd Edition** H. Ryan (Editor)
Volume 67 **Multicore Simulation of Power System Transients** F.M. Uriate
Volume 68 **Distribution System Analysis and Automation** J. Gers
Volume 69 **The Lightening Flash, 2nd Edition** V. Cooray (Editor)
Volume 70 **Economic Evaluation of Projects in the Electricity Supply Industry, 3rd Edition** H. Khatib
Volume 72 **Control Circuits in Power Electronics: Practical issues in design and implementation** M. Castilla (Editor)
Volume 73 **Wide Area Monitoring, Protection and Control Systems: The enabler for smarter grids** A. Vaccaro and A. Zobaa (Editors)
Volume 74 **Power Electronic Converters and Systems: Frontiers and applications** A.M. Trzynadlowski (Editor)
Volume 75 **Power Distribution Automation** B. Das (Editor)
Volume 76 **Power System Stability: Modelling, analysis and control** B. Om P. Malik
Volume 78 **Numerical Analysis of Power System Transients and Dynamics** A. Ametani (Editor)
Volume 79 **Vehicle-to-Grid: Linking electric vehicles to the smart grid** J. Lu and J. Hossain (Editors)
Volume 81 **Cyber-Physical-Social Systems and Constructs in Electric Power Engineering** S. Suryanarayanan, R. Roche and T.M. Hansen (Editors)
Volume 82 **Periodic Control of Power Electronic Converters** F. Blaabjerg, K. Zhou, D. Wang and Y. Yang
Volume 86 **Advances in Power System Modelling, Control and Stability Analysis** F. Milano (Editor)
Volume 87 **Cogeneration: Technologies, optimisation and implementation** C.A. Frangopoulos (Editor)
Volume 88 **Smarter Energy: From smart metering to the smart grid** H. Sun, N. Hatziargyriou, H.V. Poor, L. Carpanini and M.A. Sánchez Fornié (Editors)
Volume 89 **Hydrogen Production, Separation and Purification for Energy** A. Basile, F. Dalena, J. Tong and T.N. Veziroğlu (Editors)
Volume 90 **Clean Energy Microgrids** S. Obara and J. Morel (Editors)
Volume 91 **Fuzzy Logic Control in Energy Systems with Design Applications in MATLAB®/Simulink®** İ.H. Altaş
Volume 92 **Power Quality in Future Electrical Power Systems** A.F. Zobaa and S.H.E.A. Aleem (Editors)
Volume 93 **Cogeneration and District Energy Systems: Modelling, analysis and optimization** M.A. Rosen and S. Koohi-Fayegh
Volume 94 **Introduction to the Smart Grid: Concepts, technologies and evolution** S.K. Salman
Volume 95 **Communication, Control and Security Challenges for the Smart Grid** S.M. Muyeen and S. Rahman (Editors)
Volume 96 **Industrial Power Systems with Distributed and Embedded Generation** R. Belu
Volume 97 **Synchronized Phasor Measurements for Smart Grids** M.J.B. Reddy and D.K. Mohanta (Editors)
Volume 98 **Large Scale Grid Integration of Renewable Energy Sources** A. Moreno-Munoz (Editor)
Volume 100 **Modeling and Dynamic Behaviour of Hydropower Plants** N. Kishor and J. Fraile-Ardanuy (Editors)
Volume 101 **Methane and Hydrogen for Energy Storage** R. Carriveau and D.S.-K. Ting
Volume 104 **Power Transformer Condition Monitoring and Diagnosis** A. Abu-Siada (Editor)
Volume 106 **Surface Passivation of Industrial Crystalline Silicon Solar Cells** J. John (Editor)
Volume 107 **Bifacial Photovoltaics: Technology, applications and economics** J. Libal and R. Kopecek (Editors)

Volume 108	**Fault Diagnosis of Induction Motors** J. Faiz, V. Ghorbanian and G. Joksimović
Volume 110	**High Voltage Power Network Construction** K. Harker
Volume 111	**Energy Storage at Different Voltage Levels: Technology, integration, and market aspects** A.F. Zobaa, P.F. Ribeiro, S.H.A. Aleem and S.N. Afifi (Editors)
Volume 112	**Wireless Power Transfer: Theory, technology and application** N. Shinohara
Volume 115	**DC Distribution Systems and Microgrids** T. Dragičević, F. Blaabjerg and P. Wheeler
Volume 117	**Structural Control and Fault Detection of Wind Turbine Systems** H.R. Karimi
Volume 119	**Thermal Power Plant Control and Instrumentation: The control of boilers and HRSGs, 2nd Edition** D. Lindsley, J. Grist and D. Parker
Volume 120	**Fault Diagnosis for Robust Inverter Power Drives** A. Ginart (Editor)
Volume 123	**Power Systems Electromagnetic Transients Simulation, 2nd Edition** N. Watson and J. Arrillaga
Volume 124	**Power Market Transformation** B. Murray
Volume 126	**Diagnosis and Fault Tolerance of Electrical Machines, Power Electronics and Drives** A.J.M. Cardoso
Volume 128	**Characterization of Wide Bandgap Power Semiconductor Devices** F. Wang, Z. Zhang and E.A. Jones
Volume 130	**Wind and Solar Based Energy Systems for Communities** R. Carriveau and D.S.-K. Ting (Editors)
Volume 131	**Metaheuristic Optimization in Power Engineering** J. Radosavljević
Volume 132	**Power Line Communication Systems for Smart Grids** I.R.S Casella and A. Anpalagan
Volume 155	**Energy Generation and Efficiency Technologies for Green Residential Buildings** D. Ting and R. Carriveau (Editors)
Volume 157	**Electrical Steels, 2 Volumes** A. Moses, K. Jenkins, Philip Anderson and H. Stanbury
Volume 905	**Power System Protection, 4 volumes**

Variability, Scalability and Stability of Microgrids

Edited by
S.M. Muyeen, Syed Mofizul Islam and Frede Blaabjerg

The Institution of Engineering and Technology

Published by The Institution of Engineering and Technology, London, United Kingdom

The Institution of Engineering and Technology is registered as a Charity in England & Wales (no. 211014) and Scotland (no. SC038698).

First published 2019

The Institution of Engineering and Technology
Michael Faraday House
Six Hills Way, Stevenage
Herts, SG1 2AY, United Kingdom

www.theiet.org

British Library Cataloguing in Publication Data
A catalogue record for this product is available from the British Library

ISBN 978-1-78561-693-8 (hardback)
ISBN 978-1-78561-694-5 (PDF)

Typeset in India by MPS Limited
Printed in the UK by CPI Group (UK) Ltd, Croydon

This book is dedicated to
Lipy and Arisha
Masuma, Muntasser and Rashmi
Anja, Jakob and Ina

Contents

Preface xix
Contributors xxi

1 Introduction **1**
S.M. Muyeen, Syed Islam, and Frede Blaabjerg

1.1 Microgrid fundamentals and its anatomy 1
1.2 Microgrid technical aspects 2
1.2.1 Microgrid control issues 2
1.2.2 Power electronics in microgrid 3
1.2.3 Addressing power electronics reliability in microgrid 4
1.2.4 Use of energy storage systems in microgrid 5
1.2.5 Microgrid information and communication technology 6
1.2.6 Stability and protection issues of microgrid 7
1.3 Microgrid future form 8
1.3.1 Addressing scalability and variability 8
1.3.2 Transformation of microgrid to virtual power plant 8
1.3.3 Future trends of power electronics and its adaptation in microgrid 9
1.3.4 Future trends of energy storage technology 10
1.3.5 Future form of microgrid communication 10
1.4 What is in this book? 11
1.5 Conclusions 13
References 14

2 Microgrid control overview **15**
S. Ali Pourmousavi Kani, Farhad Shahnia, M Imran Azim, Md Asaduzzaman Shoeb, and GM Shafiullah

2.1 Introduction 16
2.2 Uncertainty of the generation and demand 17
2.2.1 Application of grid-tied MGs 18
2.3 MG control hierarchy 20
2.3.1 Primary control 20
2.3.2 Secondary control 22
2.3.3 Tertiary control 25

2.4 Case studies 25
2.4.1 Droop-based power control 25
2.4.2 Demand-side primary frequency control 34
2.4.3 Centralised secondary control 42
2.5 Conclusion 63
References 64

3 Requirements analysis in transactive energy management 73
Sreenithya Sumesh, Aneesh Krishna, and Chitra M Subramanian

3.1 Introduction 73
3.2 Transactive energy management 75
3.3 Application of requirements engineering approaches in transactive energy management 78
3.3.1 The *i** goal modelling 81
3.4 Requirements analysis and modelling of the TEM system 84
3.4.1 Goal modelling of the TEM system 84
3.4.2 Methodology 86
3.4.3 Formalisation of multi-objective optimisation functions of the *i** goal model 86
3.5 Conclusion 94
References 94

4 Transformation of microgrid to virtual power plant 99
Robert Lis and Robert Czechowski

4.1 Introduction 99
4.2 Evolution of electricity – the case of Polish electricity sector 100
4.3 Liberalization of the energy markets 102
4.3.1 Future problem identification 102
4.4 Microgrid turns to virtual power plant 106
4.4.1 MGs structure and application 106
4.5 Microgrid configuration 107
4.6 Microsource controller 109
4.6.1 Virtual power plant general concept 109
4.7 Types of Virtual Power Plants 111
4.7.1 An area-based approach to virtual power plants 111
4.7.2 Grid support and ancillary services 113
4.7.3 VPP model and algorithms 117
4.8 Difference between microgrid and VPP 118
4.9 Information communication technologies 120
4.9.1 RSTP grid mechanism 121
4.9.2 SHP grid mechanism 122
4.9.3 HSR grid mechanism 122
4.9.4 PRP grid mechanism 122

4.9.5 Microgrid/VPP cybersecurity 123
4.9.6 Energy management system 125
4.9.7 Supervision control and data acquisition 127
4.9.8 Control system operation and states 127
4.9.9 Databases 128
4.9.10 Database management process 129
4.9.11 Distribution and dispatching centre 130
4.10 Case study: regulation of VPP and MGs 132
4.11 Conclusion 138
References 138

5 Operations of a clustered microgrid 143
Munira Batool, Syed Islam, and Farhad Shahnia

5.1 Overview of clustered microgrid 143
5.2 Modeling of clustered microgrid 147
5.3 Control and operation of clustered microgrid 152
5.3.1 Droop-regulated strategy 152
5.3.2 Optimization solver 155
5.3.3 Modeling of non-dispatchable DERs 156
5.4 Optimization problem formulation and technical constraints 157
5.5 Case studies 160
5.5.1 Study case I (an overloaded MG with primary and secondary actions only) 162
5.5.2 Study case II (an overloaded MG with all actions) 162
5.5.3 Study case III (an overloaded MG with primary and tertiary actions only) 165
5.5.4 Study case IV (an overgenerating MG with primary and secondary actions only) 165
5.5.5 Study case V (an overgenerating MG with all actions) 166
5.5.6 Study case VI (an overgenerating MG with primary and tertiary actions only) 166
5.5.7 Study case VII (multiple PMGs and HMGs with all actions) 167
5.6 Concluding remarks 168
Nomenclature 168
References 169

6 Distributed energy network using nanogrid 175
Xiaofeng Sun, Wei Zhao, and Lei Qi

6.1 Overview of nanogrid 175
6.1.1 Concept of nanogrid 175
6.1.2 Architecture of nanogrid 176
6.1.3 Converters used in nanogrid 179

6.2 Energy management in nanogrid 181
6.2.1 Battery-mastered control of a simple photovoltaic/battery system 181
6.2.2 Decentralized control for multiple battery-based nanogrid 182
6.2.3 Decentralized control for multiple distributed generation units based nanogrid 183
6.2.4 Decentralized control for multiple energy storage units based nanogrid 195
6.2.5 Parameter design for a centralized hierarchical control for AC nanogrid 204
6.3 Case study 211
6.3.1 Large-scaled intelligent nanogrid 211
6.3.2 Small-scaled intelligent nanogrid 213
6.3.3 Nanogrid installed in remote villages 213
6.3.4 Nanogrid based on cogeneration system 217
6.4 Conclusion 217
References 218

7 Sizing of microgrid components 221
Ghulam Mohy-ud-din, Kashem M. Muttaqi, and Danny Sutanto

7.1 Microgrid components 221
7.2 Microgrid sizing and profit maximization 222
7.3 Models of distributed energy resources 226
7.3.1 Probabilistic wind power output model 226
7.3.2 Probabilistic photovoltaic power output model 228
7.3.3 Dynamic battery energy storage power output model 232
7.3.4 Micro-turbine power output model 233
7.4 Optimal sizing of microgrid components 234
7.4.1 Mathematical formulation 235
7.4.2 Backtracking search optimization (BSO) algorithm 237
7.4.3 Solution approach 240
7.5 Case studies 241
7.5.1 Case study 1 241
7.5.2 Case study 2 254
7.6 Summary 259
References 260

8 Optimal sizing of energy storage system 263
Kamran Jalilpoor, Rahmat Khezri, Amin Mahmoudi, and Arman Oshnoei

8.1 Introduction 263
8.2 Energy storage technologies in microgrids: types and characteristics 264
8.2.1 Battery energy storage systems 265

8.2.2 Flywheel 268
8.2.3 Fuel cell 268
8.2.4 Superconducting magnetic energy storage 269
8.2.5 Supercapacitor 269
8.2.6 Technology comparison 270
8.3 Necessity of energy storage in microgrids 273
8.3.1 Frequency regulation 274
8.3.2 Voltage support 274
8.3.3 Reliability enhancement 274
8.3.4 Demand shifting and peak shaving 274
8.3.5 Power smoothing 274
8.3.6 Black start 275
8.3.7 Storage trades/arbitrage 275
8.3.8 Non-spinning reserve 275
8.4 Case study 276
8.4.1 System description and input data 277
8.4.2 Uncertainty modelling 278
8.4.3 Problem formulation 279
8.4.4 Numerical results 282
8.5 Conclusions 286
Nomenclature 286
References 288

9 Microgrid communications – protocols and standards 291
Shantanu Kumar, Syed Islam, and Alireza Jolfaei

9.1 Introduction 291
9.2 Communication objectives and requirements 294
9.3 Communication layer 295
9.3.1 Home automation network 298
9.3.2 Building automation network 298
9.3.3 Neighbourhood area network 299
9.3.4 Local area network 299
9.3.5 Field area network 300
9.3.6 Wide area network 300
9.4 Communication infrastructure 301
9.4.1 Wired communication 301
9.4.2 Wireless communication 303
9.5 Communication protocols 305
9.5.1 Internet communications protocol suite 306
9.5.2 Modbus 308
9.5.3 Distributed Network Protocol version 3.3 309
9.5.4 IEC 61850 310
9.6 Importance of communication technology in microgrid control 311
9.7 Case study 315

9.8 Conclusion 320
Nomenclature 320
References 322

10 Voltage stability of microgrids **327**
Nasser Hosseinzadeh, Saheb Khanabdal, Yousuf Al-Jabri, Rashid Al-Abri, Amer Al-Hinai, and Mahdi Banejad

10.1 Introduction 328
10.1.1 Concept of voltage stability 328
10.1.2 Voltage stability issues of microgrid 328
10.1.3 Microgrid voltage stability assessment 329
10.2 Small-signal model of a microgrid for voltage stability analysis 335
10.3 Voltage stability enhancement 335
10.4 Case studies 336
10.4.1 Case study 1 336
10.4.2 Case study 2 343
10.4.3 Case study 3 348
10.4.4 Case study 4 350
10.5 Concluding remarks 369
References 369
Further reading 374

11 Frequency stability and synthetic inertia **377**
Nasim Ullah, Anwar Ali, Haider Ali, and Khalid Mahmood

11.1 Frequency stability issues of microgrid 377
11.2 Effect of low inertia on the frequency stability of microgrid 379
11.3 Frequency stability enhancement 380
11.3.1 Synchronous generator (SG) model-based topologies 381
11.3.2 Swing equation based 383
11.3.3 Frequency–power-response-based topologies 384
11.3.4 Droop-based approach 385
11.4 Case study 386
11.5 Concluding remarks 391
References 391

12 Microgrid protection **395**
Robert M. Cuzner, Siavash Beheshtaein, and Farzad Banihashemi

12.1 Protective system design objectives 396
12.2 Conventional protective system design practice 398
12.2.1 Fault characterization 400
12.2.2 Protective equipment and scheme components 401
12.2.3 Fault coordination analysis and protective relaying 402

12.3 Microgrid protection challenges 408
12.3.1 Impact of distributed energy resources on power flow 411
12.3.2 Impact of distributed energy resources on fault current magnitude 411
12.3.3 Impact of microgrid connection modes and changing configurations 412
12.3.4 Earthing considerations 415
12.3.5 Cyberattacks 420
12.4 Promising solutions for microgrid protection 420
12.4.1 Limiting maximum DER capacity 421
12.4.2 Evolving communication standards 421
12.4.3 Fault current limiters 423
12.4.4 Utilization of the ESS for fault discrimination 423
12.4.5 Distributed generation control modifications 424
12.4.6 Protective system design process for microgrids 424
12.4.7 Addressing cybersecurity 430
12.5 DC microgrid considerations 433
12.5.1 DC fault characteristics 434
12.5.2 DC protective system approaches 438
12.5.3 DC protective devices 445
12.5.4 DC system grounding 450
12.6 Conclusion: future of microgrid protection 451
References 453

13 Black start and islanding operations of microgrid 463
Clara Gouveia, Carlos Moreira, André G. Madureira, José Gouveia, Diego Issicaba, and Joáo Abel Peças Lopes

13.1 Microgrid operational modes 463
13.1.1 The microgrid 464
13.1.2 Microgrid hierarchical control for emergency operation 468
13.1.3 Extending the concept – the multi-microgrid 469
13.2 Microgrid islanding and reconnection 471
13.2.1 Microgrid primary frequency and voltage control 471
13.2.2 Electric vehicles contribution to primary frequency support 472
13.2.3 Secondary control and emergency dispatch strategies 473
13.2.4 Black start strategies in multi microgrids 476
13.2.5 Black start procedure 478
13.3 Case study 481
13.3.1 Microgrid islanding case study 481
13.3.2 Multi Microgrid black start case study 485
13.4 Concluding remarks 491
References 492

14 Microgrid feasibility study and economics **497**
Alessandra Parisio, Luigi Glielmo and Evangelos Rikos

14.1 Overview 497
14.1.1 Outline of the chapter 500
14.2 Theoretical background 500
14.2.1 Model-predictive control 500
14.2.2 Two-stage stochastic programming 501
14.3 Microgrid component modelling and constraints 502
14.3.1 Nomenclature 503
14.3.2 Loads 503
14.3.3 Distributed generators 505
14.3.4 Energy storage systems 505
14.3.5 Multi-energy components 507
14.3.6 Electrical and thermal balance 507
14.3.7 Interaction with the utility grid 508
14.4 Microgrid operational strategies 508
14.4.1 MPC-based energy-management system for operational optimization 508
14.4.2 MPC-based multi-objective AC optimal power flow 513
14.5 Feasibility study aspects 514
14.5.1 Design and operation 515
14.5.2 Components and topology 515
14.5.3 Active and reactive control strategies 516
14.5.4 Data collection and processing 517
14.5.5 Costing of microgrid components 518
14.6 Case studies 519
14.6.1 Experimental evaluation in Athens, Greece 519
14.6.2 Steinkjer microgrid 525
14.7 Conclusions 528
Appendix A 528
A.1 Matrices 528
References 529

15 Power electronics—microgrid interfacing **533**
Saeed Peyghami, Mohammed Alhasheem, and Frede Blaabjerg

15.1 Importance of power electronics in a microgrid 533
15.2 Classifications of microgrids 535
15.2.1 AC microgrids 535
15.2.2 DC microgrids 535
15.3 Power electronic converters 540
15.3.1 General power conversation concept 540
15.3.2 DC–DC converters 541
15.3.3 DC–AC converters 544

15.4 Power converter switching schemes 547
15.4.1 Pulse width modulation 547
15.4.2 Carrier-based pulse width modulation 547
15.4.3 Zero-sequence injection 548
15.4.4 Space vector modulation 549
15.5 Power converter basic control schemes 550
15.5.1 Electrical model of converters 550
15.5.2 Control of converters in ac grids 553
15.5.3 Control of converters in dc grids 556
15.6 Filters for power converters—active and passive 558
15.6.1 Passive filters 559
15.6.2 Active filters 562
15.7 Case studies 564
15.7.1 Case I: MPC-controlled converters in ac microgrids 564
15.7.2 Case II: Power-sharing control in a dc grid 566
15.8 Conclusions 569
References 570

Index **573**

Preface

The core theme and foundation of the traditional power system is going through a major transformation, nowadays. Small, medium, and large-scale renewable sources, often called as wind farm or photovoltaic park, are becoming the part and parcel of modern power system and distributed in a scattered way all over the power network. This introduces the term 'microgrid' which is a group of distributed energy resources and interconnected loads within a defined electrical boundary, appearing as a single controllable entity with or without being connected to the grid. This transformation needs to address many different technical, tactical, and political challenges which we need to handle collectively and carefully. The technical challenges are multifold; therefore, researchers from electrical, electronic, computer, communication, mechanical, aerospace engineering, and many other science disciplines are contributing in this domain, both from academia and industry, and writing the scripts of microgrid success.

In this book, three different mainstream technical challenges of microgrid are addressed – variability, scalability, and stability. With the term 'variability', the voltage and frequency fluctuations inside and outside microgrid boundaries are referred. On the other hand, 'stability' term includes voltage and frequency instabilities but also covers low voltage or zero voltage ride through problems. The 'scalability' part, in general, covers the optimization aspects of microgrid. The present development status and future trends of microgrid covering from generation, transmission, and distribution are presented based on the contributions from well-known researchers and academics from various disciplines. On this occasion, we the Editors sincerely acknowledge the cordial supports from all the chapter authors in this book along with their valuable contributions.

A general overview and essence of the different chapters available in this book can be obtained from introductory chapter. The microgrid topologies, its hierarchical control schemes, control of its various components along with optimal sizing, and location of microgrid components are presented in different chapters of the book. The power electronics are a mandatory component used by various components of the microgrid. This book presents various power electronic topologies used in microgrid and discusses its control and reliability issues. The microgrid protection and reliability features, black starts, economic aspects, and operations are presented in detail. The recent transformation of microgrid into the virtual power plant is another salient feature of this book.

The renewable sources of microgrid create many power system challenges when interconnected with the main grid. The challenges reach to another level when the penetration level of the renewable sources increases and the distribution system strength weakens further. This book covers all variability related issues of a microgrid, provides solutions on how to handle the scalability problems, and also discusses microgrid protection and stability augmentation methods. The Editors hope that the book will be useful for students, researchers, and engineering practitioners.

Editors
S M Muyeen
The Department of Electrical and Computer Engineering
Faculty of Science & Engineering, Curtin University
Bentley, Australia

Syed Islam
School of Science, Engineering and Information Technology
Federation University
Ballarat, Australia

Frede Blaabjerg
Department of Energy Technology
Aalborg University
Aalborg, Denmark
December 15, 2018

Contributors

S.M. Muyeen received his B.Sc. Eng. degree from Rajshahi University of Engineering and Technology (RUET), Bangladesh formerly known as Rajshahi Institute of Technology, in 2000 and M. Eng. and Ph.D. degrees from Kitami Institute of Technology, Japan, in 2005 and 2008, respectively, all in Electrical and Electronic Engineering. At the present, he is working as an associate professor in the Electrical and Computer Engineering Department at Curtin University, Perth, Australia. He is serving as Editor/Associate Editor for many prestigious Journals from IEEE, IET, and other publishers, e.g., IEEE Transactions of Sustainable Energy, IEEE Power Engineering Letters, IET Renewable Power Generation, and IET Generation, Transmission & Distribution. He is the Editor-in-Chief for Smart Grid Section of Frontier in Energy Research. He has served as guest editor-in-chief/leading editor for many special issues. He was the recipient of many awards including the Petroleum Institute Research/Scholarship Award 2012, which was the only research award for the entire university until 2013. He is the author/coauthor of about 200 scientific articles including 80+ journals and 6 books as an author/editor. In his short carrier, he has secured many prestigious research grant at national and international levels. He has given many keynotes and been invited for speeches to international conferences. His research interests are renewable energy, smart grid, and power system stability. Muyeen is the senior member of IEEE and Fellow of Engineers Australia (FIEAust).

Syed Mofizul Islam received the B.Sc. degree in electrical engineering from Bangladesh University of Engineering and Technology, Bangladesh in 1979, the M.Sc. and Ph.D. degree in electrical power engineering from the King Fahd University of Petroleum and Minerals, Dhahran, Saudi Arabia, in 1983, and 1988 respectively. He is currently the Executive Dean for the School of Science Engineering and Information Technology at Federation University Australia. Prior to joining Federation University, he was the John Curtin Distinguished Professor in Electrical Power Engineering and the Director of Centre for Smart Grid and Sustainable Power Systems at Curtin University, Perth, Australia. He has published over 270 technical papers in his area of expertise. His research interests are in condition monitoring of transformers, wind energy conversion, and smart power systems. He has been a keynote speaker and invited speaker at many international workshops and conferences. Islam was also the Dean International for the Faculty of Science and Engineering at Curtin University (2011–18). He is a Fellow of the Engineers Australia and an Engineering Executive, a Fellow of the IEEE, a Fellow of the IET and a chartered engineer in the United Kingdom, and a chartered professional Engineer in Australia. He is a founding editor of the IEEE Transaction on Sustainable Energy and an associate editor of the IET Renewable Power Generation.

Frede Blaabjerg was with ABB-Scandia, Randers, Denmark, from 1987 to 1988. From 1988 to 1992, he got the Ph.D. degree in electrical engineering at Aalborg University in 1995. He became an assistant professor in 1992, an associate professor in 1996, and a full professor of power electronics and drives in 1998. From 2017, he became a Villum Investigator. He is honoris causa at University Politehnica Timisoara (UPT), Romania and Tallinn Technical University (TTU) in Estonia. His current research interests include power electronics and its applications such as in wind turbines, PV systems, reliability, harmonics, and adjustable speed drives. He has published more than 600 journal papers in the fields of power electronics and its applications. He is the coauthor of four monographs and editor of ten books in power electronics and its applications. He has received 29 IEEE Prize Paper Awards, the IEEE PELS Distinguished Service Award in 2009, the EPE-PEMC Council Award in 2010, the IEEE William E. Newell Power Electronics Award 2014 and the Villum Kann Rasmussen Research Award 2014. He was the Editor-in-Chief of the IEEE TRANSACTIONS ON POWER ELECTRONICS from 2006 to 2012. He has been the Distinguished Lecturer for the IEEE Power Electronics Society from 2005 to 2007 and for the IEEE Industry Applications Society from 2010 to 2011 as well as 2017 to 2018. In 2019–20, he serves as President of IEEE Power Electronics Society. He serves as the vice president of the Danish Academy of Technical Sciences. He is nominated in 2014, 2015, 2016, and 2017 by Thomson Reuters to be between the most 250 cited researchers in engineering in the world. Prof. Blaabjerg is laureate of the Global Energy Prize, 2019.

Seyyed Ali Pourmousavi Kani received the B.Sc., M.Sc., and Ph.D. degrees with honors in 2005, 2008, and 2014, respectively, in electrical engineering. He worked for California ISO (CAISO), NEC Laboratories America Inc. (NECLA), and Denmark Technical University (DTU) from 2014 to 2017. He is currently a research fellow at the University of Queensland (UQ), Brisbane, Australia. He (co)authored 30+ journal and conference papers, and 5 U.S. patents and applications. His current research interests include battery integration to the grid for different applications, control-based ancillary services, and microgrids' energy-management systems.

Farhad Shahnia received his Ph.D. in Electrical Engineering from Queensland University of Technology (QUT), Brisbane, in 2012. He is currently a Senior Lecturer at Murdoch University. Before that, he was a Lecturer at Curtin University (2012–15), a research scholar at QUT (2008-11), and an R&D engineer at the Eastern Azarbayjan Electric Power Distribution Company, Iran (2005–08). He is currently a senior member of IEEE, National Council Member of the Electric Energy Society of Australia and the member of the Australasian Association for Engineering Education. Shahnia's research falls under distribution networks, smart grids, and microgrid concepts. He has authored 1 book and 11 book chapters and 100+ scholarly articles in international conferences and journals, as well as editing 6 books.

M Imran Azim has completed Bachelor of Science in Electrical and Electronic Engineering from Rajshahi University of Engineering and Technology, Bangladesh in 2013 with first class honors. He has also achieved Master of Engineering in Electrical Engineering from the University of New South Wales, Sydney, Australia in 2017 with Research Satisfactory grade. Prior to commencing Ph.D., he worked as a Graduate Electrical Engineer at RCR Tomlinson Limited. Currently, he is pursuing Ph.D. in Electrical Engineering at the University of Queensland, Brisbane, Australia. He is passionate about power and energy systems, and his research interest includes microgrids, renewable energy management, and solar PV systems.

Md Asaduzzaman Shoeb received the Master of Science degree jointly from Royal Institute of Technology, Sweden and Eindhoven University of Technology, Netherlands, in 2013. He is the recipient of the prestigious Erasmus Mundus Category-A scholarship by the European Commission and Education for Sustainable Energy Development Scholarship in 2011. He was a lecturer at Stamford University, Bangladesh (2009–11) and an assistant professor at the American International University Bangladesh (2013–15). Currently, he is a Ph.D. student at Murdoch University. His research interest includes optimal operation of microgrid and renewable energy integration.

G.M. Shafiullah received his Ph.D. in Electrical Engineering from the Central Queensland University, Australia. After completing Ph.D., he joined as a postdoctoral research fellow to Deakin University, Australia. He is currently a Senior Lecturer at Murdoch University, Australia. His research interests include power systems, smart grid, renewable energy, and its enabling technologies. He is the author of 90+ book chapters, journal articles, and conference papers. GM is a senior member of IEEE and the member of the Australasian Association for Engineering Education.

Sreenithya Sumesh graduated Master's degree in Faculty of Science and Engineering in Computing from the Curtin University and currently doing Ph.D. in Computing from the same university. Her main areas of research include data mining, grid computing, smart grids, networking, power distribution, requirements engineering.

Aneesh Krishna is currently an associate professor with the School of Electrical Engineering, Computing and Mathematical Sciences, Curtin University, Australia. He holds a Ph.D. in computer science from the University of Wollongong, Australia. His research interests include software engineering, requirements engineering, conceptual modeling, agent systems, formal methods, data-driven software engineering, data mining, bioinformatics, and renewable energy systems. He has published more than 120 articles in different journals and international conferences. His research is (or has been) funded by the Australian Research Council (ARC) and various Australian government agencies (like NSW State Emergency Service) as well as companies such as Woodside Energy, Amristar Solutions, Autism West Support Incorporated, BW Solar Australia, Dementia Training Australia, and Andrew Corporation (attracted over $1.2 million in research funding in Australia). He serves as an assessor (Ozreader) for the ARC. He has been on the organising committee, served as invited technical program committee member of many conferences and workshop in the areas related to his research.

Chitra M. Subramanian is currently an adjunct research fellow and casual academic in the Department of Computing, Curtin University, Australia. She holds a Ph.D. in Computing from Curtin University, Australia, an M.E. degree in computer science and engineering from Anna University, India and a B.E. degree in computer science and engineering from Madurai Kamaraj University, India. Her research interests include software engineering, requirements engineering, and agent systems.

Robert Lis received the Ph.D. and D.Sc. Eng. degrees in electrical engineering (electric power system) from Wroclaw University of Science and Technology (WUST) in 1996 and 2014, respectively. Since 1996, he has been with Faculty of Electrical Engineering, Department of Electrical Power Engineering, WUST as an assistant professor to 2014, and as an associate professor since 2015. His research interests include analysis and modeling of electrical power system, integration of a large number of decentralized renewable energy sources into the electric power system, power system wide-area monitoring, and control. He has published over 90 scientific articles in journal and international conferences. He has successfully accomplished several research projects at national and international levels.

Robert Czechowski received the Ph.D. degrees in electrical engineering (electric power system) in 2018. In 2011, he graduated postgraduate studies—security management of information systems on Computer Science and Management Department of WUST. He received his M.Sc. degree in computer science from the WUST in 2009 and also received the Eng. degree and Rector's award in 2007 for the best engineering thesis with a topic in the specialization: computer engineering. Presently, he is an assistant professor from 2018. His research interest are Communication and Security of Smart Grid, Data Flow in Smart Metering, ICT system dedicated Power Systems, automatics using AVR Microcontrollers and Optoelectronics. Actually he Cooperating within Polish Smart Power Grids Section. Author of many publications on the field of Smart Grids Cybersecurity.

Munira Batool received the B.Sc. and MS degrees in electrical engineering from Bahauddin Zakariya University (BZU), Multan, Pakistan and University of Engineering and Technology (UET), Taxila, Pakistan in 2007 and 2012, respectively.

Currently she is pursuing her Ph.D. degree in electrical engineering from Curtin University, Perth Australia. She is also working as Sessional Academic in the Department of Electrical Engineering of Curtin University, Australia. Before that, she was a lecturer at UET Taxila Pakistan (2013-15), lead Electrical Engineer in Power System Lab of UET Taxila (2009–13) and Demonstrator in Wah Engineering College, Pakistan (2008). Her research interest includes power system operation and microgrid system optimization.

Xiaofeng Sun M'11 received the B.S. degree in electrical engineering from Northeast Heavy Machinery Institute in 1993, Heilongjiang, China, and the M.S. and Ph.D. degrees in power electronics from Yanshan University, Hebei, China in 1999 and 2005, respectively. From 2003 to 2007, he was an associate professor with Yanshan University, where since 2008, he has been a professor and also the Director at the Key Laboratory of Power Electronics for Energy Conservation and Motor Drive of Hebei Province. He has authored or coauthored more than 70 transactions and conference papers. His current research interests include dc–dc converters, multiple-input converters, hybrid electric vehicles, microgrids, and power-quality control.

Wei Zhao received the B.S. degree and the M.S. degree in electrical engineering and Power Electronics and Power Drives from Yanshan University, Qinhuangdao, China, in 2006 and 2009. He was a Lecturer with Yanshan University, where he is currently working toward the Ph.D. degree in power electronics. His current research interests include the stability analysis of microgrid and power quality.

Lei Qi received the B.S. degree and the M.S. degree in electrical engineering from Yanshan University, Qinhuangdao, China, in 2014 and 2017. He was an Assistant Lecturer with Yanshan University, where he is currently working toward the Ph.D. degree in power electronics. His current research interests include the energy management, nanogrids, and stability analysis.

Ghulam Mohy-ud-din S'17 received the B.Sc. and M.Sc. degree in electrical engineering from the University of Engineering and Technology, Taxila, Pakistan, in 2013 and 2015, respectively. He is currently pursuing the Ph.D. degree in electrical engineering at University of Wollongong, New South Wales, Australia. He is also a lecturer with COMSATS Institute of Information Technology, Pakistan. His research interests are power system planning and operation with renewable energy resources.

Kashem M. Muttaqi M'01, SM'05 received the B.Sc. degree in electrical and electronic engineering from Bangladesh University of Engineering and Technology (BUET), Dhaka, Bangladesh in 1993, the M.Eng.Sc. degree in electrical engineering from University of Malaya, Kuala Lumpur, Malaysia in 1996 and the Ph.D. degree in electrical engineering from Multimedia University, Selangor, Malaysia in 2001. Currently, he is a professor at the School of Electrical, Computer, and Telecommunications Engineering, and member of Australian Power

Quality and Reliability (APQRC) at the University of Wollongong, Wollongong, Australia. He was associated with the University of Tasmania, Hobart, Australia as a research fellow/lecturer/senior lecturer from 2002 to 2007, and with the Queensland University of Technology, Brisbane, Australia as a research fellow from 2000 to 2002. Previously, he also worked for Multimedia University as a Lecturer for 3 years. He has more than 21 years of academic experience and authored or coauthored 300 papers in international journals and conference proceedings. His research interests include distributed generation, renewable energy, electrical vehicles, smart-grid, power system planning, and emergency control.

Danny Sutanto SM'89 received the B.Eng. (Hons.) and Ph.D. degrees from the University of Western Australia, Perth, W.A., Australia, in 1978 and 1981, respectively. He is currently a professor of power engineering with the University of Wollongong, Wollongong, N.S.W., Australia. His research interests include power system planning, power system emergency, analysis and harmonics, flexible alternating current transmission system, and battery energy storage systems. He was the IEEE Industry Applications Society Area Chair for Region 10 (Asia Pacific) from 2014 to 2017.

Kamran Jalilpoor received the B.S. degree in from Urmia University of Technology (UUT), Urmia, Iran in 2016, and the M.S. degree in from the Shahid Beheshti University (SBU), Tehran, Iran in 2018. His research interests include microgrids planning, power system resilience, power distribution systems, and optimization theories.

Rahmat Khezri received his B.Sc. degree in electrical engineering from Urmia University, Iran and M.Sc. degree in electrical engineering from University of Kurdistan, Iran. He is currently pursuing his Ph.D. degree in College of Science and Engineering at Flinders University, Adelaide, Australia. He is the recipient of Australian Government Research Training Program Scholarship (AGRTPS), Flinders University, Australia (2018). His research interest includes wind-farm optimization, battery-storage integration in renewable energy systems, load frequency control, power system, stability and intelligent control applications.

Amin Mahmoudi S'11–M'13 received his bachelor degree in electrical engineering from Shiraz University, Shiraz, Iran, in 2005, Master degree in Electrical Power Engineering from Amirkabir University of Technology, Tehran, Iran, in 2008, and the Ph.D. degree from the University of Malaya, Kuala Lumpur, Malaysia, in 2013. He is interested in research areas where energy conversion and transmission play a major role, such as hybrid power networks, renewable energy systems, transmission and distribution networks, electrical machines and drives. He currently is working as a lecturer at Flinders University. Mahmoudi is a Chartered Engineer and member of the Institution of Engineering and Technology (CEng) and Engineers Australia (CPEng).

Arman Oshnoei received the B.Sc. degree from Urmia University of Technology, Iran in 2015 and M.Sc. degree from Tabriz University, Iran in 2017 both in electrical engineering. He is currently pursuing his Ph.D. in Shahid Beheshti University, Iran. His research interest includes power system control, automatic generation control, and probabilistic load flow.

Shantanu Kumar received bachelor of engineering degree in electrical from Bangalore University, India in 1990. He obtained his MBA degree from Indore University, India in 1996 and completed his postgraduate research in power engineering leading to M.Sc.Eng. from The University of Western Australia, Perth, Australia in 2014. Currently, he is pursuing his postgraduate research work on IEC 61850 in protection, control and automation at Curtin University, Perth, WA, Australia. He has over 29 years of experience as a power engineer in diversified utilities, resources and heavy industries spanning from Asia to Australia. Currently, he is working as a power system engineer in a large consultancy in WA. He has a track record of successfully designing, engineering, and commissioning many utilities and resources HV substation projects in Asia and Australia. He is a fellow and chartered engineer of Engineers Australia (EA) and is in the interview panel to assess potential power engineering candidates to achieve CP Eng. status of EA. He also has a CAMA certificate in Asset Management Council of Australia and regularly contributes research papers on automation and control in reputed journals, conferences, and symposiums based on IEC 61850.

Alireza Jolfaei received the Ph.D. degree in Applied Cryptography from Griffith University, Gold Coast, Australia. He is a lecturer in Cyber Security at Macquarie University, Sydney, Australia. Prior to this appointment, he worked as a lecturer in Cyber Security at Federation University Australia and as an assistant professor of Computer Science at Temple University in Philadelphia, USA. His current research areas include cyber security, cyber physical systems security, AI and machine learning for cyber security. He has authored over 40 peer-reviewed articles on topics related to cyber security. He has received multiple awards for Academic Excellence, University Contribution, and Inclusion and Diversity Support. He received the prestigious IEEE Australian council award for his research paper published in the *IEEE Transactions on Information Forensics and Security*. He received a recognition diploma with cash award from the IEEE Industrial Electronics Society for his publication at the 2019 IEEE IES International Conference on Industrial Technology. He is a founding member of IEEE Northern Territory Section and Federation University IEEE Student Branch. He served as the Chairman of Computational Intelligence Society in IEEE Victoria Section and also as the Chairman of Professional and Career Activities for IEEE Queensland Section. He has served as the guest associate editor of IEEE journals and transactions, including *IEEE Internet of Things Journal* and *IEEE Transactions on Industrial Applications*. He has served over 10 conferences in leadership capacities including program co-chair, track chair, session chair, and technical program committee member, including *IEEE TrustCom* and *DependSys*. He is a senior member of the IEEE.

Nasser Hosseinzadeh received his B.Sc. degree in electrical and electronics engineering from Shiraz University in 1986, M.Sc. degree from Iran University of Science and Technology in Electronics in 1992, and his Ph.D. degree in electrical engineering from Victoria University, Melbourne, Australia, in 1998. He worked as a faculty member at Shiraz University in Iran, Monash University, Malaysia, Central Queensland University and Swinburne University of Technology, Australia, consecutively, during 1998–2011 before moving to Sultan Qaboos University in Oman. He served as the discipline leader of electrical engineering from 2005 to 2006, Head of Department of Systems from 2007 to 2008 at CQuniversity, and Head of Department of Electrical and Computer Engineering from 2014 to 2018. At SQU, he is the theme leader of Integrated Energy Systems, microgrids, and smart grid. He is also an advocate for student-centered, cooperative, and active-learning methods in engineering education. Hosseinzadeh is a senior member of IEEE. Previously, he was a member of CIGRE Australia and worked with the panel on power system developments and economics.

Saheb Khanabdal received his B.Sc. degree from Khaje Nasir Toosi University of Technology (KNTU), Tehran, Iran, in 2011 and the M.Sc. degree from University of Tabriz, Tabriz, Iran in 2013, both in electrical engineering. He is currently working toward his Ph.D. degree with the Department of Electrical Engineering, Shahrood University of Technology, Shahrood, Iran.

From June 2018 to September 2018, he was with Sustainable Energy Research Center at Sultan Qaboos University, as a research assistant in Muscat, Oman.

His research interest areas include control of microgrid and power converters, electric vehicles, and fault current limiters.

Yousuf AL-Jabri was born in Rustaq, Oman, in 1986.

Received the B.Eng. degree in electrical and computer engineering—power systems and energy at Sultan Qaboos University, Muscat, Sultanate of Oman, in 2009. He received his Master degree in power system stability at Sultan Qaboos University in 2015. He is currently a power system concept engineer at Petroleum Development Oman (PDO).

His research interests include integration of renewable energy with microgrids and stability issues related to those fields. In addition, his practical interest is on power system dynamic testing and related stability studies.

Rashid Al-Abri received the B.Sc. in electrical engineering from Sultan Qaboos University, Oman, in 2002 and M.Sc. in electrical engineering from Curtin University of Technology, Western Australia, in 2004. Then, he completed the Ph.D. degree in the Department of Electrical and Computer Engineering, University of Waterloo, Waterloo, ON, Canada in 2012. Currently, Rashid is assistant professor at Sultan Qaboos University, at ECE department. His research interests are power electronics application, renewable energy, power quality, power systems and smart-grid application, power system stability.

Amer Al-Hinai is the director of the Sustainable Energy Research Center (SERC) and associate professor of electrical engineering at Sultan Qaboos University. He has carried out more than 33 industry-funded research projects, with total funds exceeding 3 million USD, related to energy savings, power system analysis, power system quality, and transient stability of power systems. His research output has been recognized as a value-added research by the industry, engineering societies, and the academia. This is proved by the continuity of research funding from industry and academia, the awards received, and professional appointments. During 2012–16, Amer did his sabbatical leave followed by secondment to Masdar Institute (MI). The Institute is a postgraduate and research academic institution focused on sustainability, water, and renewable energy resources. In 2011, Amer was appointed as Authority for Electricity Regulation (AER) board member and then the Chairman of AER during 2014–17. The authority is an independent electricity regulatory body in Oman. Amer published over 80 articles in reputable journals and refereed international conferences and book chapters. He received several awards such as His Majesty Trust Fund research award, "One of the Pioneers in the Engineering Practice in the Gulf," Fulbright Research Scholarship, and first prize for the technical competence paper at the 39th IECON. Amer Al-Hinai is an IEEE senior member and a former Chairman of IEEE Oman Section.

Mahdi Banejad received B.Eng. degree from Ferdowsi University in Mashhad, Iran, in 1989; M.Sc. degree from Tarbiat Modarres University, Tehran, Iran in 1994; and Ph.D. degree from Queensland University of Technology, Australia in 2004 all in Electrical Engineering. After finishing his Ph.D., he undertook a research program for 1 year at QUT (QUT). He was the manager of the Section of the Relation between University and Industry of Shahrood University of Technology in 2006–08. Currently, he is a senior member of IEEE and associate professor of faculty of Electrical and Robotic Engineering at Shahrood University of Technology, Iran. His main research interests are voltage and frequency control of microgrids, decentralized state estimation in distribution system, and small signal stability of microgrids.

Nasim Ullah received the Ph.D. degree in mechatronic engineering from Beihang University, Beijing, China, in 2013. From 2006 to 2010, he was a senior design engineer with IICS, Pakistan. He is currently an associate professor of electrical engineering with the CECOS University of Emerging Science and Information Technology, Peshawar, Pakistan. His research interests include renewable energy, flight control systems, integer and fractional order modeling of dynamic systems, integer/fractional order adaptive robust control methods, fuzzy/NN, hydraulic and electrical servos, epidemic, and vaccination control strategies.

Anwar Ali was born in Mardan, Pakistan, in 1981. He received his B.E. degree in electronics engineering from NED UET Karachi, Pakistan, in 2004. He completed his M.S. degree in electronics engineering and Ph.D. degree in electronics and communication engineering from Politecnico di Torino, Italy in 2010 and 2014, respectively. From 2014 to 2017, he was an assistant professor with the

Electrical Engineering Department, Foundation for Advancement of Science and Technology (FAST) NUCES, Peshawar, Pakistan. Since August 2017, he has been an assistant professor with the Electrical Engineering Technology Department, University of Technology (UoT), Nowshera, Pakistan. His research interests include design and development of power management, attitude determination and control subsystems of small satellites. He is also working in the area of thermal analysis and thermal modeling of aerospace systems, power electronics applications and renewable energy systems.

Haider Ali was born in 1984. He completed his B.Sc. degree in telecom engineering from NUCES, Pakistan, in 2007. He received his doctorate and an M.S. degree in electronics engineering from Politecnico Di Torino, Italy, in 2010 and completed his Ph.D. in electronics and communication engineering also from there in 2014. He is currently working as an assistant professor at the department of Electrical Engineering, and Technology, University of Technology, Nowshera, Pakistan. His research interests include data-acquisition systems, power electronics systems, design and development of antenna, radio frequency (RF) front end and telecommunication subsystem for small satellites.

Khalid Mahmood received the B.S. degree in electrical engineering from UET Peshawar, Pakistan, the M.S. degree from University of Western Ontario Canada, and Ph.D. degree in electrical engineering from De Montfort University, Leicester, UK, in 1992, 2008, and 2014, respectively. Currently, he is head of department of electrical and electronics engineering technology Nowshera Pakistan. His research interests include adaptive filtering, wireless communications, and signal processing for communications.

Robert M. Cuzner received the B.S. degree from Brigham Young University, Provo, UT, USA, and M.S. and Ph.D. degrees from the University of Wisconsin–Madison, Madison, WI, USA, all in electrical and computer engineering. In 1990, his professional work began with Miller Electric Manufacturing Company, Appleton, WI, USA, designing generators for engine-driven welders. He worked at Eaton Corporation, Milwaukee, WI, USA, from 1993 to 2002 and then DRS Power and Control Technologies, Inc., from 2002 to 2014 as a designer of power conversion systems for Navy shipboard applications. He is presently associate professor in the Department of Electrical Engineering and Computer Science at the University of Wisconsin–Milwaukee. He has over 25 years of experience working in power generation, power conversion and power distribution of both military and industrial applications. A principal focus of his work has been shipboard electrification, with focus on achieving energy secure systems. His interests include microgrid protection, distributed generation, power electronics for power distribution and drive systems, low- and medium-voltage power conversion system design, high power-density packaging of power electronics, and electric machine design.

Siavash Beheshtaein received the B.Sc. and M.Sc. degrees from Shiraz University, Iran, and his Ph.D. degree from Aalborg University, Denmark, in 2011, 2013, and 2018, respectively, all in Electrical Engineering. He has also worked as a visiting scholar and post-doctoral research fellow at the University of Wisconsin–Milwaukee where he developed protective relaying approaches for microgrids, medium voltage hybrid solid-state circuit breaker for 12–35 kV systems and extreme charging stations for electric vehicles. He has also developed artificial intelligence schemes for the improvement of power quality and resilience of microgrids and grid connected power electronic converters, and various approaches to fault current limiting, fault detection and discrimination and protective relaying in microgrids based upon harmonic injection and machine learning techniques. His research interests include microgrid protection, adaptive protection, solid-state transformer, and DC circuit breakers.

Farzad Banihashemi is a Ph.D. student at University of Wisconsin—Milwaukee. He received his B.Sc. degree from the University of Guilan, Iran in Electronics in 2007. He holds an M.Sc. degree from University of Tehran, Iran and graduated in 2010, in Power Systems and High Voltage. His work was an optimization of the location and size of the distributed generation in meshed and redial AC systems. He is an expert in control and protection systems. He had been working in high voltage and medium voltage systems for 6 years. He dealt with protection design, setting calculation and relay mapping and configuration. He also has the experience of LV systems design. He moved to the US in 2017 to pursue his study at UWM. His main research focus is on protection of AC microgrids. He is developing a novel method of protection scheme, independent coordination, for meshed AC microgrids using commercial devices. His research interests also include power electronics, grid converters, and motor drives.

Clara Gouveia received her M.Sc. and Ph.D. degrees in electrical engineering from the Faculty of Engineering, University of Porto (FEUP) in 2008 and 2015 respectively. Since 2011 she is a member of the Centre for Power and Energy Systems of INESC TEC—Instituto de Engenharia de Sistemas e Computadores, Tecnologia e Ciência, where she currently holds a Senior Researcher position. She is also the leader of DMS/EMS and network automation. Since 2015, she has been involved in H2020 European projects SENSIBLE and UpGRid where her work is focused on the fields of distributed electrical energy storage control for supporting the operation of distribution networks, namely, in islanded conditions, considering deployment microgrid concepts. Her research interests are focused on the development of the microgrid concept in the context of smart grids integrating plugged-in electrical vehicles, distributed storage, and microgeneration units. She published several papers in international scientific journals.

Carlos Moreira received the licentiate degree (5-year program) in electrical engineering at the Faculty of Engineering of the University of Porto (FEUP) in 2003 and completed his Ph.D. in Power Systems in November 2008 also at the

University of Porto. He is currently a Senior Researcher and Area Leader at the Centre for Power and Energy Systems of INESC Technology and Science (INESC TEC). In February 2009 he joined the Department of Electrical Engineering at FEUP as assistant professor. He lectures classes the integrated master program in Electrical Engineering as well as in the Doctoral Program in Sustainable Energy Systems. He supervised about 30 M.Sc. dissertation and currently he is supervising 5 Ph.D. students. He is coauthor of more than 70 papers published in international journals and conferences with peer reviewing, author of a book (Ph.D. thesis edition) and coauthor of five book chapters. He has been involved in more than 15 projects involving consultancy activities related to renewables integration into electric power systems. He has been also involved in European Union funded projects MICROGRIDS, MORE-MICROGRIDS, TWENTIES, HYPERBOLE, SENSIBLE and EU-SysFLEX, where he developed activities related to the integration of renewable energy systems in electrical power systems at distribution and transmission mainly from the respective of electric power system dynamic and transient stability.

André G. Madureira was born in Oporto, Portugal, in 1980. He earned his licentiate degree (5-year program), an M.Sc. (2-year program) and a Ph.D. in electrical and computer engineering from the Faculty of Engineering of the University of Porto, Portugal in 2003, 2005, and 2010, respectively. He is currently a senior researcher at the Centre for Power and Energy Systems of INESC Technology and Science (INESC TEC) and assistant professor at the Lusófona University of Porto in Oporto, Portugal. He is author of more than 45 papers published in international journals and conferences with peer reviewing, as well as author of a book (Ph.D. thesis edition) and coauthor of four book chapters. His research interests have been directed towards the integration of distributed generation in distribution grids as well as to the development of advanced functionalities for smart grids involving renewable energy Sources, storage devices, and demand response. More recently, he has been working in energy efficiency topics.

José Gouveia received her M.Sc. completed in electrical engineering from the Faculty of Engineering, University of Porto (FEUP) in 2015 and is currently pursuing the Ph.D. degree in sustainable energy systems from University of Porto, under the MIT Portugal Program. He is currently a Researcher at the Centre for Power and Energy Systems of the INESC Technology and Science (INESC TEC), where he has been involved in scientific projects and consulting in the area of integration of renewable sources in the electrical system as well as in the dynamic analysis of electrical systems. From the work developed during its master thesis, he has won an honorable mention in the PREMIO REN 2016.

Diego Issicaba M'07 received the Ph.D. degree on sustainable energy systems from the Faculty of Engineering, University of Porto, Portugal, in association with the MIT Portugal Doctoral Program, in 2013. His research activities were hosted by the Institute for Systems and Computer Engineering of Porto (INESC Porto) where he worked on several projects financed by the EU Commission and Industry. During his doctoral studies, he was also hosted by the Department of Computational Sciences of the Pontifical Catholic University of Rio Grande do Sul (PUCRS), Rio Grande do Sul, Brazil. He received the B.S. and M.S. degrees in electrical engineering from the Federal University of Santa Catarina, Santa Catarina, Brazil, in 2006 and 2008, respectively. His research interests include self-healing grids, smart grid operation and control, multi-agent technology, and power system reliability. He is also interested in applied mathematics, distribution system planning, and machine learning.

Joáo Abel Peças Lopes is full professor at the Faculty of Engineering of Porto University (FEUP) where he teaches in the graduation and postgraduation areas. He is presently the associate director of INESC TEC, one of the largest R&D interface institutions of the University of Porto. His main domains of research are related with large-scale integration of renewable power sources, power system dynamics, microgeneration and microgrids, smart metering and electric vehicle grid integration. He is author or coauthor of more than 400 papers and coeditor and coauthor of the book "Electric Vehicle Integration into Modern Power Networks" edited by Springer. He is fellow from IEEE.

Alessandra Parisio SM'18 received her Ph.D. in automatic control from the University of Sannio, Italy, which included a year at Swiss Federal Institute of Technology (ETH), Switzerland, where she worked on building climate control within the research project "Use of weather and occupancy forecasts for optimal building climate control (OptiControl)." She undertook postdoctoral research at the Automatic Control Laboratory at the Royal Institute of Technology (KTH), Sweden, where she led the KTH-EES Smart Building Lab project and coordinated the European project EIT ICT Labs "Microgrid Operation and ICT Solutions." Since September 2015, she is a lecturer in the School of Electrical and Electronic Engineering at The University of Manchester, United Kingdom, where she is an investigator in two innovate UK projects and one H2020 EU project, focusing on energy-management systems for intelligent buildings including battery storage systems and large-scale control of multiple-distributed storage systems. Her research interests include the areas of large-scale energy-management systems and stochastic constrained control, where she has over 30 publications.

Luigi Glielmo SM'05 received the Laurea degree in electronic engineering and the Research Doctorate degree in automatic control, both from Universita di Napoli Federico II, in 1986 and 1990, respectively. He taught at the University of Palermo, the University of Naples Federico II, and the University of Sannio, Benevento, Italy, where he is currently a professor of automatic control. From 2001 to 2007, he was the head of the Department of Engineering, University of Sannio, where he is currently the rector's delegate for technology transfer and the coordinator of the Ph.D. course on information technologies for engineering. He coauthored more than 130 papers on international archival journals or proceedings of international conferences, coedited two books, and holds three patents. His research interests over the years have included singular perturbation methods, Lyapunov-based methods, model-predictive control methods, automotive controls, deep brain stimulation modeling and control, and energy-grid and water-grid control. Glielmo is on the Editorial Boards of archival journals of the area, such as the IEEE TRANSACTIONS ON AUTOMATIC CONTROL. He is an associate editor of Control Systems Letters. He is the chair of the IEEE Control Systems Society Technical Committee on Automotive Controls and the general cochair of the European Control Conference 2019.

Evangelos Rikos received his Dipl.-Eng. and Ph.D. degrees in electrical and computer engineering, from the University of Patras, Greece, in 1998 and 2005, respectively. He has been working with the Centre for Renewable Energy Sources and Saving, Department of Photovoltaics and DG since 2007. He also worked as lecturer at the University of Patras during the academic year 2006–07. His research interests are focused on the fields of renewable energy sources and especially photovoltaics, distributed generation, microgrids, energy efficiency, power electronics and electromotion systems in electric vehicles. He has participated in several EU funded projects such as ELECTRA IRP, ERIGrid, DERri, MIRABEL, EU-DEEP, MoreMicrogrids, SEESGEN-ICT, SmartGrids-ERANet. Rikos is the author or coauthor of over 40 scientific publications in international journals and conferences.

Saeed Peyghami received the B.Sc., M.Sc., and Ph.D. degrees all in electrical engineering from the Department of Electrical Engineering, Sharif University of Technology, Tehran, Iran, in 2010, 2012, 2017, respectively. He was a visiting Ph.D. Scholar with the Department of Energy Technology, Aalborg University, Denmark in 2015–16, where he is currently a postdoctoral researcher. His research interests include control, stability and reliability of power-electronic-based power systems.

Mohammed Alhasheem received the B.Sc. and M.Sc. degrees in electrical and control engineering from the Department of Electrical and Control Engineering, Arab Academy for Science, Technology and Maritime Transport, Cairo, Egypt, in 2012 and 2015, respectively. He is pursuing a Ph.D. degree at the Energy Department, Aalborg University, Aalborg, Denmark, where he is currently a visiting Ph.D. scholar with the Department of Information Engineering, Padua, Italy. His current research interests include predictive control for the power converter, microgrids, renewable energy.

Chapter 1
Introduction

S.M. Muyeen[1], *Syed Islam*[2], *and Frede Blaabjerg*[3]

Microgrids are a collection of loads and local generations that are also normally connected to the legacy grid. However, the grid connection may be simply for reliability reason, if the local generations and loads match. The grid may just serve as a trading link depending on the mutual need. Considering its many other benefits, microgrid market is growing more rapidly; nowadays, more than 1,840 projects that represent a power capacity of almost 19,280 MW are under development worldwide. According to a research report by the market research and strategy consulting firm, Global Market Insights, Inc, the microgrid market will exceed $19 billion by 2024 [1]. This chapter presents an overview, technical challenges, and future important issues of microgrid developments.

1.1 Microgrid fundamentals and its anatomy

The point of connection through an inverter that is capable of synchronization and protection enabled that is compatible with connection standards. This is a necessary feature of these power converters to be able to island the microgrid from the main grid if an emergency situation arises such as network faults. Not all microgrids have to be connected to the main grid; in remote locations, microgrids can be thought of a stand-alone system where the loads are met by local generation only and can be vulnerable to curtailment of loads, deviation from standard voltage and frequency. A microgrid thus can be either grid connected or without grid connection. The way a microgrid is connected to the main grid is demonstrated in Figure 1.1.

A microgrid can also be either ac or dc or a mixture of both. Microgrids may be of various sizes and may also act as a virtual power plant (VPP). The atomic version of a microgrid or a nanogrid is the modest residential home prosumer, who is both a consumer and a producer of electrical energy owning distributed energy resources (DERs) such as the rooftop photovoltaic (PV). Such small microgrids could form a cluster and either help each other or trade as a united entity.

[1]The Department of Electrical and Computer Engineering, Curtin University, Bentley, Australia
[2]School of Science, Engineering and Information Technology, Federation University, Ballarat, Australia
[3]Department of Energy Technology, Aalborg University, Aalborg, Denmark

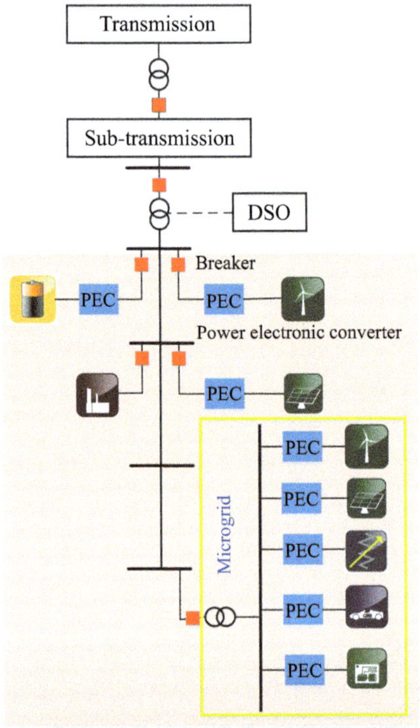

Figure 1.1 A grid-connected microgrid architecture

Commercial and industrial-scale microgrids embedded with storage offer sizeable opportunity as a dispatchable VPP for the main grid at the time of need. So far, large-scale microgrids are university campuses and commercial projects which are of sizes in excess of several tens of megawatts, in some cases.

Traditional utilities are forming partnerships with large-scale microgrid operators by investing in disruptive technologies for survivals. Such microgrid partnerships for the grid operator offer VPP opportunities to tackle few hours of extreme demand during hot summer months. This is a win–win situation for both microgrids and utilities. On the 24th of January in Victoria, Australia, the energy prices climbed to a $15,000 per MW hour from the usual a $100 per MW hour on the National Electricity Market as the mercury soared to 39°C. However, this did not last for long and for the utility, it is a good deal and a good business proposition for a microgrid who can contest in the market.

1.2 Microgrid technical aspects

1.2.1 Microgrid control issues

A microgrid can offer higher reliability of power supply compared to just feed through a grid connector. How many times, it has been witnessed that university campuses had no power due to grid failure resulting in large-scale disruption and productivity losses! During the recent blackouts in South Australia, while most of the state was without power for several days, a brewery was operating normally because it operated on an islanded microgrid mode. Residential nanogrids when operated as a microgrid can improve their reliability of supply through mutual agreement without having to purchase power from the main grid. Such clustered microgrids can also trade excess outputs or make money at times of high market prices by selling to the internet of energy operators a new business concept. These IoT operators gathering excess or opportunistic energy from microgrid clusters can bid in the national electricity market as a VPP. The advent of home area networks (HANs), local area networks, and wide-area networks (WANs) working in harmony with smart meters and communication gateways enabling demand management then can serve as a reliability value to the grid-connected customers which otherwise may face load shedding due to momentary peak demand. An important aspect of microgrids is their controls. These control measures range from real-time volt/var to frequency control to steady-state energy sources optimization within the microgrid. Microgrids are regulated through industry-accepted standards and protocols.

A key operational aspect is also the reliability of power supply. Continuity of supply in the case of either grid disturbances or disturbances within the microgrid is a key goal of the operator. For grid-connected microgrids, islanding of the microgrid ensures that the DERs of the microgrid do not absorb reactive power, and the power conversion equipment are not disconnected. Intentional islanding and noncritical load curtailment may be required to minimize widespread blackouts. For isolated or remote microgrids, reliability indices could be negotiable and may vary from grid standards to ensure that power quality issues do not cause complete blackouts. Voltage and frequency standards may be sacrificed slightly to ensure the continuity of power supply. It is apparent from the above that control measures are of extreme importance in the operation of a microgrid.

1.2.2 Power electronics in microgrid

Microgrids are systems having multiple electrical power sources and loads connected to a common grid which is either ac or dc—the sources can be conventional power sources like synchronous generators, but most often they have renewable energy-based sources like solar and wind, as well as energy-storage systems (ESSs). Here the power electronics technology is used to handle the source (power generation) as well as interconnection to the microgrid. The power electronics is the power-processing interface, and it is the system which can make the microgrid to operate—both in terms of milliseconds, seconds as well as long term. For the low-power level, it can be a single-phase system, while for the higher power system, it is a three-phase grid system.

The general demands of the power electronics are high efficiency, compactness if needed, low cost and high reliability in order to lower the operation and maintenance cost when connected to the microgrid. The typical power electronic interface will be a two-stage conversion process where the converters are voltage sourced. For the solar power conversion, it will be a dc/dc and a dc/ac power conversion, while for the wind power, it will be ac/dc and dc/ac power conversion if the microgrid is an ac microgrid. ESS is also a system, which might be interfaced by power electronics having the possibility to charge and discharge the storage systems and thereby be a buffer between the power production and the load. In such a case, the system will be dc/dc conversion and dc/ac power conversion.

At the load side, more and more load systems are interfaced by power electronics like adjustable speed drives (ac/dc/ac), power supplies to computers (ac/dc/dc), lighting (ac/dc/dc), electrical cooking (ac/ac), and so on. In the past, many systems were very little power electronics based, but due to the gain in energy efficiency—many more load systems are power electronics based. However, in microgrids, examples of having 100% power electronics based loads are seen. Seen from a power system operational perspective, this opens up for large flexibility in control of the overall microgrid—both at the generator side as well as the load side, e.g., by dispatching the loads. The power electronics controlled system makes it possible to implement a hierarchical control structure in the large power systems having a control structure using primary, secondary, and tertiary control. Typical in the primary control, current loops are used in order to protect the power converters in operation.

1.2.3 Addressing power electronics reliability in microgrid

In a microgrid, the power electronics technology (equipment) is assumed to operate hour by hour without any interruption independent on loads. Therefore, specific demands are present at the equipment level and system level in order to ensure that the electricity is available on demand. At the system level, this might be ensured by that the grid system always is able to continue to operate if one of the power sources is disappearing—which is the $N-1$ robustness (where N is the number of electrical power sources) operation or to be able to reduce the loads by using the power electronics technology [2].

At the equipment level, a trend is much better to be able to predict end of life on the most critical parts in the equipment using design software based on physics of failure models and loading models of the equipment. Here it is also important to take into account the variation of the components individually. The design is typically based on an assumed operational profile (mission profile), how the environment is (e.g., temperature and humidity), the power electronic design, and then use detailed lifetime models of the most critical components in the equipment. Having such information gives a solid information about the expected failure rates for the equipment operating in the microgrid with a given mission profile.

One way to increase the reliability in the equipment is to do condition monitoring of the power electronics during operation and detect if some malfunction is expected to happen soon and then handle the situation before it breaks down at the system level. A second method is to integrate fault-tolerant operation of the equipment if, e.g., one switch breaks down—a solution, which is applied in very demanding applications. One of the challenges in the future is to access microgrids systematically in the same manner as the power electronic equipment are assessed. It is not only hardware dependent, but also it is software dependent—e.g., can poor controller algorithms cause that the power converters interact with each other and will increase the probability of failures. Further, the hierarchical control structure can also be a reason to have instability.

If a failure happens and the microgrid loses control including power-supply interruption, then it is important to have mechanisms which can start the microgrid up again (black start). One safe source for energizing the grid is to have storage in the system and let that system first feed the grid and then connect the other power sources and at last connect the loads to the grid system.

1.2.4 Use of energy storage systems in microgrid

ESS is being utilized in a wide variety of applications including power system reliability improvement, backup energy supply during outage, storing excess energy, and backing up off-grid system. It has also found many applications in renewable energy sectors; therefore, it is a very important component of modern microgrid. The huge penetration of DERs to the existing power network requires revised regulations, additional spinning reserves, and load ramping capabilities to minimize random and stochastic power-generation patterns. Otherwise, grid sees voltage and frequency fluctuations. ESS is able to handle these fluctuations in a smart way by decoupling the generation and demand with the energy buffer zone.

Various types of ESS technologies are available in the market, and still researchers are looking for more reliable, economical, and efficient systems. Pumped storage is the oldest and most economical ESS technology, even used in modern power systems. It can be deployed on a large scale, and the majority of the biggest electricity storage plants have adopted this technology. This technology also supports the growing renewable energy integration. In doing that and addressing issues of the variable renewable energy generation, the operations of some pumped storage facilities have been adjusted, e.g., in Scotland, where pumped storage plants have gone from 4 cycles to over 60 cycles per day to complement the variability in wind power generation [3]. Many modern pumped storage facilities are adopting variable speed turbine technology to facilitate the modulation operation.

In fact, most of the modern ESSs are equipped with power electronic converter technology, which enables adjustable operation in controlling real and reactive power. These include energy capacitor system (ECS), superconducting magnetic energy storage (SMES), flywheel ESSs (FESSs), and battery ESSs [4]. In general, in a power electronics-based ESS, there exists one dc–dc converter and one dc–ac converter. The SMES is an excellent ESS device which has very fast charging and discharging capabilities. However, because of heavy investment cost, continuous operational loss, and cooling system requirement, this technology could not survive well and have limited applications. The ECS also has very fast charging/discharging capabilities like SMES. It is not that expensive like SMES and found many applications including renewable energy and microgrid. The supercapacitor in ECS has virtually unlimited life cycle, which also made the technology attractive as it last long. The FESS had initially a big standby loss; by the development of superconducting-bearing technology, FESS has overcome that challenge and, therefore, it is becoming a popular choice in power system as a storage device. In New York, Beacon Power, USA has successful commissioned a 20 MW FESS in the mid-2015. In Australia, Marble Bar microgrid from ABB is the world's first high penetration solar PV diesel power stations where a 1,000 kW flywheel system is being used.

The most popular storage device nowadays is the battery-based ESS systems. The sodium sulfide (NaS) battery is a popular choice for large renewable-based power plants, e.g., in Awamori, Japan, a 30 MW class NaS unit was commissioned successfully. Deep-cycle flooded lead acid batteries and deep-cycle valve-regulated lead acid batteries are popular for solar power system. Recently, Nickel–metal hydride and lithium-ion batteries find applications in modern electric vehicles and renewable energy sector. In Ausnet grid energy storage trial project in Australia, 1 MW lithium-ion battery is being used. In King Island, Tasmania, Australia microgrid, lead acid battery is adopted.

1.2.5 Microgrid information and communication technology

There is a good similarity between modern information and communication technology (ICT) and power system workflow. In modern ICT, there are web, messaging, and application servers at the top end, followed by information transmission/communication links, i.e., submarine cable, optical fiber link, and satellite link. At the end, there exist various service providers. Similar to this, in power system, there exist three stages, namely, generation, transmission, and distribution. Microgrid is usually connected at the distribution end of power system when it is not isolated. In microgrid architecture, ICT plays a very important role and depending on the size of the microgrid and its distance from the control center, different subnetworks such as a HAN, neighborhood area network (NAN), or WAN are being applied.

These HAN, NAN, and WAN support various wired (power line communication, optical fiber, DSL, Ethernet, etc.) and wireless (WPAN, WLAN, WiMax, cellular, satellite, Z-wave, etc.) communication media. HAN is suitable for short distance, e.g., a few tens of meter. NAN is used for a distance in between 100 m and 10 km NAN. For a distance of more than 10 km, WAN is effective to use. Different types of protocols/standards are available for these communication technologies such as IEEE P1901.2, IEEE 1901, IEEE 802.3, IEEE 802.15.4, IEEE 802.16m, ADSL, VDSL, 3G, 4G, and Z-wave 400.

There are various smart sensors used in microgrid which measures voltage, current, temperature, and other parameters, and these information is being communicated with the central control center through an ICT infrastructure which ensures normal operation of the microgrid, even in faulty conditions. Voltage sensor, current sensor, speed sensor, temperature sensor, partial discharge sensor, magnetic flux sensor, vibration sensor, pressure sensor, level sensor, medium density sensor, blade pitch angle sensor, wind speed sensor, wind direction sensor, dust level sensor, axis angle sensor, irradiance sensor, lightning sensor, conductor motion sensor, insulation sensor, energy meter, power meter, power quality meter, phasor measurement unit (PMU), etc., are very common type sensors used in microgrid. Apart from that, many smart metering infrastructures are going to be used in microgrid platform which enables peer-to-peer energy trading.

1.2.6 Stability and protection issues of microgrid

Microgrid has some typical stability issues named voltage and frequency stability. Voltage and frequency stability of a microgrid could be a local or global phenomenon, based on how the microgrid architecture is designed. The variability of renewable sources may cause any kind of local fluctuation in voltage and frequency. Apart from that, in the case of grid-connected microgrid, the instability may be caused by any event happened in the other part of the network where microgrid is connected to.

Many of the voltage fluctuations and voltage stability issues in a microgrid can be addressed well by controlling the power inverters available with renewable sources and ESS devices. Different types of control techniques have been tested to improve the voltage stability of the microgrid, including local control, hierarchical control, droop-based controls. Frequency stability is a more challenging task, and microgrid energy management can help in frequency restoration, partly or fully, depending on the size of the microgrid and associated control of microgrid components. Synthetic inertia support from wind turbines and other power electronic converters can help to restore frequency dip up to some level. In more severe cases, other flexible ac transmission systems devices can be used.

Another option could be to apply wide-area control (WAC) which modifies the reference signals of the local microgrid controller, taking a feedback from all parts of the network using PMUs. Supervisory control and data acquisition (SCADA) is a popular method for WAC, but PMU offers very high sampling rate compared to SCADA. IEEE 137.118 protocol is a popular standard in WAC.

Microgrid consists of various renewable sources, distribution line, transformer, reactors, inverters, capacitors, and loads. These microgrid components and various equipment should be protected against faults and abnormal grid conditions. There are various protection systems available based on current, voltage, impedance, and frequency. The protection schemes should be highly reliable and require improved design in microgrid architecture. The protection scheme is composed of various protection equipment that provide required protective functions, and it heavily relies on sensing and measuring instruments, advanced communications architecture, and other hardware (current transformer, potential transformer, circuit breakers, relays, and contactors) and software. SCADA, PMUs, Global Position System are other components required, based on the level of protection scheme.

1.3 Microgrid future form

1.3.1 Addressing scalability and variability

Microgrids are generally set up by entrepreneurial effort, and a business case is often required before a decision is made to invest in a microgrid. Hence, a detailed optimization process under normal, emergency, and islanded mode is carried out prior to set up a microgrid. As more often than not a collection of DERs are involved in microgrid generation, an expected criterion is the full utilization of renewable power generation. Batteries, DERs, diesel generators are key components of grid-connected and isolated microgrids. The state of charge of batteries, battery cycling, battery operating temperature, loading on diesel generators, and available output of DERs are the key variables alongside with controllable/curtailable demand, real-time tariff, and network performance measures which must be taken into considerations for the optimum operation of a microgrid. Through a rigorous optimization study, the scalability issues of microgrid can be addressed properly. Such optimization process requires network sensors, enabling communication and hardware functionalities for clever decision-making. Along with many heuristics and metaheuristics optimization algorithms, many other advanced methods, including machine-learning algorithms, were found to be very effective in solving this scalability issue. The issue of addressing variability can be achieved basically through proper demand management, and it appears that the use of ESS is the ultimate solution.

1.3.2 Transformation of microgrid to virtual power plant

As massive scale renewable energy and microgrid penetrations cause many power system operational problems such as power quality, efficiency, stability, and reliability, the concept of VPP has been introduced to make this integration smooth without compromising the grid stability and reliability along with offering many other techno-economic benefits. A VPP connects different decentralized energy production and/or consumption units using advanced communication technology. It is used for aggregation of DERs, so that they can serve as a fully dispatchable unit managing information from numerous entities such as wind, solar PVs, energy storage, and market operators and utilities. A conceptual view of centralized VPP is shown in Figure 1.2 [5].

Important features of the VPP include, but limited to, the followings:

- It connects various types of DERs in the microgrid and smart grid including storage devices.
- An efficient way to optimize grid and energy trading.
- It optimizes volume of communication data.
- It allows multidimensional communication among all parties.
- Allows forecast-based intelligent energy trading.
- It reduces grid expansion and infrastructure-replacement costs.

Microgrid can work in islanded and grid-connected modes; still many of the microgrid will be transformed into a VPP which mainly is connected to the grid for

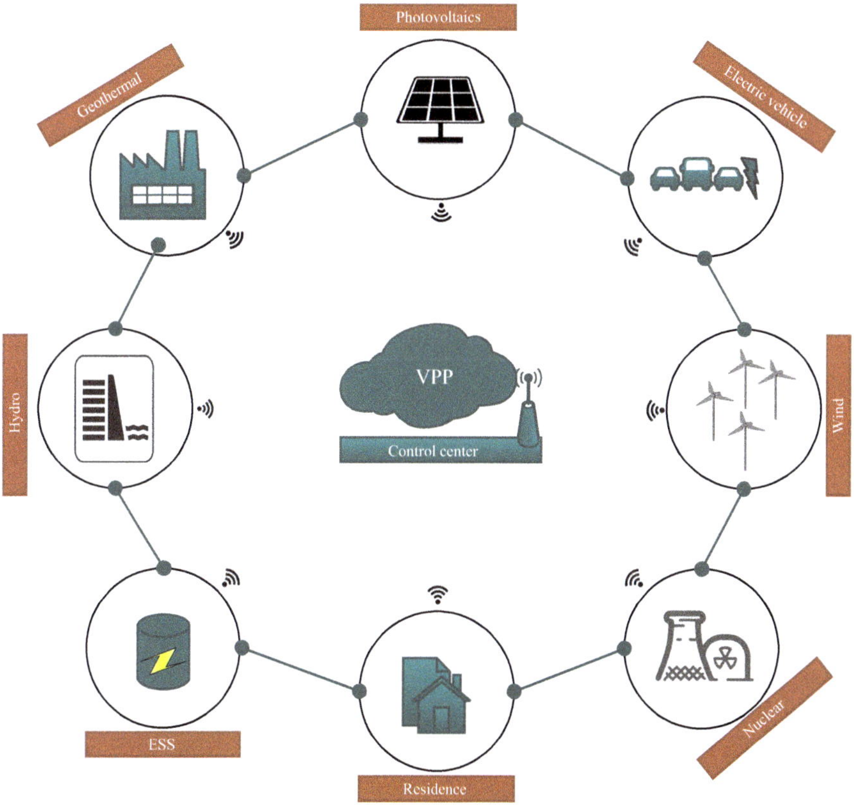

Figure 1.2 Conceptual view of centralized VPP

better flexibility. The presence of energy storage device is not mandatory for VPP as that purpose can be solved in many other ways within VPP architecture; however, without having ESS, it is difficult to imagine a microgrid. The success of VPP technology success extensively depends on smart metering and ICT; however, microgrid requires more development in inverter and switching technology. VPPs can combine a wide variety of resources in large geographic regions, but microgrids include a fixed set of resources within a fixed geographical region. Unlike microgrid which normally trades as a retail distributor, the VPP connects the wholesale market enabling forecasting-based energy trading.

1.3.3 Future trends of power electronics and its adaptation in microgrid

The future microgrids are expected to be almost 100% power electronics dominated. In terms of power electronics component technology, more power electronic converters will be based on wide bandgap power components. They are able to switch at least ten times faster than today. This will give less volume of the power converters (equipment) but also faster control dynamics in the control loops mentioned above. Such feature will give more precise control and also better system control performance as the serious interaction between the power converters in the microgrid is expected to be at a higher frequency and better damped in the grid infrastructure—the control delay is one of the root causes of microgrids instability, and the control delay will be reduced by a factor of 10 when the switching/sampling frequency is increased by a factor of 10.

A 100% power-electronic-based grid will also demand very robust hardware, which is able to cope with transients in the system, and there is a need for robust hardware protection strategies, especially overvoltage, as this is one of the more difficult hardware protection.

If the microgrid is 100% power electronics based, it will mean that each power converter can be treated as a sensor too—and many sensors will be available in the microgrid—which opens up for more sophisticated control strategies (e.g., efficiency optimization or highest possible reliability of the system) as well as enabling advanced system monitoring and give the possibility to react in due time—if, e.g., a failure can be foreseen. In such systems, the communication should of course be reliable. If that is not working—then a fallback operation mode should exist so the electricity is always supplied to the customer.

The autonomous control of microgrids can be done by droop control in order to balance power sources and load. Here voltage and frequency can vary. In the future, the source operation can be voltage sourced (grid forming) for some of the units, while others can be a current source connected where active and reactive power is used a controller reference. Whether such system is more robust compared to the control systems mentioned above—this is an issue to be addressed in the future. The power electronics technology itself is able to operate both modes.

1.3.4 Future trends of energy storage technology

The use of ESS will gradually increase more. The global utility-scale ESS capacity has reached to 158.9 GW, as per the *Renewables 2018 – Global Status report* by ren21, among which 153 GW is from pumped storage, 1.3 GW from electromechanical type, 2.3 GW from electrochemical, and 2.3 GW from thermal. This indicates that apart from the pumped hydro which is mainly for bigger storage, the battery market has a very bright future. A carbon-foam-based Absorbent Glass Mat (AGM) battery by Firefly is the latest edition in electrochemical-type energy storage technology and catching the eyes of researchers greatly, as it offers an inherently supercapacitor characteristic apart from battery characteristics. More cost-effective and efficient battery technology will be coming up soon. It is expected that microgrid, in general, will be dominated by electrochemical and electromechanical-type energy storage technology.

1.3.5 Future form of microgrid communication

ICT is changing its color very quickly, and accordingly the communication structure of microgrid will be changing too. As microgrid is rapidly transforming to

VPP, the communication technology adaptation is important. WAC is very much essential for VPP, and presently, IEEE 137.118 is an acceptable protocol considered there. However, it is noted that there are various controllers and intelligent electronic devices (IEDs) present in VPP architecture which are always from various vendors. Therefore, it is important that all controllers and IEDs are able to talk to each other for better control, operation, and flexibility. This issue is known as "interoperability" and should be handled appropriately and carefully; otherwise, the concept of VPP will never see the sunlight. Considering this, IEC61850 is also found as a promising technology, in recent years, which can handle interoperability issues of microgrid, but it requires much more research and development.

1.4 What is in this book?

The key issues addressed in each chapter are highlighted briefly in the following sections:

Chapter 2 presents an overview of different control techniques applied to microgrid for power and energy management, voltage and frequency restoration purposes. The chapter includes many interesting topics including hierarchical control, demand-side management, and multi-agent-based control. Especially, the microgrid control hierarchy is presented in detail which includes primary, secondary, and tertiary controls. Various simulation-based case studies, including results, are demonstrated too for droop-based control, demand-side primary frequency control, and centralized secondary control.

In Chapter 3, a transactive control architecture requirement analysis is presented which can tie up the market transactions at the higher levels with the inter area. An effective functioning of the electric power system necessitates the requirement engineering in transactive energy management; hence, it should be properly investigated. Therefore, the authors have defined a goal-oriented requirements engineering approach, architecting an economical transactive energy-management system, where the fundamental goal is achieving desired outcomes for decision makers or market operators. The details of the requirement analysis, modeling has also been incorporated.

Chapter 4 presents an overview of VPP giving an insight into its structure, control methods, way of interconnecting DERs. One of the problems of DERs is that usually the operational level of individual components is done in a scattered way without maintaining a bidirectional communication with control centers, which eventually affects the power-system reliability. The advantage of VPP over microgrid is that it is always grid connected, does not operate as an isolated island, and, therefore, keeps communication with other parts of the power systems. It is explained here how VPP can strengthen and diversify generation structure by adopting decentralized dispatching and distribution energy networks. The classifications of VPP are discussed from different angles, (i) area wise and (ii) ancillary service wise. The three-layer model of VPP, communication technology used in VPP, cybersecurity of VPP, SCADA, energy-management strategy, etc., are another interesting topic of this chapter. Finally, an outline is defined to show the difference between a microgrid and VPP.

Chapter 5 demonstrates the operational feasibility of clustered or networked microgrids in electrical distribution systems and shows its effectiveness in normal and emergencies. It is shown that authors proposed scheme is able to regulate the power transaction among the coupled microgrids when power generations are from various dispatchable units. The modeling of clustered microgrid is presented in detail. The droop-regulated control scheme, in the case of coupled microgrid, is discussed. The way of formulating the optimization problem to determine the most economical solution for overcoming, under/over voltage/frequency issues caused by overloading or over-generation also discusses in depth.

In Chapter 6, the architecture and control aspects of nanogrid are presented. Theoretically, a nanogrid is like an automobile or aircraft, which has its own isolated power network using batteries to support internal lightings, electric apparatus, and auxiliaries. In remote areas and mountainsides, many DERs and distributed energy storage units are available which form self-controlled islanded microgrid, known as nanogrid. In the case of several nanogrids, the power and energy management issues become complex. This chapter demonstrates various ac and dc-based design of modern nanogrid as well as some technological challenges when the size of the nanogrid becomes bigger.

Optimal sizing problem of all the microgrid components, including renewable sources, energy storage, dynamic loading, are the main attractions of Chapter 7. Through optimization study, the economic and reliability benefits to electricity consumers and microgrid owners can be achieved, i.e., the scalability issue of microgrid can be addressed properly. However, addressing scalability issues of microgrid is a challenging task because of the uncertainties of microgrid generation sources, like wind and PV systems. Authors of Chapter 7 have given great efforts in formulating the optimization problems of microgrid components in the presence of ESS including some case studies.

Chapter 8 presents first the applications of ESSs in solving various problems of microgrid, e.g., smoothing the power of renewable energy sources, peak load mitigation, voltage control, frequency regulation, and reliability enhancement. This chapter also briefly discusses various ESSs available to date and used widely in microgrid. Then the sizing details of the microgrid ESS are presented in detail, including a case study where the optimization problem formulation part is emphasized as well.

In Chapter 9, the microgrid communication technologies, standards, and protocols are presented elaborately. Various wired and wireless communications technologies, e.g., power line communication, cellular network, IP networks, ZigBee, Wi-Fi, and WiMAX, are discussed. The microgrid communication network is discussed in the form of HAN, NAN, and WAN. One of the interesting part of this chapter is the IEC61850 protocol, which has found application in distributed generations. The architecture and control issues of IEC61850 in light of microgrid are discussed in detail.

Voltage fluctuation is a very common phenomenon in microgrid which has renewable power sources. The main focus of Chapter 10 is the voltage stability and reactive power sharing in inverter populated microgrids including droop control.

The authors also have shown some control schemes to maintain voltages in microgrid busses in normal and faulty conditions. Therefore, voltage stability enhancement is another feature of this chapter.

When the penetration of renewable energy increases, the power system inertia increases significantly which raises the frequency stability issue. Chapter 11 focuses on frequency stability of a grid-connected microgrid including synthetic inertial support from power sources equipped with power electronic converters. The chapter also includes discussions on the appropriate control loop used in frequency restoration. Some good case studies are conducted as well, which is helpful for students and researchers.

Because of the renewable energy penetration into the grid, the existing practices for microgrid protection system design, including hardware and controls, are found as insufficient. Chapter 12 presents a thorough discussion on both ac and dc microgrids. Fault characterization of microgrid is discussed in detail along with protective equipment. Protective relaying and under voltage/frequency protection is discussed as well. Other interesting studies presented in this chapter are dc microgrid protection, earthing protection, and protection challenges overview.

Chapter 13 describes intentional and automatic islanding of microgrid including a local black start strategy and its reconnection scheme to the upstream distribution network. In particular, black start is presented in light of medium-voltage and low-voltage microgrids including multi-microgrid (MMG). The microgrid synchronization issues, the black start procedure, sequences of actions, etc., are presented in detail. Two different blackout case studies are included in this chapter for microgrid and MMG.

Efficient optimization and control of operation is very much essential for modern microgrid. Chapter 14 presents the technical, economical, and environmental potential of microgrids in light of control and optimization techniques. Various feasibility studies, including uncertainty of renewable sources available in a microgrid, are presented giving focus on microgrid control and operational aspects. The chapter is supported with many simulation and experimental results to demonstrate the potential economic and environmental benefits.

Chapter 15 is the last chapter in this book and serves as the reference for many other chapters as it discusses the power-electronic interfacing issues of a modern microgrid. Power converters are mandatory components of any microgrid which is being used by renewable power sources and energy-storage devices. This chapter includes discussions on different microgrid topologies, power converter structures, and control and operation principles of power electronic converters. Two case studies are also included to demonstrate the importance of power electronics in modern microgrid. This chapter also discusses the technical challenges encountering modern power-electronic converters used in microgrid.

1.5 Conclusions

Microgrid is now almost in a matured stage, but many new technologies are going to be interlinked with it in future. Variability, scalability, and stability are the main challenges of the present and future microgrids. The development of reliable and smart power inverters, advanced control strategies, and better and cheaper energy storage technologies can handle both issues—the variability and stability. On the other hand, scalability issues require handling the optimization issues carefully. On top of that, the ICT will be playing a vital role on future microgrid and VPP control, IEC61850 and other advanced protocols require more attention which can handle interoperability issues. Policies and regulations are to be microgrid friendly, so any steps taken at the government level in wide-spreading the distributed control is critical and should be carefully observed.

References

[1] Dvorak Paul, "Microgrid Market to Reach $19 Billion by 2024", Windpower Engineering and Development, [Online]. Available: https://www.windpowerengineering.com/business-news-projects/uncategorized/microgrid-market-to-reach-19-billion-by-2024/, 2018.

[2] Henry Shu-hung Chung, Huai Wang, Frede Blaabjerg, and Michael Pecht, Reliability of Power Electronic Converter Systems (Energy Engineering), The Institutions of Engineering and Technology, London, UK, 2016.

[3] Renewables 2018 – Global Status Report, [Online]. Available: http://www.ren21.net/gsr-2018/.

[4] S. M. Muyeen, J. Tamura, and T. Murata, Stability Augmentation of a Grid-Connected Wind Farm, Springer-Verlag, London, October 2008. ISBN 978-1-84800-315-6.

[5] Levent Yavuz, Ahmet Önen, S. M. Muyeen, and Innocent Kamwa, "Transformation of Microgrid to Virtual Power Plant – A Comprehensive Review," Accepted for Publications in IET Generation, Transmission, and Distribution, December 2018, 10.1049/iet-gtd.2018.5649.

Chapter 2

Microgrid control overview

S. Ali Pourmousavi Kani[1], Farhad Shahnia[2], M Imran Azim[1], Md Asaduzzaman Shoeb[2], and GM Shafiullah[2]

A microgrid (MG) is always prone to the uncertainties of its demand variation and generation of its non-dispatchable renewable sources, particularly when operating in the islanded mode. Such events can push voltage and/or frequency of the MG beyond their desired range of operation. This chapter reviews the control and management techniques to retain the voltage and frequency of such MGs within a predefined safe zone. Suitable real time, corrective, and preventive controllers are discussed on the generation and demand side, which aim to satisfy various objectives at different time instances. First, the necessity of such controllers and mechanisms is explained in both grid-tied and islanded modes and during the transition between these modes. Then, islanding detection and its impact on MG management are briefly discussed. Afterwards, the MG's control architecture is outlined, and the existing approaches in the literature are described briefly. Finally, three case studies on different aspects of MG control are reported to show the applicability and criticality of such services for MG operation. The emphasis of the case studies is on the islanded MG operation because frequency and voltage issues are more pronounced for those types of MGs. In particular, a new generalised droop-based controller is explained in Section 2.4.1 as an example of advanced power-sharing strategies for voltage and frequency regulation with the plug-and-play feature. In Section 2.4.2, the primary frequency control problem is tackled from the demand control perspective, where demand response (DR) resources are altered to provide frequency and voltage regulation within a short period of time. Finally, a corrective and preventive controller is outlined and explained in Section 2.4.3. The corrective controller takes action immediately after the occurrence of an event that violates the voltage or frequency by defining the least cost solution among available options. In the preventive controller, generation and load demand forecast are used to predict unexpected events in very short horizons that can lead to voltage/frequency violations and take suitable actions beforehand.

[1]School of Information Technology and Electrical Engineering, the University of Queensland, St Lucia, Australia
[2]Discipline of Engineering and Energy, Murdoch University, Perth, Australia

2.1 Introduction

A MG refers to a small-scale electricity grid, located at the medium- or low-voltage networks, where a cluster of loads are supplied locally by a few conventional and/or renewable-based distributed generators (DGs), and/or different types of energy-storage systems (such as batteries), and/or DR resources. An MG can operate in grid-connected (in the case of grid availability) or islanded mode (when a grid does not exist or is temporarily unavailable) [1]. Either way, an MG should have the capability to operate the entire fleet of loads, generators, and storage devices within its territory by maintaining the supply and demand balance in real time at the least operational cost [2–4]. In recent years, standalone MGs are considered as a promising solution to supply electricity to the consumers at the edge of the grid and remote areas with limited or no access to the main grid [5–7]. Also, grid-connected MGs are shown to improve the operational cost and to provide self-healing, resiliency, and reliability of the power-system operation [8,9]. In an islanded MG, maintaining a balance between the generation and demand in real time is the responsibility of the MG operator. In this case, there is no external and perhaps larger grid to absorb all the instantaneous variations in the generation and demand during real-time operation. Since the security of the power system operation depends largely on the balance between the generation and demand in spite of continuous variations and unexpected component outages, it is regarded as a fundamental task in the islanded MG operation. Moreover, islanded MGs should be able to safely ride through various credible faults and contingencies that occur in the MG's territory in real time [10]. In a grid-tied MG, where the system is connected to a larger electricity network typically through a single point of common coupling (PCC), maintaining a balance between the generation and demand in real time is not an operation objective, as it is taken care of by the larger grid.

Besides the technical challenges, cost-effective operation of an MG is a challenging task for the MG operator in both grid-tied and islanded modes. MG components, including conventional and renewable energy resources, storage devices, and DR tools, are relatively expensive [11]. In most cases, an MG is not yet a profitable solution if it is not subsidised by the government [11]. Therefore, it is of paramount importance to optimally operate MGs to minimise their overall operational cost or maximise their return on investment. Optimal operation algorithms for the generation and demand resources can achieve this goal in the presence of uncertainty through MG energy management systems (EMSs). They are designed to maximise the economic benefit of the entire system, minimise greenhouse gas emissions, and to support the upper grid when the MG is operating in grid-tied mode. Since optimal EMS involves solving an optimisation problem by considering the future operation of the system, they are updated in the order of a couple of minutes or so, and they can only consider a limited number of steps ahead in the calculation. The algorithms designed for EMSs could be complicated when the uncertainty of the generation and demand is taken into account. Different multi-level and stochastic optimisation algorithms are developed to address the operational uncertainties, which are reviewed in [11]. Such algorithms, in turn, increase the computation cost and time, which necessitates large intervals and shorter horizon in the optimisation problem. In islanded MGs, cost minimisation is the main goal and can be realised by optimally operating various generation resources, storage devices, and DR resources. In the grid-tied MG, however, it is possible to sell energy to the grid or provide ancillary services to the upper network to gain profit. In this case, the objective function (OF) involves cost and profit terms in the optimisation problem, which has to be solved by the EMS. The existing EMS algorithms will be reviewed in more detail in Section 2.3.2 as secondary control methods within the MG control hierarchy.

EMSs cannot guarantee the stability of the MG operation because of continuous and rapid changes in generation, demand, and system topology within an energy management interval. To address the system stability, power control algorithms (PCAs) are developed, where the general direction of the EMS is followed as long as they do not compromise the stability of the entire system. Typically, PCAs are intentionally designed to be simple, requiring low communication and computational resources, and are more local rather than central [12]. While the cost-effective operation of the MG is not guaranteed (and might even be neglected) by the PCAs, the integrity of the MG operation and frequency/voltage stability are preferred and provided to the maximum possible extent. PCAs will be further explored in Section 2.3.1, as primary control algorithms within the MG control hierarchy.

In some cases, an MG is expected to be able to switch from one operation mode to another based on the upper grid's conditions. It further necessitates complicated and multilayer control algorithms as well as islanding detection techniques to fulfil transition smoothly. In addition, an MG should have the capability of operating in both grid-tied and islanded modes. Section 2.2.1 will present more detail about MG's dual-mode operation and islanding detection. In conclusion, it is obvious that MG operation can be very complex based on the requirements and objectives by involving several tiers of control and management algorithms in different timeframes.

2.2 Uncertainty of the generation and demand

Typically, renewable-based DGs, such as photovoltaic (PV) and wind, are preferred for MGs, because of being more sustainable generation resources. Although realising a 100 per cent penetration of renewable resources is the ultimate goal for MG developers, it seems to be inevitable not to use small-scale diesel or gas-fired generators if installation and operation costs are concerned with current technological advancement [12,13]. Renewable-based DGs rely primarily on varying ambient conditions and, therefore, they may not be controllable (without enough storage devices for buffering) [14,15] or predictable accurately (in various time horizons) [16,17]. On the demand side, the unpredictable behaviour of consumers creates substantial variations in the realised demand, which requires even more balancing services. In practice, the uncertainties of load demand and the unpredictable generation of variable renewable resources along with the high cost of energy storage devices in the face of tight frequency and voltage standards are among the biggest obstacles for the application of 100 per cent renewable-powered islanded MGs. These uncertainties should be addressed properly in the MG design, planning, and real-time operation.

The uncontrolled DGs also referred to as non-dispatchable DGs (NDDGs), typically operate in a grid-following mode providing maximum available power or nominal power. Controllable DGs (such as diesel/gas-driven generators or those renewable-based resources coupled with an appropriate size of power smoothing battery) are referred to as dispatchable DGs (DDGs). They usually operate in the grid-forming mode, when the MG is islanded, i.e. they are responsible for controlling the MGs' voltage/frequency while supplying its electrical demand [18]. The unpredictable nature of the NDDGs production and load demand is a big hurdle in the MG operation, where prediction algorithms are needed for estimation. In one hand, day-ahead MG operation planning is preferred as the load, and some NDDGs (such as PV) have a daily cycle [11]. On the other hand, the forecast error is typically high in day-ahead prediction [19]. Therefore, MG operation optimisation in various time horizons seems to be necessary in order to account for uncertainty in production and load demand. Due to the same issue, different MG operation algorithms are needed at different time scales to cope with the quick variations in the output of NDDGs. Generally, algorithms with longer time horizons have a larger updating interval, and their primary objective is to optimise the MG's operation based on the cost, referred to as EMS. In contrary, algorithms with shorter time horizons take care of the residual load and generation from the predicted values in order to maintain the generation and demand balance in real time, called PCA. Therefore, they have shorter intervals and update more frequently. PCAs are very important for islanded MG operation since there is no upper grid connection to compensate for the deviations in the generation and demand from the expected values of the planning stage in the EMS. Overall, proper MG operation necessitates complicated and intelligent algorithms, with various resolutions and horizons, to maximise the economic and technical benefits. These algorithms should complement each other for MG's effective and comprehensive operation.

2.2.1 Application of grid-tied MGs

While the benefits of islanded MGs are largely known and well justified for remote areas in many cases, the application of grid-tied MGs (while being able to operate in islanded mode) requires further explanation. From a conceptual standpoint, an MG offers promising opportunities for larger integration of small-scale DGs by placing the generators close to the end users. They reduce the losses in the transmission and distribution systems, which improves the overall efficiency of the power system operation and results in lower prices for the end users [20].

Achieving larger penetration of MGs also can help to defer investment in building new and/or upgrading transmission and distribution systems to some extent, which consequently reduces the incurred cost of the grid operation [21]. This way, utilities, as well as the end users, benefit from MG application directly and indirectly. Moreover, the grid-tied MGs that are able to operate in islanded mode can improve the resilience of the grid operation during disasters and unexpected outages [21]. When fast intelligent islanding detection algorithms are employed, an MG (capable of operating in both islanded and grid-tied modes) can disconnect from the main grid in a reasonable amount of time and operate on its own for some time depending on the available generation, storage, and DR resources within its territory. This is an incredible opportunity for critical loads such as police stations and hospitals during natural and man-caused disasters to operate safely. Therefore, the concept of MG not only brings proven benefits for the remote communities' electrification but can also improve the technical, economic, and resilience aspects of the power system operation as a whole.

A grid-tied MG should be able to work in both islanded and grid-tied modes seamlessly to improve the power system's resilience in the face of upper grid outages, due to natural disasters, and physical and cybersecurity attacks. To do so, an MG should be able to detect the major network issues in a reasonable amount of time, and the EMS and PCA of the MG should be capable of working in both modes. Islanding detection is the key to switch between the two modes successfully, as it signals to the EMS to

- connect to or disconnect from the upper grid, to avoid fault propagation to the MG, and choose the right strategies and controls to operate in the respective mode and
- synchronise MG's voltage and frequency with the upper grid when reconnecting to the upper grid or forming the voltage and frequency for the islanded operation.

In the last decade, researchers have developed various islanding-detection algorithms by measuring real-time parameters of the main grid at the MG's PCC [22]. These algorithms can be classified broadly as local and remote [22]. Local methods are based on local measurement of some parameters at the PCC, which can be further divided into passive and active detection techniques [22]. As the names suggest, passive methods are based on measuring and real-time evaluation of some parameters, such as voltage, current, frequency, and phase, while active methods inject an intentional disturbance to detect whether it affects the voltage, frequency, power, or impedance parameters of the system, seen from the PCC's viewpoint. This is another level of complexity for MG operation, which justifies the necessity of a hierarchy of sophisticated and intelligent control algorithms at a different level of time and space for MG's control efficient.

The rest of this chapter is organised as follows: In Section 2.3, different EMSs and PCAs are explained categorically and their differences are highlighted. In particular, primary control (i.e. PCAs) is explained in Section 2.3.1; secondary (i.e. EMS) and tertiary control are described in Sections 2.3.2 and 2.3.3. Simulation studies are presented in Section 2.4, where a droop-based power-sharing control, a demand-side primary frequency control, and a centralised secondary controller integrated with a primary control functionality are introduced in Sections 2.4.1, 2.4.2, and 2.4.3, respectively. Simulation results are discussed in detail in these sections.

2.3 MG control hierarchy

As extensively explained in the Introduction, there are prominent challenges for MGs operation that should be resolved. Some of the known challenges are stability issues, low inertia, the uncertainty of DG production, power balance, and economic dispatch under uncertainty, as well as the transition between grid-connected and islanded modes of operation. In order to deal with these challenges, advanced intelligent algorithms in different time and space scales (i.e. the hierarchy of controls) are required. Based on the operational timeframe and communication requirements, MG control hierarchy can be divided into three levels: primary, secondary, and tertiary [1], as shown in Figure 2.1. These control levels are explained in detail in the rest of this section.

2.3.1 Primary control

Primary control is the lowest level of MG control hierarchy and has the highest time resolution. It operates without communication by entirely depending on the local information, as it should respond to the sudden changes without being affected by the communication delay. Thus, it is also referred to as the local or internal control. Fast response is the key feature of the primary control [12] as the purpose of

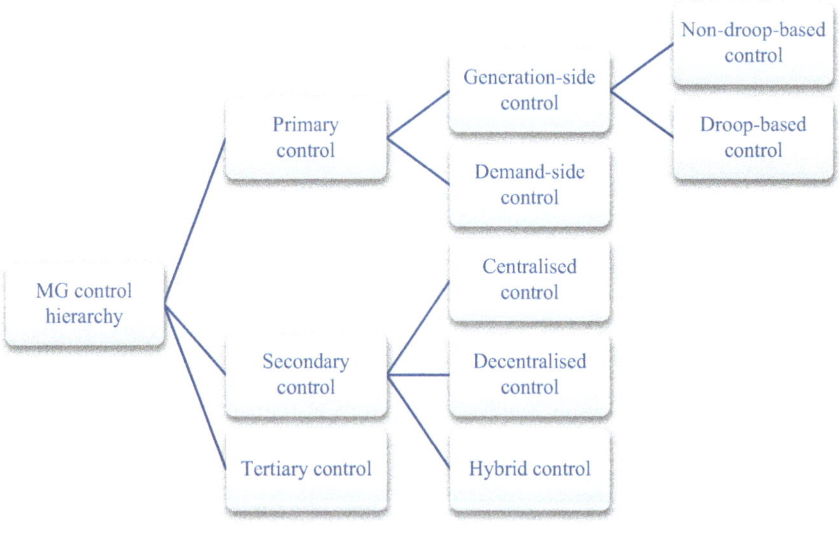

Figure 2.1 Flowchart of the MG control hierarchy

primary control is to maintain the voltage and frequency stability, power sharing among DGs in response to the rapid changes within the network, and plug-and-play capability of DGs [2,13]. As shown in Figure 2.1, primary control can be classified as generation-side and demand-side controls. The former ensures power supply in the MG based on the demand, where the generation is adjusted dynamically to cope with the changes in the load demand (known as the load-following strategy). Various generation-side primary control schemes are summarised in [3], from which two categories of control methods can be identified: non-droop-based control and droop-based control. Non-droop-based control methods, such as concentrated and master/slave control, are superior at regulating the voltage and frequency and sharing power in the steady-state condition. However, they are known to be unreliable during transient periods [4]. As a shortcoming, a dominant DG is required in the master/slave control.

To address the issues of non-droop algorithms, alternative primary controllers have been proposed by the researchers. Droop-based control is a well-established generation-side control method for islanded MGs [23]. This method is developed based on the frequency droop control in the conventional power system. There are many variations of the droop controllers for the MG frequency and voltage regulation. In the forward droop control (FDC), the active power depends on the frequency and reactive power varies proportionally with the voltage [24]. However, changing frequency from its nominal value during active power/frequency droop could result in frequency imbalance. Therefore, an additional controller needs to be introduced to readjust the frequency value. This problem has been solved in [25] by representing frequency droop as the rate-of-change of angle droop. In [26], a compensating approach has been included in the droop algorithm to cope with the slight changes in the renewable generation. A derivative droop control scheme has been proposed in [27,28] to control the start-up transients. Other approaches, such as adaptive, intelligent, and cost-based, have been added to the conventional FDC in [29] to improve the performance of the droop controller. It is shown in [30] that the traditional FDC is only effective for medium-voltage MGs and performs poorly in the low-voltage grids, where distribution lines are resistive in nature. To address this issue, [14] used virtual output inductance-based droop technique to diminish the impact of high resistance on the performance of the controller. As a side effect, however, this technique makes MGs less flexible by not supporting plug-and-play operation [15]. A robust servo control mechanism has been proposed in [16] to deal with this issue. However, the proposed algorithm does not work in multi-source MGs. To address all these limitations, the reverse droop control (RDC) scheme has been introduced in [30], in which the active power varies with the voltage and the reactive power is controlled by the frequency. Unfortunately, the RDC scheme cannot be implemented in the medium-voltage MGs. As a result, two different algorithms are required depending on the nature of the distribution lines. A generalised droop-based control algorithm is presented in [17,18], which is independent of the nature of the distribution lines, i.e. inductive or resistive.

Proportional power sharing is an important aspect of the droop-controlled MGs [31], which enables DGs to supply power based on their power ratings [32].

However, the generalised droop algorithm is unable to share power proportionally among DGs. Therefore, a linearized active and reactive power sharing approach is developed in [17]. Since linearized modelling of MGs considers the distribution line parameters, proportional power sharing cannot be guaranteed after the change in the value of the load or the distribution line parameters [33]. In order to ensure proportional power sharing, integral argument strategy is reported in [34], which has been verified for islanded MGs in [35]. Another crucial factor for droop-controlled MGs is the selection of appropriate droop parameters. Authors in [36] described a probabilistic approach to select the droop gains so that the operational cost of the MG can be minimised while a dynamic phasor analysis-based approach has been used in [37] to determine the droop coefficients considering the stability limits of the MG, as they are affected by the assumed droop coefficients for the DGs [38–43]. Despite all the advancement on this topic, there are still prominent challenges that need to be addressed. As such, Section 2.4.1 will review another droop-based PCA along with simulation results that evaluate its performance.

Although droop-based control satisfies the proportional power sharing and stability criteria in islanded MGs, flexible resources, such as energy storage devices, are required in real-world applications to regulate the frequency and voltage due to the highly unpredictable nature of NDDGs and load demand. As a result, additional costs incur for MG implementation to provide enough flexibility resources. To address the issue to some extent, demand-side control has been evolved in recent years for primary control. In principle, a demand-side controller tries to adjust responsive loads based on the available renewable generation and non-responsive loads. This is a paradigm shift in the frequency and voltage regulation, where the generation-following strategy is preferred to the well-established load-following strategy. Although it may seem likely that the individual demand resources may have a high failure rate when called to provide services on a short notice, the aggregation of many small resources makes it more probable that the desired response will be achieved. This makes the DR resources more reliable than conventional generation, where the failure of one generator to start can cause the loss of considerable spinning reserve capacity [44,45].

Demand-side control has been used to regulate frequency in [46–49] and voltage in [50]. These frequency and voltage-regulation algorithms are mostly for transmission and distribution systems operation (e.g. grid-tied MGs), but not necessarily for islanded MGs. However, the same principles are applicable to the islanded MGs, or grid-tied MGs that provide ancillary services of the same kind to the upper grid. In [47], a two-step algorithm is proposed for frequency and voltage control within an islanded MG. The algorithm along with the results of several simulation studies will be explained in Section 2.4.2.

2.3.2 Secondary control

Both generation-side and demand-side primary control schemes have some prominent limitations, such as steady-state errors in frequency and voltage, lack of decision-making process based on cost and uncertainty, and interoperability of a cluster of MGs or the upper grid [4]. Therefore, another level of control is necessary for the ultimate stability of MGs by returning the frequency to the nominal value, and the cost-effective operation by using the cheapest resources. These fall under the secondary control in the MG's control hierarchy. It is also known as the supervisory control or EMS of MGs, in which a variety of functionalities are fulfilled, such as monitoring, analysing, and forecasting of power generation of NDDGs, load consumption, energy and ancillary market prices for grid-tied MGs, and meteorological factors. These functionalities help the EMS in optimising MG operation while satisfying the technical constraints of the network and equipment. Because of the EMS operation, the voltage and frequency fluctuations, caused by the local control, are restored [1]. The EMS is also responsible for power quality control, power exchange among DGs, harmonic compensation, and uncertainty reduction while guaranteeing cost-effective operation of the MG [11,51,52]. The secondary control is intentionally designed to be slower than the primary control to allocate adequate time for the complex calculations while avoiding unnecessary actions on the equipment and interference with the lower level controls.

The secondary control's approaches can be broadly classified as decentral, central, and hierarchical [1]. A long list of papers is compiled in [11] for each category. In the decentralised approach, different generators and loads are given full autonomy to solve the energy management problem locally. Because of dividing the global optimisation problem into several sub-problems to be solved locally, computational requirements can be lessened for the individual problems. More importantly, new DGs can be incorporated easily without changing the secondary controller settings, and mutual interactions among them can be considered with minimum data sharing. Along with flexibility, it also offers resilient and reliable control of MG, by avoiding a single point of failure [53]. Another advantage of decentralised control is that private and confidential information is protected within the central unit, enhancing MG security [51]. Consensus-based decentralised MG voltage and frequency control is reported in [54,55]. However, the performance of decentralised methods shows a slow convergence rate, which consequently results in system instability. Multi-agent systems (MAS) are widely used in decentralised EMS. Other distributed control paradigms (such as gossiping and dual decomposition) are employed for MGs EMS in [56,57]. As a drawback, the implementation of decentralised methods in practical applications is difficult, especially in islanded MGs, where the PCA and EMS have to deal with the continuous changes in the generation and demand. In addition, global optimality of the solutions in decentral EMS is not achievable most of the time, since the optimisation sub-problems are solved by different local controllers with limited communication with each other.

In the centralised approaches, all DGs and responsive loads within the MG are operated by a central control entity through one- or two-way communication links. In this framework, global optimal solutions, complex intelligent supervision, and holistic system control are possible conveniently. The centralised secondary control strategies are recommended for islanded MGs in [1], where balancing demand and supply is supposed to be carried out by the supervisory control unit at all times. A central rule-based power management algorithm is proposed in [58], where the instantaneous operation of the MG was concerned. This idea is extended to an intelligent central power management system in [59], which is further expanded in [60] to also include DR and battery energy storage. The key difference between power- and energy management concept is that in the former, only one-time step ahead is concerned during the decision-making process. In the EMS, however, the operation of the system in the next time step is affected by the different scenarios, which might happen in various future time steps. It helps to make better decisions by considering the future generation and consumption patterns. In [61], the authors proposed a theoretical framework for real-time energy management of MGs and introduced a comprehensive list of technical and economic constraints and objectives relevant to the MG operation. A central analytical approach has been presented in [62] to evaluate the set points for droop controlled MGs and to keep the voltage and frequency within acceptable limits. Optimum power-sharing problem of MGs has been solved in a centralised fashion in [63]. Other criteria, including reserve capacity, reliability, renewable contribution, voltage and frequency deviation, losses, generation forecasting, and load prediction, have been added as different factors of the decision-making process during the optimal operation of MGs in [64,65]. In Section 2.4.3, a centralised secondary controller is introduced, where the corrective (as a primary control mechanism) and preventive (as a secondary control algorithm) controllers with different objectives operate the MG. A comprehensive set of simulation studies are carried out for a sample MG, and the system operation under different conditions is evaluated thoroughly.

While the decentralised EMSs provide scalability, reliability, and autonomy, central approaches deliver global optimal solutions, observability of the physical network, and interoperability with the upper grid or a cluster of adjacent MGs. In hierarchical EMSs, as the third category of techniques, the goal is to achieve the benefits of the central and decentral EMSs in a single framework. As another advantage, the hierarchical EMSs provide a framework for multi-timescale optimisation of multi-carrier energy systems (i.e. electricity, gas, and heat). Typically in a hierarchical topology, the higher control layer guarantees the least cost operation and interoperability, while the lower layer fulfils other objectives by following the commands of the higher level. It also supports plug-and-play operation, as well as maintaining communications to the minimum possible. A list of research studies on the hierarchical EMSs is reviewed in [11], where most of the studies developed algorithms for the grid-tied MGs or a cluster of MGs within a community. A multi-objective hierarchical MAS-based EMS is proposed in [66] for a grid-tied MG. The goal was to minimise its operational and emission costs (at the highest level), voltage regulation (at the middle level), and line losses (through f/V and PQ-based control strategies of DGs) at the local level. Optimal EMS of a grid-tied MG together with the vehicle-to-grid operation is offered in [67] through a two-level structure. The economic operation is considered in the lower level, while the higher level ensures the stability of the MG. In [68], a grid-tied MG operation cost and energy exchange fluctuations are minimised using a two-level hierarchical

EMS. Energy fluctuations are minimised in the lower level controller, and day-ahead economic dispatch problem is solved in the higher stage to minimise the cost. A hierarchal EMS for a standalone MG is introduced in [69], which aims at improving economic performance (at the higher level) and reliability (at the lower level) of the MG. Another two-level EMS is proposed in [70] for a cluster of MGs that cooperate with each other. The lower level focuses on an individual MG operation, while the upper level is responsible for managing the MGs and MG community-level devices. In [71], a hierarchical EMS is developed for the community of MGs in cooperative and competitive environments. The objective of the former multi-MG system is to reduce the operation cost of the entire network, while in the latter, the objective of each MG is to maximise its own profit. As it can be seen from the research studies above, most of the existing hierarchical EMS strategies are applicable for grid-tied MGs or interoperability of a community of MGs. More research is needed to advance hierarchical EMSs for islanded MGs to solve the issues of energy and power management in different timescales so that the cost-effectiveness and reliability/security can be guaranteed in different scenarios and credible contingencies.

2.3.3 Tertiary control

Apart from the primary and secondary controls, another level of control is required for MGs (particularly grid-connected ones or multi-MG systems) to coordinate between the MG operation and the upper power grid or among MGs in an MG cluster. This level of control is called tertiary control. A few algorithms are developed in [72,73] to eliminate the voltage and frequency deviations during the coupling of MG into the upper grid, under the MG tertiary control. In general, the researchers overlooked the tertiary control for managing challenges within MGs. We expect growth in research studies in the coming years to bridge the gap in knowledge and to leverage the potential of the tertiary control layer for enhanced MG operation.

2.4 Case studies

In this section, three case studies are presented for different parts of the MG control hierarchy. In the first section, a novel droop-based control approach based on the generalised droop model is explained in detail. In Section 2.4.2, a primary control algorithm will be explored based on DR for frequency and voltage regulation within an islanded MG. Finally, a central secondary controller will be explained in Section 2.4.3, where a cost- and reliability-based corrective and preventive controller is introduced.

2.4.1 Droop-based power control

This section proposes a new type of generalised droop-based control scheme for proportional power sharing in islanded MGs, which has been developed and presented in [33,35]. This method does not depend on the nature of the network lines, i.e. resistive or inductive. Using this approach, both active and reactive power is shared among DGs proportional to the droop gains. The controller gains are chosen based on the eigenvalue analysis to guarantee the system stability. Figure 2.2 shows the single-line diagram of a sample system, where buses 1–4 have converter-based DGs and bus 5 is a load bus. Line impedances among buses are represented by Z_1, Z_2, Z_3, Z_4, and Z_5 is the load impedance.

2.4.1.1 Formulation of control algorithms

The aim of developing the proposed decentralised control algorithm is to distribute generation and load-demand unbalance among the converter-interfaced DGs, according to their ratings. If the changes in phase angles are small such that $\sin(\Delta\delta) \approx \Delta\delta$ and $\cos(\Delta\delta) \approx 1$, the generalised droop equations can be expressed by [17]:

$$\Delta\dot{\delta}_i = -k_{\text{pf}_i}\frac{X_i}{|Z_i|}\left(P_{m_i} - P_i^0\right) + k_{\text{qf}_i}\frac{R_i}{|Z_i|}\left(Q_{m_i} - Q_i^0\right) \tag{2.1}$$

$$\Delta\dot{V}_i = k_{\text{pv}_i}\frac{R_i}{|Z_i|}\left(P_{m_i} - P_i^0\right) - k_{\text{qv}_i}\frac{X_i}{|Z_i|}\left(Q_{m_i} - Q_i^0\right) \tag{2.2}$$

where δ represents the voltage phase angle at the DG's PCC; ΔV denotes the change in voltage magnitude at the DG's PCC; P and Q are the active and reactive power injected by the DG to PCC while subscript m stands for measured values; superscript '0' shows the reference values; and i denotes the DG's index ($i \in \{1, \ldots, 4\}$ in Figure 2.2). The control parameters are denoted by $k_{\text{pf}_i}, k_{\text{qf}_i}, k_{\text{pv}_i}, k_{\text{qv}_i}$, while R, X, and Z are the line's resistances, inductances, and impedances, respectively, where $|Z| = \sqrt{R^2 + X^2}$. From (2.2), it can be seen that if $R \gg X$, the active power transfer is proportional to the voltage difference (ΔV), while ΔV becomes dependent on the reactive power transfer when $X \gg R$. However, (2.1) and (2.2) fail to share power proportionally as they are dependent on the network

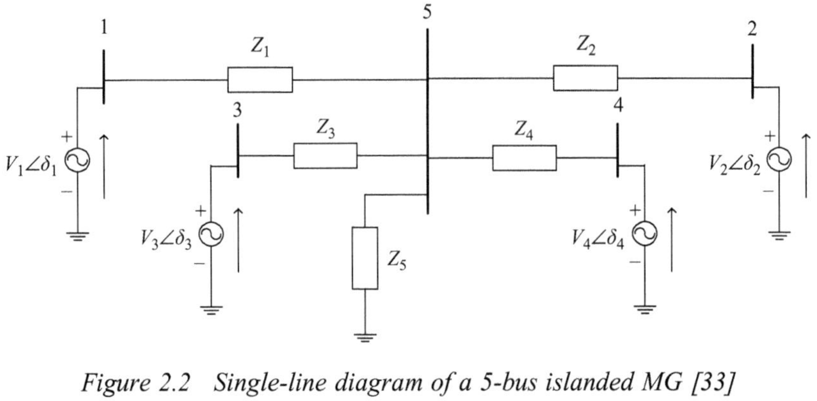

Figure 2.2 Single-line diagram of a 5-bus islanded MG [33]

parameters. To address this issue, (2.2) is replaced with the voltage control law from [34]. The voltage control law can be expressed as

$$\Delta V_i = -k_{\mathrm{p}_i}\left(\left(V_i^0 - V_{\mathrm{ref}_i}\right) - V_{\mathrm{com}}\right) + k_{\mathrm{i}_i}\int\left(\left(V_i^0 - V_{\mathrm{ref}_i}\right) - V_{\mathrm{com}}\right)dt \tag{2.3}$$

According to the droop-based control, the value of V_{ref} should change based on the MG structure using (2.4) and (2.5). For resistive-type lines, we can write:

$$V_{\mathrm{ref}_i} = D_{\mathrm{p}_i}\left(P_{m_i} - P_i^0\right) \tag{2.4}$$

Similarly, for inductive-type lines, we have

$$V_{\mathrm{ref}_i} = D_{q_i}\left(Q_{m_i} - Q_i^0\right) \tag{2.5}$$

Then, the voltage-based droop equation for resistive and inductive (RL-type) lines can be obtained by substituting (2.4) and (2.5) into (2.3), as

$$\begin{aligned}\Delta V_i = {} & k_{\mathrm{pp}_i}\left(V_i^0 - D_{\mathrm{p}_i}\left(P_{m_i} - P_i^0\right) - V_{\mathrm{com}}\right)\\ & + k_{\mathrm{pi}_i}\int\left(V_i^0 - D_{\mathrm{p}_i}\left(P_{m_i} - P_i^0\right) - V_{\mathrm{com}}\right)dt\\ & + k_{\mathrm{qp}_i}\left(V_i^0 - D_{\mathrm{q}_i}\left(Q_{m_i} - Q_i^0\right) - V_{\mathrm{com}}\right)\\ & + k_{\mathrm{qp}_i}\int\left(V_i^0 - D_{\mathrm{q}_i}\left(Q_{m_i} - Q_i^0\right) - V_{\mathrm{com}}\right)dt\end{aligned} \tag{2.6}$$

Equations (2.1) and (2.6) are the new control laws in which k_{pp} and k_{qp} are proportional control gains, k_{p} and k_{q} are the integral control gains determining the speed of the system response, D_{p} and D_{q} represent droop gains for the active and reactive power, respectively, while V_{com} is the voltage magnitude at the PCC in the MG. For the MG under study in Figure 2.2, load bus is connected to other buses such that the load voltage is the voltage at the PCC (i.e. $V_{\mathrm{com}} = V_5$).

The active and reactive power measurement sensors are assumed to have first-order dynamics and their transfer functions can be represented as [23]:

$$\begin{aligned}\frac{P'_{m_i}(s)}{P'_i(s)} &= \frac{w_{\mathrm{f}}}{s + w_{\mathrm{f}}}\\ \frac{Q'_{m_i}(s)}{Q'_i(s)} &= \frac{w_{\mathrm{f}}}{s + w_{\mathrm{f}}}\end{aligned} \tag{2.7}$$

where $P'_{m_i} = (X_i/|Z_i|)P_{m_i} - (R_i/|Z_i|)Q_{m_i}$; $Q'_{m_i} = (R_i/|Z_i|)P_{m_i} + (X_i/|Z_i|)Q_{m_i}$; $P'_i = (X_i/|Z_i|)P_i - (R_i/|Z_i|)Q_i$; and $Q'_i = (R_i/|Z_i|)P_i + (X_i/|Z_i|)Q_i$ while w_{f} is the time constant.

The overall control philosophy is demonstrated in Figure 2.3, which shows that it operates as a droop control and adjusts the voltage magnitude and rate-of-change of the voltage angle in the voltage-source converter when the load changes from its reference value. In the diagram, $k_1 = -k_{\mathrm{pf}_i}(X_i/|Z_i|)$ and $k_2 = k_{\mathrm{qf}_i}(R_i/|Z_i|)$ while a

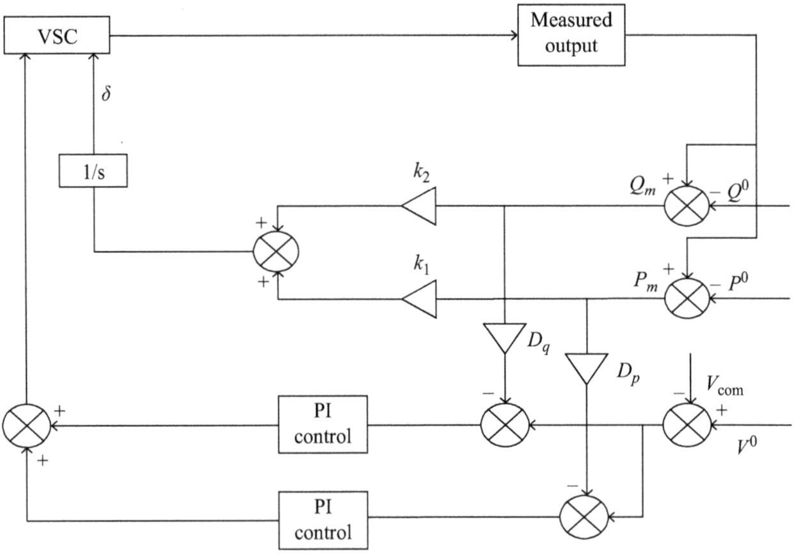

Figure 2.3 Schematic block diagram of the control scheme for each DG [35]

proportional-integral (PI) controller is included in the droop control mechanism for proportional power sharing.

2.4.1.2 Active power sharing

For a stable system with a PI controller, the argument of the first integral term in (2.6) should be zero, i.e.:

$$V_i^0 - D_{p_i} \Delta P_i - V_{com} = 0 \tag{2.8}$$

where $\Delta P = P_m - P^0$. If V^0 is the same for all DGs, it can be concluded from (2.8) that the active power is shared inversely proportional to the active droop gain (i.e. D_p).

2.4.1.3 Reactive power sharing

Similarly, the argument of the second integral term in (2.6) is zero during stable operation. Therefore:

$$V_i^0 - D_{q_i} \Delta Q_i - V_{com} = 0 \tag{2.9}$$

where $\Delta Q = Q_m - Q^0$. Equation (2.9) similarly suggests that the reactive power can be shared inversely proportional to the reactive droop gain (i.e. D_q).

2.4.1.4 Selection of the control gains

The main purpose of selecting the suitable control parameters is to ensure proportional power sharing within the MG. Accurate control gain selection is also important to guarantee well-damped responses in the MG voltages, currents, and

Table 2.1 Gain values of the controller

Parameters	Values
$k_{\mathrm{pf}_1}, k_{\mathrm{pf}_2}, k_{\mathrm{pf}_3}, k_{\mathrm{pf}_4}$	1, 5, 1, 5
$k_{\mathrm{qf}_1}, k_{\mathrm{qf}_2}, k_{\mathrm{qf}_3}, k_{\mathrm{qf}_4}$	1, 5, 1, 5
$D_{\mathrm{p}_1}, D_{\mathrm{p}_2}, D_{\mathrm{p}_3}, D_{\mathrm{p}_4}$	0.001, 0.005, 0.001, 0.005
$D_{\mathrm{q}_1}, D_{\mathrm{q}_2}, D_{\mathrm{q}_3}, D_{\mathrm{q}_4}$	0.001, 0.005, 0.001, 0.005
$k_{\mathrm{pp}_1}, k_{\mathrm{pp}_2}, k_{\mathrm{pp}_3}, k_{\mathrm{pp}_4}$	0.001, 0.005, 0.001, 0.005
$k_{\mathrm{qp}_1}, k_{\mathrm{qp}_2}, k_{\mathrm{qp}_3}, k_{\mathrm{qp}_4}$	0.001, 0.005, 0.001, 0.005
$k_{\mathrm{pi}_1}, k_{\mathrm{pi}_2}, k_{\mathrm{pi}_3}, k_{\mathrm{pi}_4}$	50, 50, 50, 50
$k_{\mathrm{qi}_1}, k_{\mathrm{qi}_2}, k_{\mathrm{qi}_3}, k_{\mathrm{qi}_4}$	50, 50, 50, 50

powers. The control gains are evaluated by successive approximation mentioned below, and their values are illustrated in Table 2.1.

The system dynamics can be represented by using (2.1) and (2.7) as follows:

$$\Delta\dot{\delta}_i = -k_{\mathrm{f}_i}\frac{1}{|Z_i|}X_i\Delta P_{m_i} + k_{\mathrm{f}_i}\frac{1}{|Z_i|}R_i\Delta Q_{m_i} \tag{2.10}$$

$$X_i\Delta\dot{P}_{m_i} - R_i\Delta\dot{Q}_{m_i} = -w_{\mathrm{f}}(X_i\Delta P_{m_i} - R_i\Delta Q_{m_i}) + w_{\mathrm{f}}X_i\Delta P_i - w_{\mathrm{f}}R_i\Delta Q_i \tag{2.11}$$

$$R_i\Delta\dot{P}_{m_i} + X_i\Delta\dot{Q}_{m_i} = -w_{\mathrm{f}}(R_i\Delta P_{m_i} + X_i\Delta Q_{m_i}) + w_{\mathrm{f}}R_i\Delta P_i + w_{\mathrm{f}}X_i\Delta Q_i \tag{2.12}$$

where $k_{\mathrm{f}} = k_{\mathrm{pf}} = k_{\mathrm{qf}}$; $\Delta P = \Delta P^0_{\delta_j}\Delta\delta_j + \Delta P^0_{V_j}\Delta V_j$; and $\Delta Q = \Delta Q^0_{\delta_j}\Delta\delta_j + \Delta Q^0_{V_j}\Delta V_j$, while the constants parameters of $P^0_{\delta_j}, P^0_{V_j}, Q^0_{\delta_j}$, and $Q^0_{V_j}$ can be obtained from the following power flow equations:

$$P_i = \sum_{j=1}^{N} V_iV_j\left(G_{ij}\cos\delta_{ij} + B_{ij}\sin\delta_{ij}\right) \tag{2.13}$$

$$Q_i = \sum_{j=1}^{N} V_iV_j\left(G_{ij}\sin\delta_{ij} + B_{ij}\cos\delta_{ij}\right) \tag{2.14}$$

where N denotes the number of buses with DG.

The Taylor series expansion for P_i and Q_i can be written as

$$\begin{aligned}P_i = P_i^0 + \Delta P_i = P_i^0 &+ \sum_{j=1}^{N} V_j^0\left(G_{ij}\cos\delta_{ij}^0 + B_{ij}\sin\delta_{ij}^0\right)\Delta V_i \\ &+ \sum_{j=1}^{N} V_i^0\left(G_{ij}\cos\delta_{ij}^0 + B_{ij}\sin\delta_{ij}^0\right)\Delta V_j \\ &+ \sum_{j=1}^{N} V_i^0V_j^0\left(-G_{ij}\sin\delta_{ij}^0 + B_{ij}\cos\delta_{ij}^0\right)\Delta\delta_{ij}\end{aligned} \tag{2.15}$$

$$
\begin{aligned}
Q_i = Q_i^0 + \Delta Q_i = Q_i^0 &+ \sum_{j=1}^{N} V_j^0 \left(G_{ij} \sin \delta_{ij}^0 - B_{ij} \cos \delta_{ij}^0 \right) \Delta V_i \\
&+ \sum_{j=1}^{N} V_i^0 \left(G_{ij} \sin \delta_{ij}^0 - B_{ij} \cos \delta_{ij}^0 \right) \Delta V_j \\
&+ \sum_{j=1}^{N} V_i^0 V_j^0 \left(G_{ij} \cos \delta_{ij}^0 + B_{ij} \sin \delta_{ij}^0 \right) \Delta \delta_{ij}
\end{aligned}
\tag{2.16}
$$

where $\Delta\delta_{ij} = \Delta\delta_i - \Delta\delta_j$ and $\Delta\delta_i = \delta_i - \delta_i{}^0$. In (2.15) and (2.16), the network admittance matrix of the MG of Figure 2.2 can be presented as

$$
Y_{\text{bus}} = \begin{bmatrix}
\frac{1}{Z_1} & 0 & 0 & 0 & -\frac{1}{Z_1} \\
0 & \frac{1}{Z_2} & 0 & 0 & -\frac{1}{Z_2} \\
0 & 0 & \frac{1}{Z_3} & 0 & -\frac{1}{Z_3} \\
0 & 0 & 0 & \frac{1}{Z_4} & -\frac{1}{Z_4} \\
-\frac{1}{Z_1} & -\frac{1}{Z_2} & -\frac{1}{Z_3} & -\frac{1}{Z_4} & \frac{1}{Z_1} + \frac{1}{Z_2} + \frac{1}{Z_3} + \frac{1}{Z_4} + \frac{1}{Z_5}
\end{bmatrix}
\tag{2.17}
$$

To simplify the calculations, Y_{bus} can be reduced by combining DG buses in 'group 1' and the load bus in 'group 4', where it is also assumed that $Y_{\text{bus-reduced}_{ij}} = G_{ij} + B_{ij}$. The different parameters of the MG are given in Table 2.2. Then, the system is linearized around its equilibrium point after the load change, and the eigenvalues of the MG are calculated accordingly. As seen from Figure 2.4, all eigenvalues are on the left half plane, where the real part is negative. Therefore, it can be concluded that the system is stable for the given MG under a specific loading condition.

2.4.1.5 Simulation results

Let us consider an example to visualise the effects of resistance and inductance of the overhead lines within the MG. The initial value of the load impedance is considered 1 pu and it is changed to $0.287 + 0.957j$ at rate of 1 s. The angle differences

Table 2.2 Parameter specifications for the given MG

Parameters	**Values**
Z_1, Z_2, Z_3, Z_4	$0.25{+}0.25j$, $0.2{+}0.2j$, $0.15{+}0.15j$, $0.1{+}0.1j$ (pu)
$V_1^0, V_2^0, V_3^0, V_4^0$	1, 1, 1, 1 (pu)
w_f	12 rad/s

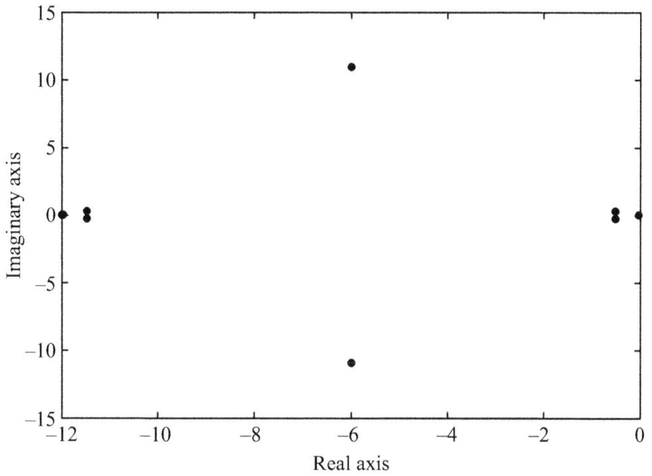

Figure 2.4 Eigenvalues of the MG under study

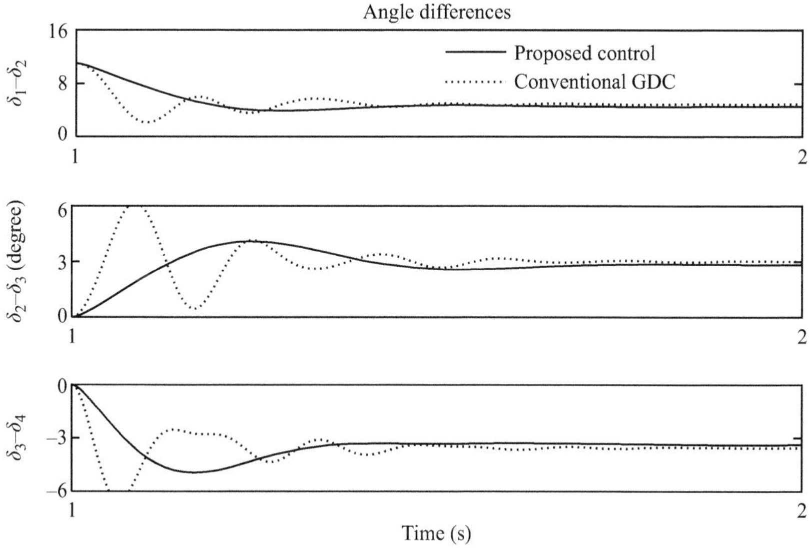

Figure 2.5 Angle differences among the DGs (solid line-proposed controller, dotted line-droop controller)

among all DGs, obtained from the operation of this droop-based controller, are shown in Figure 2.5. It can be seen that the system reaches a stable condition within 2 s. It can also be seen that this control (shown by solid line) outperforms the conventional RL-droop control (shown by dotted line). It is noticeable that the proposed controller starts with a few initial oscillations, because of the integral term in the controller. However, the oscillations do not evolve in time, whereas they are prominent in the RL-droop control.

At the steady state, the angle differences are constant in Figure 2.5. It shows that the rate-of-change of angles (i.e. frequency) is the same everywhere in the system, as expected.

The voltage waveforms of the DGs are provided in Figure 2.6. Immediately after the load change, a significant increase in the voltages is observed due to adding reactive power component to the load impedance. Figures 2.7 and 2.8 show the output powers of the DGs. From Figure 2.7, it can be seen that DG-1 is supplying more active power compared to other DGs, followed by DG-3, DG-2, and DG-4. Figure 2.8 suggests that the DG-1 and DG-4 are sinking reactive power from their respective interconnected buses, whereas the other two DGs are injecting reactive power into their PCC, in order to maintain the voltage stability of the system. Particularly, DG-3 is providing a significant amount of reactive power in this case.

Figure 2.9 illustrates the characteristics of the load. It can be seen that the reactive power consumption of the load is approximately 35 per cent of its power rating. It can be observed that the load voltage variations are contained very well within a narrow range (between 0.99 and 1 pu) at the load bus despite load variations.

Through the simulation studies, it is shown that a variation of the generalised droop-based control can be effective as a primary frequency and voltage controller

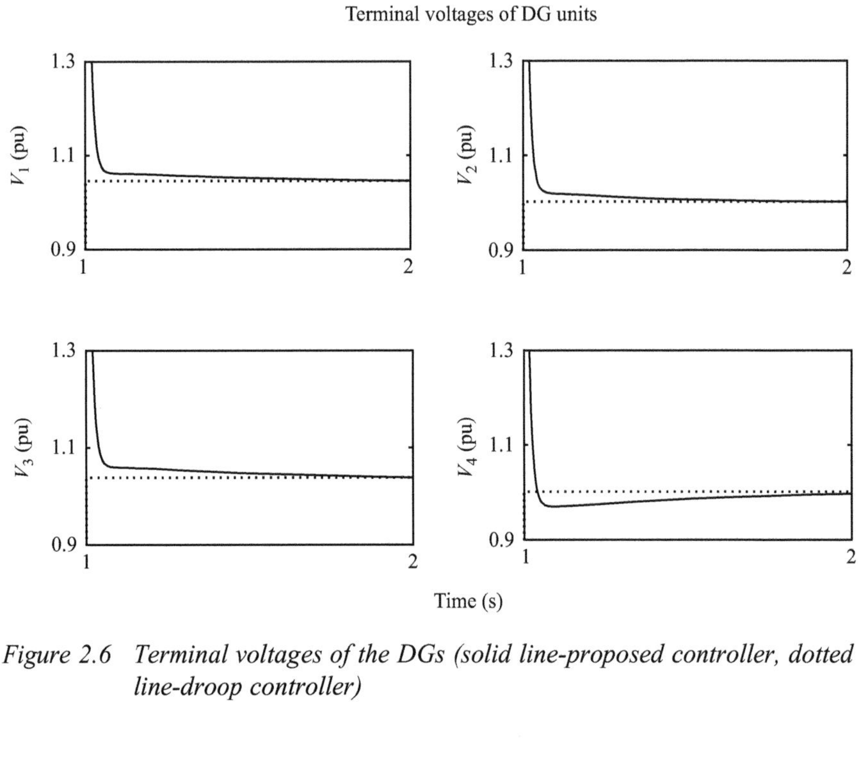

Figure 2.6 Terminal voltages of the DGs (solid line-proposed controller, dotted line-droop controller)

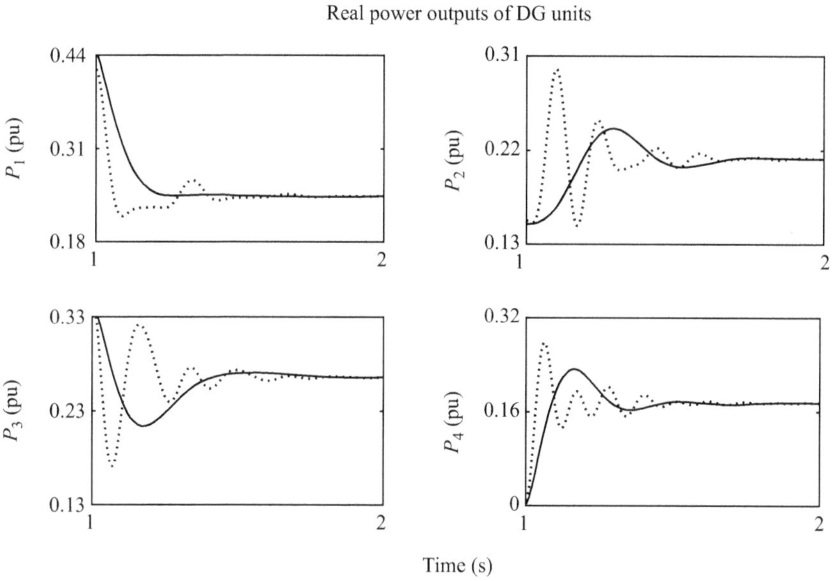

Figure 2.7 Active power outputs of the DGs (solid line-proposed controller, dotted line-droop controller)

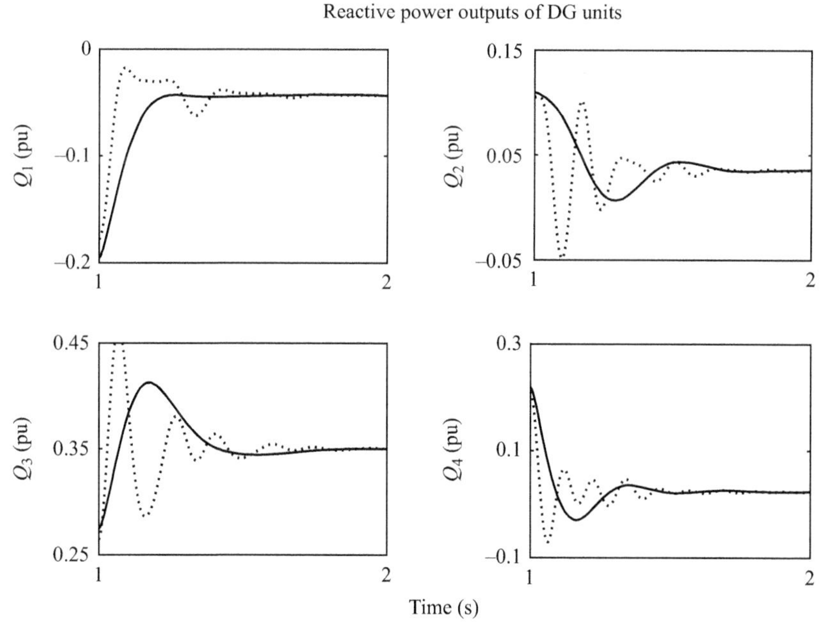

Figure 2.8 Reactive power outputs of the DGs (solid line-proposed controller, dotted line-droop controller)

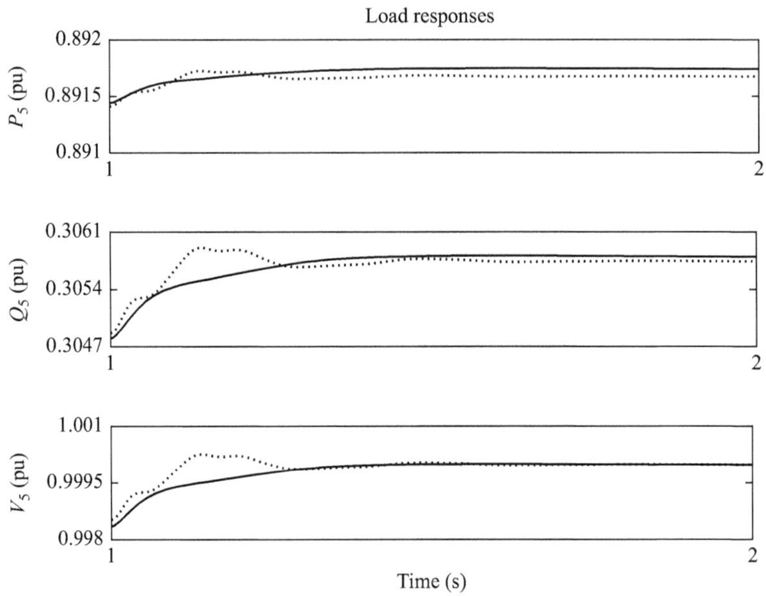

Figure 2.9 Active, reactive, and voltage of load (solid line-proposed controller, dotted line-droop controller)

in an islanded MG. The proposed algorithm has enhanced stability characteristics while being independent of the network's line parameters.

2.4.2 Demand-side primary frequency control

In this section, a comprehensive DR strategy is presented as an effective strategy to continuously balance the generation and demand in the MG, in real time, using an adaptive hill climbing (AHC) strategy [47]. The AHC control is then replaced with a step-by-step (SBS) controller to reduce the amount of manipulated load to a minimum level at steady state [47]. The proposed controller works based on a decision tree to accurately determine the operation mode of the controller.

The proposed DR strategy, referred to as a central direct load control, consists of three operation modes depending on the magnitude of the system frequency deviation (Δf), as shown in Figure 2.10. The figure shows a hypothetical Δf profile vs. time, where three operation modes (0, 1, and 2) are shown as follows:

- MODE 0: Normal operation (i.e. no control is needed).
- MODE 1: In this mode, Δf exceeds the desired range, and the load control is applied to bring Δf within the range with maximum effort, as quickly as possible. The AHC control is applied in this operation mode.
- MODE 2: In this mode, the frequency has already returned to the normal range, and the SBS load manipulation strategy is used.

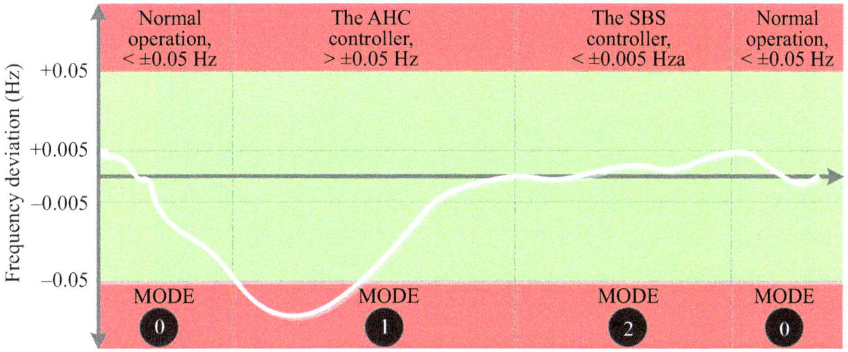

Figure 2.10 Conceptual diagram of the proposed control strategy

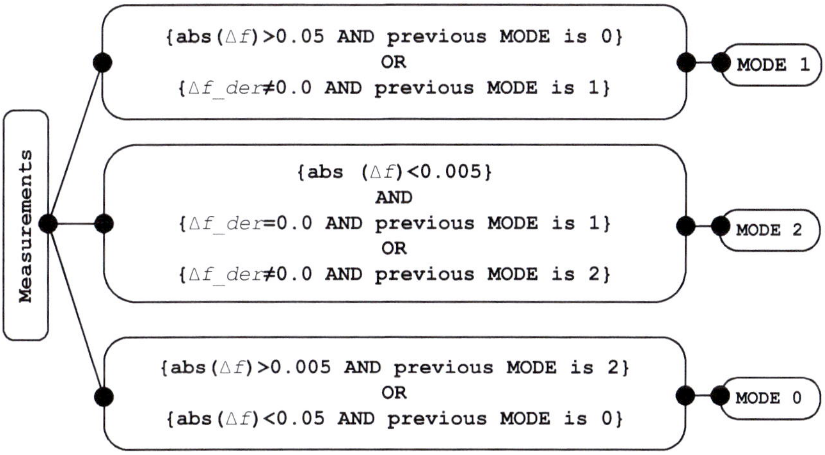

Figure 2.11 Decision tree for determining correct mode of operation

The combination of the above three steps is referred to as the comprehensive control in the remainder of this section.

The conditions for the three operation modes discussed above are given in the decision tree shown in Figure 2.11. In this figure, Δf_{der} is the first-order derivative of frequency deviation with respect to time, '`abs`' is the absolute operator (also denoted by $|.|$), and '`AND/OR`' is the logical and/or operators, respectively. The decision tree outlines the required conditions in which an operation mode will be selected. Checking the derivative of the system frequency deviation in MODEs 1 and 2 is required to assure stable operation of the system by precisely determining the appropriate time for transition from one MODE to the other. MODE 1 will begin once the MG's frequency deviates beyond 0.05 Hz, and if the MG has been previously in MODE 0. MODE 1 will continue as long as $\Delta f_{\text{der}} \neq 0$. MODE 2 will take over the control of the system if the following three conditions hold true:

1. $|\Delta f| < 0.005$ Hz,
2. $\Delta f_{\text{der}} = 0$, and
3. The MG has been in MODE 1 initially.

This mode of operation will remain active as long as $\Delta f_{\text{der}} \neq 0$ and $|\Delta f| < 0.005$ Hz if the MG was initially in MODE 2. MODE 0 takes over if $|\Delta f| > 0.005$ and the MG has been operating in MODE 2 previously. This mode will continue as long as $|\Delta f| < 0.05$ Hz and the MG is in MODE 0 previously. Different frequency thresholds are selected based on [74].

2.4.2.1 AHC control

The flowchart of the AHC controller (operating in MODE 1) is shown in Figure 2.12. The frequency, measured at the PCC of the MG, is the only input variable to the controller. At each time step, if the frequency deviation violates the dead-band (i.e. if $|\Delta f| > 0.05$), the percentage of the change in the responsive load (that will turn on or off) is computed as follows:

$$\%Load^{(k)} = \%Load^{(k-1)} + \Delta f \times M \tag{2.18}$$

where $\%Load^{(k)}$ is the percentage of the manipulated load at time step k and $\Delta f \times M$ is a perturbation parameter. M is the difference controller gain, which is calculated by trial and error (e.g. in this study, $M = 0.1$ as it gives the most satisfactory results). More detail on the AHC technique, its effectiveness, and the necessity of the follow-up controller are available in [47].

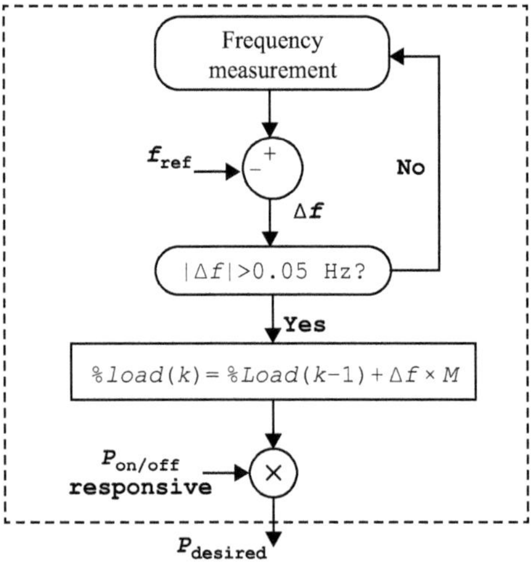

Figure 2.12 The flowchart of the AHC controller

2.4.2.2 SBS control

In order to assure that the minimum required amount of responsive load is activated at steady state, the SBS control (MODE 2), introduced in Figure 2.10, is used. Once the AHC controller stabilises the frequency, the SBS controller will start to minimise the amount of manipulated responsive loads. In order to guarantee the safe operation of the MG, the SBS will only operate within a small frequency dead band, i.e. $|\Delta f| < 0.005$ Hz, as shown in Figure 2.10. Using the SBS controller, the manipulated responsive load will be decreased by

$$\%Load^{(k)} = 0.95 \times \%Load^{(k-1)} \tag{2.19}$$

i.e. 5 per cent at 1 s time step, until the frequency starts to exceed the desired dead band.

This control strategy minimises the amount of responsive loads that need to be manipulated at the steady state to keep the system frequency within the desired range of ± 0.05 Hz. The advantage of the SBS strategy is that a lower percentage of responsive loads is manipulated at the steady state, which improves customers' quality of service. Also, a higher percentage of customer loads will be available for future control, if needed.

2.4.2.3 Simulation results

Let us consider an IEEE standard 13-bus distribution network [75], shown in Figure 2.13, as the MG system under consideration. In order to be able to observe the frequency behaviour of the system as an islanded MG, the infinite bus in the original model is replaced with a 15 MW diesel generator and a variable load which can be adjusted to set the diesel generator's operating point (e.g. to light or heavy loading). The diesel generator is equipped with a speed governor and excitation system the parameters of which are given in [47]. The dump load is only enabled in light loading conditions (in the no-control case) to prevent the frequency deviation from exceeding 0.5 Hz. A 2 MW wind turbine (about 30 per cent of the system load), shown in Figure 2.13, is added to the MG to study the effect of high penetration of wind power. To observe the system frequency deviation, the maximum mechanical torque of the diesel generator has been limited to a certain point for each study case. The network and the diesel generator's parameters are available in [47]. Also, a dynamic load model is developed based on the *d–q* frame, which is explained in detail in [47].

$L_1 - L_5$, shown in Figure 2.13, are aggregated loads with DR capability. Each aggregated load is divided into three different types: responsive, critical, and non-responsive (as depicted in Figure 2.14). The responsive loads are variable active loads, the amount of which is determined by the DR algorithm. The critical and nonresponsive loads are modelled as constant impedance loads.

Scenario I: light loading-decrease in load

In this case, Load-5 (0.9 MW) is disconnected at $t = 7$ s, and as a result, the system frequency increases, resulting in $\Delta f > 0.05$ Hz. The lower set point of the diesel

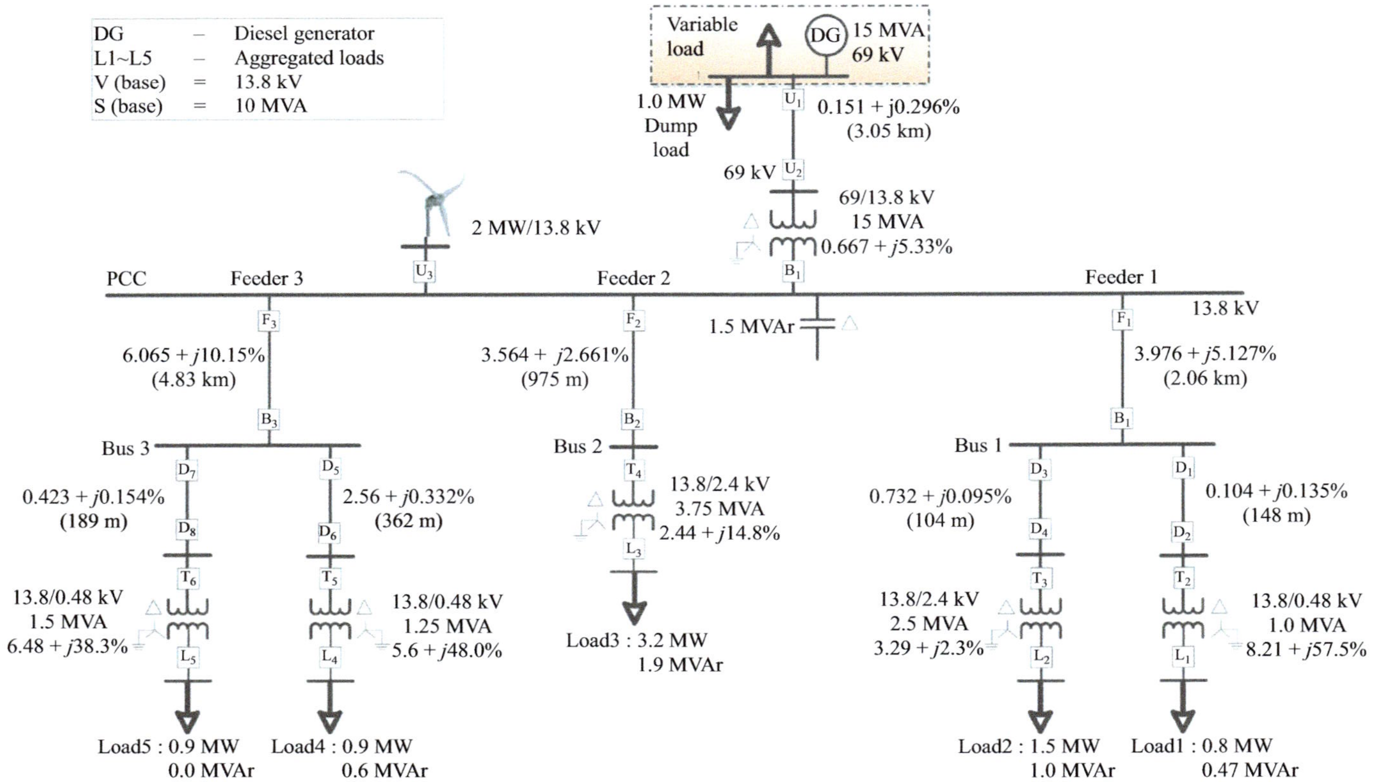

Figure 2.13 Modified IEEE 13-bus standard distribution system schematic diagram

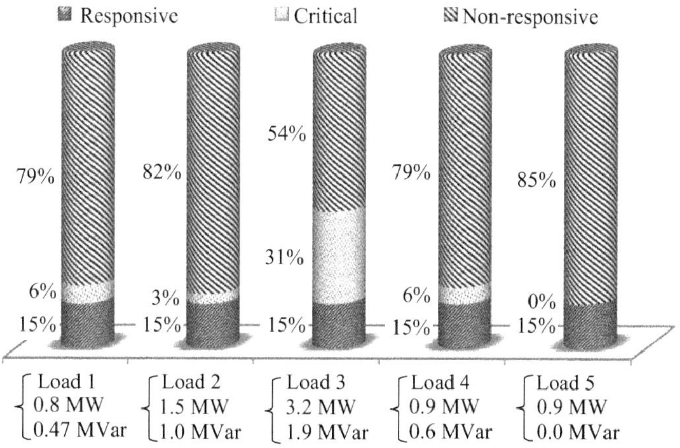

Figure 2.14 The technical characteristic of the loads in the test system

generator's torque is limited to 0.65 pu to prevent it from lowering the torque further to correct the frequency deviation under light loading.

Figure 2.15 depicts the frequency variation, voltage, and total power demand for the two DR approaches when compared with the no-control case. Figure 2.15(a) shows that both DR approaches (AHC and comprehensive) bring the frequency to the desired range faster than the no-control case. The dump load is utilised to regulate the system frequency at around 7.8 s in the no-control case. This happens when the frequency deviation exceeds 0.5 Hz. When the SBS controller is active, the frequency fluctuation is more in the range of ± 0.005 Hz. These variations are shown in Figure 2.15(b). This figure also shows the time at which the SBS controller lowers the amount of the manipulated load to a minimum, which occurs approximately at 19 s when operation at MODE 2 ends, and the MG goes to normal operation mode (i.e. MODE 0). Because of the frequency regulation, the MG's voltage is also stabilised faster, as shown in Figure 2.15(c). In addition, less voltage fluctuation is observed during the transient period, when the comprehensive controller and the AHC controller (alone) are applied. It can be noticed from the figure that the voltage remains in the acceptable range at the end of operation of the SBS controller at $t \approx 19$ s.

Figure 2.15(d) illustrates the total demand of the network for the different cases. It can be seen that a smaller percentage of the responsive loads has been manipulated at the steady state in the comprehensive approach compared to the AHC approach. Under the comprehensive control, the total load demand at the steady-state is 6.142 MW, which means that 77.38 per cent of the responsive loads is manipulated. For the AHC controller, however, the total load at the steady state is 6.238 MW, which means that 91.77 per cent of the responsive load is manipulated. This difference between the manipulated loads in the two cases corresponds to 14.39 per cent savings, which of course results in a higher quality of service for the customers.

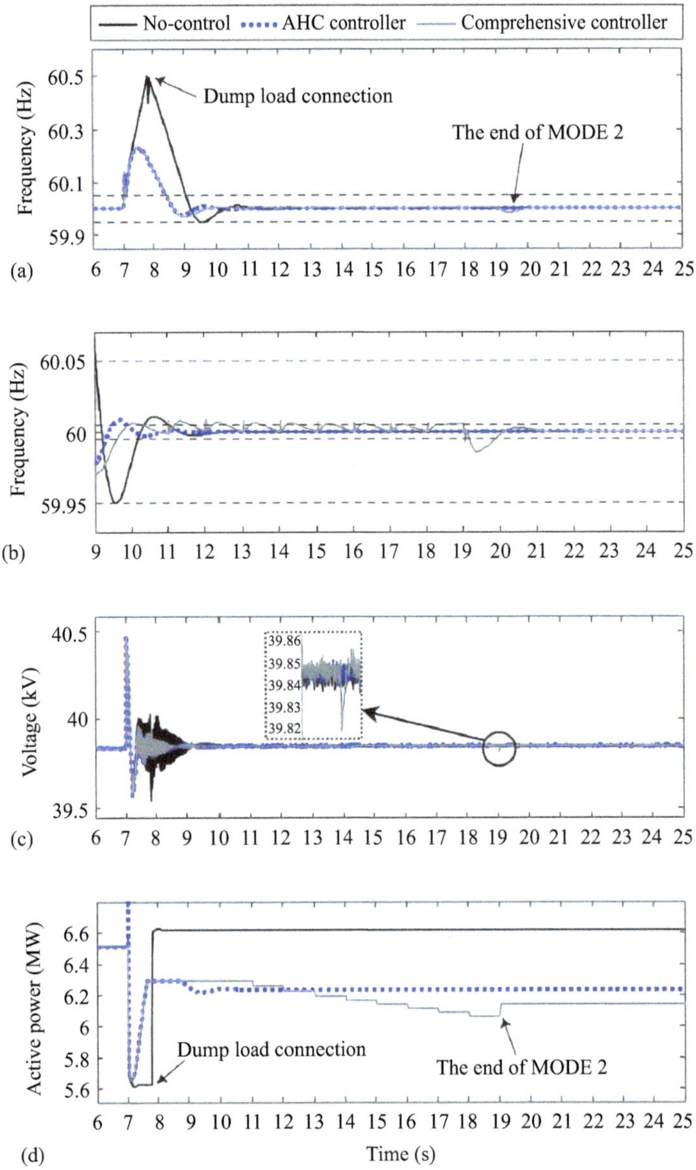

Figure 2.15 Frequency, voltage, and total load demand: light loading with no wind power generation. (a) System frequency, (b) zoomed-in system frequency, (c) voltage, and (d) total load demand

Scenario II: excess wind power generation

In this case, a 2 MW wind turbine is connected to the MG, as shown in Figure 2.13. The MG's frequency, voltage, and power demand are shown in Figure 2.16.

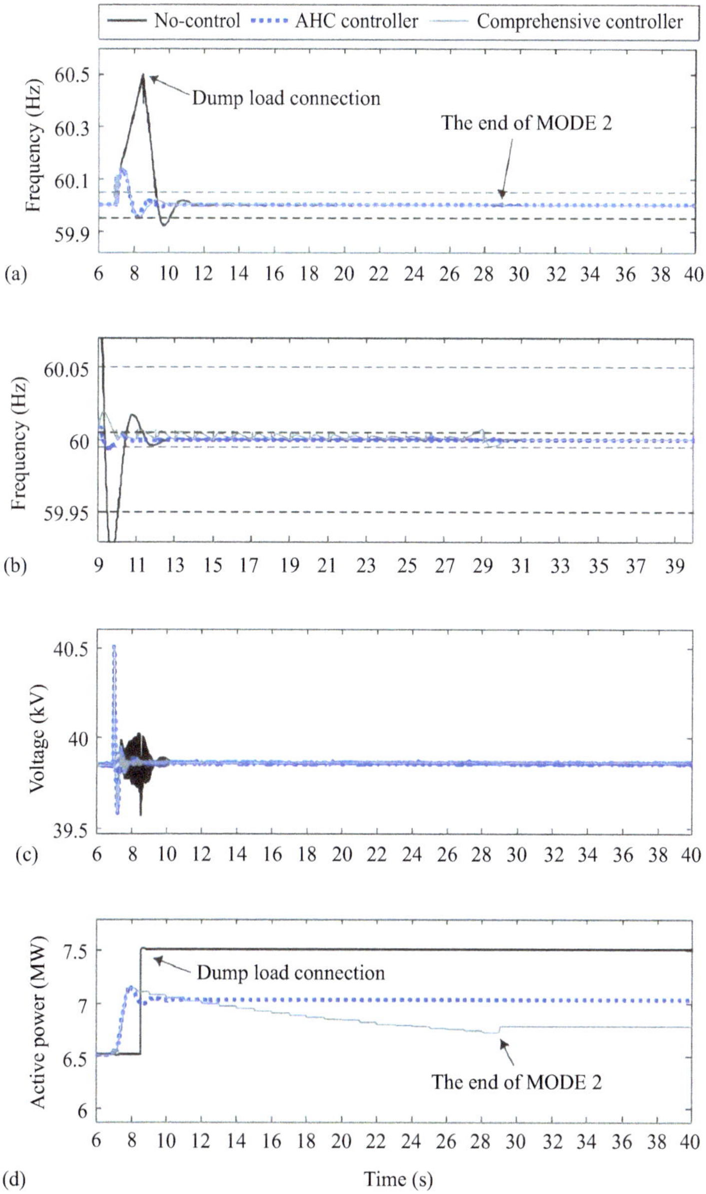

Figure 2.16 Frequency, voltage, and total load demand: light loading with 2 MW wind farm. (a) System frequency, (b) zoomed-in system frequency, (c) voltage, (d) total load demand

At $t = 7$ s, the wind power suddenly increases from 0.6 to 1.3 MW. In the no-control case, a 1 MW dump load is activated at the PCC (increasing the total system load to almost 7.5 MW) once the system frequency reaches 60.5 Hz at approximately $t = 8.3$ s, as seen from Figure 2.16(a). As shown in Figure 2.16(d), the total amount of load under the AHC and comprehensive cases, at the steady state, are, respectively, 7.037 and 6.778 MW. The percentages of manipulated responsive loads are, respectively, 65.59 and 33.04 per cent. This example, again, shows the effectiveness of the comprehensive controller in manipulating (in this case activating) a smaller percentage of the responsive loads, which translates to improved quality of service for the customers. In order to keep the system frequency at 60 Hz at the steady-state, 32.55 per cent more of the responsive loads have been saved by using the comprehensive controller compared to when the AHC controller is used. Figure 2.16(b) shows the zoomed-in frequency variation. In the case of a comprehensive controller, the frequency varies within the acceptable range of 60 ± 0.005 Hz until around $t = 29$ s. At this time, the SBS controller has accomplished the minimisation of the manipulated responsive loads (i.e. the end of MODE 2). As shown in Figure 2.16(c), because of frequency stabilisation, the MG's voltage is also stabilised in all cases.

The impact of DR communication latency on the frequency regulation performance

In an actual system, there are two possibilities of delays in the operation of the system; one is related to the dynamic response of the loads and the other related to the delay in communication. Both delays should be considered in the DR control algorithms to prevent unnecessary switching of the responsive loads. Purely resistive loads (such as electric water heaters) respond instantly to the variations in their input voltage. Therefore, it can be assumed that there is no delay in their response to the changes in the input signal. However, communication delay, often referred to as latency, should be considered. Latency is the length of time from when a request is made by a control entity to when the load receives the request and acts on it. Figure 2.17 shows the frequency variation in the MG under light loading, assuming latencies between 20 and 500 ms. It is clear from this figure that the controller is able to regulate the frequency successfully for latencies of up to 300 ms. At the latency of 500 ms, the MG becomes unstable.

2.4.3 Centralised secondary control

This section presents a corrective and a preventive controller for standalone MGs within a central controller [76–79]. Optimisation techniques are used to maintain the voltage and frequency of the MG within predefined limits during continuous variation of load and renewable generation at the minimum operational cost. More particularly, this approach minimises the total cost of energy for the MG, as well as the cost of the dynamic adequacy of supply, losses, emission, switching of reconfigurable lines, DR, and renewable curtailment. This method is a central one, where the system information is received by the central controller, and appropriate decisions are communicated to the corresponding local controllers.

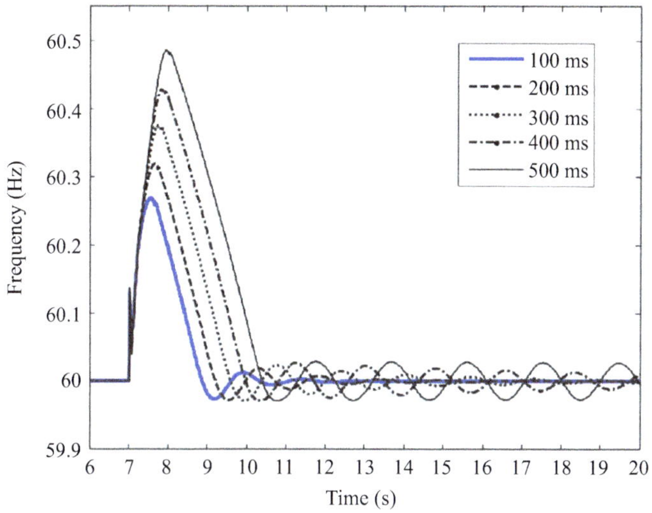

Figure 2.17 System frequency with different latencies, light loading with comprehensive controller

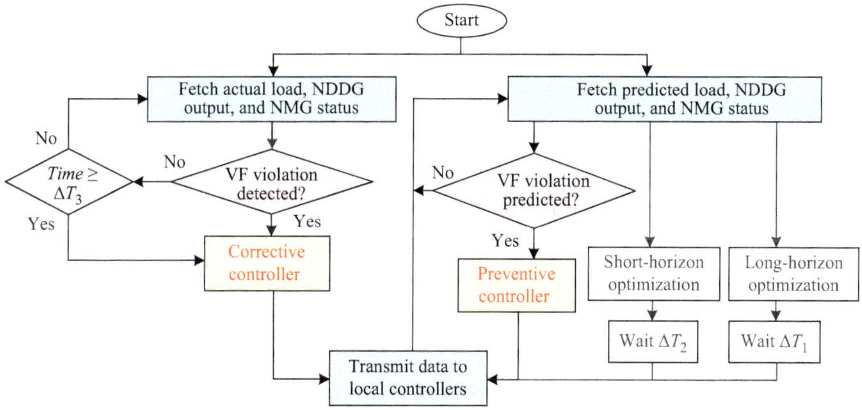

Figure 2.18 Flowchart of the long-term and short-term MG optimisation, as well as very short-term preventive and immediate corrective controllers (VF = voltage/frequency)

2.4.3.1 Corrective controller

In general, an MG operator requires to predict the load demand and NDDGs' generation in various time horizons (e.g. day, month, season ahead) to be able to operate the MG optimally, within the realm of long-term planning, as illustrated schematically in Figure 2.18. In addition, the MG operator can further improve the network performance by using short-term operational schemes (e.g. 30 and 60 min).

Consider an MG consisting of N^{DG} DGs (including N^{DDG} DDGs and N^{NDDG} NDDGs), N^{BES} batteries and N^{load} loads with DR capability interconnected through N^{bus} buses and N^{line} lines among which N^{RL} lines are reconfigurable. Also, assume that the MG is operated in standalone mode under normal operational conditions to maintain its autonomy. However, the MG is able to provisionally couple to N^{EE} external entities (if available) to exchange power during emergency conditions (such as load or generation surge) [80–82] based on the command signal received from the MG central controller and considering the power exchange cost. In this hypothetical MG, the batteries are assumed to have primary controllers that adjust their charging and discharging power based on its state-of-charge (SOC) and the command signal received from the MG central controller. In addition, the DR program (DRP) is considered to be employed at the load interconnection to the MG to facilitate shedding non-essential loads or adding future-planned loads based on the command signal from the MG central controller [83]. The MG is able to reconfigure the network using the reconfigurable lines (i.e. by turning on or off the lines according to the command signal from the MG central controller).

The NDDGs are considered as renewable-based resources with intermittent nature (e.g. PV or wind turbine [84]) unless they are accompanied by power-smoothing batteries [85]. They operate in the grid-following mode by injecting the maximum available power to the MG. Their output power is assumed curtailable if the MG central controller decides to do so. On the other hand, the DDGs are considered as resources with droop-control and grid-forming capabilities when the MG operates in standalone mode. In this case, the voltage and frequency at the output of DDGs are given by the droop equations [86]:

$$\begin{aligned} f &= f^{\max} - m^{\mathrm{DDG}} P^{\mathrm{DDG}} \\ V &= V^{\max} - n^{\mathrm{DDG}} Q^{\mathrm{DDG}} \end{aligned} \tag{2.20}$$

where P^{DDG} and Q^{DDG} are the active and reactive power injected by the DDG, respectively; f and V are the frequency and voltage magnitude at the output of the DDG, respectively; and $f^{\max}$ and $V^{\max}$ are the droop values during normal operation. In (2.20), m and n are the P–f and Q–V droop coefficients, derived from:

$$\begin{aligned} m^{\mathrm{DDG}} &= \frac{\Delta f^{\max}}{(P^{\mathrm{DDG}})^{\max}} \\ n^{\mathrm{DDG}} &= \frac{\Delta V^{\max}}{2(Q^{\mathrm{DDG}})^{\max}} \end{aligned} \tag{2.21}$$

where $\Delta f^{\max}$ and $\Delta V^{\max}$ are the maximum voltage and frequency deviations in the MG, respectively.

The voltage and frequency of the MG are supposed to be limited within $\pm V_1$ and $\pm f_1$ around the nominal values of V_{nom} and f_{nom}, as shown in Figure 2.19. While the voltage and frequency are well within the range, no change is required by the droop control. Therefore, droop control can help the MG to minimise its operation costs, where the corrective controller runs discretely in ΔT_3 intervals (e.g. 10 min).

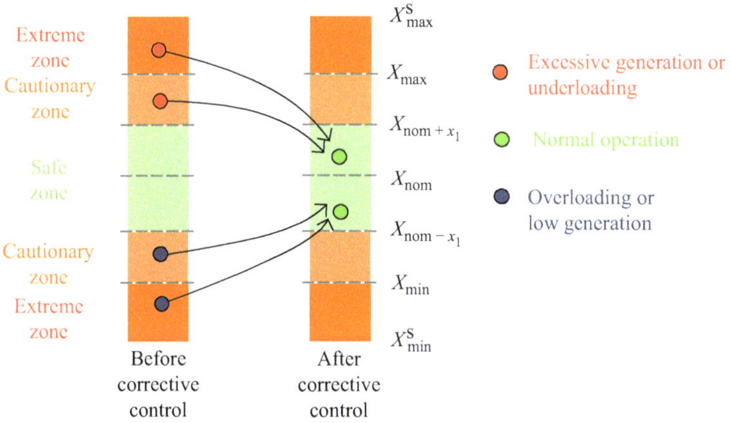

Figure 2.19 Safe, cautionary, and extreme zones illustrating the voltage and frequency violations from nominal values

The corrective controller determines the appropriate set points for the DDGs. While the operation of DDGs is local, the droop parameters of m^{DDG}, n^{DDG}, f^{max}, and V^{max} can be changed dynamically by the MG central controller. These parameters can be adjusted in real time to regulate the frequency in an islanded MG by active power sharing among DDGs. Similarly, the voltage magnitude can be controlled by modifying n^{DDG} of the DDGs. Nevertheless, suitable local control mechanisms (e.g. the virtual impedance technique [87]) should be employed to guarantee a proper reactive power sharing among the DDGs. As a result, the reactive power sharing among DDGs (i.e. modifying n^{DDG}) is not a control variable in the MG central controller.

Due to the variability of the load and NDDGs, frequency and voltage of an MG may exceed the standard range, i.e. $V_{nom} + V_1 \leq V \leq V^{st}_{max}$, $V^{st}_{min} \leq V \leq V_{nom} - V_1$, $f_{nom} + f_1 \leq f \leq f^{st}_{max}$, and $f^{st}_{min} \leq f \leq f_{nom} - f_1$, as shown in Figure 2.19, where superscript 'st' denotes the MG's stability margin of the voltage and frequency deviation. When the voltage or frequency falls outside of the safe zone, the MG observes a violation (denoted by cautionary zone). The MG operation in the cautionary zone is not desired but is permissible. If the voltage/frequency continues dropping beyond the limits of the cautionary zone, then the MG encounters an extreme violation of voltage/frequency (denoted by the extreme zone). As soon as the voltage/frequency violation falls in either of the cautionary or extreme zones, the MG needs to be recovered immediately. In such conditions, the corrective controller analyses the voltage/frequency violation and takes proper actions to return it to the safe zone, as depicted in Figure 2.19. Therefore, the controller should operate in two types of conditions:

- The event-triggered operation for incidents that cause the voltage/frequency of the MG to fall outside of the safe zone (i.e. abrupt changes in load or generation of NDDGs).

- Time-triggered operation at ΔT_3 intervals, when the voltage/frequency has been maintained within the safe zone for longer than ΔT_3, as given in Figure 2.18.

The proposed controller is an operation-stage algorithm as it collects data of the output power of the NDDGs and load consumption, which are assumed to be measured at the proper physical location using sensors with acceptable accuracies. The controller examines the current state of the MG and communicates optimal set points to the local controllers of each component.

Certainly, the processing and communication time incurred to resolve the voltage/frequency problem must be smaller than the operation time of the under/over voltage/frequency protective relays that operate following an extreme condition, as schematically depicted in Figure 2.20. The protective relays typically have a predefined operation time. In large power systems, this period is usually set to be very small (e.g. less than a second to a few seconds). However, in the case of a local and small MG in a remote area where most of the loads are expected to be residential, larger time period (e.g. less than a minute) can be allowed as far as the voltage/frequency of the MG is within the stable operation region.

The corrective controller chooses the best sets of actions that retain voltage/frequency within the safe zone based on the cost of different actions [88]. The actions are divided into four layers based on their operation cost and other hidden costs for the MG operator. The four layers are schematically shown in Figure 2.21 and are defined as follows:

- Layer-1: Adjusting the droop parameters (i.e. different droop coefficient, m^{DDG}, for each DDG and the same voltage/frequency set points of $V^{\max}$ and $f^{\max}$ for all DDGs), as well as the on/off status of the switches of each reconfigurable line (sw^{RL}).
- Layer-2: Adjusting the level of power to be imported from or exported to each available external entity (P^{ex}).
- Layer-3: Adjusting the charge/discharge power of each battery (P^{BES}).
- Layer-4: Adjusting the amount of generation curtailment of each NDDG (ΔP^{RC}) and the DR level of each load (ΔP^{DR}).

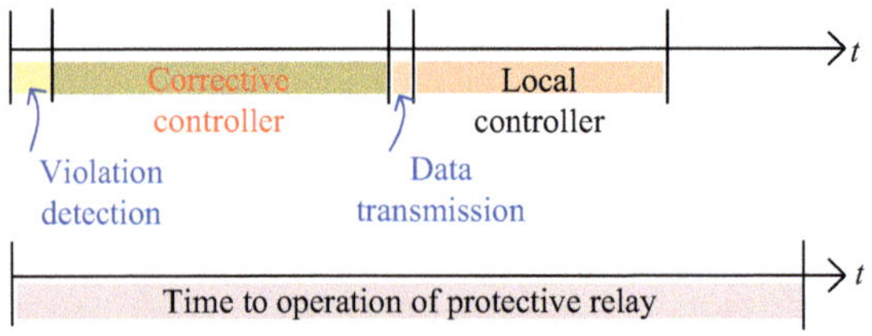

Figure 2.20 Time-sequence of actions to resolve a voltage/frequency issue in an MG by the corrective controller

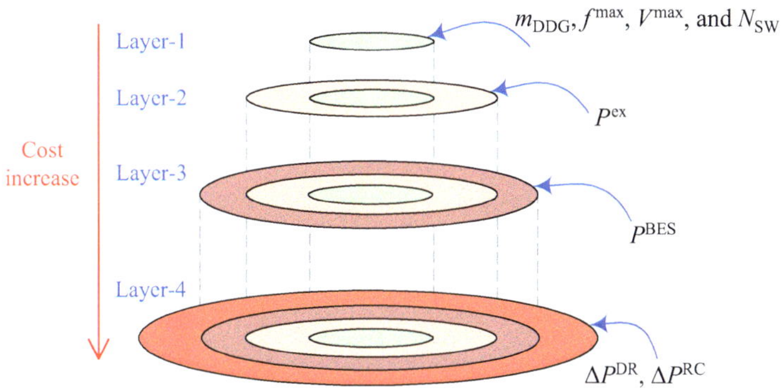

Figure 2.21 Four layers of actions with corresponding decision variables and costs in the corrective controller

Although this technique is based on a multilayer scheme, it does not operate in series according to a defined sequence. It considers all actions first and then aims to minimise the overall cost by selecting actions with lower costs. This is achieved through an optimisation problem that is explained in the following section and is solved by a modified particle swarm optimisation (MPSO) approach [89]. The corrective controller always picks less expensive layers to resolve the voltage/frequency violation when sufficient. In other words, the proposed mechanism guarantees that the controller fixes the voltage/frequency violation by the least cost actions, e.g. adjusting the droop parameters of DDGs when possible rather than selecting expensive options such as power exchange with an external entity or charge/discharge of the batteries, renewable curtailment, or DR. In this study, the cost of power exchange with an external entity is assumed to be cheaper than that of a battery because more frequent charging and discharging of a battery will reduce its lifetime and, thus, incur more operation cost for the MG operator. Similarly, it is assumed that renewable curtailment and DR cost more than power exchange with a battery unit.

For the corrective controller to operate, two-way communication links between the MG central controller and the MG devices (such as DDGs, NDDGs, batteries, loads, switches of reconfigurable lines, and external entities) are required. Figure 2.22 schematically shows the required communication links for successful operation of this method. The communication system is preferred to be a point-to-multipoint wireless connection with a bandwidth of 1 Mbps and a maximum latency of 1 s, for which reliable and secure solutions based on the IEEE 802.11n standard are currently available in the market [90]. Please note that the communication failure will not stop MG operation because of the local controllers, which improves the resilience of the MG. More information about such wireless communications, developing a suitable data pay code and locations of transmitters/receivers along with their impacts on the MG stability, are presented in [91–93].

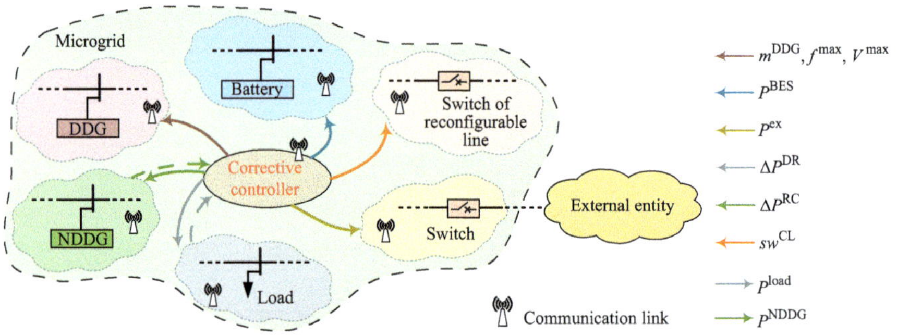

Figure 2.22 Data transmission between the corrective controller at the MG central controller and various sensors/local controllers within the MG territory

2.4.3.2 Problem formulation

As discussed above, for the corrective controller to obtain optimal solutions (e.g. droop of the DDGs, status of the reconfigurable lines switches, power exchange with external entities, battery power, renewable curtailment and DR), an optimisation problem is formulated in the form of a mixed-integer non-linear problem with an OF in the form of

$$OF = \omega_1\, OF_{\text{tech}} + \omega_2\, OF_{\text{op}} + \omega_3\, OF_{\text{sus}} + \omega_4\, OF_{\text{adq}} \\ \text{s.t:}\ \omega_i \neq 0, \sum \omega_i = 1 \tag{2.22}$$

where OF_{tech}, OF_{op}, OF_{sus}, and OF_{adq} are the technical, operation, sustainability, and dynamic adequacy costs, respectively. In (2.22), ω_1 to ω_4 are the corresponding weighting coefficients of each OF, representing the importance of each criterion in the final decision. In general, the MG operator needs to determine proper weightings to satisfy the key performance indices such as minimising the operation cost and improving technical indices or reliability. Since there is no mathematical approach to define them for a complex power system, an acceptable method can be a census from the experts or MG operators based on their experience and outlook, as discussed in [94].

The optimisation problem is solved to minimise the total operation cost and the environmental impact with maximum supply adequacy and to maintain the voltage/frequency within the safe zone. Figure 2.23 illustrates every OF and their components.

2.4.3.3 Technical objective function

OF_{tech} not only eliminates those sets of actions that violate the constraints, introduced in the rest of this section, but also aims to define a set of actions that cause the least voltage/frequency deviation when multiple actions have equal operation costs. It is derived as

$$OF_{\text{tech}} = VDI + FDI + Penalty \tag{2.23}$$

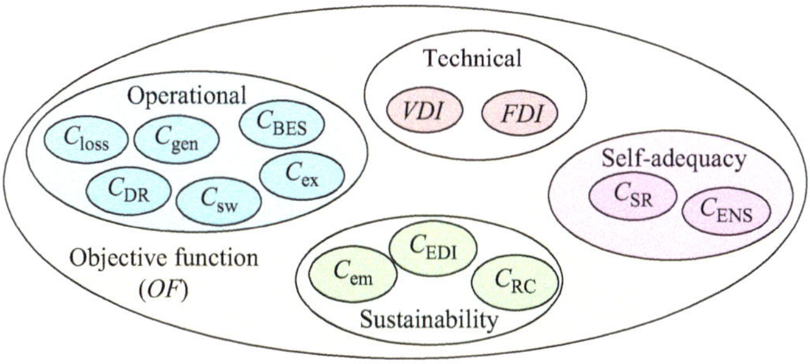

Figure 2.23 Schematic illustration of different objective functions considered in the corrective controller

where *VDI* and *FDI* are the voltage and frequency deviation indices, respectively, and *Penalty* is a penalty factor which is non-zero when at least one constraint is violated. The penalty is defined to be large enough to eliminate the actions that do not satisfy all constraints in (2.23) and are determined by

$$\begin{aligned} VDI &= \max(|V^{nom} - V_i|) \qquad \forall i \in \mathbf{BUS} && \text{(a)} \\ FDI &= |f^{nom} - f| && \text{(b)} \\ Penalty &= \begin{cases} 10^6 & \text{if any constraint is unsatisfied} \\ 0 & \text{otherwise} \end{cases} && \text{(c)} \end{aligned} \tag{2.24}$$

where **BUS** is a vector representing all buses in the MG.

2.4.3.4 Operation objective function

OF_{op} is the operational cost of the MG and is defined as

$$OF_{op} = C_{gen} + C_{BES} + C_{ex} + C_{sw} + C_{loss} + C_{DR} \tag{2.25}$$

where C_{gen} is the power generation cost by the DGs, C_{BES} is the degradation cost of the batteries, C_{ex} is the energy trading cost with external entities, C_{sw} is the cost incurred from switching reconfigurable lines, C_{loss} is the cost corresponding to power losses in the MG, and C_{DR} is the cost of sacrificing customer comfort by controlling their loads under DRPs. The following equations are developed to estimate different cost component in (2.25).

- C_{gen} is calculated as

$$C_{gen} = \sum \left(P_i^{DG} fuel_i C_i^{fuel} + C_i^{O\&M} + C_i^{life}\right)\Delta T + C_i^{start} \quad \forall i \in \mathbf{DG} \tag{2.26}$$

where **DG** is the vector representing all DGs in the MG, P^{DG} is the output power of a DG over the period of ΔT, *fuel* and C^{fuel} are DG's fuel consumption (in L/kWh) and corresponding cost (in \$/L), respectively, $C^{O\&M}$ denotes operation and maintenance cost of the DG (in \$/h), and C^{life} and C^{start} represent the depreciation and start-up cost of DGs, respectively (if applicable), formulated as

$$\begin{aligned} C_i^{\text{life}} &= \frac{\left(C_{\text{cap}}^{\text{DG}}\right)_i}{T_i^{\text{op}}} \qquad \forall i \in \mathbf{DG} \; (a) \\ C_i^{\text{start}} &= Cost_i^{\text{ST}} x_i \qquad (b) \end{aligned} \tag{2.27}$$

where T_i^{op} and $C_{\text{cap}}^{\text{DG}}$ are the total operation lifetime of the DG (in h) and its capital cost (in \$), respectively, and $Cost_i^{\text{ST}}$ is the start-up cost of a DG (in \$). $x = 1$ if a DG starts up; otherwise, it is zero.

- C_{BES} is considered as [92–94]:

$$C_{\text{BES}} = \sum \frac{\left(C_{\text{cap}}^{\text{BES}}\right)_i \lambda_i P_i^{\text{BES}}}{E_i^{\text{life}}} \Delta T \qquad \forall i \in \mathbf{BES} \tag{2.28}$$

in which **BES** is the vector representing all batteries within the MG while the charging or discharging power of the battery (in kW), total cumulative throughput energy in its lifetime (in kW h), and its capital cost (in \$/kW h) are denoted by P^{BES}, E^{life}, and $C_{\text{cap}}^{\text{BES}}$, respectively. Coefficient λ is used to model the impact of the battery SOC, as provided in [95].

- C_{ex} is assumed as

$$C_{\text{ex}} = \left(\sum \left|\left(P_i^{\text{ex}}\right)^{\text{imp}}\right| \left(Cost_i^{\text{ex}}\right)^{\text{imp}} - \left|\left(P_j^{\text{ex}}\right)^{\text{exp}}\right| \left(Cost_j^{\text{ex}}\right)^{\text{exp}}\right) \Delta T$$
$$\forall i \in \mathbf{EE}^{\text{imp}}, \forall j \in \mathbf{EE}^{\text{exp}} \tag{2.29}$$

where $\mathbf{EE}^{\text{imp}}$ and $\mathbf{EE}^{\text{exp}}$ are two vectors representing the external entities that the MG is importing power from or exporting power to, respectively, and $Cost^{\text{ex}}$ is the power exchange cost (in \$/kW h).

- C_{sw} is derived from

$$C_{\text{sw}} = N_{\text{sw}} \, Cost^{\text{sw}} \tag{2.30}$$

where N_{sw} and $Cost^{\text{sw}}$ are the total number of switchings (i.e. connecting a line to or disconnecting it from the MG) and its corresponding cost (in \$/switching), respectively. In this study, it is assumed that a connection or disconnection of a reconfigurable line involves closing or opening of two switches at the two ends of the line.

- C_{loss} is defined as

$$C_{\text{loss}} = P_{\text{loss}} \, Cost^{\text{loss}} \, \Delta T \tag{2.31}$$

in which P_{loss} is the total power loss in the MG (in kW) calculated by the power flow analysis and $Cost^{\text{loss}}$ denotes the cost of losses (in \$/kW h).

- C_{DR} is calculated as

$$C_{\mathrm{DR}} = \sum \left(\Delta P_i^{\mathrm{DR}}\right)^{\mathrm{shed}} \left(Cost_i^{\mathrm{DR}}\right)^{\mathrm{shed}} + \sum \left(\Delta P_j^{\mathrm{DR}}\right)^{\mathrm{future}} \left(Cost_j^{\mathrm{DR}}\right)^{\mathrm{future}} \quad \forall i,j \in \mathrm{DR} \tag{2.32}$$

where **DR** is the vector representing the responsive loads in the MG and ΔP^{DR} is the amount of loads modified under the DRPs (in kW). Superscripts 'shed' and 'future', respectively, denote the nonessential loads that have been shed and the upcoming additional loads in the future that are turned on. $Cost^{\mathrm{DR}}$ represents the corresponding cost of DR (in \$/kW h).

2.4.3.5 Sustainability objective function

OF_{sus} considers the cost associated with the sustainable operation of an MG. It is formulated as

$$OF_{\mathrm{sus}} = C_{\mathrm{RC}} + C_{\mathrm{em}} + C_{\mathrm{EDI}} \tag{2.33}$$

in which C_{RC} is the cost of curtailing renewable-based NDDGs, C_{em} is the emission cost of DGs, and C_{EDI} is the cost of MG's dependency on the external entities. These costs are introduced next.

- C_{RC} is considered as

$$C_{\mathrm{RC}} = \sum \Delta P_i^{\mathrm{RC}} Cost_i^{\mathrm{RC}} \Delta T \qquad \forall i \in \mathbf{NDDG} \tag{2.34}$$

where **NDDG** is the vector representing the existing NDDGs, and ΔP_i^{RC} and $Cost_i^{\mathrm{RC}}$ are the amount of curtailment (in kW) and associated cost (in \$/kW h) for unit i, respectively.

- C_{em} is given by

$$C_{\mathrm{em}} = \sum P_i^{\mathrm{DG}} Em_i^{\mathrm{DG}} Cost_i^{\mathrm{em}} \Delta T \qquad \forall i \in \mathbf{DG} \tag{2.35}$$

in which Em_i^{DG} and $Cost_i^{\mathrm{em}}$ are the emission level of a DG for electricity generation (in kg/kWh) and corresponding cost (in \$/kg), respectively.

- C_{EDI} is calculated as

$$C_{\mathrm{EDI}} = Cost_i^{\mathrm{EDI}} EDI \tag{2.36}$$

in which EDI is an index representing the MG dependency on the external entities and is formulated as

$$EDI = \frac{\sum |S_i^{\mathrm{ex}}|}{\sum S_j^{\mathrm{load}}} \qquad \forall i \in \mathbf{EE},\ \forall j \in \mathbf{LOAD} \tag{2.37}$$

where S_i^{ex} is the apparent power that the MG exchanges with all $\mathbf{EE^{imp}}$ and $\mathbf{EE^{exp}}$ (denoted by the **EE** vector), S_j^{load} is the apparent power consumed by the loads within the MG (denoted by the **LOAD** vector), and $Cost^{\mathrm{EDI}}$ in (2.36) is its corresponding cost (in \$).

2.4.3.6 Dynamic adequacy objective function

OF_{adq}, as shown in Figure 2.23, considers the dynamic supply adequacy of the MG and reflects the probability of unsatisfied demand within the MG and its capability to survive through sudden changes in the generation and demand or unexpected loss of any DGs, referred to as spinning reserve. It is considered as

$$OF_{\text{adq}} = C_{\text{ENS}} + C_{\text{SR}} \tag{2.38}$$

In (2.38), C_{ENS} and C_{SR} are the cost of energy that has not been supplied and the cost of low spinning reserve, respectively.

- C_{ENS} is assumed as

  $$C_{\text{ENS}} = ENS\ Cost^{\text{ENS}}\ \Delta T \tag{2.39}$$

  where $Cost^{\text{ENS}}$ is its corresponding cost (in \$/kW h) and ENS is the amount of unmet power (in kW), which is derived from the given availability (Av) of DGs, batteries, and external entities from:

  $$ENS = \sum (1 - Av_i)|P_i| \qquad \forall i \in \textbf{DG, BES, EE} \tag{2.40}$$

- C_{SR} is defined as

  $$C_{\text{SR}} = SRI\ Cost^{\text{SR}}\ \Delta T \tag{2.41}$$

  in which SRI represents the readily available spinning reserve within the MG, calculated from:

  $$SRI = 1 - \frac{\sum P_i^{\text{DDG}}}{\sum \left(P_i^{\text{DDG}}\right)^{\max}} \qquad \forall i \in \textbf{ROT-DDG} \tag{2.42}$$

  where **ROT-DDG** is the vector representing the MG's rotating DDGs and $Cost^{\text{SR}}$ is the corresponding cost of spinning reserve (in \$/kW).

The OF in (2.22) has to be minimised subject to the technical constraints given as follows:

$$\begin{aligned}
&|\Delta V_i| \leq |\Delta V|_i^{\max} \qquad \forall i \in \textbf{BUS} && \text{(a)}\\
&|\Delta f| \leq \Delta f_i^{\max} && \text{(b)}\\
&I_i \leq I_i^{\max} \qquad \forall i \in \textbf{LINE} && \text{(c)}\\
&SRI \geq SRI_i^{\max} && \text{(d)}\\
&RCI \leq RCI_i^{\max} && \text{(e)}\\
&EDI \leq EDI_i^{\max} && \text{(f)}\\
&P_{\text{loss}} \leq P_{\text{loss}}^{\max} && \text{(g)}\\
&\alpha_i^{\text{DDG}}\left(P_i^{\text{DDG}}\right)^{\max} \leq P_i^{\text{DDG}} \leq \left(P_i^{\text{DDG}}\right)^{\max} \qquad \forall i \in \textbf{DDG} && \text{(h)}\\
&\begin{cases} P^{\text{DDG}} = 0 & \text{if } t_{\text{DDG}}^{\text{off}} < (T_{\text{DDG}})^{\text{off}} \\ P^{\text{DDG}} > 0 & \text{if } t_{\text{DDG}}^{\text{on}} < (T_{\text{DDG}})^{\text{on}} \end{cases} && \text{(i)}\\
&\left|P_t^{\text{DDG}} - P_{t-\Delta t}^{\text{DDG}}\right| \leq \begin{cases} \left(RR_{\text{DDG}}^{\max}\right)^{\text{up}} & \text{if } P_t^{\text{DDG}} > P_{t-\Delta t}^{\text{DDG}} \\ \left(RR_{\text{DDG}}^{\max}\right)^{\text{down}} & \text{if } P_t^{\text{DDG}} < P_{t-\Delta t}^{\text{DDG}} \end{cases} && \text{(j)}
\end{aligned} \tag{2.43}$$

in which (2.43(a) and (b)) represents the maximum allowed deviations of frequency (Δf) and voltage magnitude ($|\Delta V|$) at all buses; (2.43(c)) enforces maximum thermal limits of the lines (denoted by the **LINE** vector), (2.43(d)) defines the maximum loading of the rotating DDGs to satisfy its minimum spinning reserve, while (2.43(e)) represents the maximum allowable contribution from NDDGs, denoted by RCI and obtained from:

$$RCI = \frac{\sum P_i^{\text{NDDG}} + \sum P_j^{\text{BES}}}{\sum P_k^{\text{load}}} \qquad \forall i \in \textbf{NDDG}, \forall j \in \textbf{BES}, \forall k \in \textbf{LOAD} \qquad (2.44)$$

Equation (2.43(f)) defines the MG's maximum acceptable dependency on the external entities, while the maximum allowable loss is set by (2.43(g)). The dispatching constraints of the DDGs (i.e. their minimum loading, minimum downtime, minimum uptime, maximum and maximum ramp-down rates) are given in (2.43(h)–(j)). In (2.43(h)–(j)), α_{DDG} is the percentage of minimum loading of a DDG based on its efficiency constraints (defined in the manufacturer's specifications), $(T_{\text{DDG}})^{\text{on}}$ and $(T_{\text{DDG}})^{\text{off}}$ denote minimum up and down time of the DDGs, and $RR_{\text{DDG}}^{\max}$ represents the maximum ramp up/down of a DDG. It is to be noted that a DDG may be a rotating or a converter-based DG. Hence, the parameters of T_{DDG} and $RR_{\text{DDG}}^{\max}$ can be different from one DDG to another based on their type and operation principles.

The boundaries of the assumed control variables (if applicable) can be defined by

$$\begin{aligned}
&-\left(P_i^{\text{charge}}\right)^{\max} \le P_i^{\text{BES}} \le \left(P_i^{\text{discharge}}\right)^{\max} && \forall i \in \textbf{BES} && \text{(a)}\\
&-\left(P_i^{\text{imp}}\right)^{\max} \le P_i^{\text{ex}} \le \left(P_i^{\text{exp}}\right)^{\max} && \forall i \in \textbf{EE} && \text{(b)}\\
&-\sqrt{(S_i^{\text{ex}})^2 - (P_i^{\text{ex}})^2} \le Q_i^{\text{ex}} \le \sqrt{(S_i^{\text{ex}})^2 - (P_i^{\text{ex}})^2} && \forall i \in \textbf{EE} && \text{(c)}\\
&0 \le \Delta P_i^{\text{RC}} \le \alpha^{\text{RC}} P_i^{\text{NDDG}} && \forall i \in \textbf{NDDG} && \text{(d)}\\
&0 \le \left(\Delta P_i^{\text{DR}}\right)^{\text{shed}} \le \alpha^{\text{shed}} \left(P_i^{\text{load}}\right)^{\text{shed}} && \forall i \in \textbf{DR} && \text{(e)}\\
&0 \le \left(\Delta P_i^{\text{DR}}\right)^{\text{future}} \le \alpha^{\text{future}} \left(P_i^{\text{load}}\right)^{\text{future}} && \forall i \in \textbf{DR} && \text{(f)}
\end{aligned} \qquad (2.45)$$

where charging and discharging power limits of the battery is denoted by (2.45(a)); maximum active and reactive power exchanges with the external entities are enforced by (2.45(b) and (c)). The boundary of renewable curtailment is described in (2.45(d)) in which α^{RC} is the limit of curtailment of renewables. The amount of load shedding, α^{shed}, and starting responsive loads under DRPs, α^{future}, are presented by (2.45(e) and (f)). In (2.45), the superscript 'max' denotes the maximum allowable limit.

As *OF* in (2.22) is a non-convex and non-linear problem with a large number of variables and constraints, exact mathematical optimisation methods cannot be used efficiently. Moreover, as the computational effort increases exponentially with the size of the problem, an effective method with reasonable computational time is required. Therefore, an MPSO-based optimisation approach is preferred in this case. The MPSO optimisation method includes initialisation, finding individual and global best particle, and updating the velocity and location of particles until convergence to the optimal solution is achieved. In this study, the structure of the MPSO particle is shown in Figure 2.24.

2.4.3.7 Simulation results

Let us consider the 10-bus MG shown in Figure 2.25 with three DDGs, two NDDGs, two batteries, two reconfigurable lines (i.e. line-1 and line-2), and an interconnection to a neighbouring MG as the available external entity through bus 7. DDG-1 and DDG-2 are assumed to be biomass technology and diesel synchronous generators, respectively, whereas DDG-3 is considered as a solar PV converter-based DG (with a power smoothing storage that makes it dispatchable). Therefore, DDG-1 and DDG-3 are the only renewable-based DDGs in the MG. NDDG-1 and 2 are small-scale converter-based wind turbines operating under maximum power-tracking scheme and thus are non-dispatchable. The desired limits (safe zone) of voltage/frequency are supposed to be 1 ± 0.05 pu and 50 ± 0.5 Hz, while $V_{\max}^{\rm st} = 1.1$ pu, $V_{\min}^{\rm st} = 0.9$ pu, $f_{\max}^{\rm st} = 51$ Hz, and $f_{\min}^{\rm st} = 49$ Hz. Furthermore, it is assumed that $\omega_1 = \omega_2 = \omega_3 = \omega_4 = 0.25$. The network lines are assumed to have impedances as those of given in [96]. Table 2.3 provides the technical parameters and coefficients for the numerical analyses as well as different cost components of the devices and services.

An under-frequency event

Case-1 is an example when the network is heavily loaded (i.e. above 30 per cent increase of the active and reactive power consumption, see Table 2.4), which

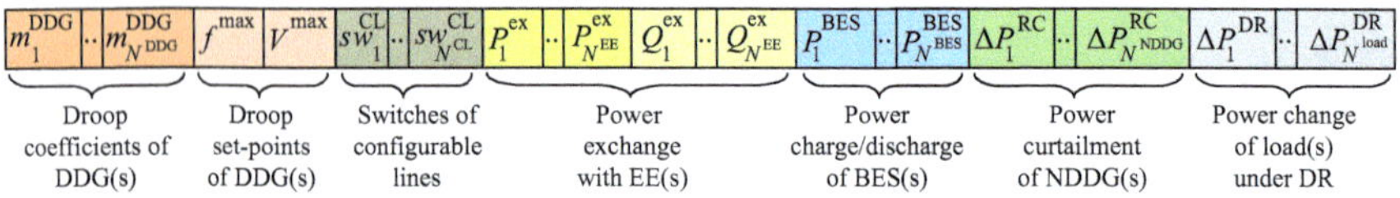

Figure 2.24 Decision variables representation in a particle of the MPSO method

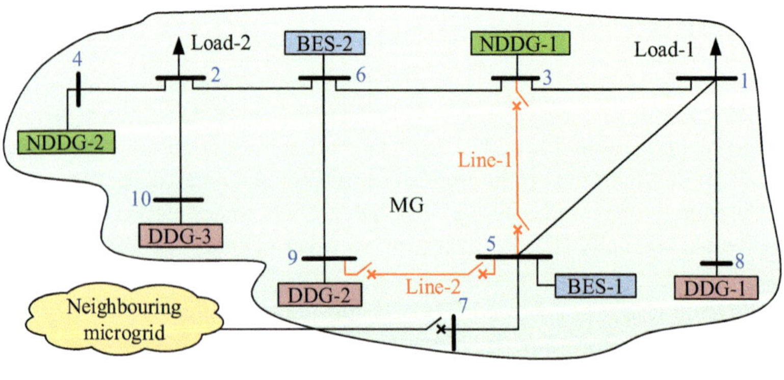

Figure 2.25 The topology of the MG under study for numerical analyses

Table 2.3 Technical parameters, costs, and coefficients used in the numerical analyses

Microgrid's DDGs	DDG-1	DDG-2	DDG-3	Costs	
α_{DDG} (%)	10	10	0	$\left(Cost_i^{ex}\right)^{imp}$	0.3 \$/kW h
$\left(P^{DDG}\right)^{max}$ (kW)	7	7	7	$\left(Cost_i^{ex}\right)^{exp}$	0.3 \$/kW h
$(T_{DDG})^{off}$ (min)	5	5	0.5	$Cost^{loss}$	0.04 \$/kW h
$(T_{DDG})^{on}$ (min)	15	15	0.5	$Cost_i^{RC}$	25 \$/kW h
$\left(RR_{DDG}^{max}\right)^{up}$ (kW/s)	10	10	60	$Cost^{em}$	0.037 \$/kg
$\left(RR_{DDG}^{max}\right)^{down}$ (kW/s)	10	10	60	$Cost^{ENS}$	20 \$/kW h
fuel (L/kW h)	0.2	0.25		$Cost^{SR}$	30 \$/kW h
C_i^{fuel} (\$/L)	0.8	1		$\left(Cost_i^{DR}\right)^{shed}$	3 \$/kW h
$C_i^{O\&M}$ (\$/h)	0.05	0.05	0.01	$\left(Cost_i^{DR}\right)^{future}$	2 \$/kW h
C_{cap}^{DG} (\$/kW)	400	500	700	C_{cap}^{BES}	800 \$/kW h
T_i^{op} (h)	12,000	15,000		$Cost^{sw}$	0.1 \$/switching
Em_i^{DG} (kg/kW h)	0	0.014	0	$Cost^{EDI}$	20 \$
C_i^{st} (\$)	7	8	0		

$Av_{NDDG} = 0.7$; $Av_{BES} = 0.9$; $Av_{EE} = 0.85$; $Av_{DDG\text{-}1} = 0.85$; $Av_{DDG\text{-}2} = 0.85$; $Av_{DDG\text{-}3} = 0.7$; $E^{life} =$ 1,950 kW h; $I^{max} = 25$ A; $SRI^{min} = 10$ per cent of load; $RCI^{max} = 50$ per cent; $EDI^{max} = 20$ per cent; $P_{loss}^{max} = 10$ per cent of load; $\alpha^{RC} = 50$ per cent; $\alpha^{future} = 30$ per cent; $\alpha^{shed} = 50$ per cent.

Table 2.4 Events leading to voltage/frequency violations in Case-1 to Case-4

Example	Load		*f* (Hz)	VDI (pu)
	Active power	Reactive power		
Base case	0.67 pu	0.22 pu	49.8	0.049
Case-1	32 per cent increase	35 per cent decrease	49.35	0.048
Case-2	7 per cent decrease	74 per cent increase	49.9	0.058
Case-3	53 per cent decrease	56 per cent decrease	50.55	0.015
Case-4	3.5 per cent increase	90 per cent decrease	49.75	0.062

triggers an under-frequency event (i.e. frequency drops from 49.8 to 49.35 Hz, see Table 2.4). The doughnut charts in Figure 2.26(a) show the contribution of each device within the MG at the beginning of the event (i.e. initial event) and after applying the corrective controller. In this case, rescheduling the DDGs and importing energy as much as 1 per cent of the demand from the external MG was enough to mitigate the frequency violation. Therefore, the corrective controller avoided expensive options (such as a battery, renewable curtailment, or DR). This figure also shows the voltage magnitudes at all buses of the MG under consideration and its frequency in both conditions. The corrective controller has recovered

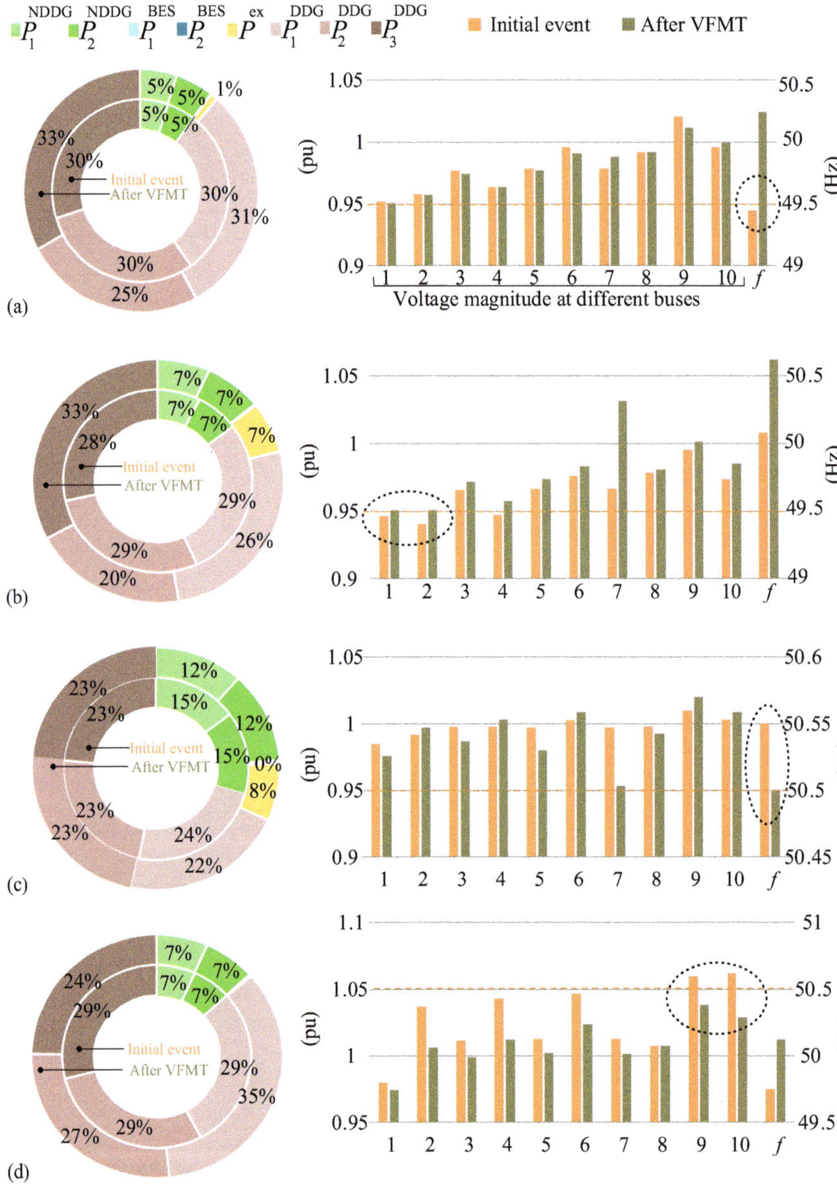

Figure 2.26 Impact of the corrective controller on the contribution of each device, as well as the voltage magnitude at different buses and the MG frequency in (a) Case-1, (b) Case-2, (c) Case-3, and (d) Case-4

the system frequency successfully, while voltage magnitudes have always been within the range.

An under-voltage event

Consider an event in which the reactive power consumption within the MG increases suddenly by 74 per cent in Case-2 (see Table 2.4). As a result, the voltage magnitude at buses 1, 2, and 4 drops below 0.95 pu, while the frequency is within the safe limit (see Figure 2.26(b)). Due to the voltage violation, the corrective controller immediately reacts by importing power from the external entity to supply 7 per cent of the demand. In addition, it reduces the contribution of the rotating DDGs by 12 per cent, while increases the contribution of solar-based DDG by 5 per cent (because of the available energy in its power smoothing storage) to improve the sustainability and performance of the MG. In this case, actions from layer-1 and layer-2 (shown in Figure 2.21) are sufficient to eliminate the voltage violation in the MG.

An over-frequency event

Assume an event in which the active and reactive power consumption within the MG drops by more than 50 per cent in Case-3 (see Table 2.4). Meanwhile, the output power of the NDDGs increases abruptly, and thus, the frequency rises to 50.55 Hz beyond the safe limits (see Figure 2.26(c)). Here, the corrective controller acted promptly and decided to export 0.03 pu power to the external entity (i.e. 8 per cent of the total power generated by all DGs within the MG), while a small amount of power (i.e. 0.002 pu) of NDDGs is curtailed to decrease the system frequency and maintain the voltages at all buses within the safe range. The corrective controller used a combination of layer-2 and layer-4 since the actions from layer-1 failed to eliminate the frequency violation.

An over-voltage event

In Case-4, the reactive power demand of the MG reduced unexpectedly by 90 per cent (see Table 2.4). Consequently, the voltage magnitudes at buses 9 and 10 increased to 1.059 and 1.062 pu, respectively, which is beyond the safe zone (see Figure 2.26(d)). By the actions taken by the corrective controller, the voltages are recovered using only the action from layer-1. It adjusted production from the DDGs (i.e. by increasing the contribution of the DDG-1 by 6 per cent while reducing the output power of DDG-2 and DDG-3 connected to these two buses by 2 and 5 per cent, respectively). As a result, the overall contribution of the DDGs remained almost the same and no other changes were required.

2.4.3.8 Preventive controller

This section discusses a preventive controller, which analyses the predicted state of the MG (using short-term prediction of demand and NDDGs' generation) and plans for the future operation of the MG (as illustrated schematically in Figure 2.18). The optimal operating set points will then be communicated to the local controllers of each device to act accordingly, as denoted in Figure 2.22. The preventive controller tries to choose the best set of actions that prevent voltage/frequency violation. Different actions can be taken by the preventive controller, based on cost such as

- readjusting droop parameters (i.e. different coefficients of m^{DDG} for each DDG and the voltage/frequency set points of $V^{\max}$ and $f^{\max}$ for all DDGs) and/or
- reconfigure network lines and/or
- power exchange with available external entities and/or
- charge/discharge power of each battery and/or
- renewable curtailment and DR.

Similar to Section 2.4.3.1, battery operation is more expensive than purchasing from an external entity, and it is cheaper than renewable curtailment and DR. The preventive controller minimises the overall cost by selecting an optimal combination of actions. This is achieved by solving a non-linear optimisation problem using the MPSO method. As a result, less-expensive actions are always preferred to resolve voltage/frequency issues. As the cost of renewable curtailment and DR are assumed very high, it is expected that their contribution will be limited in the controller.

2.4.3.9 Problem formulation

The preventive controller consists of a mixed-integer non-linear optimisation problem with an OF formulated as

$$OF = \sum \pi^s OF^s \qquad \forall s \in \mathbf{S} \tag{2.46}$$

where

$$OF^s = \omega_1\, OF^s_{\mathrm{tech}} + \omega_2\, OF^s_{\mathrm{op}} + \omega_3\, OF^s_{\mathrm{sup}} + \omega_4\, OF^s_{\mathrm{rel}} + \omega_5\, OF^s_{\mathrm{env}} + Penalty \tag{2.47}$$

The OFs in (2.47), as illustrated schematically in Figure 2.27, can be calculated as

$$\begin{aligned}
OF_{\mathrm{tech}} &= VDI + FDI + LLI + SRI + RCI + EDI && \text{(a)}\\
OF_{\mathrm{op}} &= C_{\mathrm{gen}} + C_{\mathrm{loss}} && \text{(b)}\\
OF_{\mathrm{sup}} &= C_{\mathrm{ex}} + C_{\mathrm{BES}} + C_{\mathrm{sw}} + C_{\mathrm{DR}} && \text{(c)}\\
OF_{\mathrm{rel}} &= C_{\mathrm{ENS}} && \text{(d)}\\
OF_{\mathrm{env}} &= C_{\mathrm{em}} + C_{\mathrm{RC}} && \text{(e)}
\end{aligned} \tag{2.48}$$

The MPSO particle for the preventive controller is the same as that of Figure 2.24.

All parameters in (2.48) were introduced in Sections 2.4.3.1–2.4.3.6, except *LLI* which represents the thermal capacity limit of the lines and is formulated as

$$LLI = \begin{cases} \left|1 - \dfrac{I_i}{I_i^{\max}}\right| & \exists\, I_i > I_i^{\max} \\ 0 & \text{otherwise} \end{cases} \qquad \forall i \in \mathbf{LINE} \tag{2.49}$$

Equation (2.49) allows temporary line overloading in the case of emergency. In order to consider the error in predicted values, the preventive controller assumes

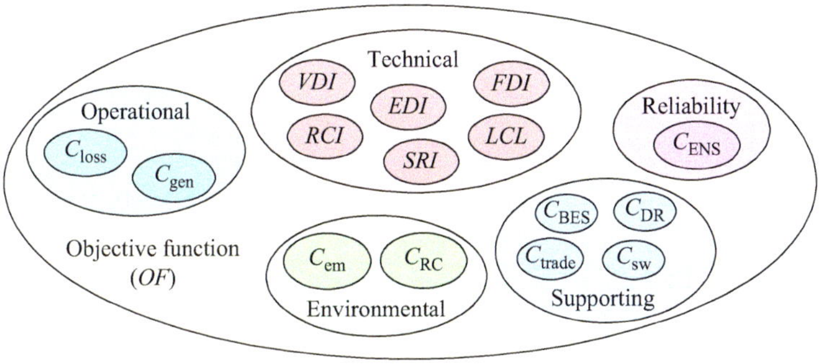

Figure 2.27 Different objective functions considered in the preventive controller

Table 2.5 Different scenarios for the simulation study

s	1	2	3	4	5	6	7	8	9
$\sum P_i^{load}$	X	$X+e$	$X-e$	$X+e$	$X-e$	$X+e$	$X-e$	X	X
$\sum P_i^{NDDG}$	Y	$Y+e$	$Y-e$	$Y-e$	$Y+e$	Y	Y	$Y+e$	$Y-e$

X = Predicted demand; Y = predicted NDDGs' generation; e = prediction error.

multiple scenarios of the predicted variables in (2.46). Considering these scenarios allows MG operator to take the stochasticity of the demand and generation into account during the decision-making process. Table 2.5 lists the different scenarios considered in the simulation study. Based on (2.46), each scenario is denoted by s and with a probability of π^s while $\mathbf{S}$ is the vector representing all scenarios.

2.4.3.10 Simulation results

Consider the MG in Figure 2.25 with the technical parameters given in Table 2.3. It is assumed that all scenarios in Table 2.5 have the same probability (i.e. $\pi^1 = \ldots = \pi^9$) with 15 per cent prediction error. Also, the weightings in (2.47) are $\omega_1 = \omega_4 = 0.1$; $\omega_2 = \omega_3 = 0.3$; $\omega_5 = 0.2$ for the simulation studies in this section.

Time-series study

Let us first envisage the operation of the MG over a sample 60-min period with variations in demand and generation of NDDGs, as shown in Figure 2.28(a). In this study, the forecast horizon is assumed at 5-min. As seen in Figure 2.28(b) and (c), a frequency rise beyond the safe zone is forecasted for t = 18 min at t = 13 min. Therefore, the preventive controller reacts promptly to prevent the prospective frequency deviation. To do so, the controller adjusts the droop coefficients of the DDGs to a new optimal level (see Figure 2.28(d)) to reduce the output power of DDG-2 and DDG-3. Also, the frequency droop set point, i.e. f^{max}, is slightly reduced (see Figure 2.28(e)), which resulted in retaining the voltage/frequency

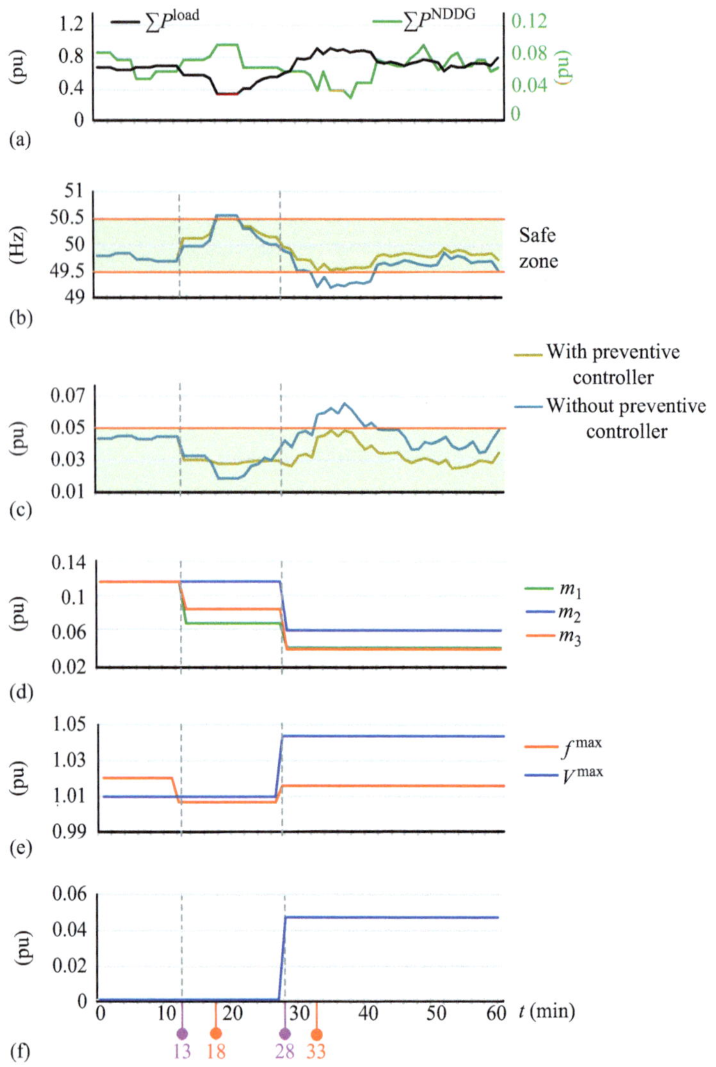

Figure 2.28 Performance of the MG operation under the preventive controller over 60-min of operation: (a) demand and NDDGs' generation, (b) frequency, (c) VDI, (d) droop coefficients, (e) droop set-points, and (f) power transaction with external entity

within the safe zone (see Figure 2.28(b) and (c)). Likewise, at $t = 28$ min, an increase in load and drop in NDDG's power are forecasted for $t = 33$ min, which will result in voltage/frequency drop beyond the safe zone (see Figure 2.28(b) and (c)). To avoid the incident, the preventive controller adjusts the power output of the DDGs to new optimal levels by changing their corresponding droop coefficients while increasing voltage/frequency droop set points (see Figure 2.28(d) and (e)). Nevertheless, the pre-emptive actions were not enough to prevent the voltage/frequency violation. In addition, as depicted in Figure 2.28(f), 0.015 pu active power was imported from the external entity. In this case, the preventive controller does not suggest expensive solutions such as exchanging power with the batteries, DR, or renewable curtailment. This example illustrates the effectiveness of the preventive controller in retaining the voltage/frequency of the MG within the safe zone with the least cost.

Performance under prediction errors

It is assumed that the preventive controller uses predicted values of different parameters from forecasting tools. Therefore, prediction error always exists and its impact should be evaluated properly. To visualise the impact of prediction error on the performance of the preventive controller, another simulation study is conducted in this section, where the active demand or generation is slightly different (below 15 per cent error) from predicted values. Figure 2.29 illustrates eight sample cases in which prediction errors are considered in the preventive controller operation. Case-1 considers four instances, where the actual values of load/NDDG power are lower than the predicted ones, while Case-2 considers another four scenarios in which the actual values are larger than the predicted ones. The predicted and observed values for the eight cases are summarised in Table 2.6.

In Case-1a at $t = 5$ min, a decrease in demand is predicted for $t = 10$ min, which will lead to a frequency rise beyond the safe limits. Therefore, the preventive controller optimises the control variables for the upcoming event. However, at $t = 10$ min, the observed load demand found to be 15 per cent higher than the predicted value. Figure 2.29(Case-1a) shows that despite the error in prediction, the voltage/frequency is maintained within the safe range by the preventive controller actions. In Case-1b, an overloading was predicted, which indicated a prospective voltage/frequency drop beyond the standard range. In reality, however, the actual demand was 9 per cent lower than the predicted value. Therefore, the system frequency is slightly higher in real-time operation, but it is still within the safe zone (see Figure 2.29(Case-1b)). Similarly, Case-1c and Case-1d represent scenarios in which 9 and 15 per cent errors are realised in the generation of NDDGs, respectively. However, in both cases, the voltage/frequency of the MG is impacted marginally by the prediction error (see Figure 2.29(Case-1c) and (Case-1d)).

Case-2a represents a scenario in which the actual demand drops 6 per cent more than the predicted value. Despite the error in prediction, the MG operates without unacceptable voltage/frequency deviation, thanks to the actions taken by the preventive controller (see Figure 2.29(Case-2a)). Case-2b shows an instance in which the actual demand rises 4 per cent more than the predicted value. In this case, it is observed in Figure 2.29(Case-2b) that the inaccuracy has a very little effect on the MG voltage/frequency. Case-2c mimics a scenario in which the actual NDDG power is 4 per cent lower than the predicted value. While the frequency drops slightly and the voltage deviation increases beyond predicted values, both are within the safe range (see Figure 2.29(Case-2c)). Finally, in Case-2d, the actual

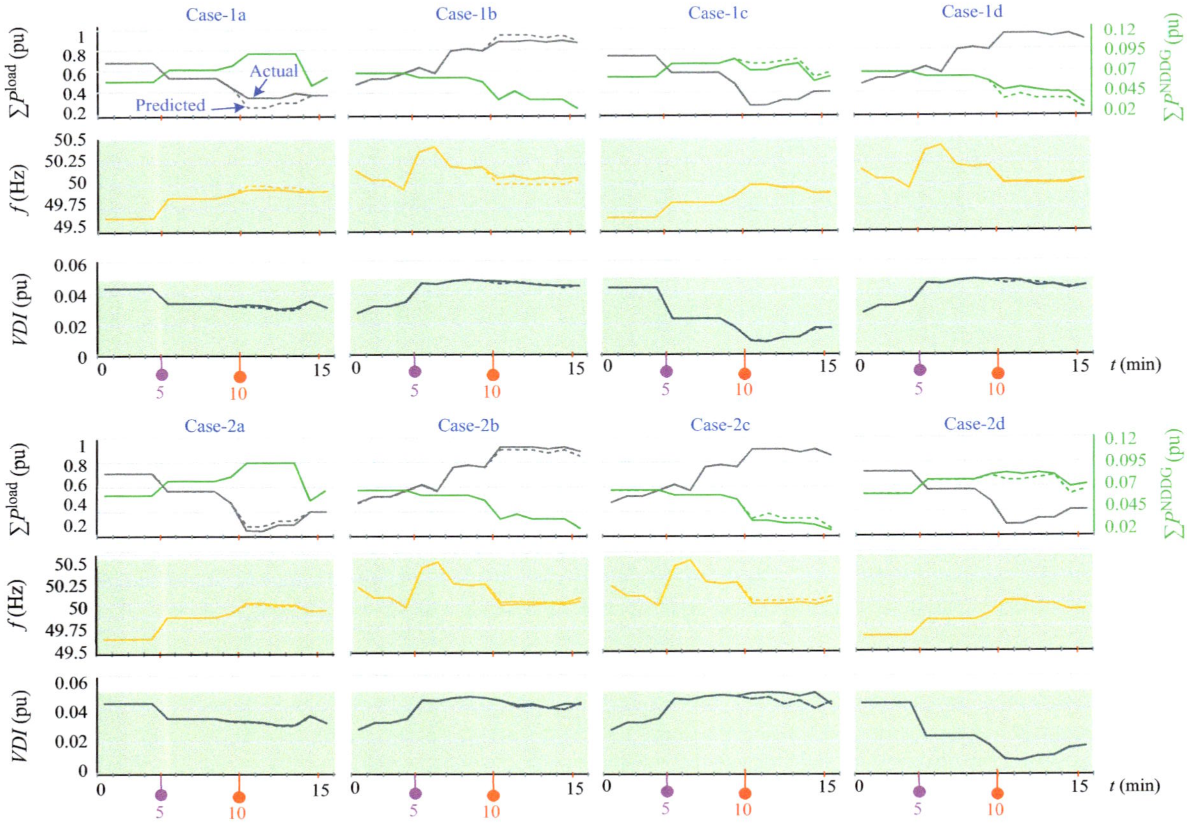

Figure 2.29 The actual and predicted demand, NDDGs' generation and the resulting frequency and VDI under the preventive controller for different study cases in Table 2.6

Table 2.6 Difference between predicted and actual load demand and NDDGs' generation between t = 5 and 10 min in different cases

	Case	1a	1b	1c	1d	2a	2b	2c	2d
$\sum P_i^{load}$	Initial (pu)	0.57	0.62	0.57	0.62	0.57	0.62	0.57	0.62
	Prediction (%)	−50	+46	−50	+46	−50	+46	−50	+46
	Actual (%)	−33	+37	−50	+46	−56	+50	−50	+46
$\sum P_i^{NDDG}$	Initial (pu)	0.08	0.06	0.08	0.06	0.08	0.06	0.06	0.08
	Prediction (%)	25	−38	0	−38	25	−38	−38	0
	Actual (%)	25	−38	−9	−23	25	38	−42	+9

NDDGs' generation is 9 per cent higher than the predicted value. However, the inaccuracy did not deteriorate the voltage/control frequency of the MG (see Figure 2.29(Case-2d)).

2.5 Conclusion

This chapter has discussed the necessity of modern control and management algorithms for islanded and grid-tied MGs operation from different aspects. Various control techniques in different time and space are classified and reviewed briefly. It is obvious from the discussion that cost-effective operation should always play an important role in the MG operation in order to facilitate applicability and suitability of the solution. In addition, a comprehensive MG management and control algorithm should account for sudden variations in the generation and demand and should be able to move smoothly from grid-tied to the islanded operation and vice versa. Another major concern that needs to be properly addressed is the capability of MGs to operate in a plug-and-play fashion. That will require distributed or hierarchical management and control methodologies to offer uninterrupted operation in different circumstances. Moreover, the high level of stochasticity and uncertainty related to renewable generation and load demand requires tools and mechanisms to effectively account for them in the decision-making processes. MG EMSs and control hierarchy also take the DR into account in a comprehensive manner for frequency and voltage support, particularly, as a fast-response ancillary service.

Three advanced algorithms for frequency and voltage regulation, as well as the cost-effective operation of MGs, are presented in this chapter. It is shown that with proper algorithms, different generation, storage, and DR resources can be utilised for secure operation of an islanded MGs. In addition, it has been shown that with preventive control, it is possible to quickly respond to unexpected events in a timely manner in order to prevent outages and cascaded faults in the MG to some extent.

Finally, it seems that the interoperability of MGs with the upper grid or with other MGs in the neighbourhood is a topic that has been overlooked in the past, where different aspects of operation from a market basis to dynamic stability are yet to be addressed in the literature.

References

[1] Olivares D.E., Sani A.M., Etemadi A.H., *et al.* 'Trends in microgrid control'. *IEEE Transactions on Smart Grid.* 2014; 5(4):1905–1919.

[2] Shahnia F., Chandrasena R.P.S., Rajakaruna S., and Ghosh A. 'Primary control level of parallel distributed energy resources converters in system of multiple interconnected autonomous microgrids within self-healing networks'. *IET Generation, Transmission & Distribution.* 2014; 8(2):203–222.

[3] Vandoorn T.L., De Kooning J.D.M., Meersman B., and Vandevelde L. 'Review of primary control strategies for islanded microgrids with power-electronic interfaces'. *Renewable and Sustainable Energy Reviews.* 2013; 19:613–628.

[4] Han H., Hou X., Yang J., *et al.* 'Review of power sharing control strategies for islanded operation of AC microgrids'. *IEEE Transactions on Smart Grid.* 2016; 7(1):200–215.

[5] Ali L., and Shahnia F. 'Determination of an economically-suitable and sustainable standalone power system for an off-grid town in Western Australia'. *Renewable Energy.* 2017; 106:243–254.

[6] Fornarelli R., Shahnia F., Anda M., *et al.* 'Selecting an economically suitable and sustainable solution for a renewable energy-powered water desalination system: A rural Australian case study'. *Desalination*, 2018; 435:128–139.

[7] Shahnia F., Moghbel M., Arefi A., *et al.* 'Levelized cost of energy and cash flow for a hybrid solar-wind-diesel microgrid on Rottnest Island'. 27th Australasian Universities Power Engineering Conference (AUPEC); Melbourne, 2017, pp. 1–6.

[8] Shahnia F., Chandrasena R.P.S., Rajakaruna S., and Ghosh A. 'Autonomous operation of multiple interconnected microgrids with self-healing capability'. IEEE Power and Energy Society General Meeting (PES); Vancouver, 2013, pp. 1–5.

[9] Khodaei A. 'Resiliency-oriented microgrid optimal scheduling'. *IEEE Transactions on Smart Grid.* 2014; 5(4):1584–1591.

[10] Kamel R.M. 'Three fault ride through controllers for wind systems running in isolated micro-grid and Effects of fault type on their performance: A review and comparative study'. *Renewable and Sustainable Energy Reviews.* 2014; 37:698–714.

[11] Zia M.F., Elbouchikhi E., and Benbouzid M. 'Microgrids energy management systems: A critical review on methods, solutions, and prospects'. *Applied Energy.* 2018; 222:1033–1055.

[12] Begum M., Li L., and Zhu J., 'Distributed control techniques for autonomous AC microgrid – A brief review'. IEEE Region 10 Humanitarian Technology Conference; Dhaka, Bangladesh, Dec 21–23, 2017. pp. 357–362.

[13] Bidram A., and Davoudi A. 'Hierarchical structure of microgrids control system'. *IEEE Transactions on Smart Grid*. 2012; 3(4):1963–1976.
[14] Li Y.W., and Kao C.N. 'An accurate power control strategy for power electronics-interfaced distributed generation units operating in a low-voltage multi-bus microgrid'. *IEEE Transactions on Power Electronics*. 2009; 24(12):2977–2988.
[15] Pogaku N., Prodanovis M., and Green T.C. 'Modeling, analysis and testing of autonomous operation of an inverter-based microgrid'. *IEEE Transactions on Power Electronics*. 2007; 22(2):613–625.
[16] Karimi H., Davison E., and Iravani R. 'Multivariable servomechanism controller for autonomous operation of a distributed generation unit: Design and performance evaluation'. *IEEE Transactions on Power Systems*. 2009; 25(2):853–865.
[17] Azim M.I., Hossain M.J., and Pota H.R. 'Design of a general droop control for islanded microgrids'. 25th Australasian Universities Power Engineering Conference; Wollongong, Australia, Sep 27–30, 2015, pp. 1–5.
[18] Bevrani H., and Shokoohi S. 'An intelligent droop control for simultaneous voltage and frequency regulation in islanded microgrids'. *IEEE Transactions on Smart Grid*. 2013; 4(3):1505–1513.
[19] Agüera-Pérez A., Palomares-Salas J.C., González de la Rosa J.J., and Florencias-Oliveros O. 'Weather forecasts for microgrid energy management: Review, discussion and recommendations'. *Applied Energy*. 2018; 228:265–278.
[20] Basu, A.K., Chowdhury S.P., Chowdhury S., and Paul S. 'Microgrids: Energy management by strategic deployment of DERs—A comprehensive survey'. *Renewable and Sustainable Energy Reviews*. 2011; 15(9):4348–4356.
[21] Hirsch A., Parag Y., and Guerrero J. 'Microgrids: A review of technologies, key drivers, and outstanding issues'. *Renewable and Sustainable Energy Reviews*. 2018; 90:402–411.
[22] Li C., Cao C., Cao Y., Kuang Y., Zeng L., and Fang B. 'A review of islanding detection methods for microgrid'. *Renewable and Sustainable Energy Reviews*. 2014; 35:211–220.
[23] Vandoorn T.L., De Kooning J.D.M., Meersman B., and Vandevelde L. 'Voltage-based droop control of renewables to avoid on-off oscillations caused by overvoltages'. *IEEE Transactions on Power Delivery*. 2013; 28(3):2545–2553.
[24] Vandoorn T.L., De Kooning J.D.M., Meersman B., *et al.* 'Automatic power sharing modification of P/V droop controllers in low-voltage resistive microgrids'. *IEEE Transactions on Power Delivery*. 2012; 27(3):2318–2324.
[25] Pota H.R. 'Droop control for islanded microgrids'. IEEE PES General Meeting; Vancouver, Canada, Jul 21–25, 2013, pp. 1–4.
[26] Azim M.I., Hossain M.A., Hossain M.J., and Pota H.R. 'Droop control for islanded microgrids with compensating approach'. 25th Australasian Universities Power Engineering Conference; Wollongong, Australia, Sep 27–30, 2015, pp. 1–6.
[27] Mohamed Y., and El-Saadany E. 'Adaptive decentralised droop controller to preserve power sharing stability of paralleled inverters in distributed

generation microgrids'. *IEEE Transactions on Power Electronics*. 2008; 23(6):2806–2816.
[28] Azim M.I., Hossain M.J., Rafi F.H.M., and Pota H.R. 'An improved droop control scheme for islanded microgrids'. 5th Australian Control Conference; Gold Coast, Australia, Nov 5–6, 2015, pp. 1–5.
[29] Han Y., Li H., Shen P., Coelho E.A.A., and Guerrero J.M., 'Review of active and reactive power sharing strategies in hierarchical controlled microgrids'. *IEEE Transactions on Power Electronics*. 2017; 32(3):2427–2451.
[30] Azim M.I., Hossain M.J., and Pota H.R. 'Design of a controller for active power sharing in a highly-resistive microgrids'. IFAC Symposium on Power and Energy Systems; Delhi, India, Dec 9–11, 2015, pp. 1–6.
[31] Shahnia F. 'Stability and eigenanalysis of a sustainable remote area micro-grid with a transforming structure.' *Sustainable Energy, Grids & Networks*. 2016; 8:37–50.
[32] Azim M.I., Hossain M.A., Hossain M.J., and Pota H.R. 'Effective power sharing approach for islanded microgrids'. IEEE PES ISGT Asia; Bangkok, Thailand, Nov 4–6, 2015, pp. 1–4.
[33] Azim M.I., Hossain M.A., Mohiuddin S.M., Hossain M.J., and Pota H.R. 'Proportional reactive power sharing approach for islanded microgrids'. IEEE Conference on Industrial Electronics and Applications; Hefei, 5–7, 2016, pp. 1–6.
[34] Johnson B., Davoudi A., Chapman P., and Sauer P. 'A unified dynamic characterization framework for microgrid systems'. *Electric Power Components and Systems*. 2011; 40(1):93–111.
[35] Azim M.I., Hossain M.J., and Pota H.R. 'Design of droop-based control for power management in islanded RL-type microgrids'. *International Journal of Emerging Electric Power Systems*. 2017; 18(2):1–13.
[36] Abdelaziz M.M.A., and EL-Saadany E.F. 'Economic droop parameter selection for autonomous microgrids including wind turbines'. *Renewable Energy*. 2015; 82:108–113.
[37] Azim M.I., Mollah K.U.Z., and Pota H.R. 'Design of a dynamic phasors-based droop controller for PV-based islanded microgrids'. *International Transactions on Electrical Energy Systems*. 2018; 28:1–16.
[38] Shahnia F., and Arefi A. 'Eigenanalysis-based small signal stability of the system of coupled sustainable microgrids'. *International Journal of Electrical Power & Energy Systems*. 2017; 91:42–60.
[39] Shahnia F. 'Eigenvalues and their participation factors for the system of coupled adjacent microgrids'. 2nd IEEE Annual Southern Power Electronics Conference (SPEC); Auckland, 2016, pp. 1–6.
[40] Shahnia F., Susanto J., and Mahmud M.A. 'Evaluating the stability impact of the interconnecting tie-line between coupled neighboring microgrids'. IEEE Power Electronics, Drives and Energy Systems Conference (PEDES); Chennai, 2018, pp. 1–6.
[41] Kunwar A., Shahnia F., and Bansal R.C. 'Impact of the capacity and number of inertial and non-inertial distributed energy resources within a microgrid

on its stability margins'. 27th IEEE International Symposium on Industrial Electronics (ISIE); Cairns, 2018, pp. 1–6.

[42] Shahnia F. 'Stability of a sustainable remote area microgrid'. IEEE Region 10 Conference (TENCON); Singapore, 2016, pp. 1–4.

[43] Susanto J., Shahnia F., and Arefi A. 'Effects of network characteristics and topology on the stability of converter-dominated microgrids'. 27th Australasian Universities Power Engineering Conference (AUPEC); Melbourne, 2017, pp. 1–6.

[44] U.S. Dept. of Energy. "Benefits of Demand Response in Electricity Markets and Recommendations for Achieving them," *A report to U.S. Congress*, Feb 2006.

[45] Federal Energy Regulatory Commission. "Assessment of Demand Response and Advanced metering," *A report to U.S. Congress*, Dec 2008. Web address: http://www.ferc.gov/legal/staff-reports/demandresponse.pdf.

[46] Molina-García A., Bouffard F., and Kirschen D.S. 'Decentralized demand side contribution to primary frequency control'. *IEEE Transactions on Power Systems*. 2011; 26(1):411–419.

[47] Pourmousavi S.A., and Nehrir M.H. 'Real-time central demand response for primary frequency regulation in microgrids'. *IEEE Transactions on Smart Grid*. 2012; 3(4):1988–1996.

[48] Klem A., Nehrir M.H., and Dehghanpour K. Frequency stabilization of an islanded microgrid using droop control and demand response'. 48th North American Power Symposium; 2016, pp. 1–6.

[49] Klem A., Dehghanpour K., and Nehrir H. 'Primary frequency regulation in islandec microgrids through droop-based generation and demand control'. 19th IEEE International Conference on Intelligent System Application to Power Systems (ISAP); 2017.

[50] Nisar A., and Thomas M.S. 'Comprehensive control for microgrid autonomous operation with demand response'. *IEEE Transactions on Smart Grid*. 2017; 8(5):2081–2089.

[51] Meng L., Sanseverino E.R., Luna A., Dragicevic T., Vasquez J.C., and Guerrero J.M., 'Microgrid supervisory controllers and energy management systems: A literature review'. *Renewable and Sustainable Energy Reviews*. 2012; 60:1263–1273.

[52] Khan A.A., Naeem M., Iqbal M., Qaisar S., and Anpalagan A. 'A compendium of optimization objectives, constraints, tools and algorithms for energy management in microgrids'. *Renewable and Sustainable Energy Reviews*. 2016; 58:1664–1683.

[53] Plytaria K.E.A., Lysikatos I.N.K., Georgilakis P.S., and Hatziargyriou N.D. 'Distributed and decentralised voltage control of smart distribution networks: Models, methods, and future research'. *IEEE Transactions on Smart Grid*. 2017; 8(6):2999–3008.

[54] Chen F., Chen M., Li Q., Meng K., Guerrero J.M., and Abbott D. 'Multiagent-based reactive power sharing and control model for islanded microgrids'. *IEEE Transactions on Sustainable Energy*. 2016; 7(3):1232–1244.

[55] Begum M., Abuhilaleh N., Li L., and Zhu J. 'Distributed secondary voltage regulation for autonomous microgrid'. 20th International Conference on Electrical Machines and Systems; Sydney, Australia, Aug 11–14, 2017, pp. 1–6.

[56] Engels J., Almasalma H., and Deconinck G. 'A distributed gossip-based voltage control algorithm for Peer-to-Peer microgrids'. IEEE International Conference on Smart Grid Communications; Sydney, Australia, Nov 6–9, 2016, pp. 1–6.

[57] Almasalma H., Engels J., and Deconinck G. 'Dual decomposition-based peer-to-peer voltage control for distributed networks'. 24th International Conference & Exhibition on Electricity Distribution; Jun 12–15, 2017, pp. 1718–1721.

[58] Wang C., and Nehrir M.H. 'Power management of a stand-alone wind/photovoltaic/fuel cell energy system'. *IEEE Transactions on Energy Conversion*. 2008; 23(3):957–967.

[59] Pourmousavi S.A., Nehrir M.H., Colson C.M., and Wang C. 'Real-time energy management of a stand-alone hybrid wind-microturbine energy system using particle swarm optimization'. *IEEE Transactions on Sustainable Energy*. 2010; 1(3):193–201.

[60] Pourmousavi S.A., Nehrir M.H., and Sharma R.K. 'Multi-timescale power management for islanded microgrids including storage and demand response'. *IEEE Transactions on Smart Grid*. 2015; 6(3):1185–1195.

[61] Colson C.M., Nehrir M.H., and Pourmousavi S.A. 'Towards real-time microgrid power management using computational intelligence methods'. Power and Energy Society General Meeting, 2010 IEEE. IEEE; 2010.

[62] Chandrasena R.P.S., Shahnia F., Ghosh A., and Rajakaruna S. 'Secondary control in microgrids for dynamic power sharing and voltage/frequency adjustment'. 24th Australasian Universities Power Engineering Conference; Perth, Australia, Sep 28–Oct 1, 2014, pp. 1–8.

[63] Moradi M.H., Abedini M., and Hosseinian S.M. 'Optimal operation of autonomous microgrid using HS-GA'. *International Journal of Electrical Power & Energy Systems*. 2016; 77:210–220.

[64] Bourbour S., and Shahnia F. 'Impact of the weightings of different criteria in selecting the suitable microgrids when forming a system of coupled microgrids'. IEEE PES Innovative Smart Grid Technologies Asian Conference; Melbourne, Australia, Nov 28–Dec 1, 2016, pp. 1–6.

[65] Bourbour S., Shahnia F., and Ghosh A. 'Selection of a suitable microgrid to couple with an overloaded neighboring microgrid based on decision making'. 47th IEEE North American Power Symposium; Charlotte, USA, Oct 1–4, 2015, pp. 1–6.

[66] Dou C-X, and Liu B. 'Multi-agent based hierarchical hybrid control for smart microgrid'. *IEEE Transactions on Smart Grid*. 2013; 4(2):771–778.

[67] Mendes PR, Isorna LV, Bordons C, and Normey-Rico JE. 'Energy management of an experimental microgrid coupled to a V2G system'. *Journal of Power Sources*. 2016; 327:702–713.

[68] Jin X., Wu J., Mu Y., Wang M., Xu X., and Jia H. 'Hierarchical microgrid energy management in an office building'. *Applied Energy*. 2017; 208:480–494.
[69] Wang C., Liu Y., Li X., Guo L., Qiao L., and Lu H. 'Energy management system for standalone diesel-wind-biomass microgrid with energy storage system'. *Energy*. 2016; 97:90–104.
[70] Tian P., Xiao X., Wang K., and Ding R. 'A hierarchical energy management system based on hierarchical optimization for microgrid community economic operation'. *IEEE Transactions on Smart Grid*. 2016; 7(5): 2230–2241.
[71] Bui V.H., Hussain A., and Kim H.M. 'A multiagent-based hierarchical energy management strategy for multi-microgrids considering adjustable power and demand response'. *IEEE Transactions on Smart Grid*. 2018; 9(2): 1323–1333.
[72] Shahnia F., Bourbour S., and Ghosh A. 'Coupling neighbouring microgrids for overload management based on dynamic multi-criteria decision-making'. *IEEE Transactions on Smart Grid*. 2017; 8(2):969–983.
[73] Arefi A., and Shahnia F. 'Tertiary controller-based optimal voltage and frequency management technique for multi-microgrid systems of large remote towns'. *IEEE Transactions on Smart Grid*. 2018; 9(6):5962–5974.
[74] Time Error Correction, Std. BAL-004-0. *North American Electric Reliability Corporation*, Apr 2005 [Online]. Available: http://www.nerc.com/files/BAL-004-0.pdf.
[75] Katiraei F., Iravani M.R., and Lehn P.W. 'Micro-grid autonomous operation during and subsequent to islanding process'. *IEEE Transactions on Power Delivery*. 2005; 20(1):248–257.
[76] Shoeb M.A., Shahnia F., and Shafiullah G. 'A multilayer and event-triggered voltage and frequency management technique for microgrid's central controller considering operational and sustainability aspects'. *IEEE Transactions on Smart Grid*. 2018, doi:10.1109/TSG.2018.2877999.
[77] Shoeb M.A., Shahnia F., and Shafiullah G. 'A multilayer optimization scheme to retain the voltage and frequency in standalone microgrids'. IEEE PES Innovative Smart Grid Technologies Asian Conference (ISGT-Asia); Auckland, 2017, pp. 1–6.
[78] Shoeb M.A., Shahnia F., and Shafiullah G. 'Sensitivity of prediction error on the performance of a preventive controller for microgrids'. IEEE International Conference on Industrial Technology (ICIT); Melbourne, 2019.
[79] Shoeb M.A., Shahnia F., and Shafiullah G. 'A multilayer preventive control to regulate voltage and frequency in autonomous microgrids'. 28th Australasian Universities Power Engineering Conference (AUPEC); Auckland, 2018, pp. 1–6.
[80] Pashajavid E., Shahnia F., and Ghosh A. 'Development of a self-healing strategy for resilience enhancement of islanded microgrids'. *IEEE Transaction on Smart Grid*. 2017; 8(2):868–880.
[81] Pashajavid E., Shahnia F., and Ghosh A., 'Provisional internal and external power exchange to support remote sustainable microgrids in the course of

power deficiency'. *IET Generation, Transmission & Distribution*. 2017; 11(1):246–260.

[82] Batool M., Shahnia F., and Islam S.M. 'Power transaction management amongst coupled microgrids in remote areas'. IEEE PES Innovative Smart Grid Technologies Asian Conference (ISGT-Asia); Auckland, 2017, pp. 1–6.

[83] Wishart M.T., Shahnia F., Ghosh A., and Ledwich G., 'Multi objective decision making method for demand side management of LV residential distribution networks with plug-in electric vehicles'. IEEE Power and Energy Society General Meeting; 2011, pp. 1–8.

[84] Fan Y., Goyal M., Ghosh A., and Shahnia F. 'Integration of wind energy conversion system with microgrid and utility'. 24th Australasian Universities Power Engineering Conference; Perth, Australia, Sep 28–Oct 1, 2014, pp. 1–6.

[85] Jabalameli N., Masoum M.A.S., Shahnia F., and Mehr T.H. 'Impact of battery rating on performance of rooftop PV supporting household loads, regulating PCC voltage and providing constant output power to grid'. 23rd Australasian Universities Power Engineering Conference (AUPEC); 2013, pp. 1–6.

[86] Chandrasena R.P.S, Shahnia F., Rajakaruna S., and Ghosh A. 'Control, operation and power sharing among parallel converter-interfaced DERs in a microgrid in the presence of unbalanced and harmonic loads'. 23rd Australasian Universities Power Engineering Conference; Hobart, Australia, Sep 29–Oct 3, 2013, pp. 1–6.

[87] Mehr T.H., Shahnia F., Chandrasena R.P.S., and Ghosh A. 'Dynamic power sharing control among converter-interfaced DERs in an autonomous microgrid'. IEEE PowerTech Conference; Eindhoven, Netherlands, Jun 29–Jul 2, 2015, pp. 1–6.

[88] Batool M., Shahnia F., and Islam S.M. 'A multi-level supervisory emergency control for the operation of remote area microgrid clusters'. *Journal of Modern Power Systems and Clear Energy*. 2018, doi.org/10.1007/s40565-018-0481-6.

[89] Shahnia F., and Arefi A., 'Defining the suitable adjacent microgrids to form a temporary system of coupled microgrids'. IEEE Region 10 Conference (Tencon); Singapore, 2016, pp. 1–4.

[90] Marzal S., Salas R., Gonzalez-Medina R., *et al.* 'Current challenges and future trends in the field of communication architectures for microgrids'. *Renewable Sustainable Energy Reviews*, 2018; 82:3610–3622.

[91] Setiawan M.A., Shahnia F., Rajakaruna S., and Ghosh A. 'ZigBee-based communication system for data transfer within future microgrids'. *IEEE Transactions on Smart Grid*, 2015; 6:2343–2355.

[92] Setiawan M.A., Shahnia F., Ghosh A., and Rajakaruna S. 'Developing the ZigBee based data payload coding for data communication in microgrids'. 24th Australasian Universities Power Engineering Conference; Perth, Australia, Sep 28–Oct 1, 2014, pp. 1–6.

[93] Setiawan M.A., Shahnia F., Chandrasena R.P.S., and Ghosh A. 'Data communication network and its delay effect on the dynamic operation of distributed generation units in a microgrid'. 6th IEEE PES Asia-Pacific Power and Energy Engineering Conference; Hong Kong, China, Dec 7–10, 2014.

[94] Jamal T., Urmee T., Shafiullah G., and Shahnia F. 'Using experts' opinions and multi-criteria decision analysis to determine the weighing of criteria employed in planning remote area microgrids'. IEEE International Conference on Green Energy for Sustainable Development (ICUE), Phuket, 2018.

[95] Zhao B., Zhang X., Chen J., Wang C., and Guo L. 'Operation optimization of standalone microgrids considering lifetime characteristics of battery energy storage system'. *IEEE Transactions on Sustainable Energy*. 2013; 4:934–943.

[96] Mumtaz F., Syed M.H., Hosani M.A., and Zeineldin H.H. 'A novel approach to solve power flow for islanded microgrids using modified Newton Raphson with droop control of DG'. *IEEE Transactions on Sustainable Energy*. 2016; 7:493–503.

Chapter 3

Requirements analysis in transactive energy management

Sreenithya Sumesh[1], Aneesh Krishna[1], and Chitra M Subramanian[1]

This chapter focusses on the effective usage of transactive energy (TE) and the importance of developing an economical TE-management (TEM) process. TE is a concept that can play a vital role in improving the efficiency and reliability of a power system. This notion is promising for the energy industry in providing an intelligent and interactive future. This concept initiates various requirements for power distribution and transmission that works efficiently and is totally reliable. This leads to the exploration of requirements engineering (RE) approaches which can play a vital role in the development of TE and management process. This chapter explains the usage of RE models in relation to micro-grid and smart grid development. The wide-ranging development of smart grid systems demands supplementary software models so that its full potential can be explored and utilised. It only makes sense that consolidation of extensive usage of distributed energy and renewable energy sources is important in relation to the future of smart grid to bring about an economical and reliable functioning of a power system. An innovative approach in the form of TE towards the future smart grid is highly beneficial for the power-system operations. This novel approach has been extensively researched in recent years around the world. Within this chapter, we are outlining a goal-oriented RE (GORE) approach to structure TEM system. The main objective of this chapter is to perform reasoning and impact of nonfunctional requirements (NFRs) on the TEM. This reasoning will help decision makers in getting the desired outcomes from an efficient and reliable power system.

3.1 Introduction

A shift is taking place within the electricity generation systems to use more renewable energy that is collected from renewable sources. This energy is becoming the focus of primary energy source, thus replacing old generators. The reasons being

[1]School of Electrical Engineering, Computing and Mathematical Sciences, Curtin University, Bentley, Australia

reduced long-term costs and also less environmental regulation. The intermittent generation characteristics of this system mostly cause added energy storage into the micro-grid installation. This can be challenging in the functioning of the grid. It can affect the efficiency of the energy-storing devices and balancing of energy in the process of scheduling non-critical loads. The good news with the use of renewable energy is that there is improved control of individual loads. This is becoming possible through the class of devices belonging to the internet of things [1].

First, let us focus on the various challenges faced by the evolving electricity system. We have witnessed a transformation of the electricity system from a highly centralised structure to a distributed system managed with a diversity of disseminated supply, storage and responsive demand assets [2]. One of the biggest challenges faced and discussed is how to optimise the usage of distributed energy resources (DER) to develop and maintain the efficiency and reliability of the power-supply system.

When we use renewable energy as the primary source of energy and to make this efficient and reliable, forecasting becomes inevitable. Foresight and forecast in the planning of the operation becomes crucial to maintain the reliability of the grid as well as efficiency of the power system. This forecasting process involves the installation of complex customised controls. These types of control are in use more often today [2]. This is demonstrated by the usage of TEM to increase the efficiency and reliability of the smart grid resources [2]. This TEM methodology incorporates the valuation of both efficiency and reliability objectives. These change in the requirements of the power system at a fine-grained nodal level. This has been prevalent in practice for the past decade [2]. The TEM system most efficiently employs DER to get the desired outcomes for both business and operational goals [2]. It helps in balancing renewable intermittency, thus aiding to drive better energy-management process. The benefit of the TEM system is that it improves the reliability and efficiency of the electric systems. It provides a way for various energy-related parties to interact and interoperate. TE systems (TESs) can help users like electric grid owners, regulators and consumers/users to manage the DER. TE is proposed to be a vital component of the future electric power system. This energy can play an important role in but not limited to supporting the expanding numbers of DER. TE can be an innovative approach for a bright future of the industry. This can be used as a regulatory model for electricity industry to become a very sustainable business. While some perceive TE as a smart grid application, others point out that the grid cannot be smart grid without it being transactive [2,3].

An explanation of the TE concept is provided in this chapter. The application and reasoning of NFRs for implementing TEM that decision makers can consider is also explained and discussed. A smart grid system based on TEM conforms to not only the functional requirements of system features (and functions) but also the NFRs such as security, reliability, interoperability, efficiency. Not a lot of focus has been accorded to the software development life cycle (SDLC) [4] before the design and development stages of the aspects of RE are carried out. There is also a connection between these notions and safety-critical software systems like smart grids [1]. A breakdown or failing in the RE process can be drastic. The Standish Group chaos report [5] identified incomplete requirements, lack of user involvement, lack of resources, unrealistic expectation, lack of executive support, changing requirements and specifications, lack of planning, etc. as causes of many failed projects. The Standish Group Report also revealed that three of the top ten reasons for challenged projects or outright project failure were lack of user involvement, unstable requirements and poor project-management practises. The RMS (Royal Mail Steamer) Titanic ship sinking in 1914 as a result of incorrect steel plating [6] or the market decline of Fords Edsel automobiles owing to the inability to capture end users' needs are examples that resulted in drastic unsafe consequences [1]. Thus, the emphasis on RE needs to follow best practices [7]. Some works in RE for smart grids have been proposed in the literature. In [8], Information and Communications Technology architectures are studied in connection with smart grids. In [9], case studies have been conducted by authors in connection with security needs of smart grids. Some of the significant cyber security and communication requirements in relation to smart grids are listed by Ericsson *et al.* [10]. Case tools [11] are used by certain researchers with regards to areas of industrial automation software in RE. In [12], modelling and checking the reliability of safety requirements were conducted by formal modelling and verification techniques. Nevertheless, there is a lack of detailed investigation in regards to the importance of NFRs [13] for the TEM smart grid systems.

In this chapter, a framework for the TES in maintaining grid performance and stability has been created to assist energy service providers, equipment suppliers, regulators and complex/sophisticated users. This framework also helps in harnessing flexibility to offset variability and for enabling value-based relationships in the electric power systems.

In the next section, we will explain what TE is; how it works; whether, and if so, how it is different from the smart grid; and why some experts think we need TE. In Section 3.2, we will provide an overview of some of the key non-functional requirements, such as scalability, security, inter-operability, reliability and efficiency in the case of transactive power system. We also discuss the notion of goal-oriented requirements engineering with respect to this new paradigm in Section 3.3. In Section 3.4, we apply requirements analysis to TEM and will then close with some concluding remarks in Section 3.5.

3.2 Transactive energy management

In TEM, value is used as a key operational parameter to establish an even distribution of supply and demand across the entire electrical infrastructure. This is made possible by the use of a set of economic and control processes. Software applications that use economic signals and operational information are used in these scenarios. These applications are used to integrate and control the gadgets for manufacture and/or usage of electricity in the grid [2].

The TE system can be perceived as a notion where an integration of retail and wholesale markets knowledge is studied and explored. Market signals are combined to form a single platform by blending retail and wholesale markets. This is achieved by employing forward and spot transactions. Also, by managing investment and operating decisions [2]. The TE system comprises three categories of players; one comprising customers, prosumers, storage, owners, producers, etc. that form part of the energy services; second comprising transmission and distribution owners forming the transport services and third comprising exchanges, market makers, system operators, etc. that forms the intermediaries [2]. In contradiction to this notion is the traditional classification of customer types: residential, commercial and industrial. The TE perception constitutes all categories comprising advanced energy-management systems and/or third party assistance. This is beneficial in the optimum use of energy as well as production build on value and grid constraints. When not interested in market exchanges, fixed-price subscriptions on a forward-looking basis can be acquired. Transactions are not risk free. The operation of a TE network is straightforward. Producers offer sell tenders to consumers indicating they have the energy to sell. The consumers then decide whether they want to purchase energy at the suggested price, wait for another offer or not use energy. For example, if a producer advertises to sell 10 MW of electricity at a plant in his city, between a specific time frame. If the selling price is 50 per MW h, let us assume a consumer wants to buy 5 MW h of the 10 MW h tender; then there is a recorded transaction between the producer and the consumer. A brief comparison between traditional grid and smart grid is shown in Table 3.1 as ready reference.

Few attributes make a smart grid transactive, and they are enabling faster transmission of information across the grid, that includes prices; empowering consumers by enabling their active participation; accommodating all new generation devices needed for a functional decentralised supply model; accommodating two-way power flows. Residential customers are having energy-management systems and/or third-party assistance. This means forward and spot transactions at the retail level being made possible. Retail customers are utilising energy to the maximum potential. All this differentiates TE from smart grid [2]. The nature of the power grid is ever changing and increasing complexity. There is an improvement in environmental goals and deeper exploration of renewable resources. This in turn is

Table 3.1 Comparison between traditional grid and smart grid

Traditional grid	Smart grid
Centralised generation	Distributed generation
Electromechanical	Digital
Few sensors	Sensors throughout
One-way communication	Two-way communication
Manual monitoring	Self-monitoring
Failures and blackouts	Adaptive and is landing
Manual restoration	Self-healing
Few customer choices	Many customer choices
Limited control	Pervasive control

motivating stakeholders to be more proactive and create solutions to foreseen problems. Some renewable resources are variable and unpredictable. Owing to this, greater flexibility and reliable customer resources are sought. With the new demands on the grid, it is becoming difficult for utilities and independent system operators to manage this complex system. Also, these systems are controlled by consumers and can be difficult to monitor in real time. Distributed systems with a hierarchy of control layers designed comprising commonly understood sets of data with sufficient interchange of information could enable a steady and universal optimisation in the way of local action. In other words, with the right design, a TE network can create a perfect market with totally rational decisions, benefiting the producers, consumers as well as the grid.

Figure 3.1 illustrates how TE resources allow for more flexibility and reliability by allowing exchanges up to the real time, as opposed to traditional demand–response resources. The expectations are high for the new TE grid. Some of the benefits of TE are that optimal usage of DER is viable. This helps in meeting both business and operational goals [2]. The TE increases the soundness and productivity of the electric power systems. The requirements for circulating reserves to bring about an even distribution of restored intermittency are controlled. Consumers are empowered by active participation. Innovation is initiated, and jobs are created. TE enables all parties to transact in the same platform in a transparent environment. This, in turn, increases the efficiency of the market as well as the power distribution network.

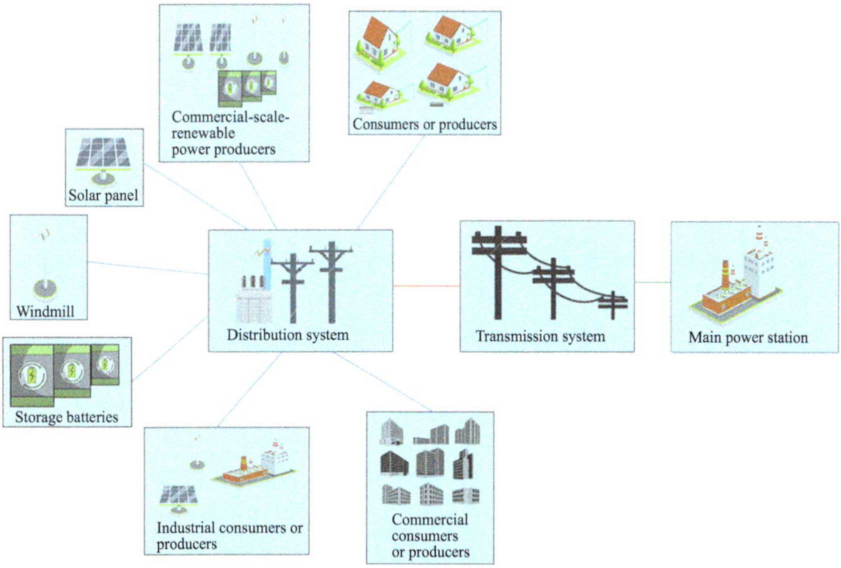

Figure 3.1 Transactive energy player model

3.3 Application of requirements engineering approaches in transactive energy management

There is a need for applying RE approaches in TEM. The smart grid system is highly complex with regards to organisational and technological aspects. Several organisations and engineering domains are involved in the design and development of the system. Thus, the TEM faces an important challenge with regards to integration of these efforts. This affects elements involved in the electrical energy consumption, generation, transportation, distribution, storage and the supporting information systems and its application. With the TEM, the functionalities and interfaces of its artefacts must be determined in advance [14]. All engineering activities are based on requirements as the decisive factor. Hence, a methodology that fits well in the description and management of requirements is necessary, that should support detectability amidst design decisions and system requirements. This also aids the association between stakeholders by allocating responsibilities. It also helps to derive the system structure with regards to software and hardware artefacts. This then assists in the trial of the implementation in comparison to the specification of the system.

One of the important phases in software engineering is RE [15,16]. Activities in RE involve requirements elicitation, modelling, analysis, negotiation and validation [17–21]. Effective RE practises lead to the improvement in software and system-development artefacts. The decisions derived using RE techniques aid in identifying customer problems to detailed specification. RE thus helps solve the problem and avoid catastrophic mistakes that could be made during the implementation stage. RE mistakes discovered in later stages could be very expensive to fix. According to Nuseibeh [18], RE represents a series of decisions that helps in recognising customer problems to the extent of detailed specification and, therefore, helps solve the problem [22]. The primary question behind RE activities is as follows: how can companies successfully arrive to an effective decision to start developing a new system, a subsystem, or a feature. The past two decades has perceived RE as a vital aspect of the SDLC. The most important and initial phase of RE is the elicitation of requirements. Elicitation helps in determining the tasks of the system and the goals that need to be met. This process helps to determine the right stakeholders as well. The requirement analyst is involved in analysing the information received from the stakeholders. After requirements elaborating, the stakeholder goals are determined. The system performs its required function based on the stakeholder goals (captured as 'hardgoals'). The goals analysed by the requirement analyst helps in building an effective software system. The requirement analyst also explores the possibility of implementing an effective system design option for a high-level alternative system [23].

With regards to an already established power system, numerous actors or stakeholders are involved with the management, operation and business aspects of the system. Also, various elements necessary for the generation, transportation and distribution of electrical energy is already established. Failures in the performance of the TEM system especially in its structural and functional planning end up being costly. An appropriate goal model must be used to model the requirements of the smart grid and traditional grid system. This will help in implementing a well-structured TEM process. This can be achieved through GORE model which can play a significant role in the TEM system. GORE characterises and models the objectives and stakeholders of the TEM systems, together with their relationships like decompositions, contributions and dependencies. Goal models can help to get the knowledge of who needs what and help to perform analyses to determine the satisfaction of goals [21,24–26].

In GORE, the requirements of the system are modelled by using goals. This technique involves eliciting, specifying, structuring, elaborating, analysing, documenting, negotiating and modifying requirements (based on goals) [17–20]. Goals play the most important role in GORE. Goals aid in establishing a domain and identify the intention of the stakeholder [24]. Goals can be established at a different abstraction levels, capturing strategic concerns to technical matters. Hence, goals are strategically planned as a very significant artefact in the early phases of RE [23]. A multi-view model or goal model illustrates the way in which goals, actors, states, objects, tasks and domain properties are interrelated in the given system [27]. This work focusses on a well-known and popular goal modelling framework known as the *i** model [20]. The *i** goal model [22] supports the essential processes of modelling organisations and sociotechnical systems. Some of the other widely used goal models are Knowledge Acquisition in Automated Space Model [28], NFRs model [24], Attributed Goal-Oriented Requirements Analysis Model [29], Tropos Model [30] and Goal-Oriented Requirement Language [31] Model.

The *i** goal model supports the essential processes of modelling organisational and sociotechnical systems. Hence, it can be used to address all the requirements of TEM system. Thus, it provides a suitable framework for the elicitation and management of TEM-based smart grid NFRs framework. In goal models, top softgoals are utilised as assessment criteria in existing quantitative [32–34] and qualitative approaches [24]. The process of goal model analysis involves generation of qualitative or quantitative values in the form of either forward propagation from the bottom softgoals to the top softgoals or backward propagation from the top softgoals to the bottom softgoals. The selected design alternative determines the satisfaction levels of softgoals. The top softgoal design that provides supreme fulfilment is chosen. Qualitative labels cause conflict in the proposed propagation algorithm. These labels are captured as *denied, partially denied, satisfied, partially satisfied and unknown* in the model. The qualitative approach has a downside to it as it delivers vagueness in the decision-making process. It leads to confusion when two alternatives have the same label or when a goal receives an unspecified conflict label. Quantitative approach as the name suggests uses crisp numbers. To express a requirement quantitatively, a stakeholder may use phrases which are practical but still vague in nature. These vague, uncertain, inappropriate or conflicting requirements are then expressed in linguistic terms with the representation of fuzzy numbers. Fuzzy numbers and fuzzy values are widely used in this chapter due to their appropriateness for expressing uncertainty [33,34].

Applications in the TEM system requires high accessibility, reliability, efficiency, interoperability and security as well as administrative consistency, adaptability and serviceability. Applications may be exposed to uncertainty and variability. Due to the interconnection with the real-world notions, TEM system requires more systematic techniques for capturing and reasoning about its framework. The early identification of goals helps the stakeholders inside and outside of the TEM power system. It enables reasoning with the specific NFRs (qualities) of the TEM systems as discussed earlier. With goals being determined and finalised early, it helps in understanding and provides a better knowledge based for the proposed system. The decision makers use goal modelling to determine whether the processes and tools are aligned with the goals of the TEM system. By systematically determining, elaborating and structuring TEM requirements goal reasoning can be performed.

The system performance or elements indicates its functional (or behavioural) requirements. The principle for checking the system performance is based on the NFRs rather than the specific behaviour. NFRs consist of the system qualities such as accessibility, reliability, interoperability and security of the system [15]. In comparison to the functional requirements, the NFRs like reliability, efficiency, interoperability and security have more influence on TEM system. NFR details criteria that can be used to analyse the performance of a system, rather than specific behaviours.

NFRs complements the functional requirements of a system and are often a result of the analysis of the efficiency, reliability, etc. But also other aspects such as security, active communication, accurate forecast, scalability, interoperability, efficiency, reliability may be involved. Many NFRs are important for the implementation of a successful TEM system. In this chapter, we have considered only a few significant NFRs like security, active communication, accurate forecast, scalability, interoperability, efficiency and reliability. The interoperability aids to combine various assets and applications into one operational system. A suitable structure has been developed by the GridWise Architecture Council [35] to assist the elicitation and management of interoperability requirements. It includes eight layers of various interoperability issues, as shown in Figure 3.2. For more details on interoperability issues, readers are directed to [35]. Another significant NFR is security. This ensures the security and privacy of the TEM system. Security should cover all aspects of the TEM system like information security, software security, physical security, hardware security, network and communication security and cloud services security. The key elements that should be considered while providing security are listed in Table 3.2 as ready reference. For providing reliable and efficient power management, the TEM system has to be able to scale to a significant number of households to provide an aggregated demand–response. This represents the scalability of the TEM system. The NFRs concerning the scalability of the TEM system is illustrated in Table 3.3.

An effective NFRs elicitation and understanding about the system helps to perform reasoning with the specific NFRs (qualities) of the TEM systems. In the next section, a detailed description of the requirements analysis process is presented for developing an optimised model of the TEM system.

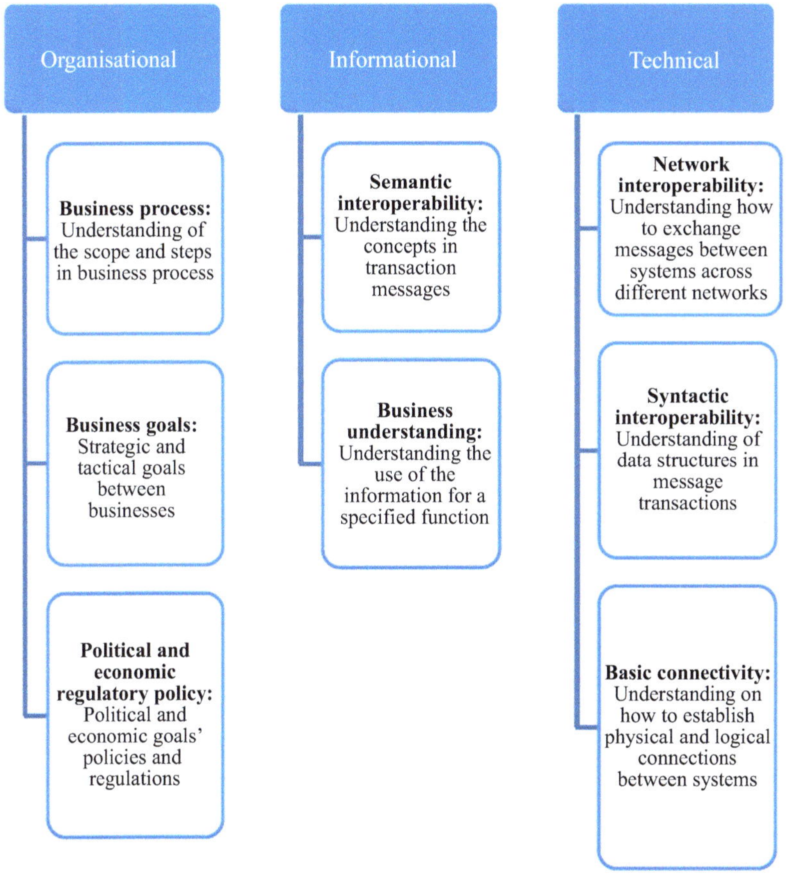

Figure 3.2 Interoperability levels

Table 3.2 Key elements in security

Key elements	Description
Authentication	Use an authentication mechanism
Trust	Earn or give, but never assume trust
Validation	Ensure all data are explicitly validated
Sensitivity	Identify sensitive data and use effective methods to handle it

3.3.1 The i* *goal modelling*

The *i** goal model within the GORE framework is an efficient tool utilised for modelling and analysing the dependencies between all the elements in a socio-economic community environment [16,19,25,36,37]. Hence, *i** framework is the

Table 3.3 Nonfunctional requirements concerned with scalability

Key elements	Description
Availability	Integration and verification of various devices to ensure 24/7 operation
Performance	System should have a real-time demand–response within less than 5 min response time
Scalability	1. Development of aggregation, forecasting and scheduling algorithms capable of managing large number of households 2. Development of a large-scale simulator to emulate the behaviour of large number of households 3. The complete electrical system and the numerous collection of appliances should be managed by the conception of an architecture that can be parallelisable, scalable, sound, productive and sturdy

preferred approach for modelling social relationships in the TEM system. In this proposed approach, the intentional strategic actor is modelled as the central unit. There are several characteristics that represent the intentional aspects of an actor such as goals, beliefs, ability and commitment [19]. An actor's intention is to successfully and strategically attain the goal. The structural relationship of an actor in comparison to other actors in sharing resources or accomplishing goals by performing some tasks is also a significant aspect. Among functional goals, in other words, 'hardgoals', some preferred behaviours are captured by the non-functional goals also known as 'softgoals' [27]. The i^* model aims at determining alternative choices through task-decomposition (OR-decomposition) reasoning. This is possible with explicit clear representation of the goals in the i^* model. The i^* goal modelling comprises two models representing the socio-economic systems: the strategic dependency (SD) model and the strategic rationale (SR) model [16,19,36,37].

The SD model displays a high-level explanation of a process or a system in the form of a graph. In a graph representation, this model exhibits the actor's dependency through behavioural goals or softgoals, tasks and resources. Figure 3.3 demonstrates an example of SD model where actors are depicted as circles, hardgoals as ovals, softgoals as cloud, resources as rectangles and tasks as hexagonal shapes. In the illustration SD model (Figure 3.3), an actor A depends upon actor B for achieving softgoal $SD1$. Actor B has a task dependency on actor D. Actor C in turn depends on actor C for a softgoal. Actor A has a goal dependency on actor A. Actor C has a resource dependency with actor A. Thus, the SD framework captures the dependency between various actors and hence captures the organisational context.

The SR model plays the role of capturing and displaying the internal modelling and analysis of all actors in the framework. This is achieved based on the actors internal intentional inter-dependencies. Nonfunctional goals or softgoals form intended qualities of the system. The SR model like the SD model is also depicted

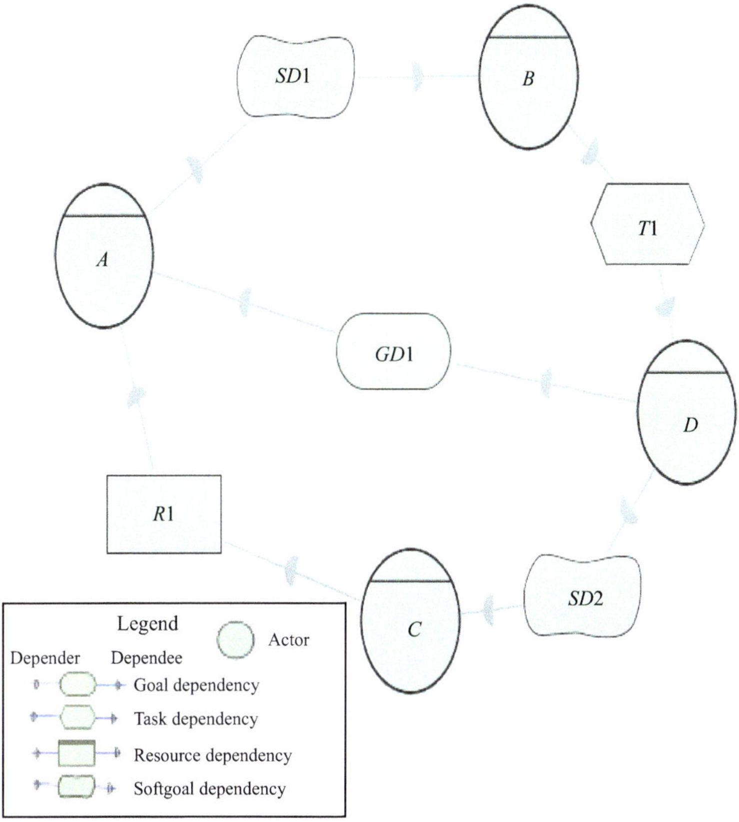

Figure 3.3 An example SD model

in the form of a graph. Through the graph, nodes are represented as goals or tasks or resources or softgoals that are interconnected by means-end links or task-decomposition links or contribution links [16,37]. The goals are connected to one or more tasks through *AND* (decomposition links) or *OR* (means-end links) relationships for accomplishing it. The contribution links can be *Make, Break, Help, Hurt, Some+, Some−*. These notions describe various types of contributions to various softgoals. This in turn leads to the satisfaction of softgoals [16,20,37]. Figure 3.4 demonstrates an example of an SR model.

The *i** model uses a top-down approach to identify the goals of each actors. This is achieved by breaking down the primary goals or hardgoals into a group of subgoals or tasks. Reasoning is performed by answering questions as follows: How to achieve? or What to achieve? By answering, How to achieve?, the softgoal is decomposed further. This decomposition is repeated till each softgoal is atomic in nature represented as operationalisation of softgoals. The following section displays how the goals and softgoals of an actor are analysed in the TEM system.

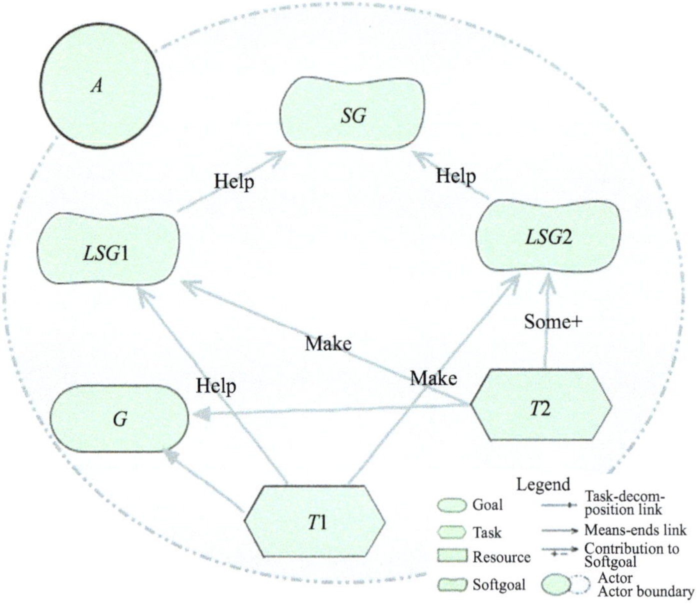

Figure 3.4 An example SR model

3.4 Requirements analysis and modelling of the TEM system

Requirement analysis and modelling is performed to determine an alternative design that provides an optimum fulfilment of the NFRs that are represented as softgoals [38–41]. This is achieved by defining a multi-objective linear optimisation function. This helps in determining the maximum satisfaction scores of the top softgoals for each alternative. Linear weighted sum method is used to integrate these multi-objective functions. The integrated linear function is solved for goal analysis, and an optimal design strategy is determined. For this purpose, IBM CPLEX optimisation tool is used [42]. An *i** goal model for TEM system is used with an actor through hierarchy of softgoals, goals and tasks. The proposed optimisation model is based upon the satisfaction scores of the top softgoals of the given *i** framework [43]. This is achieved by taking into consideration other softgoals within the hierarchy. Optimal alternative selection [39] is made on the basis of propagation of values through the entire hierarchy of softgoals. For the above process of requirements analysis, an *i** goal model is illustrated for developing an optimisation model of the TEM system.

3.4.1 Goal modelling of the TEM system

To model a generalised *i** framework for the TEM system in terms of softgoals, goals and tasks, let us consider a directed graph, G (N, A) where N as a set of nodes and A as set of arcs (Figure 3.5) [40,41,44,45]. The intentional elements such as softgoals, goals and tasks form the nodes of graph G and the means-end, task-decomposition and contribution links form the arcs of the graph G. An objective function for the optimisation model is formed in terms of the elements of the graph.

Based on the directed graph in Figure 3.5, an *i** goal model for the TEM system, to achieve an effective power energy management, is developed as shown in Figure 3.6. The developed goal model shows an actor, *power system*, that is considerably simplified but nevertheless requires some type of reasoning, namely identification and the exploration of alternatives. The aim of this system is to opt for the best alternative option on the basis of its influence on each of the softgoals. The framework has two alternatives, *traditional grid* and *smart grid*. The task of the requirement analyst is to choose an alternative that brings about maximum satisfaction to the NFRs that are represented by softgoals. The top softgoals can be thought of as nonfunctional objectives of the system, and therefore the problem can be viewed as a multi-objective optimisation problem. The selected alternative needs to maximise the satisfaction of the top objectives and hence forms the basis for maximisation optimisation problem [40,41]. A multi-objective maximisation optimisation problem is represented mathematically as follows:

$$\mathrm{Max}[F_1(\omega), F_2(\omega), F_3(\omega), \ldots, F_n(\omega)] \tag{3.1}$$

where $F_1, F_2, F_3, \ldots, F_n$ are scalar functions, ω is an element of X where X is the set of constraints.

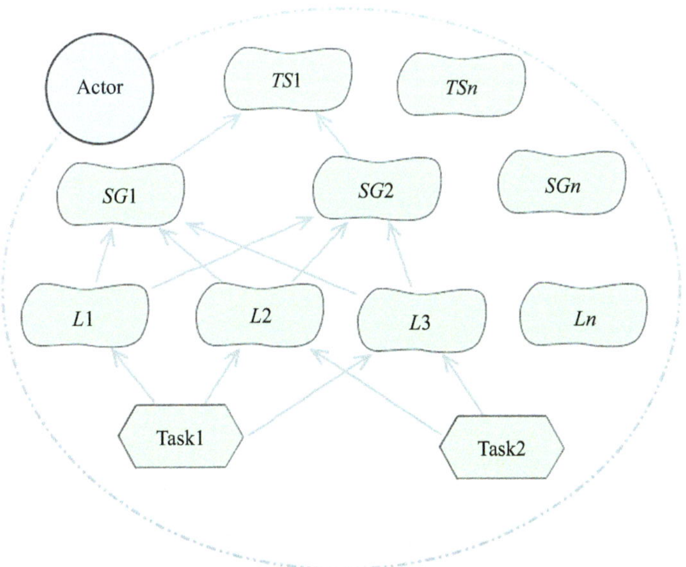

Figure 3.5 Directed graph representation of SR model for an actor with dependency

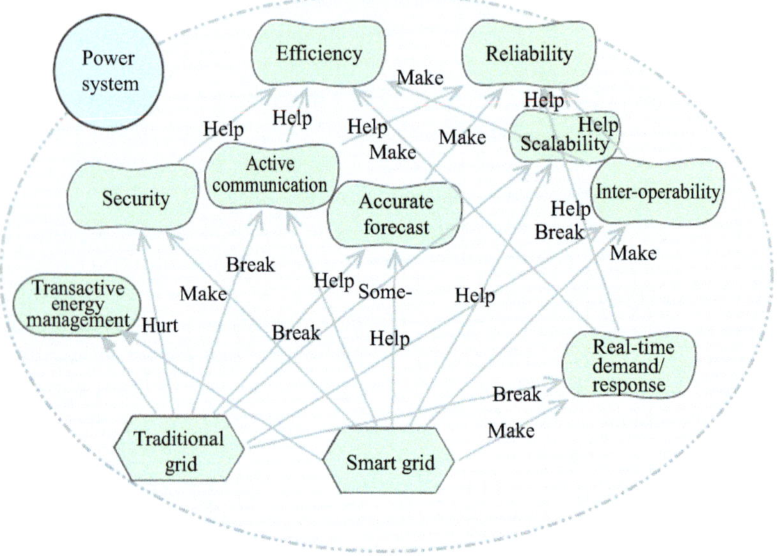

Figure 3.6 The i model of the proposed TEM system*

3.4.2 Methodology

A methodology has been proposed in this chapter to obtain an optimal strategy for the TEM system having multiple objectives.

The proposed method is presented as below:

- *Step 1:* Evaluating the scores of top softgoals based on different alternatives in the goal model.
- *Step 2:* Determine multi-objective optimisation functions based on the scores of top softgoals with respect to different alternatives.
- *Step 3:* Scalarisation of multi-objective functions using linear weighted sum optimisation method.
- *Step 4:* Applying linear programming model to obtain optimal strategy and decision-making.

3.4.3 Formalisation of multi-objective optimisation functions of the i* goal model

A generalised complete structure of the TEM *i** framework is modelled in this section by formalising the multiple objective functions in terms of softgoals, goals and tasks.

For easy understanding of the formalisation of our approach, consider a simple directed graph, as shown in Figure 3.5, whose nodes are goals (G) or leaf softgoals ($LS_1, LS_2, \ldots, LS_n$) or intermediary softgoals ($SG_1, SG_2, \ldots, SG_n$) or top softgoals

$(TS_1, TS_2, \ldots, TS_n)$. The leaf softgoals (the softgoals that are lower in the hierarchy) are assigned values as weights based on their relative importance in percentage. Let the weights of leaf softgoals LS_1 and LS_2 be ω_{L_1} and ω_{L_2}, respectively. The goal (G) can be achieved by either of the two tasks (alternatives) (A_1 and A_2). The contributions of goals or tasks to softgoals are represented by impacts that indicates the extent to which an alternative design option fulfils the leaf softgoal. Impacts are *Make, Help, Hurt, Break, Some−, Some+* which are represented as fuzzy numbers that indicates the extent to which an alternative option fulfils the leaf softgoal [33]. Both triangular and trapezoidal fuzzy numbers can be executed easily and can be quickly calculated. In this chapter, triangular fuzzy numbers are used, as to begin with a triangular membership function is the easiest way. Besides, triangular fuzzy numbers denote fuzzy numbers, whereas trapezoidal fuzzy numbers denote fuzzy intervals.

The effects of the softgoal preferences are disseminated to the top softgoals. This helps in finding the extent of satisfaction or scores of the top softgoals. The leaf softgoal scores are transmitted backwards to find the scores of the softgoals that are higher in the hierarchy. Softgoals receive numerous contribution links. For more details on generating scores, readers are directed to [33,34]. The proposed approach in this chapter is an updated version of [40].

In the proposed approach, initially each top softgoal's scores are calculated based on its inter-actor dependency under each alternative.

Consider a node that represents a leaf softgoal in the i^* model. Let $\omega_{L_{ik}}$ represents the weight of ith leaf softgoal of jth actor. From Figure 3.5, $I_{L_{ijk}}$ means the impact on ith leaf softgoal of jth alternative. Let $\omega_{L_{ijk}}$ represent the weight of ith leaf softgoal for actor k at level zero. Then the score of ith leaf softgoal for jth alternative for the kth actor is as follows:

$$score_{L_{ijk}} = I_{L_{ijk}} \times \omega_{L_{ijk}} + \sum_{d_{L_i}=1}^{n_{d_i}} \left(score_{d_{L_i}} \times I_{d_{L_i}} \right) \tag{3.2}$$

where $score_{d_{L_i}}$ is the score of d_Lth dependent for the ith leaf softgoal, $I_{d_{L_i}}$ is the d_Lth dependent impact for the ith leaf softgoal and n_{d_i} is the number of dependencies for the ith leaf softgoal (i.e. at level zero).

The approach is illustrated with the proposed TEM system as running example shown in Figure 3.6. The decision maker's task is to choose an ideal design (alternative) option from the presented choices. An objective function is generated for each alternative based on the elements of the graph.

In the proposed TEM system, each leaf softgoal is pre-assigned a unique weight that can help to select the best optimal design option for achieving the top softgoal. Let the individual weights of leaf softgoals such as *security, active communication, accurate forecast, scalability, interoperability, real-time demand/ response* be $\omega_1, \omega_2, \omega_3, \omega_4, \omega_5$ and ω_6, respectively. For improving the readability of the paper, certain terms in the TEM system are abbreviated as shown in Table 3.4.

Table 3.4 Abbreviation of terms in TEM system

Terms	**Abbreviation**
Transactive energy management	*TEM*
Security	*SR*
Active communication	*AC*
Accurate Forecast	*AF*
Scalability	*SB*
Interoperability	*IO*
Real-time demand/response	*RDR*

The scores of the leaf softgoals for actor, *power system*, under the alternative, *traditional grid*, are calculated as follows:

$$\begin{aligned} score_{SR_{TraditionalGrid}} &= Hurt \times \omega_1 \\ score_{AC_{TraditionalGrid}} &= Break \times \omega_2 \\ score_{AF_{TraditionalGrid}} &= Break \times \omega_3 \\ score_{SB_{TraditionalGrid}} &= Some- \times \omega_4 \\ score_{IO_{TraditionalGrid}} &= Break \times \omega_5 \\ score_{RDR_{TraditionalGrid}} &= Break \times \omega_6 \end{aligned}$$

Similarly, the scores for the leaf softgoals under the alternative, *smart grid* are calculated and is represented as follows:

$$\begin{aligned} score_{SR_{smart-grid}} &= Make \times \omega_1 \\ score_{AC_{smart-grid}} &= Help \times \omega_2 \\ score_{AF_{smart-grid}} &= Help \times \omega_3 \\ score_{SB_{smart-grid}} &= Help \times \omega_4 \\ score_{IO_{smart-grid}} &= Make \times \omega_5 \\ score_{RDR_{smart-grid}} &= Make \times \omega_6 \end{aligned}$$

Let us assume that there are t hierarchy levels in the directed graph. All leaf softgoals are defined at level zero. Then, at level $t = 1$, the score of the ith softgoal for jth alternative for actor k is defined in the following equation:

$$score_{SG_{i_1jk}} = \sum_{x=1}^{n_c} \left(I_x \times score_{L_{xjk}} \right) + \sum_{d_{i1}=1}^{n_i} \left(score_{d_{i1}} \times I_{d_{i1}} \right) \tag{3.3}$$

where the number of children is represented as n_c for each ith softgoal at level one and the number of dependencies at level one for ith softgoal is represented as n_i.

The score of softgoals at level one depends on the score of its leaf softgoal; (3.3) can be rewritten as

$$\begin{aligned} score_{SG_{i_1jk}} &= I_1 \times score_{L_{1jk}} + I_2 \times score_{L_{2jk}} + \cdots \\ &\quad + I_{n_c} \times score_{L_{n_cjk}} + \sum_{d_{i1}=1}^{n_i} \left(score_{d_{i1}} \times I_{d_{i1}} \right) \end{aligned} \tag{3.4}$$

Substituting with (3.3), (3.4) becomes

$$
\begin{aligned}
score_{SG_{i_1jk}} = I_1 \times \left(I_{L_{1jk}} \times \omega_{L_{1jk}} + \sum_{d_{L_1}=1}^{n_{d_1}} \left(score_{d_{L_1}} \times I_{d_{L_1}} \right) \right) \\
+ I_2 \times \left(I_{L_{2jk}} \times \omega_{L_{2jk}} + \sum_{d_{L_2}=1}^{n_{d_2}} \left(score_{d_{L_2}} \times I_{d_{L_2}} \right) \right) + \cdots \\
+ I_{n_c} \times \left(I_{L_{n_c}jk} \times \omega_{L_{n_c}jk} + \sum_{d_{L_{nc}}=1}^{n_{d_{nc}}} \left(score_{d_{L_{nc}}} \times I_{d_{L_{nc}}} \right) \right) \\
+ \sum_{d_{i1}=1}^{n_i} \left(score_{d_{i1}} \times I_{d_{i1}} \right)
\end{aligned}
\tag{3.5}
$$

Thus, it propagates upwards. Let us consider that there are m number of softgoals, n_c children and n_d dependencies for the ith softgoal. Then the score of any softgoal at level $t > 1$ is determined by calculating the product of each child score and its impact.

Hence, for an actor at level t, the softgoal's score with a dependency relationship is formalised as follows:

$$
\begin{aligned}
score_{SG_{i_tjk}} = \Pi_{l=1}^{t} I_{ijl} \sum_{i=1}^{m} \left[\sum_{d=1}^{n_c} \left(I_{d_{ij}} \times I_{d_{L_{ijk}}} \times \omega_{d_{L_{ijk}}} \right) \right. \\
\left. + \sum_{y=1}^{n_c} \left(\sum_{b=1}^{n_d} \left(score_{id_{by}} \times I_{id_{by}} \right) \right) + \sum_{b=1}^{n_d} \left(score_{id_b} \times I_{id_b} \right) \right]
\end{aligned}
\tag{3.6}
$$

In the TEM system, for actor *power system*, the score of top softgoals, *efficiency* and *reliability* under both alternatives *traditional grid* and *smart grid* are calculated. For a straightforward calculation, defuzzification is used to translate the effects that are represented in fuzzy numbers to quantifiable values [46]. These defuzzified values as shown in Table 3.5 are used to evaluate the scores of each

Table 3.5 Defuzzified impact values

Impacts	**Fuzzy value**	**Defuzzified value**
Make	(0.64, 0.8, 1)	0.8
Hurt	(0, 0.16, 0.32)	0.16
Some+	(0.32, 0.48, 0.64)	0.48
Break	(0, 0, 0.16)	0
Help	(0.48, 0.64, 0.80)	0.64
Some−	(0.16, 0.32, 0.48)	0.32

softgoal. Therefore, for actor *power system*, the score of top softgoals, *Efficiency* and *Reliability* under both alternatives are calculated as

$$\begin{aligned}
score_{Efficiency_{TraditionalGrid}} &= Help \times score_{SR_{TraditionalGrid}} + Help \times score_{AC_{TraditionalGrid}} \\
&\quad + Make \times score_{RDR_{TraditionalGrid}} + Make \times score_{IO_{TraditionalGrid}} \\
&= 0.64 \times (0.16 \times \omega_1) + 0.64 \times (0 \times \omega_2) + 0.8 \times (0 \times \omega_5) \\
&\quad + 0.8 \times (0 \times \omega_6) \\
&= 0.1024 \times \omega_1
\end{aligned}$$

$$\begin{aligned}
score_{Reliability_{TraditionalGrid}} &= Help \times score_{AC_{TraditionalGrid}} + Make \times score_{AF_{TraditionalGrid}} \\
&\quad + Help \times score_{SB_{TraditionalGrid}} + Help \times score_{IO_{TraditionalGrid}} \\
&\quad + Help \times score_{RDR_{TraditionalGrid}} \\
&= 0.64 \times (0 \times \omega_2) + 0.8 \times (0 \times \omega_3) + 0.64 \times (0.32 \times \omega_4) \\
&\quad + 0.64 \times (0 \times \omega_5) + 0.64 \times (0 \times \omega_6) \\
&= 0.2048 \times \omega_4
\end{aligned}$$

$$\begin{aligned}
score_{Efficiency_{smart-grid}} &= Help \times score_{SR_{smart-grid}} + Help \times score_{AC_{smart-grid}} \\
&\quad + Make \times score_{RDR_{smart-grid}} + Make \times score_{IO_{smart-grid}} \\
&= 0.64 \times (0.8 \times \omega_1) + 0.64 \times (0.64 \times \omega_2) \\
&\quad + 0.8 \times (0.8 \times \omega_5) + 0.8 \times (0.8 \times \omega_6) \\
&= 0.512 \times \omega_1 + 0.4096 \times \omega_2 + 0.64 \times \omega_5 + 0.64 \times \omega_6
\end{aligned}$$

$$\begin{aligned}
score_{Reliability_{smart-grid}} &= Help \times score_{AC_{smart-grid}} + Make \times score_{AF_{smart-grid}} \\
&\quad + Help \times score_{SB_{smart-grid}} + Help \times score_{IO_{smart-grid}} \\
&\quad + Help \times score_{RDR_{smart-grid}} \\
&= 0.64 \times (0.64 \times \omega_2) + 0.8 \times (0.64 \times \omega_3) \\
&\quad + 0.64 \times (0.64 \times \omega_4) + 0.64 \times (0.8 \times \omega_5) + 0.64 \times (0.8 \times \omega_6) \\
&= 0.4096 \times \omega_2 + 0.512 \times \omega_3 + 0.4096 \times \omega_4 \\
&\quad + 0.512 \times \omega_5 + 0.512 \times \omega_6
\end{aligned}$$

Then the objective function of top softgoals under each alternative for an actor are formalised from the scores as shown in (3.6) with the assumption that no inter-actor dependency relationship is considered. Let us consider that for an actor in the *i** model has *n* number of alternative design options, then there is a need to define *n* number of objective functions (maximum or minimum) for each top softgoal. Hence, to obtain the maximum score for the top softgoal under each alternative, the n maximum objective functions are defined as follows:

$$\begin{aligned}
F_i(\omega_1) &= score_{SG_{i1k}} \\
&= \text{Max } \Pi_{l=1}^{t} I_{i1l} \sum_{i=1}^{m} \left[\sum_{d=1}^{n_c} \left(I_{d_{i1}} \times I_{d_{L_{i1k}}} \times \omega_{d_{L_{i1k}}} \right) \right. \\
&\quad + \sum_{y=1}^{n_c} \left(\sum_{b=1}^{n_d} \left(score_{id_{by}} \times I_{id_{by}} \right) \right) \\
&\quad \left. + \sum_{b=1}^{n_d} \left(score_{id_b} \times I_{id_b} \right) \right] \\
F_i(\omega_2) &= score_{SG_{i2k}} \\
&= \text{Max } \Pi_{l=1}^{t} I_{i2l} \sum_{i=1}^{m} \left[\sum_{d=1}^{n_c} \left(I_{d_{i2}} \times I_{d_{L_{i2k}}} \times \omega_{d_{L_{i2k}}} \right) \right. \\
&\quad + \sum_{y=1}^{n_c} \left(\sum_{b=1}^{n_d} \left(score_{id_{by}} \times I_{id_{by}} \right) \right) \\
&\quad \left. + \sum_{b=1}^{n_d} \left(score_{id_b} \times I_{id_b} \right) \right] \\
&\quad \ldots \\
&\quad \ldots \\
&\quad \ldots \\
F_i(\omega_n) &= score_{SG_{ink}} \\
&= \text{Max } \Pi_{l=1}^{t} I_{inl} \sum_{i=1}^{m} \left[\sum_{d=1}^{n_c} \left(I_{d_{in}} \times I_{d_{L_{ink}}} \times \omega_{d_{L_{ink}}} \right) \right. \\
&\quad + \sum_{y=1}^{n_c} \left(\sum_{b=1}^{n_d} \left(score_{id_{by}} \times I_{id_{by}} \right) \right) \\
&\quad \left. + \sum_{b=1}^{n_d} \left(score_{id_b} \times I_{id_b} \right) \right]
\end{aligned} \tag{3.7}$$

Such that

$$0 \leq \omega_{d_{L_{jk}}} \leq 100 \, for \; d = 1 \; to \; n_c$$

Based on the objective functions defined in (3.7), objective functions that have to be maximised for actor *power system* in the TEM system are represented as follows:

$$F_{Efficiency}(\omega)_{TraditionalGrid} = \text{Max}\left(score_{Efficiency_{TraditionalGrid}}\right)$$

$$= \text{Max}\,(0.1024 \times \omega_1)$$

$$F_{Efficiency}(\omega)_{smart-grid} = \text{Max}\left(score_{Efficiency_{smart-grid}}\right)$$

$$= \text{Max}\,(0.512 \times \omega_1 + 0.4096 \times \omega_2 + 0.64 \times \omega_5 + 0.64 \times \omega_6)$$

$$F_{Reliability}(\omega)_{TraditionalGrid} = \text{Max}\left(score_{ReliabilityTraditionalGrid}\right)$$

$$= \text{Max}\,(0.2048 \times \omega_4)$$

$$F_{Reliability}(\omega)_{smart-grid} = \text{Max}\left(score_{Reliabilitysmart-grid}\right)$$

$$= \text{Max}(0.4096 \times \omega_2 + 0.512 \times \omega_3 + 0.4096 \times \omega_4$$

$$+ 0.512 \times \omega_5 + 0.512 \times \omega_6)$$

In the next step of formalisation, the obtained multi-objective functions of top softgoals, that has to be maximised, are integrated to a single objective function, for optimisation, using linear weighted sum method. By applying linear weighted sum method, the set of multi-objective functions are scalarised into a single objective. This is performed by adding each objective functions pre-multiplied by a user provided weight (W). The weight of an objective is predefined in proportion to its relative importance among all top softgoals in the *i** model. The linear weighted sum scalarisation process is defined as follows:

$$\text{Max } F(\omega) = \sum_{i,j=1}^{n,k} W_i \times F_i(\omega)_{A_j} \tag{3.8}$$

$$W_i \geq 0 \;\; for \;\; i = 1 \;\; to \;\; n$$

$$A_j \geq 0 \;\; for \;\; j = 1 \;\; to \;\; k$$

where W_i denotes the weight assigned to each objective function F_i under each alternative A_j.

For obtaining the optimal reasoning for the TEM system, a more subjective preference is given to *efficiency* than *reliability*; therefore, different weighting values are set to both objectives. Let *s* consider that the weights assigned to both top softgoals, *efficiency* and *reliability*, are W_1 and W_2, respectively. Also assume that the objective *efficiency* is three times more important than the objective *reliability*. Hence, the scalarised single objective function $F(\omega)$ that needs to be maximised is represented as follows:

$$\begin{aligned}\text{Max } F(\omega) &= W_1 \times F_{Efficiency}(\omega) + W_2 \times F_{Reliability}(\omega)\\ &= 3 \times F_{Efficiency}(\omega) + 1 \times F_{Reliability}(\omega)\end{aligned}$$

Therefore, the linearly integrated objective function under the alternative *traditional grid* is represented as follows:

$$\begin{aligned}\text{Max } F_{TraditionalGrid}(\omega) &= W_1 \times F_{Efficiency}(\omega)_{TraditionalGrid}\\ &\quad + W_2 \times F_{Reliability}(\omega)_{TraditionalGrid}\\ &= 3 \times F_{Efficiency}(\omega)_{TraditionalGrid}\\ &\quad + 1 \times F_{Reliability}(\omega)_{TraditionalGrid}\end{aligned}$$

Similarly, the linearly integrated objective function under the alternative *smart grid* is represented as follows:

$$\begin{aligned}\text{Max } F_{smart-grid}(\omega) &= W_1 \times F_{Efficiency}(\omega)_{smart-grid} + W_2 \times F_{Reliability}(\omega)_{smart-grid}\\ &= 3 \times F_{Efficiency}(\omega)_{smart-grid} + 1 \times F_{Reliability}(\omega)_{smart-grid}\end{aligned}$$

To evaluate the proposed reasoning method based on the *i** goal model, IBM ILOG CPLEX optimisation tool is incorporated. The robust optimisation problems are solved by using the optimisation package CPLEX, implemented in Java code, which guarantees global optimality for mixed integer programming. The optimal values of the unified objective functions optimised by the IBM ILOG CPLEX optimisation tool are provided in Table 3.6 as ready reference.

The results indicate that the alternative *smart grid* has a higher value than *traditional grid*. This means that by choosing the *smart grid* strategy, the TEM system can achieve the top softgoals in the *i** goal model reciprocally. This case study on TEM system illustrates that the proposed approach can be scaled up and applied to a reasonably complex scenario in practice.

Table 3.6 Maximum objective function values for the TEM system

Maximum objective functions	Traditional grid	Smart grid
$F_{Efficiency}(\omega)$	12.4	64
$F_{Reliability}(\omega)$	0	51.2

3.5 Conclusion

This chapter covers TEM and the importance of RE in TEM, applied to micro-grid and smart grid. The efficient and reliable functioning of the electric power system by using RE in TEM has been discussed and analysed. The aim is to help decision-makers to achieve higher efficiency and reliability of the power system. With this aim in mind, an approach to analyse the NFRs for the TEM system has been proposed. The *i** goal modelling approach has been applied to formalise the multi-objective functions of the NFRs. These requirements may be incomplete in some of the settings. In requirements analysis, new requirements may arise or existing requirements may become more or less prominent. The prominence of the requirements is based on the characteristics of a particular domain. This is the main drawback to this approach. In the future, empirical validation of the proposed approach will be conducted. Plan to develop a tool that can perform optimal goal analysis will also be explored.

References

[1] Sinha R, Patil S, Pang C, *et al.* Requirements engineering of industrial automation systems: Adapting the CESAR requirements meta model for safety-critical smart grid software. In: Industrial Electronics Society, IECON 2015 – 41st Annual Conference of the IEEE. IEEE; 2015. p. 002172–002177.

[2] Atamturk N, and Zafar M. Transactive energy: A surreal vision or a necessary and feasible solution to grid problems. San Francisco, CA: California Public Utilities Commission Policy & Planning Division; 2014.

[3] Sumesh S, Potdar V, and Krishna A. Clustered prosumer penalty framework on smart grid. In: Advances in Smart Grid and Renewable Energy. Singapore: Springer; 2018. p. 1–12.

[4] Jacobson I. The unified software development process. Chennai: Pearson Education India; 1999.

[5] Clancy T. The Standish Group CHAOS Report. Project Smart; 2014.

[6] Felkins K, Leigh H, and Jankovic A. The royal mail ship Titanic: Did a metallurgical failure cause a night to remember? JOM. 1998;50(1):12–18.

[7] Krishna A, Ghose AK, and Vilkomir SA. Co-evolution of complementary formal and informal requirements. In: Software Evolution, 2004. Proceedings. 7th International Workshop on Principles of. IEEE; 2004. p. 159–164.

[8] Rohjans S, Dänekas C, and Uslar M. Requirements for smart grid ICT-architectures. In: Innovative Smart Grid Technologies (ISGT Europe), 2012 3rd IEEE PES International Conference and Exhibition on. IEEE; 2012. p. 1–8.

[9] Zafar N, Arnautovic E, Diabat A, *et al.* System security requirements analysis: A smart grid case study. Systems Engineering. 2014;17(1):77–88.

[10] Ericsson GN. Cyber security and power system communication essential parts of a smart grid infrastructure. IEEE Transactions on Power Delivery. 2010;25(3):1501–1507.

[11] Case AF. Computer-aided software engineering (CASE): Technology for improving software development productivity. ACM SIGMIS Database: the DATABASE for Advances in Information Systems. 1985;17(1):35–43.
[12] Bitsch F. Safety patterns the key to formal specification of safety requirements. In: International Conference on Computer Safety, Reliability, and Security. Springer; 2001. p. 176–189.
[13] Affleck A, Krishna A, and Achuthan NR. Non-functional requirements framework: A mathematical programming approach. The Computer Journal. 2014;58(5):1122–1139.
[14] Uslar M, Specht M, Dänekas C, *et al.* Standardization in smart grids: Introduction to IT-related methodologies, architectures and standards. Heidelberg: Springer Science & Business Media; 2012.
[15] Pohl K. Requirements engineering: Fundamentals, principles, and techniques. New York: Springer Publishing Company, Incorporated; 2010.
[16] Yu E, and Mylopoulos J. Why goal-oriented requirements engineering. In: Proceedings of the 4th International Workshop on Requirements Engineering: Foundations of Software Quality. vol. 15; 1998. p. 15–22.
[17] Hofmann HF, and Lehner F. Requirements engineering as a success factor in software projects. IEEE Software. 2001;18(4):58.
[18] Nuseibeh B, and Easterbrook S. Requirements engineering: A roadmap. In: Proceedings of the Conference on the Future of Software Engineering. ACM; 2000. p. 35–46.
[19] Horkoff J, and Yu E. Interactive goal model analysis for early requirements engineering. Requirements Engineering. 2016;21(1):29–61.
[20] Yu ES, and Mylopoulos J. From ER to AR: Modelling strategic actor relationships for business process re-engineering. International Journal of Cooperative Information Systems. 1995;4(02n03):125–144.
[21] Vilkomir SA, Ghose AK, and Krishna A. Combining agent-oriented conceptual modelling with formal methods. In: Software Engineering Conference, 2004. Proceedings. 2004 Australian. IEEE; 2004. p. 147–155.
[22] Requirements Working Group, International Council on Systems Engineering (INCOSE). Guide for writing requirements. INCOSE; 2012. https://tcsd.instructure.com/files/99427/download?download_frd=1; accessed May 2019.
[23] Franch X, López L, Cares C, *et al.* The *i** framework for goal-oriented modeling. In: Domain-Specific Conceptual Modeling. Cham: Springer; 2016. p. 485–506.
[24] Mylopoulos J, Chung L, and Yu E. From object-oriented to goal-oriented requirements analysis. Communications of the ACM. 1999;42(1):31–37.
[25] Salim F, Chang C, Krishna A, *et al.* Towards executable specification: combining *i** and AgentSpeak (L); 2005.
[26] Subramanian CM, Krishna A, and Kaur A. Game Theory-based requirements analysis in the *i** framework. The Computer Journal. 2017;61(3):427–446.
[27] van Lamsweerde A. Goal-oriented requirements engineering: A roundtrip from research to practice [engineering read engineering]. In: Requirements

Engineering Conference, 2004. Proceedings. 12th IEEE International. IEEE; 2004. p. 4–7.
[28] Dardenne A, Fickas S, and van Lamsweerde A. Goal-directed concept acquisition in requirements elicitation. In: Proceedings of the 6th international Workshop on Software Specification and Design. IEEE Computer Society Press; 1991. p. 14–21.
[29] Kaiya H, Horai H, and Saeki M. AGORA: Attributed goal-oriented requirements analysis method. In: Requirements Engineering, 2002. Proceedings. IEEE Joint International Conference on. IEEE; 2002. p. 13–22.
[30] Bresciani P, Perini A, Giorgini P, *et al.* Tropos: An agent-oriented software development methodology. Autonomous Agents and Multi-Agent Systems. 2004;8(3):203–236.
[31] Amyot D, Ghanavati S, Horkoff J, *et al.* Evaluating goal models within the goal-oriented requirement language. International Journal of Intelligent Systems. 2010;25(8):841–877.
[32] Affleck A, and Krishna A. Supporting quantitative reasoning of non-functional requirements: A process-oriented approach. In: Proceedings of the International Conference on Software and System Process. IEEE Press; 2012. p. 88–92.
[33] Subramanian C, Krishna A, and Gopalan R. Quantitative reasoning of goal satisfaction in the *i** framework. In: SEKE; 2015. p. 666–669.
[34] Subramanian C, Krishna A, and Kaur A. Reasoning about goal satisfaction for early requirements engineering in the *i** framework using inter-actor dependency. In: PACIS; 2015. p. 89.
[35] Council GA. GridWise transactive energy framework: Version 1.0. Pacific Northwest National Laboratory, PNNL-22946 Ver1 0; 2015.
[36] Yu ES. Towards modelling and reasoning support for early-phase requirements engineering. In: Requirements Engineering, 1997. Proceedings of the Third IEEE International Symposium on. IEEE; 1997. p. 226–235.
[37] Yu E. Agent-oriented modelling software versus the world. In: International Workshop on Agent-Oriented Software Engineering. New York: Springer; 2001. p. 206–225.
[38] Burgess C, Krishna A, and Jiang L. Towards optimising non-functional requirements. In: Quality Software, 2009. QSIC'09. 9th International Conference on. IEEE; 2009. p. 269–277.
[39] Affleck A, Krishna A, and Achuthan NR. Optimal selection of operationalizations for non-functional requirements. In: Proceedings of the Ninth Asia-Pacific Conference on Conceptual Modelling-Volume 143. Australian Computer Society, Inc.; 2013. p. 69–78.
[40] Subramanian C, Krishna A, and Kaur A. Optimal goal programming of softgoals in goal-oriented requirements engineering. In: PACIS; 2016. p. 202.
[41] Subramanian C, Krishna A, and Kaur A. Optimal reasoning of goals in the *i** framework. In: Software Engineering Conference (APSEC), 2015 Asia-Pacific. IEEE; 2015. p. 346–353.

[42] Lima R. IBM ILOG CPLEX-What is inside of the box? In: Proc. 2010 EWO Seminar; 2010.
[43] Subramanian CM, Krishna A, and Kaur A. Sensitivity analysis of the *i** optimisation model. JSW. 2016;11(1):10–26.
[44] Sumesh S, Krishna A, and Subramanian C. CEA based reasoning with the I — Framework. In: PACIS; 2018.
[45] Sumesh S, Krishna A, and Subramanian CM. Optimal reasoning of opposing non-functional requirements based on game theory. In: ISD; 2018.
[46] Chou SY, Chang YH, and Shen CY. A fuzzy simple additive weighting system under group decision-making for facility location selection with objective/subjective attributes. European Journal of Operational Research. 2008;189(1):132–145.

Chapter 4

Transformation of microgrid to virtual power plant

Robert Lis[1] *and Robert Czechowski*[1]

In this chapter, the problem of microgrid (MG) and virtual power plant (VPP) concepts are considered as the most promising solutions to integrate distributed generation resources into the electric power system. Currently, the main drawback of distributed energy resource (DER) is that many of them are operating individually without any coordination with other electrical power sources, which leads to decrease of main grid reliability. MG, as a group of DER and interconnected loads, can connect and disconnect from the grid to enable it to operate in both grid-connected or standalone/isolated mode. With emerging of advanced communication and information technology, the small DER units would be aggregated then monitored, controlled and treated as a single large power plant. Bidirectional power flows, stability issues, uncertainty are the challenges to MGs. These can be overcome by integrating MGs as VPP. This concept can strengthen and diversified generation structure through decentralized dispatching and distribution energy networks. Additionally, a DER prosumers grouped together, controlled and coordinated by VPP may become significant part of the wholesale market. There are many approaches to the VPP concept. In this chapter, the concept of VPP for DER integration with power system is introduced and discussed. The focus will be paid to the modelling of proposed concept by MATLAB® Simulink™ software and the results of the simulation in different operation scenarios have been presented.

4.1 Introduction

As a result of climate changes, greenhouse gas emissions and ineffectiveness of power system with their spectacular breakdowns, the world of energy have changed for last years. The new technology development related to distributed generation (DG) and liberalization of the energy market have become a significant reason of the electrical structure transformation. In the face of conventional energy sources

[1]Department of Electrical Power Engineering, Wroclaw University of Science and Technology, Wrocław, Poland

depletion, the renewable energy sources (RES) are worthy of attention in order to meet with challenges.

The DG can be defined as energy generation close to the load as much as possible, so the reduction of energy transmission and better matching of customers' needs are main advantages. The continuous formation of solar, wind and biogas power plants are creating the foundation for new energy power structure. New emerging technologies like fuel cells, microturbines, flywheels, superconductors, highest energy density ultracapacitors, gas engines, or Stirling engine may find application in DG. Above-mentioned technologies can cause that energy transmission will become more decentralized [1].

The next step in evolution of power energy structure will be connection of different small DG to the common grid, where they work and cooperate in one large-scale energy system controlled by energy system management. Such a system is often called virtual power plant (VPP). Development of the VPP concept is possible through evolution of information and telecommunication technologies (e.g. intelligent metering and control) [2].

4.2 Evolution of electricity – the case of Polish electricity sector

Alternator discovered by Siemens in 1866 initiated the development of the power supply. In a dozen years later the first small dc power plant emerged. First larger scale transmission line with length of 50 km was built in 1882. The drawbacks of this transmission line were enormous power losses and voltage drops. Energy transfer over longer distances become possible with the move to alternating voltages, and its transformation to different parameters. In the nineties of the last century after the invention of the Doliwo–Dobrowolski three-phase generator starts with the rapid development of the power plant, which entails the need to build more and more lines. The first overhead three-phase line was built in 1891 in Frankfurt. This line worked on the voltage of 25 kV and had quite high efficiency (approx. 75%). Energy development is closely linked with the economic development of the region and the country. Especially modern industrial demand, high-reliability requirements, and the relevant parameters supplied electricity. In the 1980s observed dynamic urban development. Also observed an increase in the number of recipients of agro-breeding with significant power requirements, which require high reliability of supply. Political and social changes taking place in the country in the late 1880s caused a partial collapse of the economy or other of its orientation. Socialized factories began to privatize or had been liquidated. These developments resulted in a reduction in the demand for electricity. There was a gradual stabilization of the economy, especially the private sector. The increase in power demand for rural consumers created some problems in meeting those needs. Massive rural electrification in the post-World War II years has been directed mainly for lighting purposes. Electric motors up to 2.8 kW have been installed occasionally, because relatively few transformer stations of medium voltage (MV)/low voltage (LV) were built, and LV circuits were prolonged. The 1970s followed changes in the farms. Treadmills for exercising horses turned threshing and other activities have begun replacing electric motors. Then the problem arised that transformer stations MV/LV power are insufficient and too rarely integrated into the grid, the rural farmers have to agree among themselves about the days and hours with and without electricity supply.

After 1975, the electric power demand in the country began to grow, and existing transformer stations and LV networks were insufficient. Thus, began the so-called 're-electrification of rural areas', especially the village from the first years of mass electrification. In this regard, was developed a new government programme for the years 1985–95. The financial expenditure on the project had come from the Provincial Offices and the Ministry of Energy. An important factor limiting the connection of new customers or increasing their power consumption of the existing network became the Regulation of the Minister of Industry and Trade of 8 October 1990 and Polish 5009 Series Standards. Regulation and standards introduced significant tightening in relation to the applicable terms and conditions of protection against prior year 1990. Being the transformer stations have to meet with new conditions made it necessary to reduce the electrical circuits and increasing cross-wires.

In this case, a part of the LV network, upgraded before 1990, did not meet the new conditions and for the connection of the recipient was necessary to reupgrade. There was also a change in supply reliability LV network to increase the reliability of this operation is used in overhead lines, insulated cables, in particular wooded areas. Connections newly built insulated conductors are made to eliminate the recipients breaks caused short circuits bare conductors. New stations pole 15/0.4 kV is carried out on columns of circular 'whirling' because of their difficult accessibility to the public, as well as to reduce the occupied area. It introduces a simplification of the station pole, which is to apply new solutions arising from the practice in the field of surge protection, as well as pages 15 kV insulation. This reduces significantly the number of failures of fuses. In the 15 kV network, to avoid clipping of trees, if you need to apply a voltage of 15 kV, transformer station is used instead of overhead lines, which made wire naked, cables or overhead lines isolated. The densely built-up areas, especially in the cities, the construction of transformer stations and small-sized container, using the latest developments in the field of circuit terminals medium and LV, are envisaged (Figure 4.1).

At present, energy technologies have experienced noticeable improvement such as

- improvement the efficiency of solar cells from 15% up to 20%–24%;
- development of micro-, bio- and multifuel combined heat and power units;
- highly efficient fuel-cell technologies' development;
- wind power generator capacity increased from few kW up to several MW;
- increase of efficiency and capacity of storage devices and
- small hydrogenerators, tidal generators, etc. incorporated as new RES.

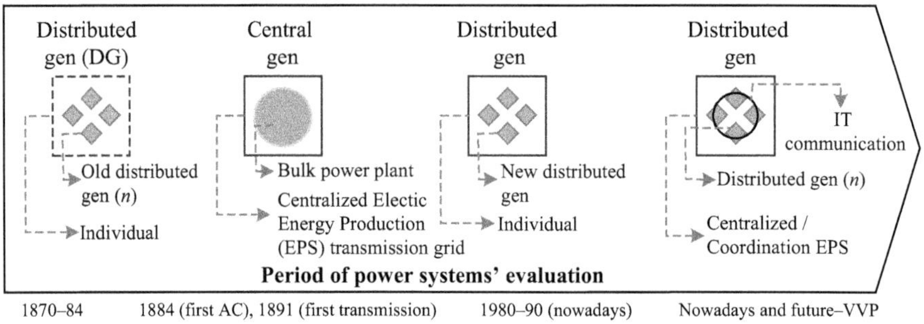

Figure 4.1 Timeline of major events in electric power system (MGs)

The observable development of power technologies aims to help reach up to 20% of energy supply from renewable resources in the EU countries by 2020 [3–5].

4.3 Liberalization of the energy markets

During the 1990s, when most of the national electricity and natural gas markets were still monopolized the European Union and the Member States decided to open these markets to competition gradually. In particular, the European Union decided to [3,6]

- distinguish clearly between competitive parts of the industry (e.g. supply to customers) and noncompetitive parts (e.g. operation of the networks);
- oblige the operators of the noncompetitive parts of the industry (e.g. the networks and other infrastructure) to allow third parties to have access to the infrastructure;
- free up the supply side of the market (e.g. remove barriers preventing alternative suppliers from importing or producing energy);
- remove gradually any restrictions on customers from changing their supplier and
- introduce independent regulators to monitor the sector.

The purpose of liberalization of an energy market is made in order to create more competitive market structures. The changeover of the monopoly market attracts more attention [7,8].

The separation between generation and transmission in liberalized electricity markets will end up with the situation where the VPP operator is positioned next to the network operator [9].

4.3.1 Future problem identification

In the context of liberalization of energy markets and increase concerns about environmental impact of conventional power plants, running a great number of different distributed energy resource (DER) units under new market conditions is inevitable, which include the following future challenges that will be met:

- Involvement of small modular power sources, storage technologies and controllable loads.
- Power output fluctuations due to weather dependent sources. Also, it causes economic penalties associated with unexpected unbalances and limits their contribution to the grid operations.
- Unification of standards for communication and execution devices.
- Many units are working isolated due to their different ownerships. Cooperation and communication often lack between neighbouring DER units. The unification of standards for communication and execution devices can help to satisfy the needs of whole grid instead of the local users.
- Combined DERs can also support the transmission system with reactive power to maintain voltage levels at transmission and distribution interfaces. Each single DER can contribute to maintain voltage level of distribution system busses and lines.

Nowadays, the way power system functions are changing drastically at all voltage levels of present structure – the digitalization of the power system takes place.

The discontinuous output of DG resources and evaluation of active power control strategies will have to be taken into account in order to facilitate the increase of DGs. Otherwise, continuing integration of DG in the distribution network in line with traditional control strategy will lead to different operational problems. In a distribution network that is supervised in a totally passive-like manner, the rise of DG supply will inevitably lead to the simultaneous voltage elevation, eventually resulting in very likely equipment impairments on the insulation background and hence possible harm to the customers on the receiving side.

One way to solve coming issues is to aggregate a numerous DER at different locations and size into a VPP, which then can be deployed like a power plant which is connected to the transmission system. This integrated approach enables DERs not only to face with upcoming challenges but participate in energy trade, but also supply sector of tertiary energy reserve. If tertiary reserve can be activated more quickly, the VPP operator will not need to call to independent transmission system operators (TSO) after the energy shortfall.

4.3.1.1 Voltage control issues

It is imperative to determine voltage decline at every given spot in the network in order to be capable of managing the necessary voltage level of the end customers. The voltage deviation is limited according to international standards like IEC 50160 in Europe [3,10,11]. If the deviation limits are represented in V_{max} and V_{min} for the illustrated network in Figure 4.2, the range of voltage variations in this network can be explained through three different demand and supply scenarios.

When the load demand is much larger than DG supply, the maximum load caused by the LV end users will decrease the voltage level at the primary side of MV/LV transformer. The built-in tap alters the HV/MV transformer, for the

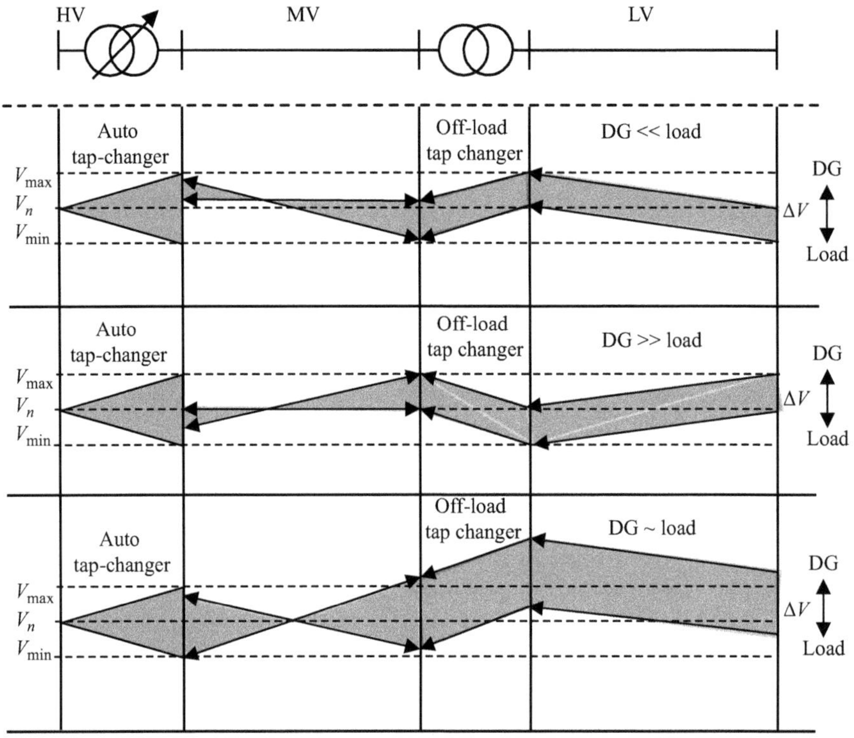

Figure 4.2 Voltage variations due to load and generation

counterbalancing purposes, will switch to a higher set-point. Any additional elevations of the LV load will eventually result in under voltages occurring in the outermost regions of MV and LV feeders. As the transformer stays intact (within the regulator's range), the regulator of the HV/MV transformer will cause this.

Due to a given constant set-point, the voltage level at the minor side of the MV/LV transformer will reach its maximum in situations involving minimum load. The built-in tap alter of HV/MV transformer will, in turn, switch to its minimum set-point to counterbalance the change. As a result, any consequent drops in the load spurred by the elevations in DG supply will overexceed the tolerance limits of V_{max}. In a situation like that, the end customers who find themselves nearby the MV/LV transformer may be a subject to some accidents caused by the flawed insulation.

The alternative to the above-mentioned scenario aiming to maintain the necessary voltage level could be the network operator fixing the set-point of MV/LV a little bit lower in cases where DG supply exceeds the load demand. In this scenario, the deficiency in DG supply will reduce the voltage level at the minor MV/LV transformer's side reaching its minimum level, V_{min}, which, in turn, is a subject to overrun caused by any of further decline in DG supply or surge of load demand. Even in cases where the network operator is capable of establishing an equilibrium between DG supply and load, a discontinuous nature of both supply and demand will nonetheless bring about a spectrum of voltage variations as depicted in the third scenario. The limits will, again, be subject to overrun, provided significant enough imbalances between supply and demand [12].

The above-mentioned scenarios testify the idea that distribution networks with DG controlled in a passive-like manner are consequently resultant of inadequate voltage control as both DG supply and load demand are of intermittent nature. One way to reduce the risks associated with the regulated voltage limits being exceeded could be, for instance, changing the structure and performance of this network, a switch from radial to meshed. By virtue of tap alter of most MV/LV transformers in the distribution network having a fixed set-point, the performance of both the load and DG at LV level are determinant of the voltage variations at LV as well as MV level [1].

The automatic tap-changer of the HV/MV transformer will attempt to keep the MV level within the tolerated limits, as shown through the arrows in Figure 4.3. Therefore, in such a distribution network, the regulator of the HV/MV transformer is also assigned to compensate load and supply changes at LV level. Since this compensation is limited by the range of regulation of the regulator, large differences from the designed load and supply ratio may move the MV and LV levels beyond the tolerated limits V_{max} and V_{min}.

4.3.1.2 Short-circuit protection issue

While the traditional power system can be seen as top-down supply system, nowadays DG will lead to bidirectional power supply. In case of top-down power supply, short-circuit currents in the distribution network will be driven from one source, centralized generation at HV level, while in bidirectional system, DG will provide multiple source contribution to the fault current as shown in Figure 4.3. In some cases, the bottom-up contribution of DG (I_c) reduces the top-down current (I_b) below the tripping level of the protection [8,13,14].

When minor impedances occur simultaneously with the surge in DG capacity, the flawed current augmentation (I_d) may trigger the mechanism protecting feeder 2, which thing in turn brings about some interruptions in the power supply to end customers utilizing feeder 2. If the voltage level oversteps the minimum frame of tolerance, the DG units will automatically switch off. Additionally, any voltage

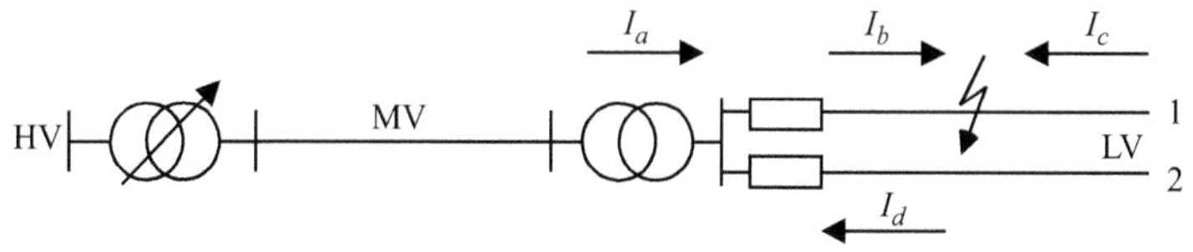

Figure 4.3 Contribution to short circuit current

declines triggered by the short-circuit flaws anywhere else in the distribution network may be responsible for the activation of protective mechanism of DG units provided the voltage limits within tolerance frames are cautiously set. It cannot be accounted, however, for single-phase flaws to an indirectly earthed MV feeder as the flaws may not be detected by the protective mechanisms of DG in the LV network. If, for instance, the MV feeder is somehow switched off by the protective mechanisms and DG supply at LV level has a value big enough at that given time then the LV network switches to operating as self-sufficient island, which situation may result in some unpredictable situations. In a nutshell, the benefits coming from the utilization of DGs greatly overweigh the inherent challenges that must be overcome in order to fully acknowledge them.

4.4 Microgrid turns to virtual power plant

Electricity networks are in the era of major transition from stable passive distribution networks with unidirectional electricity transportation to active distribution networks with bidirectional electricity transportation. Distribution networks without any DG units are passive since the electrical power is supplied by the national grid system to the customers embedded in the distribution networks. It becomes active when DG units are added to the distribution system leading to bidirectional power flows in the networks. Modern microgrids (MGs) may comprise sources of power varying from several hundred kilowatts (microsources) to several megawatts, namely [15]

- small Combined Heat and Power generation (CHP) plants (micro-CHP – μCHP) with synchronous generators, which are capable of operating in both active and reactive power control mode (when operating parallel with the system), and whose frequency and voltage can be controlled (when operating autonomously);
- small hydroelectric power plants with synchronous and asynchronous generators, capable of having generated active power remotely controlled;
- small photovoltaic (PV) power plants connected through inverters, which operate according to predefined frequency and voltage characteristics, and whose active power can be remotely controlled;
- energy storage systems connected through inverters, which can be quickly switched between charge/discharge operating mode and
- small power plants (PV, hydro, wind), whose generation is not controlled.

4.4.1 MGs structure and application

MGs are small-scale, LV combined heat and power (LVCHP) supply networks designed to supply electrical and heat loads for a small community, such as a housing estate or a suburban locality, or an academic or public community such as a university or school, a commercial area, an industrial site, a trading estate or a municipal region. MG is essentially an active distribution network because it is the conglomerate of DG systems and different loads at distribution voltage level.

The generators or microsources employed in an MGs are usually renewable/nonconventional DERs integrated together to generate power at distribution voltage. From operational point of view, the microsources must be equipped with power electronic interfaces (PEIs) and controls to provide the required flexibility to ensure operation as a single aggregated system and to maintain the specified power quality and energy output. This control flexibility would allow the MGs to present itself to the main utility power system as a single controlled unit that meets local energy needs for reliability and security [16–18].

The technical features of an MGs make it suitable for supplying power to remote areas of a country where supply from the national grid system is either difficult to avail due to the topology or frequently disrupted due to severe climatic conditions or man-made disturbances. From grid point of view, the main advantage of an MGs is that it is treated as a controlled entity within the power system. It can be operated as a single aggregated load. This ascertains its easy controllability and compliance with grid rules and regulations without hampering the reliability and security of the power utility. From customers' point of view, MGs are beneficial for locally meeting their electrical/heat requirements. They can supply uninterruptible power, improve local reliability, reduce feeder losses and provide local voltage support. From environmental point of view, MGs reduce environmental pollution and global warming through utilization of low-carbon technology. However, to achieve a stable and secure operation, a number of technical, regulatory and economic issues have to be resolved before MGs can become commonplace. Some problem areas that would require due attention are the intermittent and climate-dependent nature of generation of the DERs, low energy content of the fuels and lack of standards and regulations for operating the MGs in synchronism with the power utility. The study of such issues would require extensive real-time and off line research, which can be taken up by the leading engineering and research institutes across the globe.

4.5 Microgrid configuration

A typical MGs configuration is shown in Figure 4.4. It consists of electrical/heat loads and microsources connected through an LV distribution network. The loads and the sources are placed close together to minimize heat loss during heat transmission. The microsources have plug-and-play features. They are provided with PEIs to implement the control, metering and protection functions during stand-alone and grid-connected modes of operation. These features also help seamless transition of MGs from one mode to another. The MG consists of three radial feeders (A, B, C) to supply the electrical and heat loads. It also has two LVCHP and two non-CHP microsources and energy storage systems.

Microsources and storage devices are connected to feeders A and C through microsource controllers (MCs). Some loads on feeders A and C are assumed to be priority loads (i.e. requiring uninterrupted power supply), while others are non-priority loads. Feeder B contains only nonpriority electrical loads. The MG is

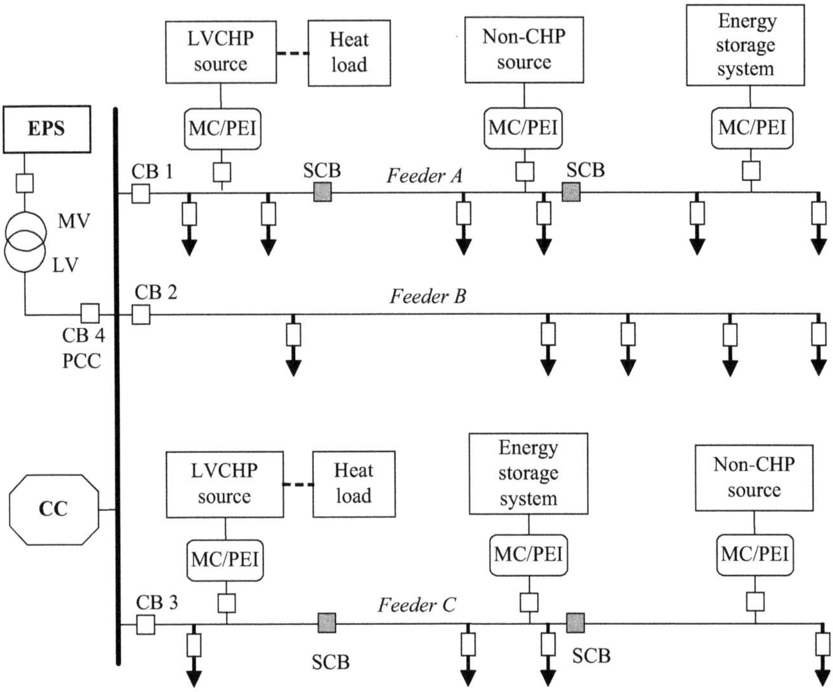

Figure 4.4 A typical microgrid configuration: LVCHP, low-voltage combined heat and power; MC, microsource controller; PEI, power electronic interface; SCB, sectionalizing circuit breaker; CC, central controller; CB, circuit breaker

coupled with the main MV utility grid through the point of common coupling – circuit breaker CB4 as per standard interface regulations. CB4 is operated to connect and disconnect the entire MGs from the main grid as per the selected mode of operation. Feeders A, B and C can however be connected and disconnected by operating breakers CB1, CB2 and CB3, respectively. The microsources on feeders A and C are placed quite apart from the MGs bus to ensure reduction in line losses, good voltage profile and optimal use of waste heat. Although the control of power flow and voltage profile along radial feeders is quite complicated when several microsources are connected to a common radial feeder and not to a common generator bus, this configuration is necessary to avail the plug-and-play feature of the microsources.

The MG is operated in two modes:

- Grid-connected
- Standalone/isolated

In grid-connected mode, the MG remains connected to the main grid either totally or partially, and imports or exports power from or to the main grid. In case of any disturbance in the main grid, the MG switches over to stand-alone mode while still feeding power to the priority loads. Either can achieve this:

1. Disconnecting the entire MGs by opening CB4,
2. Disconnecting feeders A and C by opening CB1 and CB3.

For option (1), the MGs will operate as an autonomous system with all the microsources feeding all the loads in feeders A, B and C, whereas for option (2), feeders A and C will supply only the priority loads while feeder B will be left to ride through the disturbance.

The operation and management of MGs in different modes is controlled and coordinated through local MCs and the central controller (CC).

4.6 Microsource controller

The main function of MC is to independently control the power flow and load-end voltage profile of the microsource in response to any disturbance and load changes. Here 'independently' implies without any communications from the CC. MCs also participate in economic generation scheduling, load tracking/management and demand-side management by controlling the storage devices. It must also ensure that each microsource rapidly picks up its generation to supply its share of load in stand-alone mode and automatically comes back to the grid-connected mode with the help of CC. The most significant aspect of MC is its quickness in responding to the locally monitored voltages and currents irrespective of the data from the neighbouring MCs. This control feature enables microsources to act as plug-and-play devices and facilitates the addition of new microsources at any point of MG without affecting the control and protection of the existing units. Two other key features are that an MC will not interact independently with other MCs in the MG and that it will override the CC directives that may seem dangerous for its microsource.

Central controller – The CC executes the overall control of MG operation and protection through the MCs. Its objectives are

- To maintain specified voltage and frequency at the load end through power–frequency (P–f) and voltage control.
- To ensure energy optimization for the MG. The CC also performs protection coordination and provides the power dispatch and voltage set points for all the MCs. CC is designed to operate in automatic mode with provision for manual intervention as and when necessary. Two main functional modules of CC are energy management module and protection coordination module [19].

4.6.1 Virtual power plant general concept

Actually, VPP is still a new concept, and there is not a unique standardized definition for the framework of VPP in the literature. The origin of the terminology of 'virtual power plant' can be tracked back to 1997, when Shimon Awerbuch, in his book, 'The Virtual Utility: Accounting, Technology and Competitive aspects of the Emerging Industry' defined the virtual utility as a flexible collaboration of independent, market-driven entities that provide efficient energy service demanded by consumers without necessarily owning the corresponding assets [2,20–23]. A VPP typically controls a group of generally small DG units. For instance, wind, solar and hydroelectric power generation units are internetworked. Controlling these together allows more effectivity. In spite of that, the related researches briefly listed below may depict range of tasks which are lined as VPP projects.

4.6.1.1 Overview of virtual power plants

- The Danish EDISON VPP project was launched in 2009 and corresponds to investigation happened in the Bornholm Island, which provide supportive services to already existing energy market players, like generation companies, in order to achieve effective use of the DERs [24].
- Fenix project is developed for improving DERs contribution to the electric network by aggregation them into large-scale virtual power plants. Project is based on the cooperation of DG owners, energy companies, research institutes and universities: EDF, Areva, Siemens, ECRO SRL, Imperial College London, University of Manchester and University of Amsterdam among others. The project consisted of three phases [25]. The first phase is preparing two scenarios – northern and southern, which is described in this subsection. Analysis included DER contribution to the network, its strengths and weaknesses. The second phase covered design of communication and control system between DER in form of VPP. The last phase concerned validation of previously prepared analysis and systems through realization of building 2 installations – in the United Kingdom and Spain [26,27]. Near the Bilbao City, Spain, is located FENIX Southern Demonstration VPP integrated in Iberdrola network. This installation aggregates many different DER like CHP, wind turbines, hydropower, PV and CHP-biomass installation, with total capacity of 168 MW. The distribution area, where this project is located, has a peak demand of 320 MW. It works like typical VPP, which means that information from each connected DER are processed in main control system – Distributed Energy Resource Management System (DEMS). Information is sent through an intelligent interface – FENIX Boxes that are connected to each DG. These intelligent electronic devices (IEDs) provide sufficient control and implementation of communication protocols, using wireless communication methods between DER and DEMS, based on virtual private network (VPN). In the DEMS module, portfolio of the VPP is created with characteristic of each aggregated unit, wherein revenue potential is optimized by making contracts on the markets. The VPP owned by Vattenfall Swedish power company is currently running in Germany with the control room in Berlin [10,28]. Their VPP unites CHP plants, wind and solar sources as well as heat pumps to create an interconnected, flexible system with centralized control.
- Battery storage VPP in Australia. In South Australia, largest demonstration project of VPP aggregating many small-scale batteries and PV is conducted.

Total combined capacity of these units is 5 MW/7 MW h. All these systems are connected to the central monitoring and management utility in form of VPP. The main objective of this project is to ease local network constrains, stabilize electricity prices and support RES. South Australia has great possibilities to use renewable energy – more than 40% of the generated power comes from wind farms or rooftop PV installations [29].

- Smart Power Hamburg. The project of Smart Power Hamburg aims at design VPP which aggregates variable load and CHP units in the city of Hamburg, Germany. Created virtual utility uses existing urban infrastructure: CHP units, heat storage systems and building with demand side management, to present new approaches in electricity generation and covering heat demand of the city. Combination of CHP and heat storage systems allows to lead generation unit by electrical demand, not heating needs which increase flexibility of the unit. Project shows possibility of the integration of storage potential of bunkers, swimming pools and heating networks, all controlled by VPP [30].

For the purposes of this chapter, the definition of VPP is summarized as aggregation control of a number of DG units, grid connected and installed near loads (end users). The aggregation control can be centralized or decentralized system supported by logic control algorithm and communication infrastructure then treated as a single large power plant [31]. Also, the contribution of VPP to the energy market is described in this chapter [32].

4.7 Types of Virtual Power Plants

In this work, the authors propose the purely formal division of VPPs due to the area of its operation and due to the function performed in the EPS.

The basic element of a VPP, apart from DG sources, is the Information and Communication Technologies (ICT) infrastructure. The construction of a complex information technology (IT) system is a prerequisite for the proper functioning of the VPP. It seems reasonable to use computer network technology for this purpose, developed for several decades, thoroughly tested and proven practically, hence the proposed division of the logical structure of connections in a VPP to computer networks.

4.7.1 An area-based approach to virtual power plants

Therefore, VPP in the area approach are

- Local VPP (LVPP),
- Metropolitan VPP (MVPP) and
- Centralized controlled VPP (CVPP).

LVPP: The area on which dispersed generation sources are distributed along with the ICT infrastructure that forms a VPP may include the LVPP, depending on the demand for electricity and population density, the area of a small commune, town, housing estate or large urban agglomerations single buildings. Figure 4.5 shows the

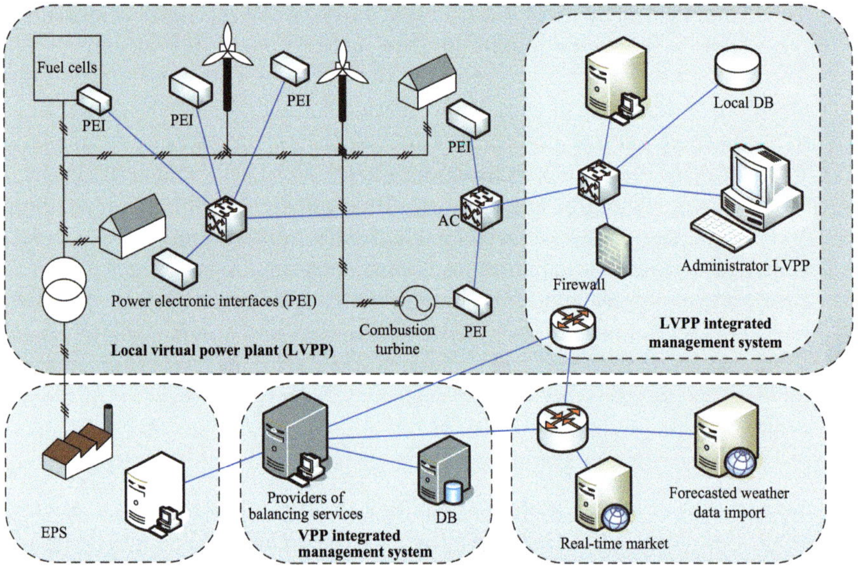

Figure 4.5 Energy flow diagram of a local virtual power plant (LVPP)

model of a LVPP. In this model, one can distinguish the power network with attached dispersed generation sources placed near to the customers, an IT network connecting logically sources and customers with the management system. Depending on the location of the LVPP, different generation units may be used to generate electricity. For a municipality or a city, these can be wind farms or single windmills supported by fossil fuel sources. In areas where rivers are located, it is beneficial to obtain energy from small hydroelectric plants. For the housing estates and large buildings, it is justified to use cogeneration systems due to the possibility of efficient heat transport and improving the security of electricity supply. Fuel cells and μCHP are perfect for use in individual buildings.

The LVPP integrated management system provides control and control of sources and receivers. The role of the system manager is performed by specialized software, monitoring the current status of LVPP and responding to the changing conditions. For reasons of cyber security, a system of user authorization, encryption of transmitted data as well as security in the form of a firewall protecting against access from external networks is necessary in the system. Data collected from the area of the LVPP are collected in the local database and are used to conduct analyses of operating modes, estimate operating costs and search for more effective ways of using available resources, taking into account meteorological conditions, the energy market and promotion payment by renewable energy acts. The forecast algorithm is a piecewise linear transformation usually of two weather variables to the expected power output according to a given transformation matrix (e.g. wind speed and direction for wind power units, light intensity and ambient temperature for PV systems).

The LVPP management system has the ability to communicate with the distribution system operator through the VPP integrated management system at level of MVPP or CVPP.

MVPP: In this model, the key rule is the hierarchy of VPP in different levels. MVPP consists of few smaller LVPP which supervise DG, but more important decisions are carried out upward to decision-making VPP on higher level in this hierarchical structure. Price signals are popular approach in this design [32,33]. This model is best suitable for big VPP which helps in cooperation between individual LVPP and better dispatch responsibilities between them.

CVPP: CVPP carries all control function and elements in one central place. It consists of limited number of DG which are controlled centrally. Its operation is based mainly on the operation of LVPP and MVPP without interfering with DG single sources. CVPP is primarily a teleinformation infrastructure and IT systems that provide data and tools for

- load forecasting taking into account the increased a higher share of DG in EPS;
- control of a whole power system (similar to the TSO function);
- modelling of EPS dynamic, e.g. voltage stability analysis with significant DG share;
- diagnostics of long-time-operated DG units planned for flexible operation;
- accommodating the relatively less controllable and predictable output of the new technologies, driving changes in the composition and operation of the entire power grid;
- determining the reliability of the system;
- analysis of the effects of integration and connection to the network of DG sources and
- power system security assessment.

4.7.2 Grid support and ancillary services

The functioning of a VPP is strongly dependent on market mechanisms and the conditions prevailing in the MGs. The use of integrated technologies covering the energy market and to provide regulation services (ancillary services for power system) in response to transmission-level requests. This allows the division of VPP due to their functionality in terms of service. The creation of a VPP is associated with a change in the demand for ancillary services. Therefore, VPPs in terms of service are

- VPP regulation (VPPR),
- peak intervention VPP (PVPP) and
- backup VPP (BVPP).

4.7.2.1 Virtual power plant for regulation

An energy trading

VPPR can cover the local power demand resulting from local load changes. In this power plant, apart from dispersed generation sources, e.g. microhydroelectric plants and small hydroelectric plants, it is proposed to introduce a mechanism for influencing the end user, who requested had the right to be supplied by its local distributor at the integral regulated tariff. This mechanism, supported by appropriate agreements between the supplier and the end user, would consist in encouraging the recipient to disconnect receivers, e.g. air condition load, storage heater or in some cases remotely switching them off by the supplier during the increased demand for electricity and switching them on when the demand decreases. The effect of this procedure will helps the energy utilities to decrease the over load demand and reshape the load curve. In addition, for each unit of saved energy, investment costs are needed, which are called 'NegaWatt cost', and it is presented as currency/kW. Final price depends on variables like initial costs, maintenance costs and life span of the unit with number of generated NegaWatts. NegaWatts can be traded on the market – power consumer commits to reduce their own energy needs for benefits from system operator. However, reduction of power demand is only dependent on the customer and its own profit. Because of that, this method requires appropriate incentives for clients for optimal operation of the system [34,35].

Load-frequency control and balancing consumption and generation

The concept of the so-called VPPR seems to be a particularly interesting idea to improve the load supply reliability. The increase in the number of distributed energy sources, as well as their technological development, supports the assumption that in both medium and LV networks, it will be possible to create structures locally balanced in terms of power generation and demand. These will be small power systems that normally operate parallel to external power system but are also capable of autonomous operation. This will result in a new situation, different from the present one, in which small electrical power sources are automatically shut down after connection with external system is lost. Such practice is irrational and attempts should be made to create such EPS that would allow for intentional switching from parallel operation to island operation and reversely. Isolation of a power balance area will be performed by opening a specific switch controlled by a controller, which performs functions related to protection, measurement and synchronization with external network. An important question arises, whether such sources will be able to maintain required voltage and frequency levels in VPPR, as load and generation change. The requirements regarding frequency may be less restrictive, however if the deviations are short-term and the frequency is kept within 47–52-Hz range.

Synthetic inertia

With increasing share of DG sources in the VPPR structure, the system's inertia is decreasing, i.e. kinetic energy in masses of rotating conventional units, which naturally covers the power loss in the EPS. Therefore, in case of loss of frequency balance in such conditions, it is faster frequency dip. An additional problem is geographical heterogeneity (and therefore the grid) distribution of nonintroductory sources inertia to the frequency response of EPS. A source of artificial inertia in the power system treats all kinds of devices that in a way are 'artificial' and can replace the natural inertia of EPS, i.e. in the initial period after the occurrence of the disturbance state (in the first moments of its duration) affect dynamics frequency changes by downloading or injecting additional active power depending on the identified one direction of frequency changes. Identification of the direction of frequency changes, and it should be done through appropriate regulation systems installed in these devices. VPPR can also become a limiting tool to the dynamics of frequency changes and – thus – contribute to reduce the depth of frequency change caused by disturbance of active power balance in EPS. It is possible, although wind sources are treated as 'restless'. To limit the dynamics of frequency changes in the first after the occurrence of the disturbance, the inertia can be used (inertia) masses of rotating wind sources. The use of a wind source to limit dynamics frequency changes requires the use of this source special control systems and regulating functions (available usually as an additional, paid option). Action of these systems should rely on a temporary change in value of the active power generated by a wind source, at what in practice is the asymmetric type of regulation, i.e. they only react to frequency reduction. For sudden changes the level of active power generation by a wind source the kinetic energy of the rotating masses of the source is used. Momentary injection/active power consumption should lead to limit the dynamics of frequency changes in the first after the occurrence of the disturbance and reduce oscillations secondary frequencies related among others with overregulation primary control systems thanks to the fast operation of systems converter.

Reactive power and voltage control

As a rule, during VPPR operation, load changes resulting from normal, continuous work of receivers should not cause a long-term voltage deviation exceeding $\pm 10\% U_n$ and frequency deviation exceeding the 49–51-Hz range. In such conditions, many customers consider it more important to maintain power supply than to keep regular standards of energy quality. During VPPR operation, there may also occur connections and disconnections of extra-high power loads, such as, for example, winding machines. In such conditions, turbine set control systems must quickly regulate the changes of voltage and frequency. The next important task is to add protection devices to the system. Special attention should be paid to short circuits in the line connecting unit with the system, which – if they happen – should be immediately switched off through the use of differential protection devices installed in such lines. Short circuits within the VPPR must also be quickly eliminated, so that the system does not become unstable. This requires changing set points for protection devices, which takes place after switching to VPPR operation (limiting the overcurrent protection settings, shortening time-delay) and often sacrificing its selective tripping. Set point thresholds for generating units' voltage and frequency protection devices must be well designed, and their proper choice may be based on test results of VPPR operation in real conditions. Preferably the change of protection devices set points is to be done by choosing an appropriate set points data bank, which is possible in modern digital relays.

Reconnection of VPPR to the power system usually first requires the generating units operating within an independent mode to be switched off, and this entails interruption of power supply to some receivers. This results from the fact that normally synchronization takes place locally, in the VPPR station, at the generator breaker. In order to avoid this inconvenience and enable resynchronization, it is necessary to use an extra 'remote' synchronization system installed at the breaker in the connecting line or in the substation, to which the VPPR station is connected.

During synchronous operation of a VPPR, the turbine controller remains in power control mode. Independent mode of a VPPR should cause the turbine controller to automatically switch to a mode, in which the priority is frequency control. The criterion for operation mode switching is usually the position of the switch on the way connecting VPPR with the main power supply station. It will be required to base the controller's operation mode switching also on the position of other breakers on the way between generating unit and the system and to use telecommunication lines to send the signals. Similarly, the generating unit's controller, when operating synchronously, remains either in power factor cosφ control mode, or in the set reactive power mode. Switching to independent operation should cause automatic (according to criteria identical to those for the turbine controller) switch to voltage control mode.

Another important issue that aids successful switch of VPPR to independent operation is ensuring uninterrupted power supply to its auxiliaries. It is not possible in a situation when power supply is provided from various sections in the power plant's switching station. The necessary thing is to group, in terms of power supply, the machinery related to generating units potentially operating within an island and to ensure to it a power supply from independent source (e.g. a generator set) in case of unsuccessful islanding. The purpose of using systems with artificial inertia is the possibility immediate use in the kinetic energy system accumulated in masses of rotating wind turbines, as in the case of rotating generating units with synchronous generators. Artificial inertia systems are based on two subsystems: the 'hidden' inertia subsystem and the subsystem controlling the position of the turbine blades.

PVPP: In the event of sudden, local load changes, it may be necessary to start sources of intervention power. This task can be implemented by PVPP, which includes sources with a short start-up time, e.g. fuel cells, diesel engines, gas engines. Such power plants should include highly urbanized and industrialized areas, and sources located at end users, e.g. fuel cells in large buildings.

BVPP: BVPP's task is to provide energy in the event of extensive system failures. In order to provide reliable power, data centres are provided with on-site generators and large backup resources. These resources are designed to provide backup power during grid failure; therefore, they are underutilized most of the time. In addition, recipients located in the area of operation of the BVPP should be granted power priorities. The highest priority should be given to institutions and plants of key importance for the functioning of the local community in crisis situations, e.g., crisis management centre, uniformed services, pumping stations, bakeries. Other recipients can be supplied with power provided that the available power reserves allow it, and there are technical conditions to deliver this capacity.

4.7.3 VPP model and algorithms

The concepts of a VPP presented above are suitable for constructing its model – the three-tier architecture of VPP that contains the following layers:

- Market model will cover production costs and their effects, e.g. energy agriculture, energy market, environmental protection.
- Management model will include
  - a strategy for the selection of DG sources adapted to the role that will be met by the VPP and its area of occurrence,
  - modelling of DG sources,
  - selection of protection automation and its settings,
  - load forecasting,
  - the security assessment of the EPS,
  - VPP control algorithms and
  - impact of weather conditions.
- Teleinformation model will be the integration of market and management models into one system, collecting and distributing data, taking into account network equipment, its parameters and distribution.

4.7.3.1 Optimal power flow (OPF) – technical aspects

Optimal power flow is the algorithm, which is used for modelling one power plant, which includes generation, loads and network limitations from aggregated characteristics of the elements of the system. OPF calculates maximum capacity of the injected active power and consumed reactive power. It is measured at the point of the connection between transmission line and distribution part of the network. All calculations are taking into account limitations of the generators, networks and demand [13].

4.7.3.2 Production planning – market aspects

Production of energy in VPP is generally planned for a longer period of time (e.g. 24 or 48 h). This is called offline method. If there is a need for adjusting generation to the changing demand or if one of the generators is corrupted, online method takes place, which means online rescheduling generation between others units. This method is used in the unexpected situation – generally VPP operates as normal plant, with contracted energy which needs to be supplied to the grid. Once planned, production scheme for specific period of time must be followed in order to keep stability in the network and avoid penalties from transmission-level or other contractors. Planning decisions are based on the forecasted load, market activities like price changes and production opportunities, if the portfolio of VPP consist mostly on the weather dependent units. After planning specified amount of energy, it is scheduled between dispersed units taking into consideration local requirements and availability of the units. μCHP needs special planning methods as well as weather dependent units. Main purpose of μCHP is to provide heat for domestic use with possibility to generate additional electricity. Depending on the heat buffer, heat generation can be shifted in time to provide more electric output. First step is forecasting heat demand for each house using μCHP, which can be done by neural network approach. After preparing production potential generation schedule is dispatched based on the local (unit owner) and global (VPP scale) needs. As was mention before, generators require control in real time in order to reschedule generation if the actual load differs from predicted one [36]. Generic method of real-time control is fully described in [32].

4.7.3.3 Load management

Load management in VPPs are mainly based on direct load control (DLC). VPP controls many controllable appliances located at the end user side like air conditioning systems, electric heaters or freezers. Load control is possible by agreements between the customer and the VPP, which allows to remote control of some devices by VPP for load reduction during requiring time periods. Device management can include disconnection of the aggravating devices for a short period of time or thermostat modifications. DLC can be used for preparing optimal strategies of the VPP operation. Literature presents many approaches to load control modelling in VPP which consist of many controllable devices. Most popular method is linear programming or dynamic programming. These methods generally base on a peak load reduction or minimization of production costs for given period of time [17].

4.8 Difference between microgrid and VPP

Both structures VPP and MG can facilitate DG integration into EPS but with different aims. MGs focus on grid operation control concerning active and reactive power using DG units existing within one local grid – mostly on medium and LV level. On the other hand, while VPP concentrate on the provision of energy and power system support services from DG units. VPPs offer lots of benefits, to end users and to the grid, but they do not offer the core MG value proposition: resilience, i.e., independence from the main grid in times of need. The future is a grid that contains hundreds of networked MGs. This kind of 'modular architecture', with multiple semi-autonomous nodes operating in parallel, is more secure and efficient than a centralized system with a few, large points of failure. Both VPPs and MGs may never eliminate the need for large utilities, power plants and transmission lines, but moving more power generation, management and consumption under local control makes everyone less dependent on them. And it makes the grid more reliable and resilient [37]. Table 4.1 compares the two mentioned concepts and highlighting their differences and similarities.

Table 4.1 Difference and similarities between microgrid and VPP

Structure	VPP	Microgrid
Different control objectives	Promotes market and grid service participation from DGs – covering the energy market and to provide regulation services (ancillary services for power system) in response to transmission-level requests (VPPR)	Implements autonomous operation of a distribution grid with many DGs, of which off-grid and island systems are of particular interest
	Providing of services to the system operator	Managing the local system for distribution system operators
	Aggregates the capacity of many diverse DGs and creates a single-operating profile from a composite of the parameters characterizing each DG to facilitate DG trading in the wholesale energy (PVPP and BVPP)	Determining of fault location
	Increases DG visibilities to transmission and distribution system operators (MVPP and CVPP)	Facilitating maintenance
Control modality	Virtual inertia and reserves	Physical inertia and reserves (generation)
Grid infrastructure	VPPs may or may not feature storage	Microgrids typically require some level of storage
	VPPs are heavily dependent upon extended sensorial devices (smart meters) with demand, phasor and power quality orientation: PMU and PDC – WAMS (synchronized)	Dependent upon hardware innovations such as inverters and smart power switches
Communication infrastructure	Advanced and dedicated IT systems and software interfaces	Advanced and dedicated IT systems and software interfaces
Demand response	Demand response management actions at both the production side and at the demand side	Demand response management actions at the production side only
Operational challenges	Supports grid services. Uncertainty management	Capable of performing automatic islanding Uncertainty management
Barriers to implement	Economic barriers:	Technological barriers:
	Pricing and metering Wholesale market access Connection charges for DG Allocation of network investment costs due to DG Intermittent nature of DG	Integrating DG will create a multidirectional power flow in parts of the distribution networks, which were not originally designed for this mode of operation System reliability: protection of the distribution grid against faults is provided by impedance relays. It is stated that DG affects the functioning of protection schemes

4.9 Information communication technologies

As smart and very modern power grids were introduced, so did, too, the importance of ensuring the energy sector's reliability and security grow. The reason for this is the rapidly developing information technologies dedicated for power systems. Therefore, discussion of the whole array of devices for controlling and commanding increasingly varied control and measurement devices should also include the security of IT systems. Adequate security should be considered as soon as at the design stage. Cybernetic security should cover not only deliberate attacks, e.g. by disgruntled employees, corporate spies or terrorists, but also accidental exposure of the IT structure caused by human errors, equipment failures or natural disasters. Such vulnerabilities of the system might allow an attacker to enter the grid, gain access to the control software and, eventually, alter the grid's load conditions in order to destabilize it. Conversion of a current grid structure into a smart grid necessitates the introduction of a number of new security solutions borrowed from already utilized ones, such as those found in banking or public administration systems.

The current degree of power systems' computerization is often limited to the necessary minimum, and the technological process is usually managed locally. Thanks to information technologies, this model of grid management has been significantly altered. The original versions of monitoring and management of production units were implemented by use of unique protocols, often individual companies' own solutions, but also grid interfaces and control systems that were incompatible with each other. As ICT systems developed, these technologies found wide use in the energy sector. Due to advanced digital solutions, an MG is today increasingly often perceived as a smart power grid (SPG). Using an MG is the solution to problems related to integration of distributed sources and a power system and belongs to a wider class of solutions referred to as smart, such as smart systems (grids), smart metering systems, smart utility.

In control, automation and supervision systems used in power stations, the basic form of data exchange between devices is digital communication based on known communication standards and protocols. The most commonly used standards are RS232 (communication standard of two electronic devices), RS485 and RS422 (communication standard of many transmitters and receivers), Ethernet (technique utilized mostly in the construction of local area networks – specification stated in the IEEE 802.3 norm) and fibre optics (communication technique utilizing fibre-optics specification stated in the IEEE 802.8 norm). The DNP protocol is used both for communication of data concentrators with devices at the station and for data transmission to supervisory systems [38]. Next to the standard serial protocol implementation, there is also a modification based on Ethernet – DNP 3.0.

Other than the aforementioned communication protocols, very popular and widely used ones are Profibus[1], OPC[2], IEC 60870-5[3] and IEC 61850[4], remote and industrial protocols Modbus[5], CANopen[6] based on manufacturer–consumer technology. Unfortunately, the problem of numerous grid standards is particularly apparent from the level of managing a large number of devices. On one hand, using different types of grids and standards defining new redundancy protocols: media redundancy protocol (MRP)[7], cross-network redundancy protocol, beacon redundancy protocol, IEC 61158 quality of service (QoS)[8], Fibre Check[9] ensure the system's greater reliability, allow for remote configuration and diagnostics and ensure the system's increased operational speed and stability. On the other hand, an end user, especially in medium-sized and large installations, who receives several types of connections (standards, protocols, grid interfaces) might face problems with data flow between them, often despite large bandwidth of the grid itself. It raises a question of whether exchange of information between so differing subsystems is necessary and if a universal 'object' connection, through ordinary inputs and outputs, is not enough. It is so in very simple and fully unrelated systems, but in case of a more complex system and with a greater amount of transferred data, other solutions are needed. All the more because for the past few years one can observe a significant increase in implementation of ERP (enterprise resource planning) class systems, i.e. effectively planning out the resources of an enterprise, for which one of the basic processed data is the state of the technological or production process.

Below are the most commonly utilized mechanisms for digital power substations [39].

4.9.1 RSTP grid mechanism

The presented solution uses the RSTP (rapid spanning tree protocol – IEEE 802.1w) protocol along with its extended RSTP implementation. During the grid's normal operation, redundant connections are blocked. In the event a transmission medium or a network switch malfunctions, RSTP allows for creation of a new

[1]process field bus – IEC 61158

[2]Open communication standard used in industrial and IT automation, OLE (object linking and embedding) for process control.

[3]Communication protocol for remote control systems and devices for tracking and controlling geographically extensive processes.

[4]Norm (standard) related to automation design for power stations.

[5]Communication protocol created by the Schneider Electric company (Modicon) aimed at communication with programmable controllers.

[6]Multicasting protocol.

[7]Based on a ring topology.

[8]QoS – prioritization of data transmission.

[9]Monitoring of optical ports.

logical grid topology, which bypasses the damaged element in a matter of milliseconds. Each of the connections has a configurable transition cost, which provides predictable communication scenarios when the grid's connections or devices are damaged.

4.9.2 SHP grid mechanism

The self-healing protocol (SHP)[10] ring-type grid architecture utilizes mechanisms allowing for dynamic reconfiguration of the connection tree in a redundant ring. SHP is a mechanism managing the communication in rings. During the grid's normal operation, information flows in a determined, basic direction. In the event of a transmission medium or a network switch malfunctions, the SHP allows for creation of a logical grid topology which bypasses the damaged element and alters the direction of data flow in a backup direction within up to one millisecond. Each of the connections has a configurable transition cost, which provides predictable communication scenarios when the grid's connections or devices are damaged.

4.9.3 HSR grid mechanism

The high-seamless redundancy (HSR)[11] ring-type grid architecture utilizes mechanisms, allowing for dynamic reconfiguration of the connection tree in a redundant ring. HSR is a mechanism managing the communication in rings. During the grid's normal operation, information flows in both directions at the same time. The receiver in every end node receives the first frame or pair of frames and ignores the duplicate. The HSR mechanism is clear to the upper protocol layers and multicast traffic, and generic object-oriented substation events messages. In the event of a transmission medium or a network switch malfunctions in one LAN network, the movement goes in both directions, and HSR sends a frame available through grid movement. Identification of an HSR marker for an HSR mechanism is at the start of the frame in order to facilitate delay-free transmission.

4.9.4 PRP grid mechanism

The parallel redundancy protocol (PRP)[12] binary star-type grid architecture utilizes two separate grids operating in parallel using the Ethernet technology Figure 4.6. PRP is a mechanism managing the communication in rings. During the grid's normal operation, information flows in parallel in both LAN networks. Each PRP node (double attached node for PRP (DANP)) has an interface for every LAN network and sends a frame to both grids simultaneously. Similarly, single attached node nodes have a single interface and are connected only to one grid. The DANP destination normally receives both frames and rejects the duplicate. In the event of

[10]Schneider Electric property.

[11]A grid protocol which ensures problem-free emergency operation before the malfunction of any grid element.

[12]Schneider Electric property

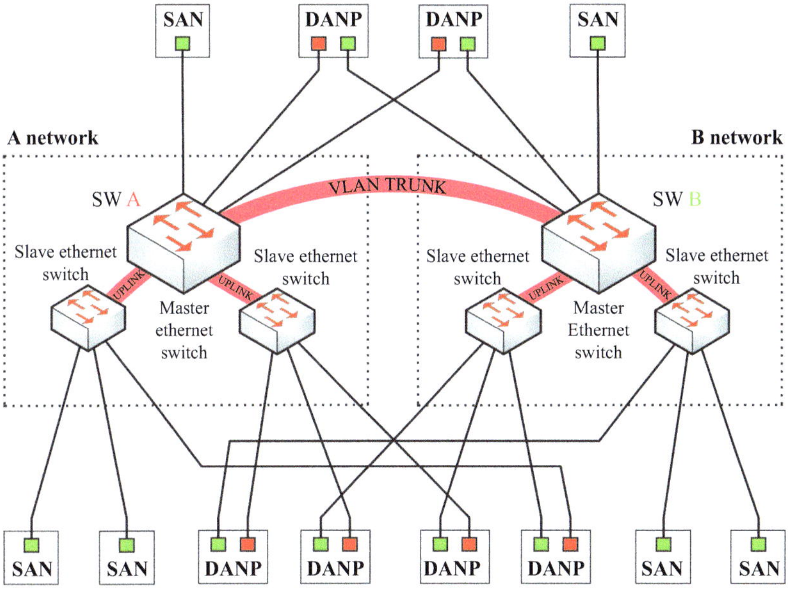

Figure 4.6 Data transmission scheme during basic network operation in the PRP dual star-type architecture

a transmission medium or a network switch malfunctions in one LAN network, the DANP destination will operate using the frame from the other LAN network. Identification of a PRP marker for a PRP mechanism is at the end of the frame in order to facilitate cooperation with every available switch.

4.9.5 Microgrid/VPP cybersecurity

Power systems are currently among the highest priority systems in the hierarchy of ensuring the stability of any country's functioning. They are referred to as critical systems. Their proper functioning, involving production and distribution of electrical energy, has to be supported by verified management solutions and, more importantly, due to implementation of the latest technical solutions, not only in terms of organizational concept of a given enterprise, but also of fitting these grids with advanced control and metering systems. Not so long ago, communication of individual devices installed in substations with the SCADA (supervisory control and data acquisition) operator's system took place entirely on the basis of serial transmission, with use of often specific and dedicated infrastructure. Interestingly, in the beginning stage of these, electronically advanced for the times, devices' expansion, many mutually incompatible standards, along with unusual communication interfaces, were made. Manufacturers used them interchangeably, which did not lead to a spread of a single leading standard and, additionally, mistakenly believed that more advanced devices using closed protocols, being the basis of their digital communications, will lead to the spread of their standard. An additional problem was often the limitations arising from the protocols themselves, physical interfaces and limited length of cords (supplied by the producer) used to connect and run those devices. Only a few years ago, it was a very common practice to move data from one system to another – most often by use external media. This phenomenon was noticed by companies specializing in production of conversers for unusual interfaces which, in a way, facilitated signal transmission through a transmission medium convenient to the administrator. A definitely better solution which, despite its simplicity, performed the function of remote administration of industrial automation, was widespread opening of backdoors in the form of VPNs. On the scale digital technology development's scale, not so long ago, it was particularly popular in industrial automation to use software called Real Virtual Network Computing [40] which, thanks to remote image capture, allowed for control of local computers and servers supporting industrial automation devices. It was widely and recklessly used by beginner administrators of power systems, which led to a decrease in the level of security of not only automation and control systems, but also ICT systems (often also database ones). It was often forgotten (in excitement about the service being implemented) that using virtual tunnels carries certain risk. On one hand, we gain remote control over distant automation supervision system, and on the other, the often unsecured system becomes a proverbial gateway for people who should not necessarily have access to our grid. Many people were forgetting to prevent communication with software from outside the local network, change the default communication ports to others or limit the communication of a station supervising technological process to its minimum, through traffic filtration or limiting the number and types of ports through which it is possible to communicate with a given network interface. The aforementioned solution was very troublesome, especially for those local enterprises which had their control and metering stations located throughout the city (often at a distance of over a dozen kilometres from each other), and integration of those systems into a single, coherent one, controlled from a single location, was a significant challenge to them.

As mentioned, an IT grid, utilized in modern VPPs, due to its topology and distributed infrastructure, is an open structure in its nature, as outside controlled grid segments, i.e. power transmission and distribution system, it also integrates distributed energy sources cooperating with the grid, and consumers and prosumers or electrical power. Such architecture, due to the lack of constant and centralized ICT protection, increases the probability of various threats even more. It is estimated that the vast majority of conducted attacks against ICT systems (which can be surprising) are ones from within the grid. Modern control information transfer systems currently require special protection against internal threats, as well as external ones. SPGs and VPPs use two kinds of subsystems in their architecture – the first are the previously described ICT information systems, and the second – operational technology industrial control systems. IT systems are mainly responsible for data processing and digital information transfer, and technological processes are related to industrial control through devices and systems rarely using IT technologies. Due to their unprecedented growth and quick integration with power systems, ICT systems have gained significant importance throughout the last years. Discussion of modern power systems' architecture cannot omit the issue of security and reliability of digital information processing systems. Initially, implementation of such solutions was only planned and considered as a potential role in power systems, but today, not only their existence but also high sophistication are largely the determining factor in operational stability and security of the entire power system. Therefore, their security is prioritized, and the lack of such solutions may lead to questionable security of such a critical and important system for the functioning of a supplied region. The administrators of VPPs' system should, already at the design stage, try to isolate the IT infrastructure and control system from each other as much as possible. This solution is related to protection of the system and IT networks but also system of visualization and control of SCADA processes and technologies. Quite often, for different reasons, separating those systems completely is not possible. An additional problem is the utilized internet protocol, which, due to many vulnerabilities [41], makes ensuring high control reliability and digital security a difficult task.

The basic and the most important element of ensuring any given enterprise's digital security is developing a security policy and putting it into practice [42]. In general sense, cyber security also includes a security policy which, next to the organizational concept of supervision over security, accounts for legal provisions, research, training, etc. All the aforementioned elements should be the pillar of every critical system's security. It should be remembered that skilful management and inclusion of modern cyber security methods in the design process, operation of ICT and power automation systems will limit the effects of deliberate attacks and accidental failures of a power system's objects, which will also contribute to increased security of its operation.

4.9.6 Energy management system

The development of smart grids has also become a new direction in the development of ICT devices which have so far not been dedicated to these grids, but also not seen as a possibility of their potential implementation in automation and control systems. The rapid development of the power sector has brought about standardization and unification of solutions applied thus far. The market currently offers highly specialized devices boasting compliance certificates with known communication protocols and popular solutions allowing for easy and compatible expansion of these systems in the future. The ongoing evolution of these devices and their implemented programmable functions currently allows for making power systems increasingly automated and reliable. Such dynamic development of this device group has also made it necessary to implement changes in the management model of the entire system. A change in approach to power system management will have a direct impact on increasing these systems' effectiveness. Current systems and power grids smartly integrate the operations of all members of the process of power generation, transmission and distribution in order to supply electrical power in an economic manner, naturally accounting for a high level of reliability and security.

The consumer of electrical energy, in the modernized VPP, is no longer just a passive receiver of electricity (Figure 4.7), but consciously and actively manages

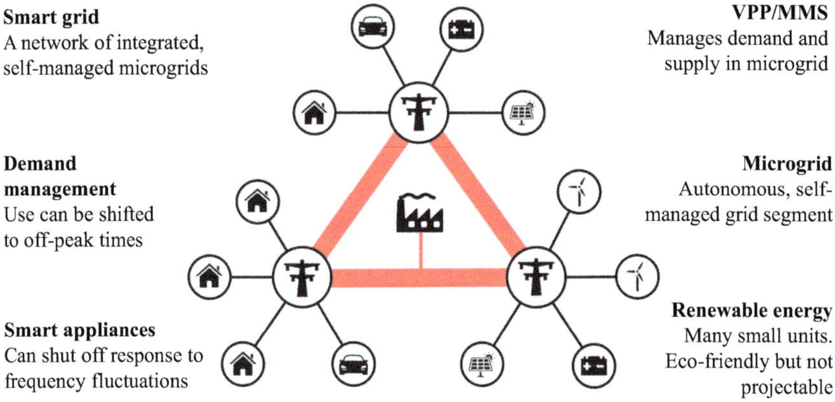

Figure 4.7 Scheme of data transmission during basic operation of a grid in the binary star PRP architecture

the energy and its use locally in their household. Power management by the consumer, and further in the process of grid transformation – prosumer, will involve conscious energy saving by use of, e.g., heating devices (with the highest power) during the evening hours or outside the peak hours. Ultimately, the consumer will become a prosumer – a producer of electric power in a microscale. Investment into knowledge about ISE and the development of the technology itself, as well as related solutions, should translate into better use of power leading to savings (and also lower electricity bills), and into opportunities for profit from selling energy – in the same of presumption and so-called DG. Even of consumers are not willing to transfer the surplus generated energy to the operator's grid, this solution will translate into a reduction of power consumed from them anyway, which will result in lowered demand during peak hours. Development and effective management of smart grids and VPPs will require an aware and active energy receiver, which will also contribute to the advent of energy-sustainable local communities (smart communities).

When considering the issues of effective and secure management from the technical standpoint, it should be remembered that smart power systems function in three environments: (1) telecommunication, encompassing grid control centres, subordinate systems, telesecurity; (2) local operator (human) and (3) process involving switching devices, transformer, auxiliary devices, etc. In each of the listed environments, the following rules should be observed:

- Each configuration of changes in the system requires their verification for compliance with the applicable security policy.
- Not adhering to the system security norms in the system should result in physical disconnection of a given component from the grid.
- The decision about connection or disconnection of the system should be ordered by authorized persons.

Moreover, one should observe the rule of granting authorization to applications supervising the grid's operation, to the grid's active devices and systems of databases in relation to the hierarchy of authorization among people managing the entire power system. The elements to be specified are also

- the level of acceptable risk,
- access control mechanisms,
- identification and access authorization mechanisms and
- registering changes made in the system: related to data configuration and modification [43].

Electric power suppliers will want quick implementation of digital technologies for production, distribution, storage and flow coordination of energy. It should be borne in mind, though, that systems of DG and VPPs, from the entire system's power security standpoint, as well as cyber security, might lead to certain problems in the event ICT means of system security are not utilized. Such units can provide wrong or falsified control information, which may, in turn, negatively impact the master system's operation. However, possible implementation of current hardware and software solutions for protection of IT networks can significantly improve the state of security. In the era of current cybernetic attack threats, it is not without significance to necessitate the implementation of proper procedures and rules of conduct in the event of a justified suspicion of an attack against the system supervising the operation of VPPs. The effective integration and growing presence of VPPs requires certain grid innovations and an unconventional approach to technical issues and ways of grid management.

4.9.7 Supervision control and data acquisition

Thanks to using and popularizing solutions from the broadly understood ICT, contemporary power systems develop towards data processing and analysis, and automated reasoning and decision-making. Additionally, the development of such disciplines as smart complex networks will allow control systems to autonomously make decisions based on current events and predefined rules [44]. Smart grids using virtual subsystems will also be a challenge, especially for specialists in computer networks and advanced server services. Especially dangerous targets for attacks are ICT systems containing important data collected in distributed databases of local subsystems and in the network operator's information processing centre. In local power systems, advanced infrastructure of control and supervision, in its basic functionality, ensures full monitoring and control of grid devices based on fully automated replication of databases.

4.9.8 Control system operation and states

One of the most important features of automated systems is utilizing a single control element to control multiple, completely different tasks – without having to use several of the same devices [39]. Smart devices, i.e. substation control system devices, IEDs, or bay control units (BCU) have the following operation modes:

1. *On-line modes:*
   (i) *Operation:* The device fully functional – all features active. In case an internal error is detected, the device can go into an error or stop mode, depending on the existing error's level of importance. It can also remain in a lowered operation mode (processing due to loss of I/O – input/output card). If possible, the error is always displayed to the operator.
   (ii) *Test:* Accessible only to BCU. All the features are operational but control commands are realized virtually; BCU simulates positive confirmation for the control sequence. What is important is that controls are still sent to IED or other BCUs.
2. *Off-line modes:*
   (i) *Service:* The device is operational, but only certain features are used, usually he supervisory ones (uploading and displaying information on database and communication states). This mode is launched at the operator's request or automatically when inconsistency of data is detected.
   (ii) *Error* (only BCU): The device is operational, but only certain features are used, usually he supervisory ones and those unrelated to managing electric processes. This mode results from the device discovering an error.
3. *Special modes:*
   (i) *Initialization (start):* Transitional mode between the device powering phase and its normal operation, error or service mode. In this mode, every device verifies its components, software and compatibility (consistency) of database. In case of inconsistency, the device goes into service mode.
   (ii) *Pause:* The device if disable from use due to a detected critical error. It is possible to program the device to be automatically restarted in such a case.

Depending on the configuration, the graphical representation has a specific colour, form or both at once (fully dependent on the designer). Detailed graphical representation is used to indicate standby states in case of redundant devices. Diagrams of states and transitions of operation modes are indicated in Figure 4.8.

4.9.9 Databases

Database of the main system contains the databases of each device in the system. Individual databases' names have unique version numbers created with each generation. It allows for identification of all changes in configuration, and control of database compatibility in all the devices of the system. Database with the version number is constantly captured on a computer disk in the system management centre and can be imported to the workstation in different ways (locally or remotely), and then uploaded to all the devices within a logical grid. The current database is an active one, used by the main application. In normal conditions, the current database in all the devices has the same version number. If not, the devices are treated as

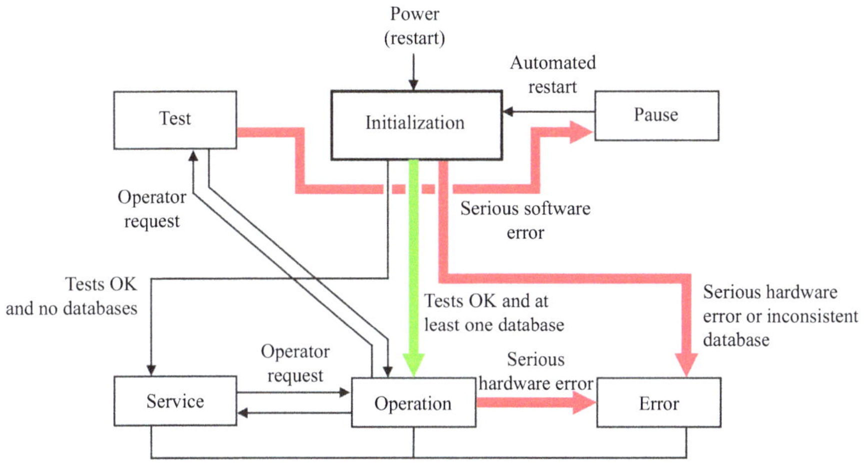

Figure 4.8 Diagrams of states and transitions of operation modes

autonomous. A backup database can be understood as a temporary one which can be set as active at any moment. Management of the current and backup database allows for optimization of global switch time (upload process does not impact the system's operation) and an easy return to previous database versions [39].

4.9.10 Database management process

Every new database is, first and foremost, uploaded to devices of a station bus (IEC61850) as a backup one. This operation can be performed on a single device or simultaneously on many. The procedure has no influence on the process station monitoring and control, as an active database operates in normal conditions. The second stage of database management is launching a backup base. This operation is referred to as switching bases. The switching process swaps the 'current' and 'backup' base. In order to conclude the switching process, all the devices are restarted. After it is complete, all the devices operate with the new database. As a result, the existing configuration is replaced with a new (generated one) [39]. The database management process and version scheme are shown in Figure 4.9.

A different issue, also being another direction for development of SPGs, is not only data analysis but also data mining [45]. Such a process will provide the electric power system operator with completely new information they would not be able to see in a traditional electric power system model. Skilful utilization of this information will allow for not only accurate knowledge and understanding of one's own network's operational logic, but also some real benefits: decreased energy consumption during peak hours, decreased losses due to automated energy balance and increased security of the energy transfer itself.

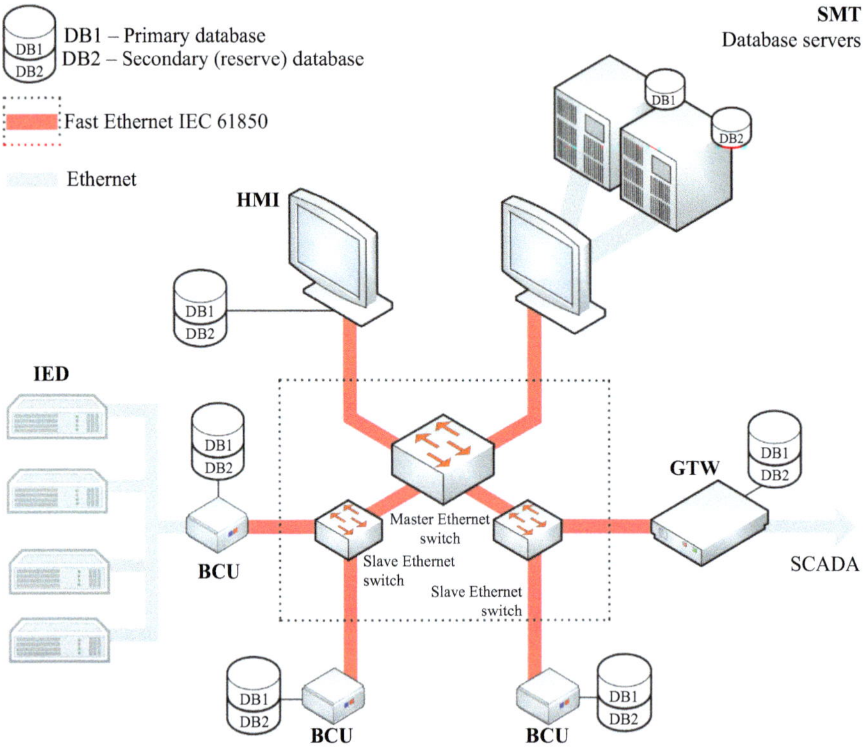

Figure 4.9 Schemes of database versions

4.9.11 Distribution and dispatching centre

Management and monitoring of expansive power systems is not an easy task; therefore, what administrators often do is divide it into individual and independent parts in order to avoid a disorienting overflow of information. With the basic digital communication security-related requirements, i.e. confidentiality, completeness and punctuality met, its key requirement is effectiveness of the communication process itself. In some cases, interrupted communication or lack of thereof might lead to disturbance in stability of the entire system. Moreover, the claim that the efficiency of information transfer depends solely on the distance between the sender and the receiver is not correct. Information transferred from node to node can get stuck somewhere along the way or in a node unable to process it quickly.

With the increasingly frequent threats in the form of cyber-attacks in mind, security should be provided for the most important resources of smart grids, i.e. access to management software, computer hardware inventory, data of the company, staff (including the list of ICT specialists) or documentation referring to telemechanic equipment and metering devices. The most important system that should be granted a high level of security is the ERP system. The most important data from the operator's standpoint is the critical data: contractor data, trade

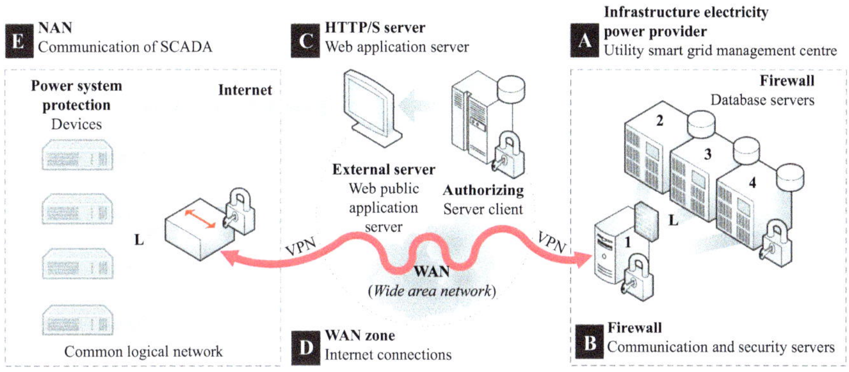

Figure 4.10 Information data flow in energy power networks

information, financial reports, which is to say all the data the loss of which puts the distributor at risk of having their positive image hurt. General smart grid model (Figure 4.10), due to the functions performed by its individual areas, can be divided as follows:

(A) *Electricity power provider infrastructure* – systems of databases and data warehouses, ICT grid administration, analysis of the process management layer and business layer.

(B) *Firewall communication and security servers* – area of grid traffic filtration, intrusion detection and prevention, mechanisms and lists of authorized external login, system logs and event logs. The acquired data, after it is filtered by the intrusion detection system and intrusion prevention system in terms of implemented traffic filtration rules, can be forwarded to the service, settlement and trade management system.

(C) *The area of electronic services provided by the client* (HTTPS server web application server) – services of Internet communication channels allowing the client (by use of a phone or computer) to remotely control their daily use of power (based on a histogram), display information on the current tariff, settle bills.

(D) *Wide area network* – composed of main information exchange buses (most often a backbone fibre-optic network), i.e.: grids connecting cities of continents. The shown WAN grid figure (Figure 4.10) presents a grid as any wide area network with an unidentified data transfer operator, unknown topology and data transfer media (cloud symbol).

(E) *The area of security and smart devices* – power system protection/IED – devices receive data from sensors and power supply equipment, able to issue control commands, e.g. activate switches if they detect anomalies of voltage, current or frequency, or increase/decrease voltage in order to keep it at a desired level. These devices provide the most interesting data whose sheer amount will provide an additional analytical challenge for the operator.

The new model of VPPs will not only revolutionize metering and the currently manual reading system, but also allows for cooperation with thus far unusual sources like PV panels, wind turbines and local power generations. These sources will be able to be placed in a residential or public building, connected (after obtaining a concession) with the grid on the plug-and-play basis to a larger VPP system.

4.10 Case study: regulation of VPP and MGs

An important question arises, whether such sources will be able to maintain required voltage and frequency levels in an MGs, as load and generation change. In this work, the problem is narrowed only to frequency control (RVPP primary frequency control) in the MGs after connection to external system is lost (standalone/isolated mode). It has been assumed here that the MGs comprises RVPP with at least one rotating power source with synchronous generator that provides a significant amount of power and has controllable frequency. Other power sources, including energy storage systems, may also participate in controlling frequency within MGs during island operation [10,11].

Let us assume that before switching to island operation, the frequency had rated value f_N. Rated frequency implies a corresponding rated kinetic energy of rotating machines currently in operation in autonomous power system (RVPP synchronous and asynchronous generators as well as asynchronous motors):

$$W_N = \frac{1}{2} J_{MGs} \omega_N^2 \tag{4.1}$$

where the W_N is rated kinetic energy of rotating masses in autonomous system, J_{MGs} is the equivalent moment of inertia of autonomous system.

A common notion used when analysing electromechanical states is inertia constant H, equal to rated kinetic energy of all machines' rotors, divided by the system's rated power:

$$H = \frac{1}{2} \frac{\omega_N^2}{P_N} J_{SEE} = \frac{W_N}{P_N} \tag{4.2}$$

where P_N is the MGs rated power equal to the sum of rated power of all installed sources, both static and rotating, $\omega_N = 2\pi f_N$ is the rated angular velocity of rotating machines' rotors in autonomous system, corresponding to frequency $f_N = 50$ Hz.

After rapid switching to island operation with N_{MW} imbalance, there will be initiated a change of angular velocity of rotors in all electric machines, which are part of the RVPP. The new value of rotating masses' kinetic energy is expressed with the following relation:

$$W = \frac{1}{2} \frac{2HP_N}{(2\pi f_N)^2} (2\pi f_{\mathrm{Hz}})^2 = W_N \left(\frac{f_{\mathrm{Hz}}}{f_N} \right)^2 \tag{4.3}$$

Present frequency in MGs is different from rated frequency by positive or negative frequency deviation.

$$f_{\mathrm{Hz}} = f_N + \Delta f_{\mathrm{Hz}} \tag{4.4}$$

Consequently it may be noted

$$W = \frac{1}{2}\frac{2HP_N}{(2\pi f_N)^2}(2\pi f_{\mathrm{Hz}})^2 = W_N\left(\frac{f_{\mathrm{Hz}}}{f_N}\right)^2$$

and then

$$\left(\frac{f_{\mathrm{Hz}}}{f_N}\right)^2 = \left(1 + \frac{\Delta f_{\mathrm{Hz}}}{f_N}\right)^2 \approx 1 + 2\frac{\Delta f_{\mathrm{Hz}}}{f_N} = 1 + 2\Delta f \tag{4.5}$$

where $\Delta f = \Delta f_{\mathrm{Hz}}/f_N$ is relative frequency increment.

For minor frequency deviations, it may be noted

$$\Delta W = W - W_N = 2W_N\Delta f \tag{4.6}$$

The speed at which rotating masses' kinetic energy changes in isolated system is approximately proportional to the speed at which frequency changes.

$$\frac{d\Delta W}{dt} = 2W_N\frac{d\Delta f}{dt} \tag{4.7}$$

After introducing inertia constant H to the preceding relation, the result is

$$\frac{d\Delta W}{dt} = 2HP_N\frac{d\Delta f}{dt} = 2HP_N\frac{d\Delta f_{\mathrm{Hz}}}{f_N dt} \tag{4.8}$$

In the first moment after MGs isolation, the derivative of kinetic energy increment in time is equal to power imbalance in autonomous power system

$$N_{MW} = \frac{d\Delta W}{dt} = 2HP_N\frac{d\Delta f}{dt} = 2HP_N\frac{d\Delta f_{\mathrm{Hz}}}{f_N dt} \tag{4.9}$$

The last equation allows to estimate the value of inertia constant of autonomous power system on the basis of power and frequency changes measured before and in the first moment after switching to island operation.

$$H = \frac{1}{2}\frac{N_{MW}}{d\Delta f_{\mathrm{Hz}}/dt}\frac{f_N}{P_N} \approx \frac{1}{2}\frac{N_{MW}}{(\Delta f_{\mathrm{Hz}}/\Delta t)}\frac{f_N}{P_N} \tag{4.10}$$

Frequency change in MGs after switching to island operation will cause a total change of loads and generation of static electrical energy sources equal to ΔP_{fGLMW}, accordingly to their frequency characteristics. For small frequency deviations, this power change is proportional to frequency change:

$$\Delta P_{fGLMW} \approx \frac{dP_{fGLMW}}{df_{\mathrm{Hz}}}\Delta f_{\mathrm{Hz}} = D_{fGLMW/\mathrm{Hz}}\Delta f_{\mathrm{Hz}} \tag{4.11}$$

where $D_{fGLMW/\mathrm{Hz}}$ is resultant frequency sensitivity of static sources and loads to frequency changes. After sudden disruption of power balance N_{MW}, active power generation individual control for each controlled source will be initiated in the MGs, which will enforce the change of generated power ΔP_{GMW}. The new power balance includes

- power imbalance N_{MW} at the moment of islanding,
- the change of kinetic energy of rotating masses in isolated system $d\Delta W/dt$,
- the change of static sources generation and loads ΔP_{fGLMW}, which results from their frequency characteristics.

As a result, the power imbalance equation is

$$\Delta P_{GMW} = N_{MW} + 2HP_N \frac{d\Delta f_{\mathrm{Hz}}}{f_N dt} + D_{fGLMW/\mathrm{Hz}} \Delta f_{\mathrm{Hz}} \tag{4.12}$$

Power imbalance equation in relative units (p.u.) is obtained by dividing both sides of preceding equation by rated power P_N of isolated system:

$$\Delta P_G - N = 2H \frac{d\Delta f}{dt} + Df_N \Delta f \tag{4.13}$$

where $\Delta P_G = \Delta P_{GMW}/P_N$ is the relative change of power generated by controlled sources, $N = N_{MW}/P_N$ is the power relative imbalance at the moment of islanding, $H = W_N/P_N$ is the equivalent inertia constant of rotating masses, $D = D_{fGLMW/\mathrm{Hz}}/P_N$ is the frequency sensitivity coefficient of static sources and loads in MGs.

After switching to island operation with N_{MW} imbalance, generated and consumed power will find balance with new frequency. This will happen due to the operation of rotating sources control systems, referred to as primary frequency control of RVPP. This process is described by

$$(\Delta P_G - N) \frac{1}{f_N D} = \frac{2H}{f_N D} \frac{d\Delta f}{dt} + \Delta f \tag{4.14}$$

Let us modify the equation by introducing new values expressing the dynamics of frequency changes in MGs, i.e. time constant $T_S = 2H/f_N D$ and amplification factor $K_S = 1/f_N D$. Finally,

$$\Delta f(s) = G_S(s)(\Delta P_G(s) - N(s)) \tag{4.15}$$

where $G_S(s) = K_S/(1 + sT_S)$ is the MGs operator transfer function.

Figure 4.11 shows a simple scheme of primary frequency control in MGs, where static sources and electric power storage system may occur alongside rotating sources. Based on this scheme, it may be possible to notate the basic relationship related to primary frequency control in MGs

$$\Delta P_G = -k_R \Delta f \tag{4.16}$$

where ΔP_G is the increase in power generated in MGs, necessary to overcome the imbalance N, which occurred after MGs lost connection with external system, k_R is

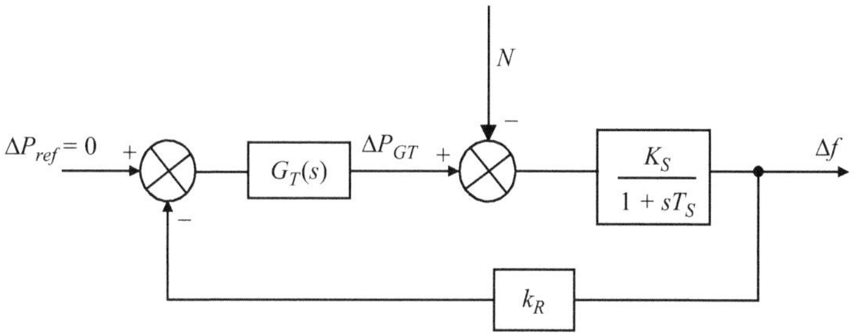

Figure 4.11 A scheme of primary frequency control in MGs

the power equivalent of MGs frequency, after switching from parallel to island operation, Δf is the change of frequency in MGs after connection to external system is lost.

Of interest here are steady-state values of changes in generated power ΔP_{GT} and in frequency Δf; therefore, the transfer functions of controller and turbine $G_T(s)$ have been omitted further in the description.

Steady-state value of frequency increment $\Delta f(t = \infty)$ caused by power balance discrete change by N after MGs islanding is

$$\Delta f = -\frac{K_S N}{1 + k_R K_S} \tag{4.17}$$

and power equivalent of frequency is expressed with the following formula

$$k_R = -\frac{K_S N + \Delta f}{K_S \Delta f} = -\left(\frac{N}{\Delta f} + \frac{1}{K_S}\right) \tag{4.18}$$

By employing preceding formula, it is possible to estimate the value of power equivalent of frequency based on power and frequency measured before and after switching MGs to island operation.

In order to illustrate the method of employing power and frequency measurements in estimating the parameters of the model here described, there were used patterns of those values, which were recorded during an experiment that consisted in islanding of a small industrial MGs, which is identical to the structure of RVPP [10].

During the simulation, there were performed three consecutive trials of switching an 11.5 MW turbo-generator to island operation with varied imbalance levels:

- Trial P1 NMW1 = −0.3 MW,
- Trial P2 NMW1 = −3.5 MW and
- Trial P3 NMW3 = −4.8 MW.

Recorded patterns of frequency changes have been included in Figure 4.12, and patterns of active power changes – in Figure 4.13. Each trial involved a different set of loads in the MGs, the loads being mainly asynchronous motors.

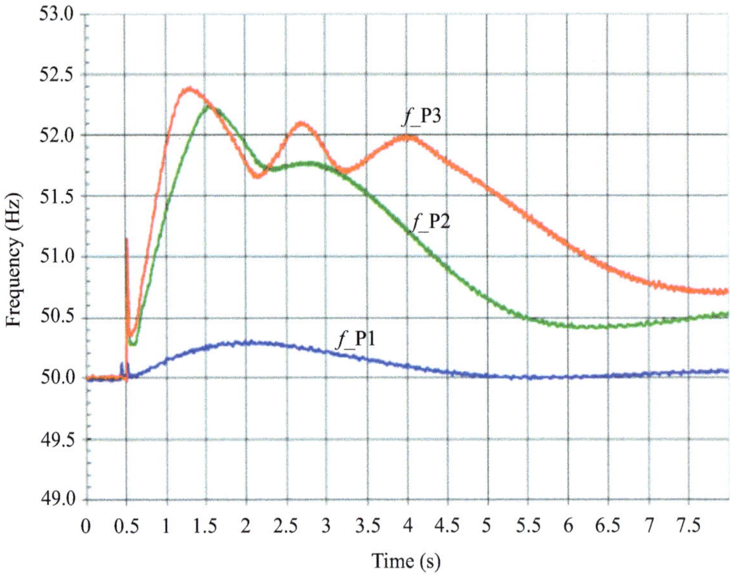

Figure 4.12 Frequency changes during switching an 11.5 MW turbine-generator set to island operation in three subsequent trials P1, P2 and P3

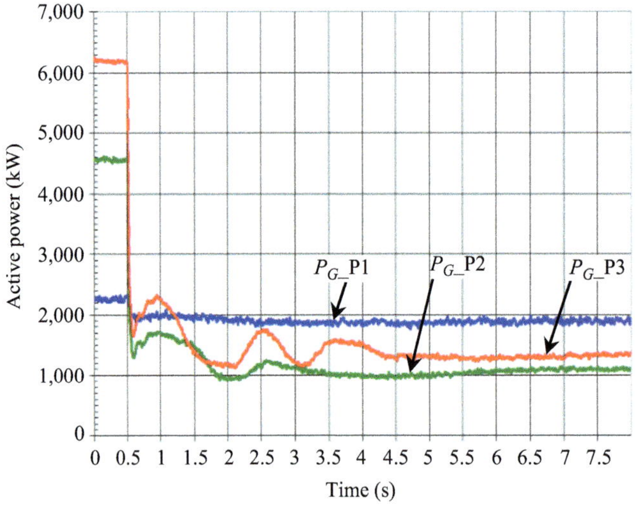

Figure 4.13 Active power changes during switching an 11.5 MW turbine-generator set to island operation in three subsequent trials P1, P2 and P3

The formulas derived earlier in this chapter served to estimate constants H and T_S, as well as factors K_S, k_R and K_{Icrit}. Transfer function parameters $G_T(s)$ of the turbine with control systems (amplification K_T and time constant T_T) were selected in a simulation by trial and error, focusing on best representation of frequency change patterns shown in Figure 4.12. All the measured and calculated parameters are shown in Table 4.2.

Due to imprecise diagram scale in Figures 4.7 and 4.13, power and frequency readings are inaccurate, this having impact on errors in estimations of the values of constants H, D, T_S, K_S, k_R and K_{Icrit}. In particular, inaccuracy affected trial P1, in which the values of power and frequency changes were small, and therefore, those results were not taken into consideration.

Figure 4.14 shows a simulation of frequency changes after switching to island operation in trial P2, with power imbalance $N_{MW} = -3.5$ MW (which corresponds

Table 4.2 Estimation of MGs model's and control system's parameters

Test –	P_N **MW**	f_{bef} **Hz**	f_{aft} **Hz**	N **MW**	df/dt **Hz/s**	H **s**	D **p.u./Hz**	T_S **s**	K_S –	k_R –	c_T –	K_T –	T_T **s**	K_{Icrit} –
1	11.5	50	50.1	−0.3										
2	11.5	50	50.5	−3.5	2.80	2.72	0.017	6.25	1.15	29.6	0.8	0.7	2.0	25
3	11.5	50	50.7	−4.8	3.80	2.75	0.025	4.42	0.81	28.6	0.8	0.7	2.0	33

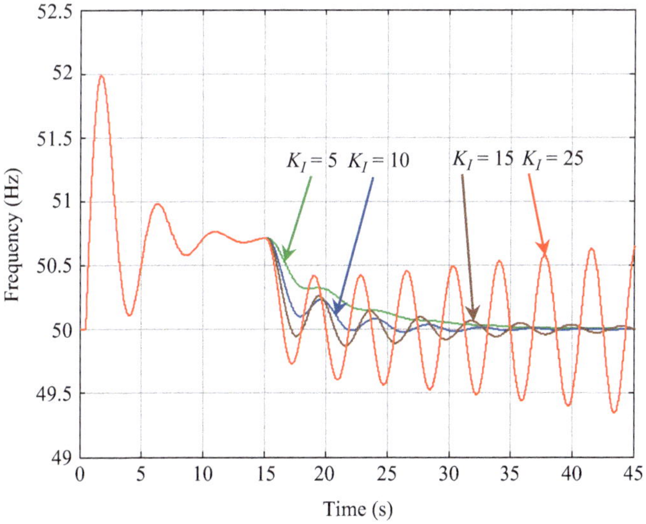

Figure 4.14 Simulation of switching an 11.5 MW turbo-generator to island operation with imbalance $N_{MW} = -3.5$ MW. Frequency changes result from primary control (t = 0.5–15 s) and secondary control (t = 15–45 s). Secondary control was simulated at various values of integrating element's amplification K_I = (5, 10, 15 and 25)

to $N = -0.3$ p.u in relative units). It has been assumed that power equivalent of frequency $k_R = 29$, while the reproduction of the dynamics of primary frequency control process was performed on the basis of the turbine's and its controller's transfer function parameters $K_T = 0.7$, $T_T = 2.0$ s. After the frequency settled, secondary frequency control was engaged, assuming that share factor for machine source is $c_T = 0.8$, while for static source, it is $c_{st} = 0.2$. Figure 4.14 shows patterns for simulations performed with the use of MATLAB Simulink toolbox. At moment $t = 0.5$ s, islanding occurs and within less than 20 s the frequency settles at about 50.7 Hz as a result of primary control. Further changes of frequency result from secondary frequency control. In Figure 4.14, they correspond to subsequent simulations, in which integrating element's amplification factor K_I equalled 5, 10, 15 and 25.

4.11 Conclusion

This chapter proposes an overview about different types of VPP, and it investigates proposed structure of a VPP applied as a grid-connected system and the optimization of the basic control structure has been made in order to minimize frequency fluctuation. Through the islanded MG and VPP concepts, individual DERs can gain access, visibility across all energy markets and benefit from VPP to optimize their performances. Both VPP and MGs can facilitate DER integration into MGs but with different aims. MGs focus on network operation control concerning active and reactive power using DER units existing within one local grid. On the other hand, while VPP concentrates on the provision of energy and power system support services from DER units. System operation can benefit from optimal use of all available capacity and increased efficiency of operation. Benefits form the VPP concept have been identified for different participants and on different levels. DERs owners may improve value of assets flexibility as well as increase value of assets through the markets, and it brings reduced financial risk through assets aggregation with improved ability to negotiate commercial conditions. VPP helps in maintaining the system secured, by using control flexibility of DER units for network management while using cost effective large-scale integration of renewable sources.

This chapter shows mathematical relations that connect frequency changes with power generated and received in islanded MGs after its switching to island operation as VPP RVPP. A method for frequency control in an autonomously operating MGs is also proposed. The parameters of frequency control system may be estimated on the basis of power and frequency measured during switching to island operation. The mathematical models shown in this paper are preliminary and may encourage further research into frequency control in autonomous power systems: MGs and VPP.

References

[1] Sundaram V.S., and Jayabarathi T. "An effect of SMES using automatic generation control in a multi area power system". *Proceedings of 2011*

International Conference on Recent Advancements in Electrical, Electronics and Control Engineering, Sivakasi, 2011, pp. 182–186. doi: 10.1109/ICONRAEeCE.2011.6129775.

[2] Awerbuch S., and Preston A. (eds.). *The Virtual Utility – Accounting, Technology & Competitive Aspects of the Emerging Industry*. Boston, MA, USA: Kluwer Academic Publishers, 1997. p. 409. https://doi.org/10.1007/978-1-4615-6167-5.

[3] Directorate-General for Research and Innovation (European Commission). *The Strategic Energy Technology (SET) Plan*, 2017. DOI 10.2777/476339.

[4] ENTSOE. *UCTE Operation Handbook*, 2017. http://www.ucte.org.

[5] Annaswamy A.M. *IEEE Smart Grid Vision for Computing: 2030 and Beyond Roadmap*, 11 January 2016, pp. 1–14. Available from http://amfarid.scripts.mit.edu/resources/Books/SPG-B02.pdf.

[6] Klessmann C., Nabe C., and Burges K. "Pros and cons of exposing renewables to electricity market risks – A comparison of the market integration approaches in Germany, Spain and the UK". *Energy Policy*, vol. 36, no. 10, pp. 3646–3661, 2008. Available from URL https://doi.org/10.1016/j.enpol.2008.06.022.

[7] Armstrong M., and Galli A. "The impact of a virtual power plant on the day-ahead market in France". Report presented at the *Winter Energy Seminar Series: "Price and Decision Support Modelling in Electricity Markets"*, Dept. of Industrial Economics & Technology Management, NTNU Trondheim, 12–13 February 2009.

[8] Invernizzi A. "Distribution systems and dispersed generation: strategy, activities and future development". *Electra*. vol. 04, pp. 22–29, 2008.

[9] Kok K. "Short-term economics of virtual power plants". *Proceedings of CIRED 2009 – 20th International Conference and Exhibition on Electricity Distribution* – Part 1, Prague, Czech Republic, 2009, pp. 1–4.

[10] Lis R., and Sobierajski M. *Integration of Distributed Resources in Power Systems*. Wrocław: Oficyna Wydawnicza Politechniki Wrocławskiej, 2011.

[11] Sobierajski M., Łabuzek M., and Lis R. *Electrical Power System Analysis in Matlab*. Wroclaw: Oficyna Wydawnicza Politechniki Wrocławskiej, 2006.

[12] Unger D., Spitalny L., and Myrzik J.M.A. "Voltage control by small hydro power plants integrated into a virtual power plant". *Proceedings of IEEE Energytech*, Cleveland, OH, 2012, pp. 1–6. doi: 10.1109/EnergyTech.2012.6304637.

[13] Pudjianto D., Ramsay C., and Strbac G. "Virtual power plant and system integration of distributed energy resources". *IET Renewable Power Generation*, vol. 1, no. 1, pp. 10–16, 2007. doi: 10.1049/iet-rpg:20060023.

[14] You S., Træholt C., and Poulsen B. *Developing Virtual Power Plant for Optimized Distributed Energy Resources Operation and Integration*. Kongens Lyngby, Denmark: Technical University of Denmark, September 2010.

[15] Hess T., and Schegner P. "Power schedule planning and operation algorithm of the Local Virtual Power Plant based on CHP-devices". *Proceedings of 2015 IEEE Power & Energy Society General Meeting*, Denver, CO, 2015, pp. 1–5. doi: 10.1109/PESGM.2015.7286261.

[16] Ogston E., and Brazier F. "Apportionment of SMEs in virtual power stations". *Proceedings of Second International Conference on Infrastructure Systems and Services: Developing 21st Century Infrastructure Networks (INFRA)*. Chennai, 2009, pp. 1–6. doi: 10.1109/INFRA.2009.5397866.

[17] Ruiz N., Cobelo I., and Oyarzabal J. "A direct load control model for virtual power plant management". *IEEE Transactions on Power Systems*, vol. 24, no. 2, pp. 959–966, 2009. doi: 10.1109/TPWRS.2009.2016607.

[18] Setiawan E.A. *Concept and Controllability of Virtual Power Plant*, Germany: University of Kassel, April 2007.

[19] Houwing M., and Ilic M. "The value of IT for virtual power plants with micro cogeneration systems". *Proceedings of 2008 IEEE International Conference on Systems, Man and Cybernetics*. Singapore, 2008, pp. 1–6. doi: 10.1109/ICSMC.2008.4811241.

[20] Nikonowicz Ł.B., and Milewski J. "Virtual power plants – general review: structure, application and optimization". *Journal of Power of Technologies*, vol. 92, no. 3, pp. 135–149, 2012.

[21] Saboori H., Mohammadi M., and Taghe R. "Virtual power plant (VPP), definition, Concept, Components and Types". *Proceedings of Asia-Pacific Power and Energy Engineering Conference*, Wuhan, 2011, pp. 1–4. doi: 10.1109/APPEEC.2011.5749026.

[22] VHPREADY. VHPREADY (Virtual Heat and Power Ready) – *Technical Requirements Specifications*, 2012. Available from http://www.vattenfall.de/de/vhp-ready.htm [Accessed October 2012].

[23] Werner T.G., and Remberg R. "Technical, economical and regulatory aspects of Virtual Power Plants". *Proceedings of 2008 Third International Conference on Electric Utility Deregulation and Restructuring and Power Technologies*, Nanjing, 2008, pp. 2427–2433. doi: 10.1109/DRPT.2008.4523819.

[24] Christensen J. *Executive Summary of Edison Project*. 2019. Available from: http://www.edison-net.dk/Project\%20Details/Executive\%20Summary.aspx. [Accessed 16 February 2017]

[25] Fuchs E., Cech H., Heher A., *et al.* "Into the new electricity age with optimal integration of decentralized energy resources – The Fenix Project". *World Energy Congress 2010*, Montreal, 12–16 September 2010. ISBN: 978-1-61738-736-4.

[26] Braun M. *Virtual Power Plants in Real Applications: Pilot Demonstrations in Spain and England as Part of the European Project FENIX*. Fraunhofer IWES, January 2009.

[27] Fenix. *Fenix – Flexible Electricity Network to Integrate the Expected "Energy Evolution"*. 2018. Available from: http://www.fenix-project.org/. [Accessed 16 February 2017].

[28] Lasseter R.H. "Certs microgrid". *Proceedings of 2007 IEEE International Conference on System of Systems Engineering*, San Antonio, TX, 2007, pp. 1–5. doi: 10.1109/SYSOSE.2007.4304248.

[29] Spence A., *'Virtual' Battery Storage Power Plant to Ease Electricity Price Spikes in Australia*. August 2016. Available from: http://www.renewableenergyworld.com/articles/2016/08.

[30] Vuthi P.P., Loreznzen P., Schäfers H., *et al.* "Smart power Hamburg: A virtual power plant for Hamburg". *International ETG Congress 2015; Die Energiewende – Blueprints for the New Energy Age*, Bonn, Germany, 2015, pp. 1–8. Available from http://ieeexplore.ieee.org/stamp/stamp.jsp?tp=\&arnumber=7388534\&isnumber= 7388454H.

[31] Lombardi P., Stötzer M., Styczynski Z., and Orths A. "Multi-criteria optimization of an energy storage system within a Virtual Power Plant architecture". *Proceedings of IEEE Power and Energy Society General Meeting*, Detroit, MI, USA, 2011, pp. 1–6. doi: 10.1109/PES.2011.6039347.

[32] Molderink A., Bakker V., and Bosman M. "Domestic energy management methodology for optimizing efficiency in Smart Grids". *IEEE Conference on Power Technology*, Bucharest, 2009. Available from https://ris.utwente.nl/ws/files/5412691/powertech.pdf.

[33] You S., Træholt C., and Poulsen B. "A market-based Virtual Power Plant". *Proceedings of 2009 International Conference on Clean Electrical Power*, Capri, 2009, pp. 460–465. doi: 10.1109/ICCEP.2009.5212012.

[34] Bakari K.E., and Kling W.L. "Virtual power plants: An answer to increasing distributed generation". *2010 Proceedings of IEEE PES Innovative Smart Grid Technologies Conference Europe* (ISGT Europe), Gothenburg, 2010, pp. 1–6. doi: 10.1109/ISGTEUROPE.2010.5638984.

[35] Zdrilić M., Pandžić H., and Kuzle I. "The mixed-integer linear optimization model of virtual power plant operation". *Proceedings of 8th International Conference on the European Energy Market (EEM)*. Zagreb, 2011, pp. 467–471. doi: 10.1109/EEM.2011.5953056.

[36] Bosman M., and Bakker M. "Production planning in a Virtual Power Plant". *ProRISC 2009, Annual Workshop on Circuits, Systems and Signal Processing*, Veldhoven, Holland, 2009. Available from https://ris.utwente.nl/ws/portalfiles/portal/ 5411833/distributed_ scheduling.pdf.

[37] Asmus P. "Microgrids, virtual power plants and our distributed energy". *The Electricity Journal*, vol. 23, no. 10, pp. 72–82, 2010. Future https://doi.org/10.1016/j.tej.2010.11.001.

[38] Czechowski R., Wicher P., and Wiecha B. "Cyber security in communication of SCADA systems using IEC 61850". *2015 Modern Electric Power Systems, MEPS'15*, Wrocław, Poland, July 06–09 2015, pp. 1–7.

[39] Schneider Electric. Technical information taken from the *Schneider Electric Energy Security Applications Catalog* of Schneider Electric Energy Poland Sp. z o.o., Book No 6 Telemechanika PACiS SN/WN, 2015. www.schneider-energy.pl.

[40] Real VNC. *Technical Documentation of Real VNC*. 2006. [online] Available from https://www.realvnc.com/products/vnc/, [Accessed 28 Sep 2018].

[41] Czechowski R. "Bad practices and most frequent mistakes in IPv4 networks implementation in smart electrical and automatic systems. Cyber security in distribution smart grid". *2016, 17th International Scientific Conference on Electric Power Engineering (EPE)*, Prague, Czech Republic. Danvers, MA: IEEE; Prague: Czech Technical University.

Faculty of Electrical Engineering. Department of Electrical Power Engineering, cop., 2016, pp. 156–160.

[42] Czechowski R. *Smart Grid Security Policies and Regulations. Communication, Control and Security Challenges for the Smart Grid, Chapter 16*. ed. by S.M. Muyeen and S. Rahman. London: The Institution of Engineering and Technology, 2017. pp. 489–528.

[43] Czechowski R. "Security policy and good practice for implementation of smart grid solutions". *Electrical Review*, vol. R. 92, no. 3, pp. 177–181, 2016. doi:10.15199/48.2016.03.42, 2016.

[44] Czechowski R. "Parameters determining the information flow in Smart Meters Complex Networks Smart Metering and Smart Grid Security". *Proceedings of the 2nd World Congress on Electrical Engineering and Computer Systems and Science, EECSS'16*, International ASET, 2016, Budapest, Hungary, 2016, p. 1–6.

[45] Zaki M.J., and Wagner J.R.M. *Data Mining and Analysis Fundamental Concepts and Algorithms*. New York: Cambridge University Press, 2014.

Chapter 5

Operations of a clustered microgrid

Munira Batool[1], *Syed Islam*[2], *and Farhad Shahnia*[3]

Microgrids (MGs) are referred to as isolated and self-sufficient electricity supply systems that well suit remote areas. These systems are generally composed of non-dispatchable and dispatchable energy resources to reduce the electricity production cost. Emergencies such as overloading, faults and shortfalls can result in difficulty for the smooth operation of MGs. The main aim of this study is to discuss the operation of MGs by presenting a power transaction management scheme. It focuses on the scenario when MGs are provisionally coupled to resolve the emergency situation and termed clustered MGs. For this purpose, power transaction is taken as an instance of purchasing or selling of electricity amongst healthy and problem MGs. The key objective of a suitable power transaction technique should then be regulating the power amongst the provisionally coupled MGs by adjusting the suitable power generation from all available dispatchable sources. An optimization problem is formulated for achieving this purpose, and its main purpose is to minimize the costs and technical impacts while focusing on the above-considered parameters. Genetic algorithm which is a heuristic optimization technique is used to solve the formulated optimization problem, and the performance of the suitable power transaction strategy is evaluated by several numerical analyses.

5.1 Overview of clustered microgrid

Due to technical and geographical limitations, it is not always possible to extend the existing transmission and distribution lines to very remote and regional areas. Thereby, utilities usually build a local power generation and distribution network at such locations. As an example, except the towns at Australia's east coast that are supplied through the National Electricity Market and those few at its southwest that are supplied through the south west interconnected system, most other towns in Australia's regional and remote areas, in which almost 31% of the population lives,

[1]School of Electrical Engineering, Computing and Mathematical Sciences, Curtin University, Perth, Australia
[2]School of Science, Engineering and Information Technology, Federation University, Ballarat, Australia
[3]Discipline of Engineering and Energy, Murdoch University, Perth, Australia

are supplied by local generators running with diesel or gas [1]. However, this type of generation is expensive; the fuel transportation is sometimes difficult because of roads' seasonal inaccessibility, and it pollutes the environment [2]. In addition to the lower reliability, the utilities also experience larger power losses due to long lines in those areas. This also results in high expenditures on supply, operation and maintenance, which are usually borne by the utilities. To reduce the overall cost of electricity generation, the utilities prefer to utilize some sort of renewable-energy-based distributed energy resources (DERs) and maximize their contribution in electricity generation [3,4]. These systems are usually designed to operate isolated and be self-sufficient and are often referred to as isolated MGs. As an example, the techno-economic analysis in [5] shows that the local utility can reduce its electricity supply cost by 70% when the rural town of Laverton in Western Australia is supplied by a group of renewable sources along with smaller sized diesel generators (Dgens).

On the other hand, it is highly probable that a large remote area accommodates multiple isolated MGs, each with a different operator (owner), with the incentives that the governments are putting forward in attracting private investors to build and operate renewable-based systems [6,7]. In such a case, to improve the reliability, resiliency and self-healing of isolated MGs in such remote areas, it is suggested in [8,9] that the MGs have some sort of physical connections amongst themselves to support each other during emergencies. Reference [10] suggested the temporary coupling of these adjacent but individually operating isolated systems under sudden emergency situations. These emergency situations can be power shortfalls, excessive generation and instantaneous faults, etc. The management of MG restoration process after faults is explained in [11], while [12–14] identify the MGs cluster with self-healing capabilities. The main aim is the enhancement of MGs' resilience against extended overloading or overgeneration events. With these objectives, the concept of clustered MG (CMG) has emerged. The key concept of the CMG is to consider a scenario in which more than one MG which is in the neighborhood of a remote area can support each other during the sudden situations of emergencies [15,16]. For this purpose, let us consider the network provided in Figure 5.1 which presents two neighboring MGs present inside a large remote area. It is shown that both MGs are connected through an interconnecting static switch (ISS) and a tie-line. It is assumed that the problem MG (PMG) is the one which is suffering from sudden emergency situation and in the similar way it can be supported by a healthy MG (HMG) available within the same remote area. To elaborate this concept, a detailed transformative architecture is presented for coupling of the neighboring MGs as described in [17], and its main focus is to enhance the resiliency of system during fault conditions. An approach based on decision-making criteria is proposed in [18], and its main task is to find the most suitable HMG(s) which can be coupled with an overloaded PMG. Furthermore, it considers various criteria, for example, electricity cost, reliability, available surplus power and the distance of the neighboring MGs on top of the technical impacts of voltage and/or frequency deviation in the CMG. While [19] proposes the key conditions which can detect the availability of excess power in the neighboring

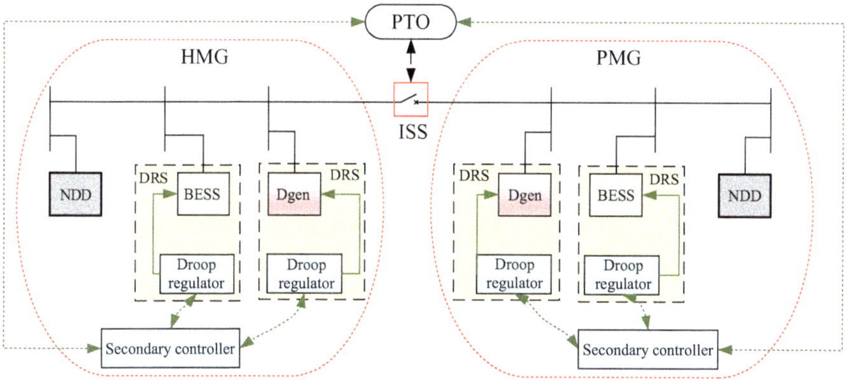

Figure 5.1 An exemplary setup of two neighboring MGs working in a cluster for achieving normal operation

HMG as well as overloading of a PMG. An interactive control of CMGs is introduced in [20] in order to guarantee system-wide stability and adequate load sharing. Reference [21] investigates the dynamic operation of DERs which are present within CMGs, whereas [22] discusses the dynamic security feature of the CMGs. While [23] describes the mutual interaction among the DERs present in MGs which are part of a CMG. The reliability aspects of a CMG is analyzed in [24] while their small signal stability as well as current and voltage controllability are analyzed in [25–27]. References [28,29] present coordinated operation of battery energy storage system (BESS) in MGs correlating with their provisional coupling. It is well explained that coupling between MGs inside CMG can be realized by the use of either back-to-back converters [30] or by incorporating ISSs [19]. To this end, the ultimate focus is that an MG present inside CMG can be interconnected to any other MG (it is important to show over here that it might not be necessarily an adjacent MG) if it is provided that a general link is available between the two to act as a power exchange highway. An optimization-solver-based technique is described in [31,32] for MGs coordination. While in [33,34], minimum operational cost calculation by the use of different optimization strategies is well discussed for CMGs. In [35,36], the focus is on the cooperative mode of operation of CMGs when high penetration of non-dispatchable DERs is present along with robust distributed control.

As discussed above, [37,38] have considered the control aspects of an MG when a voltage and/or frequency deviation is observed. In general to this end, usually an optimization problem is formulated considering different criteria or objectives. For example, studies in [39,40] propose the methods for operating the cost minimization of MGs by considering the stochastic nature of non-dispatchable resources. Cost minimization energy management techniques are presented in [41,42] while the focus was only on energy storage system optimization. Similarly, multi-MG optimization problem is formulated in [43] for achieving minimum cost by the fuel cost reduction with reserve capacity and inter-area flow limit.

Alternatively, some studies have aimed to coordinate the power exchange among MGs, load curtailment and control of the power of conventional generators. As an example, [44] has considered Dgens' fuel consumption and emission cost along with power exchange with utility grid in the formulated objective function (OF). Reference [45] discusses the impact of load curtailment in MGs by considering the sensitivities in nodal power injection and the probabilistic uncertainties of loads and renewable sources. To this end, the cost of load curtailment as well as the expense/revenue of exchanging power between the MG and a utility feeder is considered. In these studies, the main objectives are maximizing the footprint of renewable energies in supplying the demand and minimizing the contribution of conventional generators. However, the curtailment of renewable energy resources is not considered which is essential in the case of overgeneration. The voltage rise problem in MGs because of renewable-energy-based Dgens is solved in [46] by curtailing their output power using droop control. On the other hand, [47] employs an optimization technique to maximize the lifetime characteristics of BESSs in MGs when compensating the variabilities of loads and renewable sources while minimizing the power generation cost of Dgens. Alternatively, a bargaining technique is used in [48] to facilitate a proactive energy trading and fair benefit sharing among remote area interconnected MGs in which the main considered criterion is the minimization of the total operational cost. In a similar way, [49] applies demand management in remote area MGs using a cooperative power dispatching algorithm for the minimization of MG's operational cost while satisfying the load demand. References [50,51] have formulated an economic dispatch problem, which aim at minimizing the power loss on top of the costs of fuel consumption, external power sharing and BESSs. Alternatively, [52,53] had discussed the possibility of forming a provisional CMG as another resort.

The described technique operates under a power transaction scheme. The primary step, applied as the first resort, is adjusting the droop control parameters of the dispatchable distributed energy resources (DDERs) and charging/discharging control of the existing BESSs while the secondary step is the control of power exchange with one or more neighboring MGs. Finally, the third step, as the last resort, is curtailing some nonessential loads or renewable sources if the previous actions cannot eliminate the emergency. It is assumed that a low bandwidth communication is present for the transmission of the required data from sensors and MG central controller to the power transaction operator (PTO) and the decision variables from the PTO to the MG central controller and relevant local controllers. It is to be noted that the described technique considers the impact of voltage and/or frequency deviations on the power consumed by loads. Existing industrial processors by Intel® [54], National Instruments™ [55] and Analog Devices™ [56] can be effectively used when implementing the suitable technique as they satisfy the required speed.

This study recommends implementation of a scheme based on power transaction management among MGs related to CMG in the case of any sudden emergency condition. The loads and non-dispatchable DERs are assumed to be uncontrolled. On the other hand, the droop-regulated system (DRS), for example, the BESSs and the Dgen inside each MG assumes to adopt frequency and voltage droop control, and hence they play the role of the control variables. It is a known fact that adjustment of frequency is not required if the MG is operating in utility or grid-connected mode. But oppositely, the dispatchable DERs are taken responsible for the frequency regulation if a remote area is working without a utility feeder connection. The non-dispatchable DERs present inside the MGs of CMG will operate under a active reactive power (PQ) control mode with constancy. Also it is assumed that a low-bandwidth communication is available for CMG synchronization with neighboring MG(s) and to control the power flow in an MG.

The rest of the chapter is organized as follows: Section 5.2 introduces the modeling strategy for CMGs, and their operation and control is elaborated in Section 5.3 while the formulated problem and the essential constrains are discussed in details in Section 5.4. The performance evaluation of the suitable technique is illustrated through numerical analyses in Section 5.5 while the main highlights and findings of the work are summarized in the last section.

5.2 Modeling of clustered microgrid

Let us consider two neighboring MGs which can support each other by their temporary interconnection during emergency conditions for normal operation and having physical links with each other as shown in Figure 5.1. A PTO termed is considered over here with the responsibilities

- to incorporate with secondary controllers of each MG and receive information;
- identify a PMG within the considered network;
- then solving the formulated optimization problem in order to select a suitable neighboring HMG so that power exchange could take place;
- defining the approximate level of power transaction either import and/or export; and
- finally, transmit the calculated decision variables to each MG (means to its secondary controller) of considered CMG.

The output power of non-dispatchable DERs (i.e., photovoltaic (PV) and wind) is assumed to be curtailable and is dependent on the command signal from the secondary controller of MG. Moreover, non-dispatchable DERs are intermittent in nature unless they are supported by BESS which helps in power smoothening. To this end, the DRS operation is assumed to be accompanied with the following steps:

- data transmission to PTO using secondary controller of MG.
- identifying the operational set points for Dgens which are based on the information received from PTO, via the secondary controller of MG.
- the technical factors of voltage and frequency (VF) at the output of DRS can be calculated by droop equations as [39]:

$$f = f^{\max} - m^{\mathrm{DRS}} P^{\mathrm{DRS}} \tag{5.1}$$

$$V = V^{\max} - n^{\mathrm{DRS}} Q^{\mathrm{DRS}} \tag{5.2}$$

$$m^{\mathrm{DRS}} = \frac{\Delta f^{\max}}{(P^{\mathrm{DRS}})^{\max}} \tag{5.3}$$

$$n^{DRS} = \frac{\Delta V^{\max}}{(Q^{\mathrm{DRS}})^{\max}} \tag{5.4}$$

where P and Q are the active and reactive powers injected by the DRS, $f^{\max}$ and $V^{\max}$ are the set points of the droop lines for frequency and voltage. In (5.1), m and n represent P–f and Q–V droop coefficients which can be derived from (5.3) and (5.4), respectively. $\Delta f^{\max}$ and $\Delta V^{\max}$ are the maximum allowed deviations for VF in each MG of CMG. Here BESS are assumed to have primary controllers which can receive signal from secondary controller. Once it is done then charging and discharging can be adjusted by looking on state-of-charge (SoC) status. Load points are accompanied by droop-regulated algorithm for the shedding of loads to overcome the emergency situation of PMG.

An important issue to consider within CMG is the synchronization of neighboring MGs which significantly depends on the physical links (distribution lines) between them. Consider a remote area town composed of islanded MG clusters for possible coupling configurations, which are as follows:

- Topology-1: All MGs are connected to a common central node (see Figure 5.2(a)).
- Topology-2: A radial or loop line connects all MGs (see Figure 5.2(b)–(c)).
- Topology-3: An MG is connected to every other MG through multiple distribution lines (see Figure 5.2(d)).
- Topology-4: Some MGs are connected to more than one neighboring MG (see Figure 5.2(e)).

The control strategy is a multistage process in which successive layers of necessary actions are carried out to overcome the emergency situation in PMG. The step-by-step process includes

- primary step: droop parameters (i.e., droop coefficients and VF set points) and ISS are configured,
- secondary step: power transaction is decided by PTO to/from HMG(s) and
- tertiary step: non-dispatchable DERs' curtailment and/or load shedding take(s) place in the case of failure of secondary and PTO control.

The power transaction scheme first tries to achieve the minimum cost of operation by DRS control (i.e., step-I). If it is not successful to resolve the emergency situation in PMG, the developed technique will try to solve the problem by other presented actions, but each step is followed by increase in cost accordingly. This step-by-step process will guarantee that the strategy will try to operate the CMGs with the method which is low in cost rather than the curtailment of sources or loads which is the most expensive one in this case. This can be done by the development of the optimization problem formulation which is described in the next section of this chapter. Moreover, Figure 5.3(b) schematically depicts the time sequence operation of the MG network for the suitable actions explained above. If the DERs

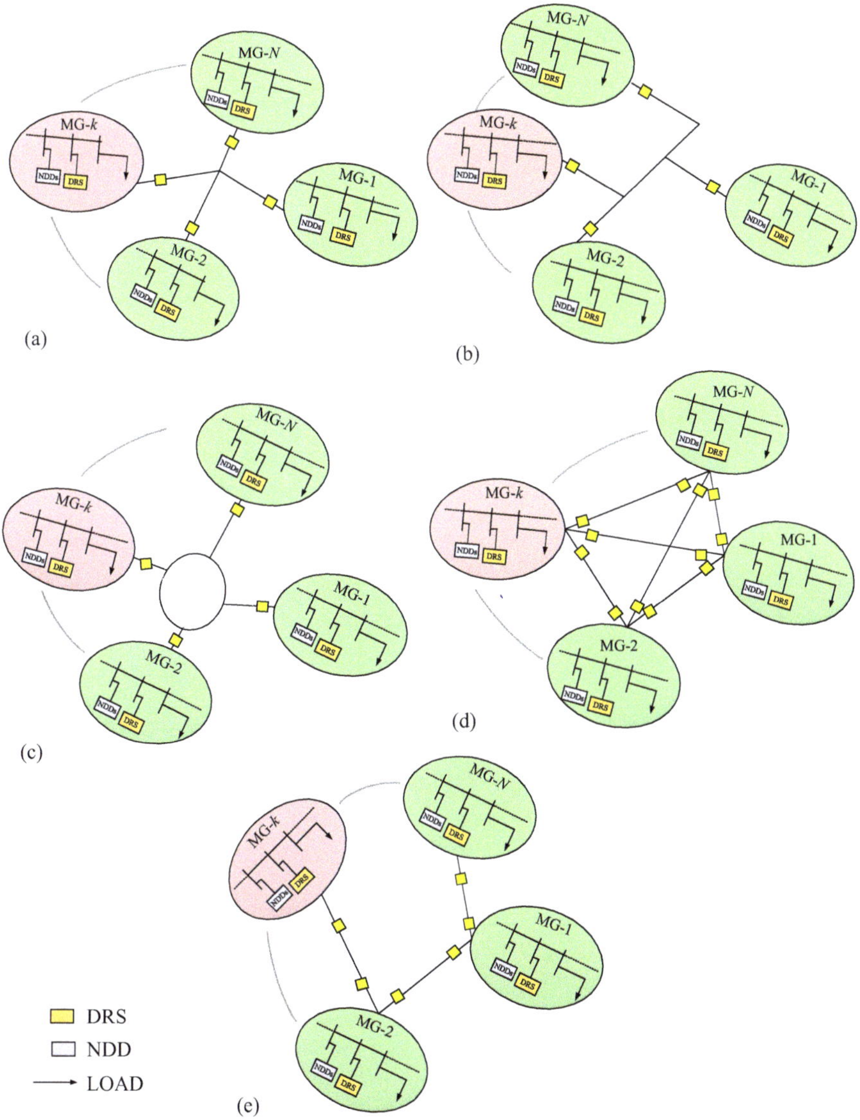

Figure 5.2 Possible physical communication links between MG(s) participating in cluster [16]

generation meets the load demand and VF lies in the safe mode, then the secondary controller will go through the normal operation as shown in Figure 5.3(a) while if the required conditions are not met, then PTO participates in the process to heal the aroused emergency situation by the applications of actions described above as described in Figure 5.3(b). All communication links among secondary, ISS and network controllers are illustrated schematically in Figure 5.4.

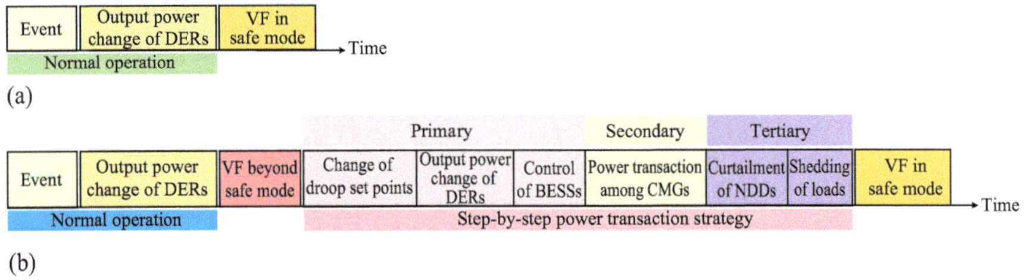

Figure 5.3 Time sequence in which actions take place to overcome the emergency situation in PMG: (a) Time sequence for normal operation of PMG and (b) Time sequence in which step-by-step power transaction strategy take place to overcome emergency in PMG

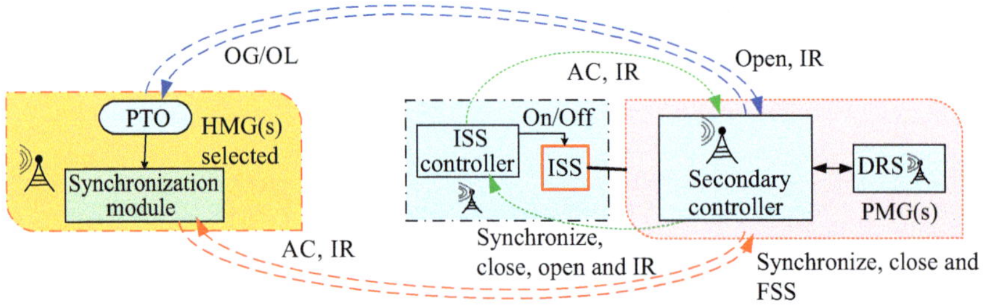

Figure 5.4 Communication protocols for the considered clustered microgrid network. AC, action complete; IR, information received; FSS, feasible solution sent; OG/OL, overgeneration/overloading

Once a PMG send information of emergencies to PTO, then PTO looks for the availability of a neighboring MG present in the healthy state. If search for an HMG is successful to support the PMG, the PTO will determine the suitable transaction of power in the tie-lines amongst the desired CMG by solving the formulated optimization problem. On contrary, if a desired HMG is not found or no cost-effective solution is searched through optimization problem-solving, the PTO will communicate a signal of curtailment to the PMG. So in this way, PMG will proceed with the curtailment of either the generation of its non-dispatchable DERs or the shedding of its nonessential loads. During the whole process, the PTO is assumed to be active and responds quickly to any changes happened inside the MGs. The system's conditions are also reevaluated by PTO in ΔT intervals. The flowchart of Figure 5.5 describes the operational principle of the PTO.

CMG which is shown in the network of Figure 5.1, one MG is defined as the power exporter, while the other is considered to be the power importer during an emergency situation. Also it is assumed that the transacted or exchanged power among MGs and denoted as P_{trans} will be negative if a power flow is happening to the HMG from the PMG and positive in the opposite situation. Hence, the difference between the planned transacted power (i.e., the output of PTO) and the actual power (before the formulated optimization) can be shown by the equation as

$$P^{\text{trans}} = P^{\text{trans}}_{\text{new}} - P^{\text{trans}}_{\text{old}} \tag{5.5}$$

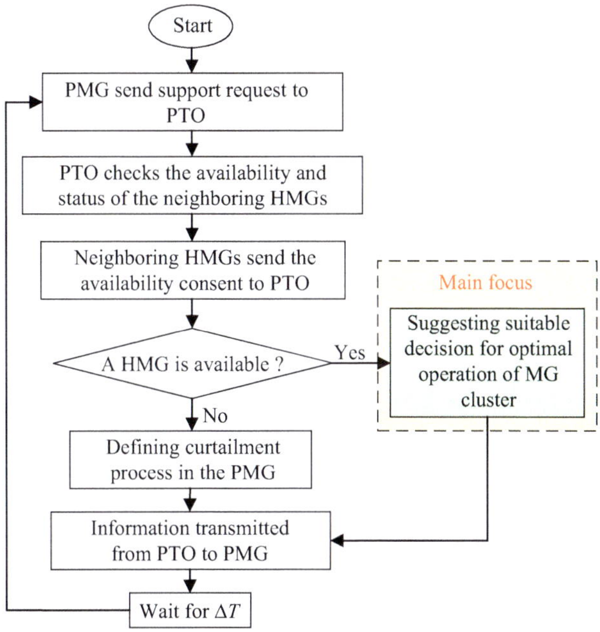

Figure 5.5 Flowchart for the suitable power transaction strategy for clustered microgrid operation

In the above equation, P^{trans} is the power required for transaction between two or more neighboring MGs present within the CMG while both subscripts "new" and "old" respectively represents the tie-line's power flow after and before the optimization application. The equal generation costs incremental principle and is considered to adjust the power transaction by each MG as described in [57]. Now let us consider that an MG becomes the PMG following a sudden increase in load demand. Therefore, the described power transaction strategy aims at increasing the flow of power from the HMG toward the PMG in order to meet the extra load demand. But it is also important that power transaction should not increase beyond a certain limit. Therefore, according to the principle of equal incremental cost, the rise of up to λ_{ix} will take place in the transacted cost for the corresponding power output of DRS as $P^{\text{HMG}}_{\text{exp-}ix}$. Therefore, the addition of variation of the power output of the BESS and DRS status will be equal to the sum of the power imported or exported from/to the HMG, i.e.:

$$\sum \Delta P_k^{\text{Dgen}} + \sum \Delta P_k^{\text{BESS}} = P_{\text{exp}} \tag{5.6}$$

The MGs which are participating in the CMG can be coupled with each other with the help of the ISS. The principle of operation and configuration of ISS is beyond the scope of this work and is discussed in [58]. Each ISS has its local controller for synchronization before closing and coupling two neighboring MGs. When one or more MG(s) are declared as PMG(s) due to overgeneration or overloading, this information is transmitted to the network tertiary controller through the secondary controller of the PMG(s). The PTO module within the network working as the tertiary controller receives this information, analyzes the system and suggests the most suitable HMG(s) to couple with the PMG(s) and form a CMG. The outcomes of the decision by the PTO are transferred to the secondary controller of the relevant PMG(s) and HMG(s), as well as the local controller of the corresponding ISS(s).

5.3 Control and operation of clustered microgrid

5.3.1 Droop-regulated strategy

When MGs are operating in either stand-alone or islanded mode, proper adjustment of both voltage magnitude and operating frequency needed to be done carefully. This can be achieved both by the control of demand as well as distributed generation, and these steps are based on network's overall demand response and mainly affected by control strategies applied to DERs (e.g., governor for diesel engines or inverter control). The presented droop-regulated strategy is based on load flow algorithm and is taken from Chapter 4 of [16]. In the available literature, generally, there are three types of buses, i.e., PV and PQ buses along with slack buses. The selection of the buses depends mainly upon the predefined parameters or quantities. In this study, it is recommended that total power, i.e., both active and reactive of all DERs buses, is either VF dependent or simply droop control buses. In general, an algorithm based on load flow strategy, the network losses are present or dumped on the slack bus. But this is not the case here as this work will be given to one of the available sources. This is done by the drop-regulated controller, i.e., PTO by looking at VF unbalance or other important adjustments, e.g., demand control. At first, nominal capacities of all sources and demand-side loads are set using stochastic modeling. Initially by using non-dispatchable DERs, rated power of BESS and load consumption and approximate output power from Dgen is estimated. Due to the fact that MG network frequency is not constant all times and it can have an effect on the reactance of lines, the admittance matrix is taken as frequency-dependent function as

$$Y^{\text{bus}}(f) = \begin{bmatrix} Y_{11}(f) & Y_{12}(f) & \cdots & Y_{1N}(f) \\ Y_{21}(f) & Y_{22}(f) & \cdots & Y_{2N}(f) \\ \vdots & \vdots & \ddots & \vdots \\ Y_{N1}(f) & Y_{N2}(f) & \cdots & Y_{NN}(f) \end{bmatrix} \tag{5.7}$$

where $Y_{xy}(f)$ denotes the mutual admittance of x and y buses, respectively. Let us suppose that bus x is connected to the load side, then non-dispatchable DERs and BESS power consumptions can be shown by the following equations:

$$S_x^{\text{load}} = P_x^{\text{load}} + Q_x^{\text{load}} \tag{5.8}$$

$$S_x^{\text{NDDs}} = -P_x^{\text{NDDs}} \tag{5.9}$$

$$S_x^{\text{BESS}} = \pm P_x^{\text{BESS}} \tag{5.10}$$

For this purpose, it is assumed that for voltages of all buses, initial values are, e.g., $1\angle 0^\circ$ pu. It is further used to calculate the current value, i.e., I_x^j drawn by each load in considered iteration. In similar way, the current injected by non-dispatchable DERs and BESS is calculated for all selected buses. Traditional power-flow analysis (PFA) is therefore applied to determine the value of voltage V_x^j of all loads, non-dispatchable DERs and BESS, respectively [20]. Acceleration factor (Ш) is then included to slightly change the bus voltages as

$$V_x^j = V_x^{j-1} + Ш\left(V_x^j - V_x^{j-1}\right) \tag{5.11}$$

MG line losses can be calculated as follows:

$$\left(S^{\text{loss}}\right)^j = \sum_{x=1}^{N_{\text{bus}}} \sum_{y=x}^{N_{\text{bus}}} -Y_{x,y}^{\text{bus}} \left(V_x^j - V_y^j\right)^2 \tag{5.12}$$

The approximate power output from Dgen will be

$$\left(\sum S^{\text{Dgen}}\right)^j = \sum S^{\text{load}} + \left(S^{\text{loss}}\right)^j - \sum S^{\text{NDDs}} \mp \sum S^{\text{BESS,rated}} \tag{5.13}$$

Powers of Dgen can be distinguished by using their predefined pu ratios:

$$\left(P_x^{\text{Dgen}}\right)^j = \text{Re}\left(\sum S^{P\text{Dgen}}\right) \times \text{Ratios}_x^{\text{Dgen}} \tag{5.14}$$

$$\left(Q_x^{\text{Dgen}}\right)^j = \text{Im}\left(\sum S^{P\text{Dgen}}\right) \times \text{Ratios}_x^{\text{Dgen}} \tag{5.15}$$

The generator powers are updated afterwards within limitations of the upper and lower bound. The considered MG's frequency can be calculated as

$$f_x^j = f_{\max} - M_x^{\text{Dgen}} \left(P_x^{\text{Dgen}}\right)^j \tag{5.16}$$

Algorithm 1: Droop-regulation-based control strategy

Generated inputs: $\boldsymbol{P}_{\textbf{DRSs}}^{\textbf{cap}}, \boldsymbol{P}_{\textbf{NDDs}}^{\textbf{cap}}, \boldsymbol{P}_{\textbf{load}}^{\textbf{cap}}, \boldsymbol{P}_{\textbf{initial}}^{\textbf{DRS}}$

i. Power transaction requirement for MG_i within the CMG is calculated,

ii. if (net available power from NDDs and Dgens is lesser than the load consumption, also the SoC of BESSs are at their lower bound) or (net available power from NDDs and Dgens is higher than the load consumption, also the SoC of BESSs are at their higher bound) then

iii. PTO declares MG_i as a PMG,

iv. else

PTO declares MG_i as an HMG,

v. end

In a similar way, the voltage magnitude of Dgen can be calculated as

$$V_x^j = V_{\text{nom}} - N_x^{\text{Dgen}} \left(Q_x^{\text{Dgen}}\right)^j \tag{5.17}$$

where V_{nom} is the system's nominal operating voltage. As slack bus is not present, the bus connected to Dgen is assumed to be working on the reference bus which has 0° angle. Similarly, rest of the angles in MG(s) can be calculated w.r.t. the selected reference bus.

Power of BESS can be updated based upon MG's frequency value. For this purpose, it is assumed that frequency difference cannot be increase than 1 Hz. For example, if nominal operating frequency f_{nom} is 50 Hz, deviation can be between lower limit of 49.5 Hz and upper limit of 50.5 Hz, respectively. Constant φ is the preset value lying in the range of 0.1–1. Therefore, if the difference between nominal and calculated frequency is $> \varphi \times \Delta f$, then the battery is assumed to be in charging mode as

$$P_{x,\text{charge}}^{\text{BESS}} = \frac{-(f_{\max} - f_x)}{M_x^{\text{BESS}}} \tag{5.18}$$

Conversely, if the difference is $< (\varphi \times \Delta f)$, the battery is in discharging mode and

$$P_{x,\text{discharge}}^{\text{BESS}} = \frac{(f_{\max} - f_x)}{M_x^{\text{BESS}}} \tag{5.19}$$

But if both charge/discharge conditions are not satisfied, then the battery assumes on standby at value $P_{\text{BSS}}^{\text{rated}}$. The load demand is assumed to be also dependent on VF and will be updated according to change in VF magnitude of the buses:

$$\left(P_x^{\text{load}}\right)^{j+1} = P_{\text{nom}} \left(\frac{V}{V_{\text{nom}}}\right)^{\alpha} (1 + K_P \Delta f) \tag{5.20}$$

$$\left(Q_x^{\text{load}}\right)^{j+1} = Q_{\text{nom}} \left(\frac{V}{V_{\text{nom}}}\right)^{\beta} (1 + K_q \Delta f) \tag{5.21}$$

$$\Delta V = |V_x^j| - 1 \tag{5.22}$$

$$\Delta f = f_x^j - f_{\text{nom}} \tag{5.23}$$

where $V_{\text{nom}}, P_{\text{nom}}$ and Q_{nom} are the nominal operating voltage, active and reactive powers on that voltage, respectively, also K_P, K_q are the frequency-dependent constants. All loads are assumed to be constant impedance (KI) type; therefore, exponent values of α and β are taken as 2. At each iteration of droop-regulated strategy, the mismatching value Δ of certain parameters will be calculated, and effort is made to make them smaller than defined tolerance value. Once DRS is deemed to be converged, then $P_{\text{load}}, P_{\text{loss}}, Y^{\text{bus}}, P_{\text{DG}}\ P_{\text{BESS}}, \Delta V, \Delta f$ will be updated accordingly. Furthermore, all the values will be converted from their original values to the recent updated ones.

5.3.2 Optimization solver

Genetic algorithm (GA) is the computational technique which is used for finding the best optimal solution for the transaction strategy. GA has proven records for solving the technical issues related to electrical distribution networks, their planning and operational analysis [59]. The main advantages of GA include an easy way to reach through the local minima, nonrestricted OF achievement and independent search direction to find out the most economical solution [60]. First, the population is initialized which is made of chromosomes. Each chromosome is coded by real number variables, and each individual in the chromosome is referred to as a gene [61,62]. Second, OF is calculated for the optimal solution along with constraints application to realize the fitness evaluation criteria. Once the parent solution pool is formed by the selection procedure, then the offsprings are created by the recombination process. Normally, crossover and mutation are considered the recombination operators which are also applied in this technique with suitable probabilities. Finally, the developed GA-based optimization technique will continue until the minimum desired cost is achieved for the considered MG's cluster to overcome the emergency situation. The full detail about the developed algorithm is explained as follows:

Algorithm 2: The GA-based optimization technique for clustered MG's operation

---

1. *for* MG-1 to MG-N
2. Define the output of secondary controller based on DRS and BESS ;
3. Declare the MG(s) as an HMG, PMG^{OL}, or a PMG^{OG} ;
4. Define the minimum and maximum bounds for NDDs/loads curtailment, Dgens, BESSs and power transaction for each MG in CMG;
5. end
6. Initialize GA parameters (initial population, individual fitness evaluation, selection and recombination)
7. Generate initial population with all selected variables and constraints along with crossover and mutation probabilities;
8. Define optimization stopping criteria;
9. *while* iterations <= Iteration ^{maximum}
10. *for* function tolerance =< 1e−6 and ΔT <= t (sec)
11. Define a number of MGs in CMG and the MG(s) which are not participating in CMG;
12. Recognize the MG(s) isolated from CMG as PMG^{OL}, or a PMG^{OG} ;
13. Call modified Gauss–Seidel-based PFA function [18] for the CMG and each PMG;
14. Identify the DRS, BESS and NDDs outputs, load demand, SoC of BESSs, frequency and voltage deviation for MG(s) in CMG and PMG(s);

15. Calculate cost of fuel, cost of emissions, BESSs life loss cost, curtailment of load/NDDs cost (if any), power loss cost, spinning reserve index (SRI) and renewable penetration level (RPL) for each MG(s);
16. Calculate the cost of power transaction (import/export) for each HMG(s) in CMG;
17. Evaluate the equality, inequality and bound constraints;
18. If the feasible solution is not converged "OR" constraints are not met then repeat lines (13 to 17) by applying
small increase in integer variables tolerance;
19. Calculate formulated objective function for optimization
20. Calculate the most optimal solution for each individual population;
21. end
22. Define the most feasible population of current iteration and identify the best individuals or chromosomes with high-rank fitness;
23. Update the values of individuals in selected population and limit them with bound constraints and tolerance;
24. end

5.3.3 Modeling of non-dispatchable DERs

The non-dispatchable distributed energy resources (NDDs) for this work are taken as PV and wind-based Dgens. The power output expected from a PV plant is calculated using a beta probability density function (PDF) to express the solar insulation level in the form of [63]:

$$f(X|a,b) = \frac{1}{B(a,b)} X^{a-1}\left(1 - X^{b-1}\right) I_{(0,1)}(X) \tag{5.24}$$

where a and b both in the range of [1,5] are beta distribution's shape parameters to yield a variety of shapes. The indicator function $I_{(0,1)}(x)$ ensures nonzero probability values of $x \in [0, 1]$ as presented in [64]. It is able to replicate the random pattern of insolation levels for any given time (h) of the day because of the fact that beta distribution has two limits or working values. Fill factor (*FF*) defined as the deviation of the current–voltage characteristic of a PV cell w.r.t. the ideal condition as

$$\mathrm{FF} = \frac{V_{\mathrm{mpp}} I_{\mathrm{mpp}}}{V_{\mathrm{oc}} I_{\mathrm{sc}}} \tag{5.25}$$

where V_{mpp} and I_{mpp} are, respectively, the voltage and current at the maximum power point [65,66], while V_{oc} is the open-circuit voltage and I_{sc} is the short-circuit current taken from PV cells of PV module. Thus, the output voltage and current of a PV module is calculated as

$$I_{\mathrm{out}} = f(X|a,b)\left[I_{\mathrm{sc}} + k_i(T - 25)\right] \tag{5.26}$$

$$V_{\mathrm{out}} = V_{\mathrm{oc}} - k_v T \tag{5.27}$$

Thereby, the total output power of a PV module is defined from:

$$P_{\text{total}} = I_{\text{out}} V_{\text{out}} FF \tag{5.28}$$

The output power of a wind-based non-dispatchable DERs is modeled in this work on the basis of wind speed (v_{wind}), which is modeled by a Rayleigh PDF (i.e., a special case of the Weibull PDF) as [67]:

$$f_{(v_{\text{wind}})} = \frac{2v_{\text{wind}}}{c_v} \times e^{-(v_{\text{wind}}/c_v)^2} \tag{5.29}$$

where $c_v \approx 1.128 v_m$ is a scale index, and v_m is the average wind speed. From (B6), the output power of the wind-based non-dispatchable DERs is calculated as

$$P_{\text{wind}} = \begin{cases} 0 & v_{\text{wind}} < v^{\text{ci}} \text{ or } v_{\text{wind}} > v^{\text{co}} \\ \dfrac{(v - v_{\text{ci}}) P_{\text{wind}}^{\max}}{v_{\text{wind}}^{\text{nominal}} - v_{\text{ci}}} & v^{\text{ci}} \leq v_{\text{wind}} < v_{\text{wind}}^{\text{nominal}} \\ P_{\text{wind}}^{\max} & v_{\text{wind}}^{\text{nominal}} \leq v_{\text{wind}} \leq v^{\text{co}} \end{cases} \tag{5.30}$$

where v^{ci} and v^{co} are the cut-in and cutout speed for wind turbine.

5.4 Optimization problem formulation and technical constraints

The optimization problem is formed to determine the most economical solution for overcoming the overloading or overgeneration issue which subsequently causes under/over VF in the PMG(s) under the presented power transaction strategy, described above. This is formulated as an OF which is solved by the PTO to yield the most feasible solution while minimizing the overall operational cost. The problem is formulated as a mixed integer linear programming problem with OF of:

$$\begin{aligned} \text{OF} = \sum_{i=1}^{n} \big(& w_1 \text{OF}_{\text{tech}} + w_2 \text{OF}_{\text{Dgen}} + w_3 \text{OF}_{\text{BESS}} + w_4 \text{OF}_{\text{curt}} \\ & + w_5 \text{OF}_{\text{trans}} + w_6 \text{OF}_{\text{loss}} \big) \text{Time} \quad \forall i \in \mathbf{MG} \end{aligned} \tag{5.31}$$

where $\text{OF}_{\text{tech}}, \text{OF}_{\text{Dgen}}, \text{OF}_{\text{BESS}}, \text{OF}_{\text{curt}}, \text{OF}_{\text{trans}}$ and OF_{loss} are respectively the OFs denoting the VF deviation in the MG followed by the operational cost of its Dgen(s), life loss cost of BESS(s), curtailment of non-dispatchable DER(s) and/or nonessential load(s), the power transaction costs for selected HMG(s) within the CMG, and the corresponding cost of power loss in the tie-line between MGs of the CMG while **MG** is a vector representing the number of MGs within the CMG or the isolated PMG(s). These costs are introduced below:

OF_{tech} is expressed as

$$\text{OF}_{\text{tech}} = |\Delta f| + \max(|\Delta V|) + \text{Violation} \times \text{Penalty} \tag{5.32}$$

in which $|\Delta V|$ and $|\Delta f|$ represent respectively the level of voltage magnitude deviation in all buses of the MGs within the CMG or the isolated PMG and the corresponding frequency deviation (in pu) while Penalty is selected as a large value (e.g., 10^8) to eliminate the sets of decisions (conditions) that yield unacceptable voltage or frequency deviation in either of the MGs of the CMG or the isolated PMG or overloading of MG lines. Thus, violation is defined as

$$\text{Violation} = V_{\text{vio}} + I_{\text{vio}} + f_{\text{vio}} + \text{Constraint}_{\text{vio}} \tag{5.33}$$

where

$$\begin{aligned}
V_{\text{vio}} &= \begin{cases} 1 & \exists|\Delta V_i| > \Delta V^{\max}, \quad i \in \mathbf{bus} \\ 0 & \text{otherwise} \end{cases} \\
I_{\text{vio}} &= \begin{cases} 1 & \exists|I_i| > I^{\max}, \quad i \in \mathbf{line} \\ 0 & \text{otherwise} \end{cases} \\
f_{\text{vio}} &= \begin{cases} 1 & |\Delta f| > \Delta f^{\max} \\ 0 & \text{otherwise} \end{cases} \\
\text{Constraint}_{\text{vio}} &= \begin{cases} 1 & \text{if a considered constraint is not met} \\ 0 & \text{otherwise} \end{cases}
\end{aligned} \tag{5.34}$$

in which $\Delta V^{\max}$, $\Delta f^{\max}$ and $I^{\max}$ are respectively the permissible limits for voltage deviation, frequency deviation and maximum line loading limit. Also, **bus** and **line** are two vectors representing the number of buses and lines of the CMG or the isolated PMG. These parameters will be calculated using a PFA by taking in account the DRS output. The other OFs are derived as

$$\text{OF}_{\text{Dgen}} = \sum\nolimits_k \left(C^{\text{fuel}} + C^{\text{cfp}}\partial_k\right) P_k^{\text{Dgen}} \quad \forall k \in \mathbf{Dgen} \tag{5.35a}$$

$$\text{OF}_{\text{BESS}} = \sum\nolimits_k C_{\text{lifeloss}}^{\text{BESS}} \left|P_k^{\text{BESS}}\right| \quad \forall k \in \mathbf{BESS} \tag{5.35b}$$

$$\text{OF}_{\text{curt}} = \sum\nolimits_k C_{\text{curt}}^{\text{NDDs}} P_{\text{curt}}^{\text{NDDs}} + \sum\nolimits_l C_{\text{curt}}^{\text{load}} P_{\text{curt}}^{\text{load}} \quad \forall k \in \mathbf{NDD},\ l \in \mathbf{load} \tag{5.35c}$$

$$\text{OF}_{\text{trans}} = \sum\nolimits_k C^{\text{trans}} \left|P_k^{\text{trans}}\right| \quad \forall k \in \mathbf{MG} \tag{5.35d}$$

$$\text{OF}_{\text{loss}} = \sum\nolimits_k C^{\text{loss}} P_k^{\text{loss}} \quad \forall k \in \mathbf{line} \tag{5.35e}$$

In (5.35a), OF_{Dgen} aims to minimize the cost of power generation by Dgens (denoted by C in \$/kW h) which includes the cost for fuel consumption and the corresponding carbon footprints (respectively denoted by "fuel" and "cfp"), where ∂_k is the emission ratio (in kg/kW h). As the BESS does not have any ongoing operational costs, only the cost of its life loss (denoted by "lifeloss") is considered in OF_{BESS} in (5.35b). Similarly, the corresponding cost of curtailing the output power of non-dispatchable DERs by $P_{\text{curt}}^{\text{NDDs}}$ and the nonessential loads by $P_{\text{curt}}^{\text{load}}$ (respectively denoted by $C_{\text{curt}}^{\text{NDDs}}$ and $C_{\text{curt}}^{\text{load}}$ in \$/kW h) is used in (5.35c) to determine OF_{curt}. The corresponding cost of power transaction ($|P^{\text{trans}}|$ in kW) over the tie-lines between the MGs of the CMG (denoted by C^{trans} in \$/kW h) is used in (5.35d) to define OF_{trans}. Equation (5.35e) aims at minimizing the power loss in the tie-lines between MGs of the CMG (P^{loss} in kW) when calculating OF_{loss} while C^{loss} is the associated power loss cost (in \$/kW h). In (5.35a)–(5.35e), **Dgen**, **BESS**, **NDD**, **load, MG** and **line** are vectors respectively representing the number of Dgens, BESSs, non-dispatchable DERs, loads and lines in the MGs within the CMG or the isolated PMG(s).

The OF of (5.31) is then solved while considering the constraints of:

$$\sum S_{\text{DRS}} + \sum S_{\text{NDD}} + \sum S_{\text{trans}} = \sum S_{\text{load}} + S_{\text{loss}} \tag{5.36a}$$

$$\begin{cases} \left(P_{\text{Dgen}}^{k}\right)^{\min} \leq P_{\text{Dgen}}^{k} \leq \left(P_{\text{Dgen}}^{k}\right)^{\max} \\ 0 \leq \left|Q_{\text{Dgen}}^{k}\right| \leq \sqrt{\left(\left(S_{\text{Dgen}}^{k}\right)^{\max}\right)^{2} - \left(P_{\text{Dgen}}^{k}\right)^{2}} \end{cases} \quad \forall k \in \mathbf{Dgen} \tag{5.36b}$$

$$\begin{cases} -\left(P_{\text{BESS}}^{k}\right)^{\max} \leq P_{\text{BESS}}^{k} \leq \left(P_{\text{BESS}}^{k}\right)^{\max} \\ \text{SoC}^{\min} \leq \text{SoC}^{k} \leq \text{SoC}^{\max} \end{cases} \quad \forall k \in \mathbf{BESS} \tag{5.36c}$$

$$V^{\min} \leq \left|V^{k}\right| \leq V^{\max} \quad \forall k \in \mathbf{bus} \tag{5.36d}$$

$$f^{\min} \leq f \leq f^{\max} \tag{5.36e}$$

$$I_i \leq I_i^{\max} \quad \forall i \in \mathbf{line} \tag{5.36f}$$

Constraint (5.36a) shows the apparent power balance equation within the CMG or the isolated PMG(s) while (5.36b) denote the active and reactive power loading of Dgens. Likewise, the active power loading and SoC limits of BESSs are given by (5.36c). The variation limits of the voltage magnitude at all buses of the CMG or isolated PMG(s) (denoted by vector **bus**) are given by (5.36d) and (5.36e), whereas (5.36f) shows the current loading of each line in those systems.

In (5.31), the Time factor is included in (5.31) to estimate the total time required for power transaction to overcome the emergency condition in PMGs (which is equal to ΔT of Figure 5.5). Also, $w_1 - w_4$ are the weighting coefficients of different considered OFs. The total cost calculation for the formulated OF highly depends on the assumed weighting coefficients values related to each cost criterion. Therefore, it is important to carefully select them. In power systems which have complex configurations, there is not a systematic method to define these weighting coefficients; however, an acceptable method is to conduct a census by the experts of the field and get their opinions based on the importance and impact of each criterion [18]. The experts may provide the importance of each in the form of either a number (i.e., 0%–100%) or linguistic (e.g., extremely/very/little big/small or neutral). Then these replies can be mapped into a digit in the range of [0,1] and then normalized. At the end, the weighting of each cost criterion will be defined as the average of all normalized values as

$$w_n = \sum_{i=1}^{N_e} \frac{w_n^i}{N_e} \tag{5.37}$$

where N_e is the number of experts participated in the census. Here, for the purpose of simplicity, it is assumed that $\sum_{n=1}^{4} w_n = 1$ (i.e., each OF cost is equally important).

5.5 Case studies

To evaluate the performance of the developed power transaction strategy for CMGs, analysis is done with the help of several case studies by using MATLAB® software, and some are described below. The data for the performed simulations has been taken from [68]. Let us consider the topology-1 (Figure 5.2(a)) with six MGs connected through a common central node (named MG-1 to MG-6). Each MG is assumed to have the configuration as described in Figure 5.6 and is composed of one PV and one wind (as the non-dispatchable DERs) connected with bus-1 and bus-2 respectively, one BESS is connected to bus-3 while one Dgen (as the droop regulated dispatchable DER for primary step) is connected to bus-5. All loads are connected to bus-4, and the tie-line for coupling with neighboring MGs is allocated to bus-6. Practically, each MG may have different configurations depending on their number of sources or may not have one or more sources included in overall generation. As for the developed technique, the internal structure of MGs is not important, so, for simplicity, all participating MGs are assumed to have the same configuration. However, the nominal load requirement of each MG, capacities of DERs and BESS and distance in kilometers from common central node are assumed to be different as described in Table 5.1. The predefined cost data for OF calculation (given in (5.31)) is also presented in Table 5.1. Also, for defining the

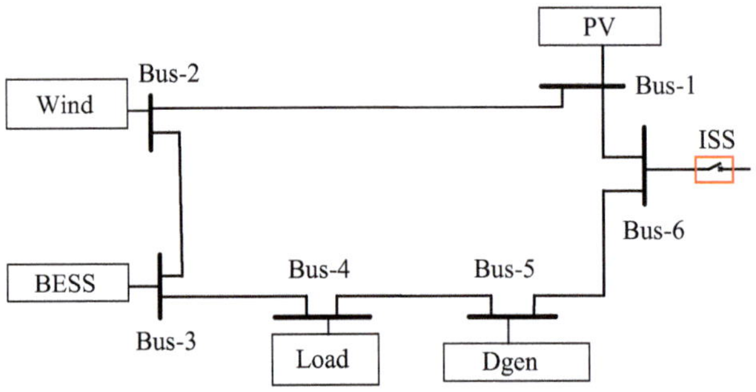

Figure 5.6 Topology of the considered MG(s) present inside the assumed cluster

Table 5.1 Considered input data for the numerical analysis

Serial no.	NDDs	Load	Dgen		BESS				Distance
	P_{NDDs}^{cap} (kW)	P_{load}^{cap} (kW)	P_{Dgen}^{min} (kW)	P_{Dgen}^{max} (kW)	Capacity (kW h)	SOC_{limit} (%)	$P_{BESS}^{ch,max}$ (kW)	$P_{BESS}^{Dch,min}$ (kW)	from central common point in km
MG-1	25	65	12	40	10	20–100	2	9	4
MG-2	25	60	13.5	45	10	20–100	2	9	6
MG-3	35	85	15	50	12	20–100	2.5	10.8	2
MG-4	20	45	9	30	8	20–100	1.6	7.2	7
MG-5	30	80	16.5	55	14	20–100	2.8	12.6	5
MG-6	25	65	12	40	10	20–100	2	9	5
			Considered costs data for the numerical analyses						
	C^{fuel}		0.31 $/kW h				C_{curt}^{load}		0.15 $/kW h
	C^{cfp}		0.02 $/kg				$C_{lifeloss}^{BESS}$		0.98 $/kW h
	C^{loss}		0.04 $/kW h				C_{curt}^{NDDs}		0.3 $/kW h
	∂		0.003 kg/kW h				C^{trans}		0.4 $/kW h

mode of operation of each MG, the acceptable limits for frequency and voltage are assumed as 50 ± 0.5 Hz and 1 ± 0.055 pu, respectively. The impedance data for all buses of MGs is taken from [18].

5.5.1 Study case I (an overloaded MG with primary and secondary actions only)

Consider an event (case study I in Table 5.2) when MG-5 is defined as a PMG^{OL} due to its working under unsafe mode of operation (a frequency drop to 49.29 Hz and a voltage drop to 0.923 pu which means both are below the acceptable limits). The load demand is 78.2 kW, out of which 23.6 kW is coming from non-dispatchable DERs while Dgen is working on 54.8 kW which is nearly equal to its nominal capacity, SRI is only 0.3% while RPL is at 30.17% to accommodate the required load. Now without the developed strategy, the only possibility is to do load curtailment of 10 kW in MG-5 so that both VF can be attained in safe mode of operation. The developed technique solves this problem by the use of primary and secondary actions, so MG-2 (an HMG with load demand of 23.3 kW, out of which 5.5 kW is supplied by non-dispatchable DERs and 18 kW by Dgen, with the minimum voltage of 1.039 pu and frequency of 50.39 Hz, SRI is 60% and RPL is 23.6%) and MG-4 (an HMG operating at 50.41 Hz frequency and minimum voltage of 1.034 pu and SRI is 58.6% with 8.9% RPL, load demand of 13.5 kW out of which 1.2 kW is contributed by non-dispatchable DERs while Dgen is operating at 12.4 kW) are found as the HMGs to be coupled with MG-5 which is PMG in this case. This is found as the most optimal solution with OF value of 9.003$. Through the application of secondary action, total 10.6 kW (out of which 6.7 kW is from MG-2 and 3.9 kW is from MG-4) is exported to MG-5 with tie-line loss of 1 kW (means 9.6 kW is imported by MG-5). In addition, primary step will make the Dgen of MG-2 and MG-4 to operate at 22.4 and 16.3 kW, respectively, which leads the MG-5's Dgen to operate at 45.2 kW. Additionally, BESS of MG-2 will discharges by 2.3 kW while BESS in MG-4 will stand by on its current situation. But as PMG is having an emergency condition of overloading, therefore BESS of MG-5 will not get charged, and all imported power goes to manage the load. As expected, CMG formation increases the bus voltages in PMG-5, while it decreases for both HMGs. Consequently, frequency settles down at 50.21 Hz with minimum voltage of 0.989 pu, which makes the MG-5 to operate under safe mode. Also, CMG SRI is 35% while accumulative RPL is 26.34%.

5.5.2 Study case II (an overloaded MG with all actions)

Consider another event (case study II in Table 5.2) in which MG-3 is detected as PMG^{OL} due to working on alarm mode of operation (with the frequency of 49.69 Hz which is within permissible limits, but minimum voltage is 0.937 pu which is below than the allowable limit). MG-3 load demand is 82 kW out of which 32.6 kW is provided by its non-dispatchable DERs while Dgen is working on the verge limit of 49.4 kW. Index of spinning reserve (SR) is only 1.2% which is very low, while RPL is relatively high as 39.7%. The developed technique proposes the

Table 5.2 Results of case studies for important MGs in the considered study cases

Case study	**I**			**II**		**III**	**IV**			**V**		**VI**	**VII**			
Observed MG(s)	MG-2	MG-4	MG-5	MG-3	MG-1	MG-2	MG-5	MG-6	MG-1	MG-4	MG-6	MG-1	MG-1	MG-3	MG-4	MG-5
Observed MG(s) State	HMG	HMG	PMG^{OL}	PMG^{OL}	HMG	PMG^{OL}	PMG^{OG}	HMG	HMG	PMG^{OG}	HMG	PMG^{OG}	PMG^{OG}	HMG	PMG^{OL}	HMG
Initial parameters																
P^{load} (kW)	23.3	13.5	78.2	82	20.6	57	25.1	15.3	32	20	30.5	34.6	13.2	46.8	43	29.6
P^{NDD} (kW)	5.5	1.2	23.6	32.6	1.7	12.8	7.4	1.9	2.3	3.7	3.8	1.2	0.4	11.6	13.2	0.4
P^{Dgen} (kW)	18	12.4	54.8	49.4	19.2	44.6	18.2	13.4	29.4	16.8	27.2	33.4	12.9	35.4	29.8	29.2
V^{max} (pu)	1.045	1.057	0.994	0.990	1.045	0.991	1.056	1.036	1.018	1.061	1.036	1.072	1.054	1.016	0.991	1.038
V^{min} (pu)	1.039	1.034	0.923	0.937	1.024	0.910	1.034	1.008	0.988	1.026	1.024	1.040	1.032	0.980	0.937	1.025
f (Hz)	50.39	50.41	49.29	49.69	50.32	49.46	50.6	49.7	49.9	50.53	50.19	50.48	50.61	51.28	49.23	49.59
CMG(s) Parameters after the application of cluster forming strategy																
HMGA[a]	Yes			Yes		No	Yes			Yes		No	Yes			
CMGF[b]	Yes			Yes		No	Yes			Yes		No	Yes			
PMGI[c]	No			No		Yes	No			No		Yes	No			
ACS[d]	Yes			Yes		Yes	Yes			Yes		Yes	Yes			
P^{trans} (kW)	6.7↑[e]	3.9↑	9.6↓[f]	5.2↓	5.6↑	[j]NA	3.2↑	2.6↓	1.9↓	4.5↑	2.3↓	NA	7.4↑	7.0↓	6.3↓	6.5↑

(Continues)

Table 5.2 (Continued)

Case study	I			II		III	IV			V		VI		VII		
$P_{curt}^{NDD_s}$ (kW)		No		No		No	No	No	No	0.8	1.2	0.9	No	3.3	No	No
P_{curt}^{load} (kW)		No		0.8	No	0.6	No	No		No		No	No	No	1.2	No
P^{BESS} (kW)	+2.3	[g]BSB	0	0	[h]+1.6	+0.8	0	[i]−2.6	−1.6	BSB	−1.1	−1.3	−3.1	−3.7	BSB	+4
P^{Dgen} (kW)	22.4	16.3	45.2	43.4	23.2	42.8	21.4	13.4	29.7	20.5	27.2	35.6	23.4	38.7	22.3	32
CMG V^{max} (pu)		1.034		1.039		1.004		1.050		1.039		1.032			1.039	
CMG V^{min} (pu)		0.989		0.987		0.955		0.993		1.017		0.998			0.987	
f (Hz)		50.21		49.87		49.61		50.26		49.88		50.21			49.77	
CMG Mode		Safe		Safe		Safe		Safe		Safe		Safe			Safe	
P^{loss} (kW)		1		0.4		–		0.6		0.3		0.2			0.7	
OF ($)		9.003		8.7065		12.2363		7.2981		10.0003		6.3246			11.4581	

[a]HMGA, healthy microgrid available.
[b]CMGF, CMG is formed.
[c]PMGI, PMG is left isolated.
[d]ACS, all constraints satisfied after actions.
[e]↑, Power export.
[f]↓, Power import.
[g]BSB, BESS Stand-By.
[h]+, BESS Discharged.
[i]−, BESS charged.
[j]NA, not applicable.

coupling of MG-1 (an HMG operating on safe mode with 50.32 Hz frequency and 1.024 pu minimum voltage, load demand is 20.6 kW out of which only 1.7 kW is coming from renewables while Dgen is working on 19.2 kW, SRI is 52% which is quite high and RPL in only 8.2% of the total generations) as the most economical solution with the OF value of 8.7065$. Through this coupling control is achieved followed by primary, secondary and tertiary actions as well. Therefore, HMG-1 will drop its frequency by increment in Dgen's output as 23.2 kW by applying primary step, then secondary action takes place by exporting 5.6 kW (also BESS of MG-1 discharges by 1.6 kW) out of which 5.2 kW is imported to PMG-3 with tie-line loss of 0.4 kW. Therefore, PMG-3 lowers down its Dgen's operational limit to 43.4 kW which result in 0.8 kW of load curtailment (as tertiary action to catch up the generation capacity). Thus, CMG formation will settle the voltage limit to 0.987 pu which within acceptable limit with nominal frequency of 49.87 Hz (also with SRI of 74% with renewable contributions of 33.4% of total generation) and make the MG-3 to operate under safe mode.

5.5.3 *Study case III (an overloaded MG with primary and tertiary actions only)*

Consider an event (case study III in Table 5.2) in which MG-2 is detected as a PMG^{OL} with its unsafe mode of operation (such that frequency is 49.46 Hz, i.e., slightly lower than the acceptable range while minimum voltage is 0.910 and not in permissible limit). In this event, MG-2 has load demand of 57 kW out of which 12.6 kW is contributed by non-dispatchable DERs, and Dgen is operating on verge point of 44.6 kW (SRI is only 0.8% which is very low and RPL in 22.4% of total generation). The developed technique proposes the direct implementation of tertiary action only in this case as the most optimal solution (with the OF value of 12.2363$) with load curtailment of 0.6 kW while the BESS will discharge by 0.8 kW. This increases the minimum voltage to 0.955 pu, and frequency reaches to 49.61 Hz, i.e., both are within acceptable range. Also the load draws 56.2 kW with Dgen operates at 42.6 kW respectively without the formation of CMG with neighboring MGs.

5.5.4 *Study case IV (an overgenerating MG with primary and secondary actions only)*

Consider an event (case study IV in Table 5.2) in which MG-5 is detected as PMG^{OG} and operating at unsafe mode (with a maximum voltage of 1.056 pu which is above the acceptable limit and frequency of 50.6 Hz, which is also exceeding beyond the applied limits). MG-1's load demand is 25.1 kW out of which renewable contribution is 7.4 kW while Dgen is operating at only 18.2 kW, SRI is relatively high with 66.9% and RPL is 29.48%, respectively. Now without the application of transaction technique, a load increase of 3 kW will result in a maximum voltage limit of 1.042 pu which is lower than the maximum limit. The technique finds the interconnection between MG-5 and MG-6 (an HMG with the operating frequency of 49.7 Hz and maximum voltage of 1.036 pu, both are within safe mode, while load demand is 15.3 kW with 1.9 kW coming from non-dispatchable DERs and 13.4 kW from Dgen, respectively, with appreciably high SRI of 66.5% and RPL is 12.415% of total generation) as the most optimal solution with OF value of 7.2981$. First, MG-6 Dgen stays at 13.4 kW, while MG-5 increases its Dgen output by 21.4 kW by primary step implementation, then by secondary action, MG-5 exports 3.2 kW out of which 2.6 kW is imported to MG-6 with the loss of 0.6 kW; therefore, BESS in MG-6 get charged by 2.6 kW. Due to the action applied by PTO, the MG-5 frequency lowers down and MG-6 frequency rises within permissible limit, so the CMG's frequency reaches to 50.26 Hz with the maximum voltage of 1.050 pu, both are attained under safe mode. This results in the accumulative SRI of 36.63% while RPL is 23.01% of total generations.

5.5.5 *Study case V (an overgenerating MG with all actions)*

Consider an event (case study V in Table 5.2) in which MG-5 is detected as PMG^{OG} operating on unsafe mode (with 20 kW of load demand with Dgen operating at 16.8 kW with reserve capacity of 44% and 3.7 kW is coming from renewables while 12.45% is RPL in total generation, and due to this, the frequency of PMG is 50.53 which is slightly higher than the acceptable limit while voltage is on the maximum limit of 1.061 pu and again higher than the higher required limit). Without the explained strategy, load increase of 4 kW can result in the maximum voltage limit of 1.047 which is within an acceptable range. With the application of described technique, the best optimal solution (with OF value of 10.0003$) is to couple MG-1 (an HMG with 32 kW load requirement out of which 2.3 kW is non-dispatchable DERs contribution and RPL is 7.18% of total generation while Dgen is working on 29.4 kW with a reserve capacity of 26.5% which is relatively low, also the maximum voltage limit of 1.018 pu and frequency of 49.9 Hz, means MG-1 is working on safe mode with both frequency and voltage are within permissible limits) and MG-6 (an HMG with 30.5 kW load demand and Dgen is operating at 27.2 kW with SRI OF 32% while power coming from renewables is 3.8 kW with RPL of 12.4% only, also VF both are in safe mode with 1.036 pu and 50.19 Hz, respectively) with MG-4. First, due to primary step, Dgen increases its active power to 20.5 kW by droop regulation, then by secondary action, 4.5 kW is exported to coupled MGs (out of which 1.9 kW is imported by MG-1, and 2.3 kW is imported by MG-6 with loss of 0.3 kW) accompanied by tertiary action which result in non-dispatchable DERs curtailment by 0.8 kW. Thus, MG-1 Dgen operates at 29.7 kW with its BESS charges by 1.6 kW, while MG-6 BESS charges by −1.1 kW with its Dgen working at 27.2 kW, respectively. Now due to the abovementioned control actions MG-4 frequency and voltage decrease and CMG's voltage reaches to 49.88 Hz with the maximum voltage of 1.039 pu, and both are within acceptable limit of safe mode, also CMG's reserve capacity is 29.63% with RPL of 9.45%, respectively, with the satisfaction of applied constraints.

5.5.6 *Study case VI (an overgenerating MG with primary and tertiary actions only)*

Consider another event (case study VI of Table 5.2) in which MG-1 is detected as a PMG^{OG} due to the alarm mode of operation (with the maximum voltage of

1.071 pu which is higher than acceptable voltage value, whereas frequency is 50.48 Hz and is still within acceptable range, 34.6 kW of load demand out of which 1.2 kW is coming from non-dispatchable DERs with RPL OF 3.46% of total generation, and Dgen is working at 33.4 kW with SRI of 74.22% which is quite high). The decision is to isolate the MG-1 (with OF value of 6.3246$). The explained technique does the optimization using primary and tertiary actions while skipping the secondary action (with non-dispatchable DERs curtailment of 0.9 kW and charging the BESS by 1.3 kW while Dgen operates at 35.6 kW) results in the maximum voltage of 1.032 pu and frequency decrease up to 50.21 Hz, consequently making the PMG to operate at safe mode of operation.

5.5.7 Study case VII (multiple PMGs and HMGs with all actions)

Consider another event (study case VII in Table 5.2) in which MG-1 is PMG^{OG} due to the alarm mode of operation (with the maximum voltage of 1.054 pu which is at verge of acceptable limit and frequency is 50.61 Hz and above than the permissible limit, also load demand of 13.2 kW out of which only 0.4 kW coming from non-dispatchable DERs with RPL of 3.03% and Dgen output of 12.9 kW with high reserve capacity index of 67.7%), and MG-4 is PMG^{OL} due to unsafe operational mode (with load demand of 43 kW while Dgen operating at 29.8 kW with SRI of only 0.66% and 13.2 kW is coming from non-dispatchable DERs with RPL of 30.69%, the minimum voltage is 0.937 pu and frequency is 49.23 Hz means both are below the acceptable range). So the suitable technique proposes the implementation of all actions with the coupling of HMGs, i.e., MG-3 (an HMG with 46.8 kW load requirement out of which 11.6 kW is non-dispatchable DERs contribution, and RPL is 24.8% of total generation while Dgen is working on 35.4 kW with reserve capacity of 29.2% which is relatively low, also the maximum voltage limit of 1.016 pu and frequency of 51.28 Hz, means MG-3 is working on safe mode with both frequency and voltage are within permissible limits) and MG-5 (an HMG with 29.6 kW load demand and Dgen is operating at 29.2 kW with SRI OF 46.90% while power coming from renewables is 0.4 kW with RPL of 1.35% only, also VF both are in safe mode with 1.038 pu and 49.59 Hz, respectively) with PMGs as the most optimal solution (with OF value of 11.4581$). Therefore, first, due to primary step MG-1 BESS charges by 3.1 kW and Dgen operates at 23.4 kW, the secondary action takes place to export 7.4 kW to MG-3 (out of which only 7 kW is imported by MG-3 with loss of 0.4 kW) and tertiary action takes place in MG-3 with non-dispatchable DERs curtailment of 3.3 kW (while MG-3's BESS charges by 3.7 kW and Dgen operates at 38.7 kW) which results in lowering down the frequency of MG-1 with slight frequency increase of MG-3. Similarly, MG-4 makes a primary step (Dgen operational limit increases up to 32 kW while BESS gets discharged by 4 kW) then as a secondary action MG-3 exports 6.5 kW to MG-4 (out of which 6.3 kW is imported by MG-4 with due to tie-line loss of 0.2 kW making the total loss in CMG as 0.7 kW in total) and in the end, 1.2 kW load curtailment takes place in MG-4 as a tertiary action (with Dgen working 22.3 kW). As the result of CMG formation, the abovementioned four MG's voltage below and above limits lie between 0.987 and 1.039 pu, respectively, while frequency is 49.77 Hz means that safe mode has been attained (with overall SRI of 66.51% and RPL is 16.8% of total generations).

5.6 Concluding remarks

This study focuses on an optimal operation of provisionally clustered neighboring MGs by proposing a power transaction strategy for a remote area which is not connected to a utility feeder. For this purpose, overgeneration from non-dispatchable DERs, overloading or the faults that can cause trouble to DRS (i.e., Dgens and BESSs in this case) are considered in emergency situations. PTO is considered for the control purpose, and it is assumed that it remains active all the time. It is also suggested that for this power transaction strategy, PTO will be able to run in specific intervals. The aim is to guarantee the CMG s operation at the lowest possible costs along with least technical deviations. The developed power flow algorithm defines the optimal operation of DRS in all MGs participating inside the CMG, power generation of non-dispatchable DERs and the curtailment of nonessential loads. Then it will send the output power information to the secondary controllers of each MG (provisionally connected) for applying them in their operative system. The performance of the explained transaction technique in terms of defining the optimal operational points has been demonstrated through exhaustive simulations by using a stochastic analysis.

Nomenclature

Abbreviations

BESS	battery energy storage systems
CMG	clustered microgrids
DER	distributed energy resource
Dgen	diesel generator
DRS	droop-regulated system
HMG	healthy microgrid
ISS	interconnecting static switch
MG	microgrid
NDD	non-dispatchable distributed energy resource
OF	objective function
OL/OG	overloaded/over-generated
PMG	problem microgrid
SoC	state of charge
PTO	power transaction operator

Parameters and variables

$C_{\text{lifeloss}}^{\text{BESS}}$	BESS lifeloss cost [$/kW h]
C^{cfp}	carbon foot print cost [$/kW h]
C^{fuel}	fuel cost [$]
C^{loss}	power loss cost [$/kW h]
$C_{\text{curt}}^{\text{load}}$	cost of load shedding [$/kW h]
$C_{\text{curt}}^{\text{NDDs}}$	NDDs curtailment cost [$/kW h]
C^{trans}	power import/export cost [$/kW h]
f	frequency [Hz]
OF_{BESS}	OF of BESS cost [$]
OF_{curt}	OF of NDDs/load curtailment cost [$]
OF_{Dgen}	OF of diesel generator cost [$]
OF_{loss}	tieline power loss cost [$]
OF_{oper}	operational cost
OF_{tech}	technical cost
OF_{trans}	power import/export cost [$]
$P_k^{\text{Dgen}}, P^{\text{NDD}}, P^{\text{load}}, P_k^{\text{BESS}}$	active power of Dgen, BESS, NDD, load [kW]
P^{trans}	power transacted between MGs [kW]
$Q_k^{\text{Dgen}}, Q^{\text{load}}$	reactive power of Dgen, load [kW]
$V^{\min}/V^{\max}$	voltage limitation [volts]
$V_{\text{vio}}, I_{\text{vio}}, f_{\text{vio}}, \text{Constraint}_{\text{vio}}$	voltage, current, frequency and constraints violation
∂	emission factor [$/kg]

References

[1] Families in regional, rural and remote Australia, Website, Retrieved 14-08-17. https://aifs.gov.au/sites/default/files/publication-documents/fs201103.pdf.

[2] L. Byrnes, C. Brown, L. Wagner and J. Foster, "Reviewing the viability of renewable energy in community electrification," *Renewable & Sustainable Energy Reviews*, vol. 59, pp. 470–481, 2016.

[3] L. Che and M. Shahidehpour, "DC microgrids: Economic operation and enhancement of resilience by hierarchical control," *IEEE Transactions on Smart Grid*, vol. 5, no. 5, pp. 2517–2526, 2014.

[4] L. Che, M. Khodayar, and M. Shahidehpour, "Only connect: Microgrids for distribution system restoration," *IEEE Power Energy Magazine*, vol. 12, no. 1, pp. 70–81, 2014.

[5] L. Ali and F. Shahnia, "Determination of an economically-suitable and sustainable standalone power system for an off-grid town in Western Australia," *Renewable Energy*, vol. 106, pp. 243–254, 2017.

[6] J. Li, Y. Liu, and L. Wu, "Optimal operation for community based multi-party microgrid in grid connected and islanded modes," *IEEE Transactions on Smart Grid*, vol. 9, no. 99, pp. 756–765, 2016.
[7] M. Batool, S. Islam, and F. Shahnia, "Selection of Sustainable Standalone Microgrid for Optimal Operation of Remote Area Town," One Curtin Post-graduate Conference (OCPC), Malaysia, 2017.
[8] M. Batool, S. Islam, and F. Shahnia, "Power Transaction Management amongst Coupled Microgrids in Remote Areas," IEEE PES Innovative Smart Grid Technologies (ISGT), Auckland, New Zealand, 2017.
[9] L. Che, M. Shahidehpour, A. Alabdulwahab, and Y. Al-Turki, "Hierarchical coordination of a community microgrid with ac and dc microgrids," *IEEE Transactions on Smart Grid*, vol. 6, no. 6, pp. 3042–3051, 2015.
[10] Z. Wang and J. Wang, "Self-healing resilient distribution systems based on sectionalization into microgrids," *IEEE Transactions on Power Systems*, vol. 30, no. 6, pp. 3139–3149, 2015.
[11] S. Conti and S.A. Rizzo, "Monte Carlo simulation by using a systematic approach to assess distribution system reliability considering intentional islanding," *IEEE Transactions on Power Delivery*, vol. 30, no. 1, pp. 64–73, 2015.
[12] S. Chanda and A.K. Srivastava, "Defining and enabling resiliency of electric distribution systems with multiple microgrids," *IEEE Transactions on Smart Grid*, vol. 7, no. 6, pp. 2859–2868, 2016.
[13] K.P. Schneider, F.K. Tuffner, M.A. Elizondo, C.-C. Liu, Y. Xu, and D. Ton, "Evaluating the feasibility to use microgrids as a resiliency resource," *IEEE Transactions on Smart Grid*, vol. 8, no. 2, pp. 687–696, 2017.
[14] C. Yuan, M.S. Illindala, and A.S. Khalsa, "Modified Viterbi algorithm based distribution system restoration strategy for grid resiliency," *IEEE Transactions on Power Delivery*, vol. 32, no. 1, pp. 310–319, 2017.
[15] R.H. Lasseter, "Smart distribution: coupled microgrids," *Proceedings of the IEEE*, vol. 99, no. 6, pp. 1074–1082, 2011.
[16] M. Batool, "Performance Optimization of Standalone and Grid Connected Microgrid Clusters," *PhD thesis submitted to Curtin University, Bentley Campus, Perth Australia*, 2018.
[17] Z. Wang and J. Wang, "Self-healing resilient distribution systems based on sectionalization into microgrids," *IEEE Transactions on Power Systems*, vol. 30, no. 6, pp. 3139–3149, 2015.
[18] F. Shahnia, S. Bourbour, and A. Ghosh, "Coupling neighboring microgrids for overload management based on dynamic multicriteria decision-making," *IEEE Transactions on Smart Grid*, vol. 8, no. 2, pp. 969–983, 2017.
[19] E. Pashajavid, F. Shahnia, and A. Ghosh, "Development of a self-healing strategy to enhance the overloading resilience of islanded microgrids," *IEEE Transactions on Smart Grid*, vol. 8, no. 2, pp. 868–880, 2017.
[20] Y. Zhang, L. Xie, and Q. Ding, "Interactive control of coupled microgrids for guaranteed system-wide small signal stability," *IEEE Transactions on Smart Grid*, vol. 7, no. 2, pp. 1088–1096, 2016.

[21] F. Shahnia, R.P.S. Chandrasena, S. Rajakaruna, and A. Ghosh, "Primary control level of parallel distributed energy resources converters in system of multiple interconnected autonomous microgrids within self-healing networks," *IET Generation, Transmission & Distribution*, vol. 8, pp. 203–222, 2014.

[22] Y. Zhang and L. Xie, "Online dynamic security assessment of microgrid interconnections in smart distribution systems," *IEEE Transactions on Power Systems*, vol. 30, no. 6, pp. 3246–3254, 2015.

[23] I.P. Nikolakakos, H.H. Zeineldin, M.S. El-Moursi, and N.D. Hatziargyriou, "Stability evaluation of interconnected multi-inverter microgrids through critical clusters," *IEEE Transactions on Power Systems*, vol. 31, no. 4, pp. 3060–3072, 2016.

[24] N. Nikmehr and S.N. Ravadanegh, "Reliability evaluation of multi-microgrids considering optimal operation of small scale energy zones under load-generation uncertainties," *International Journal of Electrical Power & Energy Systems*, vol. 78, pp. 80–87, 2016.

[25] S.A. Arefifar, M. Ordonez, and Y. Mohamed, "Voltage and current controllability in multi-microgrid smart distribution systems," *IEEE Transactions on Smart Grid*, 2018, DOI: 10.1109/TSG.2016.2568999.

[26] F. Shahnia and A. Arefi, "Eigen analysis-based small signal stability of the system of coupled sustainable microgrids," *International Journal of Electrical Power & Energy Systems*, vol. 91, pp. 42–60, 2017.

[27] F. Shahnia, "Stability and eigenanalysis of a sustainable remote area microgrid with a transforming structure," *Sustainable Energy, Grids & Networks*, vol. 8, pp. 37–50, 2016.

[28] F. Shahnia, "Semi-Decentralized Charging and Discharging Control of Floating Batteries in Microgrids," 2nd IEEE Annual Southern Power Electronics Conference (SPEC), pp. 1–6, Auckland, 2016.

[29] T.H. Mehr, A Ghosh, and F. Shahnia, "Cooperative control of battery energy storage systems in microgrids," *International Journal of Electrical Power & Energy Systems*, vol. 87, pp. 109–120, 2017.

[30] R. Majumder and G. Bag, "Parallel operation of converter interfaced multiple microgrids," *International Journal of Electrical Power & Energy Systems*, vol. 55, pp. 486–496, 2014.

[31] A. Arefi and F. Shahnia, "Tertiary controller-based optimal voltage and frequency management technique for multi-microgrid systems of large remote towns," *IEEE Transactions on Smart Grid*, 2018, DOI:10.1109/TSG.2017.2700054.

[32] F. Shahnia and A. Arefi, "Defining the Suitable Adjacent Microgrids to form a Temporary System of Coupled Microgrids," IEEE Region 10 Conference (TENCON), pp. 1–4, Singapore, 2016.

[33] J. Wang, M. Shahidehpour, and Z. Li, "Strategic generation capacity expansion planning with incomplete information," *IEEE Transactions on Power Systems*, vol. 24, no. 2, pp. 1002–1010, 2009.

[34] Y. Tohidi, F. Aminifar, and M. Fotuhi-Firuzabad, "Generation expansion and retirement planning based on the stochastic programming," *Electric Power Systems Research*, vol. 104, pp. 138–145, 2013.
[35] H. Wang and J. Huang, "Cooperative planning of renewable generations for interconnected microgrids," *IEEE Transactions on Smart Grid*, vol. 7, no. 5, pp. 2486–2496, 2016.
[36] M.J. Hossain, M.A. Mahmud, F. Milano, *et al.*, "Design of robust distributed control for interconnected microgrids," *IEEE Transactions on Smart Grid*, vol. 7, no. 6, pp. 2724–2735, 2016.
[37] H. Zhao, M. Hong, W. Lin, and K.A. Loparo, "Voltage and frequency regulation of microgrid with battery energy storage systems," *IEEE Transactions on Smart Grid*, 2019, DOI: 10.1109/TSG.2017.2741668.
[38] N. Soni, S. Doolla, and M.C. Chandorkar, "Analysis of frequency transients in isolated microgrids," *IEEE Transactions on Industry Applications*, 2016, DOI: 10.1109/TIA.2017.2746619.
[39] A. Chaouachi, R.M. Kamel, R. Andoulsi, and K. Nagasaka, "Multi-objective intelligent energy management for a microgrid," *IEEE Transactions on Industrial Electronics*, vol. 60, no. 4, pp. 1688–1699, 2013.
[40] Z. Yu, N. Gatsis, and G.B. Giannakis, "Robust energy management for microgrids with high-penetration renewables," *IEEE Transactions on Sustainable Energy*, vol. 4, no. 4, pp. 944–953, 2013.
[41] S. Conti, R. Nicolosi, S.A. Rizzo, and H.H. Zeineldin, "Optimal dispatching of distributed generators and storage systems for mv islanded microgrids," *IEEE Transactions on Power Delivery*, vol. 27, no. 3, pp. 1243–1251, 2012.
[42] C. Chen, S. Duan, T. Cai, B. Liu, and G. Hu, "Smart energy management system for optimal microgrid economic operation," *IET Renewable Power Generation*, vol. 5, no. 3, pp. 258–267, 2011.
[43] C. Yammani and V.K. Macha, "Fuel cost minimization with reserve capacity and inter-area flow limit for reliable and cost effective operation of multi microgrids," 2016, DOI: 10.1109/TENCON.2016.7848638.
[44] B. Khan and P. Singh, "Selecting a meta-heuristic technique for smart micro-grid optimization problem: a comprehensive analysis," *IEEE Access*, vol. 5, pp. 13951–13977, 2017.
[45] V. Mohan, R. Suresh, J.G. Singh, *et al.*, "Microgrid energy management combining sensitivities, interval and probabilistic uncertainties of renewable generation and loads," *IEEE Journal on Emerging and Selected Topics in Circuits and Systems*, vol. 7, no. 2, pp. 262–270, 2017.
[46] T.L. Vandoorn, J. De Kooning, B. Meersman, *et al.*, "Voltage-based droop control of renewables to avoid ON–OFF oscillations caused by over voltages," *IEEE Transactions on Power Delivery*, vol. 28, no. 2, pp. 845–854, 2013.
[47] B. Zhao, X. Zhang, J. Chen, *et al.*, "Operation optimization of standalone microgrids considering lifetime characteristics of battery energy storage system," *IEEE Transactions On Sustainable Energy*, vol. 4, no. 4, pp. 934–943, 2013.

[48] H. Wang and J. Huang, "Incentivizing energy trading for interconnected microgrids," *IEEE Transactions on Smart Grid*, vol. 9, no. 99, pp. 2647–2657, 2016.
[49] M. Fathi and H. Bevrani, "Statistical cooperative power dispatching in interconnected microgrids," *IEEE Transactions on Sustainable Energy*, vol. 4, no. 3, pp. 586–593, 2013.
[50] C. Li, F. de Bosio, F. Chen, *et al.*, "Economic dispatch for operating cost minimization under real-time pricing in droop-controlled DC microgrid," *IEEE Journal of Emerging and Selected Topics in Power Electronics*, vol. 5, no. 1, pp. 587–595, 2017.
[51] A.G. Tsikalakis and N.D. Hatziargyriou, "Centralized control for optimizing microgrids operation," *IEEE Transactions on Energy Conversion*, vol. 23, no. 1, pp. 241–248, 2008.
[52] A. Khodaei, "Provisional microgrids," *IEEE Transactions on Smart Grid*, vol. 6, no. 3, pp. 1107–1115, 2015.
[53] M. Batool, S. Islam, and F. Shahnia, "Master Control Unit based Power Exchange Strategy for Interconnected Microgrids," Australian Universities Power Engineering Conference (AUPEC), Melbourne, Australia, 2017.
[54] Intel Hardware. Software and Technologies for Industrial Automation, Intel®, 2017. https://www.intel.com.au/content/www/au/en/industrial-automation/products-and-solutions/hardware-software-technologies.html.
[55] Industrial Controllers, National Instruments™ website, 2017. http://www.ni.com/ industrial-controller.
[56] Process Control and Industrial Automation, Analogue Devices™, 2017. http://www.analog.com/en/applications/markets/process-control-and-industrial-automation.html.
[57] F. Chen, M. Chen, Q. Li, *et.al.*, "Cost-based droop schemes for economic dispatch in islanded microgrids," *IEEE Transactions on Smart Grid*, vol. 8, no. 1, pp. 63–74, 2017.
[58] E. Pashajavid, A. Ghosh, and F. Zare, "A multimode supervisory control scheme for coupling remote droop regulated microgrids," *IEEE Transaction on Smart Grid*, vol. 9, no. 99, pp. 5381–5392, 2017.
[59] M.Z. Djurovic, A. Milacic, and M. Krsulja, "A simplified model of quadratic cost function for thermal Generators," *Proceedings of DAAM International*, vol. 23, no. 1, 2012.
[60] J. Zhu, "Optimization of Power System Operation", Wiley-IEEE Press, Hoboken, NJ, eBook Chapters, 2015.
[61] S.A.-H. Soliman and A.-A.H. Mantawy, "Modern Optimization Techniques with Applications in Electric Power System", Springer, 2012.
[62] G.K. Mahanti, A. Chakraborty, and S. Das, "Floating-Point Genetic Algorithm for Design of a Reconfigurable Antenna Arrays by Phase-Only Control," 37th IEEE Conference on Software Engineering, 2005.
[63] M. Batool, S. Islam, and F. Shahnia, "Stochastic Modeling of the Output Power of Photovoltaic Generators in Various Weather Conditions," Australian Universities Power Engineering Conference (AUPEC), 2016.

[64] http://au.mathworks.com/help/stats/betapdf.html.

[65] B. Chitti Babu, T. Cermak, S. Gurjar, Z.M. Leonowicz, and L. Piegari, "Analysis of Mathematical Modeling of PV Module with MPPT Algorithm," 15th IEEE Int. Conf. on Environment and Electrical Engineering (EEEIC), pp. 1625–1630, 2015.

[66] M. Patil and A. Deshpande, "Design and Simulation of Perturb and Observe Maximum Power Point Tracking in MATLAB and Simulink," Int. Conf. on Smart Technologies and Management for Computing, Communication, Controls, Energy and Materials (ICS™), pp. 459–465, 2015.

[67] S.A. Arefifar, Y.A.I. Mohamed, and T.H.M. El-Fouly, "Supply adequacy based optimal construction of microgrids in smart distribution systems," *IEEE Transactions on Smart Grid*, vol. 3, pp. 1491–1502, 2012.

[68] M. Batool, F. Shahnia, and S. Islam, "A multi-level supervisory emergency control for the operation of remote area microgrid clusters," *Journal of Modern Power Systems and Clean Energy*, pp. 1–19, 2018.

Chapter 6

Distributed energy network using nanogrid

Xiaofeng Sun[1], *Wei Zhao*[1], *and Lei Qi*[1]

With the guidance of local governments and the concerns of the environment, more and more distributed energy resources (DERs) and distributed energy storage units (DESUs) are installed by subscribers in remote villages, outskirts and mountainous areas. The DERs and DESUs form an easy, self-controlled nanogrid. Nanogrid can be seen as smaller and technologically simpler islanding microgrids. When more and more renewable sources are interfaced to the nanogrid, a power-management issue is important for this system to supply smooth power to the customers. Therefore, some reviews for nanogrid and some representative control strategies for renewable sources in nanogrid are presented in this chapter.

6.1 Overview of nanogrid

6.1.1 Concept of nanogrid

The financial incentives of local governments, high electricity bills as well as the environmental concerns are the driving forces of the householders towards installing renewable energy-based DERs and DESUs in their premises. In the near future, there is a tendency that the majority of householders will have their own DERs which is connected to the utility grid. Hence, envision forming of a self-controlled nanoscaled power system referred as nanogrid consisted of the amalgamation of resources in neighbouring houses which are supplied by the same utility feeder would be realized. Generally speaking, the preferred sources would mainly include photovoltaic (PV) solar cells, wind generators, micro-turbines, fuel cells and local energy storages [1]. Plug-in hybrid electric vehicles (PHEVs) are also expected to become an inherent part of the future house [1,2].

In industrialized contexts, whether in developed or developing countries, the term 'nanogrid' can have a very different meaning [3]. Nanogrids can be seen as smaller and technologically simpler islanding microgrids, typically serving a single building or a single load [4]. As a local small-scale power system, compared to the microgrid, the character 'nano' in the word 'nanogrid' is referred to a smaller scale, size and energy capacity, or less elements, even simple technology demands. Up to

[1]The Electrical Engineering Institute, Yanshan University, Qinhuangdao, China

now, there is not yet remaining a standard definition or specification, especially the capacity for nanogrid. A nanogrid can cover a range less than 100 meters, and also power up to 50 households. Though it has a wide range, it is typically used to serve some buildings or loads. When connecting to a microgrid, a nanogrid could deliver up to 100 kW of power [5]. In other cases, particularly in developing countries, nanogrid refers to systems of very small scale, with fewer than 150 household customers [3].

In some other cases, it is also used to refer to distributed generation (DG) in areas which are already supplied with grid electricity [3]. The intent is to increase the use of on-site renewable generation or to improve the reliability or quality of local electric power [3]. In this sense, a nanogrid is a local collection of electricity generation and loads, typically tens of MW or less. And it is usually connected to the main grid but may be disconnected for autonomous operation to achieve very high levels of reliability compared to the grid [3].

The nanogrid can be interfaced to the higher level system through bidirectional electronic power converters [1]. Nanogrids can also be separated from a microgrid. And the function would be independent with their own voltage, phase and frequency from DC to kilohertz [5]. The interface converters help the nanogrid to be a self-control system with the self-control, protection and management features. Thus, it can be connected to microgrid, and multiple nanogrids can be interconnected to form a microgrid, and this kind of interconnection helps to increase their ranges, capacities and also the power supply security [6].

Besides the interconnection to grid, also there are some other features that could be drawn for nanogrid. A controller is used to control the load as well as manage the storage units in a centralized mode. And the system also can work in a decentralized mode featuring no central controller. Storage units can be installed internally or through as second nanogrid [5]. Power converter for the PV is most commonly unidirectional two-stage converter featuring the step-up (boost) DC–DC converter stage cascaded a single-phase or three-phase voltage source inverter stage for adequate interface with the utility grid. New converters for small wind turbines are also two-stage power converters, comprising the three-phase PWM rectifier and a voltage source inverter. Energy storage and PHEV typically require bidirectional DC–AC converters for the optimal battery utilization on one side and AC-line interface on the others. It is interesting to note that practically all electronic loads have two-stage power conversion, where the frontend converter consists of a rectifier, electromagnetic interference filter, and often a power factor correction (PFC) circuit [1].

In order to dynamically decouple the nanogrid from the rest of the power system, a full-power bidirectional converter could be used. In that case, the whole nanogrid is seen by the utility grid as a single electronic load/source, dynamically independent of the grid but dispatched by utility operator [1].

6.1.2 Architecture of nanogrid

The typical structure of the nanogird is shown in Figure 6.1. AC is the most frequently used type in the utility grid as well as the nanogrid. Meanwhile, a DC

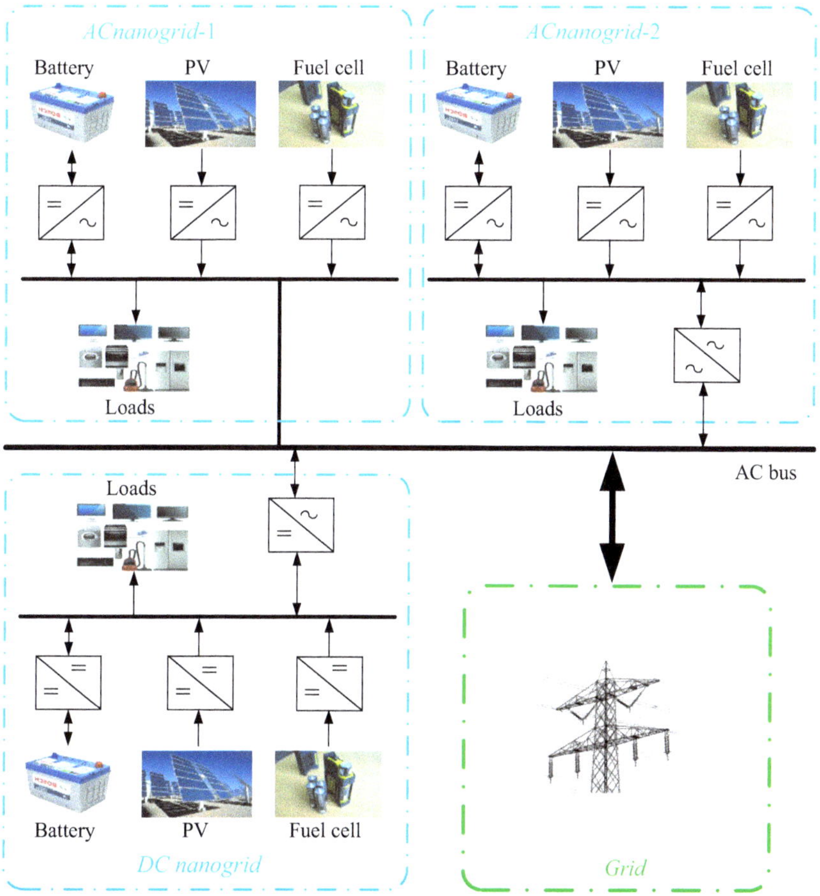

Figure 6.1 Typical structure of the nanogrid

nanogrid is preferred to deliver energy to DC customs conveniently when there are DC DERs such as PV panels or battery storage in nanogrid.

In this case, the system can feature lower costs and higher energy conversion efficiencies without DC/AC converter. But the safety issues related to the low-voltage switchgear as inherent disadvantages of DC supply system always exist in nanogrid. The bidirectional power converters as interfaces in the energy storage part are responsible to isolate the nanogrid from the utility grid in the case of a fault or other abnormal grid conditions. In AC nanogrid, the battery will perform the frequency and voltage regulation of AC-line in the stand-alone mode. Because of the existing of the DESUs, there are no load power interruptions when the nanogrid in house synchronizes and reconnects to the utility grid [1,7]. Compared to the AC nanogrid architecture, DC nanogrid brings many advantages, starting with fewer power converters, higher overall system efficiency and easier interface of renewable energy sources to a DC system [1,8]. There are no frequency stability, reactive power issues, no skin effects and AC losses [1]. Moreover, the power electronics converters within consumer electronics, electronic ballasts, LED lighting and variable speed motor drives can be more conveniently powered by DC [1].

An economic control center (ECC) is entrusted with the operation of the local renewable generation, load shedding, utilization of the battery energy, and other power-management functions, as well as nanogrid stabilization and advanced, active islanding in the event of outages or other low-frequency disturbances on the utility side. In such hierarchical grid architecture, the nanogrids are fully dynamically decoupled from grid through ECC, so that their internal architecture is completely independent and each part could have different voltages, phases and even frequencies [1].

DC nanogrid is envisioned to have two DC voltage levels: a high-voltage (380 V) DC bus powering heating, ventilation, and air-conditioning, kitchen loads, and other major home appliances; and a multitude of low-voltage (24 or 48 V) DC buses powering small tabletop appliances, computer and entertainment systems and LED lighting. The 380 V DC level is chosen to match the industry standard intermediate DC voltage in consumer electronics with the PFC circuit at the input, so that conversion from AC to DC would involve only bypassing (or eliminating) the frontend rectifier and PFC in most contemporary appliances. The 48 V DC level coincides with the standard telecom voltage to facilitate adoption, increase efficiency and provide enhanced safety when handling small appliances [1]. Besides the DC distribution systems that are already proposed for commercial and residential applications [9], DC power distribution systems are currently being considered for data centres in Japan and Europe [10–12] and are also being contemplated for PHEVs, ship and aircraft power systems [1,13]. Several manufacturers already have on the market high power-density bus control modules that supply 48 from 380 V which are intended for these applications [1].

In higher voltage DC systems, fault current interruption is a particular concern. However, if all power is fed from controllable electronic power converters which could provide active current limiting, it will reduce the need for electromechanical protection devices. The system could be even completely breakerless if all the source-converter topologies comprise serial semiconductor switches which could break in the case of catastrophic failures. This would also eliminate the need for significant oversizing of the wiring and upstream converters which are traditionally used to ensure safe clearing of the electromechanical breakers in the case of faults. Also, a power converter could transfer power between the low voltage direct current (LVDC) distribution network and the AC transmission network. The converter must possess bidirectional power flow capability to evacuate excess power to utility or to supply the loads during power deficit [1].

The application of multifunctional integrated converter is another kind of nanogrid and attracted more and more attention. Compared with multiple two-port converters, integrated structure, lower cost, flexible power flow control and high-efficiency remained in the multiport converter (MPC). Usually, a PV source, a storage battery and a load are integrated together by MPC in a stand-alone PV/battery power system. Meanwhile, the complex communication for energy management in the two-port units could be replaced by adopting easier central control for MPC in nanogrid. Therefore, complicated energy management in MPC could be realized easily [13].

6.1.3 Converters used in nanogrid

Due to the intermittent and randomness of the DERs, DESUs are necessary to maintain an uninterruptible and smooth power supply for stand-alone loads. Peak-load shifting in renewable nanogrid could be realized by installing DESUs [14,15].

Converters used in DERs of nanogrid have been fully discussed in many literatures. In order to physically implement maximum power point tracking (MPPT), a PV converter must be included in the system. As the PV power is transferred to the load or bus with a current controlled source, the bus voltage is controlled by a batter converter, and redundant energy will be stored in the battery. Interfaced by the DC/DC converter, a battery is in parallel connected to the PV panels. Also the battery charger is used to maintain the state-of-charge (SoC) of the battery and prevent overcharging or overvoltage [16].

The PV converter is typically done with simple boost, buck or buck–boost topologies, and a bidirectional conversion DC/DC converter is used to handle the charging of the battery and regulation of the load or bus voltage. Selection of the converter topology depends on the voltage ratings of the PV panels and the DC bus. In the case of very high voltages, an isolated converter or a cascading configuration of converters may be necessary to handle the high conversion ratio.

And in the energy storage part, bidirectional DC–DC converter is needed because of the bidirectional power flow. Bidirectional buck/boost converter with simple structure and easy control ability is widely used [17]. But the disadvantages of hard switching and low efficiency remained in the conventional bidirectional buck/boost converter. And the voltage ratio is also scant.

Usually, the voltage level of battery in storage system is low, and high step-up converter is most required in this part. Therefore, a novel cascaded buck/boost converter shown in Figure 6.2 is proposed to apply in energy-storage system [18,19]. Comparing to the conventional bidirectional buck/boost converter, the current ripple is reduced in current continuous mode, and the ratio of the converter is improved due to the cascaded structure.

Hard switching and limited efficiency remained in conventional converters. The losses decrease as well as operational frequency. As a result, the size and costing for magnetic elements would be increased. In order to achieve the high switching

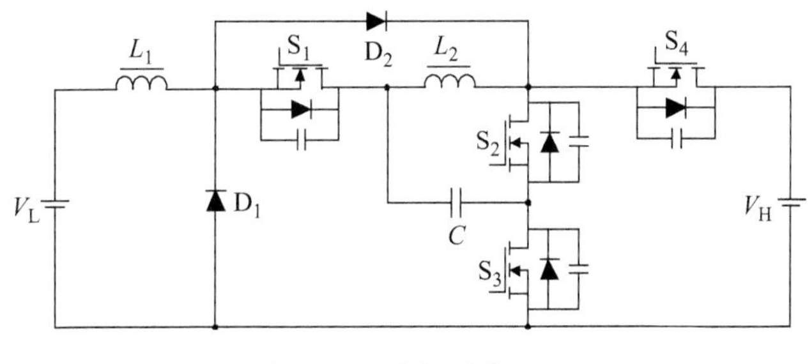

Figure 6.2 Bidirectional buck/boost type converter

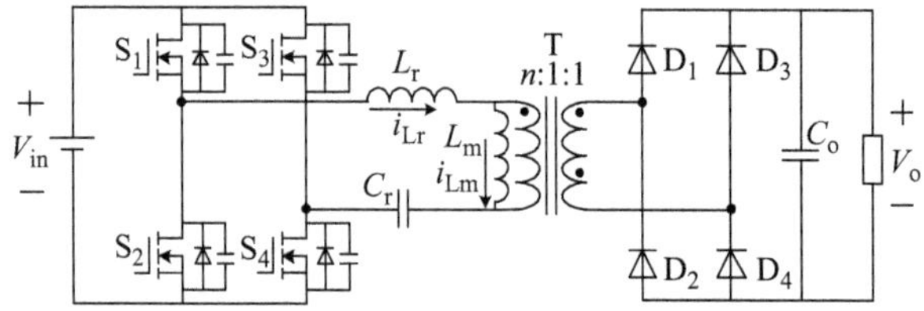

Figure 6.3 DC/DC full-bridge LLC resonant converters

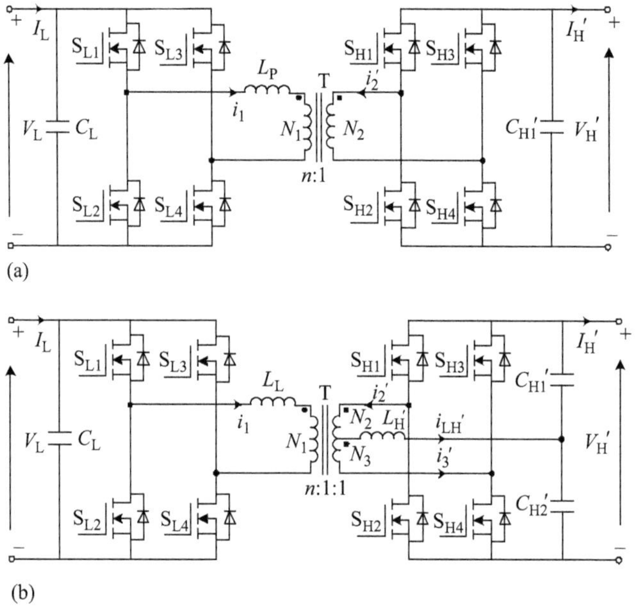

Figure 6.4 DAB type converter: (a) conventional DAB and (b) DAB proposed in [22]

frequency and the optimization of elements, isolated bidirectional converter with high voltage gain is required. DC/DC full-bridge LLC resonant converters in Figure 6.3 which utilizes magnetic inductance, resonant inductance and resonant capacitance to realize soft switches with large input/output ranges are proposed to solve the mentioned disadvantages [20]. And high efficiency could be ensured in this converter.

And in most of the isolated bidirectional converter, dual active bridge (DAB) as shown in Figure 6.4(a) is also used widely. Two full bridge circuits are connected by a transformer in dual active bridge converter, and an inductance L_p is series on one side of the transformer [21]. The direction of the power flow could be controlled by the shift phase between the two bridges. And zero voltage switching (ZVS) could be promised by controlling the shift phase. Moreover, ZVS of DAB could lead the converter to run in a high switch frequency, reduce the volume of the magnetic component and increase power density of the DAB.

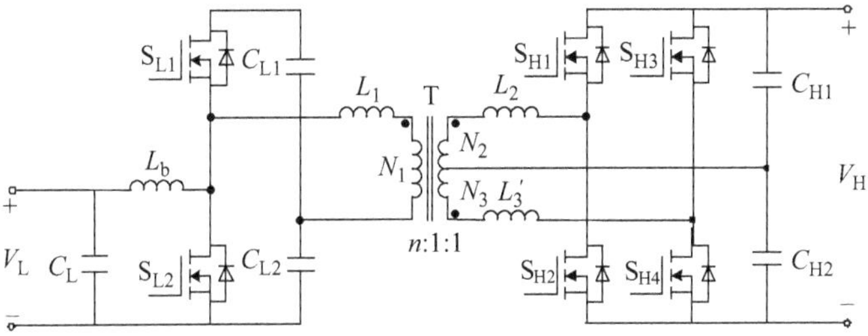

Figure 6.5 A novel current-fed bidirectional converter

But conventional DAB in light load condition, ZVS could not be realized in high voltage side, and the efficiency of the converter would be declined. Based on this basic structure, an improved DAB type shown in Figure 6.4(b) is proposed in [22]. A tapping is added in high voltage side of the transformer. ZVS could be realized in full load range by controlling the shift phase for inner bridge. And the gain of the improved DAB is twice with the conventional DAB.

Moreover, DESU usually remains the characteristic of low input voltage and large input current. Due to the large input current ripple of voltage-fed converter and its low efficiency in this case, its application has some limitations. Therefore, a novel current fed bidirectional converter shown in Figure 6.5 is proposed in [23]. An inductor is added in the low voltage side of the converter to reduce the current ripple and ZVS could be realized in full load range by phase shift control.

6.2 Energy management in nanogrid

The storage devices in a nanogrid include battery, fuel cell or gas turbine. The battery features rechargeable character, it is an important part of the PV electrical power system as it helps regulate the load voltage due to intermittent nature of solar energy. Fuel cell and gas turbine are kinds of important reserve energy, which help to supply the energy for a long time when PV is insufficient [24–26].

6.2.1 Battery-mastered control of a simple photovoltaic/battery system

During the grid-connected operation, the inverters of DERs and DESUs are controlled in the current control mode (CCM) to supply the non-dispatched active and reactive power. The nanogrid voltage is dictated by the utility grid due to its large capacity. Each DG system is initially designed to dispatch real power output of at the common base frequency of when operating in the grid-connected mode.

In the islanded operation, the DESUs which operate in the voltage control mode (VCM) are required to provide necessary voltage regulation, for the absence of utility grid. The inverters of DERs are still operating in the CCM. When the generation of DERs is deficient for the power demand, the DESUs are responsible for providing any shortage in real and reactive power within their operating limits to maintain the power balance and stability of the nanogrids.

6.2.2 Decentralized control for multiple battery-based nanogrid

Due to the intermittent and randomness of the DERs, DESUs are necessary to maintain an uninterruptible and smooth power supply for stand-alone loads in nanogrid. In a nanogrid, the output power of an inverter to grid can be controlled by varying its power angle which is decided by its internal frequency, while the flow of reactive power Q of the inverter can be controlled by changing its voltage magnitude. Thus a P–f droop hybrid Q–V loop is needed for each DG system. This kind of droop control is usually adopted by inverters in order to realize plug and play. The P–f droop control usually adopted in DESU is shown in (6.1) and (6.2). The power outputs of both DG systems are controlled in accordance with their power setting and individual droop characteristics, to supply power to the loads. This arrangement obviously allows both DG systems to share the total load demand in a predetermined manner according to their individual power ratings [24]:

$$f = f_{\mathrm{ref}} + k_{\mathrm{p}}(P_{\mathrm{ref}} - P) \tag{6.1}$$

$$V = V_{\mathrm{ref}} + k_{\mathrm{q}}(Q_{\mathrm{ref}} - Q) \tag{6.2}$$

And the P–f droop hybrid Q–E control loop is shown in Figure 6.6.

In a low-voltage nanogrid, the inverters usually adopt P–V/Q–f droop control to realize plug and play. P–V/Q–f droop control is described in the following:

$$V = V_{\mathrm{ref}} + k_{\mathrm{p}}(P_{\mathrm{ref}} - P) \tag{6.3}$$

$$f = f_{\mathrm{ref}} + k_{\mathrm{q}}(Q_{\mathrm{ref}} - Q) \tag{6.4}$$

The droop control scheme of a DC nanogrid is much simple. For a single DC converter with droop feature, when the load becomes heavy, the converter will decrease its output voltage and thus the load power might decrease. This is a proper trend to get a new power-balancing point between the supply and load. For a nanogrid or multiple converters working in parallel connected mode, the voltage reference of the converter is the original reference added to the output current times droop coefficient. The control diagram is shown in Figure 6.7.

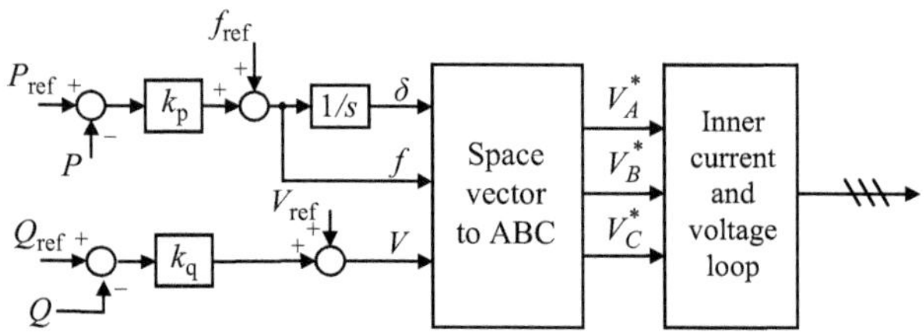

Figure 6.6 Droop control loop applied in inverter

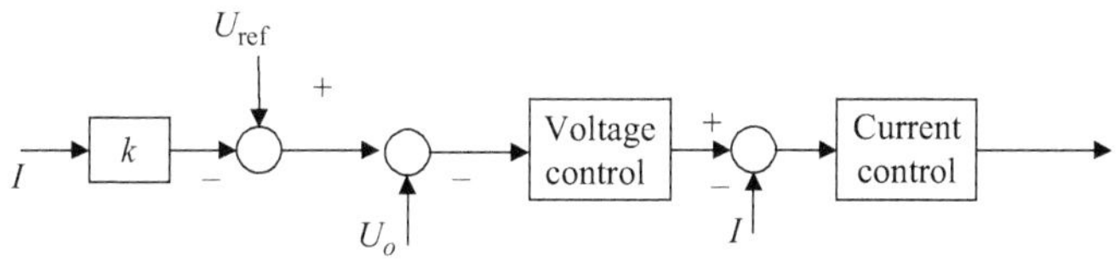

Figure 6.7 DC-droop control loop applied in converter

k is the droop coefficient characterizing the slope, which can be equivalent to a series virtual impedance of each unit converter in the control. I is the output current of the unit converter, and U_{ref} is the voltage reference of the unit converter. When the DC bus voltage is higher than the expected value U_{ref}, the bus voltage is high, the system power is excessive and the unit charge is required to store energy. When the DC bus voltage is lower than the expected value U_o, the bus voltage is low and the system needs more power. It is necessary to release energy to stabilize the bus voltage. The converter works in voltage and current dual close loop mode. With the droop-based voltage reference, the two converters can share the load equally. The greater the droop factor, the better the power sharing.

6.2.3 Decentralized control for multiple DG units based nanogrid

6.2.3.1 Decentralized control in DC nanogrid

The multiple energy storage units refer to that there are two or more kinds of storage with different properties. For example, in a nanogrid, the lead acid battery can be charged when the energy supplied by PV panel is more than the load demand, but the fuel cell and diesel turbine cannot be charged, and both of them works only at a high efficiency within a specific power range. It means that there is a minimum output power for economical operation.

Figure 6.1 is the structure of the hybrid energy system. The system is composed of two PV panels, a fuel cell unit and a battery unit. The primary source of generation for the system is PV systems and fuel cell which meet the average load demand. The battery helps to maintain a continuous supply of power in the presence of fluctuations in the sources and loads.

When the load is light, the battery could be charged according to the situation of SoC. While the power of load plus the battery charging is less than the maximum output of PV panel, the PV interface converter will work in a voltage source control mode, and the system DC bus voltage is controlled by the PV converter and battery interface converter work in a CCM.

More often, the PV systems work with MPPT, and the bus voltage is controlled by battery converter. In one of these cases, the power loads are less than PV panel, but the differences are less than the maximum battery-charging current. And the last case is that when the loads keep on increasing, the battery would discharge energy when the largest output energy of the PV system could not meet the demands of loads.

If the loads continue to increase, PV panel and the battery cannot supply enough energy to the load and the fuel cell starts to work. As the FC should work within its high-efficiency band, the battery may be charged in this case, and the bus is controlled by FC interface converter.

So the DC nanogrid is controlled by the DC-bus voltage. It also means that the bus voltage will change, and the bus will be controlled by PV converter, battery interface converter and FC interface converter independently according to the bus voltage.

According to the bus voltage and some other variable, the mode of the nanogrid could be got in Table 6.1, which the bus voltages in Table 6.1 meet the relationship V1 > V2 > V3 > V4.

If the bus voltage is V1, the nanogrid works in mode 1. If the voltage is V2, when the battery-charging current is positive, it works in mode 21. On the contrary, if the charging current is negative, it works in mode 22. If the voltage is V3, it must be in mode 3.

It must be pointed out that the bus voltage may fluctuate in a certain range, such as between Instead, it is in mode 22 (Table 6.1).

6.2.3.2 Decentralized control in AC nanogrid

For a decentralized AC nanogrid, droop control will be used to control the batteries, and the PV units will work in a CCM to carry out the MPPT. In another word, the battery inverter is always controlled to operate as voltage source with droop control in normal operation. But if there are more than two kinds of storage unit, the decentralized control scheme should be more complex. Figure 6.8 presents a

Table 6.1 Working state of the system

State		PV	Storage battery	Fuel cell	Features of power flow	Bus voltage
1		Voltage control voltage threshold: >V1	Charge with fixed current	Not work	Permitted battery charge power is less than the maximum output power of PV + loads	V1, by PV converter
2	21	MPPT voltage threshold: <V2	Charge and voltage control	Not work	The loads power is less than the maximum output power of PV	V2–V4, battery converter
	22	MPPT threshold: >V4	Discharge and voltage control	Not work	Load is greater than the MPP, less than MPP + maximum discharge power of the battery	
3		MPPT threshold: <V4	Charge	Discharge	greater than the MPP + maximum discharge power of battery	V3, fell cell converter

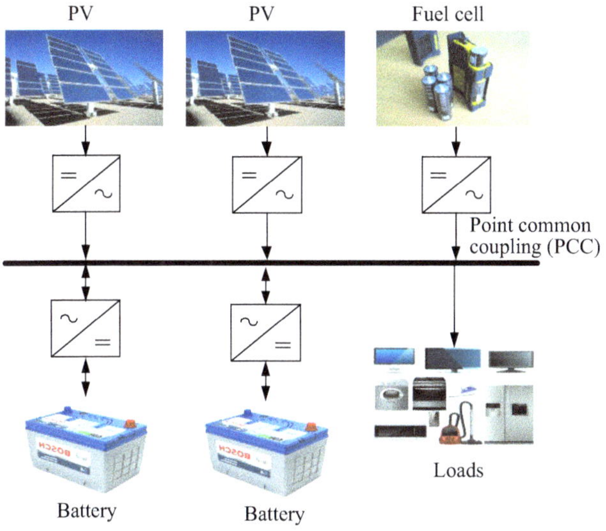

Figure 6.8 PV/battery/fuel cell AC hybrid system

simplified PV/battery/fuel cell AC hybrid system architecture, where n PV units and battery units, one fuel cell unit and some common loads are parallelly connected to the AC bus through line impedance Z_{PV}, Z_{BA} and Z_{FC}, respectively. It should be noted that the capacity of the fuel cell unit is assumed to be large enough to support the system. In this kind of hybrid energy supply system, when and how does it start up and power off the fuel cell unit is a critical issue when there are no centralized controllers. The proposed scheme is described as follows [24].

From the point of view of power management, the SoC-balancing management for the battery unit as well as the energy tracked from the PV cell and the fuel cell is mainly related to the real power. The active power management is the critical issue for the whole hybrid system. As is well known, the inverters with CCM cannot operate in islanded mode, and they need to enable VCM in islanded application. In this hybrid system, the PV and fuel cell inverters also operate with VCM, injecting appropriate power to the grid [24].

Based on conventional droop control, the frequency is related to the active power. Because the frequency has the global characteristics, the active power management based on the frequency as a communication signal could be realized. Therefore, droop control is necessary to adopt in the inverters to maintain the droop characteristics [24].

The conventional droop control could be written in formula (6.1) and (6.2). According to (6.1), the output active power P_o can be written in the following form:

$$P_o = P_{ref} + \frac{f_{ref} - f_o}{k_p} \tag{6.5}$$

The characteristics of the active power related to frequency based on (3) are shown in Figure 6.9. The initial droop curve is represented in Line1 in which the initial

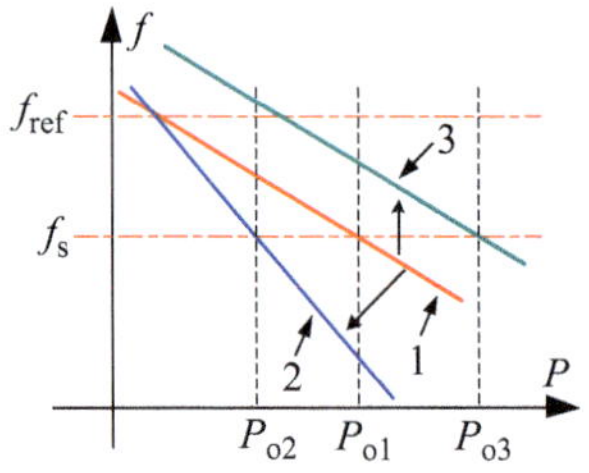

Figure 6.9 Active power/frequency droop characteristic

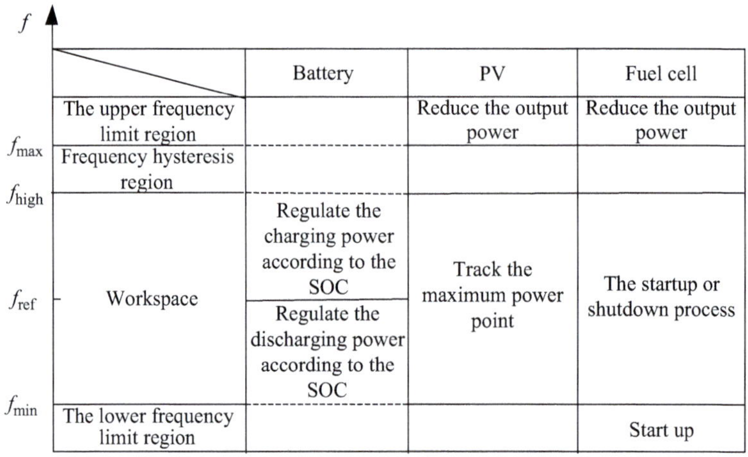

Figure 6.10 System operating status/frequency characteristics

operation point is (P_{o1}, f_o). With k_p increasing, Line1 moves to Line2 which the operation point is (P_{o2}, f_o). Also, Line1 moves to Line3 by increasing the reference power P_{ref}. As a result, the output power could be managed by adjusting the coefficients k_p and P_{ref}.

For the sake of clarity, this study is carried out for three basic conditions, and they are as follows:

1. Battery and PV units operate together under adequate illumination.
2. Battery and fuel cell units cooperate under extreme weather condition and the PV cannot output power.
3. Battery, PV and fuel cell units run together under insufficient illumination or/and heavy load.

The power-management strategy based on frequency is described in Figure 6.10. System operation status will be divided into four parts according to four frequency thresholds.

1. f_{min} is the minimum operating frequency for the system. Also, f_{min} is taken as threshold frequency for the starting of the FC. And in two cases of overload in the system and insufficient SoC of the batteries would lead the frequency down to the lower limit.
2. f_{ref} is the reference frequency of the system and is taken as a distinction of charging and discharging for battery. The output power of inverter in DESU is regulated by changing droop coefficient in transient status. Also, the droop coefficient is needed to be restored under steady state.
3. The region between f_{min} and f_{high} represents the normal operating range.
4. f_{high} and f_{max} make up the frequency hysteresis, together as the upper operating limit. During normal operation, fuel cell inverter (or/and PV inverter under adequate illumination) mainly operates at f_{high}. When the load demand is reduced or/and battery charging power is reduced, the bus frequency increases. In this case, once the system frequency is higher than f_{max}, fuel cell inverter (or/and PV inverter) will reduce output power until the system frequency is lowered to f_{high} again.

Corresponding to the system operating status/frequency characteristics described in Figure 6.10, the characteristics of the active power related to frequency are shown in Figure 6.11. For the proposed hybrid system, the method of changing reference power as showed in Figure 6.11(a) and (c) is adopted in PV and FC inverters. And the method of changing droop coefficient to regulate the active power as showed in Figure 6.11(b) is adopted in battery inverters.

Block diagram of the real power/frequency droop control for battery

Figure 6.11(b) shows the battery power/frequency characteristic curve. First, the maximum and minimum charge states SoC_{max} and SoC_{min} of the battery are set. In normal operation, the battery interface inverter operates according to the *P/f* characteristics shown in curve 1. When the inverter runs at region B with a certain power less than the maximum-rated output power $P_{B,max}$ (The instantaneous power of the battery will be much larger than this value, but the output capability of the system is

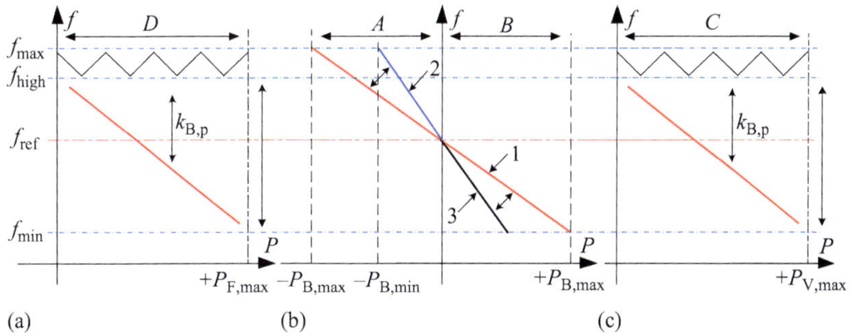

Figure 6.11 Real power/frequency characteristics: (a) Droop control applied in PV inverters; (b) Droop control applied in battery inverters and (c) Droop control applied in Fuel cell inverters

also limited by the DC/DC converter and the interface inverter. $P_{B,max}$ is the maximum output power of the DC/DC converter and the inverter), when the battery detects SoC < SoC_{min}, actively adjust the droop curve 1 (for example, to the position of curve 3 to reduce the system frequency to f_{min}, the fuel cell is activated, and the inverter is brought into the A zone where the power is absorbed, and the battery is charged). When running in area A, the inverter adjusts the droop curve 1 according to the battery SoC (for example, to the position of curve 2, reducing the charging power), thereby controlling the charging power of the battery; when the charging power reaches the floating charging power $P_{B,min}$, the drooping curve is not adjusted again [24].

The choice of droop coefficient is mainly affected by two factors: first, the charge and discharge power of the battery and second, the stability of the system. When the SoC of the battery in area B is insufficient, the fuel cell will be started by increasing the droop factor to reduce the system frequency. If the load is light at this time, it corresponds to a larger droop factor. In addition, the battery has a minimum charging power, which also corresponds to a large droop factor. According to the traditional droop control theory, a large droop coefficient will lead to instability of the system. Therefore, two factors should be considered in the selection of sag coefficient [24].

The battery adopts a three-loop control scheme based on droop control, an outer loop power droop loop, an inner loop voltage and current loop. As shown in Figure 6.12, according to the charging and discharging characteristics of the battery, the output power at rated frequency is zero, and the active power/frequency droop curve of the battery is defined as

$$f_B = f_{ref} - k_{B,p} \cdot P_{B,LPF1} \tag{6.6}$$

Among them, f_B and $P_{B,LPF1}$ are the output frequency of the battery interface inverter and the output power filtered by the low-pass filter, f_{ref} rated frequency, $k_{B,p}$ droop coefficient, respectively. Definition:

$$k_{B,p} = \frac{f_{ref} - f_{min}}{P_{B,max}} \tag{6.7}$$

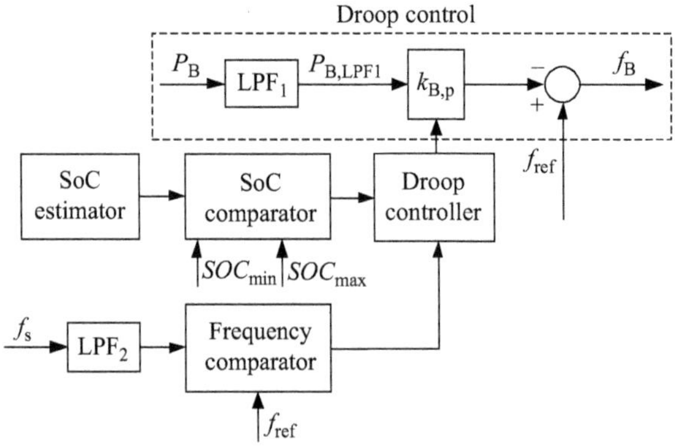

Figure 6.12 Block diagram of the real power/frequency droop control for battery

Among them, $P_{B,max}$ are the maximum output power of the battery interface inverter.

As shown in Figure 6.12, during the charging and discharging of the battery, the state of charge of the battery is first calculated by the SoC estimator, then the droop coefficient change trend is determined according to the SoC comparator and finally the droop coefficient is adjusted by the droop controller. At the same time, the phase lock loop detects the nanogrid bus frequency f_s in real time and then compares it with the reference frequency through a frequency comparator to determine the time of restoring the droop curve. Since the adjusted droop curve (as shown in curves 2 and 3 in Figure 6.12(b)) is not the normal operation curve of the inverter, the inverter needs to restore the droop curve when the nanogrid bus frequency is equal to the reference frequency. The output power of the inverter is zero at the rated frequency, and the droop factor adjustment will have minimal impact on the system at this time. Therefore, whether the battery enters the discharge area B from the charging area A or enters the area A from B, the inverter restores the droop curve at the rated frequency [24].

Since the system energy balance is mainly coordinated by the nanogrid bus frequency, the bus frequency is crucial to the normal operation of advanced energy management system (AEMS). In the actual system, the bus frequency is not a steady-state value due to load, inverter and other factors, so the frequency measured by phase-locked loop cannot be directly applied. As shown in Figure 6.12, the function of LPF_2 is to filter out high frequency components in the bus frequency. In addition, compared with the power filter LPF_1, the LPF_2 has a large time constant because the system is sensitive to frequency changes [24].

Block diagram of the real power/frequency droop control for PV

Figure 6.11(c) shows the PV cell power/frequency characteristic curve. Area C is the PV interface inverter working area, in which the PV cells are conditionally controlled by MPPT. f_{max} and f_{high} are the upper and lower thresholds of the upper limit frequency fluctuation, and the main function is to prevent the frequent-power-adjustment process of the system caused by the frequency fluctuation. After start-up, the PV battery mainly has two working states [24].

1. When the system frequency is lower than f_{high}, the PV outputs power according to MPPT power or load demand.
2. When the system frequency is greater than f_{max}, the PV cell reduces the power output and adjusts the system frequency to maintain it at f_{high}.

Unlike the battery interface inverter to adjust the droop coefficient control power, the PV interface inverter adjusts the power output by shifting the droop curve.

According to the above introduction, the interface inverter based on the PV MPPT and the constant power control strategy cannot be operated on an island, and the voltage controlled inverter is required to provide the voltage and frequency reference. Therefore, in order to improve system stability, the PV cell interface inverter adopts a hybrid control strategy based on droop control. As shown in Figure 6.13, define the PV droop control curve:

$$f_V = f_{ref} + k_{B,p}\left(P_{V,set} - P_{V,LPF1}\right) \tag{6.8}$$

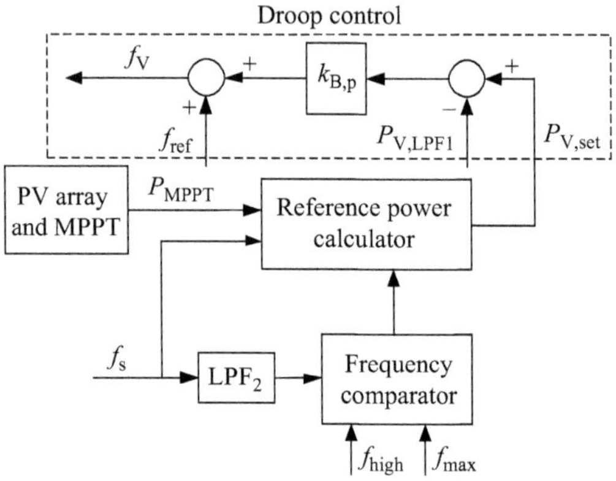

Figure 6.13 Block diagram of the real power/frequency droop control for photovoltaic

Among them, f_V and $P_{V,LPF1}$ are the output frequency of the PV cell interface inverter and the output power of the low-pass filter, respectively. Define droop control reference set power:

$$P_{V,set} = P_{MPPT} + \frac{f_s - f_{ref}}{k_{B,p}} \tag{6.9}$$

Among them, P_{MPPT} is the maximum tracking power of PV cells. Photovoltaic battery interface inverter operating status is as follows:

1. During the day, the PV cells control the output power according to the changes of the light intensity according to (6.8) and (6.9). When the system frequency satisfies the condition $f_s = f_{high}$, the droop control maximum reference set power is defined:

$$P_{V,set,max} = P_{MPPT,max} + \frac{f_{high} - f_{ref}}{k_{B,p}} \tag{6.10}$$

   Among them, $P_{MPPT,max}$ is the maximum PV power when the system frequency reaches f_{high}. At this point, redefine the droop control curve:

$$f_V = f_{ref} + k_{B,p}\left(P_{V,set,max} - P_{V,LPF1}\right) \tag{6.11}$$

   Thereafter, if the frequency increases due to the lightening of the load or the decrease in the charging power of the battery, the PV cell reduces the power on the basis of $P_{MPPT,max}$, thereby reducing the maximum reference set power $P_{V,set,max}$ and maintaining the system frequency. The interface inverter needs to refresh and record $P_{V,set,max}$ and $P_{MPPT,max}$ each time the system frequency reaches f_{high}.
2. The load power is increased. If the PV energy is sufficient, the output power is controlled according to the droop curve defined by (6.11); if the energy is insufficient, the output power is controlled according to (6.8) and (6.9).

As shown in Figure 6.13, when the reference power calculation is performed, the bus frequency f_s or the low-pass filter with a small time constant can be used to improve the system response speed, and the frequency comparator adopts a relatively large time-lapse constant. The bus frequency of the low-pass filter increases system reliability.

Block diagram of the real power/frequency droop control for fuel cell

Figure 6.11(a) is the fuel cell power/frequency characteristic curve, and the area D is the fuel cell working area. The fuel cell has the minimum operating limit frequency f_{min}. Once the system frequency is lower than the limit frequency, the fuel cell needs to start up to compensate the power of the system. In addition, the fuel cell also has the minimum output power $P_{F,min}$ limit. When the specific conditions are met, the fuel cell can stop. The same as the PV inverter, the fuel cell adjusts the power output by translating the droop curve.

There are two ways to start the fuel cell, one is to increase the load power and reduce the system frequency; the other is to charge the battery and adjust the sag coefficient to reduce the frequency. When the system frequency f_s is lower than the fuel cell starting frequency f_{min}, the fuel cell starts and continuously increases the power output until the system frequency reaches f_{high}; similar to PV, the fuel cell no longer increases the output power but maintains the system at that frequency. Thereafter, the system frequency will increase as the battery charging power decreases or the load decreases. The fuel cell maintains the system frequency f_{high} by reducing the output power. There are two cases of shutdown as well as the two cases of fuel cell start-up [24].

1. If the load power P_{load} exceeds the maximum output power $P_{B,max}$ of the battery, when the battery is charged, the output power of the fuel cell meets $P_F = P_{load} + P_{B,min} > P_{F,min}$, in which $P_{F,min} = P_{B,min} + P_{B,max}$, and the fuel cell does not shut down.
2. If the fuel cell is started by adjusting the sag coefficient of the battery, when the battery is full of electricity, the output power of the fuel cell is $P_F < P_{F,min}$. At this time, the downtime condition is satisfied and the fuel cell shuts down.

As shown in Figure 6.14, similar to PV cells, the fuel cell droop control curve and the droop control reference set power are defined first:

$$f_F = f_{ref} + k_{B,p}\left(P_{F,set} - P_{F,LPF1}\right) \tag{6.12}$$

$$P_{F,set} = P_{init,set} + \frac{f_s - f_{ref}}{k_{B,p}} \tag{6.13}$$

Among them, $P_{init,set}$ is the setting output power for fuel cell.

The operation state of the fuel cell interface inverter is as follows:

1. When $f_s \leq f_{min}$, the fuel cell starts, and the output power is controlled according to (6.12) and (6.13). Defining $P_{init,set}$ as

$$P_{init,set} = n \cdot \Delta P \tag{6.14}$$

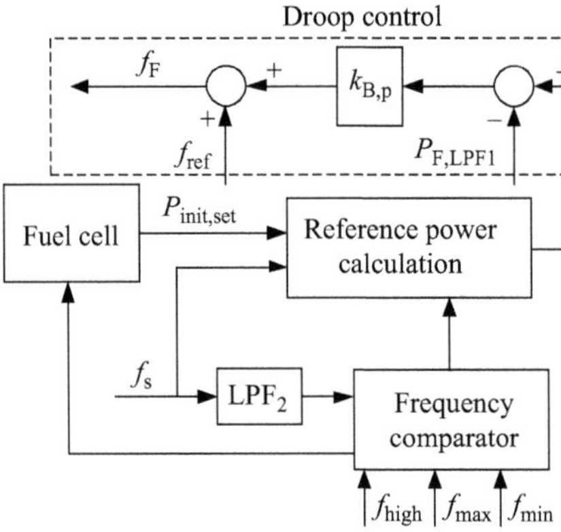

Figure 6.14 Block diagram of the real power/frequency droop control for fuel cell

Among them, ΔP is the power step and n is a function that increases linearly with time.

2. When $f_s = f_{high}$, similar to a PV cell, the droop control reference set power $P_{F,set,max}$ and the fuel cell set output power $P_{init,set,max}$ are recorded. Redefining droop control curve:

$$f_F = f_{ref} + k_{B,p}\left(P_{F,set,max} - P_{F,LPF1}\right) \tag{6.15}$$

   Thereafter, if the frequency increases due to lighter load or lower battery charging power, the fuel cell reduces power on the basis of $P_{init,set,max}$, thereby reducing the maximum reference set power $P_{F,set,max}$ and maintaining the system frequency. The interface inverter needs to refresh and record $P_{init,set,max}$ and $P_{F,set,max}$ at every time the system frequency reaches f_{high}. If $P_{init,set,max} < P_{F,min}$ is detected, the fuel cell will stop.

3. If the load increases, the output power will be controlled according to the droop curve defined in formula (6.15).

Simulation case study

According to the proposed control strategy, the analysis based on simulation software PSCAD/EMTDC in this section is carried out. Simulation parameters are shown in Table 6.2.

1. Ideal lighting condition

   The condition where only the battery and the PV work is representative of the ideal daylight conditions, and the waveform is as shown in Figure 6.15.

   Setting an initial load active power is 9 kW. At $t = 1.0$ s, the PV output power increases from zero, and the battery output power gradually decreases. When $t = 5.6$ s, the PV output power is equal to the load power, the battery output power is zero and begins to charge from the discharge state. When $t = 10.0$ s, the system frequency reaches the upper limit frequency f_{high}, the PV records the set power at this time to make a constant power output. When $t = 12.0$ s, the load

Table 6.2 System parameters

	Symbol	Value
Droop coefficient in steady state	$k_{B,p,min}/k_{V,p}/k_{F,p}$	0.05 Hz/kW
Maximum droop coefficient	$k_{B,p,max}$	0.2 Hz/kW
Minimum frequency	f_{min}	49.5 Hz
Nominal frequency	f_{ref}	50.0 Hz
Upper frequency	f_{high}	50.4 Hz
Maximum frequency	f_{max}	50.5 Hz
Time constant of LPF_1	t_1	0.01
Time constant of LPF_2	t_2	0.3
Nominal power of battery inverter	$P_{B,max}$	10 kW
Nominal power of PV inverter	$P_{V,max}$	20 kW
Nominal power of fuel cell inverter	$P_{F,max}$	large enough
Load demand under adequate illumination	P_{load1}	9/4.5 kW
Load demand under extreme weather	P_{load2}	9 kW
Load demand under heavy load	P_{load3}	12/15 kW

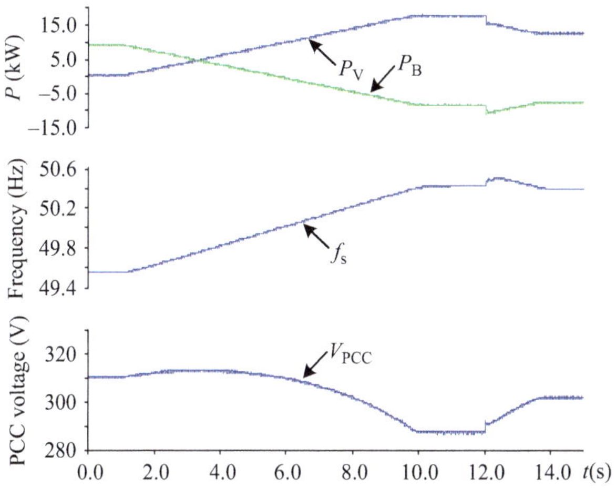

Figure 6.15 Ideal light working conditions for hybrid system

power decreases by 4.5 kW, and the system frequency naturally rises, so that the system frequency is higher than the maximum operating frequency f_{max}. Then the PV begins to adjust the output power to make the system frequency restored to f_{high} and records the set power at this time. When $t = 16.0$ s, the load power is increased by 9 kW, the output power of PV and battery are adjusted according to the droop characteristic. PV increases power output and battery reduces charging power. If the PV power is insufficient to maintain the droop control output power, the set power is adjusted according to (6.10).

To be noted that when $t = 10.0$ s, the system frequency reaches the upper limit frequency f_{high}, and when the PV stops increasing the power output, the

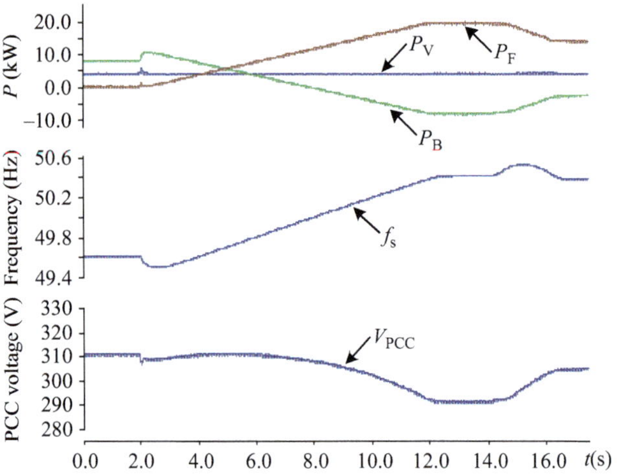

Figure 6.16 Heavy load working conditions for hybrid system

system frequency will still rise slowly for a period of time. This is because the time constant of the frequency filter is relatively large resulting in power response relatively slow. Corresponding to this, when the system frequency rises to f_{max}, the PV reduces output power to lower the frequency, and the frequency stability value will be slightly less than f_{high}. But in practical applications, the effect of the filter time constant on the actual time scale is not as obvious as in Figure 6.15 and can be ignored.

2. Heavy load working conditions (Figure 6.16)
When storage battery, PV cells and fuel cells work at the same time. It mainly represents the situation that daytime lighting conditions are not ideal and loads are heavy. To simplify the analysis, it is assumed that the output of PV is constant, as shown in Figure 6.16.

The initial active load power which is 12, 8 and 4 kW is provided by the battery and PV cells, respectively. When $t = 2.0$ s, the load power suddenly increases 3 kW, the system frequency decreases to f_{min} and fuel cells start. After that, the output powers of the PV cells remain unchanged, and fuel cells increase the output power according to (6.14). At about $t = 12.0$ s, the system frequency rises to the upper limit frequency f_{high} and fuel cells do not increase the power output. When $t = 14.0$ s, the battery charge ends and the charge power is reduced by increasing the droop coefficient until the minimum charge power $P_{B,min}$ is reached. Fuel cells detect that the system frequency reaches f_{max} at about $t = 15.2$ s and begin to reduce the power output. After that, the battery and fuel cells work together until system frequency returns to f_{high} at about $t = 18.6$ s. At this time, the output power of fuel cell $P_F > P_{B,min} + P_{B,max}$, indicating that the system load power exceeds the maximum output power of the battery and the fuel cells do not need to stop.

6.2.4 Decentralized control for multiple energy storage units based nanogrid

As aforementioned, the DESUs are dispersedly connected to common bus in nanogrid. Hence, decentralized control scheme for DESU is preferred, and droop control and its variants have been widely applied. However, the mismatches of line impedances, local loads, manufacturing tolerance, self-discharge rates, among the others will induce some problems. First, the SoC of each cell in the battery string may be unbalanced; second, the SoC of the whole DESU may be unbalanced among different DESUs; third, the reactive power cannot be properly shared in a determined manner; fourth, the voltage magnitude is deviated from the nominal value with the droop control method. As a result, overcharge or over discharge, overheat, even fire may occur and the cost of the systems is increased [25].

In order to avoid the overuse of a battery unit and prolong the service life of the DESUs, a multifunctional and wireless droop-controlled method employed into the power converter system (PCS) control of DESUs in autonomous nanogrid is introduced in this part. Accounting for the different initial SoC, mismatched line impedances and local loads, this method achieves SoC balancing and power sharing among DESUs without droop-gain scheduling and communication links in contrast to the existing control schemes proposed so far. In order to compromise the inherent trade-off between the voltage regulation and load sharing accuracy for droop method, a compensation term is added in the local controller to lead the voltage magnitude to restore an acceptable range. The structure of paralleled DESUs for smoothing the output power of the DERs is shown in Figure 6.17.

The estimation algorithm of the SoC is derived in (6.16):

$$\mathrm{SoC} = \mathrm{SoC}_0 - \frac{\int i_{\mathrm{in}} dt}{\mathrm{Ce}} \tag{6.16}$$

where SoC_0 is the initial value of SoC, Ce and i_{in} denotes the capacity and output current of DESU.

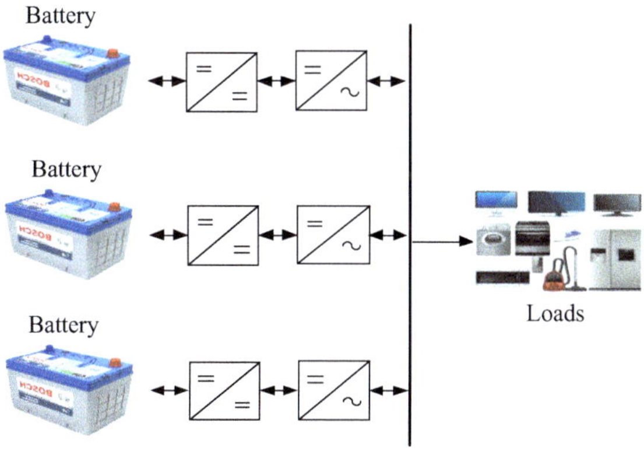

Figure 6.17 Structure of paralleled DESUs

The output voltage of the DESS can be seen as an equal constant, it is derived:

$$V_{DC1} = V_{DC2} = V_{DC3} \tag{6.17}$$

Meanwhile, ignoring the loss of the inverters yields:

$$p_{out} \approx p_{in} = V_{DC} i_{in} \tag{6.18}$$

Combining (6.16) and (6.18), it is derived:

$$\text{SoC} = \text{SoC}_0 - \frac{\int i_{in} dt}{\text{Ce}} \approx \text{SoC}_0 - \frac{\int p_{out} dt}{V_{DC}\text{Ce}} \tag{6.19}$$

From (6.19) it can be derived that the SoC of DESU is mainly determined by the SoC_0 and the active power of inverters.

Considering that the inverters output impedance in Figure 6.1 may be highly inductive when using LCL filters or due to the connection through a transformer. Then, the conventional *P–f* droop control is usually adopted in DESU which is shown in (6.20) and (6.21):

$$f = f_{ref} - k_p P \tag{6.20}$$

$$E = E_{ref} - k_q Q \tag{6.21}$$

where f and E are the frequency and the output voltage magnitude of inverter; f_{ref} and E_{ref} are the reference frequency and the magnitude of the output voltage; P and Q are the average value of output PQ of inverter; k_p and k_q are droop coefficients.

Ignoring the resistance part of the line impedance and assuming that the power angle δ is small enough, the active and reactive power flow are shown in (6.22) and (6.23), respectively:

$$P = \frac{V_{PCC} \cdot E \cdot \delta}{X} \tag{6.22}$$

$$Q = \frac{V_{PCC}(E - V_{PCC})}{X} \tag{6.23}$$

where V_{PCC} is the magnitude of PCC voltage, E is the magnitude of output voltage and X is the reactance part of impedance.

In order to achieve the SoC balancing of battery, the SoC-based *P–f* droop method is proposed in (6.24) [25]:

$$\omega = \omega_{ref} - k_p P - K_{SoC}(1 - \text{SoC}) \tag{6.24}$$

where k_p and K_{SoC} are the droop coefficients of the proposed method.

Power sharing is a crucial issue for autonomous nanogrid. However, reactive power sharing between the inverters becomes undesirable because of the mismatch of the line impedance.

The active power at t_+ is shown in (6.25):

$$P_{t_+} = \frac{V_{\mathrm{PCC}} E \int_0^{t-} (\omega_{\mathrm{ref}} - k_{\mathrm{p}} P) dt}{X} \tag{6.25}$$

where P_{t+} is the active power at t_+. The output active power of the inverter with traditional droop method adjusted by historical information of active power is described in (6.25). Similarly, the reactive power is derived:

$$Q = \frac{V_{\mathrm{PCC}}(E_{\mathrm{ref}} - V_{\mathrm{PCC}})}{X + V_{\mathrm{PCC}} k_{\mathrm{q}}} \tag{6.26}$$

Taking an example with two DESUs, the relationship of conventional *Q–E* droop in (6.26) is shown in Figure 6.18. It could be seen that the reactive power could not be shared equally under the conventional droop control.

To simulate the power flow of the traditional *P–f* droop control, the historical information of the reactive power is artificially injected into the *Q–E* drooping method, which is called the $\int Qdt$–V droop control, as shown in the following equation:

$$E_{t_+} = E_{\mathrm{ref}} - k_{\mathrm{q}} \int_0^{t-} Qdt \tag{6.27}$$

where E_{t+} is the magnitude of the output voltage reference at t_+ and k_{q} is the coefficient of the $\int Qdt$–V droop control. It could be written that:

$$Q_{t+} = \frac{V_{\mathrm{PCC}}\left(E_{\mathrm{ref}} - K_Q \int_0^{t-} Qdt - V_{\mathrm{PCC}}\right)}{X} \tag{6.28}$$

where Q_{t+} is the reactive power at t_+.

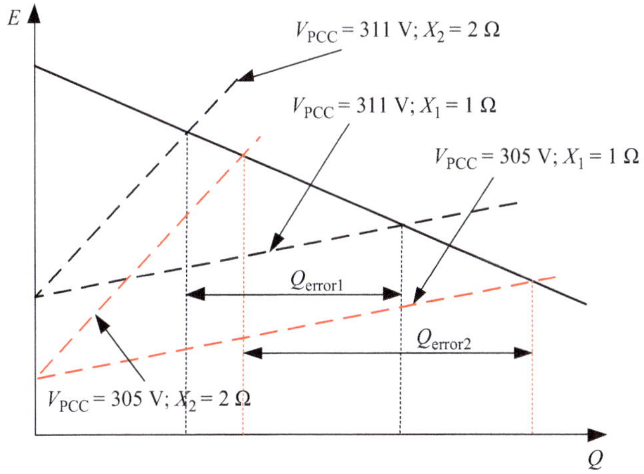

Figure 6.18 The relationship of Q–E droop and reactive power transmission

Equation (6.28) shows that the reactive power flow of the $\int Qdt$–V drooping method is strongly influenced by the historical information of reactive power. In other words, with the $\int Qdt$–V drooping method, the voltage of the DESU that provides more reactive power will be reduced more than the other voltages that provide lower reactive power. Theoretically, the transient process of the $\int Qdt$–V drooping method is shown in Figure 6.19. The intersection of the black lines means the operating point at t_1. At the same time, the intersection of the red lines represents the operating point at t_2 ($t_1 < t_2$). ΔE_1 and ΔE_2 represent the voltage drops of DESU_1 and DESU_2 from t_1 to t_2, respectively. Obviously, $\Delta E_1 > \Delta E_2$ results in a reduction in reactive power error ($Q_{\mathrm{errort2}} < Q_{\mathrm{errort1}}$) [25].

In the autonomous nanogrid, the voltage amplitude kept within the allowable range is necessary for DESUs. More specifically, there is an inherent trade-off between voltage accuracy and power sharing that exists in the proposed drooping method [25]. In order to make full use of the accurately shared active power, add the compensation term proportional to the active power integral to (6.28), which can be expressed as

$$E_{t_+} = E_{\mathrm{ref}} - K_Q \int_0^{t_-} Qdt + K_{\mathrm{PV}} \int_0^{t_-} Pdt \tag{6.29}$$

where K_{PV} is the voltage compensation coefficient.

In conjunction with (6.24) and (6.29), the overall control structure of the inverter is shown in Figure 6.20, where the different models defined by the particular dashed lines represent their respective functions in the control structure. The output variable V_{abcref} represents the three-phase reference voltage and is then sent to the internal voltage loop to track the output voltage.

In order to test the feasibility and effectiveness of the proposed droop control method, an island nanogrid consisting of three parallel DESUs (defined as DESU_1, DESU_2 and DESU_3) was studied. Table 6.2 lists the key parameters of the system. Since the line impedance of the system shown in Table 6.2 does not match, the problems mentioned in Section 6.1 cannot be avoided. This section reveals case

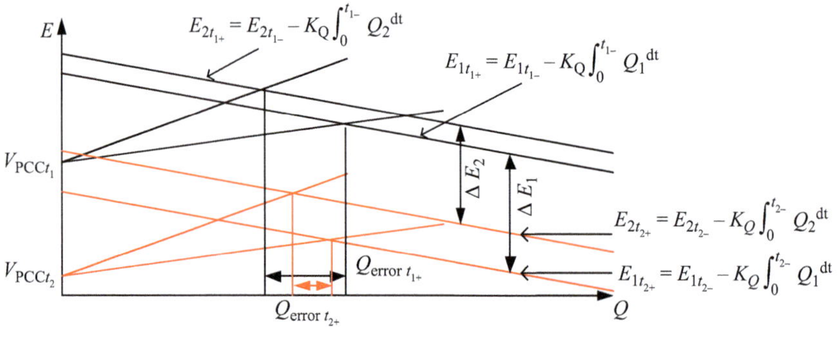

Figure 6.19 Theoretical transient process of reactive power sharing

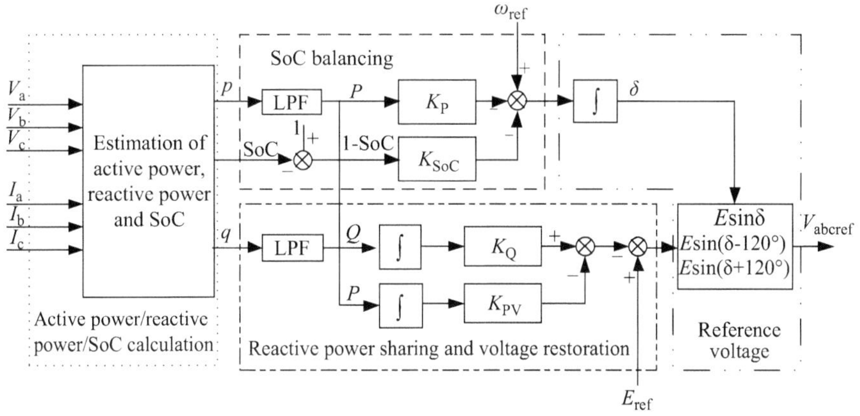

Figure 6.20 Control structure of the inverters

studies under three different conditions and proposes a drooping method to further illustrate its performance. In each case study, there are three subareas with different conditions for the initial SoC of three DESUs connected in parallel, namely $SoC_{01} = SoC_{02} = SoC_{03}$, $SoC_{01} > SoC_{02} > SoC_{03}$, and $SoC_{01} < SoC_{02} < SoC_{03}$. The response of each sub-case system is also given in each case study, which indicates that the proposed sagging method can solve the problem of different line impedance mismatch and initial SoC in the case of load step change and local load [25].

6.2.4.1 Case study I: from conventional droop control to improved droop control

In this case study, the performance of improved droop control and traditional droop control under unequal impedance and initial SoC was compared and tested. Line impedance, droop coefficient and common load are shown in Table 6.2. Figures 6.21–6.23 show the correlation results (SoC, output effective, reactive power, PCC voltage amplitude) SoC_{03} corresponding to $SoC_{01} = SoC_{02} = SoC_{03}$, $SoC_{01} > SoC_{02} > SoC_{03}$ and $SoC_{01} < SoC_{02} < SoC_{03}$, respectively. For 0–0.5 s, DESU is controlled by conventional droop control; and for 0.5–2.5 s, DESU is controlled by improved droop control. To evaluate the performance of the improved droop control, the reactive power errors of the three DESUs are defined as $Q_{error12} = Q_1 - Q_2$, $Q_{error13} = Q_1 - Q_3$ and $Q_{error23} = Q_2 - Q_3$. Similarly, the active power errors of the three DESUs are called $P_{error12} = P_1 - P_2$, $P_{error13} = P_1 - P_3$, and $P_{error23} = P_2 - P_3$.

Obviously, the balanced SoC is gradually realized as shown in Figures 6.21(a), 6.22(a) and 6.23(a). The steady-state reactive power error under conventional droop control and the improved droop control in Figures 6.21(b), 6.22(b) and 6.23(b) are listed in Table 6.3. Improved droop control can achieve the same reactive power distribution compared to conventional droop control. The voltage amplitude is limited to an acceptable range, which is ± 5% as seen from Figures 6.21(c), 6.22(c) and 6.23(c). From Figures 6.21(d), 6.22(d) and 6.23(d), it can be deduced that the active power distribution is not affected by reactive power sharing and voltage recovery.

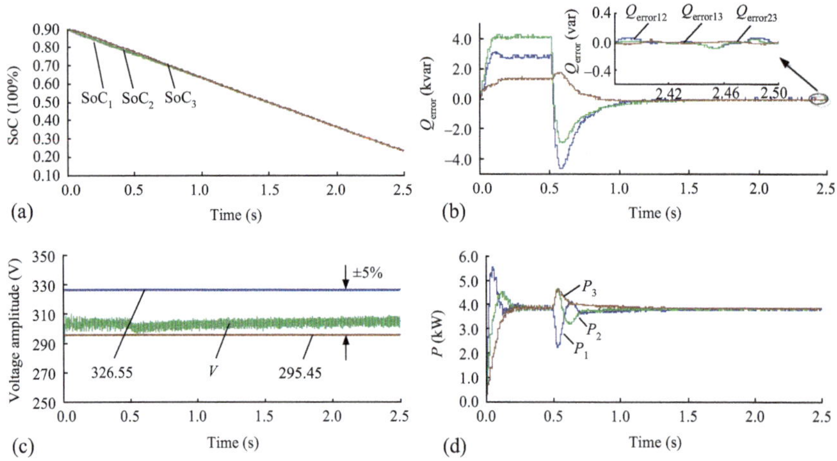

Figure 6.21 Waveforms when $SoC_{01} = SoC_{02} = SoC_{03}$: (a) SoC of the three paralleled DESUs; (b) the reactive power errors; (c) magnitude of the PCC voltage and (d) the active power errors

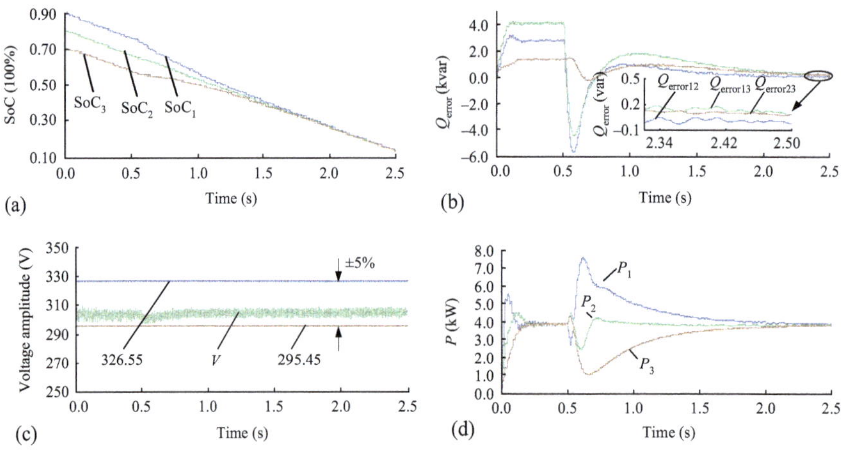

Figure 6.22 Waveforms when $SoC_{01} > SoC_{02} > SoC_{03}$: (a) SoC of the three paralleled DESUs; (b) the reactive power errors; (c) magnitude of the PCC voltage and (d) the active power error

6.2.4.2 Case study II: load step changes in active and reactive power

In this section, we will implement the drooping method proposed by the Institute to achieve SoC balance, power sharing and voltage recovery capabilities, which face load changes from active and reactive power from 4 kW+j2 kvar to 6 kW+j3 kvar.

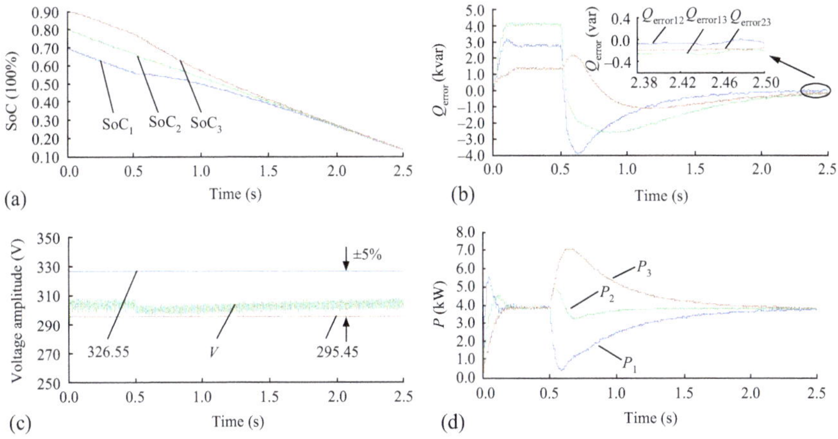

Figure 6.23 Waveforms when $SoC_{01} < SoC_{02} < SoC_{03}$: (a) SoC of the three paralleled DESUs; (b) the reactive power error; (c) magnitude of the PCC voltage and (d) the active power errors

Table 6.3 System parameters

Terms	Symbol	Simulation parameters	Experimental parameters
Line impedance	$R_1 + jX_1$	$0.15 + j0.5\ \Omega$	$0.2 + j1.0\ \Omega$
	$R_2 + jX_2$	$0.3 + j1.0\ \Omega$	
	$R_3 + jX_3$	$0.6 + j2.0\ \Omega$	$0.4 + j2.0\ \Omega$
Droop coefficient	k_p	1.88e−4 rad/W	0.0025 rad/W
	k_q	1e−6 V/var	0.00155 V/var
	K_{SoC}	6.28e−6 rad	
Compensation coefficient	K_{PV}	6e−6 V/W	0.00935 V/W

Figures 6.24, 6.25 and 6.26 show the system performance (SoC, output effective, reactive power, PCC voltage amplitude) corresponding to $SoC_{01} = SoC_{02} = SoC_{03}$, $SoC_{01} > SoC_{02} > SoC_{03} > SoC_{02} < SoC_{02}$ when active power and the reactive power is changed at 1.0 s, SoC_{03}.

The results show that with the proposed drooping method, SoC balance, power sharing and voltage recovery can be achieved regardless of the active power, reactive power or load step change of PQ.

6.2.4.3 Case study III: local load

In an autonomous nanogrid, the location of the local load will affect the power flow. Since the frequency of the nanogrid is the same at steady state, the active power

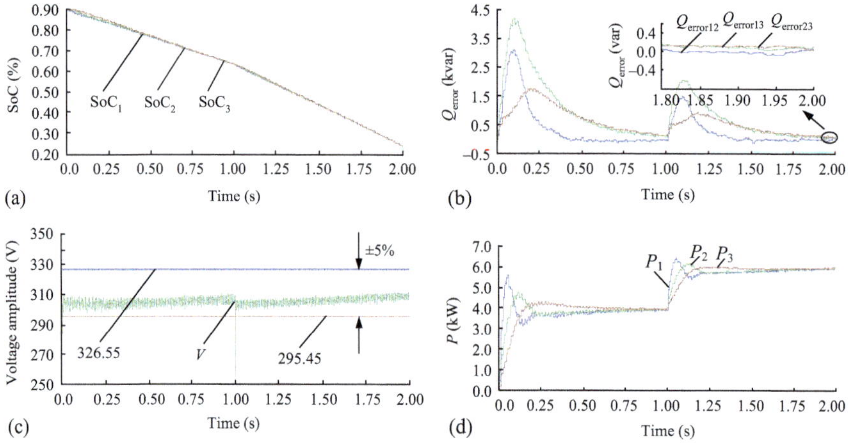

Figure 6.24 Waveforms when $SoC_{01} = SoC_{02} = SoC_{03}$: (a) SoC of the three paralleled DESUs; (b) the reactive power errors; (c) magnitude of the PCC voltage and (d) the active power error

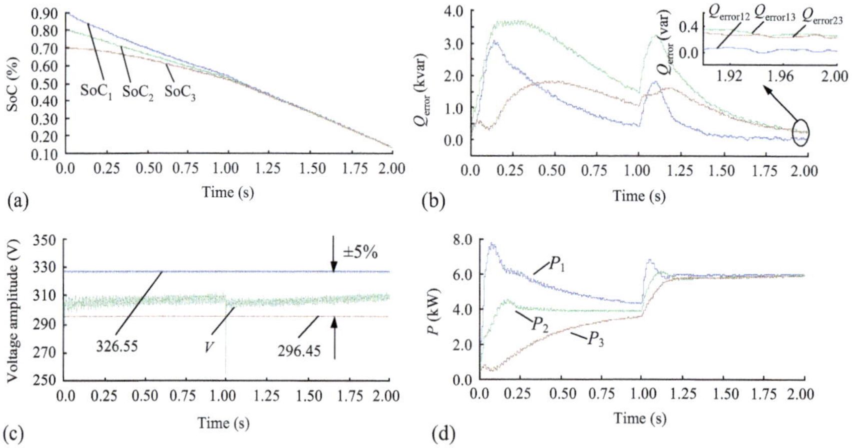

Figure 6.25 Waveforms when $SoC_{01} > SoC_{02} > SoC_{03}$: (a) SoC of the three paralleled DESUs; (b) the reactive power errors; (c) magnitude of the PCC voltage and (d) the active power errors

sharing is not affected by the local load. However, reactive power-sharing performance is strongly influenced by local loads. To further illustrate the effectiveness of the proposed drooping method, a corresponding result is displayed when there is a 2 kW+j1 kvar local load in DESU1.

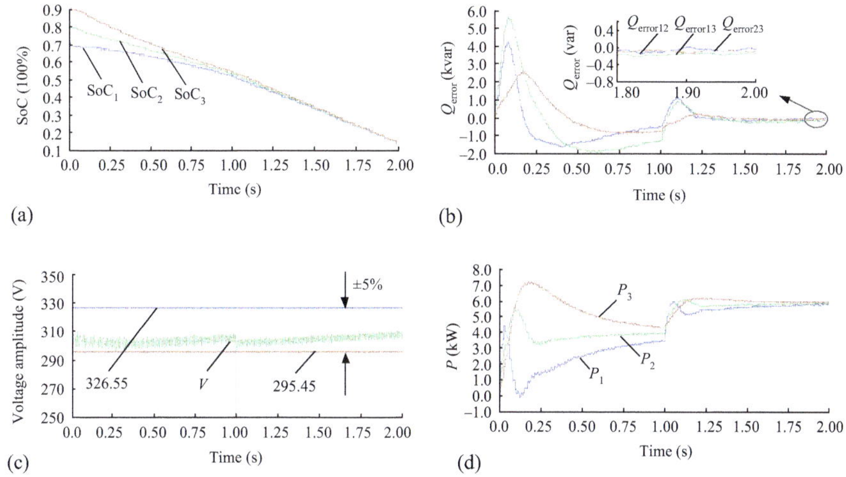

Figure 6.26 Waveforms when $SoC_{01} < SoC_{02} < SoC_{03}$: (a) SoC of the three paralleled DESUs; (b) the reactive power errors; (c) magnitude of the PCC voltage and (d) the active power errors

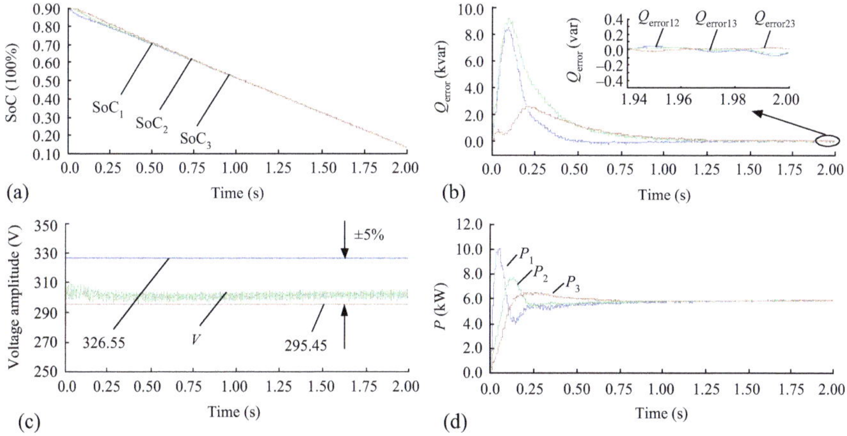

Figure 6.27 Waveforms when $SoC_{01} = SoC_{02} = SoC_{03}$: (a) SoC of the three paralleled DESUs; (b) the reactive power errors; (c) magnitude of the PCC voltage and (d) the active power errors

Figures 6.27–6.29 show that the proposed droop method can achieve SoC balance, reactive power sharing and voltage recovery in the presence of local load, which further validates the effectiveness of the proposed method.

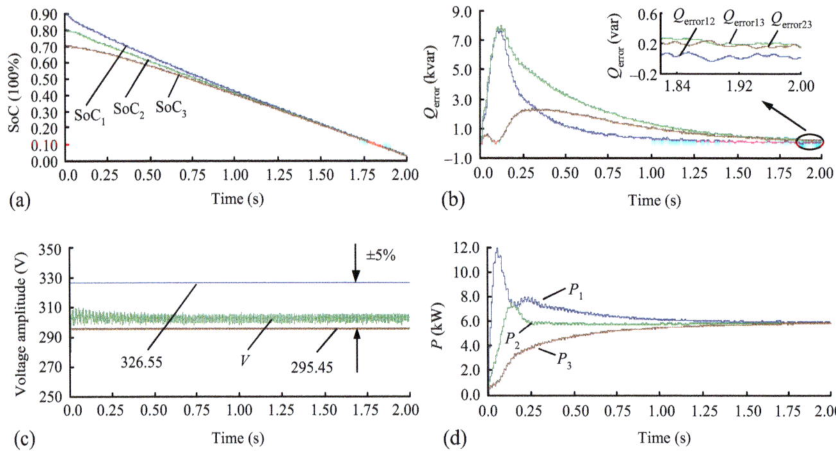

Figure 6.28 Waveforms when $SoC_{01} > SoC_{02} > SoC_{03}$: (a) SoC of the three paralleled DESUs; (b) the reactive power errors; (c) magnitude of the PCC voltage and (d) the active power errors

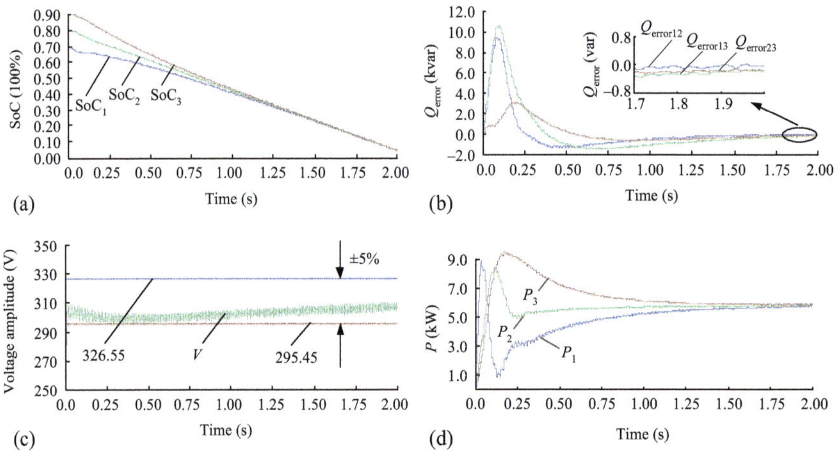

Figure 6.29 Waveforms when $SoC_{01} < SoC_{02} < SoC_{03}$: (a) SoC of the three paralleled DESUs; (b) the reactive power errors; (c) magnitude of the PCC voltage and (d) the active power errors

6.2.5 Parameter design for a centralized hierarchical control for AC nanogrid

Voltage and frequency deviations at PCC would be existed in paralleled inverters by using droop control, because droop control is a kind of proportional control. And voltage and frequency deviations at PCC are affected by droop coefficients, and voltage and frequency deviations at PCC could be decreased by adjusting droop coefficients. But, the stability of the system is related to the variable coefficients. Based on this, an optimization control scheme based on droop is proposed in Figure 6.30 [26].

Objective function optimization is an effective method to find the minimum voltage and frequency deviations at PCC. And one of the important targets in this system is voltage and frequency deviations at PCC. Therefore, considering the optimization objective in this system, a comprehensive objective function about power quality is proposed in the next:

$$F = \begin{cases} \min(\Delta V_{\mathrm{PCC}}) \\ \min(\Delta f) \end{cases} \tag{6.30}$$

In order to find the optimum solution of the multi-objective function in (6.30), weighted-sum method is applied to solve this problem. The new expression based on (6.30) is as follows. And the coefficients in (6.30) could be derived by analytic hierarchy process method, which $\alpha_1 = 0.7$ and $\alpha_2 = 0.3$:

$$F = \alpha_1 \cdot \Delta f + \alpha_2 \cdot \Delta V_{\mathrm{PCC}} \tag{6.31}$$

Under droop control, the inverters which supply a constant impedance load could be considered as a single inverter when the output power of inverters is well shared. The equivalent circuit is shown in Figure 6.31.

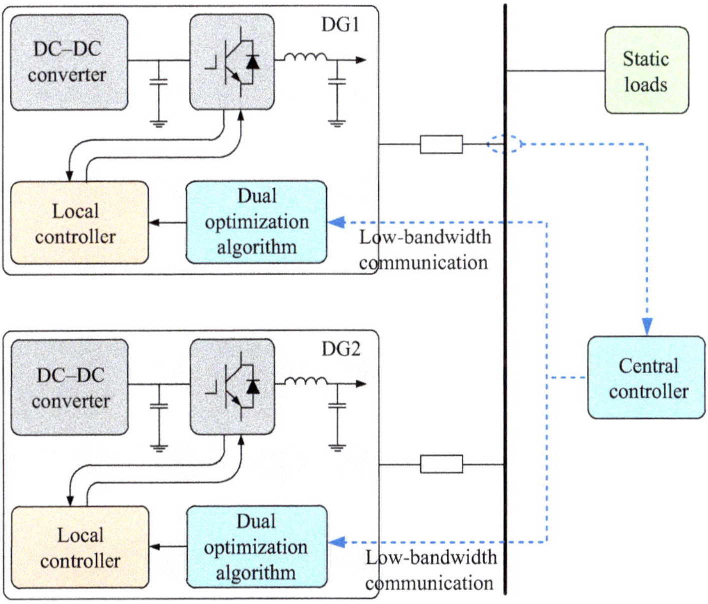

Figure 6.30 Nanogrid structure based on optimization control scheme

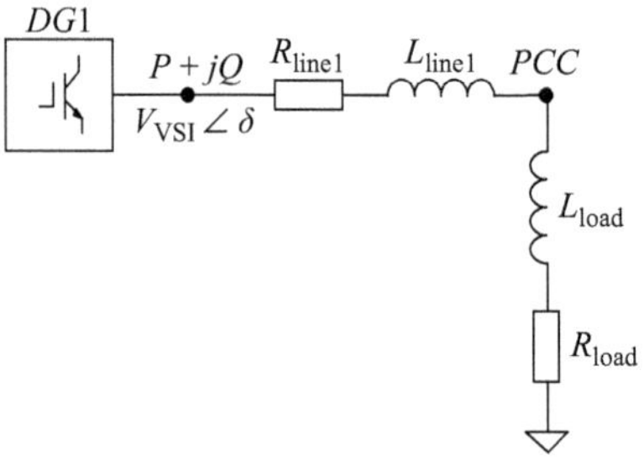

Figure 6.31 Single inverter with a constant impedance load

Based on circuitous philosophy, the apparent power of a single inverter in Figure 6.31 could be deduced in the following:

$$S = P + jQ$$
$$= \frac{V_{VSI}^2 \cdot (R_{line1} + R_{load})}{(R_{line1} + R_{load})^2 + (X_{line1} + X_{load})^2} + j \cdot \frac{V_{VSI}^2 \cdot (X_{line1} + X_{load})}{(R_{line1} + R_{load})^2 + (X_{line1} + X_{load})^2} \tag{6.32}$$

$V_{VSI}\angle\delta$ is output voltage of the inverter, $Z_{line1} = R_{line1} + jX_{line1}$ is line impedance, $Z_{load} = R_{load} + jX_{load}$ is load impedance.

Therefore, the output active and reactive power could be written in the next:

$$P = \frac{1}{3 \cdot k_p^2 \cdot (R_{line1} + R_{load})} \cdot \left(A_1 + A_2 - \sqrt{A_2^2 + 2A_1A_2}\right) \tag{6.33}$$

$$Q = \frac{(X_{line1} + X_{load})}{3 \cdot {k_p}^2 \cdot (R_{line1} + R_{load})^2} \left(A_1 + A_2 - \sqrt{A_2^2 + 2A_1A_2}\right) \tag{6.34}$$

where the coefficients A_1 and A_2 are

$$\begin{cases} A_1 = 3 \cdot k_p \cdot (V_n + k_p P_n) \cdot (R_{line1} + R_{load}) \\ A_2 = (R_{line1} + R_{load})^2 + (X_{line1} + X_{load})^2 \end{cases} \tag{6.35}$$

Combining (6.33) and (6.34) in droop control equation respectively, the characteristics of droop control considering a constant impedance load could be rewritten in the following:

$$V = V_n + k_p P_n - \frac{1}{3 \cdot k_p \cdot (R_{line1} + R_{load})} \cdot \left(A_1 + A_2 - \sqrt{A_2^2 + 2A_1A_2}\right) \tag{6.36}$$

$$f = f_n + k_q Q_n - \frac{k_q \cdot (X_{line1} + X_{load})}{3 \cdot {k_p}^2 \cdot (R_{line1} + R_{load})^2} \cdot \left(A_1 + A_2 - \sqrt{A_2^2 + 2A_1A_2} \right) \tag{6.37}$$

Based on formula (6.37), $\Delta V_{PCC} = \Delta V_{DG} + \Delta V_{line1}$. And the objective function could be written in detail [26]:

$$\begin{aligned} F &= 0.7 \cdot \Delta f + 0.3 \cdot \Delta V_{PCC} \\ &= 0.7 \times \left[k_q Q_n - \frac{k_q \cdot (X_{line1} + X_{load})}{3 \cdot {k_p}^2 \cdot (R_{line1} + R_{load})^2} \times \left(A_1 + A_2 - \sqrt{A_2^2 + 2A_1A_2} \right) \right] \\ &+ 0.3 \times \left[k_p P_n - \frac{1}{3 \cdot k_p \cdot (R_{line1} + R_{load})} \times \left(A_1 + A_2 - \sqrt{A_2^2 + 2A_1A_2} \right) \right. \\ &\left. + \frac{P \cdot R_{line1} + Q \cdot X_{line1}}{V_n} \right] \end{aligned} \tag{6.38}$$

It can be seen that the PCC voltage and frequency deviations could be decreased by adjusting droop coefficients, voltage and frequency references. For this optimization, a boundary of system stability for the objective function is necessary. The open loop transfer function of the system is shown as follows [26]:

$$G(s) = \frac{C}{s^3 + As^2 + Bs} \tag{6.39}$$

where

$$\begin{cases} A = \dfrac{1}{T} \left(2 + k_p \dfrac{\partial P}{\partial V} \right) \\ B = \dfrac{1}{T^2} \left(1 + k_p \dfrac{\partial P}{\partial V} + 2\pi T k_q \dfrac{\partial Q}{\partial \delta} \right) \\ C = \dfrac{2\pi k_q}{T^2} \left(\dfrac{\partial Q}{\partial \delta} - k_p \dfrac{\partial Q}{\partial V} \dfrac{\partial P}{\partial \delta} + k_p \dfrac{\partial P}{\partial V} \dfrac{\partial Q}{\partial \delta} \right) \end{cases} \tag{6.40}$$

When droop coefficient k_p varies from 0.0001 to 0.1 and k_q varies from -0.0001 to -0.1, the root locus of formula (6.39) could be plotted in Figures 6.32 and 6.33.

Seen from the root locus, the system would run into an unstable state with larger droop coefficients. And the system would remain stable when the droop coefficients are smaller, but the dynamic response and the power-sharing accuracy of the system would be decreased. Therefore, there is a trade-off between the dynamic response and the system stability. So, stability margin of the system is considered [26].

Based on (6.39), the stability margin of the system could be deduced in the following:

$$\gamma = F(k_p, k_q) = 180^\circ + \angle G(j\omega_\varphi) \tag{6.41}$$

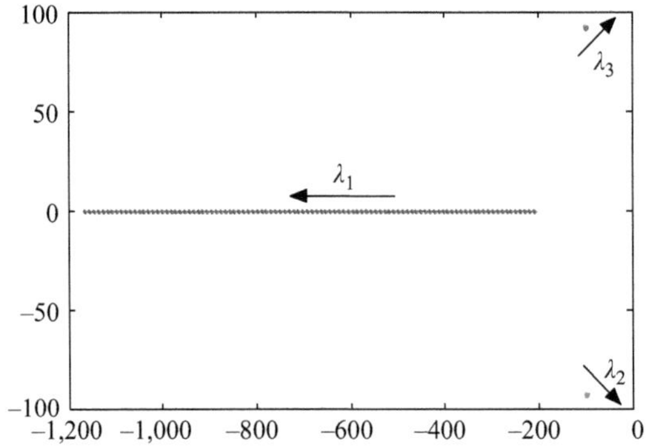

Figure 6.32 Root locus diagrams when k_p varies

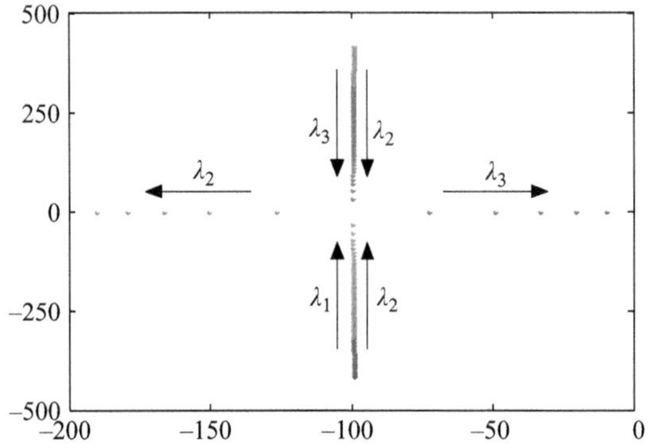

Figure 6.33 Root locus diagrams when k_q varies

Based on (6.41), the surface related to the phase margin, k_p and k_q, could be drawn in Figure 6.34.

The intersection of two planes represents the relationship between k_p and k_q when the phase margin of droop control equals 45 degrees. From (6.39), it can be seen that the simplified system is third order system, and the high order system can be approximated as a second order system in order to simplification. When the phase margin is 45 degrees, the system has the optimum damping. Therefore, the system would remain in a good performance when droop coefficients k_p and k_q satisfy the formula (6.41).

From the above analysis, a dual optimization control scheme considering droop coefficients, voltage and frequency references is proposed as shown in Figure 6.35.

Through primary optimization, the system can be promised to remain in a good stability margin while decreasing voltage and frequency deviations at PCC. When the load is heavy, primary optimization cannot decrease PCC voltage and frequency deviations completely because of proportional droop control and limitations

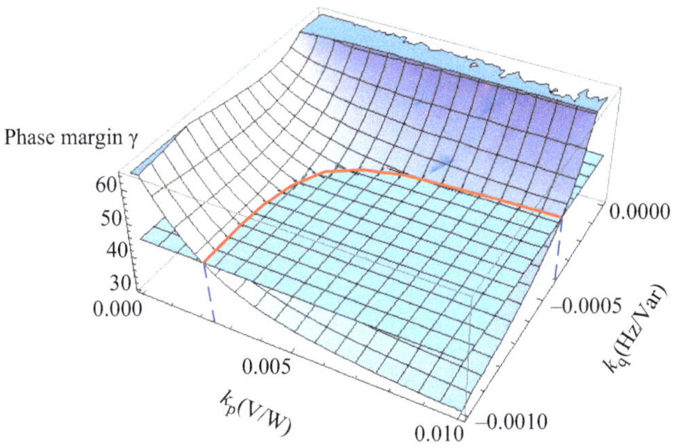

Figure 6.34 Relation between k_p and k_q

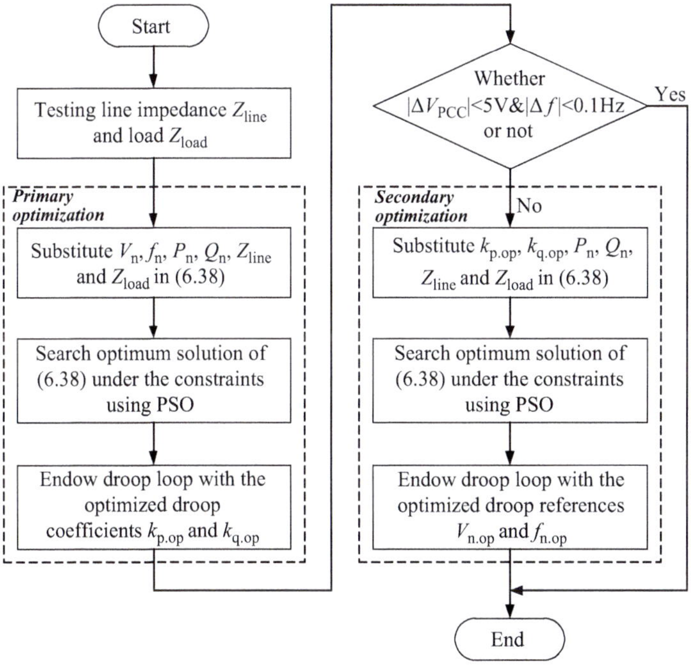

Figure 6.35 Flow chart of optimization control strategy

of stability margin of the system. Therefore, it is necessary to adopt secondary optimization to reduce PCC voltage and frequency deviations further. Following the proposed dual optimization control scheme, voltage and frequency deviations at PCC are close to zero in a good stability margin [26].

Nanogrid composed of two paralleled inverters is built, the line impedance is $Z_1 = Z_2 = 0.642 + 0.083j$, the load is $Z_{load} = 17.678 + 17.678j$. Droop coefficients are $k_p = 0.025$, $k_q = -0.005$ and optimized droop coefficients are $k_{p.op} = 0.0018576$, $k_{q.op} = -0.000883288$, respectively. And the output active power and reactive power are shown in Figure 6.36 and Figure 6.37.

From the simulation, it could be seen that the system dynamic and static performance is greatly improved by adopting the optimized droop coefficients.

When the load becomes heavy which is $Z_{load} = 14.142+14.142j$. The optimized droop coefficients ($k_{p.op}$, $k_{q.op}$) and optimized droop control references ($V_{n.op}$, $f_{n.op}$) could be achieved by the flow chart in Figure 6.35. And the PCC voltage deviation and PCC frequency deviation are shown in Figure 6.38 and Figure 6.39.

And droop control loop adopts $k_p = 0.01$ and $k_q = -0.002$ at 0 s, the optimized droop coefficients $k_{p.op} = 0.0025325$ and $k_{q.op} = -0.000580659$ at 0.5 s, and the optimized droop control references $V_{n.op} = 316.3$ V, $f_{n.op} = 49.66$ Hz at 1 s, respectively.

It can be seen from the simulation results that the PCC voltage and frequency deviations is 8.5 V and 0.9 Hz when using original droop coefficients, and the PCC voltage and frequency deviations drop to 5.5 V and 0.29 Hz when using the primary optimized droop coefficients. After using the secondary optimized droop references, the

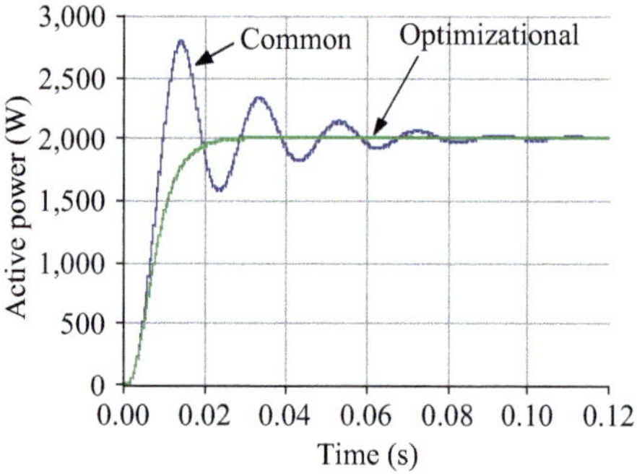

Figure 6.36 Output active power

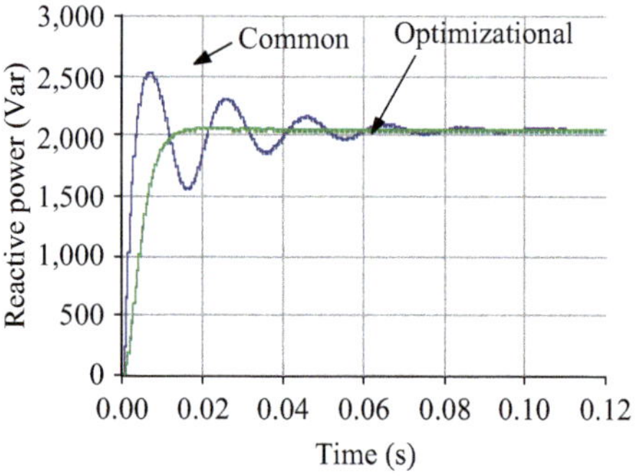

Figure 6.37 Output reactive power

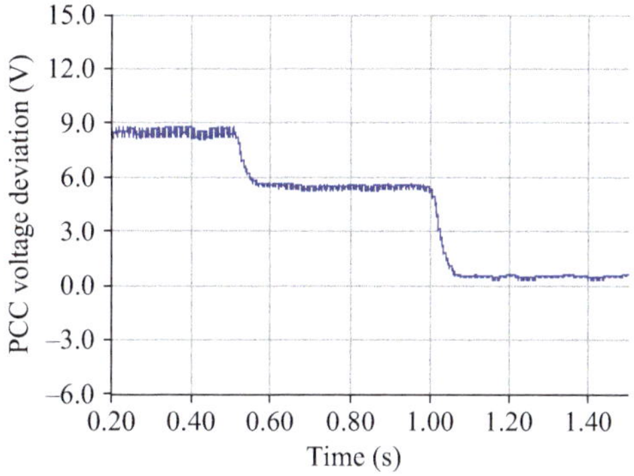

Figure 6.38 PCC voltage deviation

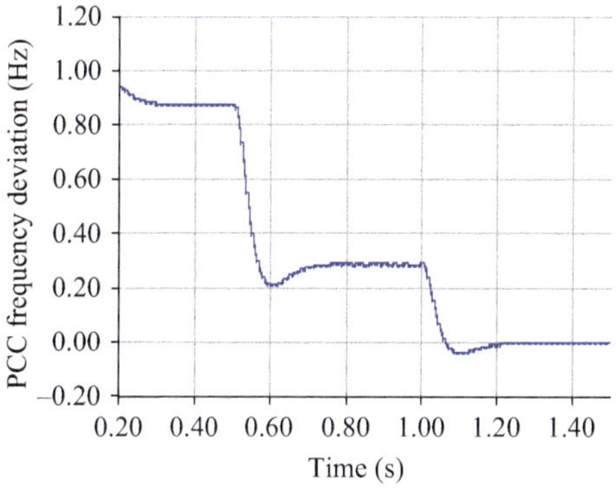

Figure 6.39 PCC frequency deviation

PCC voltage and frequency deviations decrease close to zero. Voltage and frequency deviations at PCC could be decreased by applying the proposed optimization control scheme with a good stability and performance (Figures 6.38 and 6.39).

6.3 Case study

6.3.1 Large-scaled intelligent nanogrid

A demonstration for intelligent nanogrid was established in Nantong, Jiangsu Province, China. And the nanogrid is composed of 100 kW PV, three 10 kW wind turbines, 120 kW diesel generator, 100 kW/250 kW h energy storage, active power filter and dynamic voltage regulator. The structure of the intelligent nanogrid is shown in Figure 6.40.

The penetration rate of the intermittent power supply reached 52%. The hybrid system contained wind turbines, PV and energy storage units are always in operation mode. And diesel generator would be operated in an extremely harsh environment. The switching time between the distributed system and the diesel engine is about

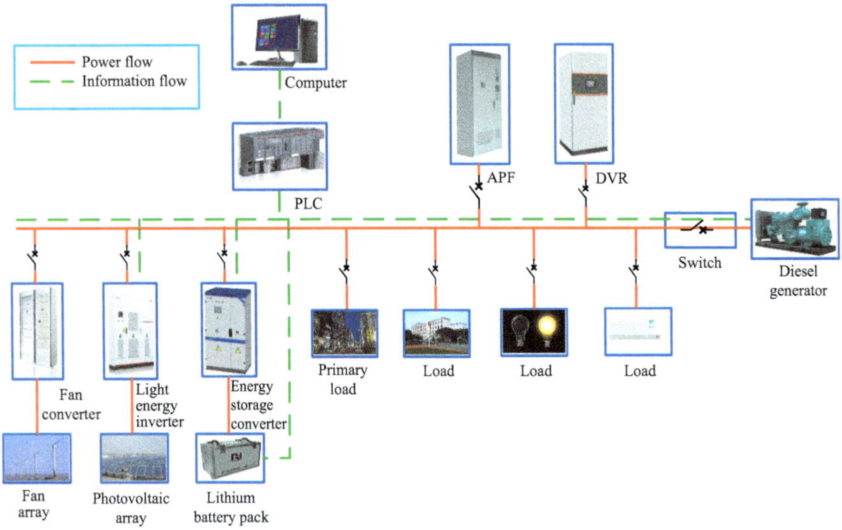

Figure 6.40 The structure of the intelligent nanogrid

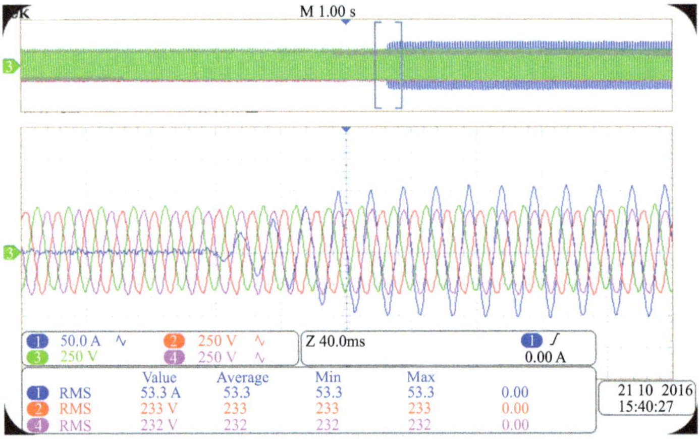

Figure 6.41 Load for energy storage converter ranges from 0 W to 50 kW

25 ms. The voltage fluctuation range of the hybrid distributed system is −10% to 7% and the frequency fluctuation is ±1%. Complete energy management and control strategy are applied in this system and some experimental results are carried out.

Experiments for dynamic response of energy-storage converter, three-phase with unbalance loads and switching between the distributed system and the diesel engine are presented as follows.

From Figure 6.41, it can be seen that the voltage and frequency remain stable when the load changes. Moreover, off-grid switching process in Figure 6.42 transfers smoothly between each state.

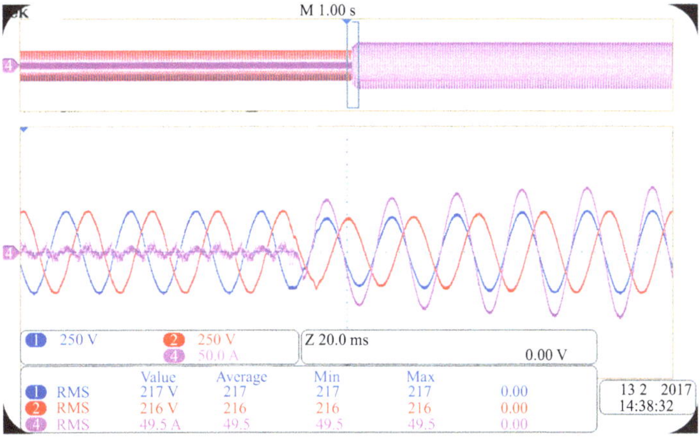

Figure 6.42 Off-grid switching process

From Figure 6.43, it can be seen that the output power in load side is smooth because of the coordinated control employed in each distributed unit. Therefore, the output power of the intelligent nanogrid could remain smooth by applying the effective energy-management strategy. And the intelligent nanogrid with complete control strategy could be installed in house, department, industrial campus, island, etc.

6.3.2 Small-scaled intelligent nanogrid

A household demonstration for intelligent nanogrid was established in our YRET Lab, Yanshan University, Hebei Province, China. The nanogrid is composed of two 3 kW PV and two 3 kW/100 kWh energy storage. This intelligent nanogrid is usually installed in the house or department. The structure of small-scaled nanogrid is shown in Figure 6.44.

In this small-scaled nanogrid, PQ control is applied in PV units to supply enough power to customers, and droop control is used in energy storage units to realize plug and play. When the solar is enough, PV units supply enough energy to the load, and the rest is stored in the energy storage units. The energy storage units would supply power to the customers when the light condition is poor or the loads are heavy. The operating waveform of the system is shown in Figure 6.45.

From Figure 6.45, it can be seen that the loads are supplied by the PV units and energy storage units proportionally. And the output power of energy storage units is shared equally by adopting droop control.

6.3.3 Nanogrid installed in remote villages

Villages in mountainous area are connected to the utility grid through long-distance transmission. In order to reduce the power transferring losses in transmission, a local nanogrid supporting local load as shown in Figure 6.46 is located in western mountainous area, Qinghai Province, China.

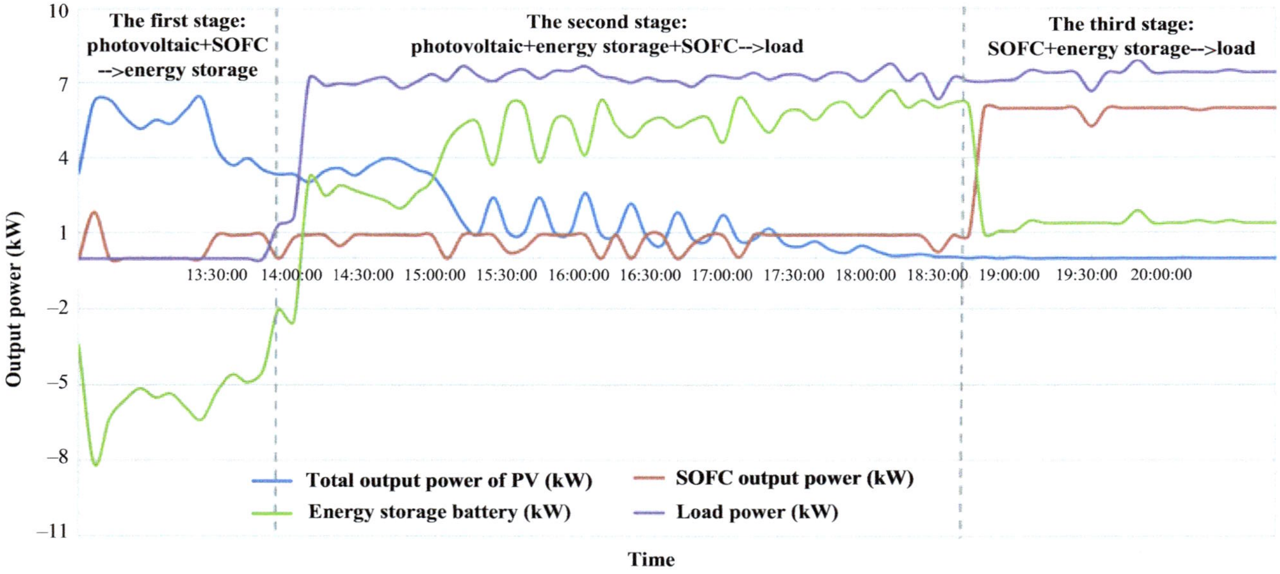

Figure 6.43 Energy management for each unit

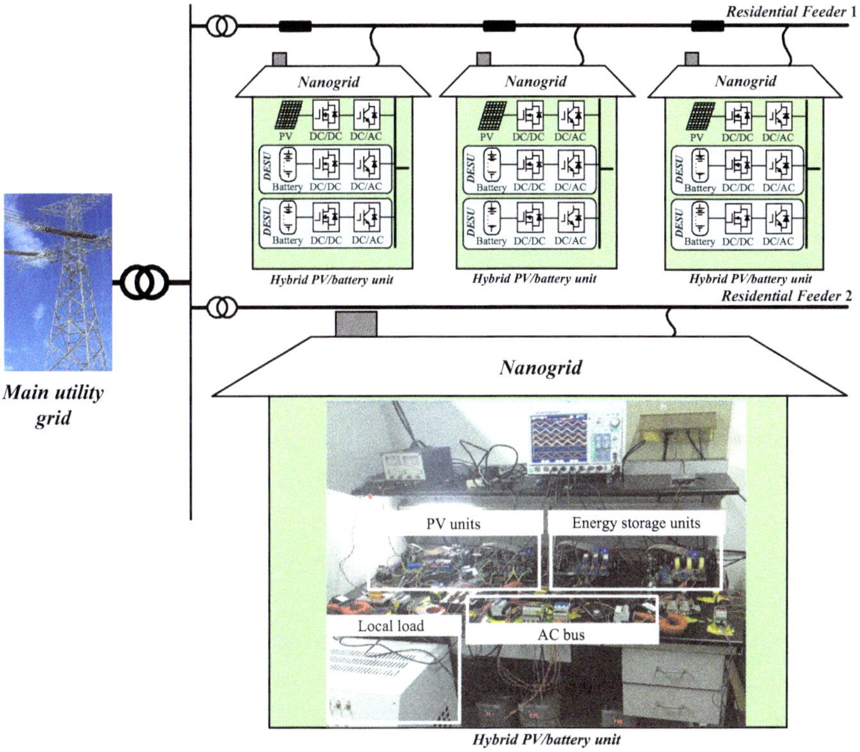

Figure 6.44 The structure of small-scaled nanogrid applied in household

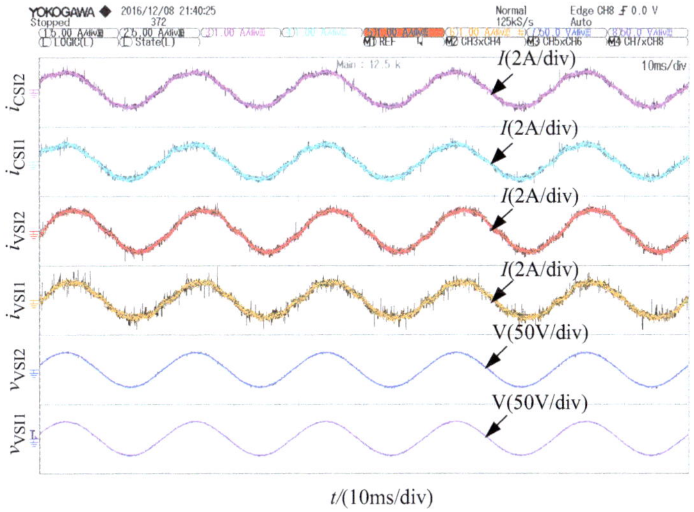

Figure 6.45 The operating waveform of the system

Figure 6.46 Nanogrid located in western mountainous area, Qinghai Province, China

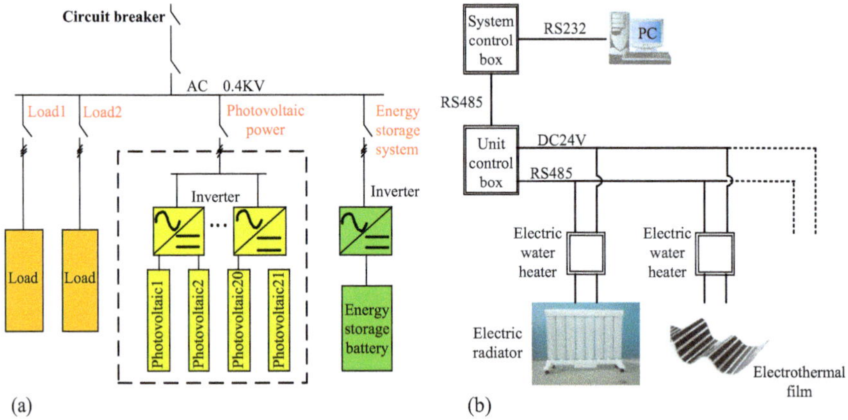

Figure 6.47 The inner structure of household: (a) renewable energy system part and (b) the heating part

There are 1,000 residents in this high-altitude village, and the climate of this western mountainous area remains cold. Considering that heating by coal burning would damage the environment, a nanogrid based on multi-energy comprehensive system is installed in this village. The inner structure of household is shown in the following.

In 50 kW PV/DESUs hybrid system shown in Figure 6.47(a), the important loads can be guaranteed to work continuously for a long time without the supporting of the utility grid. And the necessity of heating could be transferred by the PV/DESUs hybrid system. Therefore, the community energy cost could be reduced by the combined application of PV/DESUs hybrid system and heating system.

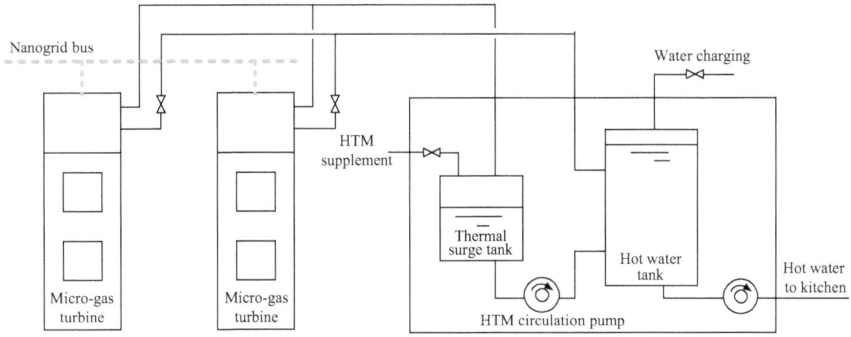

Figure 6.48 Nanogrid based on cogeneration system

6.3.4 Nanogrid based on cogeneration system

Considering the limitation of the existing energy-storage method, cogeneration system attracted more and more attention, and cogeneration system has a great significance for the energy crisis. A typical structure of nanogrid based on cogeneration system installed in Beijing, China is shown in the following.

The system in Figure 6.48 consists of two 65 kW microturbines. A turbine-driven high-speed generator is coupled with digital power electronics to produce high-quality electrical power. The output power of the microturbine could be transferred to the utility grid or connected to other renewable generation sources. The microturbine can act as a stand-alone generator for standby, backup or remote off-grid power in multiple DG units based nanogrid. Varieties of hydrocarbon gases could be used in microturbine. Dry, oxygen-rich waste gas is produced by microturbine with ultra-low emissions. And the electricity and waste heat can provide greater energy costs savings.

6.4 Conclusion

In this chapter, a framework for nanogrid and its energy management is provided. Nanogrid can be seen as smaller and technologically simpler islanding microgrids, typically serving a single building or a single load. In order to satisfy different applications, hybrid AC/DC nanogrid, AC nanogrid and DC nanogrid are widely used. Moreover, power management is a crucial issue for nanogrid, and decentralized control strategies for a simple PV/battery system, multiple battery-based nanogrid and multiple energy storage units based nanogrid are demonstrated in detail in this chapter.

In the future, more intelligent energy-management strategy considering economical efficiency, environmentally friendly and friendly human–computer interaction would be applied in nanogrid. Nanogrid with smarter and easier energy management control would become a smart dispatch unit in future high-integrated renewable grid.

References

[1] Boroyevich D., Cvetkovic I., Burgos R., *et al.* Intergrid: A Future Electronic Energy Network?. IEEE Journal of Emerging & Selected Topics in Power Electronics, 2013, 1(3):127–138.

[2] Cvetkovic I., Thacker T., Dong D., *et al.* "Future home uninterruptible renewable energy system with vehicle-to-grid technology," in Proc. IEEE ECCE, Sep. 2009, pp. 2675–2681.

[3] Rocky T. H., Islam R., and Saha U. K. "Nano solar grid (NSG): A solution for rural market power crisis," in International Conference on Green Energy and Technology. IEEE, 2014, pp. 14–17.

[4] Werth A., Kitamura N., and Tanaka K. Conceptual Study for Open Energy Systems: Distributed Energy Network Using Interconnected DC Nanogrids. IEEE Transactions on Smart Grid, 2015, 6(4):1621–1630.

[5] Castle O. D., and Shahat A. E. "Single-input – multi-output (SIMO) converter for nano-grids applications," in SoutheastCon. IEEE, 2017, pp. 1–5.

[6] Sajeeb M. M. H., Rahman A., and Arif S. "Feasibility analysis of solar DC Nano grid for off grid rural Bangladesh," in International Conference on Green Energy and Technology. IEEE, 2015, pp. 1–5.

[7] Thacker T., Burgos R., Wang F., and Boroyevich D. "Single-phase islanding detection based on phase-locked loop stability," in Proc. IEEE ECCE, Sep. 2009, pp. 3371–3377.

[8] Patterson B. T. DC, Come Home: DC Microgrids and the Birth of the 'Enernet'. IEEE Power & Energy Magazine, 2012, 10(6):60–69.

[9] Salomonsson D. and Sannino A. Low-Voltage DC Distribution System for Commercial Power Systems with Sensitive Electronic Loads. IEEE Transactions on Power Delivery, 2007, 22(3):1620–1627.

[10] Aoki T., Yamasaki M., Takeda T., Tanaka T., Harada H., and Nakamura K. "Guidelines for power-supply systems for datacom equipment in NTT," in Proc. 24th Annu. INTELEC, 2002, pp. 134–139.

[11] AlLee G. and Tschudi W. Edison Redux: 380 Vdc Brings Reliability and Efficiency to Sustainable Data Centers. IEEE Power & Energy Magazine, 2012, 10(6):50–59.

[12] Stupar A., Friedli T., Minibock J., and Kolar J. W. Towards a 99% Efficient Three-Phase Buck-Type PFC Rectifier for 400 V DC Distribution Systems. IEEE Transactions on Power Electronics, 2012, 27(4):1732–1744.

[13] Sun X., Shen Y., Li W., *et al.* A PWM and PFM Hybrid Modulated Three-Port Converter for a Standalone PV/Battery Power System. IEEE Journal of Emerging & Selected Topics in Power Electronics, 2017, 3(4):984–1000.

[14] Lucia O., Cvetkovic I., Sarnago H., *et al.* Design of Home Appliances for a DC-Based Nanogrid System: An Induction Range Study Case. IEEE Journal of Emerging & Selected Topics in Power Electronics, 2013, 1(4):315–326.

[15] Chandrasena R. P. S., Shahnia F., Rajakaruna S., *et al.* Dynamic Operation and Control of a Hybrid Nanogrid System for Future Community Houses. Generation Transmission & Distribution IET, 2015, 9(11):1168–1178.

[16] Urtasun A., and Lu D. D. Control of a Single-Switch Two-Input Buck Converter for MPPT of Two PV Strings. IEEE Transactions on Industrial Electronics, 2015, 62(11):7051–7060.
[17] Wang B., Yuan Y., Zhou Y., *et al.* "Buck/boost bidirectional converter TCM control without zero-crossing detection," in Power Electronics and Motion Control Conference. IEEE, 2016, pp. 3073–3078.
[18] Pires V. F., Foito D., and Cordeiro A. "Bidirectional boost/buck quadratic converter for distributed generation systems with electrochemical storage systems," in IEEE International Conference on Renewable Energy Research and Applications. IEEE, 2017.
[19] Pires V. F., Foito D., and Cordeiro A. A DC-DC Converter with Quadratic Gain and Bidirectional Capability for Batteries/Supercapacitors. IEEE Transactions on Industry Applications, 2018, 10(9):274–285.
[20] Cecati C., Khalid H. A., Tinari M., *et al.* DC Nanogrid for Renewable Sources with Modular DC/DC LLC Converter Building Block. IET Power Electronics, 2017, 10(5):536–544.
[21] Yaqoob M., Loo K. H., and Lai Y. M. Fully Soft-Switched Dual-Active-Bridge Converter with Switched-Impedance-Based Power Control. IEEE Transactions on Power Electronics, 2018, 10(9):9267–9281.
[22] Shen Y., Sun X., Li W., *et al.* A Modified Dual Active Bridge Converter with Hybrid Phase-Shift Control for Wide Input Voltage Range. IEEE Transactions on Power Electronics, 2016, 31(10):6884–6900.
[23] Sun X., Wu X., Shen Y., *et al.* A Current-Fed Isolated Bidirectional DC-DC Converter. IEEE Transactions on Power Electronics, 2017, 10(9):6882–6895.
[24] Sun X., Liu B., Cai Y., *et al.* Frequency-Based Power Management for Photovoltaic/Battery/Fuel Cell-Electrolyser Stand-Alone Microgrid. IET Power Electronics, 2016, 9(13):2602–2610.
[25] Sun X., Hao Y., Wu Q., *et al.* A Multifunctional and Wireless Droop Control for Distributed Energy Storage Units in Islanded AC Microgrid Applications. IEEE Transactions on Power Electronics, 2017, 32(1):736–751.
[26] Zhao W., Qi L., Sun X., *et al.* "Research on dual optimization control scheme considering voltage and frequency in islanding microgrid," in Power Electronics and Motion Control Conference. IEEE, 2016, pp. 3163–3168.

Chapter 7

Sizing of microgrid components

Ghulam Mohy-ud-din[1], Kashem M. Muttaqi[1], and Danny Sutanto[1]

A microgrid (MG) is a distinct energy system consisting of distributed energy resources (DERs) and loads having the ability to operate in parallel with, or independently from, the main power grid. MGs, which were initially introduced to ensure smooth operation and control of DERs in distribution networks, offer unprecedented economic and reliability benefits to electricity consumers with minimal carbon emission. These benefits, however, must be analysed and compared with the capital investment cost of the MG to ensure a complete return on investment and to justify the MG deployment. The biggest obstacle for the widespread and rapid deployment of MGs is the high capital investment cost of MGs. A true assessment of MGs economic benefits is a challenging task due to the significant uncertainties involved in the assessment. These uncertainties may include the intermittency of the renewable generation, the varying states of charge (SoC) of battery energy storage system (BESS), the uncertain demands, the varying market price, the probability of the MG islanding, the level of developer's risk-aversion and the unpredictably of the user preferences in the smart load management system. Moreover, some of the assessment metrics, such as the measure of reliability improvements are difficult to comprehend for consumers when represented in terms of the supply availability. Thus, efficient and optimum planning models are required to ensure the economic feasibility of MG deployments and to justify the investments based on cost-to-profit analysis under uncertain conditions. This chapter demonstrates a detailed model for the optimum sizing of MG components under the uncertainties involved in the system. The proposed model is validated with the simulation of several case studies conducted on a system depicting a similar MG in a medium-voltage (MV)-distribution system derived from electricity network of a power utility in New South Wales, Australia. The results from the case studies demonstrate the efficacy of the proposed model for the optimum sizing of the MG components to justify the MG deployment.

7.1 Microgrid components

MGs can operate autonomously in interaction with the power grid as well as in islanded operation. Therefore, the components of MG must meet the requirements

[1]Australian Power Quality and Reliability Centre, School of Electrical, Computer and Telecommunications Engineering, University of Wollongong, Wollongong, Australia

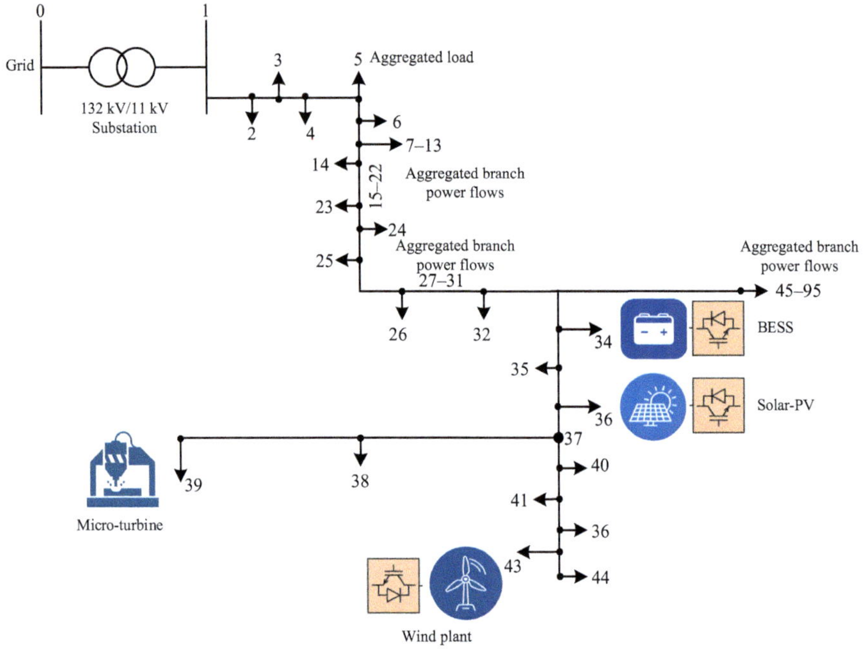

Figure 7.1 Topology of microgrid-embedded distribution system

of either modes of operation. In order to have a perfect evaluation of a MG, the factors such as the size, purpose, type of DERs, type of loads, interaction with the power grid, geographical location and spatial diversity of the network must be considered. The MG shown in Figure 7.1 is used as the study system. It is basically a community-based MG. The major components of a MG in general can be listed as follows:

- Distributed energy sources
- Storage devices
- Loads
- Control systems
- Grid interconnectors

7.2 Microgrid sizing and profit maximization

For the widespread deployment of MGs, the different aspects of the optimal sizing of the components of the MGs, such as the cost minimization, emission reduction and the profit maximization, is studied in this chapter. The size of a component can be defined by the generation capacity, peak demand, customers being served or the geographical coverage. The MGs may utilize the existing power grid infrastructure or can be physically isolated from the grid. This dynamic exchange of services within a MG and across the points of interconnection among MGs varies. This variation along with the unprecedented uncertainty associated with the renewables is the challenge in determining the optimum size of the MG components.

The recent advancements in the MGs have opened multiple venues for the future network and its energy distribution. The future MGs are envisioned as playing a greater role in the smart power grids of the developed countries, emerging economies and the evolving rural communities. Utilities across the world are highly interested in the MGs as one of the key sources to address their sustainability challenges. MGs, being locally operated within the framework of power grid, has the ability to transform the urban power grids, allowing local generation through DERs, thus resulting in higher efficiency and ecological sustainability. The intensive industrial energy consumers face the challenges, while seeking alternative sources of energy against the fluctuating fuel prices and squeezed commodity prices. They are looking for a potential solution in the form of MGs. For example, the zinc mine named *El Toqui*, that is located in a remote area of *southern Chile* has built a MG with wind, diesel and hydro, to resolve their stability problems and to reduce the energy cost [1]. The remote communities of *Canada* mostly rely on generation from diesel for their electricity. Many of these communities are thinking of implementing renewable-based MGs to reduce the expensive diesel power production. One of the remote area mines in *Canada* named *Diavik Diamond* has an operational wind-diesel MG, which has four wind turbines (WTs) of 2.3 MW, each with a total capacity of 9.2 MW, aiming to supply 17 GWh of energy per annum to decrease the dependence on the consumption of diesel and limit the resulting carbon footprint [2]. The US defence sector is also eager to achieve higher reliability and resilience for the power grid by installing the MGs. These projects include *Naval Support Facility Dahlgren, Fort Detrick, Marine Corps Air Ground Combat Centre*, and *Twenty-nine Palms* [3]. The *Hydro Tasmania*'s projects on the *King and Flinders Islands* in Australia have shown the effectiveness of renewables to reduce the diesel use in pristine remote environments while providing a reliable power for the residents and businesses [4]. The universities such as *Princeton University*, *Illinois Institute of Technology* and *British Columbia Institute of Technology* have operational applications of on-campus MGs for improving their energy distribution at the campuses [5,6].

To justify the widespread deployment and successful implementation of the MGs on industrial scale, appropriate MG sizing and profit maximization models need to be established. They involve the technical attributes of the generation resources, network and the consumer dynamics requiring a more comprehensive picture of the cost and the services that MG can provide. The expected benefits include increased resiliency to exceptional events and contingencies, reduction in the cost of energy supply from the power grid, decrease in the transmission system losses through the power grid and the reduction of the carbon footprint. The evaluation of these benefits against the investments depends on the economic efficiency of the design. Furthermore, the cost-benefit analysis provides the insight of the design models. Many of these models are detailed in the literature [7–17]. Moreover, the evolution of such MG models can also be considered from the perspective of distribution automation strategies [18,19]. The profit of the MG installation from the perspective of the regulator and policymaker is also outlined in the literature [20–26]. However, the effect of energy market price volatility and pricing strategies remains vital in the MG planning and considered as an important aspect that needs a quantification-based framework for the evaluation of the key benefits and their relationship with the entities. This chapter presents a quantification-based framework for the evaluation of the MG under the cost-to-benefit analysis, while considering the uncertainties involved in the system.

The MG installation and operation are majorly justified by the profit accrued to the MG stakeholders. Hence, the optimization strategies play key role in developing the business case for the successful MG deployments. This procedure involves the following fundamental studies as illustrated in Figure 7.2:

- The forecast of the load demand and the weather profiles for the renewable energy resources.
- The mathematical modelling of the DERs and the BESSs considering the stochastic and dynamic nature of these generation resources with respect to time.

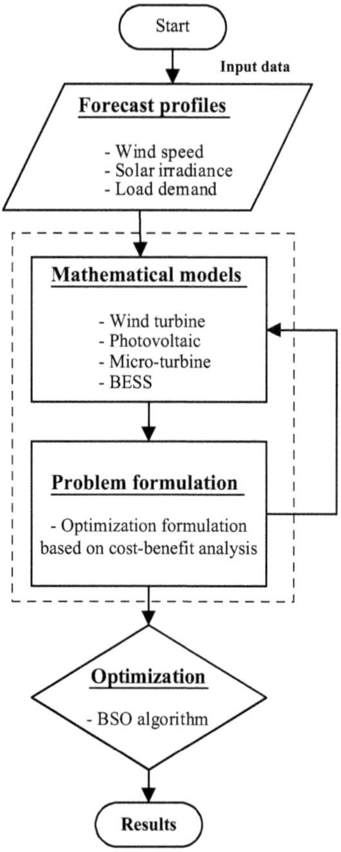

Figure 7.2 Procedure for the sizing of microgrid components

- Incorporating the mathematical models of DERs in the cost-to-benefit analysis for a generalized MG under evaluation.
- Optimization formulation of the model for the optimum solution, such as optimum size of the MG components subject, to the maximum profit.
- Optimization solution using state-of-the-art techniques and algorithms.
- Evaluation of the results.

The forecast of the load demand, wind speed and solar irradiation is very crucial for the planning of a MG, because the decisions of the sizing of MG components depend on the load demand and the generated wind and solar power profiles. The results of the load, wind speed and solar irradiation forecasts provide a reference for the regional power generation, design of electricity network, power balancing and the sharing of surplus power with the power grid. The major factors involved in the forecast are the collection of historical data and effective forecasting techniques. The forecasting time period can be long-term, medium-term and short-term depending upon the case. The medium and long-term forecast are used to plan the generation portfolio for capacity increase by enforcing the existing infrastructure with the plan of adding new assets. The short-term forecast is used for scheduling of the generation and reserves. The MG component sizing and profit maximization basically involve the medium and long-term forecast. The load forecasting is divided into total or spatial load forecast. The total load forecast involves the estimation of total energy consumption of the planning area, while the spatial forecast involves spatially diversified geographical locations of future-generation sources and loads. The common time-series models, such as ARIMA and SARIMA model, are extensively used for long-term and medium-term forecasting [27]. Similarly, the machine learning [28] and artificial neural-network-based forecasting [29] techniques have been utilized for enhanced forecasting. Since, the load forecasting is itself a wide area, it is not included in the scope of this chapter. In this chapter, given forecasts are utilized by the mathematical models of DERs for the intended model of MG component sizing.

The mathematical modelling of DERs determines the authenticity of the evaluation of the MG. The modelling of stochastic nature of renewable energy sources is crucial, as the deterministic approximations may result in false estimates, which may lead to the failure of the MG's profitability in the future. Therefore, probabilistic wind and solar PV power generation models are presented in this chapter [30]. The operation of the BESS is dynamic in nature, depending on the time of the day (peak and non-peak hours), energy price (charging and discharging price), available SoC and limitations due to battery degradation. By considering all of these factors, a dynamic BESS mathematical model is used in this study [31]. Micro-turbines (MTs) are robust as they have conventional steam turbine generators utilizing natural gas as a fuel. However, the generators have ramping limitations and prohibited operation zones (POZs) within the operating limits. Therefore, along with the generation cost, these constraints are also considered in this work.

This optimization problem is formulated as a constrained mixed integer nonlinear programming (MINLP) for higher accuracy of the results, wherein the total cost will be minimized and the profit from self-consumption and the sold power to the grid is maximized by optimally sizing the various MG components. The system states on an hourly basis are incorporated with respective time series data sets. This type of problem is normally a nondeterministic polynomial-time hard optimization problem with increased complexities imposed by power generation through renewable generation units, stochastic load demand, numerous system states and long-term planning horizon. In this chapter, a solution algorithm with the selective use of backtracking search optimization (BSO) algorithm is utilized for obtaining the optimal solution with significantly enhanced modelling task and reduced computational effort [32,33].

7.3 Models of distributed energy resources

The components of the MG are modelled and analysed from an electrical perspective, particularly, the distributed sources consisting of photovoltaic (PV) systems, WTs, BESSs, MTs and the cluster of electrical loads. In order to model the uncertainty associated with the renewable energy resources, such as wind and solar, probabilistic models for wind and PV power are presented in this chapter. The respective mathematical models are summarized in the following section.

7.3.1 Probabilistic wind power output model

The wind speed is the stochastic variable that determines the amount of wind power generated by the WT. Moreover, the type of the WT and its parameters of the power performance curve also play an important role in calculating the output wind power at a selected site. Since the speed of wind is highly intermittent in nature, it requires a long-term climatological historical data for its forecast prediction. This variability in the speed of wind, $v(\mathrm{m\,s^{-1}})$, is modelled as the Weibull distribution over a time period. The scaling factor, S_w, and the shape factor, K_w, model the characteristics of the WT. The probability density function (PDF) of the wind speed, $f_{PDF}(v)$ is given by (7.1) [30]:

$$f_{PDF}(v) = \left(\frac{K_w}{S_w}\right)\left(\frac{v}{S_w}\right)^{(k_w-1)} e^{-(v/S_w)^{k_w}} \tag{7.1}$$

The shape factor K_w and the scaling factor S_w used in (7.1) are calculated using the mean wind speed v_m and standard deviation σ as follows [30]:

$$K_w = \left(\frac{\sigma}{v_m}\right)^{-1.086} \tag{7.2}$$

$$S_w = \frac{v_m}{\Gamma(1 + (1/k_w))} \tag{7.3}$$

Once the PDF of the wind speed is determined using (7.1) for a given time interval, the wind power output can be estimated by a linear piecewise function of the wind speed given in (7.4). The rated wind speed v_r $(\mathrm{m\,s^{-1}})$, rated wind power $P_{w,r}$ (kW), cut-in wind speed v_i $(\mathrm{m\,s^{-1}})$ and cut-out wind speed v_0 $(\mathrm{m\,s^{-1}})$ are the parameters that govern the output of the WT systems. A typical WT power output curve with the rated wind speed, cut-in wind speed and cut-out wind speed is shown in Figure 7.3.

The rated wind speed is the value of the wind speed, at which the WT is able to generate the rated power output. The cut-in wind speed is the speed at which the WT starts to rotate and consequently generate power output, while the cut-off wind speed is the speed at which the WT is brought to rest in order to avoid damage from very high wind speed. Once the PDF with respect to the Weibull distribution is calculated for a specified time period, the wind power output P_w for different states of time during that time period can be calculated as follows [30]:

$$P_{w,available} = \begin{cases} 0 \leq v \leq v_i, v_0 \leq v \\ P_{w,r} \times \dfrac{(v - v_i)}{(v_r - v_i)} v_i \leq v \leq v_r \\ P_{w,r}\ v_r \leq v \leq v_0 \end{cases} \tag{7.4}$$

The PDF for the wind power output can be estimated by (7.1) and (7.4) by the application of transformation theorem given as in [30]:

$$f_w(P_w) = \begin{cases} \left(\dfrac{K_w l v_i}{C_w}\right)\left(\dfrac{(1+\rho l)v_i}{C_w}\right)^{K_w - 1} exp\left(-\left(\dfrac{(1+\rho l)v_i}{C_w}\right)^{K_w}\right) for\ 0 < P_w < P_{w,r} \\ 1 - exp\left[-\left(\dfrac{v_i}{C_w}\right)^{K_w}\right] + exp\left[-\left(\dfrac{v_0}{C_w}\right)^{K_w}\right] for\ P_w = 0 \\ exp\left[-\left(\dfrac{v_r}{C_w}\right)^{K_w}\right] - exp\left[-\left(\dfrac{v_0}{C_w}\right)^{K_w}\right] for\ P_w = P_{w,r} \end{cases} \tag{7.5}$$

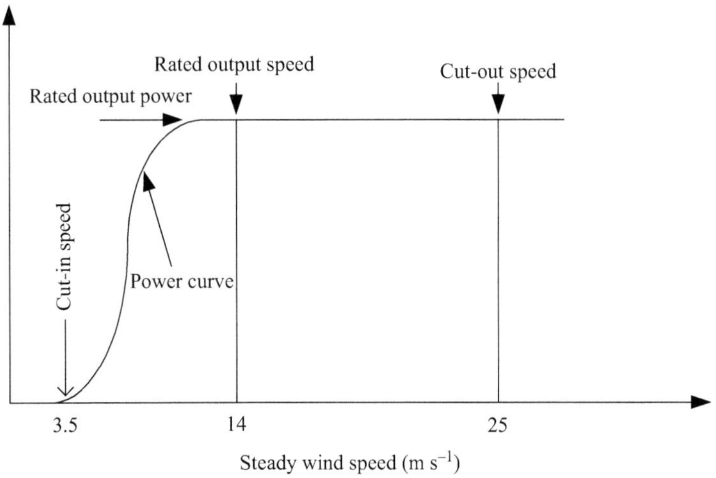

Figure 7.3 Typical wind turbine power output profile with steady wind speed

Equation (7.5) defines the probability distribution function for the wind output with respect to the Weibull distribution, where the parameters ρ and l can be calculated from the following equations, respectively:

$$\rho = \frac{P_w}{P_{w,r}} \tag{7.6}$$

$$l = \frac{v_r - v_i}{v_i} \tag{7.7}$$

7.3.2 Probabilistic photovoltaic power output model

The solar irradiation is the stochastic variable that determines the power output of a PV system. Moreover, some other factors like the day of the year, solar time, solar angle, geographical location (standard meridian, latitudes and longitudes), surface tilt, declination and the climatic condition (e.g. cloud cover, rain) also affect the output of the PV system. However, the amount of the solar irradiation that comes from the sun through the atmosphere is different from the amount of the solar irradiation that strikes the PV surface to generate electrical power. That difference is categorized by a term defined as clearness index K_t. The clearness index is basically the measure of the clearness of the atmosphere. It can be defined as the fraction of the solar irradiation that travels through the atmosphere to strike the surface of the Earth. Therefore, the clearness index instead of solar irradiation directly corresponds to the power output of the PV systems. The solar irradiation outside the earth surface in the atmosphere is called extraterrestrial solar irradiation I_{oh}(kW m^{-2}), while the amount of the solar irradiation that hits the horizontal earth surface is called terrestrial solar irradiation I_h(kW m^{-2}). The clearness index is the ratio of the terrestrial solar irradiation and the extraterrestrial solar irradiation given as follows:

$$K_t = \frac{I_h}{I_{oh}} \tag{7.8}$$

The clearness index is the variable that is used to estimate the output power of the PV system, and it is stochastic in nature because it represents a natural stochastic variable such as solar irradiation. Therefore, the variability of the PV system power output is modelled through the beta distribution of the clearness index over a time-period using historical data. The probability distribution function of the clearness index is defined as follows [30]:

$$f_{kt}(K_t) = c'\left(1 - \frac{k_t}{K_{t\max}}\right)\exp(\lambda k_t) \tag{7.9}$$

where the parameters of the beta distribution c', λ and γ are defined as follows:

$$\gamma = \frac{k_{t\max}}{k_{t\max} - k_{tm}} \tag{7.10}$$

$$\lambda = \frac{(2\gamma - 17.519\, exp(-1.3118\gamma) - 1{,}062\, exp(-5.0246\gamma))}{k_{t\max}} \tag{7.11}$$

$$c' = \frac{\lambda^2 k_{t\max}}{((\exp(\lambda k_{t\max})) - 1 - \lambda k_{t\max})} \tag{7.12}$$

where k_{tm} and $k_{t\max}$ represents the mean and maximum values of the clearness index, respectively. Sequentially, first the value of the parameter γ is determined by (7.10), then the parameter λ is calculated through (7.11) and finally, the parameter c' is calculated using (7.12). Some of the parameters are detailed in Figure 7.4.

The solar panels are normally inclined at an angle to catch the maximum possible solar irradiation. Therefore, the solar irradiance $I_{S\beta}$ on a tilted surface with an inclination angle β to the horizontal is defined as presented in [37]:

$$I_{S\beta} = \left[R_b + \left(\frac{1+\cos\beta}{2} - R_b\right)\cdot k + \rho_r \cdot \frac{1-\cos\beta}{2}\right]\cdot I_h \tag{7.13}$$

In (7.13), the parameter R_b is taken as the ratio of solar irradiation on a tilted surface to the solar irradiation on the horizontal surface calculated using (7.14). The parameter k is the fraction of solar irradiance diffused on the horizontal plane to the total solar irradiance calculated using (7.18). Here, ρ_r is the reflectance of the ground, usually in a range of 0.2–0.7 depending on the state of the ground. Here $\rho_r = 0.4$ is used. The expression for R_b is given as follows [30]:

$$R_b = \frac{\sin(L-\beta)\sin(\delta) + \cos(L-\beta)\cos(\delta)\cos(h)}{\sin(L)\sin(\delta) + \cos(L)\cos(\delta)\cos(h)} \tag{7.14}$$

In (7.14), the L represents the latitude, δ represents the declination calculated from (7.15) and h is the hour angle in degrees calculated at the midpoint of each hour given in (7.16):

$$\delta = 23.45\sin\left[\frac{360}{365}(284 + Nu)\right] \tag{7.15}$$

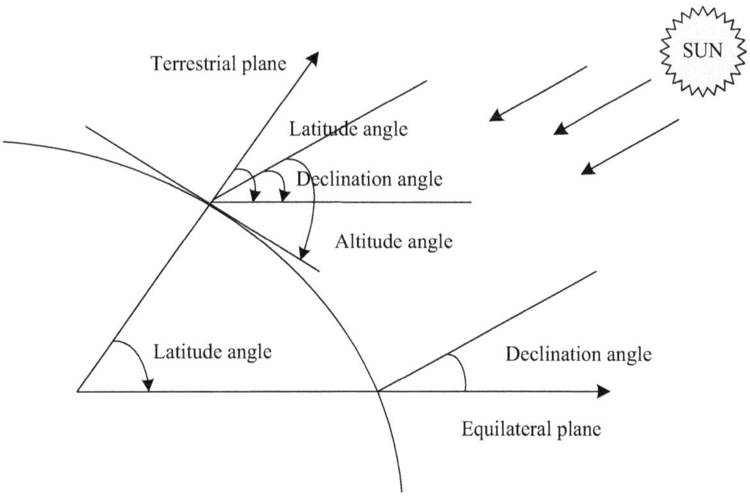

Figure 7.4 Trigonometry related to solar panel system

$$h = 0.25 \times (number\ of\ minutes\ from\ local\ solar\ noon) \tag{7.16}$$

In (7.15), the variable Nu represents the day of the year. A piecewise linear function is defined to estimate the correlation between the diffused fraction k and the clearness index K_t as given in [30]:

$$k = p - qk_t \tag{7.17}$$

where p and q are the correlation parameters of the piecewise linear function. Furthermore, the total solar irradiance I_h can be calculated as follows [30]:

$$I_h = \frac{H_o}{3{,}600} r_d K_t \tag{7.18}$$

In (7.18), r_d is the ratio of diffused irradiation in an hour to diffused irradiation per-day. It can be calculated from (7.19). H_o is the extraterrestrial irradiation on the horizontal surface in (MJ m^{-2}) for 1 day given as in (7.20):

$$r_d = \left[\frac{\pi}{24}\right] \frac{\cos(h) - \cos(h_{ss})}{\sin(h_{ss}) - \left[\frac{2\pi h_{ss}}{360}\right]\cos(h_{ss})} \tag{7.19}$$

$$H_o = \frac{24 \times 3{,}600 G_{SC}}{\pi 10^6}\left[1 + 0.033\cos\left(\frac{360Nu}{365}\right)\right] \times \left[\cos(L)\cos(\delta)\sin(h_{ss}) + \left[\frac{\pi h_{ss}}{180}\right]\sin(L)\sin(\delta)\right] \tag{7.20}$$

In (7.20), G_{SC} is the solar constant taken as 1,366.1 W m^{-2} and h_{ss} is the sunset hour in degrees calculated as follows:

$$h_{ss} = \cos^{-1}[-\tan(L)\tan(\delta)] \tag{7.21}$$

Now, the values obtained from (7.8), (7.17) and (7.18) are substituted in (7.13) to make the final form given as follows:

$$I_{S\beta} = \left[\left(R_b + \rho_r \cdot \frac{1-\cos\beta}{2}\right) + \left(\frac{1-\cos\beta}{2} - R_b\right)\cdot p\right]\cdot \frac{H_o}{3,600} r_d K_t - \left(\frac{1-\cos\beta}{2} - R_b\right)\cdot q \cdot \frac{H_o}{3{,}600} r_d k_t^2 \tag{7.22}$$

The (7.22) can be represented as

$$I_{S\beta} = \mathscr{T} K_t - \mathscr{T}' K_t^2 \tag{7.23}$$

Now, the PDF with respect to the defined beta distribution can be estimated as in [30]. The PDF has four expressions depending upon the signs of $\mathscr{T}$ and $\mathscr{T}'$, but only two of them are logically possible with the normal values of solar irradiance. They are defined as follows:

if $\mathscr{T} > 0$ *and* $\mathscr{T}' < 0$ *then*:

$$f_{pv}(P_{pv}) = \begin{cases} \dfrac{c'(k_{t\max} - 0.5(\alpha + \alpha'))}{k_{t\max} A_c \eta \mathscr{T}' \alpha'} exp \dfrac{\lambda(\alpha + \alpha')}{2} & if\ P_{pv} \in [0, P_{pv}(k_{t\max})] \\ 0\ otherwise \end{cases} \tag{7.24}$$

if $\mathscr{T} > 0$ *and* $\mathscr{T}' > 0$ *then*

$$f_{pv}(P_{pv}) = \begin{cases} \dfrac{c'(k_{t\max} - 0.5(\alpha - \alpha'))}{k_{t\max} A_c \eta \mathscr{T}' \alpha'} exp \dfrac{\lambda(\alpha - \alpha')}{2} & if\ P_{pv} \in [0, P_{pv}(k_{t\max})] \\ 0\ otherwise \end{cases} \tag{7.25}$$

where A_c is the solar panel modules array surface area in m^2 and η is the efficiency of the PV system in realistic reported conditions. Moreover, α and α' are arbitrary variables defined as follows:

$$\alpha = \frac{\mathscr{T}}{\mathscr{T}'} \tag{7.26}$$

$$\alpha' = \sqrt{\alpha^2 - 4 \cdot \frac{P_{pv}}{A_c \cdot \eta \cdot \mathscr{T}'}} \tag{7.27}$$

The solar panel PV modules power output is dependent on the solar irradiance, the ambient temperature of the site and the module characteristics. Thus, the PV power output P_{pv} considering these factors can be calculated as follows:

$$P_{pv,available} = Ni \times FF \times V \times I \tag{7.28}$$

In (7.28), Ni represents the number of cells of all the solar panel modules, and FF denotes the fill factor for the solar panel module calculated as follows:

$$FF = \frac{V_{MPP} \times I_{MPP}}{V_{oc} \times I_{sc}} \tag{7.29}$$

In (7.29), V_{MPP} and I_{MPP} are the voltage and current at maximum power point, respectively, whereas V_{OC} is the open-circuit voltage and I_{SC} is the short-circuit current of the solar panel module, respectively. The voltage V and current I of a cell of solar module can be calculated as follows:

$$V = V_{oc} - K_v \times T_c \tag{7.30}$$

$$I = I_\beta [I_{sc} + K_i(T_c - 25)] \tag{7.31}$$

where K_i represents the current per degree Celsius temperature coefficient (A °C^{-1}) and K_v represents the voltage per degree Celsius temperature coefficient (V °C^{-1}). The cell temperature T_c can be calculated as follows:

$$T_c = T_A + I_{S\beta}\frac{Not^{-20}}{0.8} \tag{7.32}$$

The variables T_C and T_A are the cell and the ambient temperatures in the degree Celsius, respectively, whereas *Not* is the cell temperature in normal operation in degree Celsius.

7.3.3 Dynamic battery energy storage power output model

The grid-level battery energy storage has evolved as one of the key sources for a reliable and resilient power grid. The grid-level energy storage can mitigate the uncertain power imbalance in the operation of power grid. Moreover, the demand and generation management, frequency regulation, flexible ramping product, Volt–Var support and the grid stabilization are some valuable features of the grid-level energy storage [34]. The energy stored in the battery storage system, usually referred as SoC, changes as the batteries are charged or discharged. The BESS can be modelled using linear expressions. It is assumed that during a time interval Δt, the charging and discharging power remains constant. The relation of energy stored in the battery storage before and after the charge or discharge process is given by the following equation:

$$E_{BESS} = E'_{BESS}(1 - \epsilon_{BESS}\Delta t) + \left[P_{BESS,C}\eta_{BESS,C} - \frac{P_{BESS,D}}{\eta_{BESS,D}}\right]\Delta t \tag{7.33}$$

In (7.33), E'_{BESS} (kW h) and E_{BESS} (kW h) are the energy stored in the BESS at the beginning and end of time step Δt, respectively. The parameter ϵ_{BESS} represents the hourly self-discharge rate for the battery storage system. The charging and discharging powers are represented by $P_{BESS,C}$ (kW) and $P_{BESS,D}$ (kW) with the corresponding efficiencies $\eta_{BESS,C}$ and $\eta_{BESS,D}$, respectively. The SoC of the BESS can be estimated from the following expression:

$$S_{BESS} = \frac{E_{BESS}}{E_{BESS,R}} \tag{7.34}$$

In (7.34), S_{BESS} represents the SoC of the BESS while, E_{BESS} (kW h) and $E_{BESS,R}$ (kW h) represents the current and rated amount of energy stored in the BESS. In the real operation of BESS, the SoC is limited within a prescribed range, and the charging and discharging power limits should also be taken into consideration. These constraints for every time step Δt are expressed as follows:

$$E_{BESS,\min} \le E_{BESS} \le E_{BESS,\max} \tag{7.35}$$

$$0 \le P_{BESS,C} \le P_{BESS,C,\max} \tag{7.36}$$

$$0 \le P_{BESS,D} \le P_{BESS,D,\max} \tag{7.37}$$

where $E_{BESS,\min}$ (kW h) and $E_{BESS,\max}$ (kW h) are the minimum and the maximum limits for the stored energy of the BESS. The parameters, $P_{BESS,C,\max}$ (kW) and $P_{BESS,D,\max}$ (kW), represent the maximum charging and discharging power limits of the BESS.

The mutual exclusiveness of the charging and discharging modes of BESSs is also considered in this model, such that the BESS cannot be in the charging and discharging mode at the same time. Therefore, a decision variable b_{BESS} is introduced, that will be 0, when the BESS is in the charging mode, otherwise it will have a value as 1, indicating that the BESS is in the discharging mode. To implement the mutual exclusiveness of the BESS charging and discharging modes, (7.36) and (7.37) are modified to incorporate the decision variable b_{BESS} such as

$$0 \leq P_{BESS,C} \leq (1 - b_{BESS})P_{BESS,C,\max} \tag{7.38}$$

$$0 \leq P_{BESS,D} \leq b_{BESS}P_{BESS,D,\max} \tag{7.39}$$

In addition, for simplicity, the battery lifetime of the BESS is assumed to be unrestrained of the discharging depths. Thus, for the calculation of BESSs's lifetime, only the total throughput is considered. The following equation represents the lifetime of the BESS:

$$l_{BESS} = \min\left[\frac{E_{BESS}^{lifetime}}{E_{BESS}^{ann}}, l_{BESS,f}\right] \tag{7.40}$$

In (7.40), l_{BESS} (years) represents the lifetime of the BESS in years. The variable, $E_{BESS}^{lifetime}$(kW h), denotes the lifetime of the BESS representing the maximum quantity of energy that can be cycled through the BESS in total. The variable, E_{BESS}^{ann} (kW h), is the annual throughput of the storage system, whereas $l_{BESS,f}$ (years) is the float life of the energy storage that refers to the maximum service life of the battery storage before it needs to be replaced.

7.3.4 Micro-turbine power output model

The consumption of fuel by a MT power generating system is normally represented by means of a quadratic function of its output power. For simplification, the consumption of fuel and power output of a MT can be expressed as a linear function, such as the consumption of the natural gas f_{mt} (m^3) during the time interval Δt of usually 1 h, can be expressed as

$$f_{mt} = b_{mt}\left(f_0P_{mt,R} + f_1P_{mt}\right)\Delta t \tag{7.41}$$

In (7.41), f_0 (m^3 kW h^{-1}) and f_1 (m^3 kW h^{-1}) represent the intercept coefficient and slope of the fuel curve, respectively, whereas $P_{mt,R}$ (kW) and P_{mt} (kW) represent the rated output power and actual output power of the MT. Here b_{mt} is the decision variable, that determines the operating state of the MT power generation system, as if the MT is off, b_{mt} takes the value of 0, which is otherwise equal to 1 indicating the system to be in operation. Equation (7.41) is further simplified by neglecting the no-load consumption of the fuel for the MT resulting in the following expression:

$$f_{mt} = b_{mt} \frac{P_{mt} \Delta t}{\gamma \eta_{mt}} \tag{7.42}$$

where γ represents the parameter of unit conversion for the consumption of natural gas. The variable η_{mt} refers to the thermal efficiency of the MT system. The safe and reliable operation of the MT power generation system is constrained by the maximum and the minimum output power, the maximum and the minimum ramp rate along with the minimum run time [30]. The output power of the MT, constrained for every time step Δt by the limits, is expressed as

$$P_{mt,\min} \leq P_{mt} \leq P_{mt,\max} \tag{7.43}$$

The MT-based generator has ramp rate constraint, which is expressed for every time step Δt as

$$P_{mt}^{t-1} - P_{mt}^{t} \leq dr \tag{7.44}$$

$$P_{mt}^{t} - P_{mt}^{t-1} \leq ur \tag{7.45}$$

where ur (kW h^{-1}) and dr (kW h^{-1}) represents the ramping-up and ramping-down limits of the MT-based generator, respectively. Also, the P_{mt}^{t} and P_{mt}^{t-1} denote the output power of the MT-based generator in tth and $(t-1)$th hour, respectively. In order to implement the ramping rate constraint, we update the output power limits of the MT-based generator in every time interval t given as

$$P_{mt,\min} = \begin{cases} P_{mt,\min} \ if \ t = 1 \\ \max\left(P_{mt,\min}, P_{mt}^{t-1} - dr\right) \ else \end{cases} \tag{7.46}$$

$$P_{mt,\max} = \begin{cases} P_{mt,\max} \ if \ t = 1 \\ \min\left(P_{mt,\max}, P_{mt}^{t-1} + ur\right) \ else \end{cases} \tag{7.47}$$

The constraint for implementing POZs of the generator is implemented as follows:

$$\begin{cases} P_{mt,\min} \leq P_{mt,t} \leq P_{mt}^{1} \\ P_{mt}^{z-1} \leq P_{mt,t} \leq P_{mt}^{z} \\ P_{mt}^{ni} \leq P_{mt,t} \leq P_{mt,\max} \end{cases} \tag{7.48}$$

The numbers of POZs of the MT generator are represented by $z \in ni$, whereas $p^{i,z-1}, p^{i,z}$ are the generating limits of the zth prohibited zone of the MT-generating unit.

7.4 Optimal sizing of microgrid components

A cost-to-profit optimization model is presented in this chapter for sizing the MG components. Both the grid-connected and islanded modes of operations can be emulated within the same formulation by considering with and without grid in/out power flows, respectively. Hence, the capital cost of the individual components comprises cost for the required controls for both the grid-connected and islanded modes of operation. It is assumed that the load demand, wind speed and solar-irradiance forecasting is accurate enough, where the planning horizon depicts the project life cycle. Moreover, it is assumed that there is no replacement cost for the devices during the project life cycle, where the net present value (NPV) of costs is considered. The objective function and the associated constraints can be formulated as given in the following section.

7.4.1 Mathematical formulation

Due to the random nature, intermittency, strict government policies and geographical dependence of renewable resources, a great effort must be made to properly size a MG. In this work, we chose to optimize the overall profit for the community being served by the MG. In particular, the profit can be considered as revenues minus expenses. The expenses include the one-time cost of the PVs, WTs and the ESs installation, the yearly maintenance cost and the energy bought from the grid. The revenues are composed of the self-consumptions-related savings and the eventual government incentives. Since our aim is to evaluate the long-term effects of an investment, we choose the financial-evaluation technique, namely the cost-benefit analysis.

This objective function minimizes the yearly costs including the capital costs, operation and maintenance costs, replacement costs and the cost of purchasing power from the grid, while maximizing the profit in terms of self-consumption and selling power to the grid by considering the NPV. The objective function for every time step Δt over a specified time horizon T with a discount rate r can be written as

$$\max f_s = -\left[C_{cpt,w} + C_{cpt,pv} + C_{cpt,B} + C_{cpt,mt}\right] + \sum_{t=0}^{T} \frac{\left[-\left(C_{om,w} + C_{om,pv} + C_{om,B} + C_{om,mt} + C_{fl,mt} + C_{em,mt} + C_{g,p}\right) + b_g\left(p_{g,s} + p_{s,c}\right)\right]}{(1+r)^t} \tag{7.49}$$

where $C_{cpt,w}, C_{cpt,pv}, C_{cpt,B}$ and $C_{cpt,mt}$ are the one time capital costs of the investment for the installation of wind power plant, PV power plant, BESS and MT, respectively. Similarly, $C_{om,w}, C_{om,pv}, C_{om,B}$ and $C_{om,mt}$ represents the annual operation and maintenance costs for wind power plant, PV power plant, BESS and MT, respectively. $C_{fl,mt}$ is the fuel cost of MT given as follows:

$$C_{fl,mt} = \alpha_{fl,mt} f_{mt} \tag{7.50}$$

In (7.50), $\alpha_{fl,mt}$ represents the cost coefficient for the fuel cost of MT. The cost related to emissions $C_{em,mt}$ from the MT is expressed as follows:

$$C_{em,mt} = a_{e,t} + b_{e,t} P_{mt} + c_{e,t} (P_{mt})^2 \tag{7.51}$$

where $a_{e,t}, b_{e,t}$ and $c_{e,t}$ represent the emission cost coefficients for the MT. The cost of the power purchased from the grid $P_{g,p}$ is represented by $C_{g,p}$, which can be expressed as follows:

$$C_{g,p} = \lambda_{g,p} P_{g,p} \tag{7.52}$$

In (7.52), $\lambda_{g,p}$ represents the grid energy price, where in (7.49), $p_{g,s}$ and $p_{s,c}$ represent the profit earned by selling energy to the grid and by self-consumption, respectively. Here, the binary variable b_g selects the grid-connected and islanded modes of operation for the MG. If the variable b_g has a value 0, it indicates the islanded operation of MG, and if the variable b_g has a value 1, it shows that the MG is operating in connection to the grid. The energy is sold to the grid at a discounted price, that is basically a fraction of the grid energy price represented by $\lambda_{g,p}{}'$. The profit earned by selling energy to the grid can be expressed as follows:

$$p_{g,s} = \lambda_{g,p}{}' P_{g,s} \tag{7.53}$$

The profit earned from self-consumption $p_{s,c}$ by the MG is expressed in terms of the grid energy price as follows:

$$p_{s,c} = \lambda_{g,p} \times \left(P_w f_w(P_w) + P_{pv} f_{pv}(P_{pv}) + P_{BESS,D} - P_{BESS,C} + P_{mt}\right) \tag{7.54}$$

The objective function is maximized subject to the following constraints for every time step Δt of the specified time horizon T:

$$P_w + P_{pv} + E'_{BESS}(1 - \epsilon_{BESS}) + P_{BESS,D}\eta_{BESS,D} - \frac{P_{BESS,C}}{\eta_{BESS,C}} + P_{mt} + b_g\left(P_{g,p} - P_{g,s}\right) = P_d \tag{7.55}$$

where E'_{BESS} is the energy stored in the BESS for the time step $t-1$, and the available energy of the BESS in the current time step t of the specified time horizon T E_{BESS} can be estimated from (7.33). The SoC of the BESS S_{BESS} for every time step Δt of the specified time horizon T can be calculated from (7.34). The constraints are listed as follows:

$$\begin{cases} P_{w,r} \leq P_{w,\max} \\ P_{pv} \leq P_{pv,\max} \\ E_{BESS} \leq E_{BESS,\max} \\ P_{mt} \leq P_{mt,\max} \\ P_{g,s} \leq P_{g,\max} \\ P_{g,p} \leq P_{g,\max} \\ P_{BESS,C} \leq (1 - b_{BESS})P_{BESS,C,\max} \\ P_{BESS,D} \leq b_{BESS}P_{BESS,D,\max} \\ S_{BESS} \leq S_{BESS,\max} \\ S_{BESS} \geq S_{BESS,\min} \\ E_{BESS} \geq E_{BESS,\min} \\ P_{mt} \geq P_{mt,\min} \\ P_w \geq 0 \\ P_{BESS,C} \geq 0 \\ P_{BESS,D} \geq 0 \\ P_{w,r} \geq 0 \\ P_{pv} \geq 0 \\ P_{g,s} \geq 0 \\ P_{g,p} \geq 0 \end{cases} \tag{7.56}$$

$$P_{mt,\min} = \begin{cases} P_{mt,\min}\ if\ t = 1 \\ \max\left(P_{mt,\min}, P_{mt}^{t-1} - dr\right)\ else \end{cases} \tag{7.57}$$

$$P_{mt,\max} = \begin{cases} P_{mt,\max}\ if\ t = 1 \\ \min\left(P_{mt,\max}, P_{mt}^{t-1} + ur\right)\ else \end{cases} \tag{7.58}$$

$$\begin{cases} P_{mt,\min} \le P_{mt,t} \le P_{mt}^{1} \\ P_{mt}^{z-1} \le P_{mt,t} \le P_{mt}^{z} \\ P_{mt}^{ni} \le P_{mt,t} \le P_{mt,\max} \end{cases} \tag{7.59}$$

The size of the decision variables like the wind power output P_w and the PV power output P_{pv} varies according to the forecasted solar irradiance and the wind speed over the time period to model the uncertainty of weather profiles associated with these technologies. Therefore, these variables are limited by their available amount of powers such as $P_{w,available}$ and $P_{pv,available}$. This optimization problem is formulated as a constrained MINLP to achieve higher accuracy of the results, wherein the total profit is maximized with the optimal sizing of the various MG components. The system states on an hourly basis are incorporated with the respective time series data sets. This type of problem is normally a nondeterministic polynomial-time hard optimization problem with increased complexities imposed by the renewable generation units, stochastic load demands and numerous system states in long-term planning horizon. In this chapter, a highly categorized BSO algorithm is utilized to obtain the optimal solution with significantly enhanced modelling task and reduced computational effort.

7.4.2 Backtracking search optimization (BSO) algorithm

The BSO is a well-known evolutionary optimization algorithm for solving the real valued constrained optimization problems [32,33]. The BSO algorithm has been proved to be better in optimization than the preceding evolutionary optimization techniques [35–40]. The BSO algorithm possesses a memory for storing past population (set of randomly selected solution) from a randomly selected previous population (previous or old historical generation) for utilizing the search-direction matrix formation. This feature of the BSO search algorithm allows the algorithm to gain advantages from the past experiences. The BSO algorithm uses two of the evolutionary operators, namely crossover and mutation. The search strategy of the BSO algorithm is very simple and effective with fast computation due to only one parameter that controls the number of elements of individual solutions among the populations involved in the crossover for generating a solution set for trial. The benefits of the BSO algorithm over the preceding evolutionary algorithms are lesser sensitivity due to only one control parameter, quick computation and efficient convergence.

The procedural steps of the BSO algorithm can be classified into five processes: initialization, Selection-I, mutation, crossover and Selection-II. These five steps are demonstrated as follows:

Step-I: Initialization

The process of BSO algorithm starts by the initialization of the population P (randomly selected set of solution) as follows:

$$\begin{aligned} P_{i,j} &= lower_j + upper_j - lower_j \times random, \\ i &= (1,2,\ldots,NP), j = (1,2,\ldots,SV) \end{aligned} \tag{7.60}$$

In (7.60), NP and SV represent the size of the populations and the number of search variables of the optimization problem, respectively. Here, random is a uniformly distributed real number between 0 and 1. $lower_j$ and $upper_j$ are the upper and lower limits of the continuous search variables in the jth element of the ith individual solution in the population, respectively.

Step-II: Selection-I

In this step, the historical population $oldP$ is initialized for the calculation of the search direction. It is initialized for the first time on a similar pattern as in the first step as follows:

$$\begin{aligned} oldP_{i,j} &= lower_j + upper_j - lower_j \times random, \\ i &= (1,2,\ldots,NP), j = (1,2,\ldots,SV) \end{aligned} \tag{7.61}$$

In every iteration, $oldP$ is expressed as follows:

$$if\, a < b\ then\ oldP := P;\ a, b \in [0, 1] \tag{7.62}$$

where := represents the update operation and a and b are the two random numbers from the uniform distribution between 0 and 1. Equation (7.62) ensures that a population in BSO algorithm can be selected randomly from old historical population. This historical old population is stored by the BSO algorithm until it is updated by using a random permutation.

Step-III: Mutation

The mutation operation is performed to generate the initial trial population as follows:

$$T = P + (oldP - P) \times F \tag{7.63}$$

In (7.63), F is a scaling factor for controlling the magnitude of the search-direction matrix $(oldP - P)$. In this paper, $F = 3 \times random$. The BSO algorithm learns from the stored previous populations to obtain the trial population.

Step-IV: Crossover

In this step, the crossover operation is performed to obtain the final trial population T. The individual solutions among the trial individuals having improved values of fitness lead the search direction for the optimization search. The crossover operator of the BSO algorithm is applied as follows:

A binary matrix named as map having size $NP \times SV$ is computed. The individual solutions of the final trial population T are generated by using the relevant individuals of the population P. If $map_{i,j} = 1$, T is updated with $T_{i,j} := P_{i,j}$.

Step-V: Selection-II

In this step, the individual solution in T_i that has better fitness corresponding to the individual in P_i is utilized to update the P_i. Moreover, when the best optimum solution P_{best} dominates the preceding global optimum value searched by the BSO algorithm, the global optimum best solution is replaced by the P_{best} and the global optimum value is also updated to the fitness value of P_{best}.

The parameters associated with the BSO algorithm in this chapter have the following values: $N = 50, Maxitr = 1{,}000, Crossate = 1$. Here ***D*** represents the dimension of the optimization problem. The pseudo code of the BSO algorithm [30] is written as

```
Input: Objfun, N, D, Maxitr, Crossate, low_{1:D}, up_{1:D}
Output: globalminimum, globalminimizer % Optimum answers
% rnd ~ U(0, 1), rndn ~ N(0, 1), w = rndint(.), rndint(.) = U(1, .)  w ∈ {1, 2, 3, ... ,.}
function bsa (Objfun, N, D, Maxcycle, low, up)
% Initialization
globalminimum = inf
for i from 1 to N do
    for j from 1 to D do
    P_{i,j} = rnd.(up_j − low_j) + low_j) % Initialization of population, P
    oldP_{i,j} = rnd. (up_j − low_j) + low_j)% Initialization of old population, oldP
    end
    fitnessP_i = Objfun(P_i)     % Initialization fitness of P
 end
 for iteration from 1 to Maxitr do
 % First Selection
 if(a < b) |a, b ~ U(0, 1) then oldP := P end
 oldP := permuting (oldP)          % Arbitrary changes in position of two
% individuals of oldP
% Generation of trial population
% Mutation
     mutant = P + 3 .rndn. (oldP − P)
% Crossover
     map_{1:N,1:D} = 1 % Initially map is an N × D matrix of ones
       if (a < b) |a, b ~ U(0, 1) then do
         for i from 1 to N do
         map_{i,u([cossrate.rnd.D])} = 0 | u = permuting(1, 2, 3, ... , D)
         end
         else
         for i from 1 to N do, map_{i,randi(D)} = 0, end
         end
     T := mutant
     for i from 1 to N do
         for j from 1 to D do
```

```
            if map_{i,j} = 1 then T_{i,j} = P_{i,j} end
            end
        end
      % Boundary control mechanism
        for i from 1 to N do
        for j from 1 to D do
            if(T_{i,j} < low_j)or(T_{i,j} > up_j) then
             T_{i,j} = rnd. (up_j − low_j) + low_j)
             end
          end
        end
    end
 % Second Selection
 fitnessT = Objfun(T)
 for i from 1 to N do
    if fitnessT_i < fitnessP_i then
    fitnessP_i := fitnessT_i
    P_i := T_i
    end
 end
 fitnessP_best = min (fitnessP) | best ∈ (1, 2, 3, ... , N)
 if (fitnessP_best < globalminimum) then
 globalminimum := fitnessP_best
 global minizer := P_best
 % Export glolbalminimum and globalminimizer
 end
end
```

7.4.3 Solution approach

The forecasted solar irradiance and the wind speed profiles are utilized to model the realistic uncertain pattern of the expected power generation through these sources in the given planning horizon. The search algorithm has to follow the variability of the forecasted load demand profile, while scheduling the generation sources including intermittent wind and PV. The size of the decision variables like the wind power output and the PV power output varies according to the forecasted solar irradiance and wind speed over the time period to model the uncertainty of the weather profiles associated with these technologies. The BESS and MT system compensates the variability of the intermittent load and generation. The BESS also utilizes the economical price hours for charging and then discharges in the peak hours to gain economy and to reduce peak load. The system states on hourly basis are incorporated with the respective time series data sets. The following steps outline the proposed solution approach:

Step-I: The forecasted load, wind and the solar irradiance profiles on hourly basis for the complete horizon are taken as input to the solution algorithm.

Step-II: The forecasted wind speed and the solar irradiance profiles are utilized by the probabilistic wind and PV models given as above to estimate the parameters of the PDF functions for the wind and the PV power outputs, respectively.

Step-III: The available amount of the wind and the PV power based on the forecast of the wind speed and the solar irradiance is calculated for comparison with the scheduled output power.

Step-IV: The BSO algorithm tends to find the solutions that are valid under all possible realizations of the uncertainties such as load, wind and PV power in a specified mode of operation (on-grid or islanding), while maximizing the NPV of the profit due to self-consumption and selling energy to the grid subject to the constraints given in Section 7.4.1.

Step-V: Based on the average values of the best schedules obtained by the BSO algorithm and considering the capacity factor (CF) of the different power plants, the optimum-rated capacities are calculated.

7.5 Case studies

To demonstrate the solution of the MG component sizing problem using the BSO algorithm, following case studies are considered:

7.5.1 Case study 1

In this case study, all the days of the year are assumed to be identical in terms of data of load demand, solar irradiance profile, wind speed and energy price for simplicity and better understanding. The parameters data associated with the MG components is given in Table 7.1, and the 1-day hourly data to be used in this example is given in Table 7.2 [30]. The discounted cash flow generated from each investment is calculated for a single day, within the period in which PV module producers guarantee at least 85% of the initial performance (CF = 0.85). We assumed that the PV system performance will degrade 20% in 25 years. By considering this fact, a derating factor of 0.5% is used for the first 8 years and then 1.0% from the 9th year. Similarly, for the wind, MT and BESS, the derating factor is assumed to be 0.5%. Also the annual maintenance cost is considered to be equal to 0.5% of the initial capital cost. The discount rate is equal to the weighted average cost of capital, and it is chosen as 5%. The CF for wind power plant is taken as 50%. The cost data is given in Table 7.3. The BSO algorithm is used to calculate the following:

1. The optimum size of the wind power plant, PV power plant, MT and the BESS.
2. The hourly savings through self-consumption and by selling energy to the grid.
3. Assuming all the days to be identical, the annual savings for this MG.
4. The profitability of the MG with the above assumption, using the given discount rate, and ignoring the derating factor for a period of 20 years.

Table 7.1 The parameters associated with the microgrid components

PV power variables	Wind power variables	Micro-turbine variables	BESS variables
$N = 34$	$C = 10$	$f_o = (\text{m}^3\ \text{kW h}^{-1})$	$\eta_{ESS,C} = 90\%$
$L = 30$	$Wr = 30$ kW	$f_1 = (\text{m}^3\ \text{kW h}^{-1})$	$\eta_{ESS,D} = 90\%$
$I_{sc} = 1.8$ A	$Vr = 15$ m s^{-1}	$P_{mt,min} = 5$ kW (5 – 25 kW)	$S_{ESS,\min} = 10\%$
$V_{OC} = 55.5$ V	$Vo = 45$ m s^{-1}	$P_{mt,\min} = 12$ W (12 – 50 kW)	$S_{ESS,\max} = 100\%$
$N_{OT} = 43$	$Vi = 5$ m s^{-1}	$ur = 2$ kW min^{-1}	$P_{ESS,C}^{\max} = 0.5P_{ESS}$
$V_{MPP} = 38$ V	$K = 2$	$dr = 2$ kW min^{-1}	$P_{ESS,D}^{\max} = 0.8P_{ESS}$
$I_{MPP} = 1.32$ A		$POZ = [7 - 10$ kW]	
$N_i = 400$			
$H_o = 800$			
$r_d = 0.102$			
$\eta = 0.5$			
$A_c = 5$ m^2			
$K_v = 194 \times 10^{-3}$			
$K_i = 1.4 \times 10^{-3}$			
$\beta = 48.5°$			
$R_b = 1.5282$			

Table 7.2 The data associated with the microgrid components

H	K_t	T_a (°C)	T_c (°C)	P_{pv} (kW)	PDF (PV)	v (m s^{-1})	P_w (kW)	PDF (Wind)	Energy price ($\lambda_{en,g}$) (\$ h^{-1})	Load (P_d) (kW)
1	0	15	15	0	0	11.75	20.25	0.589	65.10	34.3
2	0	15	15	0	0	9.650	13.95	0.497	54.70	39.7
3	0	14	14	0	0	9.250	12.75	0.437	47.17	36.3
4	0	13.5	13.5	0	0	12.90	23.70	0.602	52.42	37.9
5	0	14.1	14.1	0	0	10.50	16.50	0.432	45.43	35.4
6	0	14.15	14.15	0	0	14.52	28.56	0.377	42.75	34.5
7	0	15	15	0	0	12.75	23.25	0.609	50.69	30.1
8	0.162	16	19.77	3.4452	0.166	10.90	17.70	0.492	64.33	28.4
9	0.395	17	35.95	12.663	0.429	16.20	30.00	0.788	63.79	33.8
10	0.628	17	51.86	21.334	0.832	9.400	13.20	0.459	67.85	38.4
11	0.724	17.7	60.13	23.028	0.654	13.65	25.95	0.524	64.81	41.3
12	0.747	18.2	62.53	24.854	0.836	10.60	16.80	0.847	73.60	43.9
13	0.736	19	62.42	25.551	0.747	11.60	19.80	0.876	83.14	45.6
14	0.693	20	59.92	23.624	0.884	15.80	30.00	0.788	80.97	46.7
15	0.748	21	65.42	21.663	0.841	18.50	30.00	0.788	71.28	48.3
16	0.496	21	46.41	15.107	0.417	19.30	30.00	0.788	79.50	49.4
17	0.120	21	23.77	1.7755	0.128	13.50	25.50	0.545	89.39	49.2
18	0	21	21	0	0	17.70	30.00	0.788	94.57	50.1
19	0	20	20	0	0	9.100	12.30	0.416	105.95	51.7
20	0	19	19	0	0	16.70	30.00	0.788	102.14	50.5
21	0	18	18	0	0	11.50	19.50	0.866	79.02	45.6
22	0	17.2	17.2	0	0	9.300	12.90	0.244	67.11	40.1
23	0	17	17	0	0	11.30	18.90	0.544	66.21	38.8
24	0	16	16	0	0	11.30	18.90	0.544	60.83	36.7

Table 7.3 The cost data associated with the microgrid components

Capital costs ($ kW h^{-1})	Fuel cost ($ kW h^{-1})	Emission cost ($ kW h^{-1})
$C_{cpt,w} = 1{,}000$ $C_{cpt,pv} = 2{,}125$ $C_{cpt,BESS} = 2{,}210$ $C_{cpt,mt} = 700$	$C_{fl,mt} = 0.0451$	$C_{em,mt} = 0.020$

Table 7.4 The optimum size of the microgrid components

Wind power plant (kWp)	PV power plant (kWp)	BESS (kW h)	MT (kWp)
70	30	100	17.55

5. Optimum size of the wind power plant, PV power plant, MT and the BESS in islanded mode.

In Table 7.1, the parameter N_i represents the number of solar panels. This number is decided on the basis of the rating of a PV power plant. Here, 400 solar panels are selected for a 30 kW rating of the PV power plant.

In Table 7.2, the output power for wind and PV is calculated for a 30 kW rating. The capital costs of the MG components are given in Table 7.3.

The BSO algorithm first determines the optimum size of the MG components, then the rated peak values with reference to the corresponding CFs are estimated as shown in Table 7.4.

The NPV estimated by the objective function against the optimum size of MG components for 1 day is \$−317,062.9122. The hourly savings through self-consumption and by selling energy to the grid are given in Table 7.5.

The annual savings for this MG by assuming all the days to be identical are

$$\begin{aligned}\text{Annual Savings} &= \text{Total savings per day (\$)} \times 365 \\ &= 148.7831977 \times 365 = \$54{,}305.86716.\end{aligned}$$

The forecasted solar irradiance curve for a single day as shown in Figure 7.5 depicts the availability of the power during the hours of the day. Moreover, it necessitates the day-ahead schedule of the solar PV power to follow the availability pattern. The PDF of the scheduled PV power guides BSO algorithm to follow the expected available solar PV power as shown in Figure 7.6. Similar to the solar PV, Figure 7.7 shows the wind speed forecast during a day and the corresponding scheduled power output from BSO algorithm during that day in Figure 7.8. The load demand curve in Figure 7.9 highlights the peak hours between the hour 15 and hour 20 (evening hours). The energy price in the market follows the load demand, as the peak hours are at peak loads, the forecasted energy price for a day is shown in Figure 7.10. The energy trade as shown in Figure 7.11 reflects the unprecedented benefits associated with the on-grid operation of MG. Whenever there is excess

Table 7.5 The 1-day schedule of the microgrid components in on-grid mode

H	P_g (kW)	P_w (kW)	P_{pv} (kW)	$P_{BESS,C}$ (kW)	$P_{BESS,D}$ (kW)	P_{MT} (kW)	P_d (kW)	$\lambda_{en,g}$ ($ kW^{-1}h^{-1})	Cash inflows (savings)	
									Self-consumption ($ h^{-1})	Selling ($ h^{-1})
1	0	25.25	0	8.5	0	17.55	34.3	0.1151	2.805425	0
2	7.8	18.95	0	10	0	17.55	34.3	0.1047	1.632045	−0.81666
3	24.9	17.75	0	20.5	0	17.55	39.7	0.09717	0.295611	−2.41953
4	0	28.7	0	9.5	0	17.1	36.3	0.10242	2.604636	0
5	16.35	21.5	0	17.5	0	17.55	37.9	0.09543	0.914012	−1.56028
6	8.29	33.56	0	24	0	17.55	35.4	0.09275	1.371948	−0.7689
7	0	28.25	0	11	0	17.25	34.5	0.10069	2.35083	0
8	−33.5952	22.7	2.4452	0	21	17.55	30.1	0.11433	6.139767	3.840939
9	−35.813	35	11.663	0	0	17.55	28.4	0.11379	6.164292	4.075161
10	−52.284	18.2	20.334	0	30	17.55	33.8	0.11785	9.002494	6.161669
11	−64.128	30.95	24.028	0	30	17.55	38.4	0.11481	10.62873	7.362536
12	0	21.8	24.884	22.934	0	17.55	41.3	0.1236	3.962175	0
13	−13.831	24.8	24.381	9	0	17.55	43.9	0.13314	6.5438	1.841459
14	−17.074	35	22.624	12.5	0	17.55	45.6	0.13097	7.065909	2.236182
15	−5.513	35	24.663	25	0	17.55	46.7	0.12128	5.189888	0.668617
16	4.143	35	15.107	23.5	0	17.55	48.3	0.1295	4.575827	−0.53652
17	0	30.5	1.7755	0	0	17.125	49.4	0.13939	5.771061	0
18	0	35	0	0	0	14.2	49.2	0.14457	6.188424	0
19	−24.75	17.3	0	0	40	17.55	50.1	0.15595	10.53035	3.859763
20	−40.85	35	0	0	40	17.55	51.7	0.15214	12.93805	6.214919
21	8.45	24.5	0	0	0	17.55	50.5	0.12902	4.282786	−1.09022
22	10.15	17.9	0	0	0	17.55	45.6	0.11711	3.009045	−1.18867
23	0	23.9	0	0	0	16.2	40.1	0.11621	3.605401	0
24	0	23.9	0	0	0	14.9	38.8	0.11083	3.330214	0

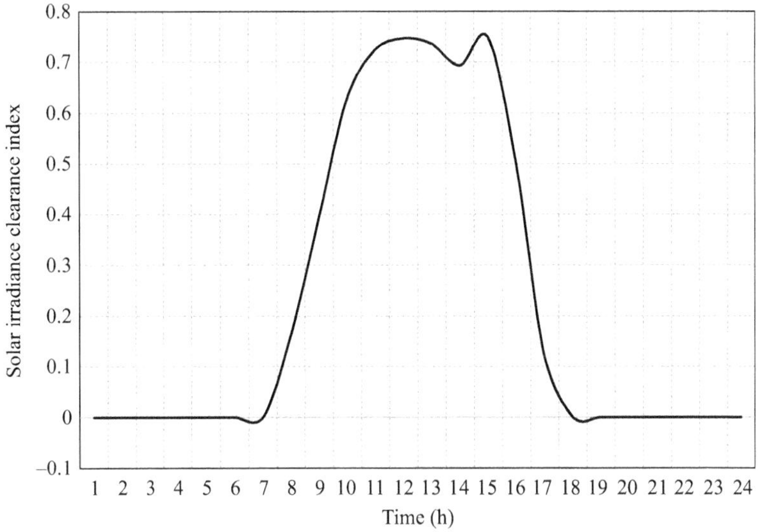

Figure 7.5 Solar irradiance clearance index for 1 day

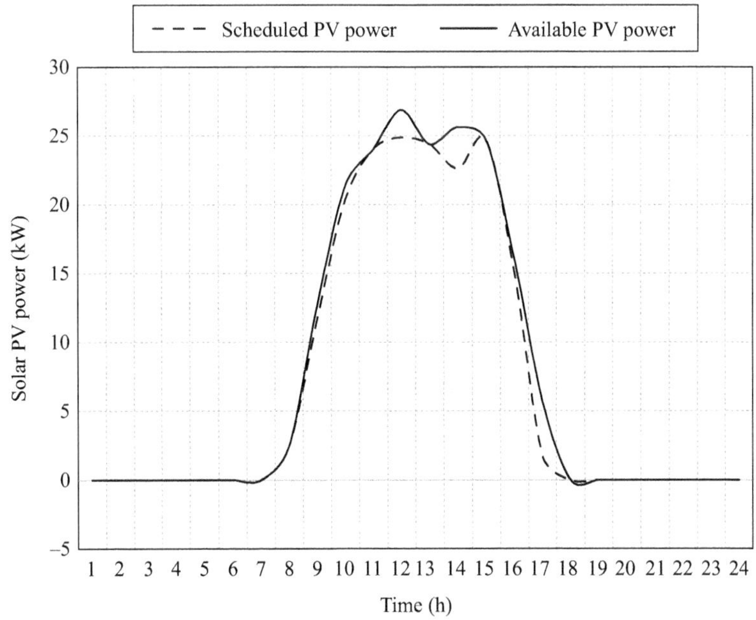

Figure 7.6 Scheduled and available solar PV power output for 1 day

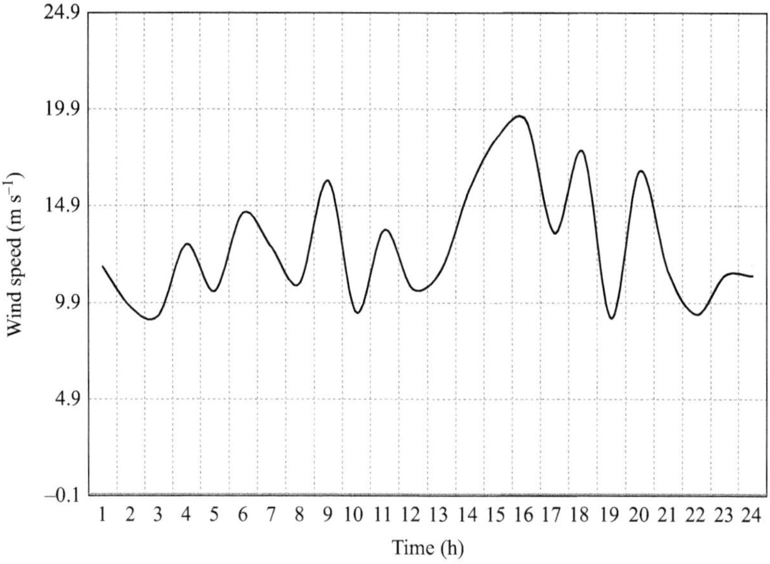

Figure 7.7 Wind speed profile for 1 day

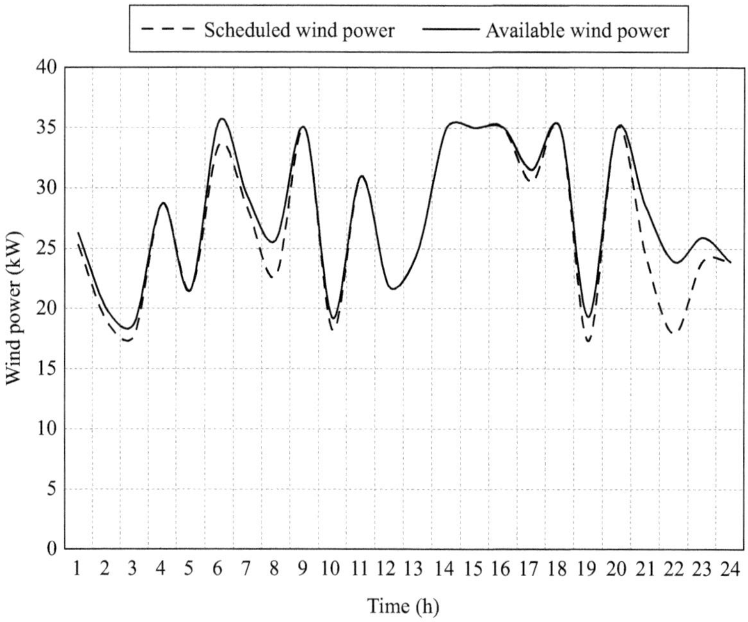

Figure 7.8 Scheduled and available wind power output for 1 day

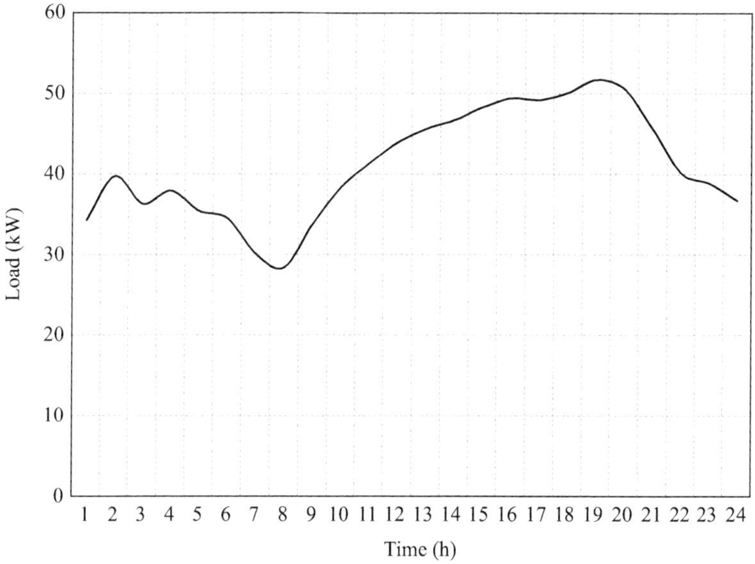

Figure 7.9 Forecasted load demand profile

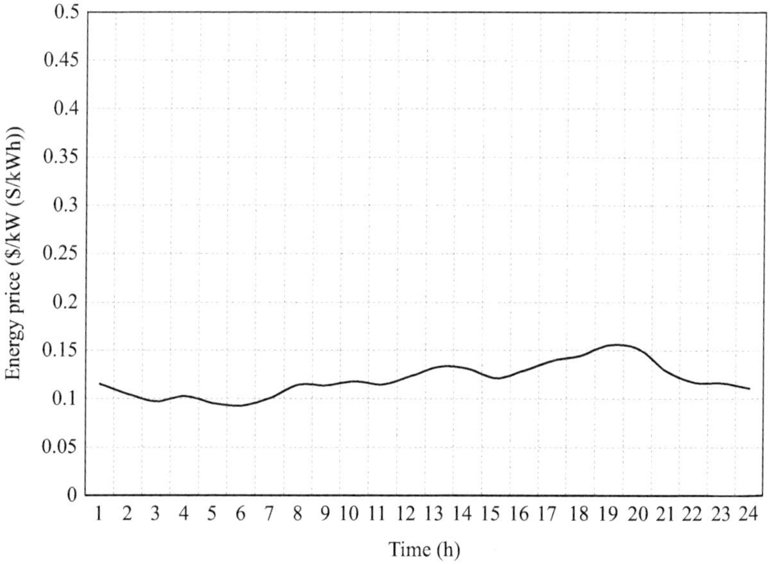

Figure 7.10 Market energy price for the day

amount of available power, it can be sold to the grid to earn profit, and whenever there is a requirement of power in the MG, the grid can rescue that possible imbalance. The MT operates quite smoothly and scheduled in most of the hours at an optimum position of 17.55 kW as shown in Figure 7.12. The BESS undergoes two

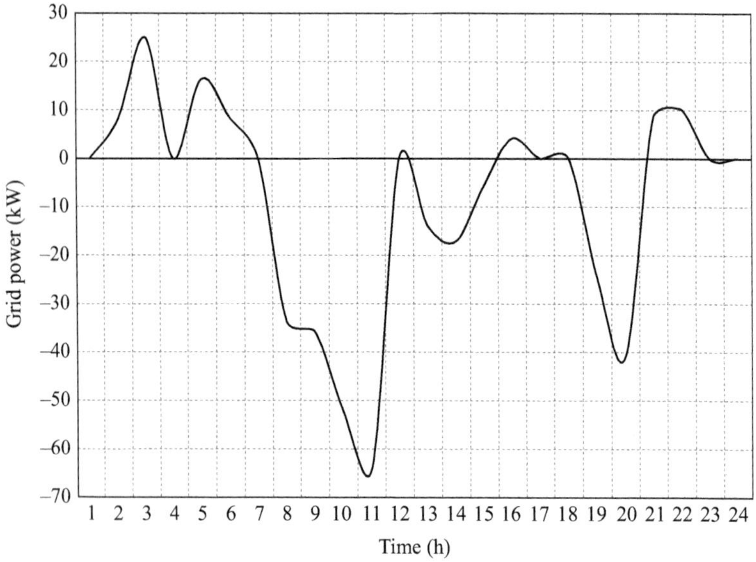

Figure 7.11 Power bought and sold to the grid

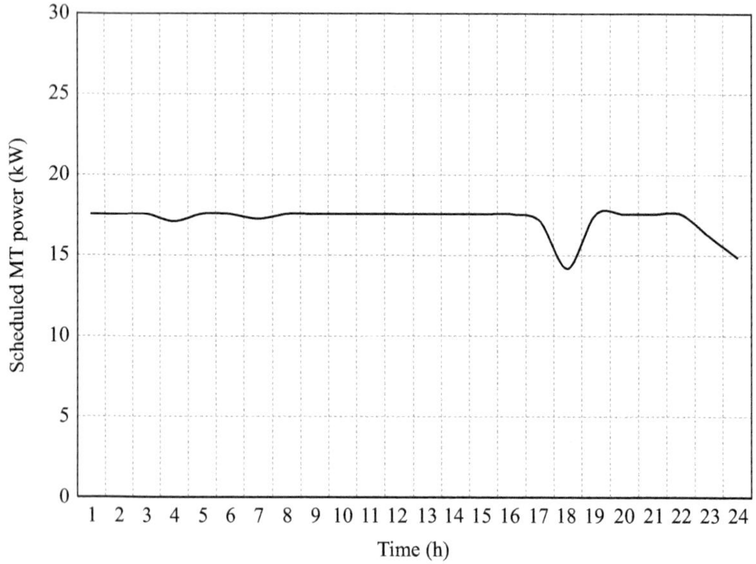

Figure 7.12 Schedule of the MT for the day

cycles of charge and discharge during the day as shown in Figure 7.13. In can be seen from the price variation that the proposed BSO algorithm has successfully utilized the low-price hours for charging the BESS and high-price hours for the discharge to earn the dual benefits of financial profit as well as the shifting of peak load. All these

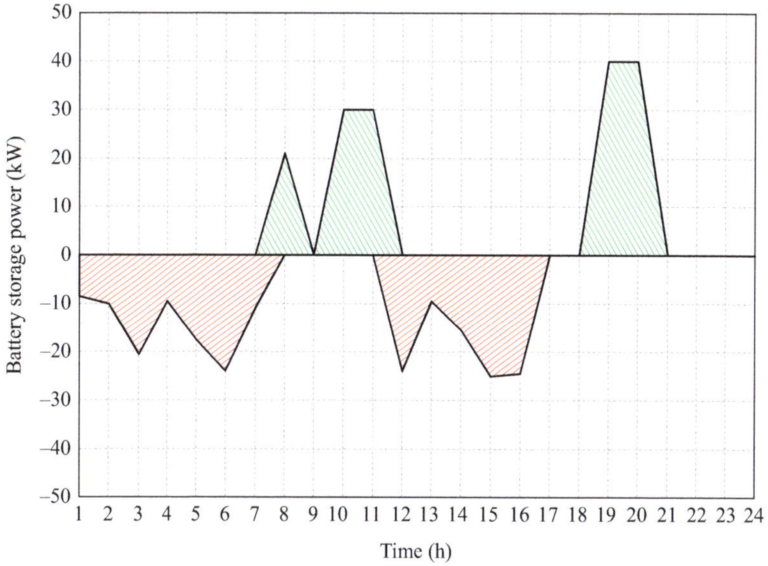

Figure 7.13 Schedule of the BESS for the day

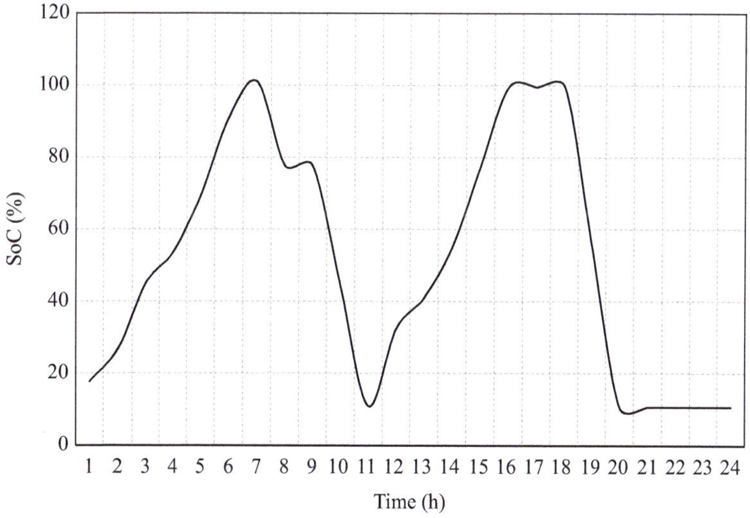

Figure 7.14 Percentage SoC of the BESS for the day

benefits are obtained while securing the SoC limits expressed in Figure 7.14. The savings through self-consumption of energy by the MG are presented in Figure 7.15, while the savings through energy trade with the grid are expressed in Figure 7.16. The self-consumption savings in the early part of day from hours 1–7 and the latter part of the day from hours 20–24 are low reflecting the non-availability of solar PV

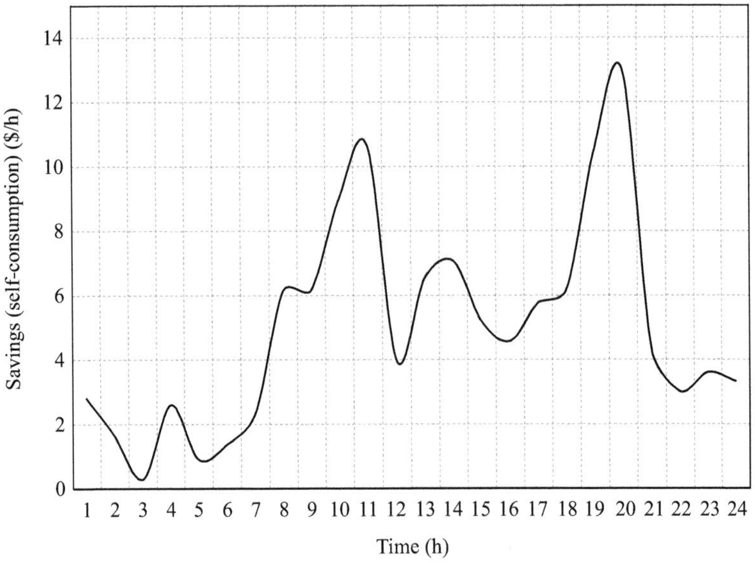

Figure 7.15. Daily savings through self-consumption

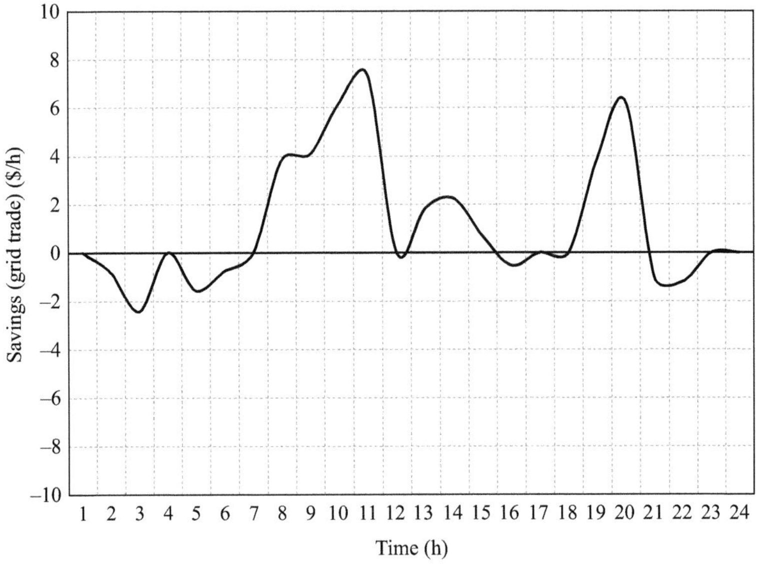

Figure 7.16 Daily savings through energy trade with the grid

power, while the energy trade-to-grid savings are negative in these hours reflecting purchase of energy from the grid. The overall peak saving hours of the day are between hour 8–12 and hour 18–20. Both these peaks coincide with the peak discharge from BESS, which shows the financial worth and corresponding importance of BESS in the MG. The convergence curve of the BSO algorithm is shown in Figure 7.17. The convergence to the optimum solution in less than 50 iterations demonstrates the robustness and fast computational attributes of the BSO algorithm.

The peak for the wind power is noted to be 35 kW; therefore, with the given CF of 50%, the rated peak wind power plant size resulted to be 70 kWp. Similarly, for the solar PV power plant with the CF of 85%, the rated peak plant size is chosen as 30 kWp. The BESS system's optimum size is 100 kWh while the optimum size for the MT is calculated as 17.55 kW by the BSO algorithm.

The capital cost for the given optimum size of the MG components is

Total capital cost = \$367,035.
The maintenance cost per annum = 0.05 Total capital cost = \$1,835.175 per year.

The objective function in this problem maximizes the NPV for the years t to T, that can be written as

$$NPV = -P_o + \sum_{t=0}^{T} \frac{P_t}{(1+r)^t} = -P_o + \frac{P_1}{(1+r)^1} + \frac{P_2}{(1+r)^2} + \ldots + \frac{P_T}{(1+r)^T} \tag{7.64}$$

In (7.64), P_o represents the first term of the objective function that is capital investment, and P_t represents the net annual profit calculated by subtracting the total annual costs from the total annual savings. Table 7.6 demonstrates the NPV for the period of 20 years.

Here, the annual O&M cost is kept constant for the period of 20 years and the derating factor is ignored. The discount rate is considered to be 5%. The yearly net

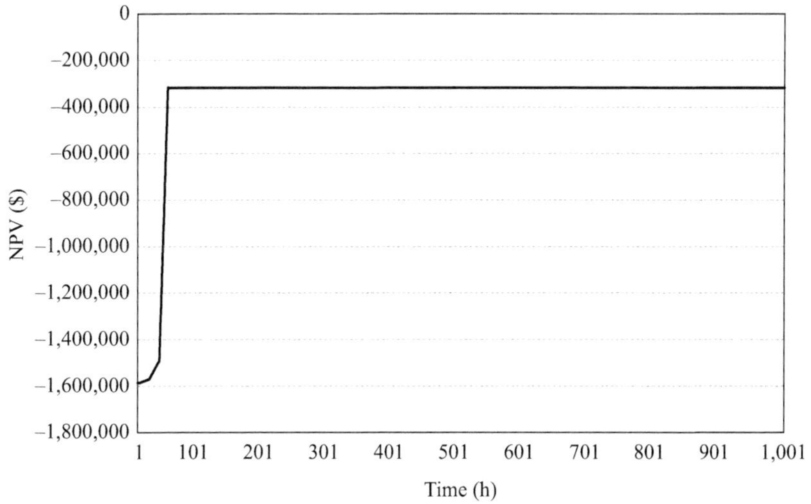

Figure 7.17 Convergence curve of the BSO algorithm for the day

Table 7.6 The savings and the NPV of the microgrid components in on-grid mode

Years	Capital cost ($)	O&M cost ($ year^{-1})	Total savings ($ year^{-1})	Net cash flow ($ year^{-1})	NPV ($)
1	367,035	1,835.18	54,305.86716	49,972.08777	−317,062.9122
2		1,835.18	54,305.86716	47,592.46455	−269,470.4477
3		1,835.18	54,305.86716	45,326.15671	−224,144.291
4		1,835.18	54,305.86716	43,167.7683	−180,976.5227
5		1,835.18	54,305.86716	41,112.16028	−139,864.3624
6		1,835.18	54,305.86716	39,154.43836	−100,709.924
7		1,835.18	54,305.86716	37,289.9413	−63,419.98274
8		1,835.18	54,305.86716	35,514.22981	−27,905.75293
9		1,835.18	54,305.86716	33,823.07601	5,917.32308
10		1,835.18	54,305.86716	32,212.45334	38,129.77642
11		1,835.18	54,305.86716	30,678.52699	68,808.30341
12		1,835.18	54,305.86716	29,217.64475	98,025.94816
13		1,835.18	54,305.86716	27,826.32834	125,852.2765
14		1,835.18	54,305.86716	26,501.26508	152,353.5416
15		1,835.18	54,305.86716	25,239.30008	177,592.8417
16		1,835.18	54,305.86716	24,037.42865	201,630.2703
17		1,835.18	54,305.86716	22,892.78919	224,523.0595
18		1,835.18	54,305.86716	21,802.65637	246,325.7159
19		1,835.18	54,305.86716	20,764.43464	267,090.1505
20		1,835.18	54,305.86716	19,775.65203	286,865.8025

cash flow due to savings is shown in Figure 7.18. Although the net cash flow is decreasing as the number of the years is progressing, still this project is profitable. The yearly NPV, as shown in Figure 7.19, demonstrates that within 9 years the MG can return its capital cost and the rest of the years will be in profit. The variation of the NPV with respect to the size of the MG components is presented in Figure 7.20 by decreasing and increasing the optimum size at a rate of 10%. The result show that the NPV decreases in both directions, whether increasing the size from optimum or decreasing the size from the optimum. Hence, the highest value of the NPV decides the optimal size of the MG components so that MG can earn maximum possible profit.

Now, if the grid is operated in the islanding mode, the BSO algorithm first determines the optimum size of the MG components, then the rated peak values with reference to the corresponding CFs are estimated as shown in Table 7.7.

The NPV estimated by the objective function against the optimum size of MG components for the 1 day is $-396,710.5333. The hourly savings through self-consumption and by selling energy to the grid are given in Table 7.8.

The annual savings for this MG by assuming all the days to be identical are

$$\begin{aligned}\text{Annual Savings} &= \text{Total savings per day (\$)} \times 365\\ &= 109.487643 \times 365 = \$39{,}962.98969.\end{aligned}$$

Table 7.9 demonstrates the NPV for the period of 20 years.

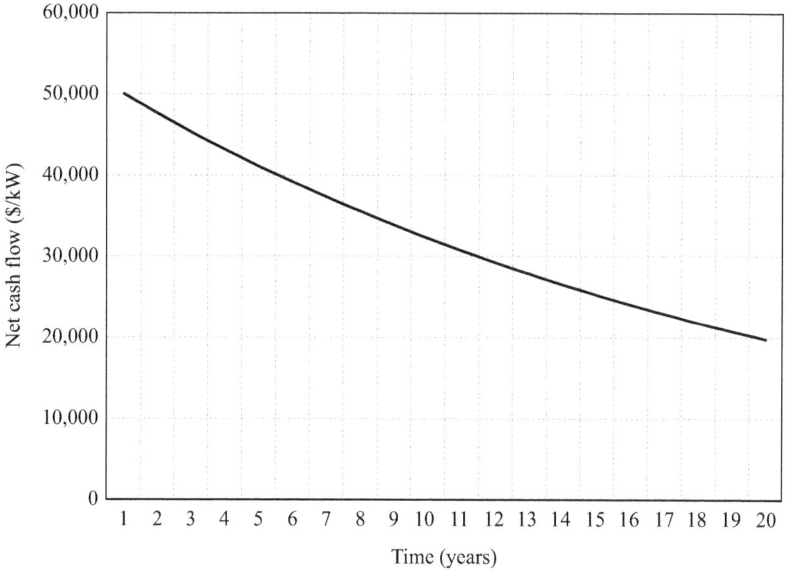

Figure 7.18 Yearly net cash flow due to savings

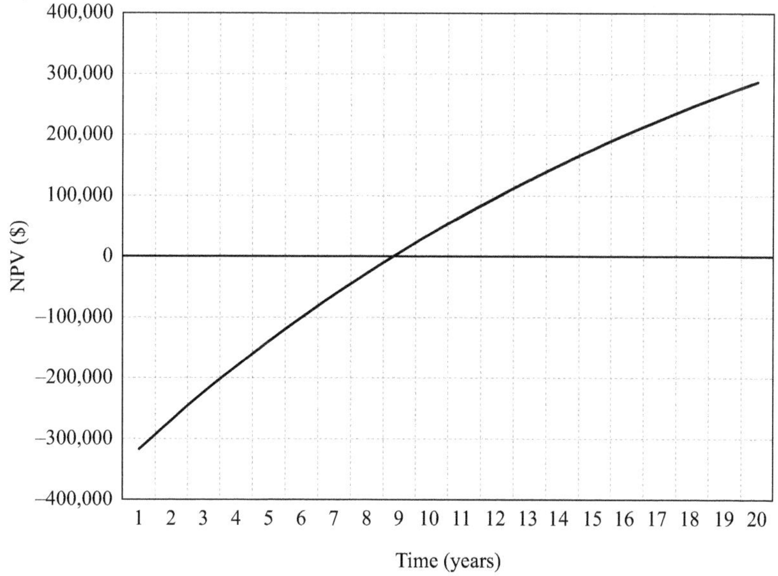

Figure 7.19 Yearly net present value

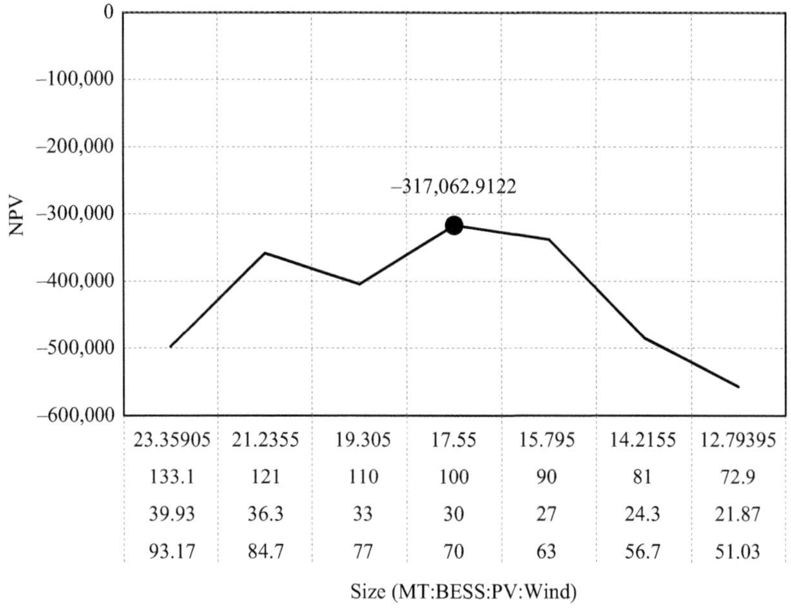

Figure 7.20 The variation of the NPV with respect to the size of the microgrid components

Table 7.7 The optimum size of the microgrid components in islanded mode

Wind power plant (kWp)	PV power plant (kWp)	BESS (kW h)	MT (kWp)
74.5	30	127	19.7

The savings have been decreased in the islanding mode as the opportunity to sell the surplus energy to the grid is no longer available. The BESS and MT operate in a fashion to compensate for the intermittency of wind and solar PV power. Early in the day, when the BESS is not available and there is no solar PV power, the MT ensures the balance. The hours with solar power have excess energy available, that is why BESS charges during these hours and discharges later when there is no solar power and the energy price is at its peak. However, the overall savings are less than on-grid mode operation. Table 7.9 shows that the MG will return the investment in 17 years if it is operated in the islanded mode.

7.5.2 Case study 2

A MG in a MV distribution system adapted from the electricity network of a power company in New South Wales, Australia, that features an average daily demand of 4,000 kWh as shown in Figure 7.1, is considered in this case study. The economic

Table 7.8 The 1 day schedule of the microgrid components in islanding mode

H	P_g (kW)	P_w (kW)	P_{pv} (kW)	$P_{BESS,C}$ (kW)	$P_{BESS,D}$ (kW)	P_{MT} (kW)	P_d (kW)	$\lambda_{en,g}$ (\$ $kW^{-1}h^{-1}$)	Cash inflows (savings)	
									Self-consumption (\$ h^{-1})	Selling (\$ h^{-1})
1	0	27.5	0	0	0	6.8	34.3	0.1151	3.50525	0
2	0	21.2	0	0	0	13.1	34.3	0.1047	2.7384	0
3	0	20	0	0	0	19.7	39.7	0.09717	2.575179	0
4	0	30.95	0	0	0	5.35	36.3	0.10242	3.066846	0
5	0	23.75	0	0	0	14.15	37.9	0.09543	2.695632	0
6	0	30.4	0	0	0	5	35.4	0.09275	2.95785	0
7	0	29.5	0	0	0	5	34.5	0.10069	2.822805	0
8	0	24.95	2.4452	2.2952	0	5	30.1	0.11433	3.115833	0
9	0	37.25	11.663	25.513	0	5	28.4	0.11379	2.906136	0
10	0	20.45	20.334	11.984	0	5	33.8	0.11785	3.65783	0
11	0	30.2	24.028	20.828	0	5	38.4	0.11481	4.083204	0
12	0	24.05	24.884	12.634	0	5	41.3	0.1236	4.77918	0
13	0	27.05	24.381	12.531	0	5	43.9	0.13314	5.519346	0
14	0	30.25	22.624	12.274	0	5	45.6	0.13097	5.646732	0
15	0	37.25	24.663	20.213	0	5	46.7	0.12128	5.338276	0
16	0	37.25	15.107	9.057	0	5	48.3	0.1295	5.92935	0
17	0	32.75	1.7755	0	9.8745	5	49.4	0.13939	6.560366	0
18	0	37.25	0	0	6.95	5	49.2	0.14457	6.787344	0
19	0	19.55	0	0	25.55	5	50.1	0.15595	7.487595	0
20	0	25.25	0	0	21.45	5	51.7	0.15214	7.540138	0
21	0	26.75	0	0	18.75	5	50.5	0.12902	6.19001	0
22	0	20.15	0	0	20.45	5	45.6	0.11711	5.014716	0
23	0	26.15	0	0	8.95	5	40.1	0.11621	4.334521	0
24	0	26.15	0	0	7.65	5	38.8	0.11083	3.974704	0

Table 7.9 The savings and the NPV of the microgrid components in islanding mode

Years	Capital cost ($)	O&M cost ($ year^{-1})	Total savings ($ year^{-1})	Net cash flow ($ year^{-1})	NPV ($)
1	−432,710	2,163.55	39,962.99	−35,999.47	−396,710
2		2,163.55	39,962.99	34,285.21	−362,425
3		2,163.55	39,962.99	32,652.58	−329,773
4		2,163.55	39,962.99	31,097.69	−298,675
5		2,163.55	39,962.99	29,616.85	−269,058
6		2,163.55	39,962.99	28,206.52	−240,852
7		2,163.55	39,962.99	26,863.36	−213,988
8		2,163.55	39,962.99	25,584.15	−188,404
9		2,163.55	39,962.99	24,365.86	−164,038
10		2,163.55	39,962.99	23,205.58	−140,833
11		2,163.55	39,962.99	22,100.55	−118,732
12		2,163.55	39,962.99	21,048.14	−97,684.1
13		2,163.55	39,962.99	20,045.85	−77,638.2
14		2,163.55	39,962.99	19,091.29	−58,546.9
15		2,163.55	39,962.99	18,182.18	−40,364.7
16		2,163.55	39,962.99	17,316.36	−23,048.4
17		2,163.55	39,962.99	16,491.77	−6,556.61
18		2,163.55	39,962.99	15,706.45	9,149.839
19		2,163.55	39,962.99	14,958.52	24,108.36
20		2,163.55	39,962.99	14,246.21	38,354.57

Table 7.10 The cost data associated with the microgrid components

Capital costs				Yearly O&M costs
WT ($ kW^{-1})	PV ($ kW^{-1})	BESS ($ kW h^{-1})	MT ($ kW^{-1})	2% of capital investment costs
3,000	2,500	1,950	500	

and technical data of the system components are given in Tables 7.10 and 7.11, respectively. The solar irradiance and wind speed forecast data is taken from [41]. While the load demand data is forecasted using built-in ARIMA model of the MATLAB® software. The degradation is considered as 2% per year. The parameters related to probabilistic wind and solar PV models are expressed in Table 7.1.

The MG component sizing is analysed in the following two different scenarios:

7.5.2.1 Scenario 1

In this scenario, it is considered that the system can utilize all the resources such as WT, solar PV, BESS and MT. Additionally, the surplus energy generated by the MG can be sold to the grid at a 20% discounted price of the purchase rate in order to account for the transmission losses. Table 7.12 shows the optimum size of the components calculated by the BSO algorithm. Since the load is variable, it is highly economical to store the energy produced by the renewable energy sources and to

Table 7.11 The parameters data associated with the microgrid components

Nominal return rate (%)	Project lifetime (years)	$P_{BESS,C}$, $P_{BESS,D}$ (kW)	$S_{BESS,\min}$ (%)	$S_{BESS,\max}$ (%)	$S_{BESS,start}$ (%)	$\eta_{BESS,C}$ (%)	$\eta_{BESS,C}$ (%)
3.75	20	50	10	95	50	90	90

Table 7.12 The optimum size of the microgrid components

Wind power plant (kW)	PV power plant (kW)	BESS (kW h)	MT (kW)
432	1,113	2,169	400

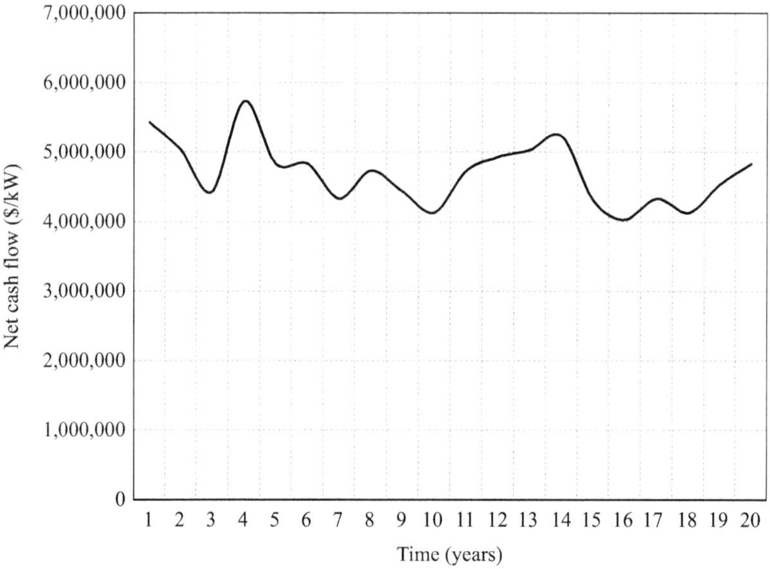

Figure 7.21 The yearly net cash flow due to savings for the 20 years

recharge the BESS when required; this needs a higher BESS size. The higher the size of BESS, the higher is the possibility of shifting the load between the solar PV and the wind as per economy. However, this feature is limited due to the variability of the renewable energy resources. Figure 7.21 shows the yearly net cash flow (profit from savings) for the 20 years. The net cash flow has an irregular pattern due to the intermittent generation sources and variable load demand. The profit is higher in the starting years, while it decreases gradually due the degradation of MG components. In the 1st year, the value of NPV resulted as $−41,218,178.59, and in the 11th year, the value of NPV became positive and resulted as $2,438,329.518.

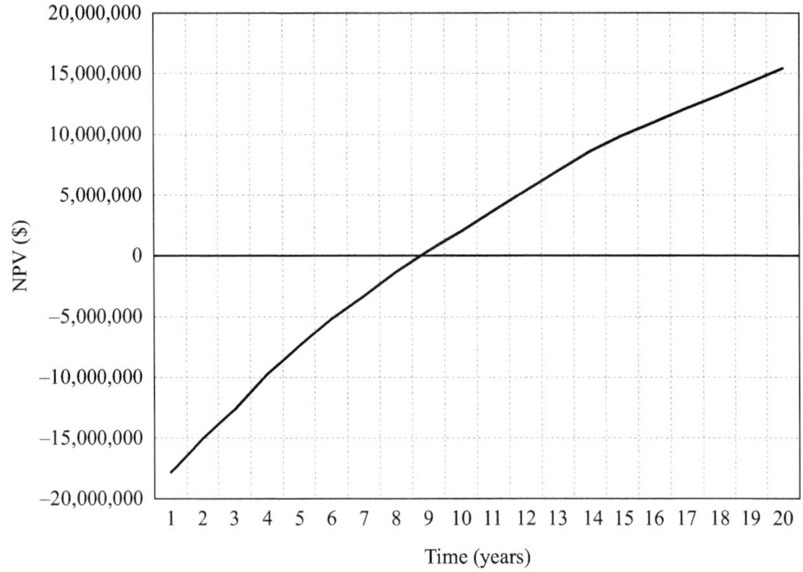

Figure 7.22 The yearly net present value for the 20 years

Table 7.13 The optimum size of the microgrid components in islanded mode

Wind power plant (kW)	PV power plant (kW)	BESS (kWh)	MT (kW)
397	863	1,500	416

Consequently, by the end of the 20th year, the NPV becomes $26,396,470.67. The NPV for the MG as shown in Figure 7.22 for the period of 20 years demonstrates the fact that it takes 10 years to recover the initial investment, although the project is in profit.

7.5.2.2 Scenario 2

In this scenario, by considering the limitations of large deployments of renewable energy sources in the real-world design cases, the focus is on the sizing of the MG components for matching the need of MG itself. Here, the savings through selling of power to the grid are neglected. Table 7.13 shows the results, where the resulted capacities are a bit smaller than scenario 1. The NPV for this scenario is shown in Figure 7.23. The BESS size in this case decreases significantly as expected due to lesser peak-shaving profits. However, the most remarkable is the decrease in the size of solar PV, whereas the size of WT does not decrease much, that is due to the lower on–off margins. Moreover, the size of the MT increases slightly, that is because of its definite nature of power supply. In the 1st year, the value of NPV resulted as $−17,820,565.01, and in the 9th year, the value of NPV became positive

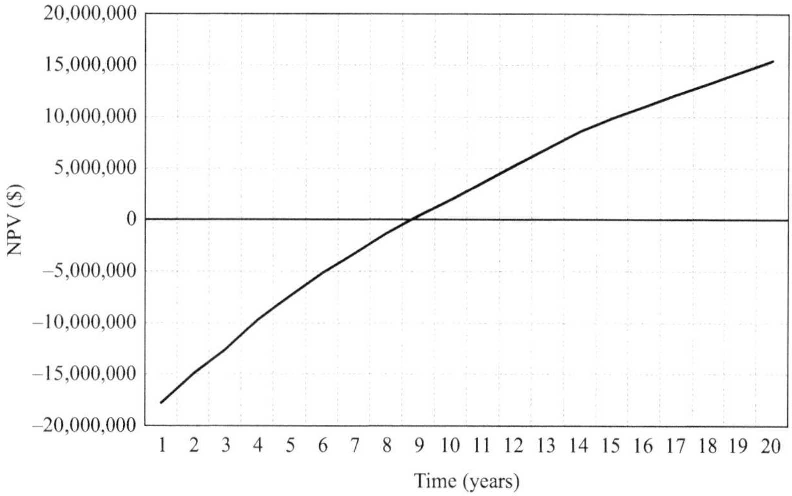

Figure 7.23 The yearly net present value

and resulted as $393,608.4392. Consequently, by the end of 20th year, the NPV becomes $15,430,924.48. The NPV for the MG as shown in Figure 7.22 for the period of 20 years demonstrates the fact that it takes 8 years to recover the initial investment. The NPV is lesser than that obtained in the first scenario, that is majorly because of negligence of the energy trade with the grid, but the size of the components is effectively decreased, which decreases the capital cost. That is why it took 8 years in this mode to return the initial investment.

7.6 Summary

In summary, this chapter details MG component sizing problem in conjunction with other chapters. It discusses the common methodologies applied for obtaining time-series data sets, deriving the planning model and formulation, and solving the associated optimization problem. The probabilistic wind and solar PV power models are important as they handle the corresponding uncertain generation profiles. The solution technique, BSO algorithm, is one of the key aspects of this chapter that derives the optimal sizing of the MG components.

It is observed in the case studies, that PV generation always coincides with price peaks, while the wind power generation is distributed throughout all time periods. Therefore, when the prices become more flat, solar PV generation loses the associated benefits and the wind generation starts becoming more effective. Furthermore, the supplied power becomes more expensive under such cases. Similarly, when the price peaks come too often, the BESS size tends to increase, as more flexibility is required to gain the benefit out of higher discharge during the peak hours. All of these resources, such as WT, solar PV and BESS, behave in a similar manner against the load demand. In the meanwhile, MT plays quite a passive role as a flexible but a definite source of energy that is mostly utilized during the non-peak hours, although the emission cost always tends to limit its supply with a higher preference for WT, solar PV and BESS as much as possible.

In conclusion, the detail contents in this chapter accompanied with case study simulations will provide a comprehensive guide for MG planners in the context of sizing MG components.

References

[1] Quashie, M., 'Optimal Planning of Advanced Microgrids with an Energy Management System', (McGill University, 2017).

[2] Arriaga Marin, M., 'Long-Term Renewable Energy Generation Planning for Off-Grid Remote Communities', 2015.

[3] Broekhoven, S.B.V., Judson, N., Nguyen, S.V.T., and Ross, W.D., 'Microgrid Study: Energy Security for DoD Installations', (Lincoln Laboratory, Massachusetts Institute of Technology, 2012).

[4] Murray-Wallace, C. and Goede, A., 'Aminostratigraphy and Electron Spin Resonance Dating of Quaternary Coastal Neotectonism in Tasmania and the Bass Strait Islands', Australian Journal of Earth Sciences, 1995, 42, (1), pp. 51–67.

[5] Abbey, C., Cornforth, D., Hatziargyriou, N., *et al.*, 'Powering through the Storm: Microgrids Operation for More Efficient Disaster Recovery', IEEE Power and Energy Magazine, 2014, 12, (3), pp. 67–76.

[6] Talei, H., Zizi, B., Abid, M.R., Essaaidi, M., Benhaddou, D., and Khalil, N., 'Smart Campus Microgrid: Advantages and the Main Architectural Components', in, Renewable and Sustainable Energy Conference (IRSEC), 2015 3rd International, (IEEE, 2015).

[7] Chen, S., Gooi, H.B., and Wang, M., 'Sizing of Energy Storage for Microgrids', IEEE Transactions on Smart Grid, 2012, 3, (1), pp. 142–151.

[8] Stadler, M., Cardoso, G., Mashayekh, S., *et al.*, 'Value Streams in Microgrids: A Literature Review', Applied Energy, 2016, 162, pp. 980–989.

[9] Sheikhi, A., Ranjbar, A., and Oraee, H., 'Financial Analysis and Optimal Size and Operation for a Multicarrier Energy System', Energy and Buildings, 2012, 48, pp. 71–78.

[10] Bahramirad, S., Reder, W., and Khodaei, A., 'Reliability-Constrained Optimal Sizing of Energy Storage System in a Microgrid', Perspectives, 2012, 1, p. 3.

[11] Kaur, A., Kaushal, J., and Basak, P., 'A Review on Microgrid Central Controller', Renewable and Sustainable Energy Reviews, 2016, 55, pp. 338–345.

[12] Gamarra, C. and Guerrero, J.M., 'Computational Optimization Techniques Applied to Microgrids Planning: A Review', Renewable and Sustainable Energy Reviews, 2015, 48, pp. 413–424.

[13] Xiao, C., Soetanto, D., Muttaqi, K., and Zhang, M., 'A Dynamic Evolutionary Strategy for Time Ahead Energy Storage Management in Microgrid', in, 2016 IEEE International Conference on Power System Technology (POWERCON), (IEEE, 2016).

[14] Basak, P., Chowdhury, S., Chowdhury, S.P., and nee Dey, S.H., 'Simulation of Microgrid in the Perspective of Integration of Distributed Energy Resources', in, Energy, Automation, and Signal (ICEAS), 2011 International Conference on, (IEEE, 2011).
[15] Bayindir, R., Hossain, E., Kabalci, E., and Perez, R., 'A Comprehensive Study on Microgrid Technology', International Journal of Renewable Energy Research, 2014, 4, (4), pp. 1094–1107.
[16] Morris, G.Y., Abbey, C., Joos, G., and Marnay, C., A Framework for the Evaluation of the Cost and Benefits of Microgrids', (Lawrence Berkeley National Lab. (LBNL), Berkeley, CA (United States), 2011).
[17] Tan, K., So, P., Chu, Y., and Chen, M., 'A Flexible Ac Distribution System Device for a Microgrid', IEEE Transactions on Energy Conversion, 2013, 28, (3), pp. 601–610.
[18] Tavakoli, A., Negnevitsky, M., and Muttaqi, K.M., 'A Decentralized Model Predictive Control for Operation of Multiple Distributed Generators in an Islanded Mode', IEEE Transactions on Industry Applications, 2017, 53, (2), pp. 1466–1475.
[19] Roy, T.K., Mahmud, M.A., Oo, A.M.T., Haque, M.E., Muttaqi, K.M., and Mendis, N., 'Nonlinear Adaptive Backstepping Controller Design for Islanded DC Microgrids', IEEE Transactions on Industry Applications, 2018, 54, (3), pp. 2857–2873.
[20] Tan, Y., Meegahapola, L., and Muttaqi, K.M., 'A Review of Technical Challenges in Planning and Operation of Remote Area Power Supply Systems', Renewable and Sustainable Energy Reviews, 2014, 38, pp. 876–889.
[21] Rabiee, A., Farahani, H.F., Khalili, M., Aghaei, J., and Muttaqi, K.M., 'Integration of Plug-in Electric Vehicles into Microgrids as Energy and Reactive Power Providers in Market Environment', IEEE Transactions on Industrial Informatics, 2016, 12, (4), pp. 1312–1320.
[22] Li, Y., Vilathgamuwa, D.M., and Loh, P.C., 'Microgrid Power Quality Enhancement Using a Three-Phase Four-Wire Grid-Interfacing Compensator', IEEE Transactions on Industry Applications, 2005, 41, (6), pp. 1707–1719.
[23] Mendis, N., Mahmud, M.A., Roy, T., Haque, M.E., and Muttaqi, K.M., 'Power Management and Control Strategies for Efficient Operation of a Solar Power Dominated Hybrid Dc Microgrid for Remote Power Applications', in, 2016 IEEE Industry Applications Society Annual Meeting, (IEEE, 2016).
[24] Hossain, M.A., Pota, H.R., Squartini, S., Zaman, F., and Muttaqi, K.M., 'Energy Management of Community Microgrids Considering Degradation Cost of Battery', Journal of Energy Storage, 2019, 22, pp. 257–269.
[25] Khederzadeh, M., 'Application of Wide-Area Protection Concepts in Microgrids', IET Conference Proceedings, 2013, pp. 0212–0212. Available: https://digital-library.theiet.org/content/conferences/10.1049/cp.2013.0612.
[26] Abdelaziz, M.M., Farag, H.E., and El-Saadany, E., 'Fuel-Saving Benefit Analysis of Islanded Microgrid Central Controllers', Electric Power Systems Research, 2015, 125, pp. 45–54.
[27] Valipour, M., 'Long-Term Runoff Study Using Sarima and Arima Models in the United States', Meteorological Applications, 2015, 22, (3), pp. 592–598.

[28] Fan, S., Chen, L., and Lee, W.-J., 'Machine Learning Based Switching Model for Electricity Load Forecasting', Energy Conversion and Management, 2008, 49, (6), pp. 1331–1344.
[29] Guo, Z., Zhao, W., Lu, H., and Wang, J., 'Multi-Step Forecasting for Wind Speed Using a Modified EMD-Based Artificial Neural Network Model', Renewable Energy, 2012, 37, (1), pp. 241–249.
[30] Mohy-ud-din, G., 'Hybrid Dynamic Economic Emission Dispatch of Thermal, Wind, and Photovoltaic Power Using the Hybrid Backtracking Search Algorithm with Sequential Quadratic Programming', Journal of Renewable and Sustainable Energy, 2017, 9, (1), p. 015502.
[31] Xiaoping, L., Ming, D., Jianghong, H., Pingping, H., and Yali, P., 'Dynamic Economic Dispatch for Microgrids Including Battery Energy Storage', in, Power Electronics for Distributed Generation Systems (PEDG), 2010 2nd IEEE International Symposium on, (IEEE, 2010).
[32] Civicioglu, P., 'Backtracking Search Optimization Algorithm for Numerical Optimization Problems', Applied Mathematics and Computation, 2013, 219, (15), pp. 8121–8144.
[33] Modiri-Delshad, M. and Rahim, N.A., 'Solving Non-Convex Economic Dispatch Problem Via Backtracking Search Algorithm', Energy, 2014, 77, pp. 372–381.
[34] Castillo, A. and Gayme, D.F., 'Grid-Scale Energy Storage Applications in Renewable Energy Integration: A Survey', Energy Conversion and Management, 2014, 87, pp. 885–894.
[35] Walters, D.C. and Sheble, G.B., 'Genetic Algorithm Solution of Economic Dispatch with Valve Point Loading', IEEE Transactions on Power Systems, 1993, 8, (3), pp. 1325–1332.
[36] Park, Y.-M., Won, J.-R., and Park, J.-B., 'A New Approach to Economic Load Dispatch Based on Improved Evolutionary Programming', Engineering Intelligent Systems for Electrical Engineering and Communications, 1998, 6, (2), pp. 103–110.
[37] Yang, H.-T., Yang, P.-C., and Huang, C.-L., 'Evolutionary Programming Based Economic Dispatch for Units with Non-Smooth Fuel Cost Functions', IEEE transactions on Power Systems, 1996, 11, (1), pp. 112–118.
[38] Park, J.-B., Lee, K.-S., Shin, J.-R., and Lee, K.Y., 'A Particle Swarm Optimization for Economic Dispatch with Nonsmooth Cost Functions', IEEE transactions on Power Systems, 2005, 20, (1), pp. 34–42.
[39] Al-Sumait, J., Al-Othman, A., and Sykulski, J., 'Application of Pattern Search Method to Power System Valve-Point Economic Load Dispatch', International Journal of Electrical Power & Energy Systems, 2007, 29, (10), pp. 720–730.
[40] Alsumait, J., Sykulski, J., and Al-Othman, A., 'A Hybrid GA–PS–SQP Method to Solve Power System Valve-Point Economic Dispatch Problems', Applied Energy, 2010, 87, (5), pp. 1773–1781.
[41] BoM, Climate information for solar energy, Bureau of Meteorology – BoM [online], 2019, Available: http://www.bom.gov.au/climate/data-services/solar-information.shtml

Chapter 8

Optimal sizing of energy storage system

Kamran Jalilpoor[1]*, Rahmat Khezri*[2]*, Amin Mahmoudi*[2]*, and Arman Oshnoei*[1]

Energy storage system (ESS) as a growing technology plays a substantial role in operation and planning of microgrids. It is a viable option in smoothing the power of renewable energy sources (RESs), peak load mitigation, voltage control, frequency regulation and reliability enhancement of microgrid. Sizing of energy storages in microgrids is still a challenge. A non-optimal small capacity of storage system cannot ensure the proper operation of a microgrid. On the other side, a non-optimal large capacity may increase the cost and power losses in the system. In this chapter, different types of energy storages with their characteristics such as capacity, cost and efficiency are introduced. The necessity of energy storages in microgrids is explained. Optimal sizing of an ESS in a microgrid is discussed. The objective function (OF), system constraints and operation conditions are addressed.

8.1 Introduction

A new concept of an energy delivery system known as the microgrid is a localised cluster of distributed energy resources. It especially includes the RES, energy storages and loads which can operate in stand-alone and grid-connected modes. Microgrids offer some unique benefits such as high reliability, clean energy by renewable sources, low power losses due to assembled infrastructure and improved grid resiliency. In contrast, microgrids have complexities and introduce new challenges in design and operation. Majority of the power is produced by intermittent and unreliable renewable energies. Spinning reserves are still a drawback as the penetration of diesel generator's production has decreased. Additionally, to work in a stand-alone mood with renewable energies, microgrid needs facilities to supply the loads all the time. A potential remedy for most of the identified challenges of microgrids is ESS.

Energy storage is an important component of microgrids. When demand power management is a concern, ESS improves the peak load mitigation in microgrids.

[1]Faculty of Electrical and Computer Engineering, Shahid Beheshti University, Tehran, Iran
[2]College of Science and Engineering, Flinders University, Adelaide, Australia

Considering RES integration in microgrids, ESS smooths the intermittent output power of renewable sources. ESS is applicable to improve voltage control and frequency regulation. In economic planning, ESS can store energy in low-cost times and release it in periods of high electricity price to the main grid in grid-connected mode. Although energy storages are potential candidature for microgrid problems, there are challenges in ESS integration. ESS should be optimally designed for microgrid to obtain a maximum economic and operation efficiency.

The sizing problem of ESS concerns both economic and operation aspects of the microgrids. Energy storage technologies are not profitable if oversized. A large capacity of energy storage may increase the power losses and costs in microgrid. Conversely, if undersized, ESS may not satisfy the desired operation and economic benefits.

This chapter is organised in five sections. Section 8.2 explains the energy storage technologies and their characteristics. Section 8.3 addresses the necessity of energy storages in microgrids and their applications. Section 8.4 evaluates and discusses the optimal sizing and placement of the ESS in a microgrid.

8.2 Energy storage technologies in microgrids: types and characteristics

ESSs reserve the electrical energy in multiple forms (chemical, electrochemical, electromagnetic, mechanical and thermal) and when needed convert it back into the electrical energy. Energy storages are developed for multiple applications. They are accessible in generation, transmission and distribution levels of power systems. Figure 8.1 shows a classification of applicable energy storage technologies for

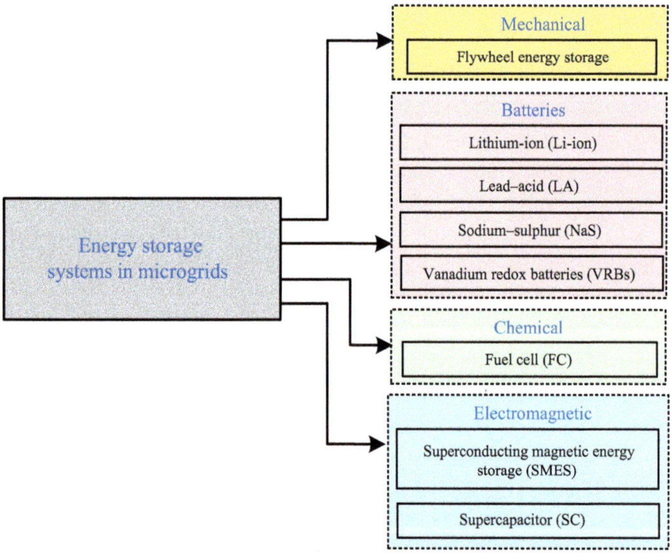

Figure 8.1 Energy storage technologies classification suitable for microgrids

microgrids based on the forms of storage. Other types of energy storages such as thermal storages are either still in technological development stage or not appropriate for the microgrids. Pumped hydro storages are the popular type of energy storages applicable in power system scales but not in microgrids. They are geographically limited and not economical in medium or small sizes. Battery energy storage technology has the most share in microgrids among the other storage technologies. In 2018, the total estimated installed energy storage around the world, without considering hydro pump energy storage, reached 5.9 GW for stationary and grid-connected energy storage capacity [1].

8.2.1 Battery energy storage systems

Battery ESSs (BESS) are one of the most applicable storage systems in microgrids. Batteries store electrical energy in the form of electrochemical energy. They are composed of multiple cells connected in series or parallel or both, to provide a desired capacity and voltage through electrochemical reactions. In the battery structure, each cell consists of an electrolyte and two electrodes (one cathode and one anode). Due to low voltage of each module, a number of modules are connected in series or parallel to provide the desired voltage and capacity in a battery. Figure 8.2 shows the schematic diagram of a BESS. The output power of a BESS is a direct current (DC) power which needs converting to alternative current (AC) to be used in the external grid. As demonstrated, BESS is connected to the external grid via bidirectional inverter which makes the possibility of converting DC to AC and vice versa.

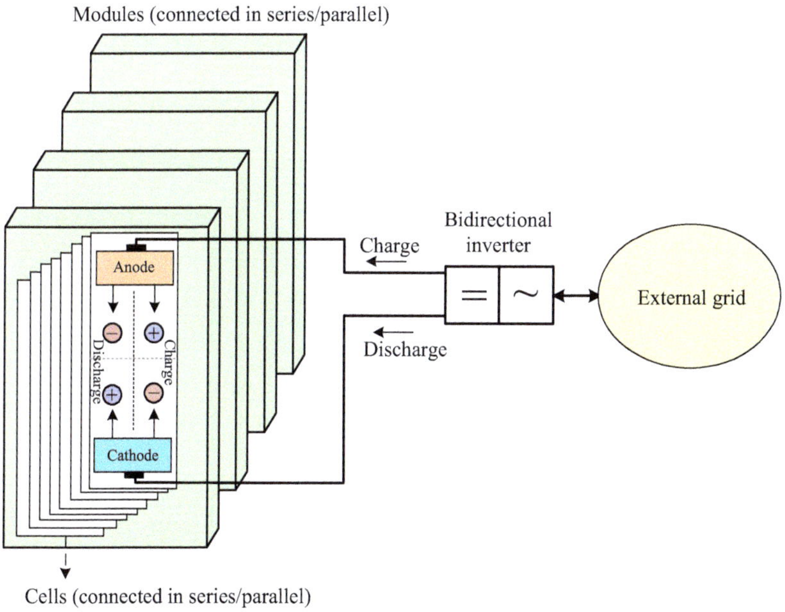

Figure 8.2 Structure of a battery storage system

Various types of battery technologies with different materials in their electrodes and electrolyte are developed recently. Some of these technologies are in the commercialising stage, and some are still in the experimental stage. This chapter only investigates the battery technologies popular in microgrids which include the lithium-ion (Li-ion), lead–acid (LA), sodium–sulphur (NaS) and VRB. Figure 8.3 demonstrates the installed capacity of different types of BESS around the world by 2016 [2]. Explanations of this chapter include the advantages, disadvantages and applications of BESS in microgrids. The application of different materials and chemical reactions in the BESS are ignored in this chapter.

Table 8.1 lists the advantages, disadvantages and applications of various battery technologies in electrical power systems. High efficiency and fast response time are the common features of the battery storage systems. Due to the fast response time, BESS contributes to short-term applications such as frequency and voltage control as well as the long-term applications such as demand shifting by high-energy density. Furthermore, battery storage systems prepare the microgrids for high level of RES integration.

8.2.1.1 Lithium-ion

Li-ion battery is one of the most popular types of batteries commercially available. It provides high power and energy densities. The 100 MW/129 MW h battery system integrated in Hornsdale Wind Farm in South Australia is the biggest commercialised Li-ion battery in the world [3]. Although this BESS is integrated in power system level, it shows that the Li-ion can be simply available in a large size for microgrid application. Other features of Li-ion battery storage system are high efficiency, long lifetime without memory effect and low self-discharging.

The Li-ion BESS is an appropriate selection for variety of applications in microgrids such as frequency regulation, voltage control, renewable resources integration and power quality improvement.

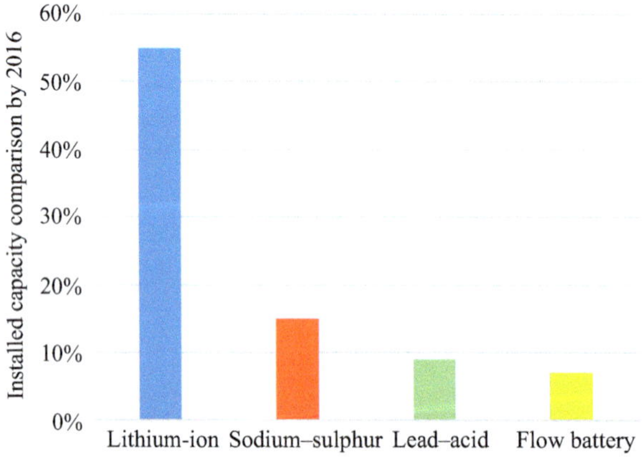

Figure 8.3 Installed BESS capacity around the world by 2016

Table 8.1 Batteries advantages, disadvantages and applications

Battery type	Advantages	Disadvantages	Applications
Li-ion	• High energy and power density • Fast charge/discharge capability • High efficiency	• Hard maintenance • High cost • Require recycling	• Demand shifting • Peak shaving • Network expansion • Voltage control • Frequency regulation
LA	• Lowest capital cost • High efficiency	• Low power and energy density • Short lifetime • Limited temperature operation • High self-discharge rate	• Energy management • Network stabilisation • Emergency backup • Power quality • End-user service • Load levelling • Renewable resources integration
NaS	• High-energy density • Long lifetime (cycles) • High efficiency and no self-discharge	• High temperature operation • High cost • Require recycling	
VRB	• High efficiency • Long lifetime (cycles) • Low self-discharging	• Low energy and power density	

8.2.1.2 Lead–acid

LA battery is the oldest rechargeable storage technology, which was invented in 1859 by Gaston Plante. This type of battery is the most economical type with the lowest capital cost for microgrid applications. However, it has some disadvantages which make it inappropriate for microgrid application. Short lifetime and high discharge rate as well as low power and energy densities have limited the application of LA batteries. Deep and/or rapid discharges decrease the LA capacity significantly. A real-case LA battery storage system (3 MW/1.6 MW h) is installed in King Island microgrid of Tasmania, Australia for the purposes of power smoothing and demand shifting [3].

8.2.1.3 Sodium–sulphur

As a recent battery technology, NaS is a promising option for high-power energy applications. The commercialised 45 MW NaS battery storage system in Japan is the largest reported capacity of NaS [4]. Zero self-discharging, long lifetime cycle and low maintenance are the important features of NaS. Despite all the advantages, high-temperature operation condition of the NaS battery is the major concern for this technology. The NaS battery operates under high temperature around 350°C [4].

8.2.1.4 Vanadium redox batteries

As a type of rechargeable flow battery, VRB is a proportionately young technology. The VRB technology uses two tanks to store two different aqueous electrolytes, an analytic reservoir and a catholyte reservoir. Fast response time and high charge/discharge cycles (more than 12,000) as well as the high-operation efficiency are the prominent features for VRB. Low energy and power densities reduce the VRB potential to be used in long-term applications.

8.2.2 Flywheel

Flywheel energy storage as a rotating mass stores the electrical energy in the form of rotational energy. When flywheel starts charging, it operates by accelerating a rotor to high speeds in order to store the electrical energy as kinetic energy. By deaccelerating or speed reduction, flywheel starts discharging and releasing electrical energy to the system. Flywheels act as an uninterruptable power supply in microgrids successfully [5]. It is considered an appropriate short-term energy storage with low cost as well as high lifetime (15–20 years) and efficiency. Flywheel has a low power rating and energy density which decrease its suitability for long-term applications. Figure 8.4 demonstrates a general schematic diagram of a flywheel. In a flywheel-based system, rotation speed and inertia determine the amount of stored energy. The output power of flywheel is an AC power. However, to use flywheel in a microgrid, it is needed to control the frequency and voltage of the flywheel. The AC/DC/AC converter in the main structure of flywheel gives the capability to control the output voltage and frequency of flywheel.

8.2.3 Fuel cell

As a type of hydrogen-based energy storage technology, fuel cell (FC) converts chemical energy into electrical energy and vice versa. The FC is an environmental friendly technology since it is a zero or low-emission pollutant storage system. High-energy density and power rating are other important advantages of the FCs. High capital cost and low efficiency of the FCs are hindrances for their viable applications in microgrid.

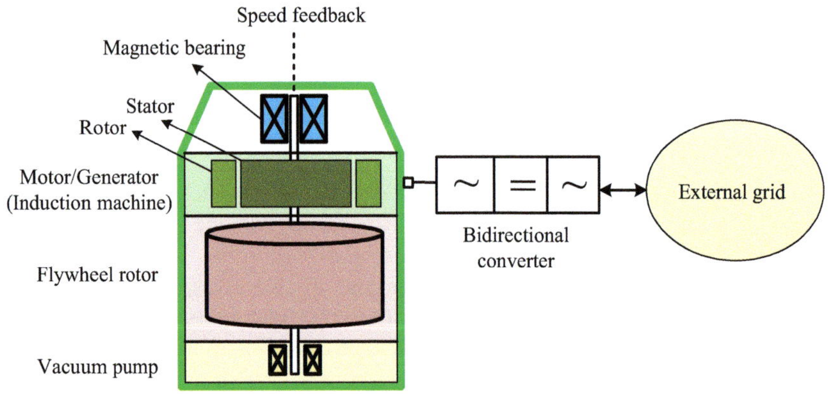

Figure 8.4 Flywheel schematic diagram

An FC storage system is shown in Figure 8.5. The main structure contains a FC system, water electrolyser and hydrogen storage. Electrolyser decomposes the water into hydrogen and oxygen. Hydrogen storage is the key component as it stores the energy.

8.2.4 Superconducting magnetic energy storage

As an almost recent technology, superconducting magnetic energy storage (SMES) stores energy in a magnetic field created by the DC current in a superconducting coil. The most important feature of the SMES is very fast charge and discharge capability with high power density. SMES is still a new technology and needs maturity to be completely applicable. Figure 8.6 illustrates an SMES structure which consists of two main components known as refrigeration system and superconducting coil. The electrical energy is stored in the magnetic field of the superconducting coil generated by the DC connection. A bidirectional inverter is used to convert the output DC power of SMES to an AC power and vice versa.

8.2.5 Supercapacitor

Capacitors are the main component in the supercapacitor (SC) energy storages. SCs have similar structure as batteries with one electrolyte and two electrodes (cathode and anode) as well as a porous membrane separator. They are known as ultracapacitors or electric double-layer capacitors. The SC benefits from delivering high power for a short period of time; actually, it has the highest power density among

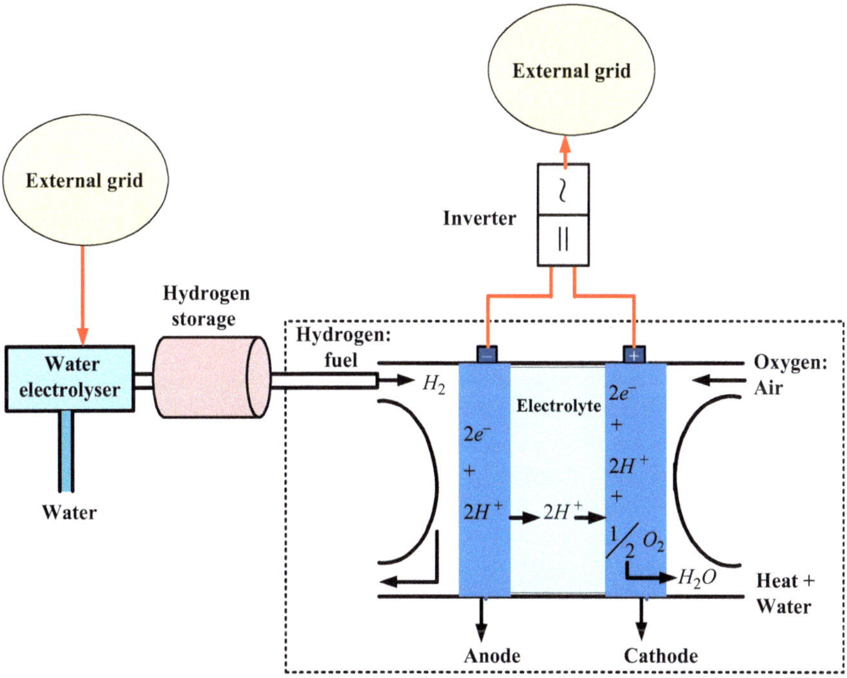

Figure 8.5 Fuel cell storage configuration

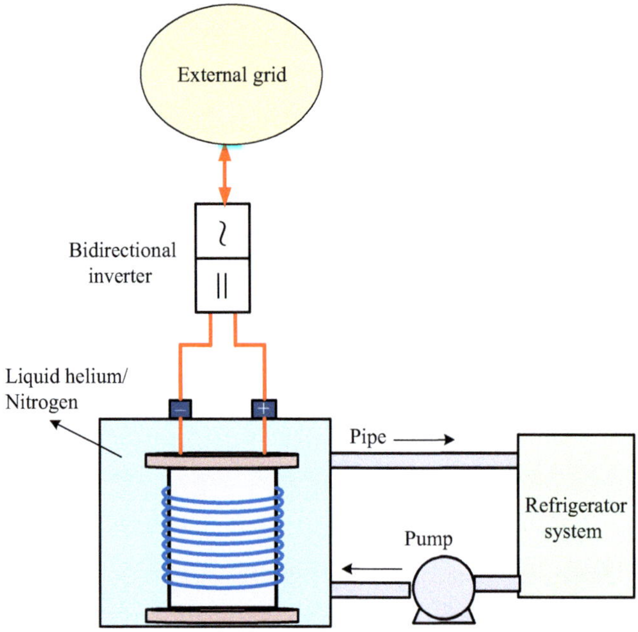

Figure 8.6 SMES schematic diagram

the other energy storage technologies. Long cycling lifetime $(\mathbf{1} \times \mathbf{10^5})$ as well as high efficiency is the other significant feature of the SCs. However, the power rating and energy density are low in an SC.

8.2.6 Technology comparison

Extracted from [6–8], Table 8.2 summarises the operation and cost characteristics of ESSs applicable in microgrids. The operation characteristics include power rating and density, energy density, discharge time, efficiency, response time and lifetime, which are extremely important considering ESS technologies. Based on the table, the Li-ion battery storages have the highest power rating and capital cost. The electromagnetic storage system including SMES and SC possess the fastest response time. Flywheel is unique for the lowest cost. FC has the highest energy density.

The application of energy storages in microgrids is related to the necessity of the microgrid. For fast response operations like frequency control and power smoothing, SMES and SC are appropriate candidates. Flywheel is a suitable selection for black start in microgrids, and FCs can be selected when the high amount of energy is required. This is while, battery storage systems offer mature and available technology in the market and hence are considered suitable choices in microgrids generally.

A proper attention is required for the point of common coupling of ESSs in microgrids. AC- and DC-coupling microgrids are the key elements for integrating renewable and distributed energy resources as well as ESSs. In AC-coupling microgrids, all the components are connected to an AC bus, while in DC-coupling

Table 8.2 Operation characteristics and approximate costs of various storage systems [8–10]

Energy storage technology	Capital cost ($/kW)	Power rating (MW)	Discharge time	Power density (W/l)	Energy density (W h/l)	Efficiency (%)	Response time	Lifetime (years)	Lifetime (cycles)
Flywheel	250–350	0–0.25	ms–15 min	5,000	20–80	80–90	<s	15–20	100–1,000
LA	300–600	0–20	s–h	90–700	50–80	75–90	<s	3–15	250–1,500
Li-ion	1,000–4,000	0–100	s–h	1,300–10,000	200–400	85–90	<s	5–20	600–1,200
NaS	1,000–3,000	0.05–40	s–h	120–160	15–300	80–90	<s	10–15	2,500–4,500
VRB	600–1,500	0.03–3	s–10 h	0.5–2	20–70	80–90	<s	5–10	$12,000^+$
FC	$10,000^+$	0–50	s–24 h	0.2–20	600	30–40	s–min	5–15	1,000
SMES	200–400	0.1–10	ms–8 s	2,600	6	80–90	<ms	10–20	1,000
SC	100–300	0.05–0.3	ms–1 h	$50,000^+$	10–30	80–90	<ms	8–20	$10,000^+$

microgrids, a DC bus is the place of integration of the components. Figure 8.7 shows the connection of energy storages in AC- and DC-coupling microgrids. The energy storages are connected via DC–DC converters in the DC-coupling to convert the output voltage of energy storage to the standard level of the microgrid. This occurs when energy storages use bidirectional inverters in AC-coupling microgrids to have a bidirectional connection with the network.

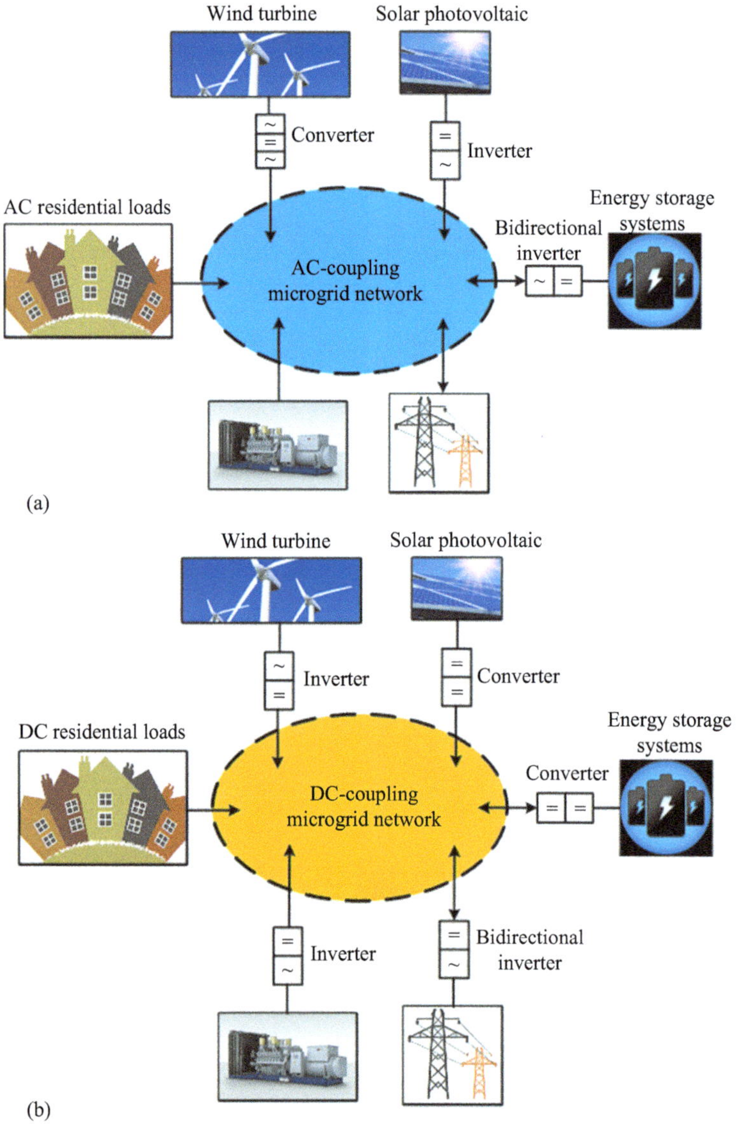

Figure 8.7 Microgrid network: (a) DC-coupling microgrid, (b) AC-coupling microgrid

8.3 Necessity of energy storage in microgrids

A microgrid is defined as a set of loads, distributed resources and ESSs which operate as a stand-alone/grid-connected system. Distributed renewable energy systems in the microgrid increase the voltage, frequency and active power fluctuations which result in imbalanced supply and demand power. The uncertainty in the output power of RES complicates the operation of microgrid [9] and should be supported by fossil-fuel-based generation units (which are expensive and air-polluter energy resources).

The output power of RESs is intermittent. Actually, an RES is more productive at some hours (such as wind power at midnight and solar energy at noon). This productivity may not match with peak load in the microgrid. These factors lead the microgrid to employ storage systems with distributed generation resources [10].

ESS in the microgrid improves the power quality and reliability. Another important advantage in microgrid planning and management is air pollution reduction using storage systems which reduces the penetration of fossil-fuel-based units. By the application of the ESS as a source of energy saving, a large benefit of microgrid can be obtained by energy exchanging with the main grid. This means that the electricity can be purchased from the main grid in low-cost times and then sold in high-price periods. Figure 8.8 shows different benefits of the ESS application in microgrids.

Key applications for ESS in microgrids can be categorised in the following subsection.

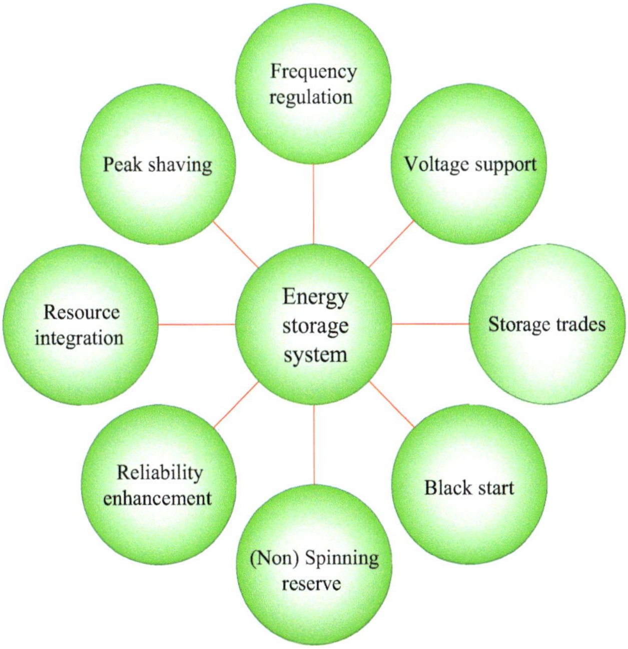

Figure 8.8 Advantages of using energy storage systems in microgrid

8.3.1 Frequency regulation

Frequency regulation is the continuous balancing of supply and demand under normal and disturbance conditions. It is automatically performed at intervals of minutes (or even shorter) by primary and secondary frequency control. High speed in charging and discharging is the main advantage of the ESS for frequency regulation. This means that ESS can respond to the power imbalances in a very short time compared to the conventional power plants. In a microgrid, shutting down of energy sources, disconnection of large loads and RES fluctuations can be concluded in frequency deviations [11]. ESS can be generally integrated in the microgrid by an appropriate strategy to improve the frequency regulation (either by active power releasing/absorbing or using a controller in its structure).

8.3.2 Voltage support

Voltage and reactive power control are widely used to increase the power quality in microgrids. Recently, the use of distributed generation based on renewable energy has been increased. This issue affects the voltage and reactive power control as well as the system performance. Random and unpredictable changes in RESs and uncertainty of the electric demand cause severe system voltage fluctuations. ESS has the capability to control the reactive power in microgrids for efficient voltage regulation [12].

8.3.3 Reliability enhancement

Reliable power supply means an uninterruptible power supply which is a significant issue in microgrids with intermittent RES. Any supply failure in a microgrid can result in performance degradation and disruption of the system. This causes serious damages to the generation and demand sides of microgrids. The average cost of a power outage or generation failure in the microgrid is very high. In such condition, the microgrid can use ESS to avoid interruptions and maximise resource efficiency. In addition to microgrid resources, ESS has stored energy as a backup which increases the reliability and stability of the microgrid [13].

8.3.4 Demand shifting and peak shaving

Energy management means optimised and rationalised power consumption so that at lower electricity costs, more electrical efficiency is achieved. As a part of the energy management, demand power can be shifted using the ESS technology to match the electricity supply. It shall make possible the integration of various RES suppliers. By demand shifting, the capacity of transmission lines can be managed. The power systems can decrease the high energy generation in peak demand that results in a longer equipment lifetime [14].

8.3.5 Power smoothing

Due to the uncertain inherent of the wind and solar energies as well as the errors in prediction methods, the amount of energy production by RES cannot be accurately calculated. Therefore, in the planning stage of microgrids, appropriate strategies are required to reduce the uncertainties to make a safe integration of renewable sources. Electricity storage technologies like batteries which operate on short time intervals (seconds, minutes and hours) can be used to maintain a more accurate balance between the power supply and demand. Installing the energy storage at the generation side smoothens the variations in generated power. Additionally, the output power of RES can directly charge the energy storage. In this case, the loads will be supplied by a smoothed and high-quality power through energy storage discharging. Energy storage is generally used to modify the variable output of supply resources. It helps to reduce the rapid and seasonal changes in the output of RES and hence improves the supply power quality [15].

8.3.6 Black start

Restoring a microgrid after blackouts or islanding is known as black start. Low speed of the conventional generation units for restoration and intermittency in RES has complicated the black start in microgrids. In most of the cases, black start is related to transient states and stability issues in microgrid [16]. Application of storage systems like flywheels with a fast response time is a remedy for black starts in microgrids to supply the loads uninterruptedly.

8.3.7 Storage trades/arbitrage

The energy storage will be very effective in the economic efficiency of microgrids connected to the main grid. With a proper planning, ESS can save energy during a period of low demand with a cheap price and release it when the demand and electricity price are high. In this way, microgrid can exchange energy with the main grid to make a higher economic profit [17]. In addition, the energy storage can purchase electricity from the main grid in low-cost periods and sell the energy in high-cost times.

8.3.8 Non-spinning reserve

The aim of reserve capacity of electricity is to keep the system equilibrium and provides quick compensation in the event of unexpected loss of the supply resources. In conventional power systems, spinning reserve is provided by conventional power plants. In microgrids, conventional generation units are usually utilised in their highest capacity. Renewable resources are considered non-reliable generation units for the reserve systems due to their unpredictable power generation. The ESS capacity in the generation side of microgrid is to maintain the system balance and power compensation in unexpected conditions.

An important aspect of the reserve capacity is discharging time of the storage for various applications. In high-power applications such as emergency power supply, it is required to discharge energy in a few seconds. In microgrid stability applications, it is required to discharge energy within minutes. For high energy applications and energy management, including arbitrage and peak reduction, a few hours of discharge time is necessary [18].

8.4 Case study

In this section, the optimal sizing of an ESS is studied for a microgrid. To determine the optimal capacity of the ESS, total cost of the microgrid should be minimised. The total cost includes the cost of the ESS investment and expected operation cost of the microgrid in the presence of ESS. Operation costs include the cost of energy generation from local generation units of microgrid, cost of imported/exported energy from/to the main grid and load shedding cost of the microgrid. In addition to determining the optimal capacity of the ESS in microgrid, optimal depth of discharge is achieved in this study. Figure 8.9 shows the trend of microgrid costs variations with increasing the capacity of ESS. As the capacity of ESS increases, the investment cost of ESS increases and the operation cost of microgrid decreases. What is important in Figure 8.9 is the total planning cost which is high for low and high capacities of ESS. In fact, the capacity of ESS should be significantly optimised in microgrids to achieve an optimal trade-off between ESS investment cost and microgrid operational cost to reduce the total planning cost of microgrid.

In summary, the main objectives of this chapter include

- modelling the demand and wind turbine generation uncertainties in a microgrid,
- modelling the energy storage in a microgrid,
- determining the battery storage optimum capacity and depth of discharge in a microgrid,
- planning the generation and consumption of energy resources in a microgrid and
- reducing the operation cost of a microgrid.

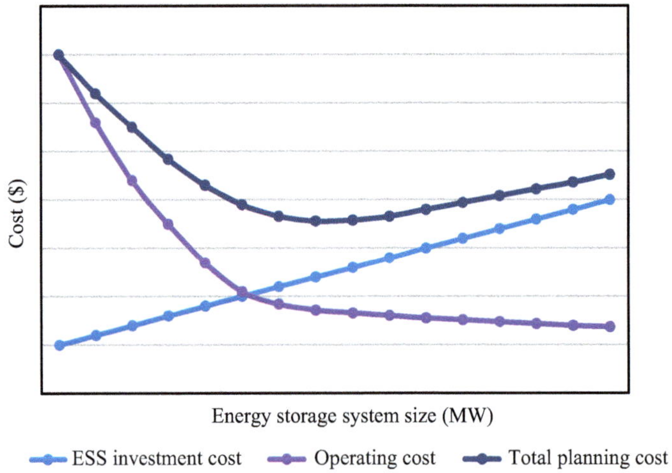

Figure 8.9 Optimal sizing of energy storage system

Objectives are considered to help designers to make the right decisions on economic operation and planning of the microgrid. A mixed integer nonlinear programming (MINLP) model has been used to formulate the problem, load management program and determine the optimal capacity of the ESS. All parameters used in this chapter are described in 'Nomenclature' section.

8.4.1 System description and input data

A sample microgrid is considered to optimise the ESS capacity. Figure 8.10 shows the microgrid under the study in this book chapter with its various components.

This microgrid has three thermal units and one wind turbine unit. Table 8.3 lists the specifications of generation units. The capacity of power line between the microgrid and main grid is 10 MW.

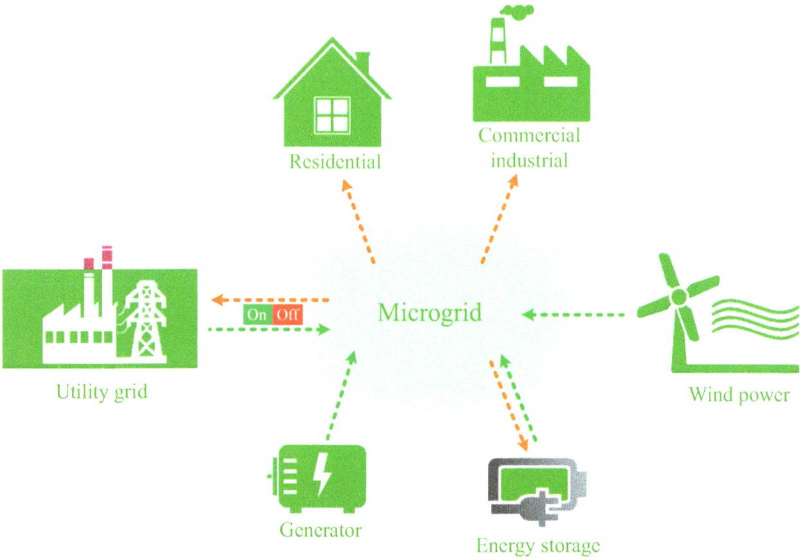

Figure 8.10 Under the study microgrid with its various components

Table 8.3 Characteristics of generating units

Unite no.	**Bus no.**	**Cost coefficient ($/MW h)**	**Minimum capacity (MW)**	**Maximum capacity (MW)**	**Ramp-up rate (MW/h)**	**Ramp-down rate (MW/h)**
1	Gas	31.2	1	5	2.5	2.5
2	Gas	44.8	1	3	2.5	2.5
3	Gas	55.6	0.8	3	3	3
4	Wind	0	0	1	–	–

Figure 8.11 shows the microgrid load for a sample day. The electricity prices on the upstream network to which the microgrid is connected are presented in Table 8.4.

No load development of the microgrid in the coming year is assumed. Therefore, the total planning is considered for only 1 year. ESS for the microgrid has an investment cost of power and energy of 40,000 \$/MW/year and 11,000 \$/MW h/year, respectively. The charge and discharge efficiency of ESS are considered 95% and 90%, respectively. The proposed system is modelled as a MINLP problem which is implemented in GAMS and solved by basic open-source nonlinear mixed integer programming (BONMIN) solver [19]. To demonstrate the effectiveness of the proposed model, three different cases including a microgrid without ESS, with a random capacity of ESS and with an optimised capacity of ESS are analysed. Then, impacts of the energy demand management program on determining the optimal size of ESS in the microgrid are reviewed.

8.4.2 Uncertainty modelling

In order to model a real problem and achieve accurate results, it is necessary to consider uncertainties. Uncertainty in the data of an optimisation problem can be

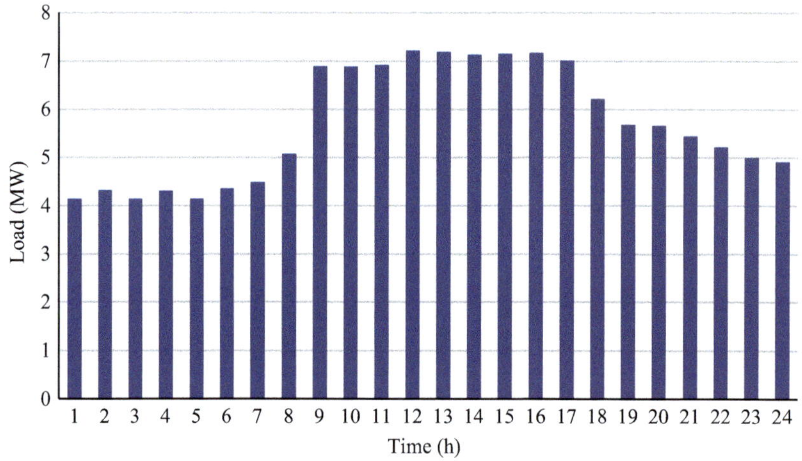

Figure 8.11 Hourly load demand for a sample day

Table 8.4 Hourly real electricity price

Hour	**1**	**2**	**3**	**4**	**5**	**6**	**7**	**8**
Price (\$/MW h)	40.23	20.97	20.51	20.36	20.71	30.30	40.18	30.83
Hour	9	10	11	12	13	14	15	16
Price (\$/MW h)	20.05	40.39	40.86	60.65	60.39	50.57	40.44	30.97
Hour	17	18	19	20	21	22	23	24
Price (\$/MW h)	45.45	60.28	80.15	120.53	90.38	60.95	50.42	60.68

because of error in prediction, measurement, implementation, etc. Electrical load demand and output power of renewable resources are the most effective and important uncertainties in the microgrid planning problem. This subsection explains the model of demand and wind turbine generation uncertainties.

8.4.2.1 Demand uncertainty

Different patterns of energy consumers make changes in the demand of microgrid. For proper planning of microgrids, uncertainty of demand requires a precise model. Normal probability density function is used to model the demand uncertainty. In the following, a formula for the probability density function of the normal distribution is given:

$$p(x) = \frac{1}{\sigma\sqrt{2\pi}} \exp\left(-\frac{(x-\mu)^2}{2\sigma^2}\right) \tag{8.1}$$

where $p(x)$ is the demand distribution function. μ and σ represent the mean and standard deviations of the distribution function. Values of μ and σ are given in [20].

8.4.2.2 Wind power generation uncertainty

In this study, in addition to thermal units, a wind turbine is used to generate the required power of the microgrid. Wind turbine power depends on the wind speed. The nature of wind speed is stochastic and can be described by probability distribution functions. A detailed explanation for the production of different wind power scenarios is presented in [21].

8.4.2.3 Combined wind and load uncertainty

By producing the appropriate number of wind power generation and load demand scenarios, total number of combined wind-load scenarios is obtained by multiplying the number of wind power and load scenarios:

$$\pi_s^T = \pi_s^{wind} \times \pi_s^{load} \tag{8.2}$$

Five scenarios for the uncertainty of wind and microgrid demand are created. Table 8.5 lists the combined wind-load scenarios [19].

8.4.3 Problem formulation

In this section, the optimisation model for microgrid planning and determining the optimal capacity of ESS is discussed. Detail model of each component is introduced. The OFs and ESS constraints are presented. Limitations of the microgrid, generation units and operating constraints are explained. The OF including the cost of investment of ESS and the expected daily operating cost of the microgrid for 1 year is as follows [22]:

$$\forall i \in N_i, \quad s \in N_s, \quad t \in N_t, \quad h \in N_h$$

$$OF = \text{Min}\,\{IC + OC\} \tag{8.3}$$

Table 8.5 Wind-load scenarios and their corresponding probabilities

Scenario number	∂_s^{wind} (%)	ς_s^{Load} (%)	π_s^T
s_1	0.00	95	0.002
s_2	0.00	97	0.009
s_3	0.00	100	0.047
s_4	0.00	103	0.009
s_5	0.00	105	0.002
s_6	12.87	95	0.005
s_7	12.87	97	0.028
s_8	12.87	100	0.139
s_9	12.87	103	0.028
s_{10}	12.87	105	0.005
s_{11}	49.37	95	0.010
s_{12}	49.37	97	0.055
s_{13}	49.37	100	0.275
s_{14}	49.37	103	0.055
s_{15}	49.37	105	0.010
s_{16}	86.83	95	0.005
s_{17}	86.83	97	0.027
s_{18}	86.83	100	0.135
s_{19}	86.83	103	0.027
s_{20}	86.83	105	0.005
s_{21}	100.00	95	0.003
s_{22}	100.00	97	0.017
s_{23}	100.00	100	0.083
s_{24}	100.00	103	0.017
s_{25}	100.00	105	0.003

$$IC = ICP \cdot P_{\max}^{soc} + ICE \cdot SOC_{\max} \tag{8.4}$$

$$OC = \sum_{s=1}^{N_s} \pi_s^T \sum_{t=1}^{N_t} \sum_{h=1}^{N_h} \sum_{i=1}^{N_i} F_i(P_{s,t,h}^i) + \sum_{s=1}^{N_s} \pi_s^T \sum_{t=1}^{N_t} \sum_{h=1}^{N_h} \omega_{t,h} \cdot P_{s,t,h}^M + \sum_{s=1}^{N_s} \pi_s^T \sum_{t=1}^{N_t} \sum_{h=1}^{N_h} \rho_{t,h} \cdot P_{s,t,h}^{lsh} \tag{8.5}$$

Initial costs are considered for each year and distributed through the life of ESS technologies. Investment cost of ESS calculated by (8.4) depends on the maximum power and energy rated of the ESS. Microgrid operational costs include the electric power production cost by local units, purchasing and selling electricity costs, exchanging energy costs with the main grid, as well as the load shedding cost of the microgrid, as shown in (8.5). In this equation, the cost of operation of fossil fuel units is considered by $F_i(P_{s,t,h}^i)$.

In general, the ESS is used in three modes of charging, discharging and non-use. The charging and discharging equations of the EES and their constraints are modelled in (8.6)–(8.13). ESS self-discharging is ignored in this study:

$$\forall s \in N_s,\ \ t \in N_t,\ \ h \in N_h$$

$$SOC_{s,t,h} = SOC_{s,t,h-1} + \left(\frac{P^c_{s,t,h}\eta_c - P^d_{s,t,h}}{\eta_d} \right) \cdot \Delta h \tag{8.6}$$

$$0 \le P^c_{s,t,h} \le P^{ESS} \tag{8.7}$$

$$0 \le P^d_{s,t,h} \le P^{ESS} \tag{8.8}$$

$$P^{ESS} = k \times SOC_{\max} \tag{8.9}$$

$$SOC_{\min} \le SOC_{s,t,h} \le SOC_{\max} \tag{8.10}$$

$$SOC_{s,t,h(\text{start})} \le SOC^0 \tag{8.11}$$

$$SOC_{s,t,h(end)} = SOC^{end} \tag{8.12}$$

$$ICP \cdot P^{ESS} + ICE \cdot SOC_{\max} \le CIF \tag{8.13}$$

The energy of ESS is determined by (8.6). The energy stored per hour is calculated with the accumulated stored energy at the previous hour and the charge or discharge energy status at the same hour. A 1-h interval ($\Delta h = 1$) is considered for the operation. Charge and discharge powers of the ESS are limited by (8.7) and (8.8). The rated power of ESS is obtained by (8.9). Equation (8.10) shows the limitations of the amount of stored energy in each hour. The ESS charge status at the start and end of each day is obtained by (8.11) and (8.12). Equation (8.13) is a constraint on the initial investment of the ESS installation by considering the power and energy of ESS.

The total power generated by local power generation units, the power exchanged between the microgrid and the main grid and the charge or discharge power of ESS should be equal to the hourly microgrid load. The balance equation for power generation and demand in the microgrid is expressed by (8.14):

$$\forall i \in N_i,\ \ r \in N_r,\ \ s \in N_s,\ \ t \in N_t,\ \ h \in N_h,\ \ \left\{ UX^i_{s,t,h}, UY^M_{s,t,h} \right\} \in \{0,1\}$$

$$\sum_{i=1}^{N_i} P^i_{s,t,h} + \sum_{r=1}^{N_r} P^r_{s,t,h} + P^d_{s,t,h} + P^M_{s,t,h} + P^{lsh}_{s,t,h} = P^L_{s,t,h} - P^d_{s,t,h} \tag{8.14}$$

$$P^i_{\min} \cdot UX^i_{s,t,h} \le P^i_{s,t,h} \le P^i_{\max} \cdot UX^i_{s,t,h} \tag{8.15}$$

$$P^i_{s,t,h} - P^i_{s,t,h-1} \le RU_i \tag{8.16}$$

$$P^{i}_{s,t,h-1} - P^{i}_{s,t,h} \leq RD_i \tag{8.17}$$

$$\left|P^{M}_{s,t,h}\right| \leq P^{M}_{\max}.UY^{M}_{s,t,h} \tag{8.18}$$

$$0 \leq P^{r}_{s,t,h} \leq \left(0.01 \times \partial^{wind}_{s}\right) \cdot P^{r}_{\max} \tag{8.19}$$

$$P^{L}_{s,t,h} = \left(0.01 \times \zeta^{Load}_{s}\right) \cdot \overline{P}^{L}_{t,h} \tag{8.20}$$

$$0 \leq P^{lsh}_{s,t,h} \leq P^{L}_{s,t,h} \tag{8.21}$$

The limitation of maximum and minimum production capacity of unit i is expressed by (8.15). Constraints (8.16) and (8.17) control the increase or decrease of production power of units between two consecutive hours. The power capacity exchanged between the microgrid and the main grid is obtained by (8.18). $P^{M}_{s,t,h}$ is positive when the power is imported from the main grid, and it is negative when selling power, and when the microgrid is separated from the grid, it is zero.

Due to the dependency on wind speed, the wind turbine generation cannot exceed the maximum wind turbine capacity in each scenario, which is indicated by (8.19). Calculation of the load and limit of microgrid load shedding are obtained by (8.20) and (8.21), respectively.

8.4.4 Numerical results

In this subsection, determining the optimal ESS capacity by using microgrid programming is analysed in three cases. To demonstrate the performance of the proposed model, the following cases are considered:

Case 1: Microgrid without ESS
Case 2: Microgrid with non-optimised ESS
Case 3: Microgrid with optimised ESS

8.4.4.1 Case 1: microgrid without ESS

Case 1 is assumed as the base case, which the microgrid is programmed without ESS. The total cost of microgrid is \$ 1,137,929. The total cost of generation is \$ 2,503,656, and the total cost of purchasing power from the upstream grid is \$ 126,781. The economic benefit of \$ 1,492,508 from the sale of power to the main grid in overproduction periods is achieved. As shown in Figure 8.12, the power transmission from the main grid to the microgrid is initiated when the price of electricity is reduced in the early hours. When the cost of electricity from upstream grid increases, the generation of thermal units inside the microgrid increases to supply the load and the surplus generated power of thermal units, if exists, is sold to the upstream grid simultaneously.

8.4.4.2 Case 2: microgrid with non-optimised ESS

In this case, an 8 MW h ESS with a rated power of 3.68 MW and discharge depth of 0.46 is added to the microgrid. Figure 8.13 shows that the storage in this microgrid

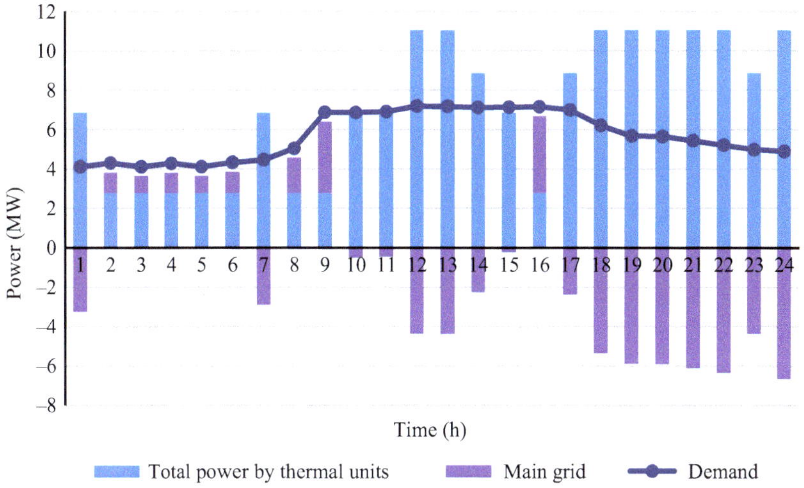

Figure 8.12 Operation planning of the microgrid and production units in case 1

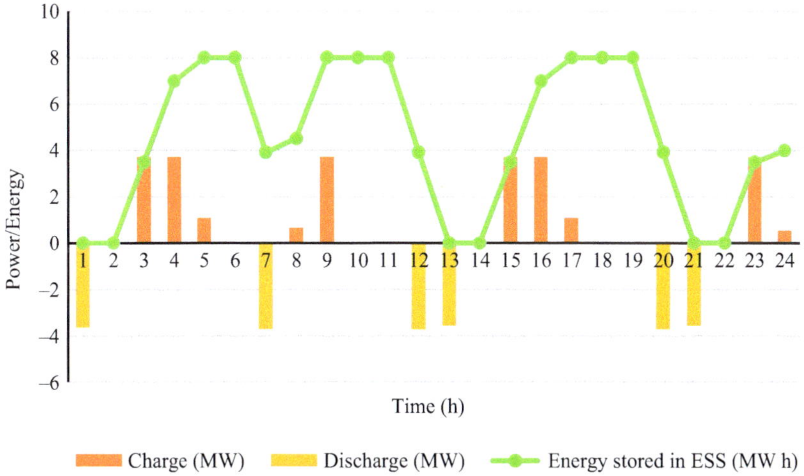

Figure 8.13 The hourly dispatch of ESS in case 2

is fully charged and reaches the maximum SOC at 5 o'clock. Usually, the ESS is charged at non-peak times when energy is cheap. At peak hours when the cost of energy is expensive, it will be discharged. The power stored at peak hours is discharged to supply demand during overload in a microgrid. The power stored at peak hours is discharged to supply the demand during overload in the microgrid or to sell the power to the main grid and increase profits.

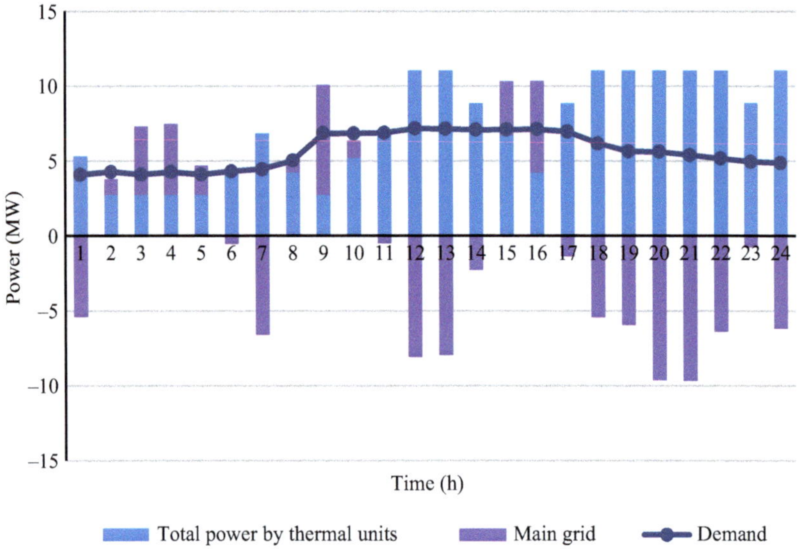

Figure 8.14 Operation planning of the microgrid and production units in case 2

The total cost of the microgrid is \$ 113,009, which includes \$ 2,520,210 of total generation cost, \$ 287,112 total cost of purchasing power, \$ 235,200 the ESS investment cost and \$ 1,130,091 obtained from power sales to main grid. Reduced total microgrid cost compared to case 1 is due to the sale of ESS power to the upstream network. Figure 8.14 shows the planning of production units and the exchange of power between the microgrid and main grid.

8.4.4.3 Case 3: microgrid with optimised ESS

Using the proposed method, the optimal size of 15.79 MW h at the discharge depth of 0.26 with the rated power of 4.15 MW is obtained for the ESS. In this case, the storage can be charged in 4 h. The ESS charging and discharging schedule are shown in Figure 8.15.

By adding ESS, the total cost of generation and the total cost of the power purchased from the main grid are \$ 2,508,809 and \$ 366,315, respectively. Also, the ESS investment cost is \$ 339,985. The microgrid income for the sale of power is \$ 2,123,216. After including these costs, the total cost of the microgrid operation will be \$ 1,130,091; this cost reduction is due to increased sales of microgrid power to the upstream grid. The planning of the production units and the exchange of power between the microgrid and the main network, in this case, are shown in Figure 8.16.

8.4.4.4 Results comparison

Table 8.6 presents a summary of results for the case studies 1–3. The optimal capacity of the ESS is determined to minimise the total cost of the microgrid during

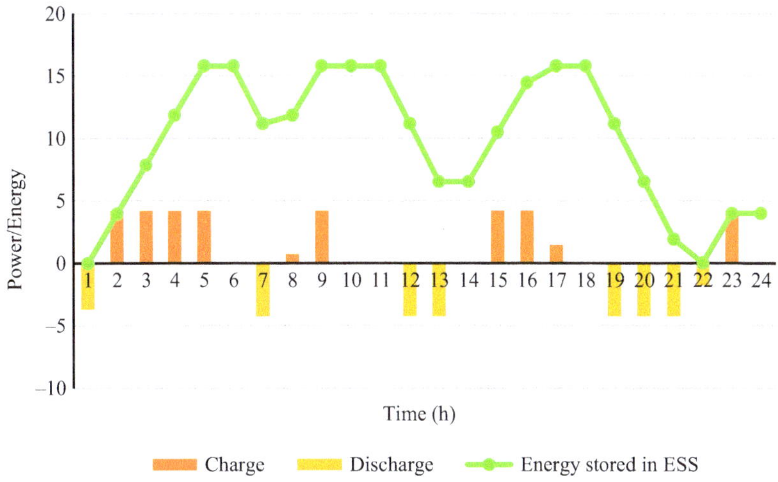

Figure 8.15 The hourly dispatch of ESS at the optimal size in case 3

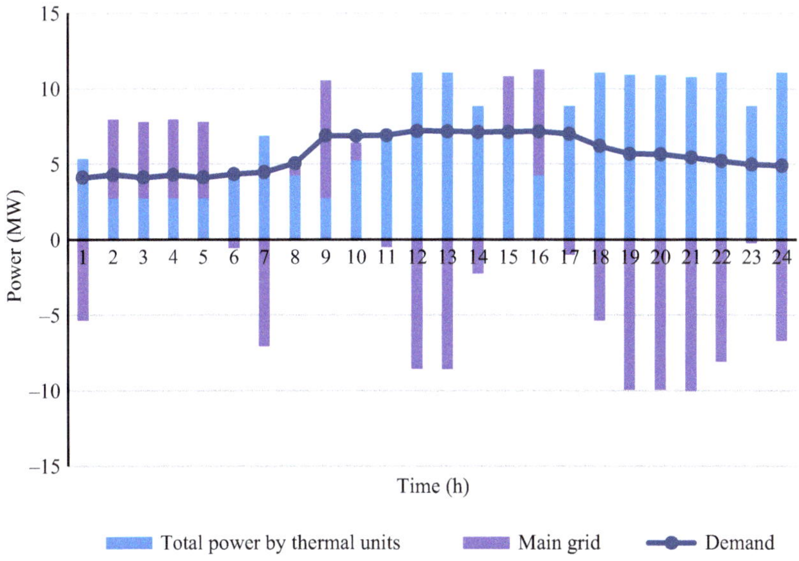

Figure 8.16 Operation planning of the microgrid and production units in case 3

the system planning. The ESS stores energy at cheap electricity prices and provides low-cost power to the microgrid loads whenever required. It reduces the need for local production or power import from the main grid.

Table 8.6 Summary of system costs related to case 1, 2 and 3

Different parameters	Case 1	Case 2	Case 3
Depth of discharge (%)	0	46	26.3
ESS rated power (MW)	0	3.68	4.15
ESS rated energy (MW h)	0	8	15.79
Generation cost ($)	2,503,656	2,520,210	2,508,809
Power import ($)	126,781	287,112	366,315
Power export ($)	−1,492,508	−1,912,431	−2,123,216
ESS investment cost ($)	0	235,200	339,985
Total cost ($)	1,137,929	1,130,091	1,091,893

8.5 Conclusions

This chapter presented a review on applicable energy storage technologies in microgrids. Different energy storage technologies were introduced and explained, and multiple characteristics of them were studied. Application and benefits of energy storages in microgrids were analysed. It was found that high-energy density of energy storage is an appropriate trait for energy arbitrage and demand shifting. High power rating and density of energy storages are excellent features for spinning reserve in microgrid. Frequency regulation and voltage control of microgrids are enhanced with fast charging/discharging feature of energy storages. In overall, BESS as a mature and commercialised technology is the most appropriate candidate for microgrids. Batteries have high-energy density, high power rating, fast charge/discharge capability and high efficiency.

It was found that the capacity of energy storages is an important index when integrated into the microgrids. Oversized energy storage lifts the costs and undersized storage deteriorates the desirable operation. The optimal sizing of ESS was done on a sample microgrid under load and wind power uncertainties. Investment cost of ESS and operation cost of microgrid were considered as the OF for minimisation. It was shown that energy storage can make the microgrid more profitable by decreasing the dependency on main grid and saving the energy in low-price periods and releasing it in high-cost periods.

Nomenclature

Indices and sets

Index of thermal unit	i
Index of renewable unit	r
Index of scenario	s
Index of day	t
Index of hour	h
Set of thermal units	N_i

Set of renewable units	N_r
Set of scenarios	N_s
Set of days	N_t
Set of hours	N_h

Parameters

Probability of combined wind-load scenarios	π_s^T
Probability of wind power generation scenarios	π_s^{wind}
Probability of load scenarios	π_s^{load}
Percentage of existing wind power in scenario s	∂_s^{wind}
Percentage of existing load demand in scenario s	ζ_s^{Load}
Electricity price at day t at hour h	$\omega_{t,h}$
Value of loss load at day t at hour h	$\rho_{t,h}$
Microgrid load demand at day t at hour h in scenario s	$P_{s,t,h}^L$
Average microgrid load demand at day t at hour h	$\overline{P}_{t,h}^L$
Minimum power generation capacity of thermal unit i	$P_{\min}^i$
Maximum power generation capacity of thermal unit i	$P_{\max}^i$
Ramp-up rate limit of thermal unit i	RU_i
Ramp-down rate limit of thermal unit i	RD_i
Maximum power generation capacity of renewable unit r	$P_{\max}^r$
Maximum power capacity exchanged between the microgrid and the main grid	$P_{\max}^M$
Maximum state-of-charge capacity	$SOC_{\max}$
Minimum state-of-charge capacity	$SOC_{\min}$
State-of-charge at the start of every day	SOC^0
State-of-charge at the end of every day	SOC^{end}
ESS charge efficiency	η_c
ESS discharge efficiency	η_d
The investment cost of ESS power rating	ICP
The investment cost of ESS energy rating	ICE
The cost of the investment funds for ESS	CIF
The standard deviation of the normal distribution	σ
The mean of the normal distribution	μ

Variables

Power generation thermal unit i at day t at hour h in scenario s	$P_{s,t,h}^i$
Power generation renewable unit r at day t at hour h in scenario s	$P_{s,t,h}^r$
Power exchanged between the microgrid and the main grid at day t at hour h in scenario s	$P_{s,t,h}^M$

Load shedding at day t at hour h in scenario s	$P^{lsh}_{s,t,h}$
Generation cost function of unit i	F_i
State-of-charge at day t at hour h in scenario s	$SOC_{s,t,h}$
ESS power charged at day t at hour h in scenario s	$P^{c}_{s,t,h}$
ESS power discharged at day t at hour h in scenario s	$P^{d}_{s,t,h}$
Depth of discharge	k
ESS rated power	P^{ESS}
Outage status of thermal unit i at day t at hour h in scenario s	$UX^{i}_{s,t,h}$
Outage status of the line connecting microgrid to the main grid at day t at hour h in scenario s	$UY^{M}_{s,t,h}$
Objective function	OF
The total cost of the microgrid investment	IC
The total cost of the microgrid operation	OC

References

[1] *Renewables global status report*, 2018. Available from http://www.ren21.net/wp-content/uploads/2018/06/17-8652_GSR2018_FullReport_web_final_.pdf.

[2] Zhang, C., W. Yi-Li, C. Peng-Fei, and L. Meng-Chang, *Energy storage system: Current studies on batteries and power condition system*. Renewable & Sustainable Energy Reviews, 2018. **82**: p. 3091–3106.

[3] May, G.J., A. Davidson, and B. Monahov, *Lead batteries for utility energy storage: A review*. Journal of Energy Storage, 2018. **15**: p. 145–157.

[4] Kawakami, N., Y. Iijima, M. Fukuhara, *et al.*, *Development and field experiences of stabilization system using 34 MW NAS batteries for a 51 MW wind farm*. in *IEEE International Symposium on Industrial Electronics (ISIE 2010)*. 2010: p. 2371–2376.

[5] Arghandeh, R., M. Pipattanasomporn, and S. Rahman, *Flywheel energy storage systems for ride-through applications in a facility microgrid*. IEEE Transactions on Smart Grid, 2012. **3**(4): p. 1955–1962.

[6] Das, C.K., O. Bass, G. Kothapalli, T.S. Mahmoud, and D. Habib, *Overview of energy storage systems in distribution networks: Placement, sizing, operation, and power quality*. Renewable & Sustainable Energy Reviews, 2018. **91**: p. 1205–1230.

[7] Diaz-Gonzalez, F., A. Sumper, O. Gomis-Bellmunt, and R. Villafáfila-Robles, *A review of energy storage technologies for wind power applications*. Renewable & Sustainable Energy Reviews, 2012. **16**(4): p. 2154–2171.

[8] Aneke, M. and M.H. Wang, *Energy storage technologies and real life applications – A state of the art review*. Applied Energy, 2016. **179**: p. 350–377.

[9] Etxeberria, A., I. Vechiu, H. Camblong, J.M. Vinassa, and H. Camblong, *Hybrid energy storage systems for renewable energy sources integration in microgrids: A review*. in *IPEC, 2010 Conference Proceedings*. 2010. IEEE.

[10] Benysek, G., M. Kazmierkowski, J. Popczyk, and R. Strzelecki, *Power electronic systems as a crucial part of smart grid infrastructure-a survey*. Bulletin of the Polish Academy of Sciences: Technical Sciences, 2011. **59**(4): p. 455–473.

[11] Li, J., R. Xiong, Q. Yang, F. Liang, M. Zhang, and W. Yuan, *Design/test of a hybrid energy storage system for primary frequency control using a dynamic droop method in an isolated microgrid power system*. Applied Energy, 2017. **201**: p. 257–269.

[12] Zeraati, M., M.E.H. Golshan, and J.M. Guerrero, *Distributed control of battery energy storage systems for voltage regulation in distribution networks with high PV penetration*. IEEE Transactions on Smart Grid, 2018. **9**(4): p. 3582–3593.

[13] Mitra, J., *Reliability-based sizing of backup storage*. IEEE Transactions on Power Systems, 2010. **25**(2): p. 1198–1199.

[14] Hossain, E., M.Z. Xahin, K.R. Islam, and M.D. Qays Akash, *Design a novel controller for stability analysis of microgrid by managing controllable load using load shaving and load shifting techniques; and optimizing cost analysis for energy storage system*. International Journal of Renewable Energy Research, 2016. **6**(3): p. 772–786.

[15] Nosratabadi, S.M., R.-A. Hooshmand, and E. Gholipour, *A comprehensive review on microgrid and virtual power plant concepts employed for distributed energy resources scheduling in power systems*. Renewable and Sustainable Energy Reviews, 2017. **67**: p. 341–363.

[16] Dang, J. and R.G. Harley, *Islanded microgrids black start procedures with wind power integration*. in *Power and Energy Society General Meeting (PES), 2013 IEEE*. 2013. IEEE.

[17] Ross, M., C. Abbey, F. Bouffard, and G. Joós, *Microgrid economic dispatch with energy storage systems*. IEEE Transactions on Smart Grid, 2018. **9**(4): p. 3039–3047.

[18] Lott, M.C., S. Kim, C. Tam, *et al.*, *Technology roadmap: energy storage*. International Energy Agency: Paris, France, 2014. Available from https://www.iea.org/publications/freepublications/publication/TechnologyRoadmap Energystorage.pdf.

[19] Bussieck, M.R. and S. Vigerske, *MINLP solver software*. 2010, Wiley: Chichester, UK. p. 1–12.

[20] Nikkhah, S., K. Jalilpoor, E. Kianmehr, and G. B. Gharehpetian *Optimal wind turbine allocation and network reconfiguration for enhancing resiliency of system after major faults caused by natural disaster considering uncertainty*. IET Renewable Power Generation, 2018. **12**(12).

[21] Atwa, Y.M. and E.F. El-Saadany, *Probabilistic approach for optimal allocation of wind-based distributed generation in distribution systems*. IET Renewable Power Generation, 2011. **5**(1): p. 79–88.

[22] Bahramirad, S., W. Reder, and A. Khodaei, *Reliability-constrained optimal sizing of energy storage system in a microgrid*. IEEE Transactions on Smart Grid, 2012. **3**(4): p. 2056–2062.

Chapter 9

Microgrid communications – protocols and standards

Shantanu Kumar[1], *Syed Islam*[2], *and Alireza Jolfaei*[3]

The recent advancements in the Internet of Things (IoT) and telecommunication infrastructure have significantly increased the reliability and effectiveness of communication protocols in microgrid environment. Nowadays, the equipment in a smart microgrid not only exchange information with one another much faster but also control, monitor and diagnose faults much faster and more reliably. Despite the benefits of new technologies, they are tailored to satisfy specific requirements, and therefore, they can be only used for certain applications. The microgrid communication model consists of a three-layer architecture, where the energy management system (EMS) sits in the top layer and controls the overall operations of the island of microgrids. The middle layer includes the local controllers (LCs) that regulate operations within the local grid. The bottom layer includes IoT devices, such as smart meters, fault recorders and protective relays, which continuously capture and transmit the stream of sensed data. Such hierarchical architecture introduces specific computation and latency requirements for each layer of microgrid communication. To meet these requirements, each layer must use different communication equipment and protocols. This chapter provides an insight into communication requirements, system architecture, standards, protocols and tools used in microgrid communications. The chapter concludes with a case study, where wireless technology is utilised for reliable and optimal operations in a microgrid.

9.1 Introduction

Microgrids are intentional islands with distributed energy resources (DERs) and associated loads formed at a facility or in an electrical distribution system [1]. Microgrids provide reliable energy, which can control the power flow when they are connected to the grid or operate autonomously in an electrical island, or in transition between these two states. A typical structure of a microgrid with key elements is shown in Figure 9.1. The figure shows a local microgrid attached to a

[1]Faculty of Science and Engineering, Curtin University, Perth, Australia
[2]School of Science, Engineering, and Information Technology, Federation University Australia, Mount Helen, Australia
[3]Department of Computing, Macquarie University, Sydney, Australia

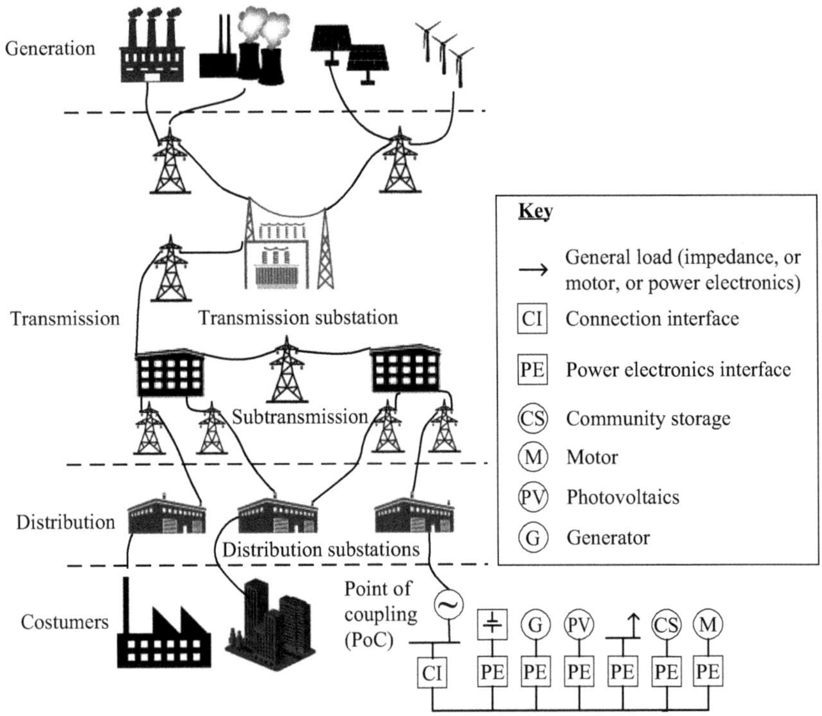

Figure 9.1 Typical structure of microgrid connected to distribution network

bigger smart grid (SG), which contains a number of bulk generation/transmission, distribution, and customer domains. Microgrid network is connected to the power grid using high-voltage (HV) equipment, such as electromechanical circuit breakers, solid-state switches or back-to-back power electronics (PE) converters. Architecture of a microgrid normally contains an energy storage (ES) to balance power loads, loads on transition to the islanded mode as a fast-acting energy absorption or injection application. ES could be necessary for the control of net power flow in the grid-connected mode bidirectional to the power grid. In this way, microgrid is capable to assist a stable network operation by providing improved power quality and voltage control. PE converters are added to load/generation units for additional control and speed of response. PE can also be used to provide continuous variable power consumption by noncritical power loads, for example, ventilation. Furthermore, to coordinate microgrid elements and to make them operate as a system, an LC regulates microgrid's state variables, such as instantaneous phase currents and voltages at the point of common coupling (PCC).

In practice, a number of microgrids may be connected to a single point of coupling, through which power is exchanged with the main grid. As shown in Figure 9.2, the communication model of microgrids follow a three-layer edge computing architecture [2], where the main controller controls a number of microgrids, and the

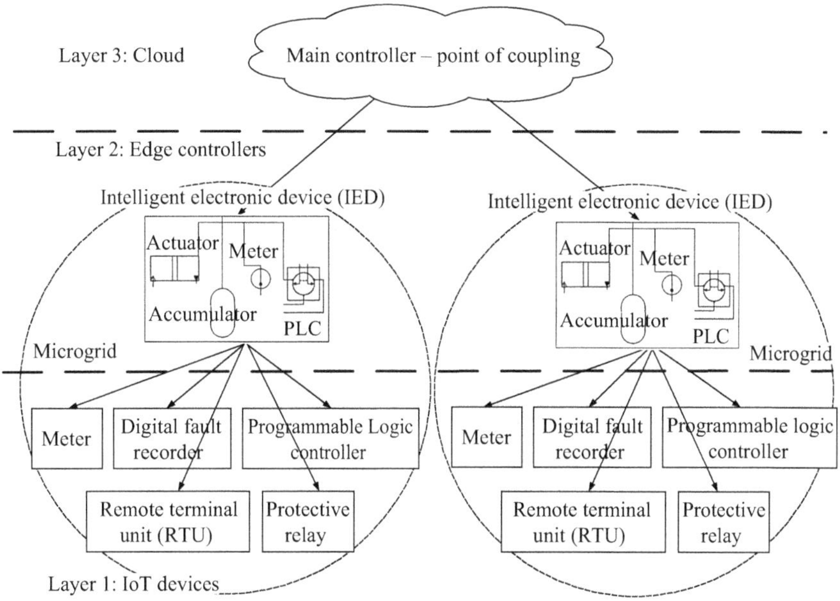

Figure 9.2 Three-layer communication architecture of microgrids

smaller microgrids regulate local operations within the local grid. The bottom layer includes IoT devices, such as meters, fault recorders, programmable logic controllers (PLCs) and protective relays, which continuously monitor the stream of sensed data. The middle layer includes a set of edge controllers, such as intelligent electronic devices (IEDs), each of which controls IoT devices in its vicinity and is responsible for real-time data analytics and temporary data storage. The top layer is a cloud service that provides longer term data storage and deep analytics.

As shown in Figure 9.2, the processing power and communication latency requirements are different across all three layers. This mandates the use of differing communication protocols in each layer, such as home area network (HAN) [3] and local area network (LAN) [4] in Layers 1 and 2, neighbourhood area network (NAN) [5] and wide area network (WAN) [6] in Layers 2 and 3. This chapter not only reviews these layers but also technologies related to microgrid communication protocols and standards, including various types of wired and wireless-communication-tiered networks. The rest of the chapter is as follows: Section 9.2 outlines the communication objectives and requirements in a microgrid environment. Section 9.3 reviews the communication layers. Section 9.4 gives an overview of the existing communication infrastructure with reference to already functioning examples. Section 9.5 highlights the commonly used communication protocols in microgrids. Section 9.6 elucidates the importance of communication technologies in microgrid control. Section 9.7 studies a microgrid communication architecture in an infrastructure project in India. Finally, Section 9.8 concludes the chapter.

9.2 Communication objectives and requirements

The implementation of microgrid technologies involves participation of the big number of DERs and implies their simultaneous functioning in real time. Stable, reliable and responsive operation of microgrid components in providing services to the utility network, such as near-instantaneous feedback on grid's injection/consumption demand, is impossible without bidirectional, fast and compound interaction. In addition, the additional islanded mode of operation and periods of transition to and from this mode make the communication more challenging. Several aspects play an important role in the design of a microgrid communication system including

- type and configuration of equipment in the system, where particular devices require different time of response, by that demanding separate arrangement of communication equipment and interfaces, along with each communication channel's bandwidth;
- number of participants in the system and their physical allocation determine the type of connection needed, data traffic volume and maintenance costs and
- control system, which can be centralised or decentralised, designates the communication architecture. The centralised communication system gives an advantage of simplification and unifying of system components, because all the processes executes by one common software without significant problems regarded to operation conflicts and schedules. On the other hand, the decentralised communication system allows independent functioning of various devices and sections and can provide high rate of redundancy to the system.

To maintain a reliable and safe operation of the microgrid, the communication network needs to preserve two main requirements: real-time performance and security by providing confidentiality, integrity and availability while respecting the real-time boundaries. These requirements are challenging since most of the digital systems in operation including the sensing devices, embedded control subsystems and the execution of control logics would impose delays in the communication network. In addition, there are many communication devices in operation that have no hardware support for cryptographic algorithms, and thus, they cannot utilise computationally expensive security algorithms [7,8].

Let us consider the following scenario where an emergency event is detected. This would trigger the protection equipment to send safety messages to a neighbouring relay to prevent cascading failures. Such safety messages would help if they arrive before the arrival of any harmonics in the electrical network. Indeed, digital massages would be racing against transient analogue voltage. In addition, when protection messages are more than one hop away, they would need to be processed by an intermediate network device, which can impose additional processing delay. Buffering delay would also be added if the network is congested.

Thee local protection equipment would take independent actions if the protection messages do not arrive before the arrival of power line's disturbance, and this would force the equipment to send additional warning messages, creating even

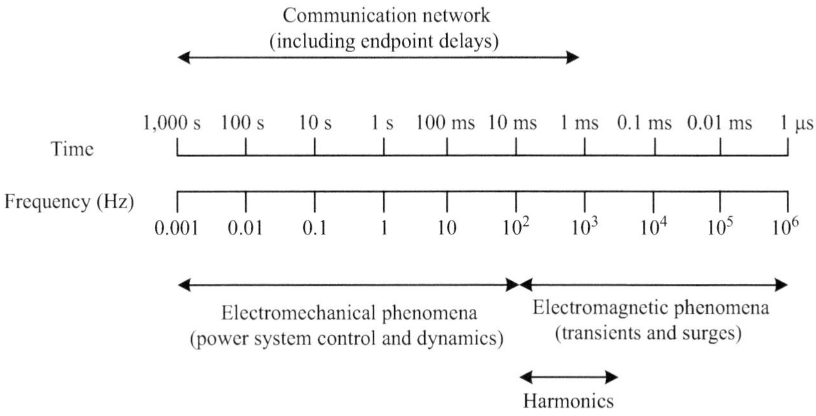

Figure 9.3 Power vs. communication network timescale

more traffic in the already congested communication network. Moreover, the problem would exacerbate with each additional protection tripping. A practicable and viable microgrid communication must ensure to properly handle such scenarios.

Figure 9.3 depicts various timescales in the communication and power networks [9]. In practice, it would be hard to prevent any electrical power phenomenon from spreading by making use of telecommunication signalling. This is mainly because the end-to-end packet delay is close to the upper time boundary of the power transient, as shown in Figure 9.3. This would be even more complicated when an additional processing delay is imposed by the encryption algorithm at each message delivery point. Furthermore, since most of the messages are broadcast and/or multicast in nature, there might be multiple packets for a single event notification. To this end, the proper delivery of all broadcast and/or multicast messages needs to be included in the reliability analysis of communication protocols.

9.3 Communication layer

The layered communication architecture of power grid is depicted in Figure 9.4. The WAN [6] serves as a backbone for SG and provides communication for the control centre with transmission and distribution substations. WAN comprises a big number of communication nodes, including smart meters, remote terminal units (RTUs) [10], phasor measurement units [11] and other sensors for remote automation purposes, EMS [12] and supervisory control and data acquisition (SCADA) systems [13]. To ensure fast, broadband and real-time communication without considerable latency required for WAN, the optical fibre communication is mostly applied. Another option is the WiMAX technology due to its relatively long distance coverage and high data flow rate.

NAN/field area network (FAN) [14] is responsible for the level of distribution substations as well as for control of information flows between WAN and customer

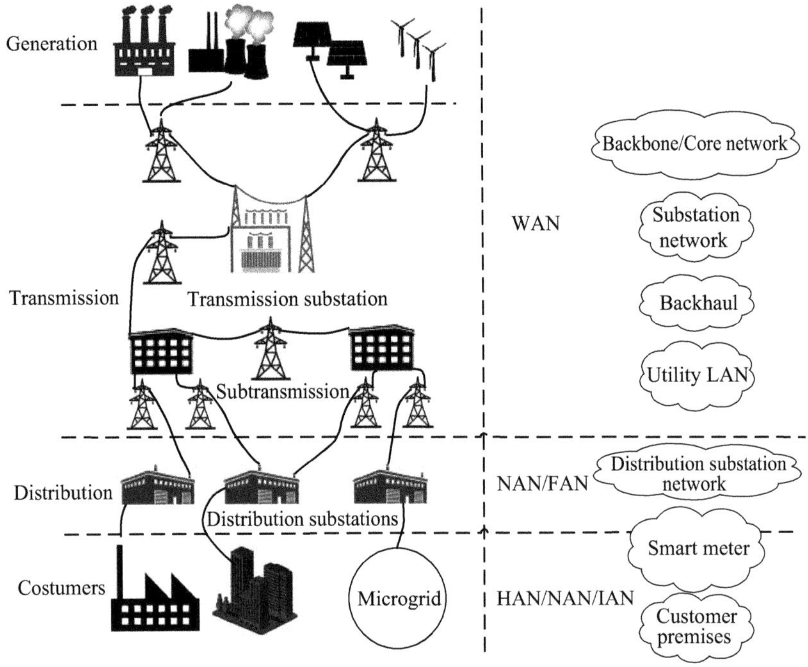

Figure 9.4 Layered communication architecture

premises networks. This type of network provides a large number of SG services, such as smart metering from customer site to control centre, load management and distribution automation. Typical data flow rate can vary from 100 kbps to 10 Mbps covering the distance between 100 m and 10 km. Various communication infrastructures can be used depending on the application of each particular network including ZigBee, wireless LAN (WLAN), PLC, twisted pair and other wired and wireless technologies dependent on funding.

There are different types of customer premises networks [14], such as HAN, NAN, industrial area network (IAN) or building area network (BAN). These networks contain various types of devices depending on microgrids applications, and they can interact with FAN and WAN. HAN mostly includes home automation appliances and smart meters. In addition, it can be made up of energy sources and storages. IAN and BAN are complicated networks with a large number of control devices and sensors for building and industrial EMS and SCADA systems. Customer premises networks require low data rate and power consumption, scalability and security of connection, which gives possibilities for implementation of various communication technologies. According to the number and type of participants, microgrids can be gathered in different types of networks including customer premises area networks as well as NANs or FANs. Figure 9.5 compares various communication technologies using their coverage range and data rates [15]. Table 9.1 provides a list of technologies used in microgrid communications.

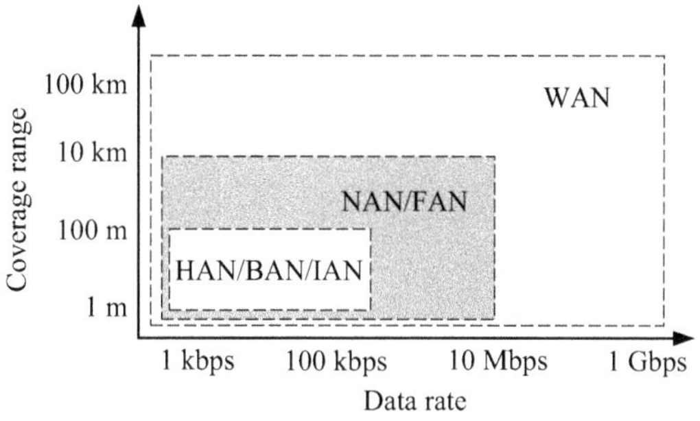

Figure 9.5 Comparison of various subnet communication networks

Table 9.1 List of technologies used in microgrid communications

Communication technology	Standards	Applications
Narrow band PLC [16]	IEEE P1901.2 [17] G3-PLC [18] PRIME [19]	HAN/NAN/WAN
Broadband PLC [20]	IEEE 1901 [21] HomePlug 1.0 [22] HomePlug Turbo [23] HomePlug AV [23] HomePlug CC [24] HomePlug Green PHY [25] HomePlug BPL [25]	HAN/BAN/NAN
DSL [26]	ADSL [27] ADSL2 ADSL2+	NAN/FAN
VDSL [27]	VDSL VDSL2	NAN/FAN
HDSL [28]	HDSL	NAN/FAN
Ethernet [29]	IEEE 802.3 [29] IEC 61850 [30]	HAN/BAN/NAN SAS [31]
Optical fibre [32]	PON [32] WDN [32] SONET/SDH [32]	WAN WAN WAN
WLAN [33]	IEEE 802.11 [33]	Wi-Fi/LAN/HAN/NAN/V2G [34]
WPAN [35]	IEEE 802.15 [35]	Bluetooth/ZigBee [43]/ HAN/NAN/V2G
WiMAX [36]	IEEE 802.16 [36]	MAN [37]/NAN/WAN
Cellular [38]	2G 2.5G 3.5G 4G	V2G GPRS /EDGE/NAN/WAN HSPA/NAN LTE-A [39]/NAN/WAN
Microwave	Direct microwave link [42]	WAN
Satellite Internet		WAN
CR networks (TVWS) [40]		Spectrum bands/WAN
Z-wave [41]		HAN

The communication topologies involved in the subnet communications are coupled either tightly or loosely. In a tightly coupled communication, all equipment must be in a highly available state, while in a loosely coupled communication, equipment are able to manage their communication independently. LAN and metering infrastructure integration systems are examples of tightly and loosely coupled topologies, respectively. In the following subsections, we describe the above mentioned layers.

9.3.1 Home area network

In an HAN, communications happen within a building or smaller network using wired or wireless technologies. This mode of communication is also named an access tier home management system, wherein the operator is able to act as an interface and control appliances within the home domain for energy management. Some of the systems that use HAN mode of communications are heating and ventilation, operation of water pumps, occupancy control system, smart meter reading and remote surveillance for home security. HAN communication equipment are usually set up in a building or local premise. The devices used in this network consume less power and have data speed approximately 100 kbps in a 100 m radius. Some of the communication protocols used in HAN are Wi-Fi, Bluetooth and ZigBee. A case study is narrated in Section 9.7, where ZigBee is used for communications.

9.3.2 Building automation network

As depicted in Figure 9.6, a BAN is a two-layer hierarchical structure with backbone and control level structures, which include sensors, actuators and controllers. The Internet plays a major role in the BAN communications. The sensors used in a BAN are able to capture the field information periodically. The communication bandwidth of the used sensors are in the order of kbps. Sensors communicate with management devices using interconnection devices (ICDs). The communication bandwidth between ICDs and the control system is in the order of Mbps. Typical examples of ICDs are routers, repeaters and protocol converters. Management devices at the top level are the intelligent equipment which monitors, logs, interrogates and archives process data values. In the top tier, management data have a human–machine operating system, which produces graphs and analysis charts for a better network-asset management such as configuration. Devices in the network interfaces are compatible and are able to communicate with one another, while in the point-to-point interface, these devices perform configuration tasks. In the process interface, data is acquired from the network and environmental conditions. Human–machine interface allows the operator to interrogate the field-level equipment. A BAN set-up can be operated by battery system and power consumption is very low. As it is housed in a building, the chances of environmental impact are less, and the cost is minimal. A BAN may use ZigBee and Ethernet protocols for its communications.

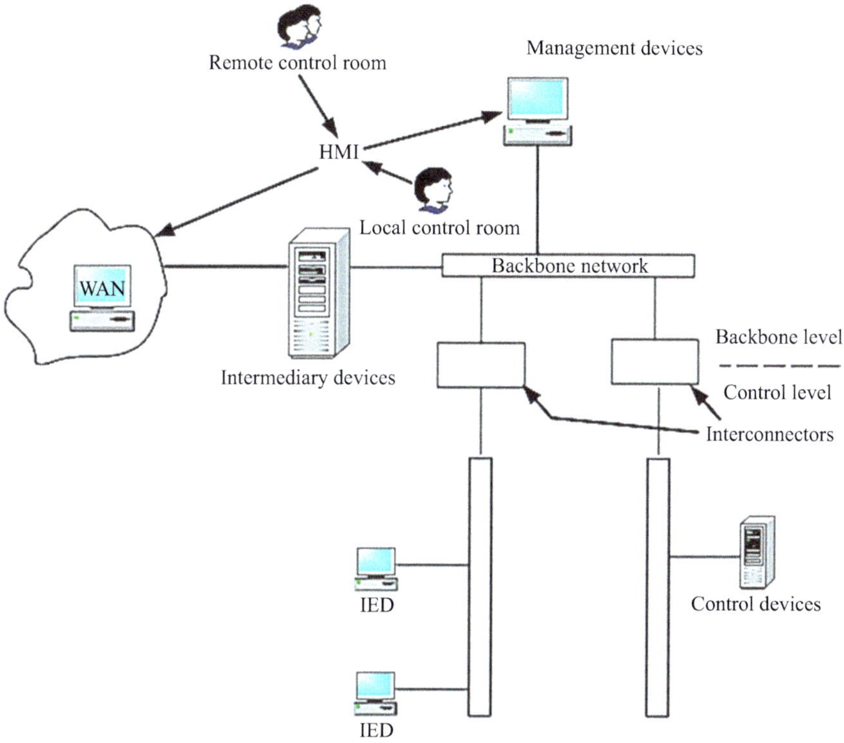

Figure 9.6 BAN network architecture

9.3.3 Neighbourhood area network

A NAN is a mode of communication, in which devices are within a smaller distance to communicate at a higher speed rate, that is 10–100 Mbps, within a radius of 10 km. In a smart microgrid, NAN communications are at the distribution level. Devices in a NAN can communicate using both wired and wireless technologies. Data collectors located in the distribution network use a NAN network to communicate with smart meters. NANs have an advantage over hotspot technology, as it offers more devices to be connected into its network than hotspots over a wider range. However, NAN's bandwidth and speed of data transfer is a matter of further investigation and research.

9.3.4 Local area network

A LAN is a group of intelligent devices, such as computers, in a small network, such as school, office, business and group of buildings. A typical LAN set-up consists of interconnected workstations, personal computers, printers, scanners and data storage devices. Communication speed amongst LAN devices is fast, and there is no requirement of leased communication lines. Ethernet and Wi-Fi are the two examples of protocols used in a LAN. A typical application of LAN is the airport internet network, where passengers access the internet and browse websites leveraging on LAN. The LAN topology can be in the form of ring, star, mesh and tree. Due to cyber security requirements, a LAN needs to have firewalls, network intrusion detection and load balancing. LANs have the ability to manage data exchanges and the quality of traffic.

9.3.5 Field area network

Figure 9.7 shows a FAN architecture, which is gaining popularity due to its ability to connect one point to multiple points in a distributed automation set-up. FAN leverages on HAN and NAN network architectures. A FAN can be used in a number of applications including asset management and diagnostics, smart metering and distributed power generation. FAN reduces the cost of the network and is able to maintain a high availability of the data and quality of service (QoS) [44].

9.3.6 Wide area network

WAN is a telecommunications network that extends over a large geographical distance. WAN are often established with leased telecommunication circuits, and it facilitates a smarter control applications, protection and monitoring in a larger geographical space. The communication rates in a WAN are from 1 Mbps to 1 Gbps [45]. In a SG communication, WAN forms the core tier communication and connects with all nodes within the network, including NANs and FANs. The communication protocols in a WAN are usually implemented in a SG system covering approximately 100 km radius. WAN can maintain a robust communication system encompassing few LANs connected together with routers at the end of the leased line.

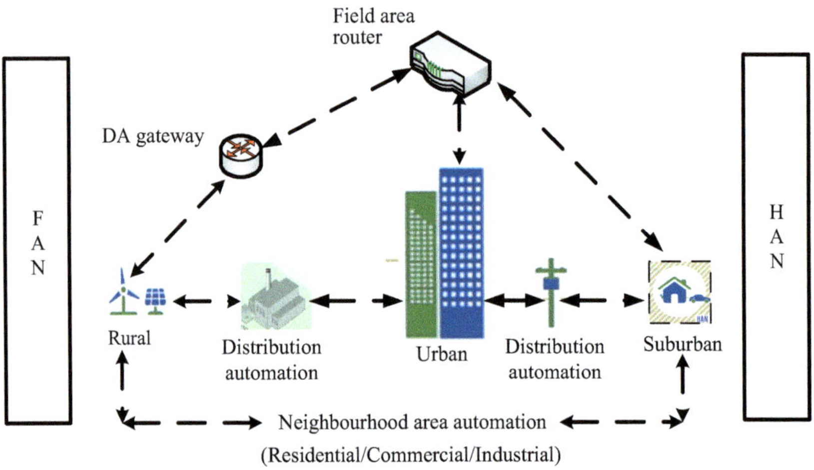

Figure 9.7 FAN architecture

WAN circuit uses two different switching methods, one is circuit switching the other is packet switching. In circuit switching method, two nodes are connected with an established dedicated channel. When a call is made at one end of the line, continuity is established until the call lasts, and circuit is complete making it point-to-point wired communication. In packet switching technique, data packets are transmitted all over the nodes and are received by IEDs. This method of communication suffers from data errors, losses and delays. Circuit switching is mainly for voice transmission from one node to another. Telephone exchange is a good example of circuit switching, while APPLENET, Advance Research Project Agency Network and Bio-Resource Network are the examples of packet switching.

9.4 Communication infrastructure

Depending on the device constraints, location and financial resources, microgrid communication network may transmit signals wired or wirelessly. The following subsections give an overview of existing communication infrastructure with reference to already functioning examples.

9.4.1 Wired communication

Wired communication is the simplest mode of communication, and it can utilise power lines, twisted-pair cables and optical fibres.

9.4.1.1 Power line

Power line communication [16] is the technology that uses electrical power lines as signal carriers. It appeared in the early 1900s as a low data rate service for remote control of power network components. From the introduction of power line communication and until present times, several frequency ranges and various signal modulation techniques were used to achieve the data rates at first from a few bps to a few kbps with audio/low-frequency bands and up to 200 Mbps with a high-frequency range (3–20 MHz). Despite the narrow applications of power line communication in the early years, the technology has become more important nowadays because of the potential in high-speed connection through medium and low voltage (LV) power lines, 15/50 kV and 110/220 V, respectively. The traditional application of power line communication technology is a last mile communication, that is the connection of end devices to the common network or backbone. Power line communication gives benefits by its low costs for implementation and total coverage, because the power lines are installed everywhere in the electrical networks.

However, the power line communication technology has the following several disadvantages:

- It is vulnerable to electromagnetic noise from electrical motors, radio signals interference and power supplies. This is mainly because power lines are not twisted and shielded, so there is considerably a high possibility of interception of critical data transmission by unauthorised parties.

- Connection interrupts when there is an open circuit on the power line section with switches, reclosers and insulators.
- Physical topology of the grid, impedance fluctuations and wave reflection at the terminal points attenuates and distorts signals, which leads to the loss of communication.

9.4.1.2 Twisted pair

The communication technologies based on twisted-pair copper conductors have been used for many years, evolving from the signal carrier of telephone lines to the basic communication medium for LANs. The structure of the cable contains from one to several pairs of conductors with plastic insulation for each wire receiving and transmitting signals in the form of electrical current. In a single pair of wires, one wire is, used for sending the signal to the receiver, and the second one is a ground reference. The difference between these two signals is used by the receiver on the other side. Depending on the cable protection, twisted-pair telecommunication cables are divided into an unshielded twisted pair, shielded and/or foiled twisted pair (FTP and S-FTP). The shield of the cable is implemented by braided mesh and metal foil, which covers each pair of conductors or the whole set of conductors. Shielded cables are used when there is a need for laying the communications in the electromagnetic interference (EMI) environment and by thus preventing the communication channel from the noise penetrations and cross talk. Twisted-pair cable is a cheap communication technology with limited distance capabilities and maximum up to 1.54 MHz channel capacity. Among the disadvantages are cable breakage and water ingress, difficulties in failure pinpointing, rise of ground potential due to power faults and lighting.

9.4.1.3 Optical fibre

The communication systems based on optical fibre technology were introduced in the 1960s and became a good alternative to conventional copper-wired cables. The general configuration of the optical fibre communication systems is as follows [32]:

- Passive optical network (PON) – It provides point-to-multipoint network structure by utilising optical splitters in order to establish the communication service for multiple customers by single wire.
- Wavelength-division multiplexing – It uses cable bandwidth capacity to transmit several data streams simultaneously through the same fibre.
- Synchronous optical networking (SONET)/synchronous digital hierarchy (SDH) – This technology is used for carrying the high-capacity traffic by the time-division multiplexing architecture.

Fibre optic cables give considerable advantages for automation in electrical systems by extremely high data flow rates (5, 10, 20 or 40 Gbps), invulnerability to EMI and radio frequency (RF) interference, long distance of data transmission with significantly smaller necessary amount of repeaters (every 100–1,000 km). Listed features make the technology ideally suitable for backbone communication to the different SG applications and communication infrastructure for HV medium in electrical substations. Despite the obvious benefits of using an optical fibre, it has one main constraint i.e. very high cost for implementation. However, because of the huge bandwidth capacity, the technology is becoming viable by recovering the expenditure for its installation with big amounts of consumers using one communication channel as a backbone. Therefore, the optical fibre infrastructure can serve as a highly reliable and fast communication.

9.4.2 Wireless communication

In congested areas where too many devices are connected in a wired mode, the wiring system may complicate the finding of fault locations. With equipment in close proximity, the wireless technology is a preferred option from the installation, modularity, and the ease of operation perspectives. Challenges in wireless technologies are interference in RF bands, line of sight issue, presence of obstacles and fluctuation of weather conditions. Wireless technologies adopted for microgrids are either short range or long range (cellular), which are described as follows.

9.4.2.1 ZigBee

ZigBee technology provides a short-range (up to 100 m and up to 1,600 m with ZigBee Pro), low-rate wireless communication for personal area networks and based on IEEE 802.15.4 standard [43]. It uses unlicensed industrial, scientific and medical (ISM) bands and performs with the following data rates depending on the frequency: 20 kbps at 868 MHz, 40 kbps at 915 MHz and 250 kbps at 2.4 GHz. The technology supports different network topologies and applies for residential, commercial and industrial buildings automation, energy monitoring and automatic meter reading (AMR) systems. The implementation costs and power consumption are quite low, which along with the high level of security make it a reliable communication solution. However, slow data rate, limited area coverage and the interference by parallel wireless networks, such as Wi-Fi, restrict deployment of ZigBee technology to in-home applications.

9.4.2.2 Wireless local area network

WLAN is a high-speed technology for wireless Internet and network communication based on IEEE 802.11 series of standards and commonly known as Wi-Fi. It provides data rates from 2 to 600 Mbps and operates on ISM frequency bands, such as 2.4, 3.6 and 5 GHz. Wi-Fi gives advantages in fast, secure and reliable connection, but short operation range, high cost for deployment and high power consumption constrain the implementation of WLAN technology to primarily residential and commercial local networks. Table 9.2 lists and compares some versions of IEEE 802.11 standards for WLANs.

9.4.2.3 Wireless mesh

Wireless mesh is an economically viable, reliable and scalable wireless network consisting of a large number of nodes including routers and mesh clients. This type of network uses dynamic routing, where every single communication point has a

Table 9.2 Comparison of WLAN standards

Standard	Data rate	Coverage	Applications
IEEE 802.11b	Up to 11 Mbps	Up to 40 m	HAN, AMI
IEEE 802.11s	Up to 54 Mbps	Up to 300 m	HAN
IEEE 802.11p	Up to 54 Mbps	Up to 1 km	Vehicular network
IEEE 802.11e	Up to 54 Mbps	Up to 300 m	QoS
IEEE 802.11n	Up to 600 Mbps	Up to 300 m	High throughput

possibility to work as a repeater, transmitting in such a way data packets from one node to another. In the case of outage of one node, all others are capable to communicate through the rest of the points because of interconnected structure of the network. Wireless mesh networks exploit various standards, such as IEEE 802.11, 802.15 and 802.16 and are able to cover large territories by using multiple radiowave reflection technology for routing. Because of the benefits in high coverage, robustness and self-healing, wireless mesh has found its application in home automation and AMR systems. Along with that, the technology has limitations in low-speed data rates and vulnerability to the interference from other wireless networks.

9.4.2.4 Z-wave

Z-wave is a low-cost, short-range and low power consumption wireless network technology. It uses 900 MHz ISM frequency and can provide 40 kbps data rate up to 30 m [41]. The technology applies in home automation for residential and commercial light utilities. Because of its low power requirements, it can be used in SG application.

9.4.2.5 Satellite communication

The satellite communication is used for remote monitoring and control of electric substations and especially for the time synchronization based on a Global Positioning System technology. In the satellite communication, the process of the data transmission includes the signal modulation and sending from the one point, amplifying by the satellite and sending back to the receiver on the earth. The technology supports the coverage between any communication points and data rate up to 1 Mbps. The advantages of technology are global coverage, which permits establishing of connection between the nodes in a wide geographical area, and fast installation in comparison to wired technologies by only deploying the necessary equipment. Among the constrains are considerably long round-trip delay dependent on the satellite orbit altitude above the earth's surface, satellite channel characteristics that can be heavily influenced by weather conditions, the effect of fading and considerably high initial investments for satellite transceivers and equipment.

9.4.2.6 Cellular communication

The cellular technology is based on the radio network with a big amount of transmitters, which create cells and reuse frequencies form both coverage and data transmission. Cellular networks work in 850, 900, 1,800 and 1,900 MHz frequencies. The common classification separates cellular communication by generations of technology development: 1G, 2G (Global System for Mobile (GSM) communications), 2.5G (general packet radio service (GPRS) and Enhanced Data rates for GSM Evolution), 3G (universal mobile telecommunications system), 3.5G (high-speed packet) and 4G (WiMAX and long-term evolution (LTE)). Cellular networks are widely employed in AMR systems worldwide by gathering the data from remote metering points through GSM/GPRS modems in offline or real-time modes. Additionally, 3G technology is regarded as one possible solution for SG applications because of its fast data rate, wide coverage, low latency and security. Constraints in cellular communication can arise due to sharing of the same network with mobile users, leading to congestions and lower network performance.

9.4.2.7 Worldwide interoperability for microwave access

Worldwide interoperability for Microwave Access (WiMAX) is a 4G generation of cellular network communication based on the IEEE 802.16 series of standards. WiMAX works in 2.3, 2.5, 3.3 and 3.5 GHz frequency bands together with unlicensed 5.8 GHz frequency giving up to 75 Mbps data rate on the 50 km operating range. Due to its small latency (10–50 ms) and the capability of providing communication to multiple users by exploiting just one base station, the technology is applicable to real-time microgrid and substations' automation applications. The WiMAX is suitable for SCADA systems in big industrial manufacturing facilities, where there is no possibility to provide wired communication. Despite its fast communication and wide coverage, the technology is costly in terms of initial investments, demands high power supply and can be influenced by severe weather conditions.

9.5 Communication protocols

The given subsection highlights the most commonly used and adopted communication protocols in microgrids. Reliable functioning of microgrid components depends on efficient information exchange, such as the current state of the power system, its historic information and its applications. All this information is provided by communication system, which makes communication standards and protocols especially important. The communication system arranges the information exchange between different grid members, such as substations' equipment, DERs and control centres through the common frame of regulations for data format and transmission. Different grid applications have different constrains in terms of necessary communication, which is resulted in various communication protocols existing for data exchange in power industry. For instance, protection devices require much faster response and real-time communication than monitoring applications. On the other hand, communication channels for customer metering infrastructure have to be more protected than substation monitoring devices. Table 9.3 provides a summary of communication protocols used in microgrids.

Table 9.3 List of communications protocols used in microgrid

Protocol	Applications
Modbus [46]	HAN/NAN/Can be used with Ethernet/fiber-optic (FO) [47]
DNP3 [48]	SCADA
PPP [49]	WAN
Improved on-demand distance vector routing protocol [50]	NAN
Geographic routing [51]	NAN
Powerline multipath routing (PMR) [50]	NAN/Narrow band PLC [16]
CSMA/CA [52]	
RT-MAC [53]	
DRX and FDRX [54]	
Max MAC	
QoS-MAC	
Tree based MAC	HAN
Rate allocation MAC	
IEEE 802.15.4 [55]	ZigBee/Tree routing/On Ad hoc On-Demand
6LoWPAN routing protocol	Distance Vector (AODV) [57]/In source routing
EQ-MAC [56]	
Z-MAC [58]	
Wireless HART [59]	Enhanced least—hop first routing [50]
ISA 100.11a [60]	
Z-wave [41]	
INSTEON [61]	HAN
IPV6 hybrid RF-PLC network [62]	BAN
Hybrid protocol for PLC-ZigBee	PLC/ZigBee network
IEEE 802.15	HAN/ZigBee/6LoWPAN/Wireless HART/ISA 100.11a
IEEE 1901	HAN/INSTEON/IPv6RF-PLC/PLC-ZigBee

9.5.1 Internet communications protocol suite

To implement a reliable and secure microgrid architecture, the integration of communication systems is required. This can be achieved by utilising a communication protocol that consists of a layered architecture, where each layer performs a functionality using one or more protocols. There are various networking models, and the most common benchmark communication architecture is the Internet model, which is derived from the Open Systems Interconnection (OSI) reference model [63]. The Internet communication model is another suite that is derived from OSI model. In the Internet communication model, each layer has more than one protocol to carry out various tasks. Both models and their layers are shown in Table 9.4.

The physical layer in the Internet model is the physical connection between the sender and receiver. The role of the physical layer is to transfer analogue/digital signals through the communication medium. The physical layer entails all the hardware devices and physical media. The type of connection and the type of

Table 9.4 Communication network layers

OSI model	Internet model	Groups of layers	Protocols
Physical layer Data link layer	Physical layer Data link layer	Hardware layer	2G, 3G, 4G, 5G, IEEE 802.3, IEEE 802.11, IEEE 802.15.4g, IEEE 802.16, IEEE 802.22
Network layer	Network layer	Internetwork layer	IPv4/v6, ICMP [64], IGMP [65], ARP, RARP
Transport layer	Transport layer		UDP, TCP
Session layer Presentation layer Application layer	Application layer	Application layer	IEC 61968 [66], IEC 61850, IEC 60870, DNP3/IP, Modbus/TCP, SSH, IPfix/Netflow, NTP, DNS, SNMP [67]

signals (electrical signals, radio waves and/or light pulses) are specified by the physical layer.

The data link layer is responsible for movements of individual bits from one hop (node) to the next. It controls the physical layer by deciding when to transmit messages over the media, formats the messages by indicating where they start and end and detects and may correct any errors that have occurred during transmission.

The network layer is responsible for routing and identifies the next communication node to which the message should be sent. The transport layer establishes end-to-end connections between the sender and receiver by linking the application layer to the network layer. The transport layer is able to detect lost messages and request retransmission of the lost messages. The application layer is the application software, which provides access to the network and controls messages that are sent over the network.

There are strong interdependencies between the communication layers of the Internet model; decisions made in one layer impose certain requirements on other layers. For example, the data link and the physical layers are closely tied together as the data link layer controls the physical layer regarding the data transmission. Due to such strong interdependencies, the physical and data link layers are grouped together as the hardware layers. Similarly, the network and transport layers are closely tied together and are sometimes named the Internetwork layer.

The Internet Protocol (IP) suite is a set of protocols for providing of the Internet services. The following protocols are mostly used in microgrid applications: Network Timing Protocol (NTP) for time synchronization [68], IP, Transmission Control Protocol (TCP), User Datagram Protocol (UDP), File Transfer Protocol and Simple Mail Transfer Protocol [69]. The protocols that supply network and transport services, such as IP, UDP and TCP, are of the highest importance. The IP protocol sends individual packets of data (datagrams) from one node to another carrying the addresses of sender and receiver in the datagram header as 32-bit fields. It fragments data into smaller packages for transmitting purposes and reassembles them at the receiver end. However, the data delivering is a concern for higher layers, and IP does not guarantee it by only checking the header sum. Besides that, IP does not establish or monitor the connection on the channel and just sends each datagram independently.

TCP utilises the functionality of IP with the establishing of communication links between each end, maintains error-free transmission and makes possible the bidirectional communication. It fragments the data into sections and conveys them to IP, which further cuts those to smaller datagrams. UDP also uses IP services additionally providing information about sender and receiver ports, and a checksum. Unlike TCP, UDP discards data packets with detected error and does not supervise reliable data delivering, and it is not connection oriented. By utilising IP broadcast addressing mode, UDP allows the transmission of data to several destinations. It applies for real-time systems where the smallest delay of connection is of the highest importance.

The IP suite is used for carrying specific automation protocols, such as Modbus, Distributed Network Protocol version 3.3 (DNP3) and IEC 61850, over a network environment by conveniently utilising Ethernet data link. These protocols are encapsulated in the TCP/IP stack and act over its layers. Another application of the IP suite is the data communication between terminal servers, data logging servers and the control centre.

9.5.2 Modbus

The Modbus protocol is originally designed for data exchange between PLCs [46]. It is openly published and from 2004 controlled and maintained by the community of vendors and users of the automation equipment Modbus-IDA. The protocol works at application level with a foundation on a client/server architecture, where the client requests server's operation. By using master multiplexing in gateway-based network for gathering diverse communication interfaces, Modbus is capable for routing of different system configurations with not only one master controlling all the slave devices. Figure 9.8 shows the layered structure of Modbus.

Modbus can be deployed over several communication interfaces, they are as follows:

TCP/IP over Ethernet: The data is framed into binary format in TCP using Ethernet protocol. Addressing and channel access control mechanisms are provided by carrier sense multiple access with collision detection (CSMA–CD).

Serial transmission: The data with serial communication can be transmitted in eight-bit binary format (Modbus RTU) or in seven-bit American standard code for information interchange format (Modbus ASCII), which requires a twice-higher data traffic. This mode can be implemented over various communication mediums, such as wired, fibre optic or radio. The most commonly used standards in wired communication for serial transmission are RS-232, RS-422 and RS-485.

Modbus Plus: The advanced technology for high data transfer with extra addressing, routing, data consistency and diagnostic features.

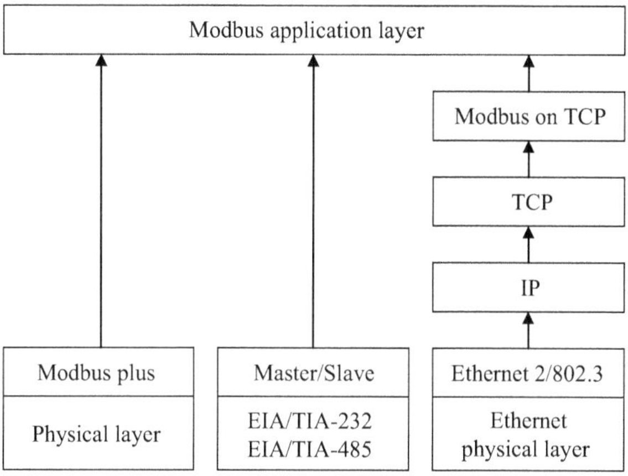

Figure 9.8 Modbus protocol layers

9.5.3 Distributed Network Protocol version 3.3

The DNP3 is an open standard for telecommunication designed for interaction between master stations, RTUs and other IEDs in electrical utilities and industrial environments, such as oil and gas, water/waste and security [48]. It was designed for SCADA systems to transmit considerably small data packets in an interruptible manner with a predefined sequence. DNP3 uses several topologies including point-to-point, multi-drop or hierarchical topologies. DNP3 messages are usually mapped to the upper three layers of the OSI model, that is data link, transport and application layers.

Point-to-point topology is a straightforward model where one central station interacts with one remote station. In multi-drop topology, one master device communicates with several slaves by polling the data. Each remote station receives the request messages but sends the data to the master only when the destination address corresponds to its address. The hierarchical layout comprises one master and several slave devices, which also can be a master device to other outstations in a lower layer. The last possible topology with utilization of DNP3 protocol is data concentrator, where several remote stations can have different protocols, and by using gateways, the central station can collect information from them.

DNP3 supports two modes of operation – poll and quiescent, or reporting by exception, meaning that master device does not send any requests to check if the communication with the remote station is stable. When the state of the channel is changed, the slave devise sends 'unsolicited response', and without any troubles the system remains in silent state. This mode of operation contributes to full bandwidth capacity utilization of the communication channel. The above described modes of operation imply the initiation of communication performed by slave devices, which is referred as balance communication. However, only master devices are able to request data or send commands to other devices in the DNP3-supported networks. DNP3 is typically deployed on serial communication (RS-232 and RS-485 [70]) over various physical links, such as twisted pair, fibre optic, radio and satellite communication, and in some cases, it is encapsulated in TCP/IP protocol on Ethernet networks.

9.5.4 IEC 61850

IEC 61850 is a set of standards of IEC Technical Committee 57 for electrical substation automation systems. The IEC 61850 defines various aspects of substation communication system along with a number of several related standards for communication between substations and the industrial automation [30]. The standard is organised in ten main sections with few subsections performed as separate IEC 61850 documents. The main communication architecture concept in IEC 61850 is the creation of data objects and services independent of any particular protocol, which allows further mapping of the data objects and services to any other protocol meeting the data/service requirements. For the purpose of building large and abstract data objects, the concept of common data classes (CDCs) was defined. The standard also determines the ways of mapping the abstract data and services into the Manufacturing Messaging Specification (MMS) standard [71] and sampled measured values into the Ethernet data frame with point-to-point and multipoint, unidirectional and bidirectional communication, respectively. For the explicit representation of the relations between automation system and substation itself, the Extensible Markup Language [72]-based Substation Configuration Language (SCL) was developed, so that each device has to provide its own SCL file with the configuration description.

Unlike legacy communication protocols, which specify the format of bytes transmitting over the wire, the IEC 61850 standard gives the extensive model for the organization of data, which is consistent with all vendors and devices in the power system. This feature reduces significantly the efforts needed for the engineers for devices configuring. It can be done simply by importing the SCL configuration file into the device, and then IEC 61850 client application can retrieve the object characteristics from the device over the network. IEC 61850 provides a variety of advantages over the proprietary protocols including interoperability of devices from different suppliers, lower installation, configuration and maintenance costs, enhanced scalability and possibility for further improvements of the automation processes of systems.

To meet the strict timing requirements of the power grid, the IEC 61850 standard designates a 4 ms latency requirement that ensures the timely delivery of interlocking trip commands and data-sampling messages. The latency requirement is derived from the time required for a protection message to travel from a logical node in the detecting IED to the delivery point in the logical node of the protection IED. Table 9.5 details the latency requirements of the protocols in the IEC 61850 standard [73]. Since the messages have a multicast nature, they need to be delivered to multiple IEDs, and hence, the timing requirements apply to all subscribers. Sampled value (SV) messages include information about system's voltage, current and frequency. Generic-Object-Oriented Substation Events (GOOSE) packets are used for tripping circuit breakers or signalling availability of

Table 9.5 IEC 61850 protocol time requirements

Protocol	Type	Latency (ms)
SV	Data sampling	4
GSSE and GOOSE	Protection	4
TimeSync and ACSI	Monitoring and control	16
ACSI	Maintenance	1,000

Table 9.6 Criticality and size of GOOSE/SV messages

Message type	Objective	Priority	Message size (byte)
Type 1	Trip	5	50
Type 2	Interlocks	3	150
Type 3	Status indication	2	200
Type 4	Sample values	4	98
Type 5	File settings	1	1,000

equipment. As shown in Table 9.6, the criticality and size of GOOSE/SV messages vary by the type of event [74].

In comparison with DNP3, IEC 61850 protocol has a number of advantages, such as a faster command in tripping, control, measurement and event recording on IEDs. In addition, data packets are sent in the form of Multimedia Specification [75] or GOOSE or sampled measurement value. This fast communication has an impact on equipment and cost saving to the end user in terms of operation, that is exchange of status, data logging, time synchronization and file transfer. IED translates the signal from power components, such as frequency, voltage, current and watts into object-oriented data representation, which facilitates the ease of operation and provides a better control.

9.6 Importance of communication technology in microgrid control

In a microgrid system, there might be a number of interconnected DERs, such as fuel cells, wind, solar and hydropowers, which can deal with a number of communication challenges. These challenges are as follows:

- Bidirectional power flow between DERs and controllers require a robust communication and voltage regulation methods.
- Application of inverters, controllers and PLCs produces undesirable harmonics.
- Uncertainty in the production of renewable generation due to varied environmental and weather conditions which may affect DER output to maintain a balanced power system;

- System integration by making use of various equipment, including DERs and IT peripherals manufactured by multiple vendors. This can be a reason for misconfiguration.

Figure 9.9 shows a typical solar farm microgrid. The overall system consists of photovoltaic (PV) cells, which acts like generators equipped with converters, inverters and filters. The presence of large-scale electronic devices is a source for harmonics. Fundamentally, the third and fifth harmonics pumped into the network cause issues with power quality and could damage insulation of the equipment over a period of time if not checked. In order to mitigate this unwanted power quality issue, suitable filters are inserted into the circuit, which mitigates harmonics.

To coordinate communications within a distributed microgrid system, a suitable control and communication protocol is required. The microcontroller communicates with different generators and governs charge flow into ES devices in an off-grid mode. Traditionally, microgrids have focused more on AC systems, but these days DC microgrids are resurging in large scale. Also, hybrid microgrids, that is a combination of DC and AC systems, are also finding wide-scale applications. DC microgrids have few advantages over AC microgrids, but they have problems with respect to protection and stability. In DC microgrids, control of DERs is achieved by introducing control loops. The two common methods available in controlling microgrid equipment are droop control and master–slave control.

A droop controller is adopted with converters to retain each voltage source by comparing it with a set characteristic, while a master–slave controller builds a link between system components and the master controller (MC). It calculates the reference of each source periodically and compares it with the MC. Employing

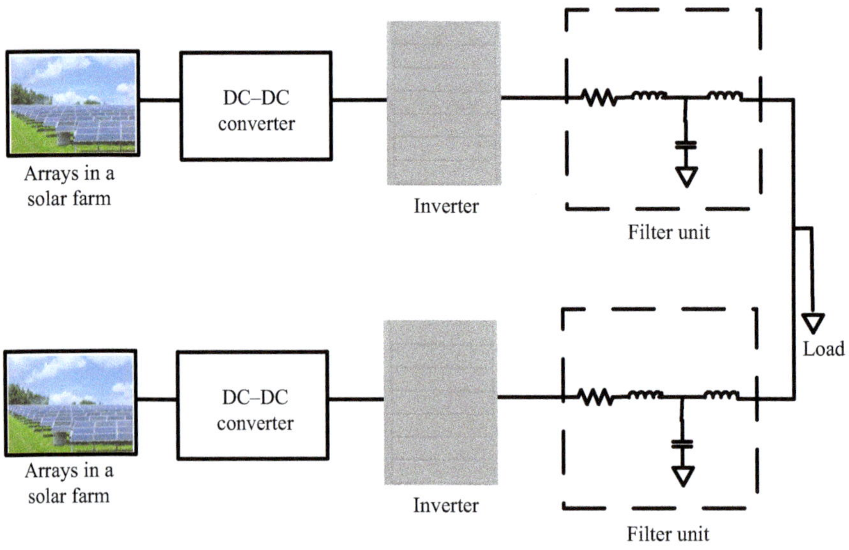

Figure 9.9 Solar farm microgrid and mitigation of harmonics using filters

these techniques, the microcontroller is able to achieve simplicity in operations, desired reliability and efficiency.

As microgrid generators are mainly weather dependent, it is important to establish an appropriate communication between generators and controllers. The microcontroller regulates bus voltage fluctuations, battery storage availability and ability to load shed and cut-off DERs whenever there is a fluctuation in power supply for stable operation. The importance of communication is emphasised by having a coordinated approach to manage grid-connected converters, ES systems, battery EMS (BEMS). The BEMS controller communicates with MC and LC of the storage cells to oversee the charging and discharging of batteries due to the non-linear nature of generation and load variation. If storage cells continue to perform in the exponential region, it reduces the battery life drastically. BEMS have set upper and lower limits for charging and discharging. The intelligence in BEMS smart model allows batteries to disconnect when it peaks at charges up to 80% and reconnects it back to power supply when the batteries are 40% discharged. Smart MCs have inbuilt intelligence to detect varying load and weather conditions to monitor the network condition. It appropriately communicates to DERs and controls its output in off-grid and grid connection mode including battery output optimisation and its operations. This is beneficial for smooth grid service and power stability. In short, proper communication between batteries and BEMS could expand life of batteries by implementing appropriate control technology to discharge and charge cycle during a power supply deviation.

Nowadays, hybrid microgrids are finding wide applications. It is worth noting that by suitable interfaces, DC microgrid power output can be converted to AC with low losses. It ensures the stability of the system and equal power-sharing management. In a hybrid state, appropriate communication between AC and DC controller ensures the maintenance of reference voltage locally and at the master control room of the microgrid operator.

In a grid connection mode, MC's role is to regulate power flow into the grid via PCC after verifying the quality, frequency and other parameters. Grid connection is not accomplished without having proper communication link between field devices and IEDs linked to MC. Large exchange of data is essential feature in a complex microgrid network deriving huge data to analyse, store and control the DERs. The microcontroller of such a complex network needs to be reliable and able handle data for the optimal operation of microgrid. In order to achieve handling of huge data, a two-way communication such as TCP/IP is usually a host-centric communication that is employed, wherein the microcontroller and field device equipment exchange data rapidly. It may be noted, due to number of DERs in the network, it is important for MC to interrogate and communicate with all DER generators.

TCP/IP communication has been widely used for communication in a traditional microgrid set-up backed up by Modbus or Profibus communication. However, these days with high usage of Internet and wireless mode in our cyber system, it is important to have a smart MC that communicates with devices to store, interrogate, command and manage data of interconnected DERs and peripherals.

Smart IEDs in the network are able to manage lesser capable DERs having short cell life, low processing capabilities, lesser memory storage units, etc. All these advanced technologies make the control and communication of microgrid system complex depending upon its size. While a simple microgrid could have an output less than 2 MW, a large utility-scale microgrid connected to a utility network could be of the order 20 MW or greater having industrial or commercial loads. Hence, it is of great interest to the operator to communicate in a complex network by smarter and intelligent way to process data.

One can have either a centralised control or decentralised control of communication in a microgrid network. A centralised control example is SCADA centre system. Centralised control schemes are easy to implement and cost-effective way to control a microgrid network. Such schemes are usually established in a smaller microgrid network. However, there are issues with the computation of data due to low flexibility, scalability, high bandwidth, single point failure, etc. On the contrary, decentralised controllers have LCs positioned at different locations in the microgrid network take faster decisions while interfacing DERs. This overcomes issues related to flexibility, bandwidth, failure of communication channels, improved communication speed, plug and play capability, etc. The most important reason for having such a decentralised communication control command is to have system expandability. However, its implementation is not so easy, and it is overcome by using multi-agent system (MAS) [76]. In a MAS set-up, multiple IEDs are positioned at strategic points having complex algorithm built within it to compute, compare, analyse and exchange data amongst microgrid smart equipment having interoperability features. IEDs within MAS set-up react to any changes to their field of operation in an abnormal condition. They are goal oriented and autonomous in operation, while fully cooperating with other intelligent devices.

Figure 9.10 shows a typical microcontroller in a conventional set-up with three layers in its network for control and communication. The first layer, that is LC, interacts with DERs, ESs and controllable loads, etc. It obtains data from field regarding frequency, voltage and current. The second layer, that is the centralised control layer (CCL), controls power deviations, frequency, voltage, power quality, etc. It interacts with main grid via PCC and exchanges data to the control room of the utility operator. This controller has the ability to decide on grid connection and islanded operation mode as applicable. The third level is a distribution network dispatch centre, which monitors the inflow of power dispatched into the grid and has the ability to command the CCL to cut itself up from the grid in the case of deviation in frequency, quality, excess demand or under voltage condition.

More recently, two apex bodies i.e. IEEE and CIGRE working groups, are studying methods to improve the control and communication set-up in the absence of a communication and control standard. Communication and control are essential in AC and DC microgrid network to determine bidirectional power flow, that is from AC to DC and DC to AC. Conventional methods of analysing power flow between AC to DC and DC to AC revolves around droop characteristics based on static and dynamic loads. The power flow is frequency dependent in the case of an AC system. When the AC bus voltage is greater than the DC bus voltage, power

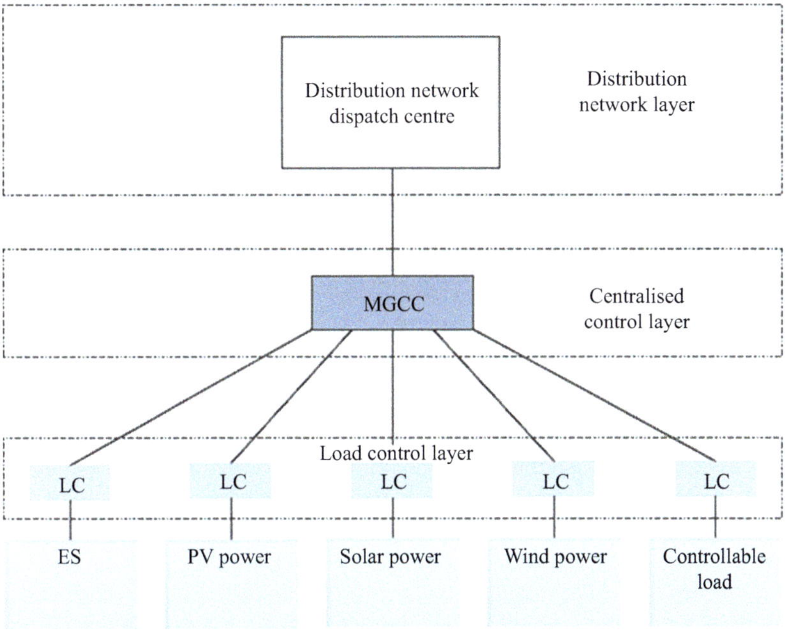

Figure 9.10 A three-layer microgrid control scheme

flows from AC to DC and vice versa. Some of the challenges encountered in the control and communication within a microgrid network are [77]

- absence of a common standard (interoperability),
- difficulty in providing protection to multiple DERs operating at the same time,
- control of power flow,
- varying weather and environmental condition and
- toggling between off grid and islanded mode due to demand and supply condition.

9.7 Case study

In this section, we study the communication protocols in a car manufacturing plant in India. The manufacturing plant includes five major areas, all of which have a number of sheds, buildings, power equipment and control devices. The equipment used in these premises are described as follows:

- *Area A* includes an HV substation (132/11 kV) switchyard with a 11 kV diesel generator (1 MW). In this premise, a local control room (LCR) is equipped with switchgears, protection equipment, a substation automation system (SAS) and AC/DC systems. The total load, that is 20 MW, is transferred to two downstream substations using two power transformers.

- *Area B* includes two 5 MW wind turbines, one 3 MW PV generation with associated switchgears, and one control building housing SAS equipment at 11 kV.
- *Area C* includes an engine shop with a small 11/415 kV substation that feeds power to 20 machines in the manufacturing plant. This area also includes a protection control room, a SAS equipment, a kitchen area, and a common room, that is accessed by 50 people. The load in this area is approximately 7 MW.
- *Area D* includes a long shed with a small 11/415 kV substation that feeds power to 15 machines, one training centre and one SAS equipment room, that is accessed by 30 people.
- *Area E* includes the main control room (MCR), a SAS room with AC and DC equipment, an admin and corporate building, a finance department and an HR department which includes 100 people. In addition, the area includes a heating, ventilation and air conditioning (HVAC) system, private automatic branch exchange (PABX) telecommunication infrastructure and firefighting equipment.

Figure 9.11 shows an HV and an LV power reticulation within the car manufacturing plant. During the normal operation, power flows from the wind turbines and PV systems (DERs) to the main receiving 11 kV substation. In the event of an inclement weather, for example, DERs are switched off by the operators and power supply flows through via two incoming 132 kV lines through the PCC. The car manufacturing company has as an agreement with the state utility to sell surplus power to the grid. The emergency diesel generator (EDG) backs up the essential loads of various areas connected to 11 kV bus. It provides essential 24-h services to SAS, MCR and admin building even if a blackout occurs. Power equipment in areas A, B, C, D and E communicate via FO and hard wired copper cables based on IEC 61850 and DNP3 communication protocols.

The communications in plant happen using both IEC 61850 and DNP3 protocols. IEC 61850 is the primary mode of communication within the plant, while DNP3 is the backup. This redundancy helps to maintain the reliability in case one of the communications fail. In the backup communication mode, DNP3 acts as an alternative communication scheme for the power equipment. DNP3 transmits analogue status to a RTU in the form of binary codes, such as 0, 1, 2 or 3; mapping the binary code to a trip, a fault, an open or a close command.

The plant aims to make the distributed generation system more dependent to microgrid's own power supply rather than the utility system. As shown in Figure 9.11, the merging units (MUs) collect sampled sensory data (e.g. currents and voltages), namely SVs, from the power equipment, DERs and EDG equipment. The SVs published by MUs are classed into five different categories, all of which have differing priorities with different payload sizes. Table 9.6 shows the criticality and size of SV messages. For example, the IED would issue a trip command when it receives a Type 1 message. A Type 5 SV message has the highest priority as it relates to protection, and once it is received the circuit breaker needs to be tripped.

The plant communication architecture is shown in Figure 9.12. IEC 61850 and DNP3 protocols complement one another by making the load management and

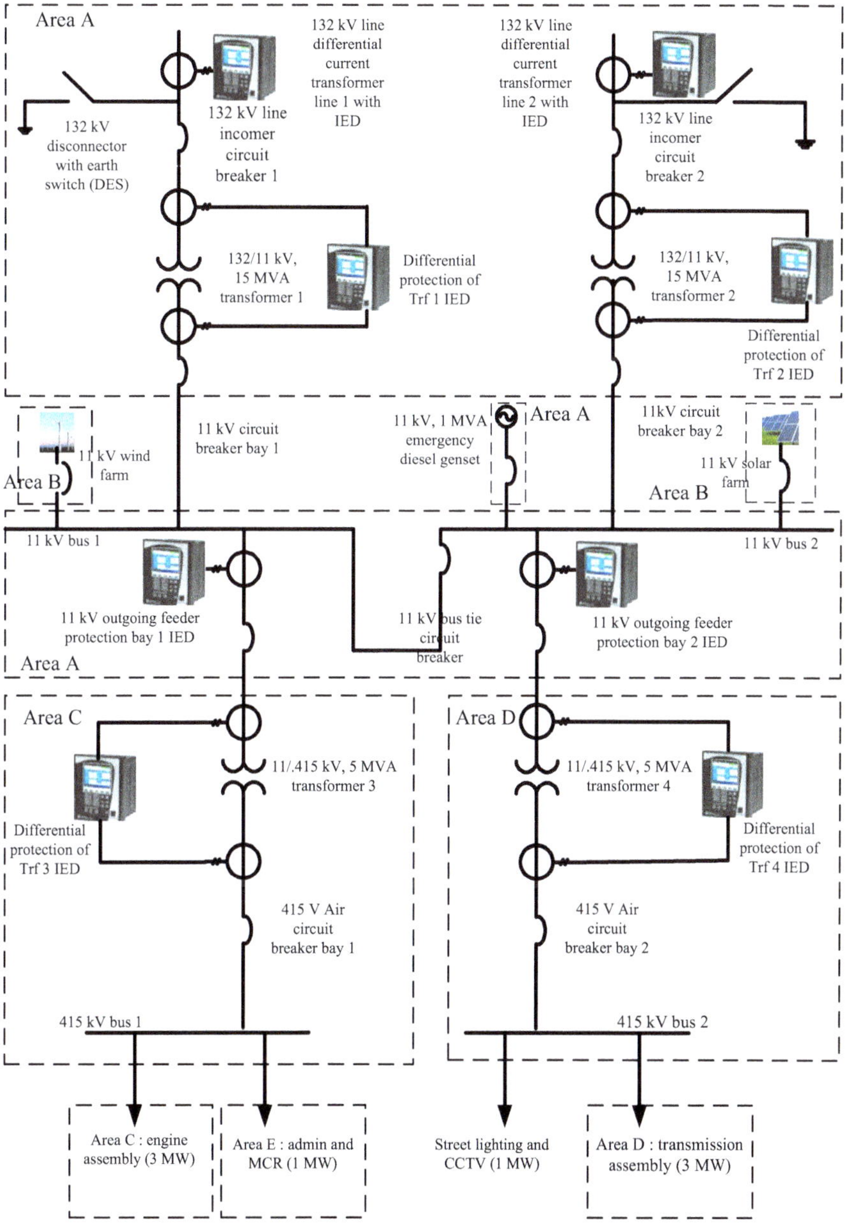

Figure 9.11 Power and SAS diagram of the case study

protection of HV/LV apparatus more efficient. From the safety and reliability point of view, all power equipment need a backup protection scheme. In this regard, in the event of malfunctioning of the main protection, the backup protection of power equipment is managed by DNP3.

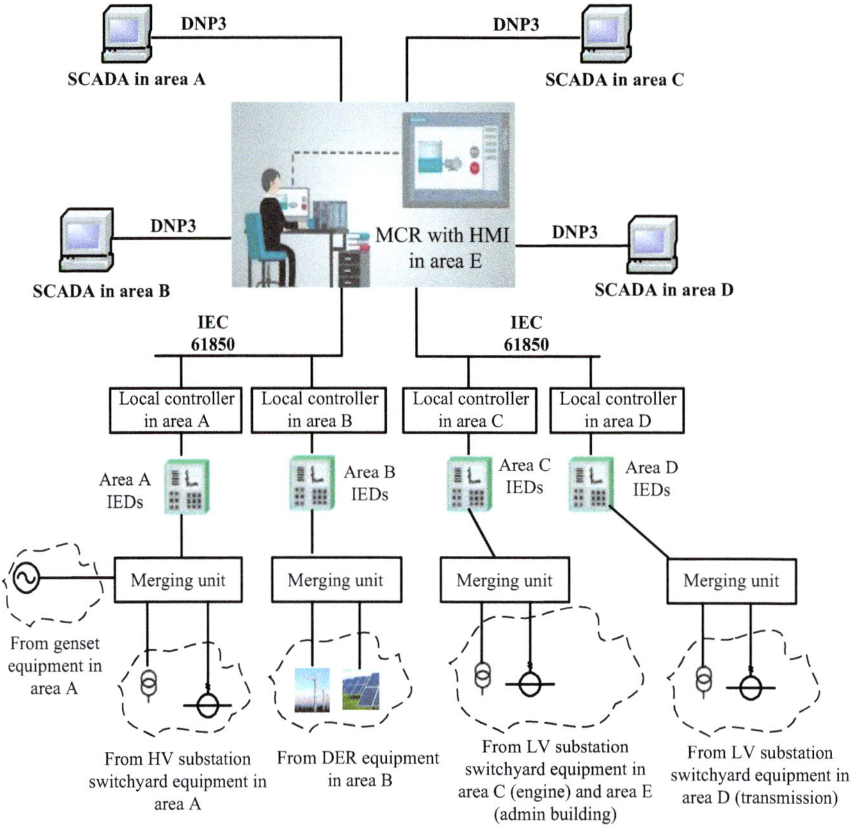

Figure 9.12 Block diagram of IEC 61850 and DNP3 communication protocols

The first protection scheme, which is based on IEC 61850 communication mode, regularly scans the network by communicating with field devices via MUs for faults and alarms. For example, when a fault occurs in a power transformer, IEDs trip the switchgear in less than 4 μs. Alternatively, if an operator wishes to shut down a plant, he or she can trip and isolate the HV device by sending digital SV messages to IEDs. However, in a digital communication system, there are issues with delays and latencies due to data clogging or congestion. This makes the data packets lost and never reach its destination as a result, the subscriber IED is not able to see the messages and understand the byte size making communication fail or trip or alarm. Under such a circumstance, redundancy in protection is essential, and the DNP3 mode of communication comes into the picture. In a DNP3 communication protocol, SCADA system monitors the state of HV equipment through hard wired copper cable. DNP3 uses a 16-bit cyclic redundancy check, which is able to detect errors in data communication.

Figure 9.13 shows the mapping of digital IEC 61850 data sets to DNP3 in Boolean form in the event of a failure of the first protection scheme. Figure 9.13

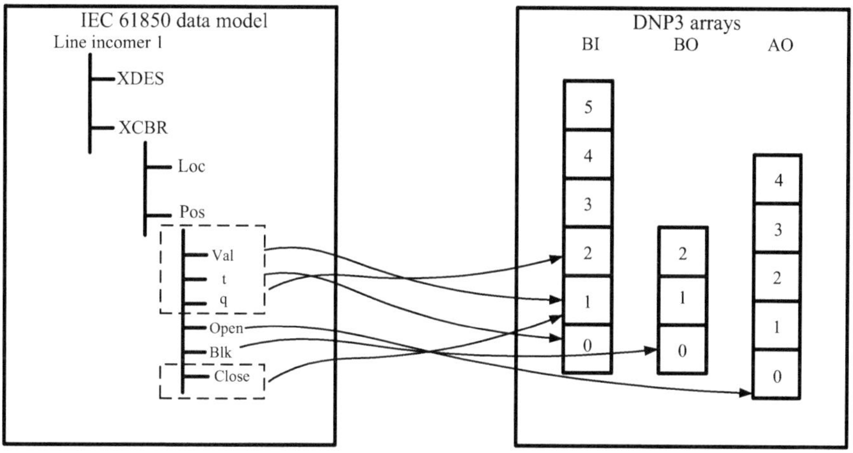

Figure 9.13 IEC 61850 attribute mapping

shows the mapping of a feeder in a particular HV primary plant. Logical nodes of the feeder have been named XDES (disconnector with earth switches) and XCBR (circuit breaker). Logical nodes are further subclassified as CDC representing an object, such as a set of data attributes value (Val), time of last change (t), quality flags (q), open, block (Blk) and close. Figure 9.13 shows how digital and analogue values are converted into one another.

The HVAC system in buildings adjusts the temperature based on environmental circumstance monitored by IoT sensors, which wirelessly communicate with the MC in area E. The load management and distributed generation system controls the utility interconnections. The data is recorded and logged every 10 s at the LCR, and it is used by the asset engineer for performance monitoring.

Area E uses a ZigBee communication network. The application of ZigBee over a small area, such as a three-storied admin building and canteen facilities, gives adequate flexibility and reliability communicate and controls LV appliances, such as dimming or increasing lux level of building lights or wirelessly communicating with sensors of building automation system. In order to mitigate adversaries who infiltrate building security system, a two or a three-layer authentication is used. The installed ZigBee network uses 2.4 GHz frequency, which is shared by RF products and Wi-Fi networks. Tablets and mobile devices are Wi-Fi enabled, which enables field technicians to carry out diagnostics on LV system, such as battery chargers and battery banks, by latching on to Wi-Fi mode.

The main issue in this set-up is the communication drop out, and this can be resolved by having a number of PLCs as backups. All nodes in a PLC set-up interface uses physical wiring connection and that has a slim chance compromising the security of the set-up. The only issue encountered in a PLC set-up is the interference due to the noise picked up by power cables reticulated around the plant. Noise is a normal phenomenon due to a large number of power apparatuses operating in the network. The EMI is mitigated by using isolated, grounded twisted cables. In this plant, the use of wired and wireless technology is decided based upon the ease of installation, costs, security, power consumption, reliability of protection and the speed of the network.

9.8 Conclusion

This chapter provides a detailed overview of communication technologies, standards and protocols used in smart microgrid applications. The chapter discusses applications of several smart and intelligent protocols, such as IEC 61850 and DNP3, and issues pertaining to the application of digital technologies, such as latency and physical link conversion. Efficiency, reliability and security of interconnected devices and systems are critical to enabling smart microgrid communication infrastructures. Interoperability is achieved while avoiding being isolated into non-competitive technical solutions. This chapter describes the translations between IEC 61850 and DNP3. Finally, a case study detailing a car manufacturing plant in India is presented in order to demonstrate the actual implementation of smart microgrid communication technologies and communication protocols.

Nomenclature

ARPANET	Advance Research Project Agency Network
ASCII	American standard code for information interchange
BAN	building area network
BEMS	battery energy management system
BRNET	Bio-Resource Network
CCL	centralised control layer
CSMA–CD	carrier sense multiple access with collision detection
DERs	distributed energy resources
DNDC	distribution network dispatch centre
DNP3	Distributed Network Protocol version 3.3
DSL	digital subscriber line
EDGE	Enhanced Data rates for GSM Evolution
EIA	Electronic Industrial Alliance
EDG	emergency diesel generator
EMS	energy management system
EMI	electromagnetic interference
FO	fiber-optic
FTP	foiled twisted pair
GPRS	general packet radio services
GOOSE	Generic-Object-Oriented Substation Event
HAN	home area network
HDSL	high-bit-rate digital subscriber line
HSPA	high-speed packet access

HV	high voltage
HVAC	heating, ventilation and air conditioning
IAN	industrial area network
ICD	interconnection device
IED	intelligent electronic device
IEEE	Institute of Electrical and Electronics Engineering
IoT	Internet of Things
LAN	local area network
LCR	local control room
LTE	long-term evolution
LV	low voltage
MAN	metropolitan area network
MAS	multi-agent system
MC	master controller
MMS	Manufacturing Messaging Specification
NAN	neighbourhood area network
PABX	private automatic branch exchange
PCC	point of common coupling
PE	power electronics
PLC	programmable logic controller
PLC	power line communication
PON	passive optical network
PMU	phasor measurement unit
PV	photovoltaic
QoT	quality of traffic
QoS	quality of service
RF	radio frequency
RTU	remote terminal unit
SAS	substation automation system
SCL	Substation Configuration Language
SCADA	supervisory control and data acquisition
SFTP	shielded foiled twisted pair
SG	smart grid
SDH	synchronous digital hierarchy
SONET	synchronous optical networking
SV	sampled value
TIA	Telecommunication Industry Standard
TC	technical committee
TCP/IP	Transmission Control Protocol/Internet protocol
WAN	wide area network

WDM	wavelength division multiplexing
WiMax	Worldwide Interoperability for Microwave Access
WLAN	wireless local area network
UTP	unshielded twisted pair
VDSL	very high-bit rate digital subscriber line
XCBR	circuit breaker
XDES	disconnector with earth switch
XML	Extensible Markup Language

References

[1] R. H. Lasseter and P. Paigi, "Microgrid: A conceptual solution," In *IEEE 35th Annual Power Electronics Specialists Conference*, vol. 6, 2004, pp. 4285–4290.

[2] A. Jolfaei and K. Kant, "Data security in multiparty edge computing environments," In *Annual GOMACTech Conference: Artificial Intelligence and Cyber Security – Challenges and Opportunities for the Government*, Albuquerque, NM, USA, 2019, pp. 1–6.

[3] M. Z. Huq and S. Islam, "Home area network technology assessment for demand response in smart grid environment," In *20th Australasian Universities Power Engineering Conference*, 2010, pp. 1–6.

[4] D. Budin, A. Herman, and C. Lanzl, "Wireless local area network communications system," U.S. Patent 5,276,703, issued January 4, 1994.

[5] H. Hu, D. Kaleshi, A. Doufexi, and L. Li, "Performance analysis of IEEE 802.11af standard based neighbourhood area network for smart grid applications," In *IEEE 81st Vehicular Technology Conference*, 2015, pp. 1–5.

[6] M. Borka and M. Begovic, "Voltage-stability protection and control using a wide-area network of phasor measurements," *IEEE Transactions on Power Systems*, vol. 18, no. 1, 2003, pp. 121–127.

[7] A. Jolfaei and K. Kant, "A lightweight integrity protection scheme for fast communications in smart grid," In *Proceedings of the 14th International Conference on Security and Cryptography*, Madrid, Spain, 2017, pp. 31–42.

[8] A. Jolfaei and K. Kant, "A lightweight integrity protection scheme for low latency smart grid applications," *Computers and Security*, 2018 Oct 13. (In press).

[9] P. Kundur, N. J. Balu, and M. G. Lauby, "Power system stability and control," vol. 7, McGraw-Hill, New York, 1994.

[10] D. C. Osburn III, "Automation Solutions Inc., Remote terminal unit," U.S. Patent 6,628,992, 2003.

[11] M. M. Eissa, M. E. Masoud, and M. M. M. Elanwar, "A novel back up wide area protection technique for power transmission grids using phasor measurement unit," *IEEE Transactions on Power Delivery*, vol. 25, no. 1, 2010, pp. 270–278.

[12] C. Chen, S. Duan, T. Cai, B. Liu, and G. Hu, "Smart energy management system for optimal microgrid economic operation," *IET Renewable Power Generation*, vol. 5, no. 3, 2011, pp. 258–267.

[13] S. A. Boyer, "SCADA: Supervisory control and data acquisition," 4th edition, *International Society of Automation*, New York, 2009.

[14] M. Kuzlu, M. Pipattanasomporn, and S. Rahman, "Communication network requirements for major smart grid applications in HAN, NAN and WAN," *Computer Networks*, vol. 67, 2014, pp. 74–88.

[15] S. Shukla, D. Yi, and L. Mili, "Construction of a microgrid communication network," In *IEEE Power Engineering Society Innovative Smart Grid Technology Conference*, February 2014, pp. 1–5.

[16] L. Lampe and A. J. H. Vinck, "On cooperative coding for narrow band PLC networks," *AEU-International Journal of Electronics and Communications*, vol. 65, no. 8, 2011, pp. 681–687.

[17] V. C. Gungor, D. Sahin, T. Kocak, *et al.*, "Smart grid technologies: Communication technologies and standards," *IEEE Transactions on Industrial Informatics*, vol. 7, no. 4, 2011, pp. 529–539.

[18] K. Razazian, M. Umari, A. Kamalizad, V. Loginov, and M. Navid. "G3-PLC specification for powerline communication: Overview, system simulation and field trial results," In *IEEE International Symposium on Power Line Communications and Its Applications*, 2010, pp. 313–318.

[19] S. Thokozani and A. J. H. Vinck, "Interleaving and nulling to combat narrow-band interference in PLC standard technologies PLC G3 and PRIME," In *17th IEEE International Symposium on Power Line Communications and Its Applications*, 2013 pp. 258–262.

[20] L. Jianming, Z. Bingzhen, G. Liang, Y. Zhou, and W. Yirong, "Communication performance of broadband PLC technologies for smart grid," In *IEEE International Symposium on Power Line Communications and Its Applications*, 2011, pp. 491–496.

[21] S. Goldfisher and S. Tanabe, "IEEE 1901 access system: An overview of its uniqueness and motivation," *IEEE Communications Magazine*, vol. 48, no. 10, 2010, pp. 150–157.

[22] M. Y. Chung, M.-H. Jung, T.-J. Lee, and Y. Lee, "Performance analysis of HomePlug 1.0 MAC with CSMA/CA," *IEEE Journal on Selected Areas in Communications*, vol. 24, no. 7, 2006, pp. 1411–1420.

[23] L. Guerrieri, E. Guerrini, D. Veronesi, P. Bisaglia, and G. Dell'Amico, "LLR-based bit-loading algorithm for the turbo coded HomePlug AV," In *IEEE Global Telecommunications Conference*, 2007, pp. 140–145.

[24] S. Bradbury, "HomePlug Command & Control (C&C) Overview White Paper," Yitran Communications Ltd. and the HomePlug Command and Control Marketing Work Group, 2008.

[25] R. R. Mohassel, A. Fung, F. Mohammadi, and K. Raahemifar, "A survey on advanced metering infrastructure," *International Journal of Electrical Power & Energy Systems*, vol. 63, 2014, pp. 473–484.

[26] R. Cendrillon, W. Yu, M. Moonen, J. Verlinden, and T. Bostoen, "Optimal multiuser spectrum balancing for digital subscriber lines," *IEEE Transactions on Communications*, vol. 54, no. 5, 2006, pp. 922–933.

[27] G. Ginis and J. M. Cioffi, "Vectored transmission for digital subscriber line systems," *IEEE Journal on Selected Areas in Communications*, vol. 20, no. 5, 2002, pp. 1085–1104.

[28] S. V. Ahamed, P. L. Gruber, and J. J. Werner, "Digital subscriber line (HDSL and ADSL) capacity of the outside loop plant," *IEEE Journal on Selected Areas in Communications*, vol. 13, no. 9, 1995, pp. 1540–1549.

[29] K. Christensen, P. Reviriego, B. Nordman, M. Bennett, M. Mostowfi, and J. A. Maestro, "IEEE 802.3az: the road to energy efficient Ethernet," *IEEE Communications Magazine*, vol. 8, no. 11, 2010, pp. 50–56.

[30] C. Brunner, "IEC 61850 for power system communication," In *IEEE/PES Transmission and Distribution Conference and Exposition*, 2008, pp. 1–6.

[31] M. A. Ko, "Layering serial attached small computer system interface (SAS) over Ethernet," International Business Machines Corp., U.S. Patent 8,140,696, 2012.

[32] G. Keiser, "Optical fibre communications," McGraw-Hill, New York, 2008.

[33] D. Jiang and L. Delgrossi, "IEEE 802.11p: Towards an international standard for wireless access in vehicular environments," In *Vehicular Technology Conference*, 2008, pp. 2036–2040.

[34] C. Guille and G. Gross, "A conceptual framework for the vehicle-to-grid (V2G) implementation," *Energy Policy*, vol. 37, no. 11, 2009, pp. 4379–4390.

[35] S. Rajagopal, R. D. Roberts, and S. K. Lim, "IEEE 802.15.7 visible light communication: Modulation schemes and dimming support," *IEEE Communications Magazine*, vol. 50, no. 3, 2012, pp. 72–82.

[36] D. Johnston and J. Walker, "Overview of IEEE 802.16 security," *IEEE Security & Privacy*, vol. 2, no. 3, 2004, pp. 40–48.

[37] Y. Xiao, "Energy saving mechanism in the IEEE 802.16e wireless MAN," *IEEE Communications Letters*, vol. 9, no. 7, 2005, pp. 595–597.

[38] A. R. Mishra, "Fundamentals of cellular network planning and optimisation: 2G/2.5G/3G... evolution to 4G," John Wiley & Sons, Chichester, 2004.

[39] A. Ghosh, R. Ratasuk, B. Mondal, N. Mangalvedhe, and T. Thomas, "LTE-advanced: next-generation wireless broadband technology," *IEEE Wireless Communications*, vol. 17, no. 3, 2010, pp. 10–22.

[40] A. Bourdena, G. Kormentzas, C. Skianis, E. Pallis, and G. Mastorakis, "Real-time TVWS trading based on a centralized CR network architecture," In *IEEE GLOBECOM Workshops*, 2011, pp. 964–969.

[41] Y. Rodriguez and B. Garcia, "Programmable multi-function Z-wave adapter for Z-wave wireless networks," Homerun Holdings Corp., U.S. Patent 8,117,362, 2012.

[42] G. Pataut, "Direct microwave modulation and demodulation device," Thomson Composants Microondes, U.S. Patent 5,122,765, 1992.

[43] J. S. Lee, Y.W. Su, and C. C. Shen, "A comparative study of wireless protocols: Bluetooth, UWB, ZigBee, and Wi-Fi," In *33rd Annual Conference of the IEEE Industrial Electronics Society*, 2007, pp. 46–51.

[44] X. Xiao and L. M. Ni, "Internet QoS: a big picture," *IEEE Network*, vol. 13, no. 2, 1999, pp. 8–18.
[45] R. Sundar, M. Aravamudan, S. A. Naqvi, P. R. Iyer, K. K. Vishwanathan, and G. U. Pai, "Method, system, and apparatus for a mobile station to sense and select a wireless local area network (WLAN) or a wide area mobile wireless network (WWAN)," Winphoria Networks Inc., U.S. Patent 7,200,112, 2007.
[46] I. D. A. Modbus, "Modbus application protocol specification v1. 1a," North Grafton, Massachusetts, 2004, Online: www.modbus.org/specs.php.
[47] E. Omerdic and D. Toal, "Smart ROVLATIS: from design concepts to test trials," *IFAC Proceedings Volumes*, vol. 42, no. 18, 2009, pp. 346–357.
[48] G. Clarke, D. Reynders, and E. Wright, "Practical modern SCADA protocols: DNP3, 60870.5 and related systems," Newnes, Oxford, 2004.
[49] J. Benoit, S. Gagnon, L. Tétreault, and A. Engineer, "Securing distribution automation," In *Western Power Delivery Automation Conference*, 2010.
[50] N. Saputro, K. Akkaya, and S. Uludag, "A survey of routing protocols for smart grid communications," *Computer Networks*, vol. 56, no. 11, 2012, pp. 2742–2771.
[51] Y. J. Kim, R. Govindan, B. Karp, and S. Shenker, "Geographic routing made practical," In *Proceedings of the 2nd USENIX conference on Symposium on Networked Systems Design & Implementation*, vol. 2, 2005, pp. 217–230.
[52] A. Koubaa, M. Alves, and E. Tovar, "A comprehensive simulation study of slotted CSMA/CA for IEEE 802.15. 4 wireless sensor networks," In *IEEE International Workshop on Factory Communication Systems*, 2006, pp. 183–192.
[53] R. O. Baldwin, I. V. Davis, S. F. Midkiff, and R. A. Raines, "Packetized voice transmission using RT-MAC, a wireless real-time medium access control protocol," *ACM SIGMOBILE Mobile Computing and Communications Review*, vol. 5, no. 3, 2001, pp. 11–25.
[54] I. Al-Anbagi, M. Erol-Kantarci, and H. T. Mouftah, "Priority-and delay-aware medium access for wireless sensor networks in the smart grid," *IEEE Systems Journal*, vol. 8, no. 2, 2014, pp. 608–618.
[55] D. M. Han and J. H. Lim, "Smart home energy management system using IEEE 802.15. 4 and Zigbee," *IEEE Transactions on Consumer Electronics*, vol. 56, no. 3, 2010, pp. 1403–1410.
[56] B. Yahya and J. Ben-Othman, "An energy efficient hybrid medium access control scheme for wireless sensor networks with quality of service guarantees," In *IEEE Global Telecommunications Conference*, 2008, pp. 1–5.
[57] C. Perkins, E. Belding-Royer, and S. Das, "Ad hoc on-demand distance vector (AODV) routing," no. RFC 3561, 2003.
[58] I. Rhee, A. Warrier, M. Aia, J. Min, and M. L. Sichitiu, "Z-MAC: a hybrid MAC for wireless sensor networks," *IEEE/ACM Transactions on Networking*, vol. 16, no. 3, 2008, pp. 511–524.
[59] D. Chen, M. Nixon, and A. Mok, "Why wirelesshart," In *WirelessHART*™, Springer, Boston, MA, 2010, pp. 195–199.
[60] I. Silva, L. A. Guedes, P. Portugal, and F. Vasques, "Reliability and availability evaluation of wireless sensor networks for industrial applications," *Sensors*, vol. 12, no. 1, 2012, pp. 806–838.

[61] C. M. Talbot, M. A. Temple, T. J. Carbino, and J. A. Betances, "Detecting rogue attacks on commercial wireless insteon home automation systems," *Computers & Security*, vol. 74, 2018, pp. 296–307.

[62] F. Lemercier and N. Montavont, "Performance evaluation of a RPL hybrid objective function for the smart grid network," In *International Conference on Ad-Hoc Networks and Wireless*, 2018, pp. 27–38.

[63] B. N. Jain and A. K. Agrawala, "Open systems interconnection: its architecture and protocols," CERN Accelerating Science, North-Holland, 1990.

[64] X. Li, C. Bao, and F. Baker, "IP/ICMP translation algorithm," no. RFC 6145, 2011.

[65] M. Christensen, K. Kimball, and F. Solensky, "Considerations for Internet group management protocol (IGMP) and multicast listener discovery (MLD) snooping switches," no. RFC 4541, 2006.

[66] M. Uslar, M. Specht, S. Rohjans, J. Trefke, and J. M. González, "The Common Information Model CIM: IEC 61968/61970 and 62325-A practical introduction to the CIM," Springer Science & Business Media, Chennai, 2012.

[67] W. Stallings, "SNMP, SNMPv2, SNMPv3, and RMON 1 and 2," Addison-Wesley Longman Publishing Co., Inc., Chennai, 1998.

[68] D. Mills, "Network time protocol," RFC 958, M/A-COM Linkabit, 1985.

[69] P. Hoffman, "SMTP service extension for secure SMTP over transport layer security," Network Working Group, ietf draft, 2002.

[70] B. Ratner, "RS-485 multipoint power line modem," ACEX Tech Inc., U.S. Patent 5,684,826, 1997.

[71] R. E. Mackiewicz, "Overview of IEC 61850 and Benefits," In *Power Systems Conference and Exposition*, 2006, pp. 623–630.

[72] T. Bray, J. Paoli, C. M. Sperberg-McQueen, E. Maler, and F. Yergeau, "Extensible markup language (XML) 1.0," W3C Recommendation Report, 2008.

[73] V. C. Gungor, D. Sahin, T. Kocak, *et al.*, "A survey on smart grid potential applications and communication requirements," *IEEE Transactions on Industrial Informatics*, vol. 9, no. 1, 2013, pp. 28–42.

[74] L. Gomes, J. Silva, P. Faria, and Z. Vale, "Microgrid Demonstration Gateway for Players Communication and Load Monitoring and Management," In *Clemson University Power Systems Conference*, 2016, pp. 1–6.

[75] C. L. Hsu, H. P. Lu, and H. H. Hsu, "Adoption of the mobile Internet: An empirical study of multimedia message service (MMS)," *Omega*, vol. 35, no. 6, 2007, pp. 715–726.

[76] A. L. Dimeas and N. D. Hatziargyriou, "Operation of a multiagent system for microgrid control," *IEEE Transactions on Power systems*, vol. 20, no. 3, 2005, pp. 1447–1455.

[77] A. Hirsch, Y. Parag, and J. Guerrero, "Microgrids: A review of technologies, key drivers, and outstanding issues," *Renewable and Sustainable Energy Reviews*, vol. 90, 2018, pp. 402–411.

Chapter 10

Voltage stability of microgrids

Nasser Hosseinzadeh[1]*, Saheb Khanabdal*[2]*, Yousuf Al-Jabri*[3]*, Rashid Al-Abri*[1]*, Amer Al-Hinai*[1]*, and Mahdi Banejad*[2]

This chapter focuses on voltage stability of microgrids. Some control schemes to maintain voltages of all the buses of a microgrid at acceptable levels in normal conditions and after being subjected to disturbances will be presented. Voltage stability enhancement and reactive power sharing in inverter-based microgrids are discussed, including droop-control techniques.

In a microgrid, a good portion of the required power is supplied by distributed energy resources or distributed generators (DGs), usually called microsources. As such, the dynamic characteristics of DGs are important in determining the dynamic behavior of the microgrid. According to the interface modes, DGs can be classified into two categories: (1) inverter-interfaced DGs, e.g., photovoltaic (PV) electricity generators, small direct-drive wind turbines, micro-gas turbines, batteries, flywheel energy storage, and super capacitors; or (2) DGs connected to the microgrid directly, e.g., doubly fed induction generator (DFIG), diesel generator, and small hydro units. To build a microgrid, renewable energy sources are usually applied as much as possible, so inverter-interfaced DGs are used widely in microgrids, which makes the operating characteristics of a microgrid quite different from the traditional grid.

When most of the microsources are connected to the grid through power electronic inverters, the system inertia is low and dynamic responses are much faster than the rotating machines used in the conventional generators. Thus, some traditional methods such as P–V and Q–V curves are not sufficient in assessing the voltage stability of microgrids. Instead, the small-signal stability analysis is commonly used for assessing the stability of microgrids.

[1]Department of Electrical and Computer Engineering, Sultan Qaboos University, Muscat, Oman
[2]Department of Electric Power, Shahrood University of Technology, Shahrud, Iran
[3]Power System Concept Engineer, Petroleum Development Oman (PDO), Muscat, Oman

10.1 Introduction

In this section, first, the concept of voltage stability in a power system is reviewed. Then, voltage stability issues of microgrids are presented, and some differences with voltage stability in conventional power systems are highlighted. Finally, some methods of assessing voltage stability of microgrids are explained.

10.1.1 Concept of voltage stability

Voltage stability in a power system is defined by a joint IEEE/CIGRE Task Force as "the ability of a power system to maintain acceptable voltages at all buses in the system under normal condition and after being subjected to a disturbance." A major factor contributing to voltage instability is the voltage drop that occurs through inductive reactances of the transmission network due to the active and reactive power flows; this limits the capability of the transmission network for power transfer and voltage support. Voltage stability is threatened when a disturbance increases the reactive power demand beyond the sustainable capacity of the available reactive power sources [1]. In general, voltage instability stems from the attempt of load dynamics to restore power consumption beyond the capability of the combined transmission and generation system [2]. Voltage instability shows itself in the form of large, uncontrollable voltage drops at a number of network buses and sometimes results in voltage collapse.

Loads are the driving force of voltage instability; however, loads are not the only players in this phenomenon. In voltage instability, both active and reactive power share a role. The decoupling between active power and phase angles on one hand and reactive power and voltage magnitudes on the other hand apply to normal operating conditions and cannot be extended to the extreme loading conditions typical of voltage instability scenarios. In other words, there is no "cause and effect" relationship between reactive power and voltage instability. In AC power systems, reactive power makes the picture much more complicated, but it is certainly not the only source of the problem [2]. Voltage stability may be classified in different ways [3]. Small-signal voltage stability refers to the ability of the system to control the voltage when small perturbations occur, such as changes in loads. Large-signal voltage stability refers to the ability of the system to control the voltage after being subjected to large disturbances such as load outages, faults, and large-step changes in the loads.

Long-term voltage stability has been many times misunderstood as a "static" problem because usually static tools (such as modified power flow programs) are acceptable for simpler and faster analysis. Voltage stability, however, is dynamic by nature, and for better simulation results, one has to resort to dynamic analysis tools (such as time-domain methods) [2].

10.1.2 Voltage stability issues of microgrid

In distribution systems with embedded DG units, the main factors that influence stability are the control strategies of the DG units, the energy storage systems, the types of load in the system, the location of faults, and the inertia constant of motors [4]. As the deployment of DGs continues, an increasing number of DC/AC power inverters will be interfaced into the electric grid. Microgrids may be more vulnerable to voltage instability due to various types of source generation injected in such systems as well as the reactive power limits, load dynamics and transformers tap changers [5].

Similar to the role of synchronous generators in traditional grids, inverter-interfaced DGs have the same important roles in microgrids. In order to classify and analyze microgrid stability more precisely, the significant differences between inverter-interfaced DGs and traditional synchronous generators, such as operation mechanism, control mode, response speed, and overcurrent capability, should be taken into account [6].

At present, the research on microgrid stability is progressing in areas such as development of mathematical models for microgrid stability analysis, methods of microgrid stability analysis, and methods of improving stability of microgrids [7,8]. A lot of attention is paid to the small signal stability analysis and transient stability simulation analysis [5]. For example, in [9], small-signal stability analysis of a single-phase grid-connected inverter is presented. Small signal stability analysis of islanded microgrids is covered in [10–28]. Influences of droop control gains, line impedance, and load fluctuations on the microgrid voltage and frequency characteristics are also discussed. By using the small signal stability analysis of microgrid, better droop control gains can be obtained.

10.1.3 Microgrid voltage stability assessment

Voltage stability can be evaluated by two different methods of analysis: static and dynamic, the details of which are presented in the following subsections.

10.1.3.1 Static voltage stability analysis

This method examines the level of stability of the system at a specific equilibrium point represented by a specified operating condition of the power system. The static method is evaluated by means of a variety of techniques such as *P–V* and *Q–V* curves, and *V–Q* sensitivity analysis.

10.1.3.2 *P–V* and *Q–V* curves

P–V and *Q–V* curves are generated by executing a large number of power flow solutions for different operating points. Generally, a power system can be modeled with nonlinear differential algebraic equations [29]:

$$\dot{X} = f(X, \lambda) \tag{10.1}$$

where X represents a state vector, including the bus voltage magnitudes (V) and angles (δ).

λ is a parameter vector that represents the real and reactive power demand at each load bus. The parameter vector λ changes due to variations in the load. Therefore, the power flow solution varies as λ varies. The power system's power flow is represented by

$$0 = f(X, \lambda) \tag{10.2}$$

As an example, [30] derives the P–V curve for a simple single-machine PV bus, which supplies a PQ load of constant power factor through a transmission line, as shown in Figure 10.1. The P–V curve of such a generator bus is shown in Figure 10.2, which demonstrates a typical P–V curve without specific scaling of voltage and power.

10.1.3.3 *V–Q* sensitivity analysis

In this method, the network is represented by a power flow equation that can be linearized, as given in the following equation [31]:

$$\begin{bmatrix} \Delta P \\ \Delta Q \end{bmatrix} = \begin{bmatrix} J_{P\theta} & J_{PV} \\ J_{Q\theta} & J_{QV} \end{bmatrix} \begin{bmatrix} \Delta\theta \\ \Delta V \end{bmatrix} \tag{10.3}$$

where ΔP is the incremental change in bus real power, ΔQ is the incremental change in bus reactive power, ΔV is the incremental change in bus voltage magnitude, $\Delta\theta$ is the incremental change in bus voltage angle, and J is the Jacobian matrix.

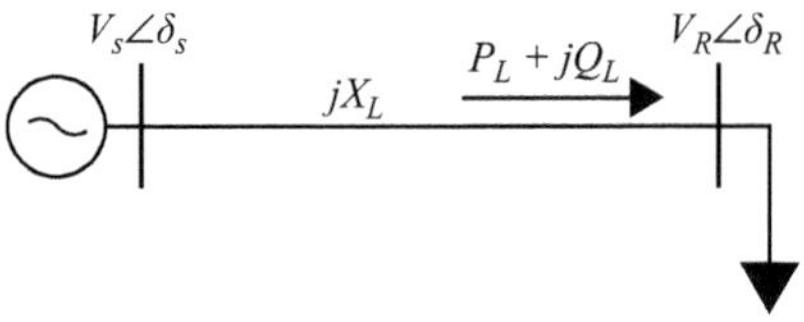

Figure 10.1 Single-machine PV bus supplying a PQ load bus

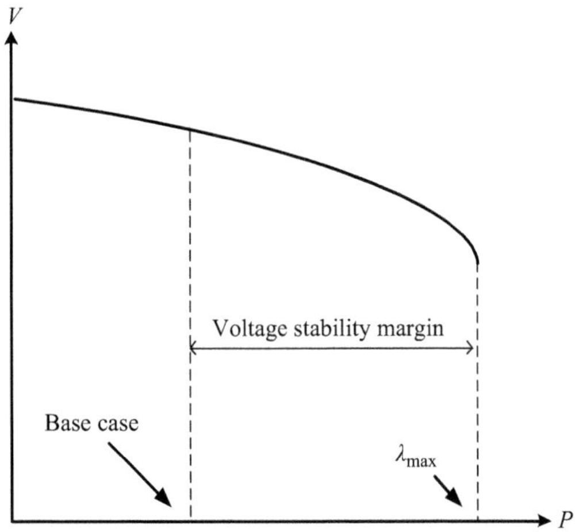

Figure 10.2 P–V curve or nose curve

With the assumption of the real load power (P) being constant, the incremental change in the bus real power ΔP equals to zero. Then, using the partial inversion of (10.3) gives:

$$\Delta Q = \left(J_{QV} - J_{Q\theta}J_{P\theta}^{-1}J_{PV}\right)\Delta V$$

or

$$\Delta V = \left(J_{QV} - J_{Q\theta}J_{P\theta}^{-1}J_{PV}\right)^{-1}\Delta Q \tag{10.4}$$

The V–Q sensitivity can be calculated by solving (10.4). The V–Q sensitivity at a bus represents the slope of the Q–V curve, Figure 10.3, at a given operating point. Note that this is a general figure to show the concept; therefore, the scaling of axes is not given. A positive V–Q sensitivity is indicative of stable operation, and a negative sensitivity is indicative of unstable operation [30]. In the stable region, as the injected reactive power is increased, the voltage of the corresponding bus also increases.

10.1.3.4 Dynamic analysis

Dynamic analysis can show the real behavior of the system and its components such as loads, DG units, and control equipment. The overall power system is represented by a set of first-order differential equations, as given in the following equation:

$$\dot{X} = f(X, V) \tag{10.5}$$

and a set of algebraic equations

$$I(X, V) = Y_N V \tag{10.6}$$

with a set of known initial conditions (X_0, V_0), where X is the state vector of the system, V is the bus voltage vector, I is the current injection vector, Y_N is the network node admittance matrix.

Equations (10.5) and (10.6) can be solved in time domain using numerical integration methods.

Based on these methods, researchers have studied voltage stability with the integration of DG units in distribution systems. The literature has covered this

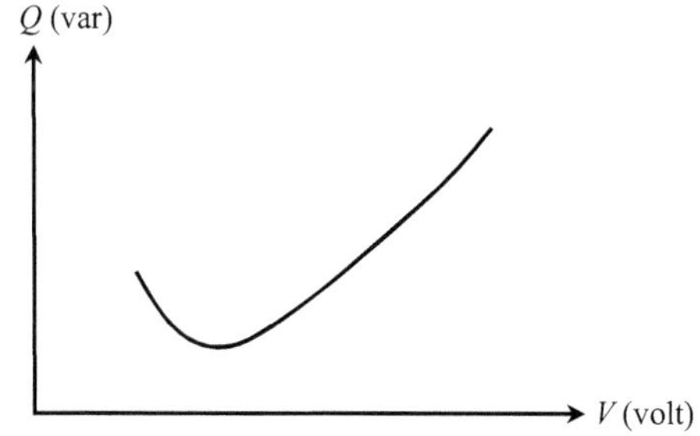

Figure 10.3 Q–V characteristic curve

impact from different points of views. For example, the authors of [31,32] investigated the impact of DG technology. The impact of induction generators due to small and large disturbances was studied in [33–36]. A practical investigation of the impacts of DG units on system stability can be found in [37]. An assessment of the impact of the DG units' size and location under a change in the loading conditions due to a contingency on unbalanced distribution systems was presented in [38]. In [39], the effect of DG units capacity and location on voltage stability enhancement of distribution networks was investigated. The DG units were allocated and sized based on minimizing overall cost. In [40,41] methods were proposed to locate distributed generation units to improve the voltage profile and voltage stability of a distribution system. Authors of [40] placed DG units at the buses most sensitive to voltage collapse, which resulted in improvement in voltage profile, as well as decline in the power losses. Authors of [41] developed the work in [40] further to maximize the loadability conditions in normal and contingency situations.

Voltage and frequency stability of a distribution system in grid-connected mode and in a stand-alone microgrid were studied in [42]. The study considered the imbalance of loads and sources and was divided into two parts: simulation with storage DG units only and then simulation with storage and renewable energy units operating in parallel. However, it was a time domain simulation study and the renewable energy DG units were assumed to supply the system with constant power.

10.1.3.5 Proximity to voltage instability

As mentioned earlier, the static technique can be analyzed by using the relation between the receiving power (P) and the voltage (V) at a certain bus in a system, which is known as a P–V curve or nose curve. As illustrated in Figure 10.2, the critical point $\lambda_{\max}$ (saddle-node bifurcation point) in the P–V curve represents the maximum loading of a system. This point corresponds to a singularity of the Jacobian power flow equations. The stability margin can be defined by the MW distance from the operating point to the critical point. The penetration of the DG units in a distribution system can increase or decrease the voltage stability margin depending on their operation at unity, lead or lag power factors as well as their location. λ may vary from zero to the maximum loading ($\lambda_{\max}$):

$$P_i = \lambda P_{0,i} \text{ and } Q_i = \lambda Q_{0,i} \tag{10.7}$$

However, static analysis cannot determine the control action and the interaction between the integrated DG units in the system. Proximity to the voltage instability method can be used to determine those issues.

Small-signal stability analysis is achieved in frequency domain using eigenvalue analysis. It is carried out by linearizing the mathematical model of the system and then solving the eigenvalues and eigenvectors of the linearized model. A small-signal model of inverter DG units in a microgrid was presented in [43]. Authors in [44] modeled microgrids with small-signal and included a mix of inverter-based and conventional DG units. In [45], the small-signal model to investigate the dynamic behavior of the system was utilized and the control parameters of the DG units were selected. In addition, they used small signal analysis to improve the control of the DG units, as well as studying the mode of control when the DG system is moved from grid-connected to a stand-alone microgrid mode.

Although the main voltage stability analysis tool in microgrids is small-signal analysis, *P–V* and *Q–V* curves can still be used. Maximum loadability of the system can be worked out from the *P–V* curves, and the required reactive power from the microgrid for maintaining voltage stability can be obtained by studying *Q–V* curves. The required reactive power from the microgrid, Q_G, may be expressed in terms of the inverter terminal voltage, V_{cf}, the load voltage, V_t, the load active power, P_L, and load reactive power, Q_L. It should be noted that the load demand for reactive power depends on the type of the load.

A generator based on a voltage-source inverter has a faster response than a synchronous generator. The production of reactive power from the generators should be distributed among them in such a way that the reactive power output of inverter-based generators does not exceed their permitted max reactive power output, or it does not create oscillations in system voltages.

Figure 10.4(a) shows a typical microgrid, with two inverter-based micro-sources and one directly connected micro-source. Figure 10.4(b) illustrates three different ways of assessing the voltage stability of a system [5]. In this figure, the stable points are shown by the letter s, and the unstable points are shown by the letter u.

Condition 1 uses a graph of total generated reactive power from the microgrid versus the total Q_L demand from the load. It states that the condition for voltage stability is that the rate of change of Q_G with respect to Q_L is positive,

$$\frac{dQ_G}{dQ_L} > 0$$

Or, in other words, the slope of Q_G versus Q_L at the point related to the system operating condition is positive.

Condition 2 uses two curves of generated reactive power versus voltage and load reactive power versus voltage. The condition for voltage stability is

$$\frac{dQ_L}{dV} > \frac{dQ_G}{dV}$$

which says when the voltage changes, for the system to remain stable, the rate of change of generated reactive power should be less than the rate of change of the load reactive power with respect to voltage change.

Condition 3 uses a curve of terminal voltage of the microgrid versus terminal voltage of the load. In this case, the condition for voltage stability is

$$\frac{dV_{cf}}{dV} > 0$$

which says the microgrid terminal voltage should change in the same direction that the load voltage changes for the system to remain stable, i.e., if the load voltage increases, the microgrid terminal voltage must also increase, and if the load voltage decreases, the microgrid terminal voltage must also decrease.

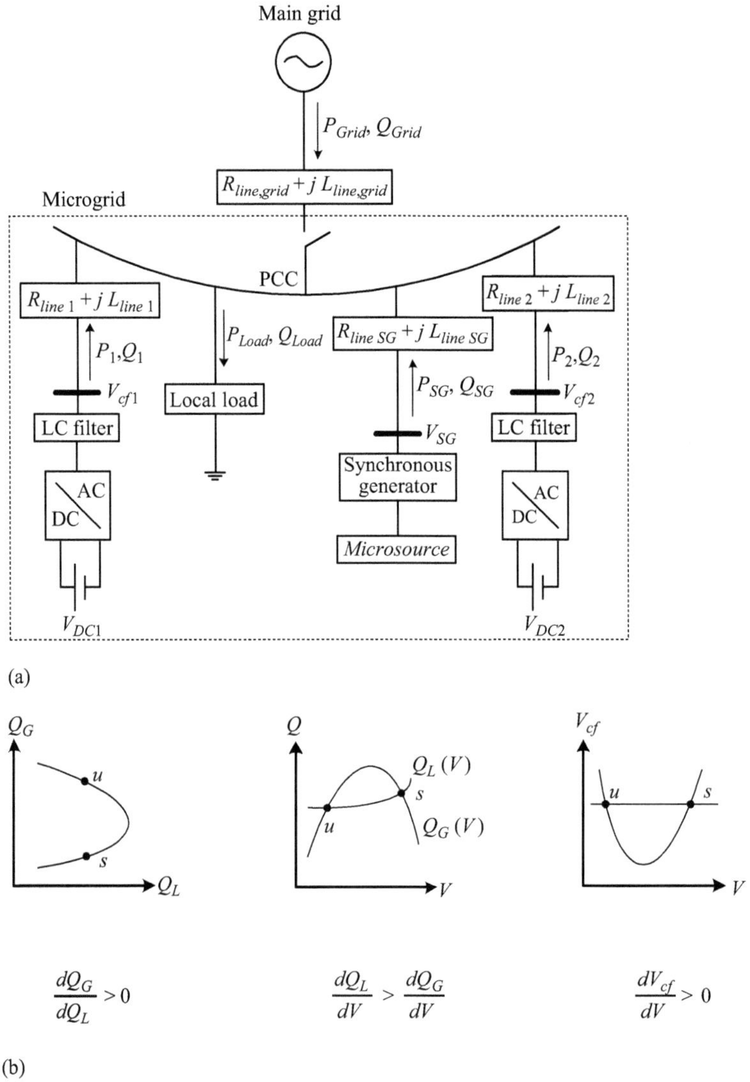

Figure 10.4 Methods of assessing voltage stability of a microgrid—static methods: (a) example of a microgrid and (b) three methods of assessing the voltage stability (Adopted from [5])

Important points to consider in the voltage stability analysis are

- reactive power control strategy,
- load characteristics,
- controlling the system to make gradual increase in reactive power demand rather than sharp changes, and
- in emergency situations, having a plan to disconnect a part of the system.

10.2 Small-signal model of a microgrid for voltage stability analysis

Unlike conventional power systems, in which generators are mainly synchronous machines, in microgrids usually various types of generators are used, such as inverter-based generators (solar PV, some types of wind energy conversion systems, and other type of generators which initially generate DC voltage), induction generators, and diesel generators using synchronous machines. Therefore, to develop a benchmark model for the voltage stability of a microgrid, all these different types of micro-sources should be included. For such a microgrid, the small-signal analysis becomes more complicated than the analysis for conventional power systems. In this section, some examples are given about developing small-signal models for microgrids that include different level of generator-type mix in their structures.

10.3 Voltage stability enhancement

Voltage stability of microgrids may be improved by regulating voltages in DGs, using reactive power compensators, load controllers, or current limiters [5].

Similar to the droop characteristics of synchronous machines, the inverter-based micro-sources have droop characteristics, which can be represented by the following equations:

$$\omega = \omega^* - m(P - P^*) \tag{10.8}$$

$$V = V^* - n(Q - Q^*) \tag{10.9}$$

where P^* is the set point of active power, Q^* is the set point of reactive power, ω^* is the nominal value for the angular frequency, and V^* is the nominal value for the voltage magnitude. ω, V, P, and Q are the operating values of the angular frequency, voltage magnitude, active power, and reactive power, respectively. m and n are the droop coefficients. Figure 10.5 shows the droop characteristics of a typical micro-source [46].

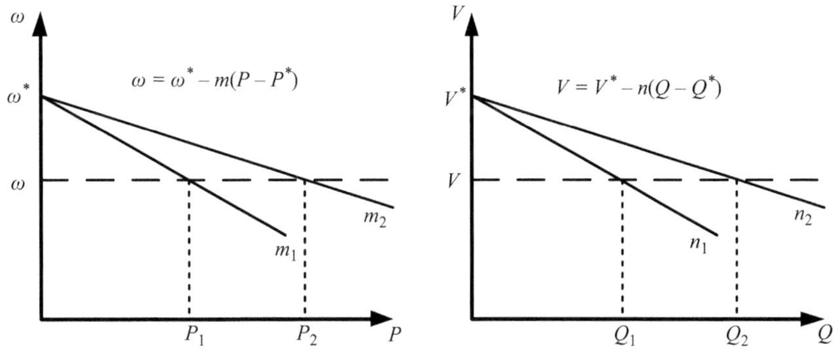

Figure 10.5 Droop characteristics for a typical inverter-based micro-source (Adopted from [46])

10.4 Case studies

Four cases will be presented in this section. The first case, studies the voltage stability of a microgrid with inverter-based micro-sources. The second case study covers the voltage stability analysis for a microgrid that has synchronous generators, such as diesel generators, in addition to the renewable energy sources. The third case adds a directly connected induction generator to the combination of a synchronous generator and inverter-based micro-source. The fourth case is the voltage stability study of a real multi-microgrid in Oman comprising interconnected microgrids powered by diesel generators and connected via a transmission system to a central gas-fired power plant.

10.4.1 Case study 1

Majority of microgrids include inverter-based renewable-energy micro-sources [47,48]. Inverter-based micro-sources have the capability of providing control on the level of power flow both active and reactive from the micro-source. Thus, they can provide reactive power for regulating voltage and enhancing voltage stability of the microgrid. On the other hand, the inverter-based micro-sources have low inertia, which is a disadvantage in terms of the system stability [49]. Studying a microgrid with inverter-based micro-sources, such as the microgrid of Figure 10.6, is then the first case study given here, which is based on [49].

In order to derive a small-signal model for such a microgrid, the models for three components are derived separately and are then aggregated as a *d–q* model in a single reference frame. The three components are micro-sources, the network, and loads. Figure 10.7 shows this concept of modeling.

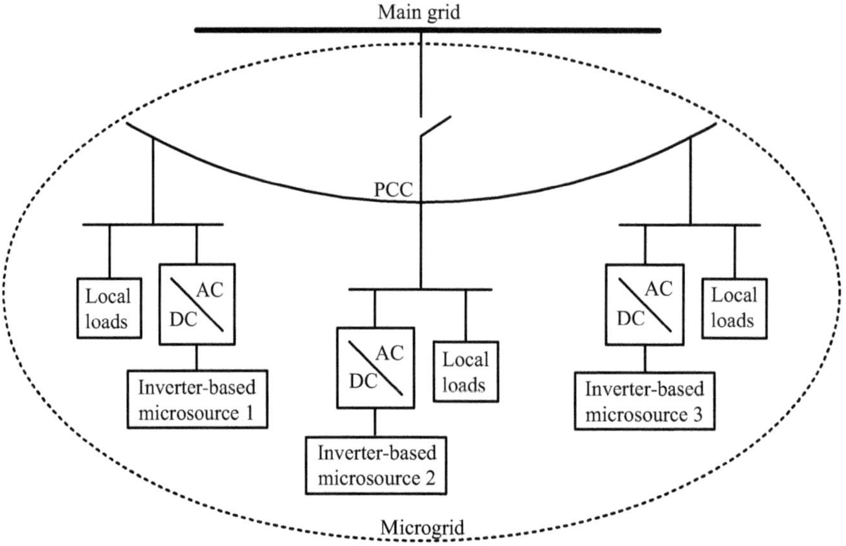

Figure 10.6 A microgrid with three inverter-based micro-sources (Adopted from [49])

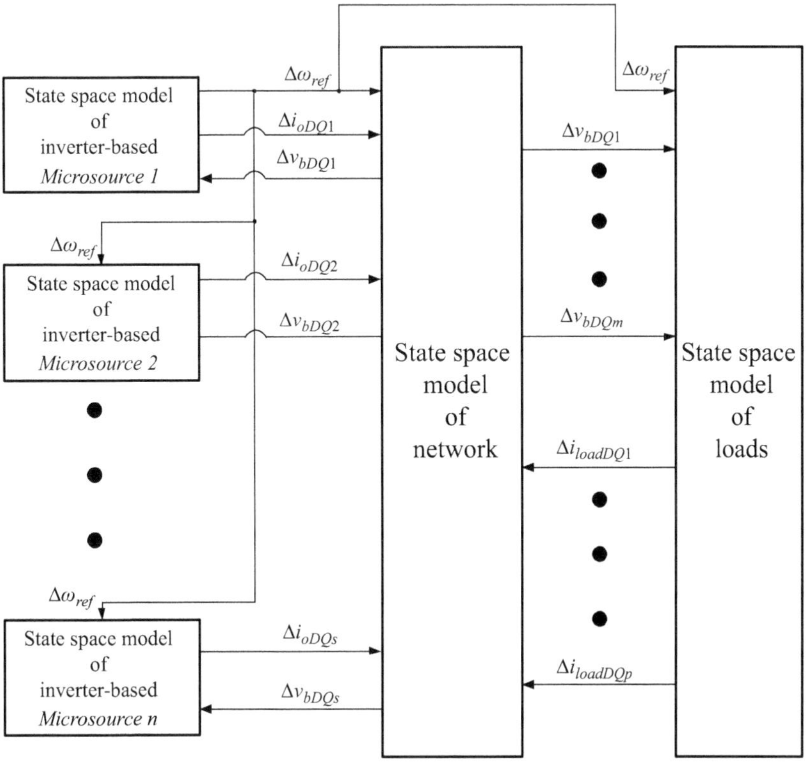

Figure 10.7 Block diagram of an inverter-based microgrid for developing its small-signal model (Adopted from [11])

Modeling of each component is done in its own *d–q* local reference frame, and then, models of all components are converted to a *D–Q* global reference frame (GRF). All differential equations will be eventually written for the integrated system in the GRF. Figure 10.8 shows how the conversion of local *d–q* frame to the *D–Q* GRF is done [11]. The reference frame of inverter number 1 can be selected as the GRF; for other inverters, a rotational speed of the local reference frame, ω_i, is calculated to move the local frame to coincide with the global one.

The following equations are used to convert the current and voltage quantities from the local frames to the GRF. The current and voltage values in the GRF are used as inputs to the model of each component:

$$[\Delta i_{oDQ}] = [T_s][\Delta i_{odq}] + [T_c][\Delta\delta] \tag{10.10}$$

$$T_s = \begin{bmatrix} \cos\delta_0 & -\sin\delta_0 \\ \sin\delta_0 & \cos\delta_0 \end{bmatrix} \tag{10.11}$$

$$T_C = \begin{bmatrix} -I_{od}\sin(\delta_0) - I_{oq}\cos(\delta_0) \\ I_{od}\cos(\delta_0) - I_{oq}\sin(\delta_0) \end{bmatrix} \tag{10.12}$$

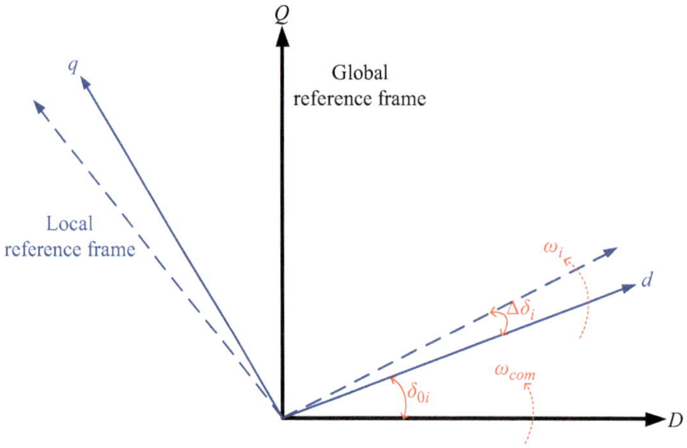

Figure 10.8 Conversion of local d–q frame to the D–Q global reference frame (GRF) (Adopted from [11])

$$[\Delta v_{bdq}] = [T_s^{-1}][\Delta v_{bDQ}] + [T_V^{-1}][\Delta\delta] \tag{10.13}$$

$$T_V^{-1} = \begin{bmatrix} -V_{bD}\sin(\delta_0) + V_{bQ}\cos(\delta_0) \\ -V_{bD}\cos(\delta_0) - V_{bQ}\sin(\delta_0) \end{bmatrix} \tag{10.14}$$

In these equations, Δi_{oDQ} is the inverter current deviation in the *D–Q* GRF, δ_0 is the angle of the inverter terminal voltage at the operating point, I_{od} and I_{oq} are the operating-point currents in the local frame, V_{bD} and V_{bQ} are the inverter terminal voltages in the *D–Q* GRF.

10.4.1.1 Small signal model of the inverter—case study 1

To derive the small-signal model of an inverter with its control system, Figure 10.9 is used in [11]. As seen from the figure, the inverter control system is comprised from three sections: power controller, voltage controller, and current controller. The power controller uses the droop control technique, and PI controllers are used for voltage and current control.

The following state equations are derived from Figure 10.9:

$$\Delta\dot{X}_{invi} = A_{INVi}[\Delta X_{invi}] + B_{INVi}[\Delta v_{DQi}] + B_{i\omega com}[\Delta\omega_{com}] \tag{10.15}$$

where

$$\Delta X_{invi} = [\Delta\delta_i \Delta P_i \Delta Q_i \Delta\phi_{dqi} \Delta\gamma_{dqi} \Delta i_{ldqi} \Delta v_{odqi} \Delta i_{odqi}]$$

$$\begin{bmatrix} \Delta\omega_i \\ \Delta i_{oDQi} \end{bmatrix} = \begin{bmatrix} C_{INV\omega i} \\ C_{INVci} \end{bmatrix} [\Delta X_{invi}] \tag{10.16}$$

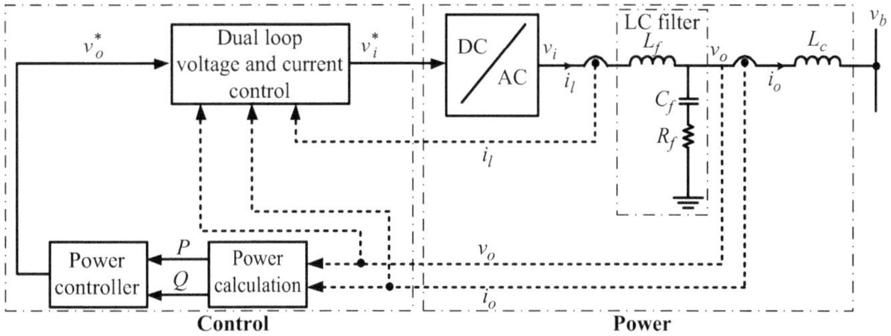

Figure 10.9 Block diagram of the inverter control system (Adopted from [11])

The coefficient matrices A_{INVi}, B_{INVi}, $B_{i\omega com}$, $C_{INV\omega i}$, and C_{INVci} are derived in [5,50]. In a system similar to the one given in Figure 10.6, the reference frame of the inverter number 1 is chosen as the GRF; therefore, $\Delta\omega_{com} = \Delta\omega_1$. The output $\Delta\omega_i$ for all other inverters will be zero. As such, the state equations related to the micro-sources can be derived as follows:

$$[\Delta x_{INV}] = A_{INV}[\Delta x_{INV}] + B_{INV}[\Delta v_{DQ}] \tag{10.17}$$

$$[\Delta i_{oDQ}] = [C_{INVc}][\Delta x_{INV}] \tag{10.18}$$

$$[\Delta x_{INV}] = [\Delta x_{inv1}\ \Delta x_{inv2} \cdots \Delta x_{invn}]^T \tag{10.19}$$

The coefficient matrices A_{INV}, B_{INV}, and C_{INVc} are given in [5,51].

10.4.1.2 Small signal model of the power network—case study 1

For modeling an arbitrary network, as shown in Figure 10.10, it is necessary to write differential equations of the network in terms of bus voltages and branch currents in the GRF:

$$\frac{di_{line\ Di}}{dt} = \frac{-r_{line\ i}}{L_{LINE\ I}} i_{line\ Di} + \omega i_{line\ Qi} + \frac{1}{L_{line\ i}} v_{bDj} - \frac{1}{L_{line\ i}} v_{bDk} \tag{10.20}$$

$$\frac{di_{line\ Qi}}{dt} = \frac{-r_{line\ i}}{L_{LINE\ I}} i_{line\ Qi} + \omega i_{line\ Di} + \frac{1}{L_{line\ i}} v_{bQj} - \frac{1}{L_{line\ i}} v_{bQk} \tag{10.21}$$

In the above equations, $i_{line\ Di}$ and $i_{line\ Qi}$ are the current d–q components in the GRF, v_{bDj} and v_{bQj} are the d–q bus voltage components in the GRF, and $r_{line\ i}$, and $L_{line\ i}$ are the resistance and inductance of line i.

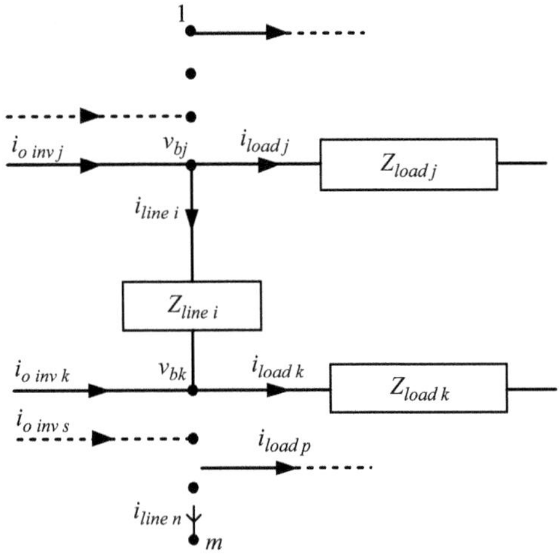

Figure 10.10 A part of the power network

Considering the line currents as the state variables, the bus voltages as the inputs, and the angular speed as the linking input between the local reference frame and the GRF, the small signal model of the network is found as follows:

$$\left[\Delta i_{line\ DQ}\right] = A_{NET}\left[\Delta i_{line\ DQ}\right] + B_{1NET}\left[\Delta v_{bDQ}\right] + B_{2NET}\Delta\omega \tag{10.22}$$

$$\left[\Delta i_{line\ DQ}\right] = \left[\Delta i_{line\ DQ\ 1}\Delta i_{line\ DQ\ 2}\cdots\Delta i_{line\ DQ\ n}\right]^{T} \tag{10.23}$$

$$\left[\Delta i_{v\ DQ}\right] = \left[\Delta v_{b\ DQ\ 1}\Delta v_{b\ DQ2}\cdots\Delta i_{b\ DQ\ m}\right]^{T} \tag{10.24}$$

$$\Delta\omega = \Delta\omega_{com} = \Delta\omega_{1} \tag{10.25}$$

$$A_{NET} = \begin{bmatrix} A_{NET\ 1} & 0 & \dots & 0 \\ 0 & A_{NET\ 2} & \dots & 0 \\ \dots & \dots & \dots & \dots \\ 0 & 0 & \dots & A_{NET\ n} \end{bmatrix}_{2n\times 2n} \tag{10.26}$$

$$B_{1NET} = \begin{bmatrix} B_{1\ NET\ 1} \\ B_{1\ NET\ 2} \\ \dots \\ \dots \\ B_{2\ NET\ n} \end{bmatrix}_{2n\times 2m} \tag{10.27}$$

$$B_{2NET} = \begin{bmatrix} B_{2\ NET\ 1} \\ B_{2\ NET\ 2} \\ \dots \\ \dots \\ B_{2\ NET\ n} \end{bmatrix}_{2n\times 1} \tag{10.28}$$

The coefficient matrices A_{NET}, B_{1NET}, and B_{2NET} are given in [50,52].

10.4.1.3 Small signal model of the load—case study 1

A constant impedance load is assumed with the following differential equations:

$$\frac{di_{load\ Di}}{dt} = \frac{-R_{load\ i}}{L_{load\ i}} i_{load\ Di} + \omega i_{load\ Qi} + \frac{1}{L_{load\ i}} v_{b\ Di} \tag{10.29}$$

$$\frac{di_{load\ Qi}}{dt} = \frac{-R_{load\ i}}{L_{load\ i}} i_{load\ Qi} - \omega i_{load\ Di} + \frac{1}{L_{load\ i}} v_{b\ Qi} \tag{10.30}$$

In the above equations, $i_{load\ Di}$ and $i_{load\ Qi}$ are the load current d–q components in the GRF, $R_{load\ i}$ and $L_{load\ i}$ are the resistance and conductance of load i. The small signal equations of the load may be written in the following form:

$$\left[\Delta i_{load\ DQ}\right] = A_{load}\left[\Delta i_{load\ DQ}\right] + B_{1\ LOAD}\left[\Delta v_{b\ DQ}\right] + B_{2\ LOAD}\,\Delta\omega \tag{10.31}$$

For integrating the loads, the following equations will be used:

$$\left[\Delta i_{load\ DQ}\right] = \left[\Delta i_{load\ DQ\ 1}\ \Delta i_{load\ DQ\ 2} \cdots \Delta i_{load\ DQ\ P}\right]^T \tag{10.32}$$

$$A_{load} = \begin{bmatrix} A_{load\ 1} & 0 & \dots & 0 \\ 0 & A_{load\ 1} & \dots & 0 \\ \dots & \dots & \dots & \dots \\ 0 & 0 & \dots & A_{load\ P} \end{bmatrix}_{2p\times 2p} \tag{10.33}$$

$$B_{1LOAD} = \begin{bmatrix} B_{1\ LOAD\ 1} \\ B_{1\ LOAD\ 2} \\ \dots \\ \dots \\ B_{2\ LOAD\ n} \end{bmatrix}_{2n\times 2m} \tag{10.34}$$

$$B_{2LOAD} = \begin{bmatrix} B_{2\ LOAD\ 1} \\ B_{2\ LOAD\ 2} \\ \dots \\ \dots \\ B_{2\ LOAD\ n} \end{bmatrix}_{2n\times 1} \tag{10.35}$$

$$A_{load\ i} = \begin{bmatrix} \dfrac{-R_{load\ i}}{L_{load\ i}} & \omega_0 \\ -\omega_0 & \dfrac{-R_{load\ i}}{L_{load\ i}} \end{bmatrix} \tag{10.36}$$

$$B_{2LOAD\ i} = \begin{bmatrix} I_{load\ Qi} \\ -I_{load\ Di} \end{bmatrix} \tag{10.37}$$

$$B_{1LOAD\ i} = \begin{bmatrix} \dots & \dfrac{1}{L_{load\ i}} & 0 & \dots \\ \dots & 0 & \dfrac{1}{L_{load\ i}} & \dots \end{bmatrix}_{2\times 2m} \tag{10.38}$$

The coefficient matrices $A_{load\ i}$, $B_{1LOAD\ i}$, and $B_{2LOAD\ i}$ are given in [50,52].

It is observed that bus voltages appear in the small signal models of all components. Therefore, it is necessary to express the bus voltages in terms of the state variables. For doing this, it is assumed that each bus is connected to the ground by a virtual resistance. The value of this resistance will be chosen big enough so that it does not affect the network dynamics much [11]:

$$\left[\Delta v_{b\ DQ}\right] = R_N\left(M_{INV}\left[\Delta i_{oDQ}\right] + M_{load}\left[\Delta i_{load\ DQ}\right] + M_{NET}\left[\Delta i_{line\ DQ}\right]\right) \tag{10.39}$$

10.4.1.4 Small signal model of the whole system—case study 1

By substituting (10.39) in (10.17), (10.22), and (10.32), the small signal equation of the whole system can be written as follows in (10.40). In this equation, A_{MG} is the system state space matrix, which can be determined from (10.41):

$$\begin{bmatrix} \Delta\dot{x}_{INV} \\ \Delta i_{line\ DQ} \\ \Delta i_{load\ DQ} \end{bmatrix} = A_{MG}\begin{bmatrix} \Delta x_{INV} \\ \Delta i_{line\ DQ} \\ \Delta i_{load\ DQ} \end{bmatrix} \tag{10.40}$$

$$A_{MG} = \begin{bmatrix} A_{INV}+B_{INV}R_NM_{INV}C_{INVc} & B_{INV}R_NM_{NET} & B_{INV}R_NM_{load} \\ B_{1NET}R_NM_{INV}C_{INVc}+B_{2NET}C_{INV\omega} & A_{NET}+B_{1NET}R_NM_{NET} & B_{1NET}R_NM_{load} \\ B_{1LOAD}R_NM_{INV}C_{INVc}+B_{2LOAD}C_{INV\omega} & B_{1LOAD}R_NM_{NET} & A_{load}+B_{1LOAD}R_NM_{load} \end{bmatrix} \tag{10.41}$$

10.4.1.5 Stability analysis—case study 1

After finding the state space equation of the whole system, the eigenvalue analysis and sensitivity analysis can be used to evaluate the voltage stability of the microgrid. Table 10.1 gives the parameters of the system under study in this case [11].

Figure 10.11 illustrates the location of the eigenvalues of the state space matrix of this microgrid. It is worth mentioning that in this system, a *Q–V* droop control

Table 10.1 Parameters of the microgrid of case study 1 (Adopted from [11])

Parameter	Value
f_s	8 kHz
L_f	1.35 mH
C_f	50 μF
r_f	0.1 Ω
L_c	0.35 mH
r_{L_c}	0.03 Ω
ω_c	31.41
m_p	9.4e^{-5}
n_q	1.3e^{-3}
K_{pv}	0.05
K_{pc}	390
K_{ic}	16e^3
F	0.75

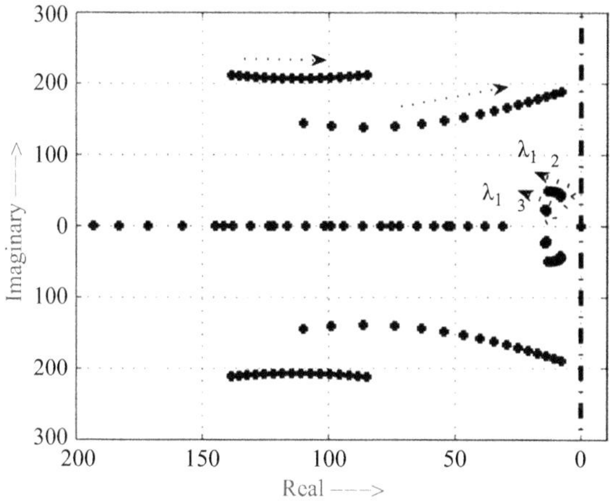

Figure 10.11 *Root locus of the roots of the characteristic equation of the microgrid in terms of change in the voltage control gain (adopted from [11])*

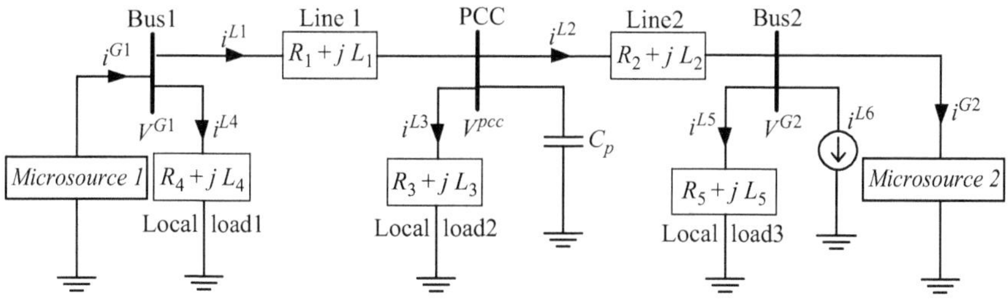

Figure 10.12 *The structure of a test system for studying a microgrid comprised a renewable energy source and a diesel generator (Adopted from [44])*

for voltage control is used. As long as the eigenvalues are located in the left hand side of the imaginary axis, i.e., all of them have negative real values, the system remains stable. Therefore, to design the voltage control system, its gain is selected such that all eigenvalues have negative real values which give enough margin for voltage stability.

10.4.2 Case study 2

This case study covers the voltage stability analysis for a microgrid that has diesel generators in addition to the renewable energy sources. Particularly, for off-grid microgrids, a hybrid system which comprises renewable energy micro-sources working in parallel to diesel generators may provide a reliable source of electrical energy. Therefore, this case study investigates the modeling and voltage stability analysis of a test microgrid that has one inverter-based generator (to represent a Renewable Energy System (RES)) and one synchronous generator (to represent diesel generator) [44]. This test-system microgrid is shown in Figure 10.12.

To study this microgrid, first, the small-signal models of four subsystems are derived, and then the model of the whole integrated system is obtained [11]. Similar to case study 1, all the models will be done in the GRF. For this purpose, it is assumed that the GRF is rotating with a specific speed. Then, according to Figure 10.13, the small-signal model of each subsection is transformed to the GRF using the following equation:

$$f^g = T_n f^n \tag{10.42}$$

In the above equation, f^n is any variable in its local reference frame and f^g is its transformed variable in the GRF. T_n is the transformation matrix, given in (10.43). The angle δ is illustrated in Figure 10.14:

$$T_n = \begin{bmatrix} \cos(\delta_n) & \sin(\delta_n) \\ -\sin(\delta_n) & \cos(\delta_n) \end{bmatrix} \tag{10.43}$$

The variables in d–q global reference from are obtained as

$$\begin{bmatrix} \Delta f_q^g \\ \Delta f_d^g \end{bmatrix} = \begin{bmatrix} \cos\left(\delta_n^o\right) & \sin\left(\delta_n^o\right) \\ -\sin\left(\delta_n^o\right) & \cos\left(\delta_n^o\right) \end{bmatrix} \begin{bmatrix} \Delta f_q^n \\ \Delta f_d^n \end{bmatrix} + \begin{bmatrix} -f_d^{g0} \\ f_q^{g0} \end{bmatrix} \Delta\delta_n \tag{10.44}$$

10.4.2.1 The synchronous generator model—case study 2

In modeling a synchronous generator connected to a network, with a good approximation, it is fine to assume the rest of the network as an infinite bus. The generator is modeled along with the excitation and governor systems. The small-signal equations are obtained as follows:

$$\Delta\dot{x}_{G1} = A_{G1}\Delta x_{G1} + B_{G1}^{v}\Delta x_{v1} + B_{G1}^{u}\Delta u_{G1} \tag{10.45}$$

$$\Delta x_{G1} = \left[\Delta i_q^{G1}\ \Delta i_d^{G1}\ \Delta i_{klql}\ \Delta i_{k2ql}\ \Delta i_{kdl}\ \Delta i_{fdl}\ \Delta\dot{\delta}_1\ \Delta\delta_1\right]^T \tag{10.46}$$

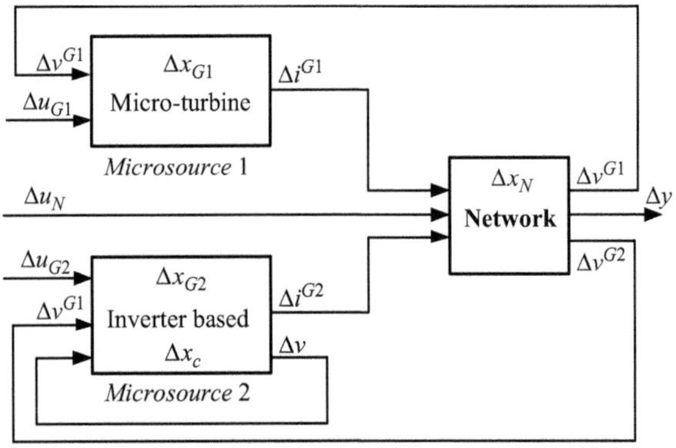

Figure 10.13 State space model for the microgrid of case study 2 (Adopted from [44])

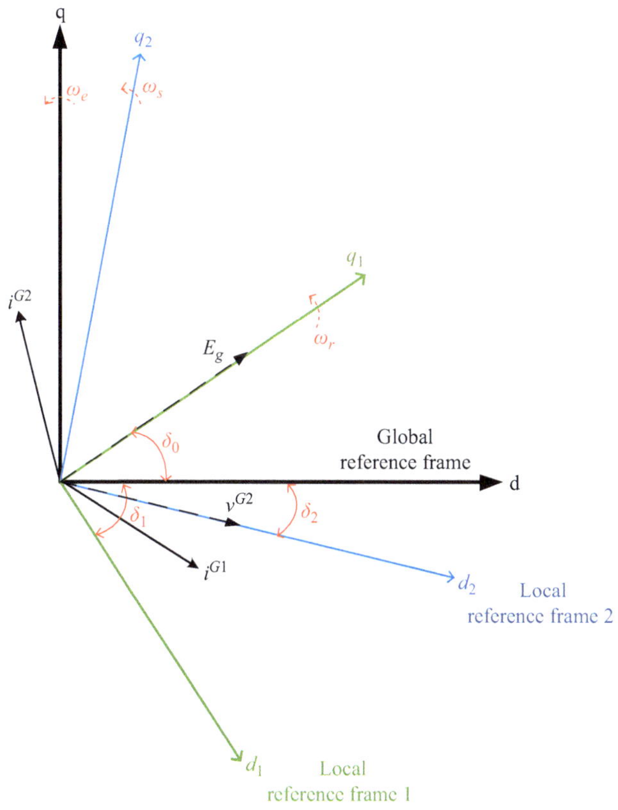

Figure 10.14 Reference frames of the microgrid (Adopted from [44])

$$\Delta v^{G1} = \left[\Delta v_q^{G1} \ \Delta v_d^{G1}\right]^T \tag{10.47}$$

$$\Delta u_{G1} = \left[\Delta v_{fd1} \quad \Delta T_m\right]^T \tag{10.48}$$

In the above equations, Δx_{G1} is the vector of state variables of the synchronous generator, Δv^{G1} is the vector of d–q components of the terminal voltage of the generator in the GRF, and Δu_{G1} is the input vector, which comprises Δv_{fd1}, the excitation system voltage, and ΔT_m, the input torque of the governor.

10.4.2.2 The inverter model–case study 2

To find the inverter model, it is assumed that the whole microgrid is an infinite bus. Then, based on Figure 10.15, the small signal model of the inverter is obtained as follows:

$$\Delta \dot{x}_{G2} = A_{G2}\Delta x_{G2} + B_{G2}^{v'}\Delta v' + B_{G2}^{v}\Delta v^{G2} + B_{G2}^{\omega}\Delta\omega_s \tag{10.49}$$

where Δx_{G2} is the vector of state variables of the synchronous generator including the d–q components of the inverter output currents, Δv^{G2} is the vector of d–q

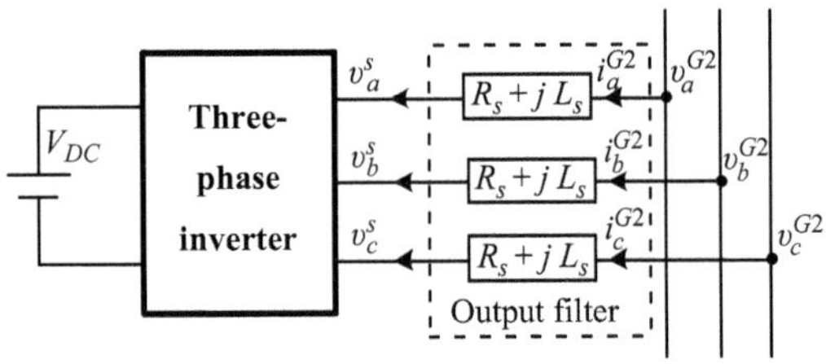

Figure 10.15 The inverter and its output filter (Adopted from [53])

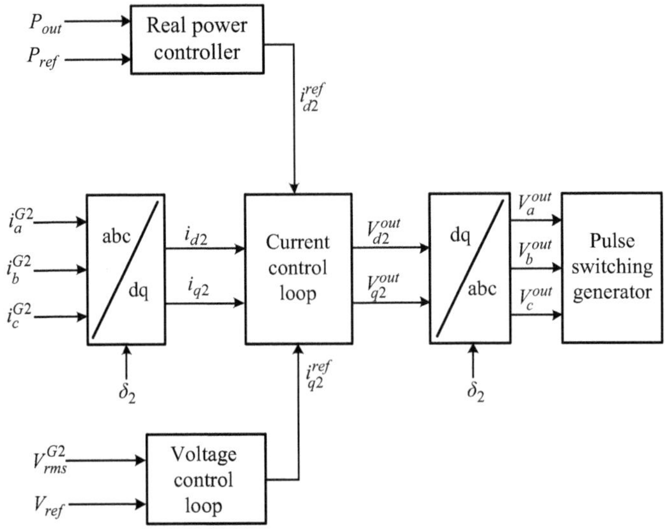

Figure 10.16 The control system of the inverter-based renewable energy micro-source (Adopted from [44])

components of the terminal voltage of the inverter, and $\Delta v'$ is the vector of the intermediate variables, all in the GRF, given in [44,45].

10.4.2.3 Model of the inverter control system—case study 2

The control system of the inverter includes three controllers for controlling the terminal voltage, current, and active power, which are shown in Figure 10.16.

The small-signal model of the control system of an inverter-based micro-source is then obtained as follows [44,45]:

$$\Delta\dot{x}_c = A_c\Delta x_c + C_{G2}\Delta x_{G2} + C_{G2}^{v}\Delta v^{G2} + C^{v}\Delta v + C_{G2}^{u}\Delta u_{G2} \tag{10.50}$$

$$\Delta x_c = \left[\Delta v_{q2'}^{G2}\Delta v_{d2'}^{G2}\ \Delta v_{q2'}\ \Delta v_{d2'}\ \Delta i_{iq2}\ \Delta i_{id2}\ \Delta i_{q2'}^{G2}\Delta i_{q2'}^{G2}\ \Delta v_{iq2}\ \Delta v_{id2}\ \Delta\omega_s\right]^T \tag{10.51}$$

$$\Delta v^{G2} = \left[\Delta v_{q2}^{G2}\ \Delta v_{d2}^{G2}\right]^T \tag{10.52}$$

$$\Delta v = \left[\Delta v_{q2}\ \Delta v_{d2}\right]^T \tag{10.53}$$

$$\Delta u_{G2} = \left[\Delta P_{ref}\ \Delta V_{rms}\right]^T \tag{10.54}$$

10.4.2.4 Network model—case study 2

Feeders and loads are modeled as *R–L* impedances. Nonlinear loads are modeled as equivalent current sources. As a result, differential equations of the network including loads are derived as follows [44]:

$$R_1 i^{L1} + L_1 \frac{d}{dt}\left(i^{L1}\right) = v^{G1} - v^{pcc} \tag{10.55}$$

$$R_2 i^{L2} + L_2 \frac{d}{dt}\left(i^{L2}\right) = v^{pcc} - v^{G2} \tag{10.56}$$

$$C_p \frac{d}{dt}\left(v^{pcc}\right) = i^{L1} - i^{L2} - i^{L3} \tag{10.57}$$

In these equations, R_i and L_i are the ith line resistance and inductance, respectively, v^x is the voltage of bus x and i^x is the current of line x. Based on these equations, the small-signal model of the network is derived. Details with the coefficient matrices, etc., are given in [44,45]:

$$\Delta \dot{x}_N = A_N \Delta x_N + B_N^{v1} \Delta v^{G1} + B_N^{v2} \Delta v^{G2} \tag{10.58}$$

$$\Delta v^{G1} = E_{11}^N \Delta x_N + E_{12}^N \Delta \dot{x}_N + D_{11}^N \Delta x_{G1} + D_{12}^N \Delta \dot{x}_{G1} \tag{10.59}$$

$$\Delta v^{G2} = E_{21}^N \Delta x_N + E_{22}^N \Delta \dot{x}_N + D_{21}^N \Delta x_{G2} + D_{22}^N \Delta \dot{x}_{G2} + B^N \Delta u_N \tag{10.60}$$

$$\Delta x_{G1} = \left[\Delta i_q^{L1}\ \Delta i_d^{L1}\ \Delta i_q^{L2}\ \Delta i_d^{L2}\ \Delta i_q^{L3}\ \Delta i_d^{L3}\ \Delta v_q^{pcc}\ \Delta v_d^{pcc}\right]^T \tag{10.61}$$

$$\Delta u_N = \left[\Delta i_q^{L6}\ \Delta i_d^{L6}\right]^T \tag{10.62}$$

10.4.2.5 Model of the whole system—case study 2

Integrating all the subsystem models, the small-signal state-space model of the whole system is given as follows. Details with the coefficient matrices are given in [44,45]:

$$\Delta \dot{x}_N = A_N \Delta x_N + N_N^{v1} \Delta v^{G1} + B_N^{v2} \Delta v^{G2} \tag{10.63}$$

10.4.2.6 Voltage stability analysis—case study 2

Using (10.57), the stability analysis of the system is carried out by finding the eigenvalues of the coefficient matrix. The nominal values of the test system are given in Table 10.2 [45].

Table 10.2 Eigenvalues of the test system in case study 2 in grid-connected and autonomous modes (Adopted from [44])

Eigen values	Mode	
	Grid connected	Autonomous
1–2	–	$-119.28 \pm j2044.26$
3–4	–	$-147.06 \pm j1293.26$
5–6	$-113.90 \pm j486.80$	$-86.54 \pm j675.74$
7–8	$-100.85 \pm j436.19$	$-90.75 \pm j430.86$
9–10	$-576.74 \pm j376.93$	$-757.54 \pm j376.41$
11–12	$-34.83 \pm j372.13$	$-517.50 \pm j371.09$
13–14	$-420.19 \pm j376.98$	$-314.46 \pm j364.13$
15–16	$-1.14 \pm j23.17$	$-1.06 \pm j10.41$
17	-139.2	$-127.9 - j8.42$
18	-138.4	$-127.9 + j8.42$
19–21	-333.33	-333.33
22	-327.11	-326.18
23	-305.48	-137.46
24	-93.27	-93.04
25	-20.99	-24.39
26	-3.09	-17.26
27	-2.57	-2.57
28	-1.28	-0.672
29	-0.429	-0.277

By comparing the eigenvalues in the two modes, it is observed that the tendency for oscillations in the system responses to possible disturbances is more in the case of autonomous mode of operation. However, since the real values of all eigenvalues in both modes of operation are negative, it is concluded that for the test system, under nominal operating conditions, is stable for small and slow disturbances.

10.4.3 Case study 3

A directly connected induction generator is added to the case study 2 to make this case study 3, as shown in Figure 10.17 [15].

Similar to the previous cases, the model of all subsystems is derived first, and then the model of the whole system in GRF is developed. To avoid repetition, only the model of the induction generator is explained, and then the model of the whole system will be presented, as given by [15].

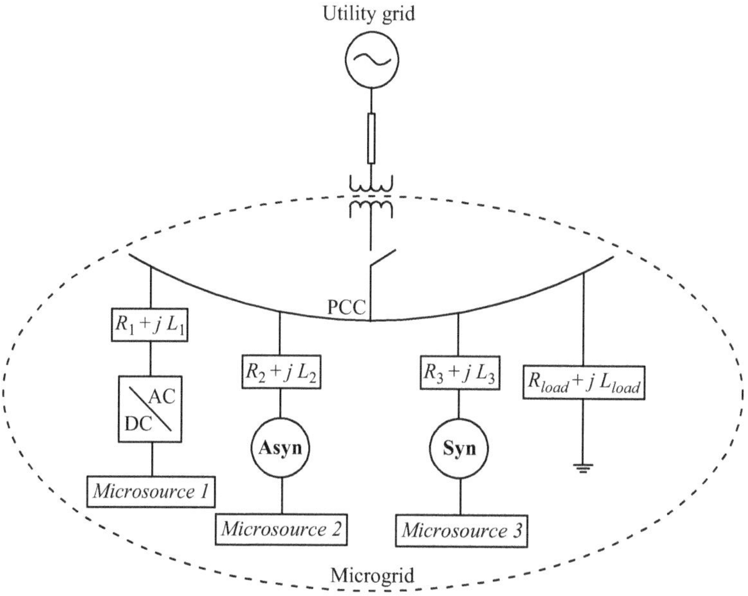

Figure 10.17 The microgrid of case study 3 comprising three micro-sources, synchronous generator, inverter-based generator, and induction generator directly connected (Adopted from [15])

10.4.3.1 Models of the induction generator and the whole system—case study 3

Some renewable energy sources such as small wind turbines use induction generators directly connected to the microgrid. The small-signal model of this type of generator is given by the following equations [15]:

$$\Delta \dot{X}_{G2} = A_{G2}\Delta X_{G2} + B_{G2}\Delta u_{G2} + C_{G2}\Delta u_{pcc} \tag{10.64}$$

$$\Delta X_{G1} = \left[\Delta u'_d \; \Delta u'_q \; \Delta\omega_g\right]^T \tag{10.65}$$

$$\Delta u_{G2} = \Delta T_m \tag{10.66}$$

The simplified coefficient matrix is found as follows, with the details in [15]:

$$A = \begin{bmatrix} B_{G1} & 0 & C_{G1}G_1 \\ 0 & B_{G2} & C_{G2}G_1 \\ 0 & 0 & B_{G3} + C_{G3}G_1 \end{bmatrix} \tag{10.67}$$

10.4.3.2 Stability analysis—case study 3

In this case, too, the eigenvalues and sensitivity analysis is used to perform stability analysis. At first, the parameters of the coefficient matrix is obtained using power

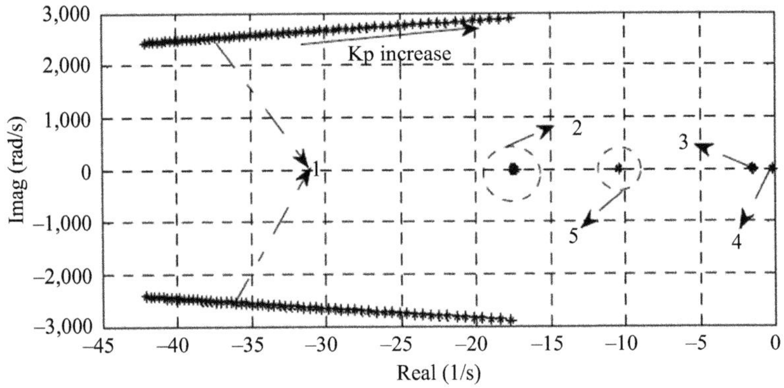

Figure 10.18 Root locus for the case study 3 (adopted from [15])

flow analysis; then, the root locus of eigenvalues, as the gains of controllers are changed, are plotted as shown in Figure 10.18 [15].

The root locus in this figure, which only shows the eigenvalues related to the induction generator and synchronous generator, shows that the system is stable. As the bus voltages are included in the vector of state variables, the voltage stability is maintained for the test system with the selected range of K_p in the PI controller of the voltage control loop in the inverter-based DG. However, as K_p increases, the stability margin of the system reduces as it is observed that the eigenvalues move toward the imaginary axis from the left side.

For a thorough stability analysis, this sensitivity analysis should be repeated for other controller gains, and possibly with combination of a few changes in the gains simultaneously. Furthermore, optimal controller gains can be worked out using the sensitivity analysis.

10.4.4 Case study 4

This case studies a real multi-microgrid in Oman, comprising interconnected microgrids powered by diesel generators and connected via a transmission system to a central gas-fired power plant. This multi-microgrid system is located in Musandam, Oman.

10.4.4.1 The context: current status and future perspective of microgrids in Oman

Oman electricity sector restructure

The electricity sector in Oman went through a major restructuring in 2005. From a vertically integrated system which was fully owned and operated by the Ministry of Housing, Electricity, and Water (MHEW), the sector has been unbundled to successor companies and an independent regulator to replace the role of MHEW [54]. The role of the government, which is limited to setting up the policy of the sector, is currently taken care of by the Public Authority for Electricity and Water. The main objective of restructuring the electricity sector and related water production was to improve the sector's efficiency, the financial transparency, and the security of electricity supply. The sector has witnessed a rapid growth in terms of the amount of energy delivered to the consumers with the new system during the first decade to reach more than 240% of the system before change [55].

Due to the natural terrain and the geographical distribution of the population in Oman, the sector has been divided into three subsystems. The Main Interconnected System (MIS) which represents 88% of the sector with several central gas-fired power generation and associated water-production plants, transmission network operated by Oman Electricity Transmission Company and three distribution and supply companies to serve the majority of seven governorates out of the ten governorates in Oman [55]. In addition, Dhofar Power System (DPS) represents about 9% of Oman electrical system and serves the southern part of Oman to cover the majority of Dhofar Governorate [55]. DPS, which is currently isolated from the MIS, is composed of three central gas-fired power plants and associated water production. The third subsystem is the Rural Areas Electricity Company (RAECO) which represents about 3% of Oman in terms of population but covers a huge area and serves Musandam Governorate, Alwosta Governorate, and portions of Dhofar, Dakhlyah, and Alsharqyah governorates that are not covered by MIS nor DPS [56]. Figure 10.19 represents the distribution of the electricity sector in Oman between the three subsystems based on the energy consumptions [55]. The next section will shed the light in more details about the RAECO and the formation of multi-microgrids in Oman.

Rural area electricity company

RAECO is the smallest portion of Oman Electrical Sector in terms of generation capacity or the demand of energy. Nevertheless, it is considered the most expensive portion due to the high subsidy paid by the Government of Oman per account or per kW h delivered to customers. This is mainly due to the formation of this system, which is mainly a small isolated power system in the rural areas operated by relatively small power generators. Each system is composed of multiple diesel

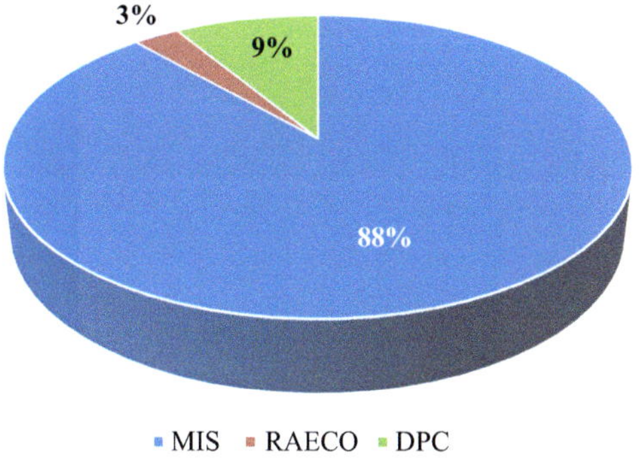

Figure 10.19 Oman Electricity Sector based on the Energy Consumption 2017 [56]

generators connected to customers through 33 kV and/or 11 kV long distance overhead lines and low-voltage 415 V circuits. This structure may be considered as a microgrid system with its own power generation, transmission and distribution system, and loads.

RAECO has three different setups of microgrids. The first type that represents the majority of RAECO system is completely isolated microgrids and powered fully by diesel generators. The second type is multi-microgrids interconnected with each other and powered by diesel generators and connected to 132 kV transmission system with a central gas-fired power plant, such as Musandam system, where the majority of the energy supplied to the customers is coming from the gas-fired power plant. The third type of RAECO microgrids is an island of customers supplied with electricity via a 33 kV and/or 11 kV overhead lines from the main Petroleum Development Oman (PDO) company grid. PDO has its own grid that serves its own needs of electricity for the exploration and production of oil and gas activities. The grid is fully owned and operated by PDO and based on central gas-fired power plants with long 132 kV transmission network and 33 kV distribution system. RAECO signed an agreement with PDO to interconnect its rural areas customers either solely from PDO grid or in a hybrid way with RAECO diesel power generators. Figure 10.20 presents the number of the diesel-microgrids and total energy supplied to customers under RAECO territory [56].

In 2014, RAECO commissioned the first solar PV power plant as an Independent Power Producer (IPP). This power plant supplies electricity to Almazyonah

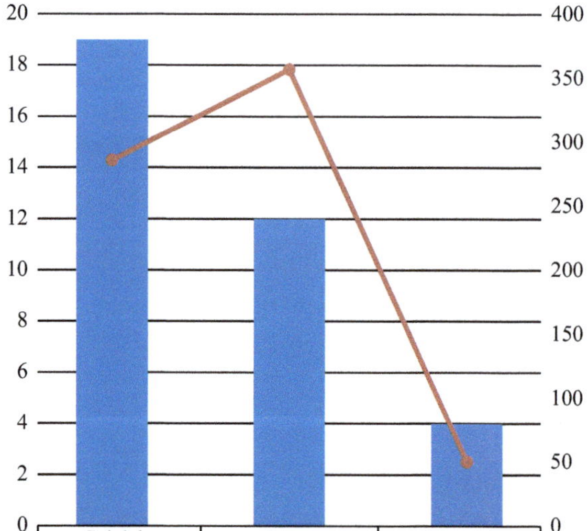

	Salalah	Westa	Musandam
Number of diesel power plants	19	12	4
GWh production from diesel	286.025	356.948	50.341

Figure 10.20 RAECO diesel microgrids power plants and energy production 2017 [56]

microgrid. The PV power plant has a peak capacity of 303 kW, which is integrated with existing 14 MW diesel generators power plant. This integration reflects the first hybrid PV-Solar and Diesel microgrid in Oman with main objective to reduce dependency on the expensive diesel fuel and improve the overall system efficiency [57,58]. The project has been set as a pilot project to understand the challenges associated with the hybrid microgrids powered by solar PV and diesel generators.

Figure 10.21 shows the contribution of power generation in RAECO microgrids based on the source of energy as per 2017. It is clear that the main source of energy in RAECO microgrids is coming from diesel generators which represent more than 70% [56].

Sahim PV-Rooftop Project

In 2017 and to promote an eco-friendly lifestyle, creating a greener future and contributing to the development and prosperity of Oman, the Authority for Electricity Regulation lunched the first version of Sahim renewable energy initiative. The project put in place the required technical and legislation to integrate solar PV system at the consumer ends. This initiative allows the electricity customers in Oman to install rooftop PV system with total peak capacity equal to 50% of their total connected load. The PV can be as rooftop system or installed within the customers' premises. The main objective is to reduce the pressure on the main grid or distribution network and to facilitate the consumer participation in producing part of their energy demand [59]. The consumers would have the right to export excess energy to the grid and get a return in cash or adjustment to their electricity bill equal to the Bulk Supply Tariff.

While the microgrid concept was limited in RAECO power system since the restructuring of the electricity sector in Oman, this was changed in 2017 upon lunching of the first version of Sahim initiative. With the presence of the solar PV

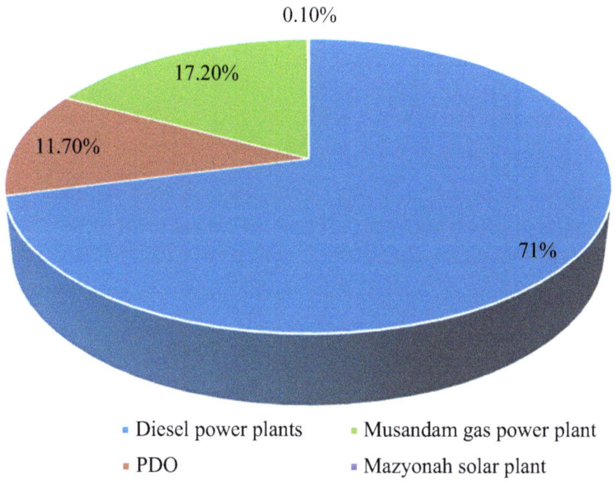

Figure 10.21 RAECO power production contribution by source 2017 [56]

system within the distribution networks of MIS, DPS, and RAECO system, Oman electricity sector changed to showcase microgrids that are of grid-connected type. The resulted microgrids are expected to improve the distribution networks efficiency, improve the reliability, reduce the carbon footprint, and increase the overall awareness within the consumers [60].

Musandam power network in Oman—case study 4

This case study [61] is about Musandam Power System, an area in the north of Oman, under the responsibility of RAECO. As RAECO was planning to expand the Musandam electrical network and add new power stations and transmission lines, it was very important to make voltage stability assessment for this project in the planning phase. In addition, since Musandam power network is not linked with the MIS of the national electricity grid, it can be treated as a microgrid case.

At the time of this study, the generation in the Musandam microgrid was based on diesel generators. In our study, a DFIG connected to a wind turbine was added to the system for the study purpose in order to have close look in how such renewable energy sources can affect the stability of micro-grids. It is expected that RAECO will implement many renewable energy power supplies in its networks, as a general plan has already been approved.

In 2015, when this case study was performed, there were two diesel-based power stations in Musandam, namely, Khasab and Dibba power stations. They were running at 33 and 11 kV and their installed capacity power were 65 and 22 MW, respectively. However, there were plans to implement a 132 kV power line to interconnect Khasab and Dibba. In addition, an IPP project was awarded to construct a 100 MW gas turbine power station in Khasab. Based on load forecasting and the expected load growth in Khasab and Dibba, some of the power stations were going to be upgraded as well. Khasab power station consisted of 8×10 MW diesel-based generators.

The system was modeled for two modes of operation: when the 132 kV line is connected and when it is disconnected. Four types of disturbances were applied to the system for their effect on the voltage stability, and also for studying the effect of DFIG wind turbine: (1) the effect of losing one transmission line was studied, (2) the effect of losing a 10 Mvar capacitor bank was studied. Disconnection of the DFIG from the microgrid and its effect was analyzed. Then, three-phase short circuits were applied to two locations, DFIG terminal bus and a synchronous generator bus. The critical fault clearing time t_c was determined. Finally, the effect of DFIG power factor was examined during each of the three-phase short circuits.

The Musandam Power System, as shown in Figure 10.22, is modeled in ETAP.

Masandum Power System, static analysis

Since the main concern and the worst case may happen for the islanded mode, Figure 10.23 shows some snapshots of the load flow solution performed using MATLAB®.

Figure 10.24 represents the voltage profile for the islanded mode of Musandam microgrid. It should be noted that the lowest voltage bus is not always the weakest in terms of voltage instability. In this case, buses 9 and 13 have lowest voltages in

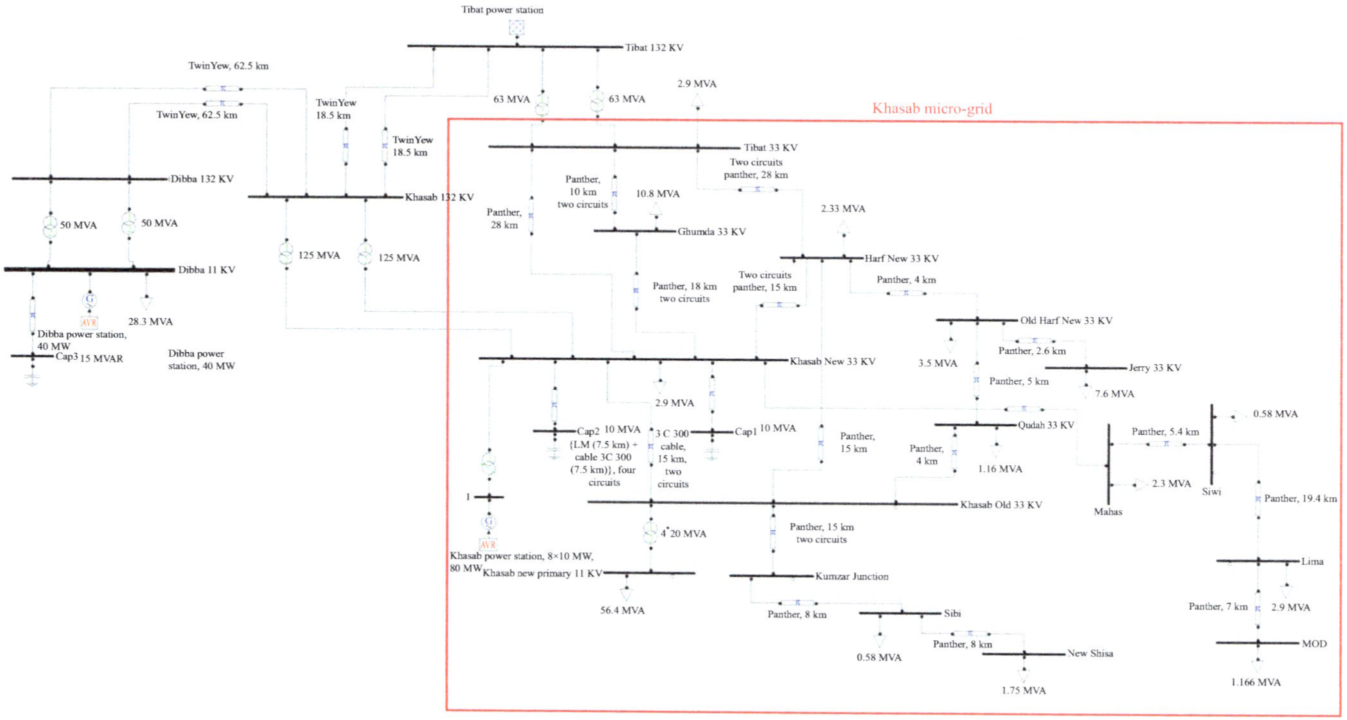

Figure 10.22 Khasab, Oman, power system model in ETAP

Total loss 1.284 1.687

Power Flow Solution by Newton–Raphson Method
Maximum Power Mismatch = 2.71575e–005
No. of Interactions = 3

Bus No.	Voltage Mag.	Angle Degree	------Load------ MW	Mvar	------Generation------ MW	Mvar	Injected Mvar
1	1.000	0.000	0.000	0.000	80.782	30.676	0.000
2	0.980	–0.577	2.500	1.530	0.000	0.000	0.000
3	0.980	–0.581	9.000	5.560	0.000	0.000	0.000
4	1.000	–0.035	2.500	1.500	0.000	0.000	20.000
5	0.977	–0.695	0.000	0.000	0.000	0.000	0.000
6	0.976	–0.736	45.000	28.000	0.000	0.000	0.000
7	0.973	–0.803	0.000	0.000	0.000	0.000	0.000
8	0.969	–0.919	0.500	0.300	0.000	0.000	0.000
9	0.966	–1.006	1.500	0.900	0.000	0.000	0.000
10	0.980	–0.594	2.000	1.200	0.000	0.000	0.000
11	0.972	–0.821	3.000	1.800	0.000	0.000	0.000
12	0.974	–0.767	1.000	0.600	0.000	0.000	0.000
13	0.968	–0.941	6.500	4.000	0.000	0.000	0.000
14	0.992	–0.242	2.000	1.200	0.000	0.000	0.000
15	0.987	–0.939	0.500	0.300	0.000	0.000	0.000
16	0.970	–0.877	2.500	1.500	0.000	0.000	0.000
17	0.968	–0.928	1.000	0.600	0.000	0.000	0.000
Total			79.500	48.990	80.782	30.676	20.000

Figure 10.23 Snapshot of load flow solution for Musandam Power System

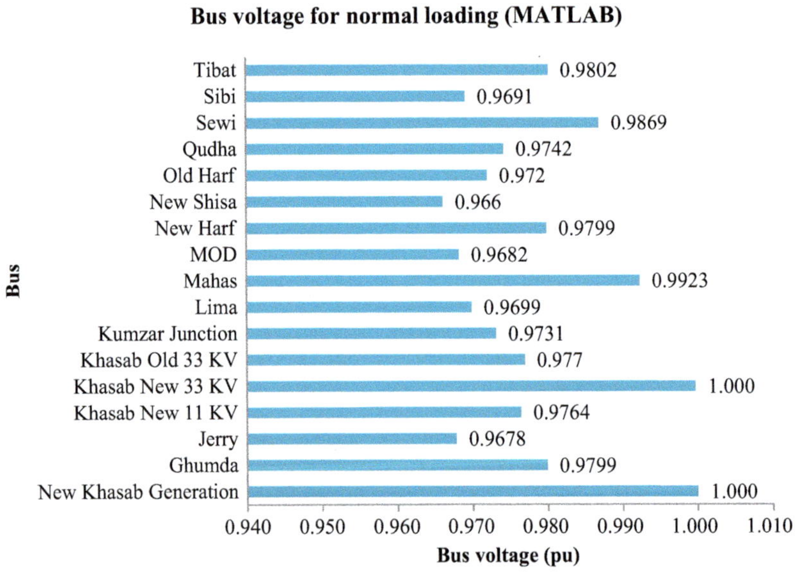

Figure 10.24 Voltage profile of Musandam Power System

the system, i.e., 0.966 and 0.968 pu, respectively. But, they are not the most critical buses as it will be seen in Figure 10.25.

By examining the location of eigenvalues, we can work out the critical buses. The critical mode that may lead to instability of the voltage is the eigenvalue 7 with magnitude of 0.7 as shown in Table 10.3, which illustrates the eigenvalues of the reduced Jacobian matrix of Musandam microgrid. Since the microgrid consists of 17 buses, then the reduced Jacobian matrix will have only (17−1 slack−0 PV = 16 buses = 16 eigenvalues).

The participation factor for the defined critical mode (eigenvalue = 0.7) is shown in Figure 10.25. It can be seen that the weakest bus is bus 17 (MOD bus) followed by bus 16 (Lima bus).

The *QV* and *PV* curves for the critical bus (MOD bus) are shown in Figures 10.26 and 10.27. It is very vital to note that for this bus, the operating *P* and *Q* are 1 and 0.6 MW, respectively. It can be seen that the stability margins for both are still large (35 MW, 30 Mvar). However, it will be worth mentioning that this microgrid cannot handle this amount of power due to limitation in the generation side which is 80 MW in the case of islanded mode.

Hence, one can see that the PV and QV analysis alone are not enough to determine the stability issues of the microgrid even though modal analysis was added to determine the weakest buses. So, time-domain analysis or dynamic analysis is required for further assessment of Musandam microgrid.

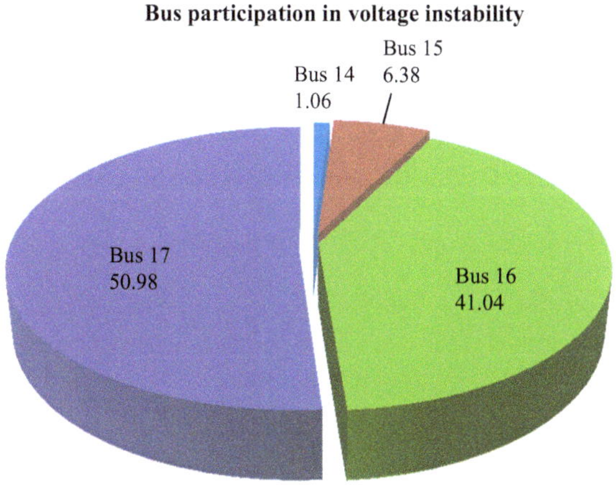

Figure 10.25 Bus participation factor for voltage instability—Musandam System [61]

Table 10.3 Eigenvalues of the J_R for Musandam microgrid system

Eigenvalue no.	1	2	3	4	5	6	7	8	9	10	11	12	13	14	15	16
Value	1,339.4	1,270.9	58.9	31.4	3.2	1.1	0.7	5.5	5.8	8.9	27.6	26.1	23.1	20.6	15.3	15.8

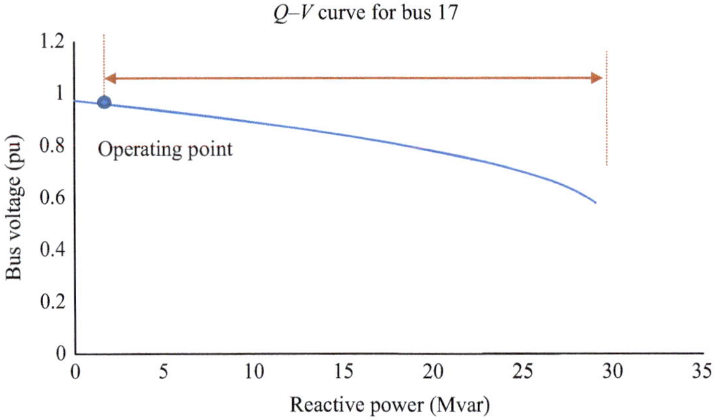

Figure 10.26 QV curve for Bus 17—Musandam microgrid

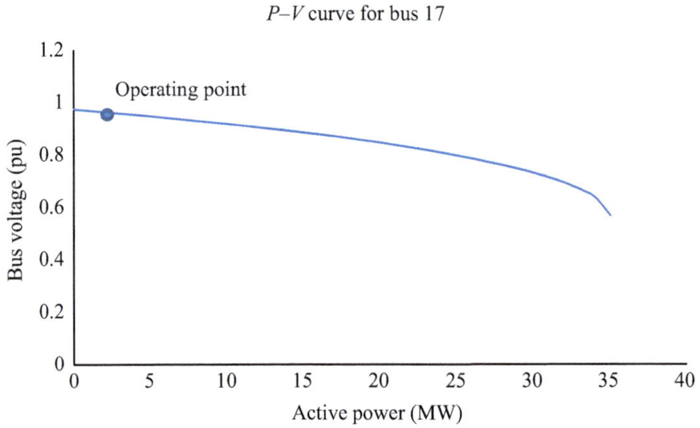

Figure 10.27 PV curve for Bus 17—Musandam microgrid

Musandam Power System dynamic analysis

For the dynamic analysis, first, the time-domain simulation was performed for the case of loss of capacitor bank (to assess long-term voltage stability). The main concern in the comparison process was between the weakest bus obtained from the modal analysis (MOD Bus) and the bus which had the lowest voltage magnitude among the others (New Shisa bus).

Loss of 10 Mvar capacitor bank from Khasab 33 kV old substation

Khasab 33 kV old substation contains capacitor banks (2×10 Mvar). Let us examine the Musandam microgrid system voltage and frequency responses when one of the 10 Mvar capacitors is lost. Figures 10.28 and 10.29 represent a comparison of the microgrid responses when it loses 10 Mvar with and without the wind turbine generator. As expected, it is worth noting that for both buses, the

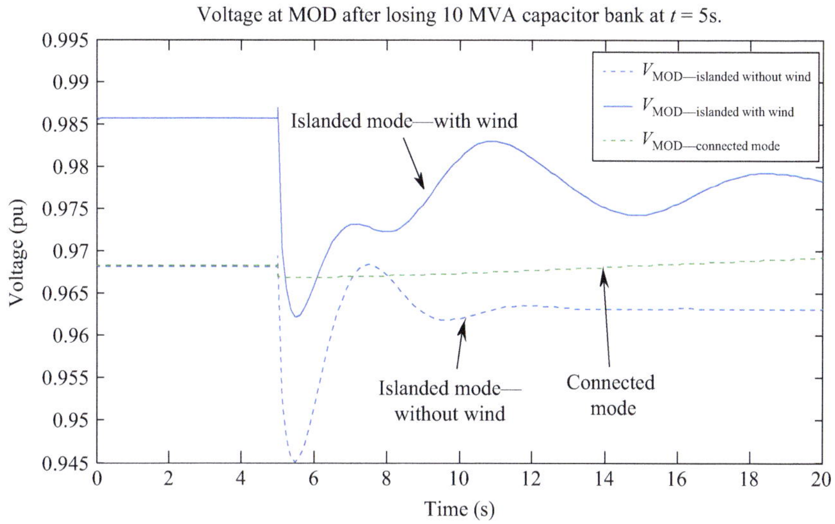

Figure 10.28 MOD voltage after losing 10 Mvar capacitor bank

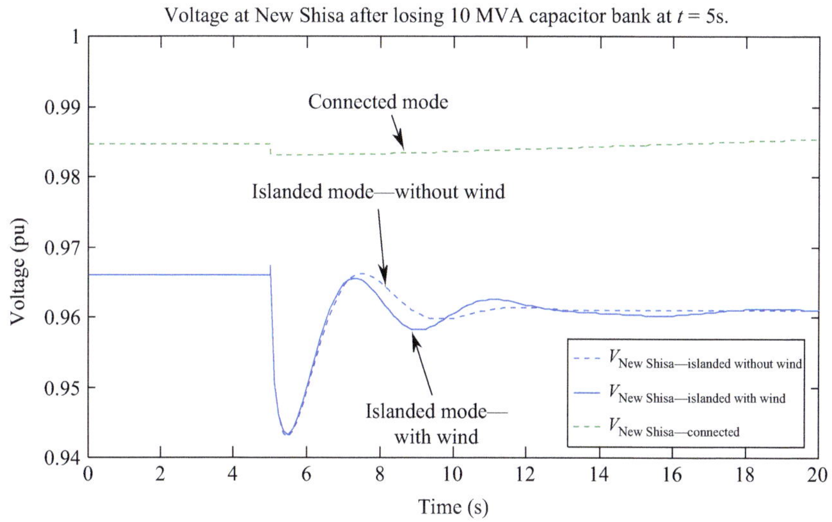

Figure 10.29 New Shisa voltage after Losing 10 Mvar capacitor bank

connected mode seems to be much better than the autonomous mode against this disturbance. The same can be extended to the frequency as seen in Figure 10.30.

Figure 10.28 shows the MOD-bus voltage after this disturbance, it is interesting to mention that with the wind turbine included in the system, the voltage is a little bit higher than that for the connected mode as well as the islanded mode with the absence of wind generation. In the former (islanded with wind generator), the

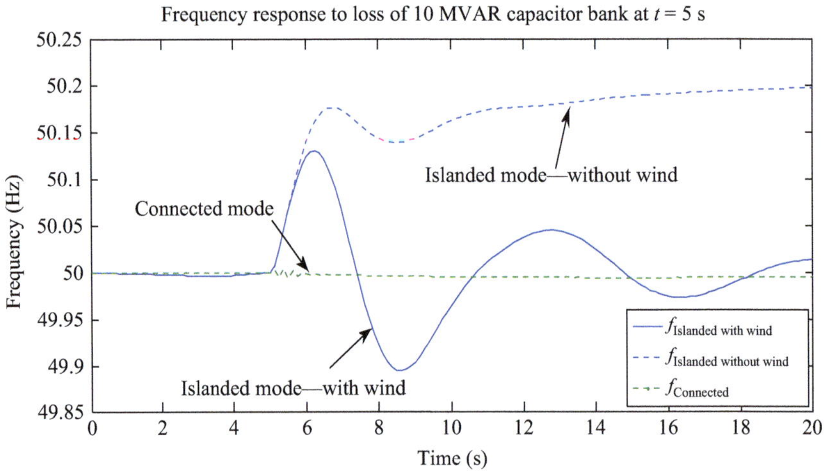

Figure 10.30 Frequency of the system after losing 10 Mvar capacitor bank

voltage starts at around 0.985 pu and then at $t = 5$ s drops and settled after a while to around 0.98 pu. In the latter case, the voltages were around 0.968 and then dropped to lower value for the islanded mode (without wind) to 0.964 pu while it returned back to its original value in the connected mode after few cycles. Small transient oscillations are observed, but they are damped after a while.

The same can be said for the New Shisa voltage at Figure 10.29. However, one can notice that there is only a slight difference between the islanded mode with wind and that without wind turbine compared to the MOD bus. This is due to the fact that the wind turbine was installed very close to MOD bus, while New Shisa bus is located around 76 km away from the place of the wind turbine.

However, Figure 10.30 shows interesting behavior of system frequency after this disturbance. It can be seen that the frequency returns back to nominal operation point for the connected mode and the islanded mode with wind turbine, while in the case of absence of the wind turbine, the frequency rose up slightly and then normalized at 50.2 Hz. This is due to the fact that, DFIG controller may be able to stabilize the frequency after some types of disturbances [62]. The DFIG inertia contribution to the system can improve the system frequency regulations taking in consideration some controller factors [63].

Loss of a transmission line

In this section, the scenario of losing one of the transmission lines will be tested. The selected line was Kumzar feeder, a feeder that fed the New Shisa area and some other areas with total length of 31 km and loaded around 2 MW. Previously, MOD bus was compared with the New Shisa bus, but in this case, since the latter bus is disconnected, the nearest feeder which is Khasab Old 33 kV substation was analyzed instead.

The same situation repeated in this disturbance, Figures 10.31 and 10.32 show MOD and Khasab Old 33 kV substation voltages, respectively, after being disturbed

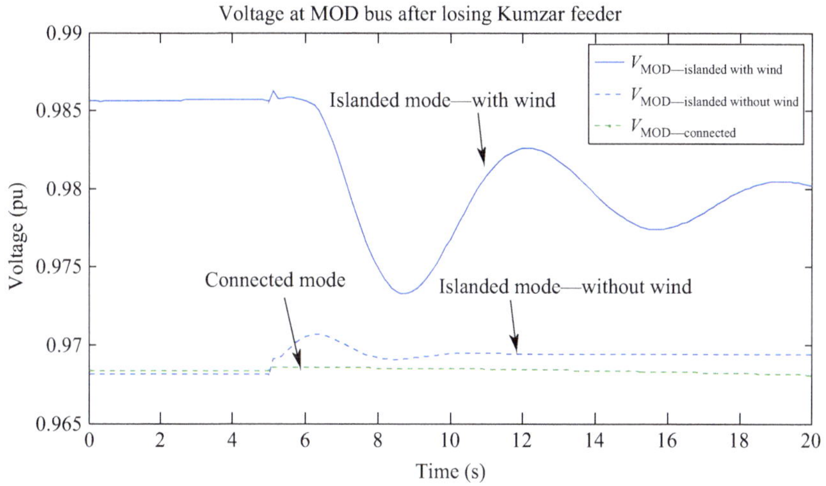

Figure 10.31 MOD voltage after losing Kumzar feeder

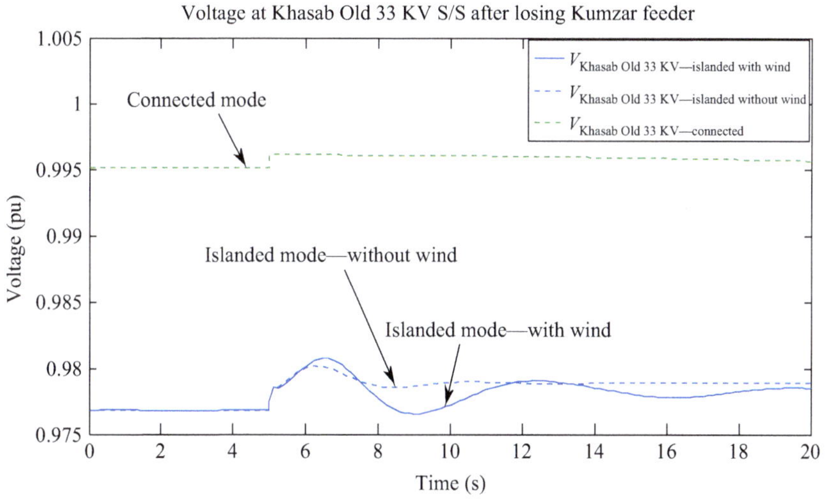

Figure 10.32 Khasab Old 33 kV S/S voltage after losing Kumzar feeder

by the loss of Kumzar feeder at $t = 5$ s. The voltages for the connected modes in both locations do not make significant changes from their original operation points, 0.968 and 0.995 pu for MOD and Khasab Old 33 kV substation, respectively.

Again for MOD in Figure 10.31, the islanded mode (without wind) in case of the voltages, behaves similar to that for the connected mode due to the long distance between the lost line and the MOD which is around 52 km. In contrast, in the isolated case associated with wind, it can be seen that the voltage jumps slightly at $t = 5$ s and then drops from 0.96 to 0.980 pu with oscillations for few cycles caused by the transient period. This may be due to the fact that the DFIG wind turbines failed to provide natural damping for normal cases and normal controllers, which then increase the possibility of having voltage oscillations [64].

However, in low wind penetration, as in Musandam microgrid case, the islanded mode connected with wind can improve the frequency of the system, as seen in Figures 10.30 and 10.33.

Effect of disconnection the DFIG wind turbine

Figures 10.34 and 10.35 illustrate the behavior of the real and reactive power generated by diesel-based synchronous generators in Khasab power station after

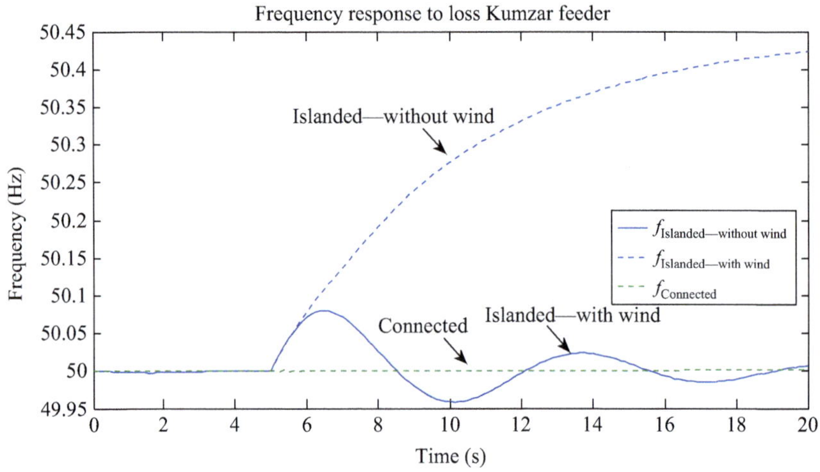

Figure 10.33 Frequency of the system after losing Kumzar feeder

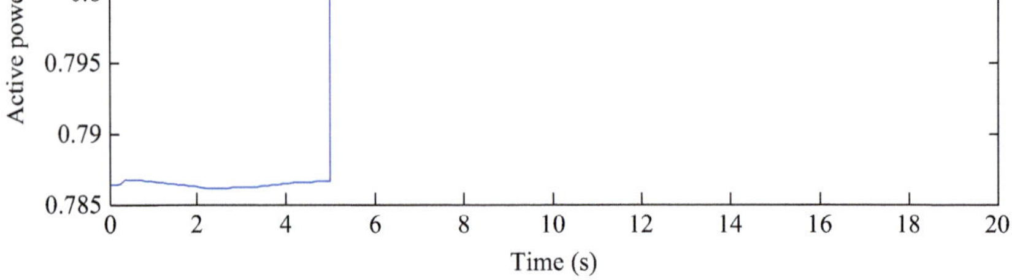

Figure 10.34 Real power of the synchronous generators after losing the DFIG wind turbine at t = 5 s

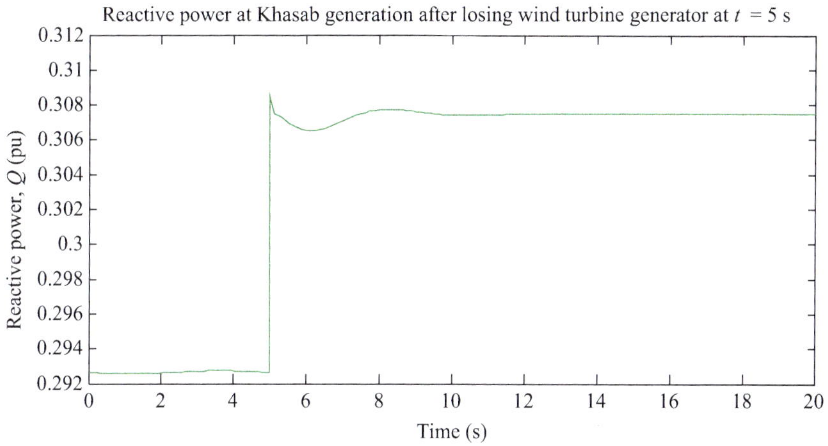

Figure 10.35 Reactive power of the synchronous generators after losing the DFIG wind turbine at $t = 5$ s

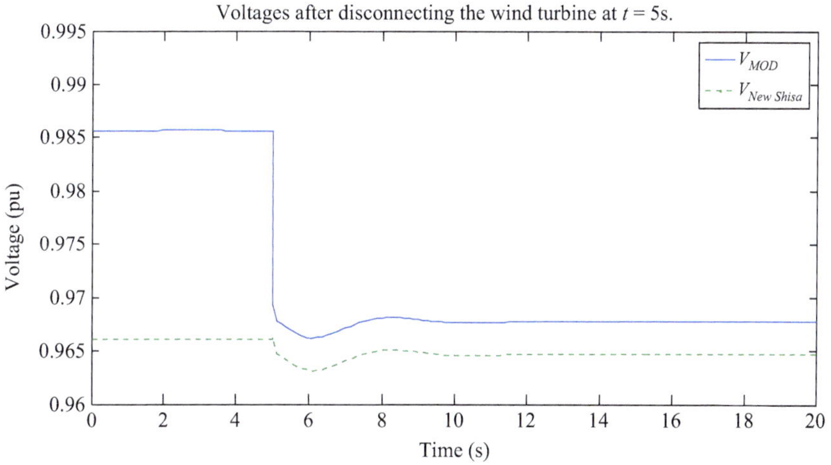

Figure 10.36 Voltages of MOD and New Shisa following disconnection of wind turbine

losing the wind turbine at $t = 5$ s. The real power jumped from 0.786 to around 0.80 pu, while the reactive power changed from 0.293 to 0.307 pu. The synchronous generators tried to recover the shortage in the power happened due to this event.

Figure 10.36 shows that MOD voltage was supported by the DFIG and it drops from 0.986 to around 0.968 pu due to the loss of the DFIG. On the other hand, since New Shisa is located very far from the DFIG, the effect of disconnection is much smaller.

The frequency, however, in Figure 10.37 becomes unstable for about 14 s and finally was stabilized close to 49.6 Hz. Since the frequency f in proportional to rotor speed n, the synchronous generators try to recover the electrical power and equalize it to the mechanical one, the speed for a while will decrease causing the frequency to do so.

Three-phase short circuit

In this scenario, three-phase short circuit fault was applied to two different locations one at a time; one is in the synchronous generators bus and the second in the DFIG wind turbine generator bus.

The amount of the fault applied was 25 kA based on the maximum fault current that can a typical 33 kV switchgear withstand (as per Oman Electrical Standard). To find out this, the following calculation was performed:

$$Since\ S_{base} = 100\ \text{MV A},\ V_{base} = 33\ \text{kV},\ then$$

$$I_{base} = 1{,}010\ \text{A},\ hence,\ I_{sc(\text{pu})} = 24.75\ \text{pu}$$

$$Then,\ Z_{sc(\text{pu})} = \frac{1}{24.75} = 0.04\ \text{pu}$$

Hence, amount of 0.04 pu short circuit impedance was applied to the test system for duration of 0.2 s.

Figures 10.38 and 10.39 show how the system is affected by the three-phase short circuit. It is observed that when the fault occurred at DFIG bus, the voltages have some oscillations. This is supported also by Figure 10.40 which represents the frequency response during the fault. Since the frequency is directly proportional to the speed of the synchronous generators, then it can be seen that the speed drops

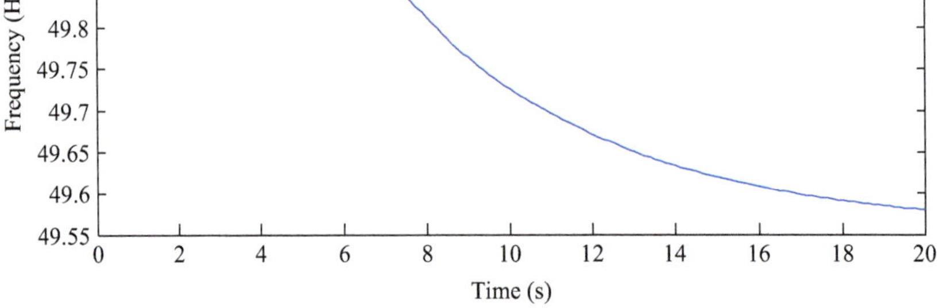

Figure 10.37 System frequency following disconnection of wind turbine

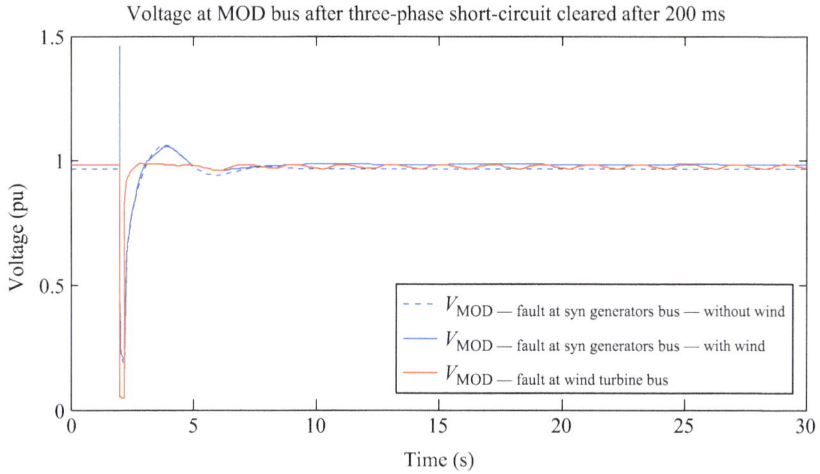

Figure 10.38 MOD voltage response after three-phase short circuit

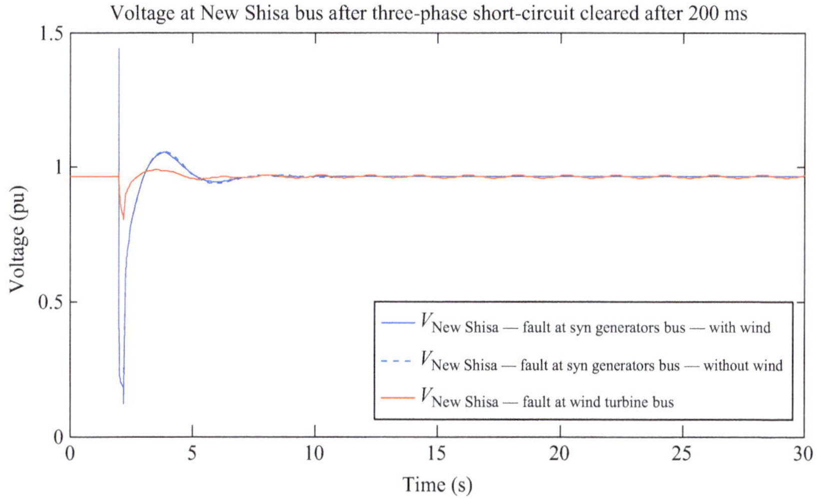

Figure 10.39 New Shisa voltage response after three-phase short circuit

down (black dashed dot line) and starts to oscillate at 49.2 Hz, and this may enable the protection scheme in the generator to operate causing tripping the system.

Figure 10.40 shows that the worst effect happens when the short circuit occurs at the DFIG bus while the wind turbine is in operation, although the system could be stable if the fault had occurred at the synchronous generator bus. In addition, if the fault occurred at the DFIG bus (but no DFIG exists, red dashed line), the system still behaves as if the fault occurred at synchronous generators terminals.

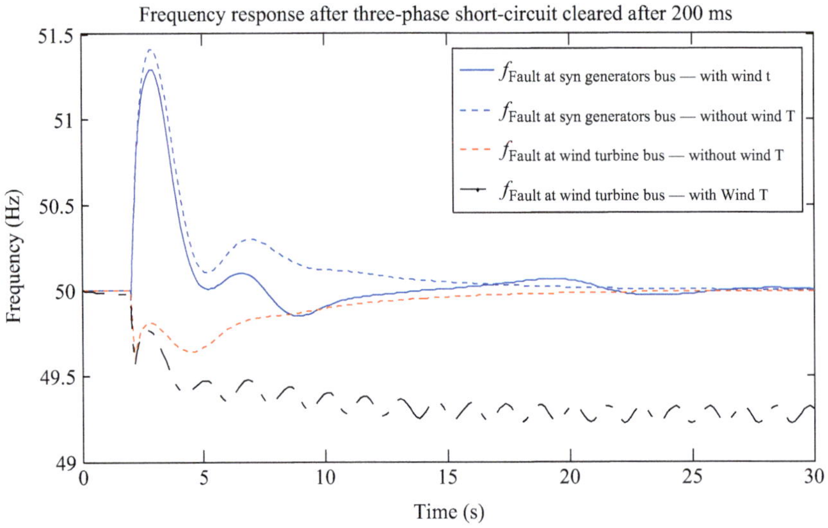

Figure 10.40 Frequency response after three-phase short circuit

It can be concluded then that in the case of DFIG fault, there is no voltage support or even reactive power supply which prevents to build up the voltage or the frequency after the fault, unlike synchronous generators [62]. The DFIG has negligible inertia during the transient disturbance of the grid because the dynamic behavior is decoupled by the power electronic devices. However, the inertia can be counted in some cases if proper controllers are added to the system [63,64].

By incrementing the fault clearing time and repeating the dynamic simulation until it halts, it was found out that the critical clearing time for this fault is about 395 ms. However, the protection scheme of the generators will operate before reaching this time, since in Figure 10.40, the frequency raised up to 51.4 Hz in case of the fault at the synchronous generator terminals.

Effect of DFIG power factor during three-phase short circuit

In this section, the power factor of the modeled DFIG will be varied and the voltage response to a fault disturbance will be examined. Figures 10.41 and 10.42 show the voltage response after the occurrence of a three-phase short circuit for two different power factors: 0.8 lagging (generator convention) and unity power factor. It should be noted that with the generator convention, lagging power factor means the generator is supplying reactive power. Figure 10.43 is the same as Figure 10.41 but zoomed in to show voltage oscillations more clearly.

It is interesting to note that for 0.8 lagging power factor, the voltage is more stable and it returned to the original operation point. The figure also shows that the situation got worst when the power factor is unity, going through prolonged voltage oscillations. This fact is also observed in the system frequency shown in Figure 10.44. It can be seen that the damping for lagging case (generator convention) is much better than that for the unity power factor.

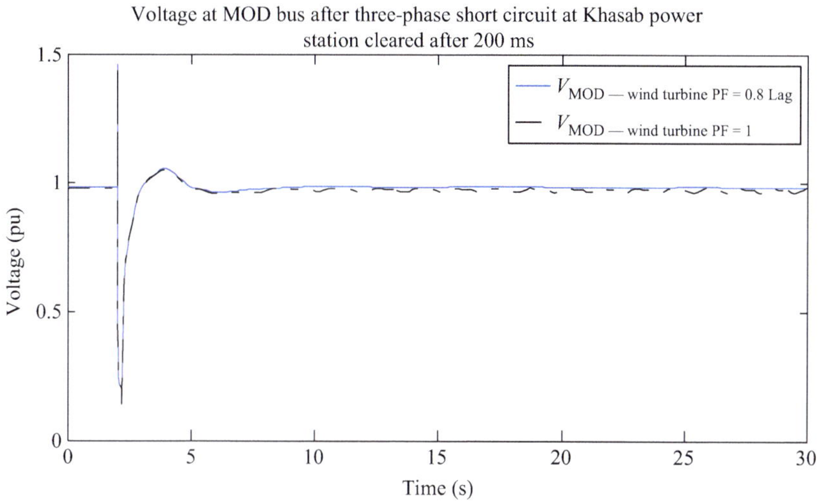

Figure 10.41 MOD voltage after three-phase SC for two DFIG Power Factors, 0.8 lagging and unity

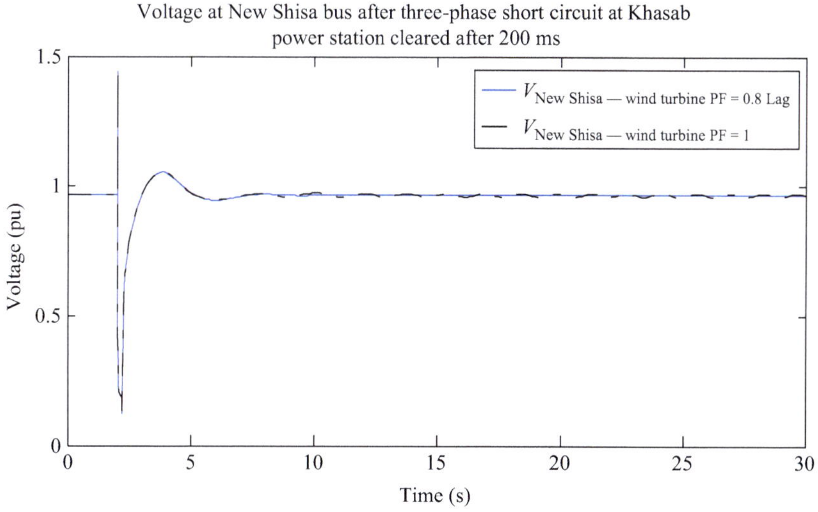

Figure 10.42 New Shisa voltage after three-phase SC for different DFIG power factors

This behavior of the DFIG comes from the fact that when DFIG wind turbine is connected to the system with unity power factor, the loading margins as well as the static damping will be worst compared to that for lagging power factor where the voltage is maintained and the reactive power is produced by the DFIG [65].

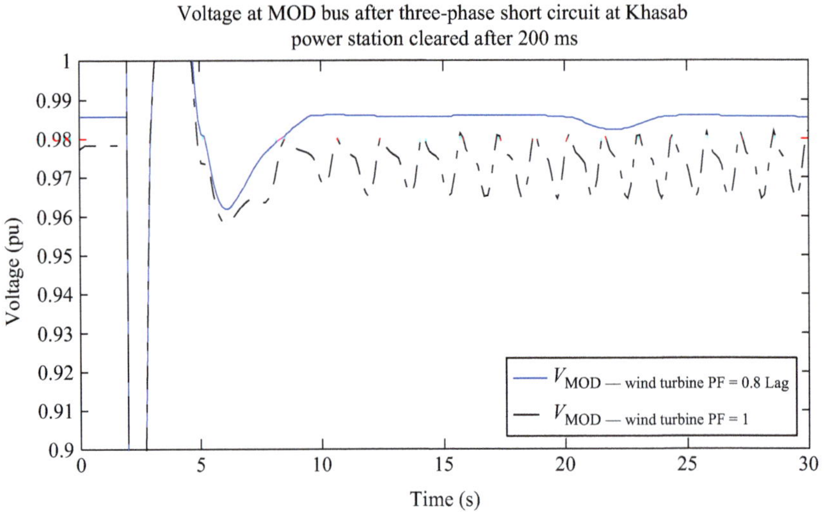

Figure 10.43 MOD voltage after three-phase SC for different DFIG power factors—magnified

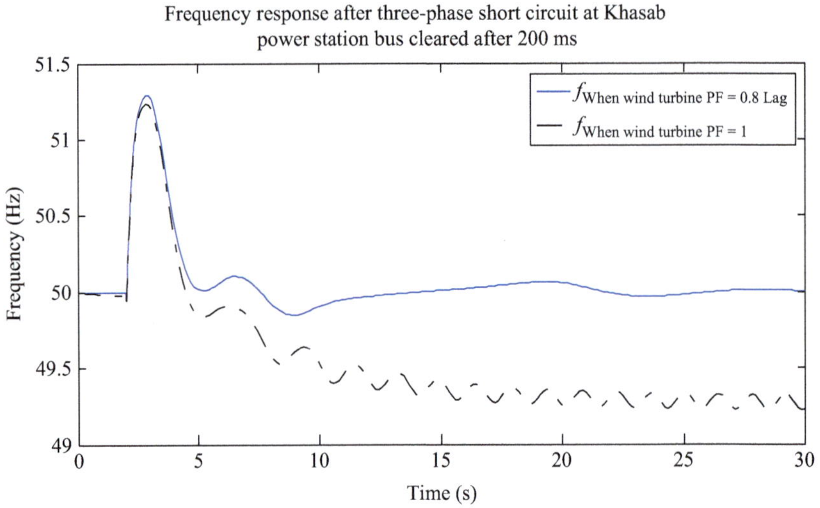

Figure 10.44 System frequency after three-phase SC for different DFIG power factors

10.4.4.2 Conclusion and recommendations for case study 4

In this research, the voltage stability of a microgrid power system is presented. The study utilized well-defined techniques to evaluate the voltage stability of a selected microgrid. *PV* and *QV* curves are generated, and the modal analysis technique is used to identify the weakest node in the systems. Musandam Power System is an isolated microgrid going under expansion plans. The loadability of the system buses and the weakest bus has been identified. Such results are very important while considering the network expansion and its future operation. The results have been validated via time-domain simulations to estimate the system behavior under different types of disturbances. The voltage stability analysis was evaluated using MATLAB load flow program and ETAP while the time-domain simulation was carried out using the powertrain system analysis toolkit (PSAT) program. It was found that the system would remain stable under some types of disturbances with short fault clearing time and certain amount of reactive power that could be suddenly lost.

It was seen that the *PV* and *QV* curves method is not enough to make serious decisions about the voltage stability of the system. This is also applicable to the modal analysis method. Time domain simulation provides a powerful support to those techniques in order to make the analysis more realistic. In this research, the time domain simulation was performed to analyze the system behaviors under certain types of disturbances. In addition, the effect of DFIG power factor during disturbance was analyzed. It was found that the system was more stable in the lagging power factor case, generator convention.

As mentioned in the literature, many studies prove the advantages of using DFIG in the micro-grids to improve the stability in general as well as the voltage stability. In this research, this was well proven, and it is recommended for such system taking in consideration other factors such as social and economic.

10.5 Concluding remarks

The most appropriate method of voltage stability analysis for microgrids is small-signal modeling and analysis. This was demonstrated by studying several case studies. Some of these case studies were borrowed from the literature, and a specific case study was presented from Oman based on an earlier study carried out by the authors.

References

[1] IEEE/CIGRE Joint Task Force on Stability Terms and Definitions, P. Kundur, J. Paserba, V. Ajjarapu, *et al.* "Definition and classification of power system stability," IEEE Trans. Power Syst., vol. 19, no. 2, pp. 1387–1401, 2004.

[2] T. van Custem, and C. Vournas. Voltage Stability of Electric Power Systems. Springer, 1998.

[3] S. M. Amelian, and R. Hooshmand. Small signal stability analysis of microgrids considering comprehensive load models – a sensitivity based approach. In: Proceedings of the Smart Grid Conference, December 2013. p. 143–149.

[4] N. Jayawarna, X. Wu, Y. Zhang, N. Jenkins, and M. Barnes, Stability of a microgrid. In: IET International Conference, Power Electronics, Machines and Drives, 2006. p. 316–320.

[5] R. Majumder. "Some aspects of stability in microgrids," IEEE Trans. Power Syst., vol. 28, no. 3, pp. 3243–3252, 2013.

[6] Z. Shuai, Y. Sun, Z. J. Shen, *et al.* "Microgrid stability: classification and a review," Renew. Sustain. Energy Rev., vol. 58, pp. 167–179, 2016.

[7] N. Jayawarna, X. Wut, Y. Zhangt, N. Jenkins, and M. Barnes. Stability of a microgrid. In: Proceedings of the International Conference on Power Electronics, Machines and Drives, April 2006. p. 316–320.

[8] Z. Zeng, H. Yang, and R. Zhao. "Study on small signal stability of microgrids: a review and a new approach," Renew. Sustain. Energy Rev., vol. 15, pp. 4818–4828, 2011.

[9] E. E. A. Coelho, P. C. Cortizo, and P. F. D. Garcia. Small signal stability for single-phase Inverter connected to stiff ac system. In: Proceedings of the IEEE Industry Applications Conference, October 1999, p. 2180–2187.

[10] E. A. A. Coelho, P. C. Cortizo, and P. F. D. Garcia. "Small-signal stability for parallel-connected inverters in stand-alone ac supply systems," IEEE Trans. Ind. Appl., vol. 38, pp. 533–542, 2002.

[11] N. Pogaku, M. Prodanovic, and T. C. Green. "Modeling, analysis and testing of autonomous operation of an inverter-based microgrid," IEEE Trans. Power Electron., vol. 22, pp. 613–625, 2007.

[12] S. S. Acevedo, and M. Molinas. Power electronics modeling fidelity: impact on stability estimate of micro-grid systems. In: Proceedings of the IEEE Innovative Smart Grid Technologies Asia, 2011. p. 1–8.

[13] L. Herrera, E. Inoa, F. Guo, J. Wang, and H. Tang. "Small-signal modeling and net-worked control of a PHEV charging facility," IEEE Trans. Ind. Appl., vol. 50, pp. 1121–1130, 2014.

[14] M. Zhu, and H. Li. Improved state-space model and analysis of islanding inverter-based microgrid. In: Proceedings of the IEEE International Symposium on Industrial Electronics, May 2013. p. 1–5.

[15] Y. Wang, Z. Lu, Y. Min, and Z. Wang. Small signal analysis of microgrid with multiple micro-sources based on reduced order model in islanding operation. In: Proceedings of the IEEE Power and Energy Society General Meeting, July 2011. p. 1–9.

[16] A. Elrayyah, Y. Sozer, and M. Elbuluk. Simplified modeling procedure for inverter-based islanded microgrid. In: Proceedings of the IEEE EnergyTech, 2012. p. 1–6.

[17] X. Tang, W. Deng, and Z. Qi. "Investigation of the dynamic stability of microgrid," IEEE Trans. Power Syst., vol. 29, pp. 698–706, 2014.

[18] J. C. Vasquez, M. Guerrero Josep, A. Luna, P. Rodríguez, and R. Teodorescu. "Adaptive droop control applied to voltage-source inverters operating in grid-connected and islanded modes," IEEE Trans. Ind. Electron., vol. 56, pp. 4088–4096, 2009.

[19] C. Marinescu, and I. Serban. Analysis of frequency stability in a residential autonomous microgrid based on a wind turbine and a microhydro power plant. In: Proceedings of the IEEE Power Electronics and Machines in Wind Applications, 2009. p. 1–5.

[20] C. N. Rowe, T. J. Summers, R. E. Betz, and D. J. Cornforth. Small signal stability analysis of arctan power frequency droop. In: Proceedings of the IEEE Ninth International Conference on Power Electronics and Drive Systems, December 2011. p. 5–8.

[21] S. Tabatabaee, H. R. Karshenas, A. Bakhshai, and P. Jain. Investigation of droop characteristics and *X*/*R* ratio on small-signal stability of autonomous microgrid. In: Proceedings of the 2nd Power Electronics, Drive Systems and Technologies Conference, February 2011. p. 223–228.

[22] E. Barklund, N. Pogaku, M. Prodanovic, C. Hernandez-Aramburo, and T. C. Green. "Energy management in autonomous microgrid using stability-constrained droop control of inverters," IEEE Trans. Power Electron., vol. 23, pp. 2346–2352, 2008.

[23] D. P. Ariyasinghe, and D. Vilathgamuwa. Stability analysis of microgrids with constant power loads. In: Proceedings of the IEEE International Conference on Sustainable Energy Technologies, November 2008. p. 279–284.

[24] A. Kahrobaeian, and Y. A. R. I. Mohamed. Stability analysis and control of medium-voltage micro-grids with dynamic loads. In: Proceedings of the IEEE Power and Energy Society General Meeting, July 2013. p. 1–5.

[25] A. K. Abbasi, M. W. B. Muatafa, and A. S. B. Mokhtar. Small signal stability analysis of rectifier-inverter fed induction motor drive for microgrid applications. In: Proceedings of the IEEE Region 10 Conference, November 2011. p. 1015–1019.

[26] G. Diaz, C. Gonzalez-Moran, J. G. Aleixandre, and A. Diez. "Composite loads in stand-alone inverter-based microgrids – modeling procedure and effects on load margin," IEEE Trans. Power Syst., vol. 25, pp. 894–905, 2010.

[27] P. Arboleya, G. Diaz, and M. Coto. "Unified AC/DC power flow for traction systems: a new concept," IEEE Trans. Veh. Technol., vol. 61, pp. 2421–2430, 2012.

[28] A. Kahrobaeian, and I. Abdel-Rady. "Analysis and mitigation of low-frequency instabilities in autonomous medium-voltage converter-based microgrids with dynamic loads," IEEE Trans. Ind. Electron., vol. 61, pp. 643–658, 2014.

[29] C. A. Cañizares. Voltage Stability Assessment: Concepts, Practices and Tools. IEEE/FES Power System Stability Subcommittee, August 2002.

[30] P. Kundur, N. J. Balu, and M. G. Lauby. Power System Stability and Control. New York: McGraw-Hill Professional, 1994.

[31] J. Machowski, J. W. Bialek, and J. R. Bumby. Power System Dynamics: Stability and Control, Second ed. Chichester: John Wiley & Sons, 2008.

[32] J. G. Slootweg, and W. L. Kling. Impacts of distributed generation on power system transient stability. In: Proc. IEEE PES Summer Meeting, Vol. 2, 2002. p. 862–867.

[33] W. Freitas, L. C. P. DaSilva, and A. Morelato. "Small-disturbance voltage stability of distribution systems with induction generators," IEEE Trans. Power Syst., vol. 20, pp. 1653–1654, 2005.

[34] W. Freitas, J. C. M. Vieira, L. C. P. da Suva, C. M. Affonso, and A. Morelato. Long-term voltage stability of distribution systems with induction generators. In: Proc. IEEE PES General Meeting, San Francisco, CA, Vol. 3, 2005. p. 2910–2913.

[35] N. Jenkins, R. Allan, P. Crossley, D. Kirschen, and G. Strbac. Embedded Generation. IET, 2000.
[36] W. Freitas, A. Morelato, X. Wilsun, and F. Sato. "Impacts of AC generators and DSTATCOM devices on the dynamic performance of distribution systems," IEEE Trans. Power Delivery, vol. 20, pp. 1493–1501, 2005.
[37] R. R. Londero, C. M. Affonso, and M. V. A. Nunes. Impact of distributed generation in steady state, voltage and transient stability—Real case. In: Proc. IEEE Power Tech, in Bucharest, 2009. p. 1–6.
[38] S. Kotamarty, S. Khushalani, and N. Schulz. "Impact of distributed generation on distribution contingency analysis," Electr. Power Syst. Res., vol. 78, pp. 1537–1545, 2008.
[39] N. G. A. Hemdan, and M. Kurrat. Distributed generation location and capacity effect on voltage stability of distribution networks. In: Proc. IEEE Annual Conference, 2008. p. 1–5.
[40] H. Hedayati, S. A. Nabaviniaki, and A. Akbarimajd. "A method for placement of DG units in distribution networks," IEEE Trans. Power Delivery, vol. 23, pp. 1620–1628, 2008.
[41] M. Alonso, and H. Amaris. Voltage stability in distribution networks with DG. In: Proc. IEEE PowerTech, in Bucharest, 2009. p. 1–6.
[42] N. L. Soultanis, S. A. Papathanasiou, and N. D. Hatziargyriou. "A stability algorithm for the dynamic analysis of inverter dominated unbalanced LV microgrids," IEEE Trans. Power Syst., vol. 22, pp. 294–304, 2007.
[43] N. Pogaku, M. Prodanovic, and T. C. Green. "Modeling, analysis and testing of autonomous operation of an inverter-based microgrid," IEEE Trans. Power Electron., vol. 22, pp. 613–625, 2007.
[44] F. Katiraei, M. R. Iravani, and P. W. Lehn. "Small-signal dynamic model of a micro-grid including conventional and electronically interfaced distributed resources," IET Gener. Transm. Distrib., vol. 31, pp. 369–378, 2007.
[45] F. Katiraei, and M. R. Iravani. "Power management strategies for a microgrid with multiple distributed generation units," IEEE Trans. Power Syst., vol. 21, pp. 1821–1831, 2006.
[46] S. Khanabdal, M. Banejad, F. Blaabjerg, and N. Hosseinzadeh. Virtual flux droop control with constant switching frequency for power sharing between parallel inverters in islanded microgrids. In: Proceedings of the Iranian Conference on Electrical Engineering, April–May 2019.
[47] Y. Li, D. M. Vilathgamuwa, and P. C. Loh. "Design, analysis, and real-time testing of a controller for multibus microgrid system," IEEE Trans. Power Electron., vol. 19, no. 5, pp. 1195–1204, 2004.
[48] C. K. Sao, and P. W. Lehn. "Control and power management of converter fed microgrids," IEEE Trans. Power Syst., vol. 23, no. 3, pp. 1088–1098, 2008.
[49] S. Leitner, M. Yazdanian, A. Mehrizi-Sani, and A. Muetze. "Small-signal stability analysis of an inverter-based microgrid with internal model-based controllers," IEEE Trans. Smart Grid, vol. 9, no. 5, 5393–5402, 2018.
[50] K. Yu, Q. Ai, S. Wang, J. Ni, and T. Lv. "Analysis and optimization of droop controller for microgrid system based on small-signal dynamic model," IEEE Trans. Smart Grid, vol. 7, no. 2, pp. 695–705, 2016.

[51] A. G. Tsikalakis, and N. D. Hatziargyriou. "Centralized control for optimizing microgrids operation," IEEE Trans. Energy Convers., vol. 23, no. 1, pp. 241–248, 2008.
[52] J. Chen, and J. Chen. "Stability analysis and parameters optimization of islanded microgrid with both ideal and dynamic constant power loads," IEEE Trans. Ind. Electron., vol. 65, no. 4, pp. 3263–3274, 2018.
[53] F. Katiraei. "Dynamic analysis and control of distributed energy resources in a micro-grid." Ph.D. dissertation, University of Toronto, Canada, 2005.
[54] M. H. Albadi. "Electricity sector in Oman after 10 years of reform: status, trends, and future perspectives," Electr. J., vol. 30, no. 7, pp. 23–30, 2017. DOI: 10.1016/j.tej.2017.07.005.
[55] AER Annual Report, Authority for Electricity Regulation (AER), Oman, 2017, accessed in May 2019: www.aer.om/pdfs/AERO_Annual_Report_2017_Eng.pdf.
[56] RAECO Annual report 2017.
[57] S. Acharya, M. S. E. Moursi, and A. Al-Hinai. "Coordinated frequency control strategy for an islanded microgrid with demand side management capability," IEEE Trans. Energy Convers., vol. 33, no. 2, pp. 639–651, 2018.
[58] S. Acharya, M. S. E. Moursi, A. A. Hinai, and M. Albadi. Coordinated control for frequency regulation in microgrids using RES, battery storage and demand response. In: 2017 IEEE Power & Energy Society General Meeting, Chicago, IL, 2017. p. 1–5.
[59] E. B. Ssekulima, and A. A. Hinai. Coordinated voltage control of solar PV with MPPT and battery storage in grid-connected and microgrid modes. In: 2016 18th Mediterranean Electrotechnical Conference (MELECON), Lemesos, 2016. p. 1–6.
[60] S. Acharya, M. S. El Moursi, A. Al-Hinai, A. S. Al-Sumaiti, and H. Zeineldin. "A control strategy for voltage unbalance mitigation in an islanded microgrid considering demand side management capability," IEEE Trans. Smart Grid. DOI: 10.1109/TSG.2018.2804954.
[61] Y. Al Jabri, N. Hosseinzadeh, R. Al Abri, and A. Al Hinai. Voltage stability assessment of a microgrid. In: 2015 IEEE 8th GCC Conference & Exhibition, Muscat, 2015. p. 1–6.
[62] M. Rizwan, N. Kumar, A. H. Quadri, and H. Ahuja. "Transient stability analysis of distribution system with DFIG based wind penetration," Special Issue of Int. J. Sustain. Dev. Green Econ. (IJSDGE), vol. V, no. 1, pp. 2315–4721, 2013.
[63] M. Kayikci, and J.V. Milanovic. "Dynamic contribution of DFIG-based wind plants to system frequency disturbances," IEEE Trans. Power Syst., vol. 24, no. 2, pp. 859–867, 2009.
[64] L. Meegahapola, D. Flynn, J. Kennedy, and T. Littler. Impact of Wind Generation Mix on Lasantha Transient Stability for an Interconnected Power System. Power & Energy Research Centre, The Queen's University of Belfast, United Kingdom, 2009.
[65] J. Muñoz, and C. Cañizares. Comparative stability analysis of DFIG-based wind farms and conventional synchronous generators. In: Power Systems Conference and Exposition (PSCE), 2011 IEEE/PES. IEEE, 2011.

Further reading

- C. Hsiao-Dong, and R. Jean-Jumeau. "Toward a practical performance index for predicting voltage collapse in electric power systems," IEEE Trans. Power Syst., vol. 10, pp. 584–592, 1995.
- L. L. Grigsby, The Electric Power Engineering Handbook. Boca Raton, FL: CRC Press/IEEE Press, 2001.
- R. B. Prada, and L. J. Souza. Voltage stability and thermal limit: constraints on the maximum loading of electrical energy distribution feeders. In: Proc. IEEE Generation, Transmission and Distribution, Vol. 145, 1998. pp. 573–577.
- C. Haiyan, C. Jinfu, S. Dongyuan, and D. Xianzhong. Power flow study and voltage stability analysis for distribution systems with distributed generation. In: Proc. IEEE PES General Meeting, 2006. p. 8.
- Z. Ruimin, M. H. J. Bollen, and Z. Jin. Harmonic resonances due to a grid-connected wind farm. In: Proc. International Conference in Harmonics and Quality of Power, 2010. p. 1–7.
- G. J. Wakileh. Power Systems Harmonics: Fundamentals, Analysis, and Filter Design. Springer, Berlin, New York, 2001.
- J. J. Mesas, and L. Sainz. "Stochastic assessment of distribution system resonance frequencies with capacitors or shunt filters," Electr. Power Syst. Res., vol. 81, pp. 35–42, 2010.
- T. E. Grebe. "Application of distribution system capacitor banks and their impact on power quality," IEEE Trans. Ind. Appl., vol. 32, pp. 714–719, 1996.
- S. A. Papathanassiou, and M. P. Papadopoulos. "Harmonic analysis in a power system with wind generation," IEEE Trans. Power Delivery, vol. 21, pp. 2006–2016, 2006.
- D. Patel, R. K. Varma, R. Seethapathy, and M. Dang. Impact of wind turbine generators on network resonance and harmonic distortion. In: Proc. Canadian Conference, Electrical and Computer Engineering, 2010. pp. 1–6.
- R. A. Walling, R. Saint, R. C. Dugan, J. Burke, and L. A. Kojovic. "Summary of distributed resources impact on power delivery systems," IEEE Trans. Power Delivery, vol. 23, pp. 1636–1644, 2008.
- H. H. Zeineldin, E. F. El-Saadany, and M. M. A. Salama. Distributed generation micro-grid operation: control and protection. In: Power Systems Conference: Advanced Metering, Protection, Control, Communication, and Distributed Resources, 2006. p. 105–111.
- T. Ackermann. Wind Power in Power Systems. John Wiley, Chichester, West Sussex, England, Hoboken NJ, 2005.
- Y. M. Atwa, E. F. El-Saadany, M. M. A. Salama, and R. Seethapathy. "Optimal renewable resources mix for distribution system energy loss minimization," IEEE Trans. Power Syst., vol. 25, pp. 360–370, 2010.
- C. Grigg, P. Wong, P. Albrecht, *et al.* "The IEEE reliability test system-1996. A report prepared by the reliability test system task force of the

application of probability methods subcommittee," IEEE Trans. Power Syst., vol. 14, pp. 1010–1020, 1999.

- Y. M. Atwa, E. F. El-Saadany, M. Salama, and R. Seethapathy. Distribution system loss minimization using optimal DG mix. In: Proc. IEEE PES General Meeting, 2009.
- Z. M. Salameh, B. S. Borowy, and A. R. A. Amin. "Photovoltaic module-site matching based on the capacity factors," IEEE Trans. Energy Convers., vol. 10, pp. 326–332, 1995.
- M. H. Albadi, and E. F. El-Saadany. Novel method for estimating the CF of variable speed wind turbines. In: Proc. IEEE PES General Meeting, 2009.
- M. H. Albadi, and E. F. El-Saadany. "Wind turbines capacity factor modeling- a novel Approach," IEEE Trans. Power Syst., vol. 24, pp. 1637–1638, 2009.
- A. M. Borbely, and J. F. Kreider. Distributed Generation: The Power Paradigm For The New Millennium. CRC Press, Boca Raton, 2001.
- P. Chiradeja, and R. Ramakumar. "An approach to quantify the technical benefits of distributed generation," IEEE Trans. Energy Convers., vol. 19, pp. 764–773, 2004.
- J. M. Uudrill. "Dynamic stability calculations for an arbitrary number of interconnected synchronous machines," IEEE Trans. Power Appl. Syst., vol. 87, pp. 835–844, 1968.
- M. Prodanovi. "Power quality and control aspects of parallel connected inverters in distributed generation." Ph.D. dissertation, University of London, UK, 2004.
- Y. A.-R. I. Mohamed. "New control algorithms for the distributed generation interface in grid-connected and micro-grid systems." Ph.D. dissertation, University of Waterloo, Canada, 2009.
- K. Jalili, and S. Bernet. "Design of LCL filters of active-front-end two-level voltage-source converters," IEEE Trans. Ind. Electron., vol. 56, pp. 1674–1689, 2009.
- J. H. R. Enslin, W. T. J. Hulshorst, A. M. S. Atmadji, *et al.* Harmonic interaction between large numbers of photovoltaic inverters and the distribution network. In: Proc. IEEE Power Tech Conference, in Bologna, Vol. 3, 2003. p. 6.
- A. Xin, L. Zhili, and C. Mingyong. A novel assessment method for harmonic environment in microgrid. In: Proc. International Conference, Critical Infrastructure, 2010. p. 1–7.
- W. Xu. Status and future directions of power system harmonic analysis. In: Proc. IEEE PES General Meeting, Vol. 2, 2003. p. 1184.
- Y. Tang, and A. A. Mahmoud. "Evaluation and reduction of harmonic distortion in power systems," Electr. Power Syst. Res., vol. 17, pp. 41–48, 1989.
- Y. A. R. I. Mohamed. "Mitigation of converter-grid resonance, grid-induced distortion, and parametric instabilities in converter-based distributed generation," IEEE Trans. Power Electron., vol. 26, pp. 983–996, 2011.

- IEEE Std 399-1997. "IEEE Recommended Practice for Industrial and Commercial Power Systems Analysis," 1998.
- M. Reza, P. H. Schavemaker, J. G. Slootweg, W. L. A. K. W. L. Kling, and L. A. v. d. S. L. van der Sluis. Impacts of distributed generation penetration levels on power systems transient stability. In: Proc. IEEE PES General Meeting, Vol. 2, 2004. p. 2150–2155.
- C. A. Canizares, and F. L. Alvarado. "Point of collapse and continuation methods for large AC/DC systems," IEEE Trans. Power Syst., vol. 8, pp. 1–8, 1993.

Chapter 11
Frequency stability and synthetic inertia

Nasim Ullah[1], *Anwar Ali*[2], *Haider Ali*[3], *and Khalid Mahmood*[2]

Due to low inertia, stochastic nature of renewable energy sources (RESs) and sudden load changes, nearly all the modern microgrids are associated with the dynamic frequency stability issues. Thus it restricts the maximum number of the renewable energy systems that can be penetrated to the microgrid. In order to increase the penetration of low-inertia sources to the microgrid, the frequency stability issues need to be addressed. The frequency stability issues of a typical microgrid are addressed by the addition of extra inertial support from the power sources using power converters and appropriate control loop. In this chapter, the readers are introduced with the basic concepts of the synthetic inertia support for the dynamic frequency stability issues of a hydro-photovoltaic (PV) microgrid. The details of the frequency–power-response-based synthetic inertial support and current control loops are elaborated using a simulation example. The deviation in system's frequency is usually compensated by sourcing or absorbing the active power by the inertial support loop, so by utilizing the concept of inertial loop, the maximum number of RESs integration can be enhanced.

11.1 Frequency stability issues of microgrid

In order to fulfill the future clean energy needs of the world's population, the RESs such as PV and wind energy are replacing the traditional generation sources, such as thermal and nuclear [1,2]. According to a survey report, the global penetrations of RESs have exceeded 600 GW by the end of 2015 [3]. A microgrid optimally handles the deployment of power from the distributed energy resources to the community centers including both in the metropolitan and rural areas. A microgrid can be operated in grid-connected or islanding mode depending on the requirement of the community. The grid-tied microgrids offer several advantages such as the

[1]Department of Electrical Engineering, CECOS University of IT and Emerging Sciences, Peshawar, Pakistan
[2]Department of Electrical Technology, University of Technology, Nowshera, Pakistan
[3]Department of Electronics Technology, University of Technology, Nowshera, Pakistan

injection of extra power capacity back to the grid and take back power from the stiff grids. Thus the main grids can act as a backup source for the microgrids, and the requirement for additional storage devices for the microgrid is minimized. All the RESs are integrated to the power network through power electronics converters, so practically, these sources do not contribute to the inertial behavior of the power network [4]. The frequency stability of a microgrid greatly depends on the balance between the active power consumed and active power sourced by different RESs. In conventional generation, the synchronous generators (SGs) with rotating parts can absorb or release kinetic energy during the imbalance of power consumed and sourced, thus it offers resistance to the deviation of system frequency. This property of the SGs is called moment of inertia [5]. The reliable operation of the microgrids is a major challenge, and it is an emerging research topic. The reliability of the microgrids with high number of RESs penetration is threatened by the stochastic nature of the RESs and load fluctuations, and it is the major cause of frequency and voltage instabilities in such power networks [6]. With low inertial behavior and the intermittent changes in RESs sources and load, the rate of change of frequency (ROCOF) is high, and as a result, frequency relays may trip. The response of the primary loop (governors) is not fast enough to stabilize the system's frequency, so an additional control loop is required with fast response [7]. The additional control loop (inertial loop) detects the imbalance of active power and is responsible for adding or absorbing the active power from the power network through power converter interface [8]. Several methods have been proposed in the literature for the frequency stabilization of microgrids. In order to have sufficient energy support for the inertial loop, an algorithm that operated the PV system below its maximum power point tracking (MPPT) has been reported in the literature [9]. The reserved energy was proposed to be used and fed to the microgrid during frequency deviations. The frequency response was improved at the cost of the poor efficiency of the PV system. In order to ensure both the efficiency and the inertial support for the PV system, battery energy storage system (ESS) (BESS) has been proposed in [10]. The integration of BESS ensures the operation of PV system at MPPT, and the additional stored energy was used as a back for the inertial support. The disadvantage of this method is the increased cost and large size of the BESS system. A method based on the ROCOF detection is proposed in [11]. In this method, the additional reference power is calculated based on the ROCOF and fed to the inverter to reflect the additional inertial part. In [12], the authors reported a synchronverter-based inertial control loop for the inverter to minimize the frequency deviations caused by the sudden changes in the load or sources. Similarly, the swing equation of the SG was used to calculate the reference mechanical phase angle and inverter output voltage, and based on these, calculations were used in the damping loop of the control system to stabilize the system's frequency [13]. To damp out the frequency oscillations, the concept of virtual oscillator control for synchronization and regulating a number of islanded power electronics converters has been reported in [14]. Recently, considerable research has been reported on the application of different control schemes for the inertia control loop [15–21]. In [15], a comparative analysis is reported between the droop control method and virtual synchronous. In [16], the droop control method is proposed to emulate the synthetic inertia in the wind-energy-based microgrid. The authors in [17] proposed a novel control method for the synthetic inertia addition in the European power system. In [18], proposal is made to utilize DC microgrid as a main backsource for inertial loop. In [19], a model predictive control technique-based inertia control loop is presented to improve the frequency stability performances of microgrids. In [20], the authors have proposed the robust virtual inertia, and in [21,22], the robust tuning methods are presented to adapt the parameters of virtual inertia control.

11.2 Effect of low inertia on the frequency stability of microgrid

In this section, basic mathematical concepts are introduced to derive the relation between the power system's frequency and the inertia. The following swing equation is used to approximate the frequency deviation of a power system in event of a disturbance [23]:

$$P_g - P_L = \frac{1}{2}J\frac{d{w_s}^2}{dt} = Jw_s\frac{dw_s}{dt} \tag{11.1}$$

From (11.1), P_g represents the generated power, P_L represents the consumed power by the load including the losses, J is the total moment inertia of the system and w_s represents the angular frequency (rad/s) of the system. Let H represents the inertia constant of the power system and the expression for H is given in the following equation:

$$H = \frac{1}{2}J\frac{{w_s}^2}{S_s} \tag{11.2}$$

In (11.2) S_s represents the apparent power of the connected generator. Using (11.2), (11.1) can be rewritten as

$$\frac{2H}{w_s}\frac{dw_s}{dt} = \frac{P_g - P_L}{S_s} \tag{11.3}$$

Equation (11.3) can be represented in terms of system's frequency (Hz) as follows:

$$\frac{2H}{f_s}\frac{df_s}{dt} = \frac{P_g - P_L}{S_s} \tag{11.4}$$

where f_s represents the system's frequency in Hz and df_s/dt represents the rate of change of frequency (ROCOF). From (11.4), it is obvious that the ROCOF is inversely proportional to the system's inertia. The effect of the virtual inertia on ROCOF and nadir frequency of a power system is given in Figure 11.1. It is obvious that the power system without inertial support reaches the lowest nadir frequency and also the ROCOF is high. The maximum limit of the frequency tripping relay is 51 Hz, and the minimum limit is 48.9 Hz. In the event of the imbalance of the real power, a power system without inertial support will cross the

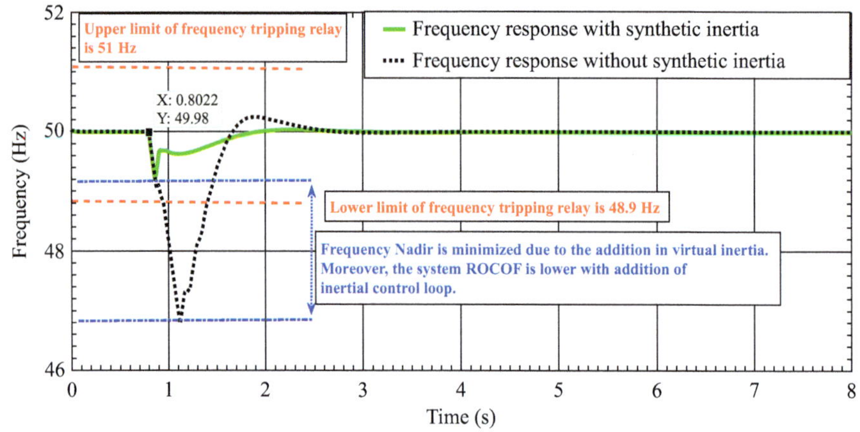

Figure 11.1 Effect of virtual inertia on ROCOF and nadir frequency of the system

lower limit, and the frequency tripping relays will be activated. In order to avoid such scenarios, the power system with additional inertial support will stay within the maximum and minim limits of the frequency as shown in Figure 11.1. In contrast, the power systems with low inertia will suffer from high ROCOF in the event of the disturbance either on source or load side. The power systems with high ROCOF will suffer from the frequency instabilities issues. Moreover, the dependency of the ROCOF on system's inertia will limit the maximum penetration of the RESs in the power grids. As discussed in the previous section, the microgrids with high penetration of RESs have low inertia due to the nonrotating behavior of these sources. To solve the issues associated with low system's inertia, frequency instability and limit imposed on maximum penetration of RESs, the concepts of synthetic or virtual inertia are introduced. To improve the stiffness of the microgrids, the virtual or synthetic inertia is virtually introduced to the microgrids using power electronics converters. The virtual or synthetic inertia concepts are used to emulate and add the inertial response to the microgrids by utilizing the combination of control algorithms, RESs, ESS and power electronics converters.

11.3 Frequency stability enhancement

This section elaborates the brief introduction of different methods reported in the literature for the frequency stabilization of the microgrids. Generally, the short-term frequency stability issue of a microgrid arises due to the insufficient response or generation reserves. The short-term frequency stability enhancement is usually achieved by adding virtual inertia to the power system through the controlled power converters and storage devices. Various techniques proposed in the literature are discussed briefly, and the details are given next.

The synthetic inertia is the combined effect of power electronics, RESs, ESSs and control algorithms [24].

RESs sources such as PV and wind turbines are integrated to the main grid through power converters, so physically, these sources add no inertia to the grid [4,25]. The inertial response is achieved by the current voltage and frequency feedback, and the power converter behaves as SG from grid point [26]. Different topologies have been reported in the literature as per different scenario and complexity. The details about different topologies are given in the following section [27].

11.3.1 Synchronous generator (SG) model-based topologies

This topology is widely utilized to replicate the dynamic response of SG. The control algorithms reported in the literature for this particular topology are independent of system's frequency and its derivative. The main disadvantage of this topology is related to its numerical instability. This topology is implemented as voltage source (https://en.wikipedia.org/wiki/Synchronverter). Figure 11.2 shows the subclassification of the SG-model-based topologies.

11.3.1.1 Synchronverters

Synchronverters utilize the inverter unit, ESS and a core control algorithm for providing the synthetic inertial support for the frequency stability of a power grid [28,29]. As shown in Figure 11.3, the synchronverter has two subunits, i.e., power unit and measurement/control unit. The power unit consists of a power converter and a filter while measurement/control unit consists of different sensors, transducers and processors [30,31]. The dynamic behavior of the synchronverter should be the same as SG. The dynamics of a SG is given as [32]:

$$T_E = Mi_f < i, \sin\alpha > \tag{11.5}$$

$$E = \dot{\alpha} i_f M \sin\alpha \tag{11.6}$$

$$Q = \dot{\alpha} i_f M < i, \cos\alpha > \tag{11.7}$$

where T_E represents the electromagnetic torque of the synchronverter, M represents the mutual inductance, i_f is the field excitation current, α is the angle, E represents

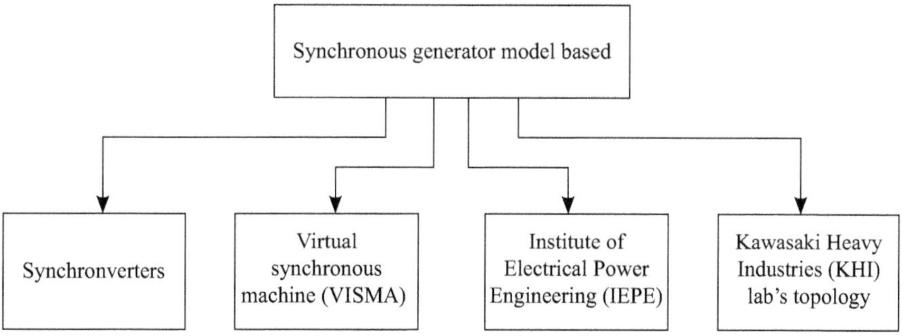

Figure 11.2 Synchronous-generator-model-based topologies

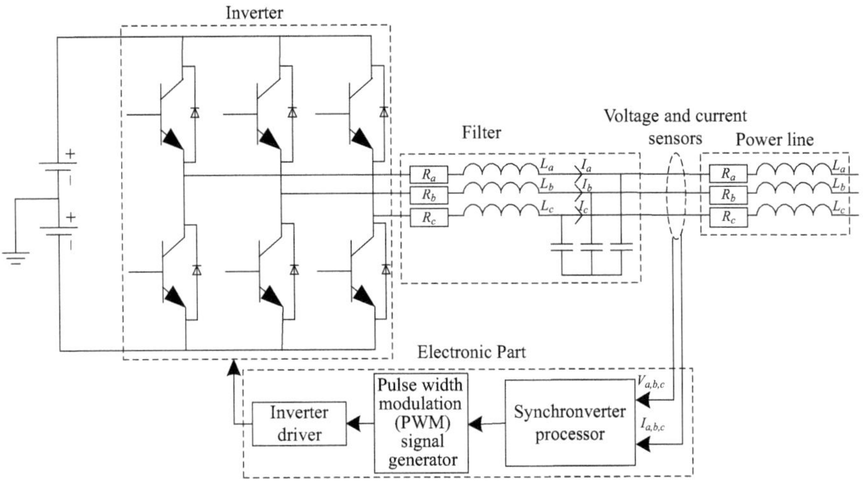

Figure 11.3 Block diagram of synchronverter topology

the no load voltage and Q is the reactive power. The three-phase stator current, i, $\sin\alpha$ and $\sin\alpha$, are defined in the following equation [32]:

$$i = \begin{bmatrix} i_a \\ i_b \\ i_c \end{bmatrix}; \ \sin\alpha = \begin{bmatrix} \sin\alpha \\ \sin(\alpha - 2\pi/3) \\ \sin(\alpha - 4\pi/3) \end{bmatrix}; \ \cos\alpha = \begin{bmatrix} \cos\alpha \\ \cos(\alpha - 2\pi/3) \\ \cos(\alpha - 4\pi/3) \end{bmatrix}$$

The above expressions given in (11.5)–(11.7) are utilized to generate the gate driver signal of the inverter unit.

11.3.1.2 Virtual synchronous machine (VISMA) topology

In the virtual synchronous machine (VISMA) topology, the power converter is embedded between a DC bus and AC power line. The dynamic response generated by VISMA is also similar the SG [33,34]. VISMA works on the basic relation of inverse proportionality between the active power and the rotor speed. Figure 11.4 shows the basic architecture of the VISMA topology.

11.3.1.3 The IEPE's topology

The IEPE's topology utilizes a simplified SG model and generates the inverter's gate signals to add inertial support to the power grid [32]. The working diagram of the IEPE's topology is shown in Figure 11.5. Similar to the VISMA topology, the inverter in the IEPE's topology is also embedded between a DC and AC power bus.

11.3.1.4 Kawasaki Heavy Industries (KHI) topology

The Kawasaki Heavy Industries (KHI) topology utilizes the phasor diagram of the SG to formulate the inertial support control loop [32]. The detailed working diagram of the KHI topology is shown in Figure 11.6.

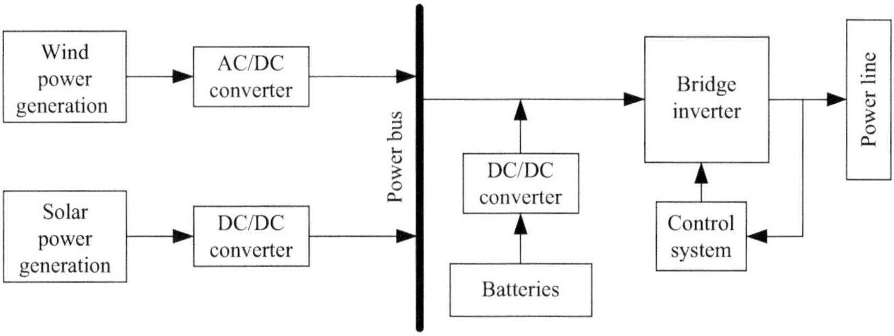

Figure 11.4 Block diagram of virtual synchronverter machine topology

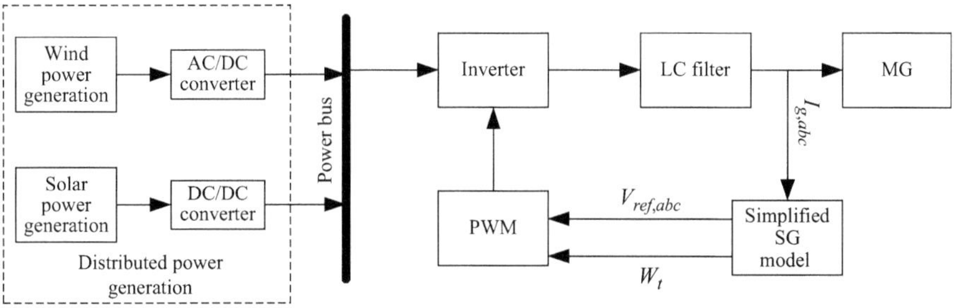

Figure 11.5 Block diagram of IEPE topology

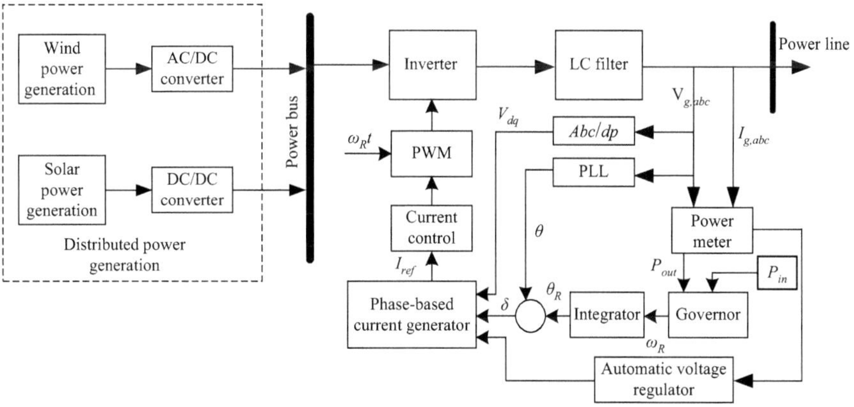

Figure 11.6 Block diagram of KHI topology

11.3.2 Swing equation based

The swing-equation-based topology utilizes relatively simple model as compared to the SG-based topologies. The swing-equation-based topology does not require the frequency derivative, and the synchronization is done using phase lock loop (PLL). A popular swing-equation-based topology is discussed in the following section [35].

11.3.2.1 ISE lab's topology

In order to emulate the inertial support through power electronics converter, this topology solves the power–frequency swing equation [32,36]. The power–frequency swing equation is given as follows:

$$P_{in} - P_{out} = J\omega_m\left(\frac{d\omega_m}{dt}\right) + D_p\Delta\omega \tag{11.8}$$

$$\Delta\omega = \omega_m - \omega_g \tag{11.9}$$

From (11.8) and (11.9), P_{in}, P_{out}, ω_m, ω_g, J and D_p represent the input power, the output power of the inverter, virtual angular frequency, grid angular frequency, moment of inertia and damping factor, respectively. Working diagram of ISE lab's topology is shown in Figure 11.7. This calculates the active and reactive powers using a power meter and based on these quantities formulate reference voltage and angle to generate the PWM signals for the power converter [37].

11.3.3 Frequency–power-response-based topologies

The frequency–power-response-based topologies require the frequency and its derivative for the inertial support control loop. These topologies are typically implemented as current sources so it offers the inherent advantage of overcurrent protection [32,38]. With the frequency derivate, noise is introduced in the control loop which can limit the performance of the closed-loop system. The frequency–power-based topologies are briefly discussed in the following sections.

11.3.3.1 VSYNC's topology

The working diagram of the VSYNC's topology is shown in Figure 11.8 [32]. The frequency and derivate of the frequency are calculated through PLL. Reference

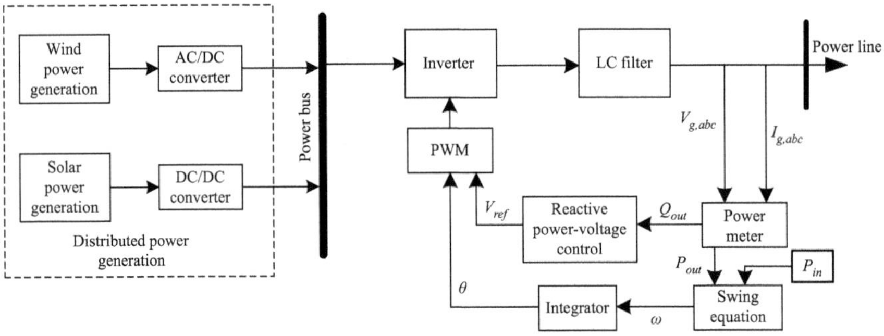

Figure 11.7 Block diagram of ISE lab's topology

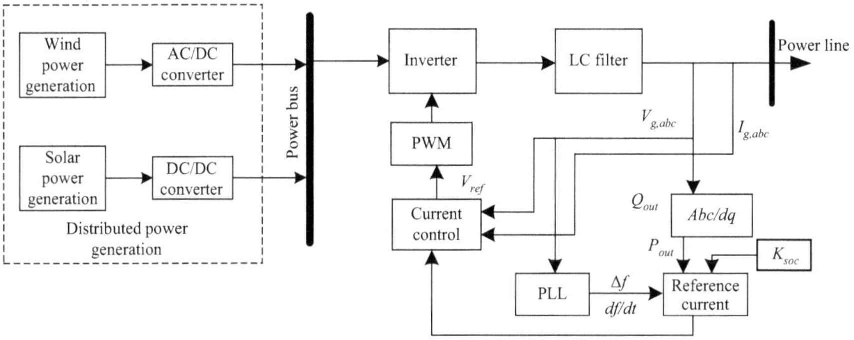

Figure 11.8 Block diagram of VSYNC's topology

current is calculated based on the reference power. The reference power balance equation is expressed in the following equation [39]:

$$P = P_{soc} + K_i\left(\frac{d\omega}{dt}\right) + K_p\Delta\omega \tag{11.10}$$

From (11.10), P_{soc} represents the power component which is calculated based on state of charge of the battery, K_i represents the inertial support gain and K_p represents the damping term gain.

11.3.3.2 Virtual synchronous generators

The virtual SG (VSG) topology requires measured frequency and its derivative to emulate the inertial response. VSG is also implemented as current source [40–42]. The reference power consists of two components, namely, the damping and inertial components (http://pedo.kp.gov.pk/page/introduction_to_pakhtunkhwa_energy_development_organization_pedo). The output power of the VSG is expressed as

$$P_{VSG} = K_i\left(\frac{d\omega}{dt}\right) + K_p\Delta\omega \tag{11.11}$$

where $\Delta\omega$ and $d\omega/dt$ represent the change in angular frequency and derivative of the angular frequency. K_i and K_p represent the inertial and the damping gains, respectively.

11.3.4 Droop-based approach

The droop-based inertial support does not require the communication link, and it is quite similar to the conventional droop control of SG. As the dynamic response of this topology is slow, so it is not very effective for transient power sharing [43–45]. With inductive impedance, the frequency droop is expressed as

$$\omega_g = \omega^* - m_p(P_{out} - P_{in}) \tag{11.12}$$

From (11.12), ω^* represents the reference frequency, ω_g is grid frequency, P_{in} is the input active power, P_{out} represents the output active power and m_p represents the active power droop gain. In the same way, the voltage droop is expressed as

$$v_g = v^* - m_p(Q_{out} - Q_{in}) \tag{11.13}$$

From (11.13), v^* represents the reference voltage, v_g is grid voltage, Q_{in} represents the reference reactive power, Q_{out} is the output reactive power and m_p is the reactive power droop control gain.

11.4 Case study

Northern areas of Pakistan have a great potential for small micro hydropower and PV generation units. Apart from such great potential and abundant sources in the northern areas, population is divided in many small communities. Therefore the supply of AC power to all of these small communities from the main grid was not economically feasible for the power generation authorities of Pakistan. In order to supply continuous and cheap power to these small communities, The Khyber Pakhtunkhwa provincial government of Pakistan has installed several micro hydropower plants (MHPP) ranging from several kilowatts to hundreds of kilowatts. However, the frequency regulation of all the installed MHPP units is achieved through mechanical governors. In the events of sudden load or flow rate changes, large frequency fluctuations are being observed, and on several occasions, the operators have to shut down several units. Moreover, with increase in the population, the number of consumers will increase, so the existing installed capacity may not be sufficient. Since in the northern areas of Pakistan, the PV energy is also available over the period of the whole year, so this chapter proposes a microgrid architecture based on the integration of PV and the existing MHPP units. For dynamic frequency regulation of the microgrid, the synthetic inertia control loop is proposed, and the inertial response is added to the microgrid through an inverter with ESS source. As a test case for the simulation purpose, a 15 kW hydropower generator unit is integrated to 3 kW PV unit. The test case is simulated in MATLAB®/Simulink™ environment. The block diagram of the test case is given in Figure 11.9.

From Figure 11.9, P_{si} represents synthetic power component, and it is calculated based on the system's ROCOF. The inner loop represents the current control scheme of the inverter in *dq* reference frame. The inner current loops are implemented using conventional proportional integral controllers. The ESS represents the energy storage system. The reference command for synthetic power is generated by calculating the system's ROCOF and by choosing appropriate gains [16]. The conceptual diagram of the synthetic inertia loop is shown in Figure 11.10.

From Figure 11.10, the reference synthetic power is expressed as

$$P_{si} = -K_{SI}\frac{d}{dt}\Delta f + K_d\Delta f \tag{11.14}$$

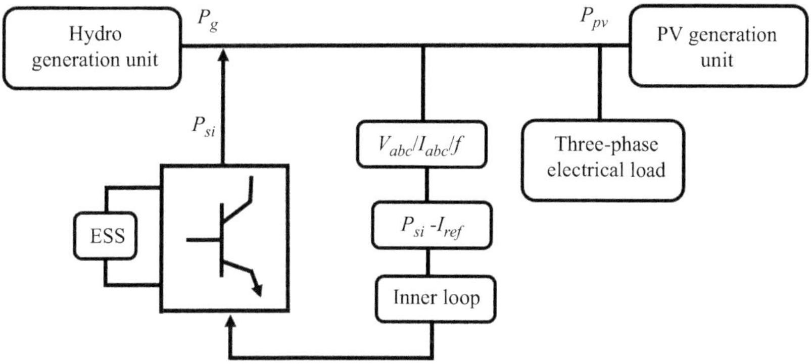

Figure 11.9 Hydro-PV microgrid with synthetic inertia loop

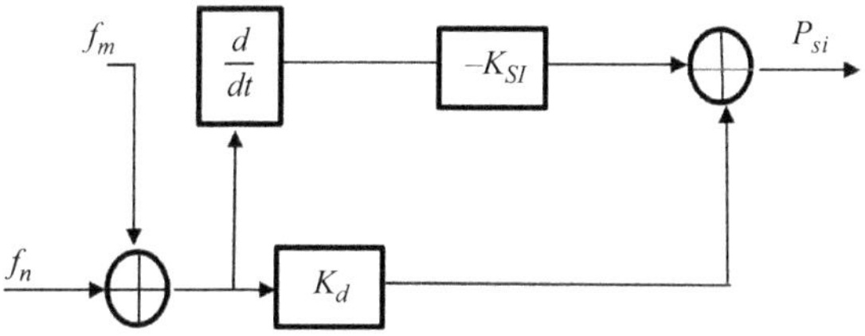

Figure 11.10 Implementation of synthetic inertial loop

From (11.14), K_{SI} represents synthetic inertial loop gain, K_d is the droop control gain, Δf is the frequency error, f_m is the measured frequency and f_n represents the reference frequency. The implementation details of the inner loop current controllers are given in Figure 11.11.

The effects of synthetic inertial loop for the frequency stabilization of the hydro-PV microgrid are investigated with sudden load and source changes. In the first scenario, the three-phase load is abruptly changed from 1.5 to 6.5 kW at time $t = 0.8$ s. In the second case, a fixed three-phase load of 5 kW is connected to the microgrid, and the mechanical power (P_m) is suddenly reduced from the nominal value to one third. For simplicity, the change in P_m is assumed to be directly proportional the decrease in the amount of volumetric flow rate of the water. For this simulation study, the nominal power of the hydro generator is 15 kW with 50 Hz frequency. The voltage levels are chosen as 330 V rms. The frequency response of the microgrid with sudden load increase and without synthetic inertia support is shown in Figure 11.12. Without synthetic inertia support, the minimum frequency of the microgrid is 46.8 Hz.

Figure 11.13 shows the comparative results of the frequency response with and without synthetic inertia support. As mentioned previously the minimum frequency reaches 46.8 Hz after the application of sudden load change. However, in the addition of the inertial support, the minimum frequency is restricted to 49.4 Hz. The synthetic inertia is added to the microgrid through power inverter which is

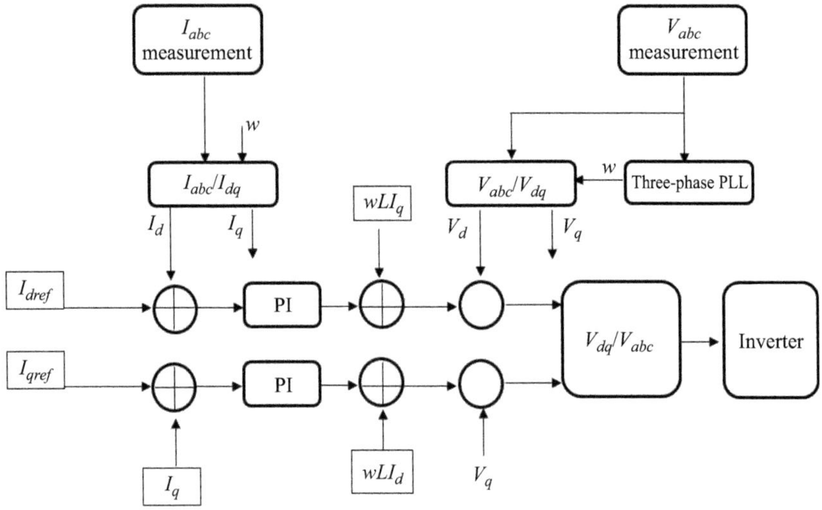

Figure 11.11 Inner current loops

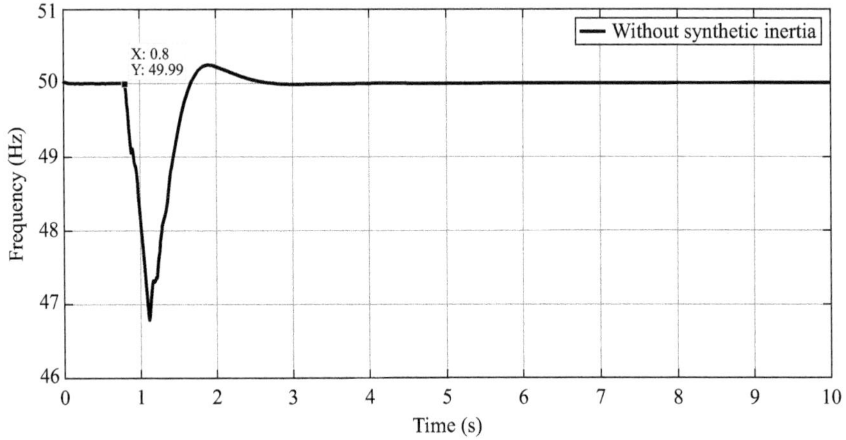

Figure 11.12 System frequency without synthetic inertia

supplying active power calculated based on the power frequency response. The active power response with synthetic inertial support is shown in Figure 11.14. From the simulation results, the maximum and minimum peak active power is recorded as 2,800 and −800 W, respectively. The peak maximum and minimum active power is sourced and absorbed in a very fast transient, and thus the synthetic inertial support restricts the minimum frequency of the microgrid to 49.4 Hz. The system returns to the nominal frequency, i.e., 50 Hz after a short interval of time. From Figure 11.14, it can be observed that the active power of the synthetic support is zero when the system's frequency stays near the nominal value, i.e., 50 Hz.

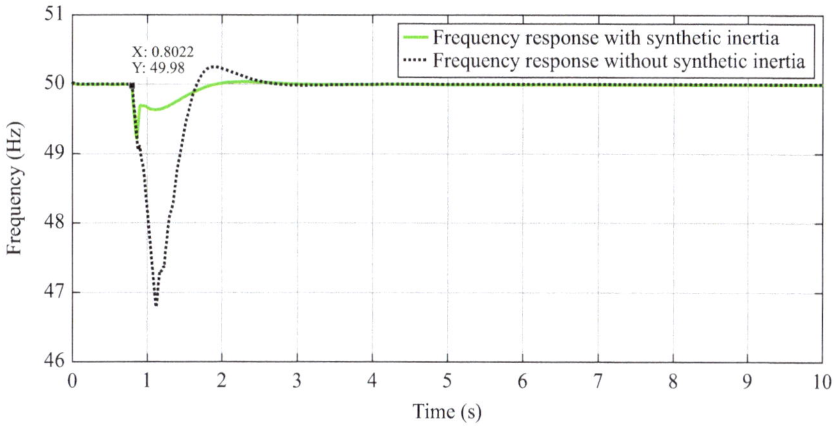

Figure 11.13 System frequency with and without synthetic inertia

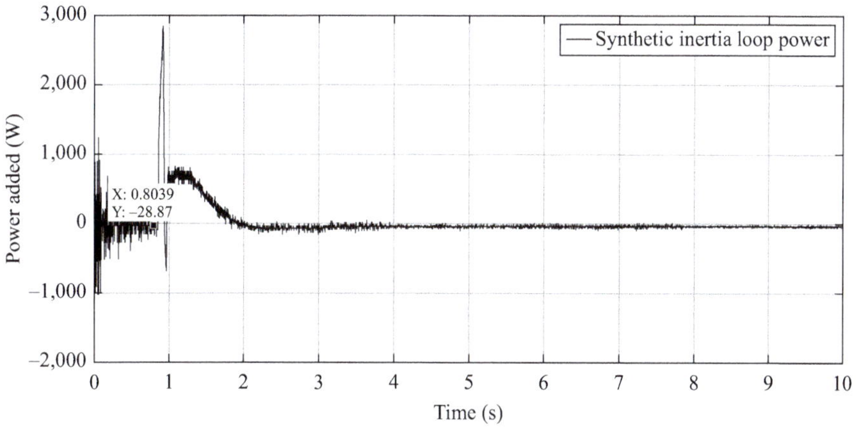

Figure 11.14 Active power with synthetic inertia

In the second scenario, the fixed three-phase electrical load of 5 kW is connected to the microgrid. The mechanical power of the hydroturbine generator is suddenly reduced from 1 to 0.5 p.u. The sudden change in the mechanical power will affect the electrical output power of the generator. Figure 11.15 shows the frequency variation of the system in response to the sudden reduction of the mechanical power of the hydro generation unit. No synthetic inertia support is provided, so the minimum peak of the frequency is recoded as 48.8 Hz. Moreover, at time $t = 5$ s, the maximum peak of the frequency is recorded as 51 Hz. To compensate the frequency deviations, the proposed synthetic inertia loop is incorporated, and the frequency response of the microgrid with synthetic loop is shown

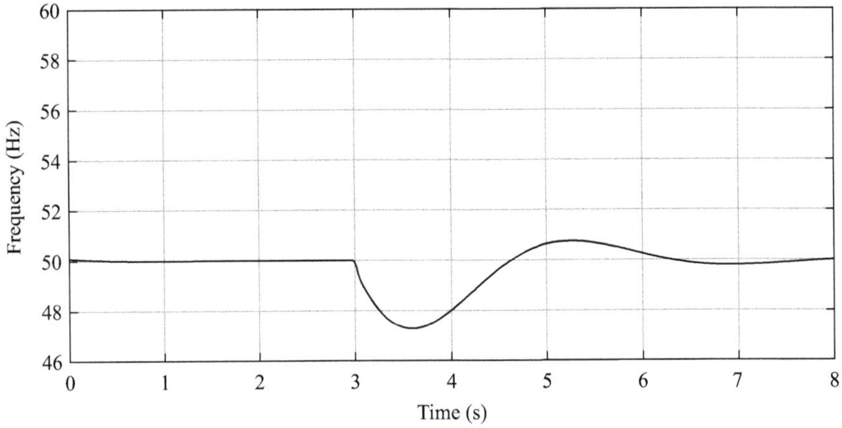

Figure 11.15 Frequency response without synthetic inertia

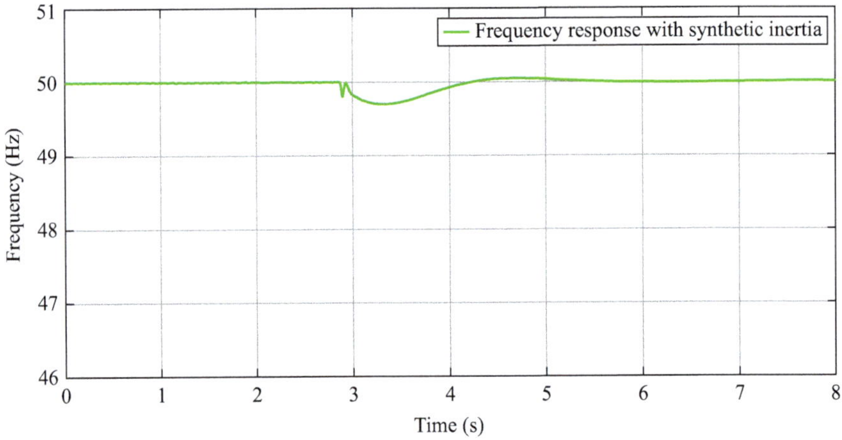

Figure 11.16 System frequency with synthetic inertia

in Figure 11.16. With the application of the synthetic inertial support, the minimum peak of the system's frequency is restricted to 49.7 Hz.

The simulation case for the active power supplied by the power inverter as a result of the application of the synthetic support is shown in Figure 11.17. From the simulation results, it is recorded that the maximum and minimum active power supplied and absorbed is around 1,100 and −400 W, respectively. Moreover, from Figure 11.16 and after time interval $t = 4$ s, since the system's frequency is above 50 Hz, so the active power level is negative (around −100 W) as shown in Figure 11.17.

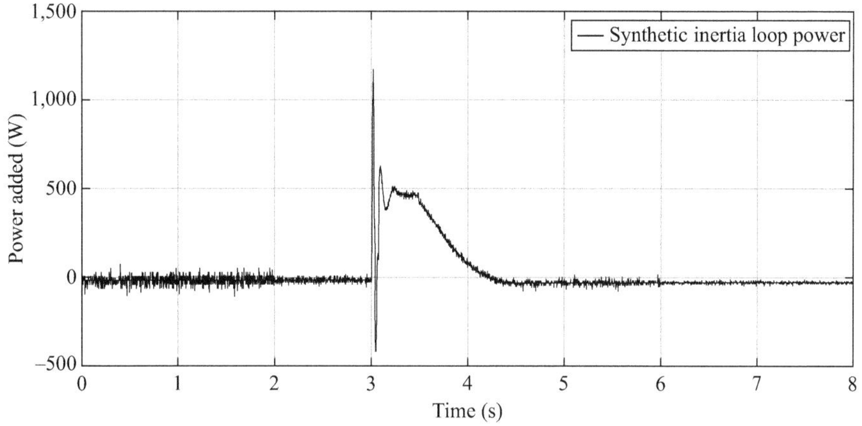

Figure 11.17 Active power with synthetic inertia

11.5 Concluding remarks

This chapter briefly introduced the basic concepts of the power-converter-based synthetic inertial support for the frequency stability of a microgrid. As a test case, a hydro-PV microgrid was tested under the source and load changes. The frequency–power-response-based topology was selected to add inertial support to the microgrid using a separate inverter and ESS. From the simulation results, it is concluded that with the addition of the synthetic inertial support, the stability of the system's frequency is enhanced.

References

[1] U.S. Energy Information Administration. Annual Energy Outlook 2017; U.S. Department of Energy: Washington, DC, USA, 2017. Available online: https://www.eia.gov/outlooks/aeo/pdf/0383 (2017).pdf (accessed on 21 June 2017).
[2] The Sunshot Initiative. Available online: https://www.energy.gov/eere/solar/sunshot-initiative.
[3] IEA PVPS. Trends in 2016 in Photovoltaic Applications; T1-30:2016; IEA PVPS: Paris, France, 2016 Available online: http://iea-pvps.org/fileadmin/dam/public/report/national/Trends_2016_-_mr.pdf (accessed on 21 June 2017).
[4] Kroposki, B.; Johnson, B.; Zhang, Y.; *et al.* Achieving a 100% renewable grid: operating electric power systems with extremely high levels of variable renewable energy. IEEE Power Energy Mag. 2017, 15, 61–73.
[5] Yan, R.; Saha, T.K.; Modi, N.; Masood, N.A.; and Mosadeghy, M. The combined effects of high penetration of wind and PV on power system frequency response. Appl. Energy 2015, 145, 320–330.

[6] Moslehi, K.; and Kumar, R. A reliability perspective of the smart grid. IEEE Trans. Smart Grid 2010, 1(1), 57–64.
[7] Gururng, A.; Galipeau, D.; Tonkoski, R.; and Tamrakar, I. "Feasibility Study of Photovoltaic-hydropower microgrids," in 5th International Conference on Power and Energy Systems (ICPS), 2014.
[8] Kerdphol, T.; Rahman, F.S.; and Mitanni, Y. Virtual inertia control application to enhance frequency stability of interconnected power systems with high renewable energy penetration. Energies 2018, 11, 981.
[9] Rahmann, C.; and Castillo, A. Fast frequency response capability of photovoltaic power plants: the necessity of new grid requirements and definitions [On the electrodynamics of moving bodies]. Energies 2014, 7(10), 6306–6322.
[10] Wang, X.; Yue, M.; and Muljadi, E. "PV generation enhancement with a virtual inertia emulator to provide inertial response to the grid," in 2014 IEEE Energy Conversion Congress and Exposition (ECCE), 2014, pp. 17–23.
[11] van Wesenbeeck, M.P.N.; de Haan, S.W.H.; Varela, P.; and Visscher, K. "Grid tied converter with virtual kinetic storage," in PowerTech, 2009 IEEE Bucharest, 2009, pp. 1–7.
[12] Zhong, Q.C.; and Hornik, T. "Synchronverters: grid-friendly inverters that mimic synchronous generators," in Control of Power Inverters in Renewable Energy and Smart Grid Integration. Wiley-IEEE Press, 2012, pp. 277–296, ISBN: 9781118481806.
[13] Hirase, Y.; Abe, k.; Noro, O.; Sugimoto, K.; and Sakimoto, K. "Stabilization effect of virtual synchronous generators in microgrids with highly penetrated renewable energies," 2016 IEEE 17th Workshop on Control and Modeling for Power Electronics (COMPEL), Trondheim, 2016, pp. 1–8. doi: 10.1109/COMPEL.2016.7556690
[14] Johnson, B.B.; Sinha, M.; Ainsworth, N.G.; Dorfler, F.; and Dhople, S.V. Synthesizing virtual oscillators to control islanded inverters. IEEE Trans. Power Electron. 2016, 31(8), 6002–6015.
[15] Liu, J.; Miura, Y.; and Ise, T. Comparison of dynamic characteristics between virtual synchronous generator and droop control in inverter-based distributed generators. IEEE Trans. Power Electron. 2016, 31, 3600–3611.
[16] Van De Vyver, J.; De Kooning, J.D.M.; Meersman, B.; Vandevelde, L.; and Vandoorn, T.L. Droop control as an alternative inertial response strategy for the synthetic inertia on wind turbines. IEEE Trans. Power Syst. 2016, 31, 1129–1138.
[17] Thiesen, H.; Jauch, C.; and Gloe, A. Design of a system substituting today's inherent inertia in the European continental synchronous area. Energies 2016, 9582.
[18] Chen, D.; Xu, Y.; and Huang, A.Q. Integration of DC microgrids as virtual synchronous machines into the AC grid. IEEE Trans. Ind. Electron. 2017, 64, 7455–7466.
[19] Kerdphol, T.; Rahman, F.S.; Mitani, Y.; Hongesombut, K.; and Kufeoğlu, S. Virtual inertia control-based model predictive control for microgrid frequency

stabilization considering high renewable energy integration. Sustainability 2017, 9, 773.

[20] Kerdphol, T.; Rahman, F.S.; Mitani, Y.; Watanabe, M.; and Kufeoglu, S. Robust virtual inertia control of an islanded microgrid considering high penetration of renewable energy. IEEE Access 2018, 6, 625–636.

[21] Bevrani, H.; Francois, B.; and Ise, T. Microgrid Dynamics and Control; John Wiley & Sons: Hoboken, NJ, USA, 2017; ISBN 0780345304.

[22] Shi, R.; Zhang, X.; and Hu, C. Self-tuning virtual synchronous generator control for improving frequency stability in autonomous photovoltaic-diesel microgrids. J. Mod. Power Syst. Clean Energy 2018, 6, 482. https://doi.org/10.1007/s40565-017-0347-3.

[23] Tielens, P.; and Hertem, D.V. The relevance of inertia in power systems. Renew. Sustain. Energy Rev. 2016, 55, 999–1009.

[24] Tamrakar, U.; Galipeau, D.; Tonkoski, R.; and Tamrakar, I. "Improving transient stability of photovoltaic-hydromicrogrids using virtual synchronous machines," in Proceedings of the IEEE Eindhoven PowerTech, Eindhoven, The Netherlands, 29 June–2 July 2015, pp. 1–6.

[25] Gonzalez-Longatt, F.; Chikuni, E.; and Rashayi, E. "Effects of the Synthetic Inertia from wind power on the total system inertia after a frequency disturbance," in Proceedings of the IEEE International Conference on Industrial Technology (ICIT), Cape Town, South Africa, 25–28 February 2013, pp. 826–832.

[26] Bevrani, H.; Ise, T.; and Miura, Y. Virtual synchronous generators: a survey and new perspectives. Int. J. Electr. Power Energy Syst. 2014, 54, 244–254.

[27] Piya, P.; and Karimi-Ghartemani, M. "A stability analysis and efficiency improvement of synchronverter," in Proceedings of the IEEE Applied Power Electronics Conference and Exposition (APEC), Long Beach, CA, USA, 20–24 March 2016, pp. 3165–3171.

[28] Fairley, P. "Can Synthetic Inertia from Wind Power Stabilize Grids?," 7 November 2016, IEEE. Retrieved 29 March 2017, Available from https://spectrum.ieee.org/energywise/energy/renewables/can-synthetic-inertia-stabilize-power-grids.

[29] Synchronverter-enabled DC power sharing approach for LVDC microgrids. IEEE Trans. Power Electron. 2016.

[30] Waffenschmidt, E.; and Hui, R.S.Y. "Virtual inertia with PV inverters using DC-link capacitors," 2016 18th European Conference on Power Electronics and Applications (EPE'16 ECCE Europe), Karlsruhe, 2016, pp. 1–10. doi: 10.1109/EPE.2016.7695607

[31] Tamrakar, U.; Shrestha, D.; Maharjan, M.; Bhattarai, B.P.; Hansen, T.M.; and Tonkoski, R. Virtual inertia: current trends and future directions. Appl. Sci. 2017, 7, 654; doi:10.3390/app7070654.

[32] Zheng, Y. Virtual Inertia Emulation in islanded microgrids with energy storage system, Master of Science Thesis, at the Delft University of Technology, August 30, 2016.

[33] Bevrani, H.; Ise, T.; and Miura, Y. Virtual synchronous generators: a survey and new perspectives. Int. J. Electr. Power Energy Syst. 2014, 54, 244.

[34] Chen, Y.; Hesse, R.; Turschner, D.; and Beck, H.-P. "Improving the grid power quality using virtual synchronous machines," in Power Engineering, Energy and Electrical Drives (POWERENG), 2011 International Conference on IEEE, 2011, pp. 1–6.

[35] Sakimoto, K.; Miura, Y.; and Ise, T. "Stabilization of a power system with a distributed generator by a virtual synchronous generator function," in Proceedings of the 8th International Conference on Power Electronics (ECCE Asia), Jeju, Korea, 30 May–3 June 2011, pp. 1498–1505.

[36] Alipoor, J.; Miura, Y.; and Ise, T. Power system stabilization using virtual synchronous generator with alternating moment of inertia. IEEE J. Emerg. Sel. Top. Power Electron. 2015, 3, 451–458.

[37] Shrestha, D.; Tamrakar, U.; Ni, Z.; and Tonkoski, R. "Experimental verification of virtual inertia in diesel generator based microgrids," in Proceedings of the 18th Annual International Conference on Industrial Technology (ICIT), Toronto, ON, Canada, 22–25 March 2017, pp. 95–100.

[38] Bevrani, H.; Ise, T.; and Miura, Y. Virtual synchronous generators: a survey and new perspectives. Int. J. Electr. Power Energy Syst. 2014, 54, 244.

[39] Torres, M.; and Lopes, L.A.C. "Virtual synchronous generator control in autonomous wind-diesel power systems," in Proceedings of the IEEE Electrical Power & Energy Conference (EPEC), Montreal, QC, Canada, 22–23 October 2009, pp. 1–6.

[40] Van Wesenbeeck, M.P.N.; de Haan, S.W.H.; Varela, P.; and Visscher, K. "Grid tied converter with virtual kinetic storage," in Proceedings of the IEEE Bucharest PowerTech, Bucharest, Romania, 28 June–2 July 2009, pp. 1–7.

[41] Van, T.V.; Visscher, K.; Diaz, J.; *et al.* "Virtual synchronous generator: an element of future grids," in Proceedings of the IEEE Innovative Smart Grid Technologies Conference Europe (ISGT Europe), Gothenburg, Sweden, 11–13 October 2010, pp. 1–7.

[42] Torres, M.; and Lopes, L.A. Virtual synchronous generator: a control strategy to improve dynamic frequency control in autonomous power systems. Energy Power Eng. 2013, 5, 2A:1–2A:7. Available online: http://file.scirp.org/Html/5-6201497_30602.htm (accessed on 29 April 2017).

[43] Katiraei, F.; and Iravani, M.R. Power management strategies for a microgrid with multiple distributed generation units. IEEE Trans. Power Syst. 2006, 21, 1821–1831.

[44] Pogaku, N.; Prodanovic, M.; and Green, T.C. Modeling, analysis and testing of autonomous operation of an inverter-based microgrid. IEEE Trans. Power Electron. 2007, 22(2), 613–625. doi:10.1109/TPEL.2006.890003.

Chapter 12

Microgrid protection

Robert M. Cuzner[1], Siavash Beheshtaein[1], and Farzad Banihashemi[1]

Power-system protection encompasses the interrelated concerns of assurance of human safety, mitigation of equipment damage and provisions for reliable transmission and distribution of electrical power to end users. For conventional power delivery systems—from national power grids to household electrical distribution—the design methods and engineering design solutions for protection have benefited from over a century of experience. However, as renewable energy penetration into the grid is increasing, long-held practices for protective system design and associated hardware and controls are proving to be inadequate. This is particularly the case for microgrids. The microgrid has emerged as an efficient means for introduction of locally supplied *renewable energy source(s)* (RESs), such as solar photovoltaic (PV) and wind, and a diversity of local generation sources, such as national gas generation, into existing grid structures. With the inclusion of energy storage systems (ESSs), the microgrid provides a means for improving grid resiliency and achieving truly *energy secure* systems. Throughout this chapter, *distributed generation* (DG) or *distributed generators* (DGs) refers to any distributed source of power, including RESs or more conventional natural gas generators and back-up supply diesel generators. The term *distributed energy resource* (DER) refers generally to DGs and ESSs. The inclusion of ESS is implied by the use of the term DER, whereas DG or DGs refer only to power generation. The microgrid concept effectively integrates and manages simultaneously generation, energy storage and load demand. The microgrid also provides a viable means for electrification of energy poor areas. Although technically not a microgrid, electrified transportation power-distribution systems share many common traits with microgrids.

From the beginning, the achievement of accurate fault discrimination within a microgrid has been an issue. In the case of AC microgrids, this protection challenge has not prohibited implementations because the microgrid can be integrated into existing protective structures, albeit with suboptimal behavior during fault events.

[1]Department of Electrical Engineering and Computer Science, University of Wisconsin-Milwaukee, Milwaukee, United States

In the case of DC microgrids, where there is potential for lowest cost and highest efficiency, protection is the greatest impediment to large-scale deployment.

The power system is composed of different components including generators, transmission lines, bus bars, transformers, reactors, capacitors and loads. All equipment must be protected against various forms of faults. Current-based, voltage-based, frequency-based or impedance-based methods are among the most well-known protection systems. In order to understand the path forward for proper design of microgrid protection systems, it will be important to understand both the established (or conventional) process for protective system design and the limitations of conventional protection systems when they are applied to microgrids. The intention is, therefore, to establish a context relating to conventional distribution systems and then determine what modifications must be made in both process and implementation to achieve levels of selectivity equal to, or even better than, conventional protection systems. This chapter addresses the distribution system protection issues associated with both AC and DC microgrids. For the AC microgrid discussion, a community microgrid provides a good basis for reference because all microgrid protection challenges can be represented. Here, most of the protection challenges exist at the low-voltage AC (LVac) interfaces, but as power electronic interfacing DERs are increasingly applied to medium-voltage AC (MVac), without intervening transformers, similar challenges will occur at the MVac interfaces. The DC microgrid discussion applies universally to a wide range of applications. Low voltage DC (LVdc) and medium-voltage DC (MVdc) fault characteristics and protective solutions will be different, so distinctions are made with respect LVdc versus MVdc.

12.1 Protective system design objectives

As a first order of business, the *taxonomy* of the protective system is identified. As a convention, important taxonomical terms will be italicized when they are first identified, described and defined after which the italics will be dropped in subsequent textual references. To begin, the conventional trade space for the design of the protective systems is defined as follows [1]:

> *Reliability:* The ability of the installed *protection equipment* and hardware associated with the *protection scheme* to predictably identify, locate and isolate the fault. High reliability includes high functionality of the installed equipment over time and the avoidance of spurious actuation (tripping) on normal operational transients or noise and avoidance of maloperation when the fault is out of zone. Reliability has two aspects including *dependability* and *security*. From a protective system standpoint, *dependability* is defined as "the degree of certainty that a relay or relay system will operate correctly" (IEEE C37.2, 2008). On the other hand, *security* is "the degree of certainty that a relay or relay system will not operate incorrectly" (IEEE C37.2, 2008).

In general, improving the *dependability* may lead to decrease in *security*, and vice versa.

Speed: After the fault is removed, normal operating voltages are restored with minimal interruption to all connected, persisting loads.

Selectivity: The protection system the protective system isolates the fault by the main relay in the shortest possible time, and if the main relay fails to operate, the protective system includes backup relay(s) to disconnect the faulty part by a specific time delay. As a result, coordination of protective relays is a necessary task to insure the *selectivity*. In a system with a high level of selectivity, the requisite *continuity of power* to, *quality of power* (QoP) and *quality of service* to the loads persists following isolation of the fault from the system is met. Another descriptor for selectivity may be the *reliability of power* during fault events and after fault removal. A system with a low level of selectivity will exhibit poor *recoverability*, i.e., a persistent and significant degradation of service as a result of the fault. A system with a low level of selectivity will also exhibit a degradation in QoP following fault recovery.

Economics: The protective system is economically viable, according to the application. Since protective elements within the distribution system are not the principal function, it will be desirable to minimize costs. The technical risk associated with new protective approaches would also fall under this category since technical risk often results in increased costs in non-recurring engineering activities.

Simplicity: The quantity of parts, zones of protection, level control decentralization needed to ensure reliability of fault discrimination do not result in excessive commissioning costs or reductions in lifetime due to the increase in probability of component failure. The protective relay has to be designed in a simplest possible structure to accomplish its protective goals. Any additional unit/component must contribute a significant improvement to the protective relay, otherwise it increases complexity and maintenance cost.

It will be shown later that it is very helpful to look at networked grids and microgrids as a power and energy delivery service to downstream users. In this way, the organizing principle of *dependability* can be applied in a similar fashion to the way this *dependability theory* is applied to communications networks [2]. According to dependability theory, a *dependable* system will have the measurable attributes of *availability*, *reliability*, *safety*, *confidentiality*, *integrity* and *maintainability*. In this context, the design trade space mentioned above could be augmented to include all of dependability attributes resulting from dependability theory, where the attribute of *reliability* is already accounted for. Achievement of a certain level of *reliability*, *safety*, etc., will be attended by corresponding levels of *economics* (cost), *simplicity*, etc. Acknowledging this point sheds some light on how conventional practices for protective system design may be improved upon when considering the evolving nature of the electrical power grid and, particularly, with respect to microgrids.

12.2 Conventional protective system design practice

According to dependability theory, the *threats* to achievement of dependability attributes fall under three categories: *faults*, *errors* and *failures*. These are interdependent system conditions and states. A *fault* is a condition (external or internal to the system) that when active produces an *error*. An *error* is a part of the system total state that may lead to a *failure*. A *failure* occurs when a part or all of the service cannot be delivered as a result of the impact of an *error* or *fault*. The purpose of the protective system is to detect the instances of errors that must be acted upon and minimize the extent of failures within the system.

The dependability theory definitions of *faults*, *errors* and *failures* work very well within the context of protective system design. The principle threats that the protective system design must deal with are faults, where faults are non-malicious or malicious events leading to equipment failures and subsequent threats to dependable delivery of the power and energy-delivery service. The principal faults that must be dealt with are inadvertent line-to-line (LL), line-to-ground (LG), multiple LG, LL-to-ground (LLG) or three-phase LLG 0 Ω connections that may occur within the distribution system or within distribution-connected equipment.

The onset of the fault, or *fault event*, is a very important consideration. For example, the fault may be a *slow inception fault* that starts from a high impedance fault and then progresses to a low impedance or short-circuit (0 Ω) path over time. Any one of the fault types identified may occur intermittently (*intermittent faults*). Of greatest concern, especially to the design of fault protection in microgrids, is the *sudden inception fault*. From an analytical perspective, this fault occurs suddenly and with a nearly infinite associated rate of change in voltage (dv/dt) or rate of change in current (di/dt). While physics will always dictate non-infinite dv/dt or di/dt, it is an established and good engineering practice to perform protective design starting with the assumption of instantaneous *sudden inception* of 0-Ω short circuit faults. The *sudden inception fault* will assume a 0 Ω short circuit fault. If this practice is followed, the *fault characterization* phase, or "Characterize Fault Response" phase, shown in the conventional protective system design process of Figure 12.1, will reveal the excitation of system resonance, capacitor discharge, over-current and over-voltage behaviors. It is then within the judgment of the protective system designer to incorporate known practical limitations in dv/dt, di/dt and short circuit fault circuit impedance in order to avoid protective system over-design. Other fault events to be considered, including high impedance, arcing, over-current, over-voltage, under-voltage, over-frequency and under-frequency.

Once again, relying upon the definitions established by dependability theory, the means for achieving dependability attributes are as follows:

1. *Fault prevention or susceptibility*
2. *Fault tolerance or vulnerability*

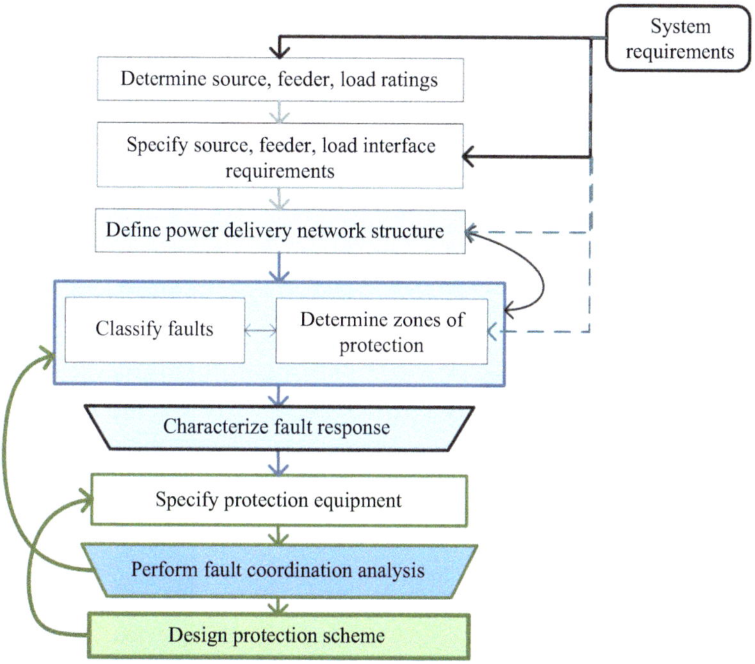

Figure 12.1 Conventional protective system design process

3. *Fault removal or recoverability*
4. *Fault forecasting*

The alternate definitions of (1)–(3) are typically utilized in the design of military equipment and are grouped under the subsuming category of *survivability*. The *survivability* of a system (typically applied to shipboard power and energy delivery systems) is a parallel but not all-encompassing organizing methodology for identifying and allocating measurable requirements to system components, when compared to *dependability*. Reconciling *survivability* with dependability theory, survivability includes the measurements of both attributes and means for achieving a dependable system. Grid *resiliency* is becoming a very important consideration in the present-day environment, where the increase of malicious threats to power grids, coupled with natural disasters, extreme weather causes/effects and human cost of unreliable and inadequate grid infrastructure, highlight the fact that the availability of electrical power is a basic human right. The achievement of resilient electrical grid infrastructures coincides directly with the means for achieving a dependable power and energy systems. In terms of the conventional design space identified above, resiliency is a function of the combined attributes of reliability, speed and selectivity. Resiliency can be improved through more networked grids, microgrid implementations with a parity of dependability in both grid-connected and grid-islanded modes.

12.2.1 Fault characterization

Protection design for utility transmission and local power-distribution systems is largely based upon well-established standards and practices that dictate the structure of the source, feeder and load network. This process is described in Figure 12.1. The conventional structure of a power system starts from centralized generation resources that provide electrical power to wide geographical areas. Power is typically generated and then transformed to high voltages so that it can be delivered to areas of usage over long transmission lines. Well-defined transition points, electrical substations, transform to lower voltages where power delivery fans out through radially distributed networks to end users.

The process shown in Figure 12.2 applies to both transmission and distribution systems design. The methodology and practices for fault characterization and follow-on protective system design are well established [1,3]. They are largely based upon well-established standards and practices that dictate the structure of the source, feeder and loads. The power transmission/distribution system starts as a network of interconnecting cables and interfacing transformers between source(s), feeds and loads. For protective system design at the transmission level, the definition of the power-delivery network structure is driven by geographic and geopolitical considerations. Zones of protection are defined by geography, weather and population. Fault classification is mostly limited to LG faults. Fault characterization is impacted principally by the cabling and the levels of transmission voltage which, for strong grids, are in the high voltage range (i.e., >100 kV).

For protective system design at the distribution level, the distribution transformers, feeder networks, grounding and feeder loads are prominent considerations. The majority of power delivery network structures are radial, meaning that power is distributed from a substation feed by fanning out to lower current rated branches

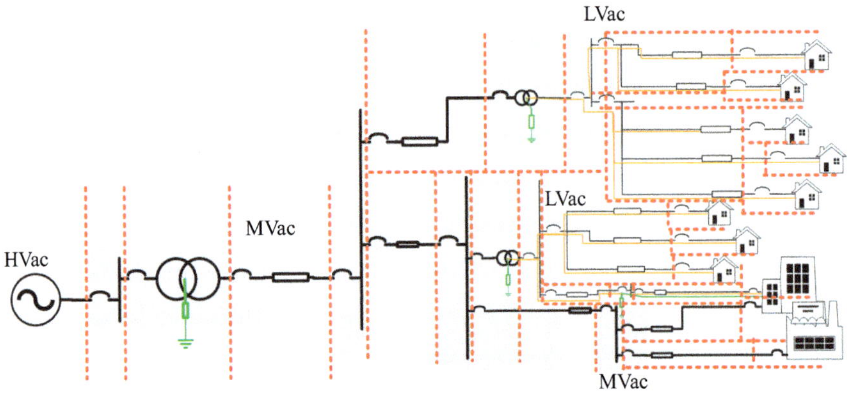

Figure 12.2 A conventional radial AC grid divided up into zones of protection

that are feeds to other parts of the system or feeds to loads at the point of end usage. Protective system design closer to the point of usage is decoupled from the impedance and time-current considerations of the upstream substation distribution feeder. Here the utility grid is treated as an infinite source of current to faults which are limited mostly by the transformer feeder impedance at points of usage. Such a system is shown in Figure 12.2. The zones of protection are defined by the power-network delivery structure from the main utility feeder downward, as shown by the division of the system between the red-dashed lines in Figure 12.2. The network structure and zones of protections are often evolutionary and dynamic. Fault classification must cover the entire range of fault types, three-phase LL/LLG, single-phase LL/LLG, LG, high to low impedance, etc. Fault characterization will be a function of voltage level (i.e., 1–100 kV for medium voltage, below 1 kV for low voltage), fault type, grounding and transformer impedance.

In distribution protective system design, particularly in systems with a weak interface to the main utility grid or with DG, certain load characteristics must also be taken into account, such as the impacts of motors sourcing currents into the fault and start-up inrush currents, principally from transformers. These latter considerations are the key to achieving a high reliability in the protective system design—as it will be important for the protective devices (PDs) to be insensitive to the effects of transients currents induced by loads.

As a minimum requirement, the protective system must reliably isolate the *instantaneous inception* near-zero ohm fault. The level of *fault discrimination* within the network, i.e., the closeness to which the fault can be isolated from the rest of the system and level of *recoverability*, with respect to non-faulted portions of the system, is traded off against *economics* and *simplicity*. The system recoverability in trades involves *speed*, *selectivity*, *economics* and *simplicity*. The extreme corner case to be considered for fault characterization is the near zero ohm impedance, *sudden inception* fault. The reliability of the protective system design may vary from the slow-inception fault, but safety must always be a primary objective.

12.2.2 Protective equipment and scheme components

The protective system ultimately consists of the *protection equipment* and the *protective scheme* specified by the protective system designer. The *protection equipment* is the backbone protection system that ensures fault current limiting and fault isolation. Protective equipment components, selected by the protective designer after the fault characterization stage, include relays, fuses and any other fail-safe device. The *protective scheme* consists of the collection of *protection equipment* providing the required protective functions aided by sensing, controls and communications hardware, software and firmware required to make the system work. The protection scheme will incorporate additional components, such as current transformers (CTs), circuit breakers (CBs), relaying coils, contactors and logic, supervising control and data acquisition, global position system and power-management unit. Per the conventional protective system design practice of Figure 12.2, the protection scheme is not complete until a fault-coordination analysis is performed involving the protection equipment and the additional hardware, software and firmware associated with the protection scheme.

If the system design and construction remain within the norms of standard practice, the tools that are available for design improvement are limited to the available subsystems and components associated with protective relaying design. Referring to Figure 12.2, the fault classification and characterization for conventional power-delivery systems are based upon practice well-established analytic principles and methodologies. Once power ratings and voltages are defined at the source of generation and step-up transformation, transmission cable lengths are determined and power ratings and voltages are defined at feeder distribution points and points of usage, then fault characterization is performed using phasor analysis and symmetrical components. Fault characteristic is a function of well-defined impedances that are supplied or can be derived from component ratings; typical information used includes generator sub-transient, transient and steady state (SS) reactances, cable impedance and load rating and power factor. Transient time-current characteristics are derived from impedances and associated sub-transient, transient or L/R time constants. Particular fault behavior will depend upon fault location within the system. The fault classification defines the symmetrical components and symmetrical component-based equivalent circuits (i.e., positive, negative, zero sequence) used in conjunction with derived impedances and component-level time-current characteristic to derive system-level, time/position-dependent fault time-current characteristics. Resultant fault time-current characteristics versus fault location and fault type are then used for protection equipment selection and protection scheme design. Fault characteristics combined with the specific protection equipment relay characteristics are then used to evaluate the coordinate time interval (CTI) between upstream and downstream breakers. This last step forms the foundation for the fault-coordination analysis, which identifies any adjustments or enhancements to the protection scheme that are required to optimize reliability, speed, selectivity, economics and simplicity against system design requirements and objectives.

Considering the selection of protective equipment, the protective relay components making up the protective equipment and protection scheme should be designed with the simplest possible structure necessary to accomplish its protective requirements. Any additional unit/component must contribute a significant improvement to the protective relay to justify the associated increases in cost and maintenance. Deviations of the well-established methodologies for fault classification and protective relaying design are driven by needs to increase grid resiliency. Such deviations will certainly occur as the grid complexity increases.

12.2.3 Fault coordination analysis and protective relaying

Figure 12.3 shows a community distribution system with a single utility feed that is designed to increase grid resiliency through cross-connections between two radial grid structures responsible for providing power to two sections of the community through multiple paths. Each community section is backed up by localized power generation. The three most common types of protective relaying schemes will be applied to such a grid: *overcurrent relays* (OCRs), *directional OCRs* (DOCRs) and *differential relays*.

12.2.3.1 Overcurrent relays

OCRs are simple and economical PDs. They are employed as the main protective relays for distribution and backup protection for transmission systems. For the distribution system of Figure 12.3, reliable protection within the radially distributed residential, commercial and industrial sub-grids can be accomplished using OCRs. The fault discrimination capability of the OCR is accomplished through electro-thermal and electromagnetic means that are inherent to the device and can be mechanically tuned to achieve a time-current trip characteristic, hence their low cost. Generally, there are three types of time-current characteristics for OCRs: (1) instantaneous; (2) definite time and (3) inverse-time. If the current amplitude exceeds a predefined value, the relay with instantaneous and definite-time characteristics will send trip signal instantly and after a definite time, respectively. For the inverse time-current characteristic, operating time is mathematically defined by IEEE C37.112 as follows:

$$T_p = TDS \cdot \left(\frac{\beta}{(I_{sh}/I_P)^{\alpha} - 1} + \gamma \right) \tag{12.1}$$

where TDS is the time-delay setting of the relay; I_P is the pick-up current; I_{sh} is the current detected by relay; α, β and γ denote the slope of the relay characteristic.

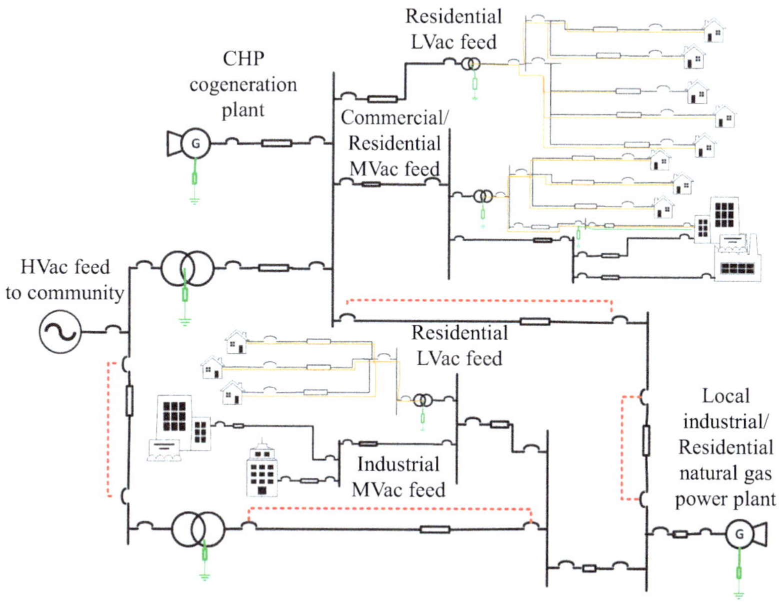

Figure 12.3 Conventional AC grid showing radial and ring bus structures

According to IEEE C37.112, moderately inverse (*MI*), very inverse (*VI*) and extremely inverse (*EI*) trip characteristics are the three classes of inverse time-current modes, which are presented in Table 12.1.

The proper OCR selectivity is maximized through selecting proper pickup currents and TDSs. Pickup currents are usually determined in proportion to maximum possible load current; however, choosing TDSs is more complicated [4]. Correct relay coordination by appropriate settings would ensure isolation of faults in the protected zone by corresponding primary relays as quickly as possible, and if they fail, the corresponding backup relays trip after a coordination time delay. The operating time of backup relay is equal to the operating time of primary relay plus the CTI. This strategy creates a safety margin between primary and backup relays to guarantee the selectivity criteria and prevent maloperation. The coordination problem can be formulated as follows:

$$\begin{aligned}
&\text{Min } T = \sum_{i=1}^{N} T_i \quad \text{such that} \\
&T_j - T_i \geq CTI \\
&\forall(i, j) \in \Omega \\
&TDS_i^{\min} \leq TDS_i \leq TDS_i^{\max} \\
&I_P^{\min} \leq I_P \leq I_P^{\max}
\end{aligned} \tag{12.2}$$

where Ω is set of the main-backup pairs of the relays. There are three ways of solving the OCR coordination problem including trial and error, curve fitting and optimization methods [5,6]. The trial and error approach requires a large number of iterations to reach an optimal result; therefore, it has slow convergence rate. In curve fitting methods, the characteristics of OCR are modeled. These methods may need large computing memory to store many points of OCR characteristic curve. Finally, optimization-based methods model OCR coordination with an objective function and several constraints, then try to solve it by different methods such as mixed integer non-linear programming (MINLP) [7], genetic algorithm [8], citer-azavi2008new\cite, particle swarm optimization [9] and bee colony algorithm [10]. Since the evolutionary algorithms are multipoint search strategies, they can ensure reaching the global optimal solutions.

Table 12.1 Parameters for three modes of time-current characteristics of OCR

Curve description	α	β	γ
MI	0.02	0.051	0.114
VI	2.0	19.61	0.491
EI	2.0	28.2	0.121

12.2.3.2 Directional overcurrent relays

The DOCR is one of the alternatives for overcoming the shortcomings of the OCRs in distribution systems where power can flow in either direction through a line. This will be the case for the hypothetical distribution system of Figure 12.3, when DGs are added, such as the combined heat and power (CHP) cogeneration plant and natural gas power plant. Under fault conditions, multiple paths for the flow of energy into the fault are possible, and, therefore, the OCR alone may not be sufficient to adequately discriminate fault location. This will also be the case as the grid structure becomes more meshed, as is the case when cross-connections are added or if a ring bus architecture is required. For these distribution systems, the DOCR addresses bidirectional power flow between parts of the grid by enabling the discrimination between power flow from different power sources into the fault. The direction of current is typically identified by using a CT and voltage transformer (VT). Figure 12.4 shows the scheme and block diagram of a typical DOCR. Additionally, the speed of protection in ring bus architectures can be enhanced by communications between DOCRs at each end of the crossties. In Figure 12.3, the cross-connections between the DERs in the lower part of the grid in Figure 12.3 effectively creates a ring bus. Communicating DOCRs on the ring bus (where communication lines are represented by the dashed red lines) can isolate a fault to a section of the bus between them.

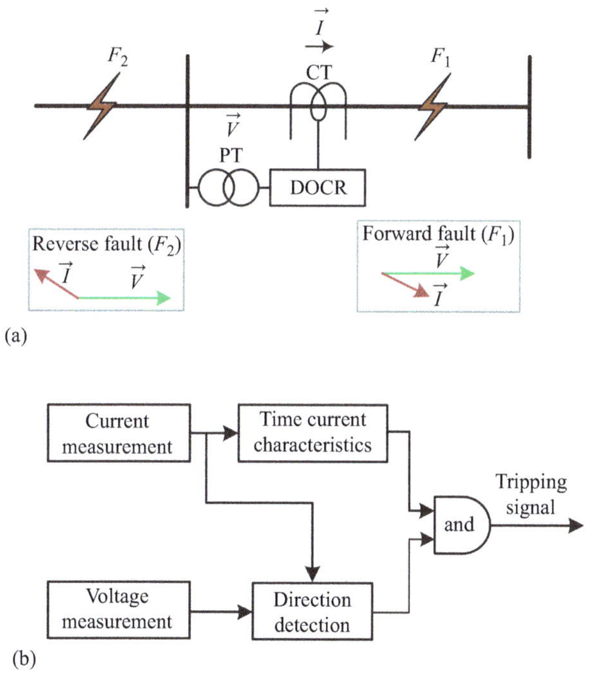

Figure 12.4 Directional overcurrent relay: (a) directional overcurrent relay scheme and (b) block diagram of directional overcurrent relay

There are potentially three measurements approaches for DOCRs: voltage reference phasors [11] derived from VTs, and pre-fault current [12] and post-fault current [13] derived from CTs. In the first approach, VTs and associated local processing calculate voltage reference phasors that are measured during pre-fault and post-fault condition. In the post-fault condition, the voltage magnitude might be very low and the phasor may be immeasurable. In this case, the voltage phasor(s) and current phasor(s) are derived from the pre-fault condition and then compared with the post-fault current phasor to determine the direction of the current, indicated the post-fault VT RMS measurement dropping below a predetermined value. Since the voltage reference requires a VT, there will be a considerable additional cost. Therefore, DOCR with VT and CT measurement is usually only considered when voltage sensing is required or available as a result of additional system functionality or requirements. Another approach, utilizing only a CT for post-fault current measurement, increases the economics of the system. This method detects changes in direction in current after the fault occurs [14]. It is applicable for the radial systems in which the power flow is single directional when a fault is not present.

DOCRs must be coordinated in order to disconnect the faulty area from the healthy parts of the grid in the shortest possible time to increase the speed attribute of system protection. In a radial distribution system, the coordination process starts with the furthest relay position from the source in a feeder, and then the relays are set one by one. The concept of break point set (BPS) is introduced to apply this method for meshed networks [15]. One of the classes of DOCR coordination is topological analysis. This method utilizes graph theory to find minimum BPS [16,17].

Other approaches can achieve the functionality of DOCRs utilizing lower cost OCRs. Information resulting from the fault characterization stage of the protective system design (see Figure 12.2) can be used to adjust the tuning of lower cost OCRs that are adjacent to each other in order to mimic the behavior of DOCR protection, by optimizing the TDS and pick-up current sensing levels. These approaches include trial and error methods [18], curve fitting [19], linear programming (LP) [20–22], MINLP [23], NLP [24,25] and heuristic-based algorithms [26,27]. In LP, the value of pickup current is selected between accepted overload and minimum fault current, and the optimal value of TDS is obtained through LP. In [8], interval analysis is integrated into LP to consider uncertainty in the network topology. Typically, the pickup current is selected near to its minimum value to increase DOCR sensitivity; however, it will not lead to the optimum values of TDS(s) and therefore will cause higher relay operating times. MINLP considers pickup current(s) and TDS(s) as discrete and continuous variables, respectively. In NLP, both pickup current and TDS are considered to be continuous variables. The continuity of these variables is consistent with the nature of modern digital relays. However, long preprocessing time and complexity during the fault coordination analysis stage of Figure 12.2 are the main disadvantages of NLP [25]. The main advantage of these heuristic algorithms is that they can search a wider space of solutions by the high number of populations and generations to find a global optimum solution.

Intercommunication between OCRs enhances speed and reliability of the protection system. The simplest approach is to provide a communication between a pair of OCRs at the both ends of the line so that they detect whether the corresponding OCR detects a fault in the opposite direction and therefore operate instantaneously without communications. There is a time delay associated with the time it takes for OCRs to independently act on their time-trip settings. This approach is applied to achieve the functionality of DOCR using low-cost OCRs. Comparing the current phase of the adjacent OCRs indicates if the currents are in the same or opposite direction. In the normal condition, the current must be of the same direction at the two ends of the line. If the fault is between the two OCRs, then opposite directions are detected.

12.2.3.3 Differential relays

Differential relays may be required to discriminate between the fault locations and to isolate internal faults within distribution components (i.e., transformers, generators), short lines and busbars. Because of their high cost, differential relays should be applied judiciously. A current differential protective relay measures and subtracts the currents flowing into and out of bus-terminal connections, a transformer or transmission lines. If the measured current difference is above a specified threshold, then the fault of the associated terminals, component or bus section is isolated from the rest of the distribution system. Using a differential relay as a line protection, the measured information of the one end typically is transferred by a pilot wire, fiber optics communication, power line carrier, or a wireless communication network to the other end of the line [28]. A typical differential scheme is shown in Figure 12.5. With the rapid increase in communication technology, implementation of the differential protection is becoming less costly. The main benefits of the differential relay include high sensitivity, selectivity and immunity to power swing and load encroachment, high selectivity to high impedance faults, and simplicity.

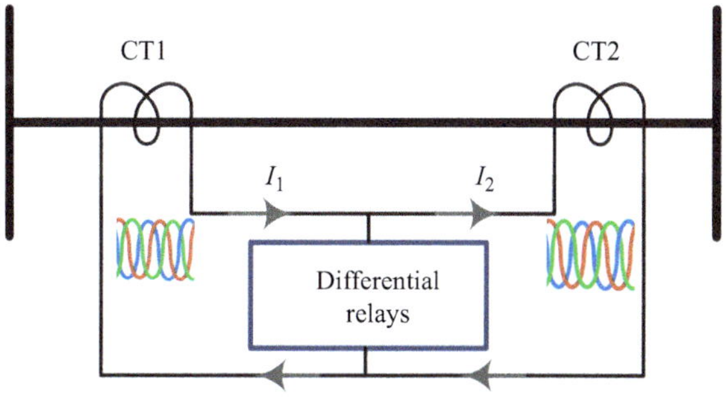

Figure 12.5 Illustrative two-terminal differential protection scheme

12.2.3.4 Under/overvoltage/frequency protection

Voltage sag usually occurs as a result of the fault, overload and starting of the large motors. On the other hand, overvoltage takes place due to many reasons such as lightning, switching, disconnection of bulk loads, ferroresonance and insulation faults. Typically, disturbances in power system disrupt the balance between consumption and generation of active and reactive powers, and as a result, voltage and frequency stability are jeopardized, simultaneously. Accordingly, under voltage relay (UVR) and overvoltage relay (OVR), under frequency relay (UFR) and over frequency relay (OFR) are four other conventional protective relays used to enhance selectivity of the system through mitigation of voltage and frequency instability aftereffects due to fault events and other extreme events within the system.

12.3 Microgrid protection challenges

Renewable energy can be introduced into the hypothetical grid of Figure 12.3 through independent installations that are residential, commercial or industrial, as shown in Figure 12.6. As a result, each independent installation interfaces independently with the main utility grid through its own distribution interfaces. Typically, the only changes to the distribution network are the addition of feeds from the renewable source (i.e., wind or solar), and the associated RES is treated as a load by the rest of the distribution system (from a distribution protection standpoint). The RES may also be considered as bulk generation to be used by the community, as will be the case for the Municipal Wind installation of Figure 12.6, which is meant to either offload the demand from the high voltage AC (HVac) feed to the community or sell excess energy to the utility through the HVac interface. This will typically be a utility-owned asset. The community wind installation of Figure 12.6 could be a collaboratively owned and managed RES within the community, meant to service only a part of the community. This method for introducing renewables into the grid places burdens on the distribution grid owner that are difficult to manage, poorly utilizes energy resources, from the overall community standpoint, and that can actually reduce grid resiliency. There have been several notable impacts to the grid, driving the need for increased utilization of UVRs, OVRs, UFRs and OFRs. For example, recently, it has been recognized that high penetration of PV systems at the distribution level causes overvoltages, caused by reverse power flow [29].

The better approach for increasing the penetration of renewables, combined with carefully managed energy storage is through the use microgrids. The microgrid enables the management DERs, both RESs and ESSs, in close coordination with the loads being serviced by the microgrid. The microgrid enables the cooperative combination of independently installed RESs through communications and controls that are enabled by the introduction of a cyberphysical structure that allows all users of the RESs to interact with loads and sources to self-optimize usage, while, at the same time, the system autonomously optimizes energy usage within the system according to objectives that are set by the stakeholders—i.e., the energy

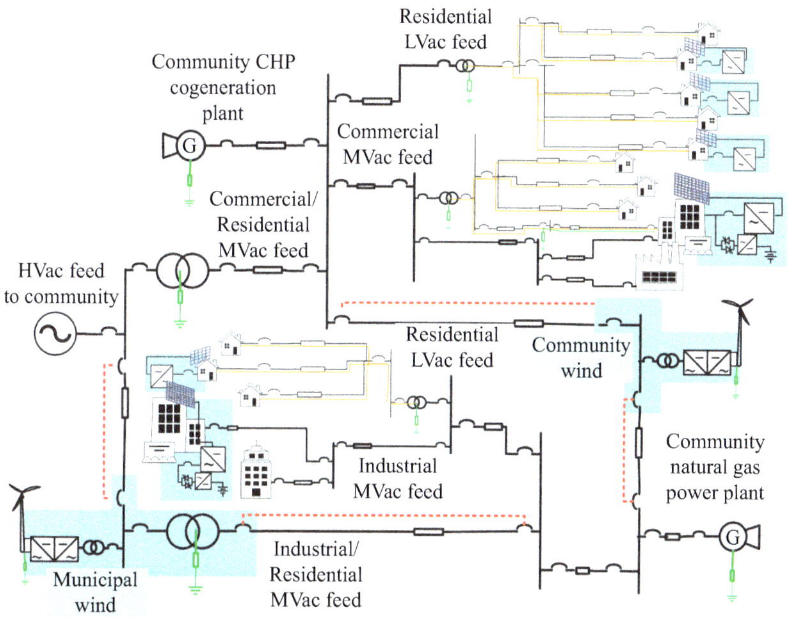

Figure 12.6 Conventional AC grid with independent renewable energy installations

consumers and the energy asset owners. Under normal operating conditions, the microgrid presents a much more attractive option for all participants, local microgrid asset owner, energy consumers and the utility. The microgrid will also increase the resiliency of the distribution system by mitigating the operation of OVR/UVR/OFR/UFR trips associated with poor regulation of uncoordinated RESs and losses of stability within the system. The hypothetical municipal/community distribution system with microgrid installations is shown in Figure 12.7.

The microgrid is largely the creation of power electronics engineers, with the main focus being upon regulation and control through power-electronic interfacing RESs and ESSs. For the most part, the distribution engineer has had little to do with the microgrid—from conception to installation. A fallacy that has resulted is that if the microgrid is installed into the existing AC distribution architecture, the key protective functionality of the distribution system (other than just ensuring the connection of parts of the power system together) is taken care of. In fact, with the introduction of smart-interconnected microgrids, as shown in Figure 12.8, there may be a temptation to allocate all distribution functions to the microgrid(s) without adequate attention to *dependability* of the microgrid(s) to assure the delivery of power and energy service to the loads.

To understand the protection challenges associated with microgrids, it will be important to understand how the introduction of power electronic converters in the fault-feeding path affects the complexity of the protective relay

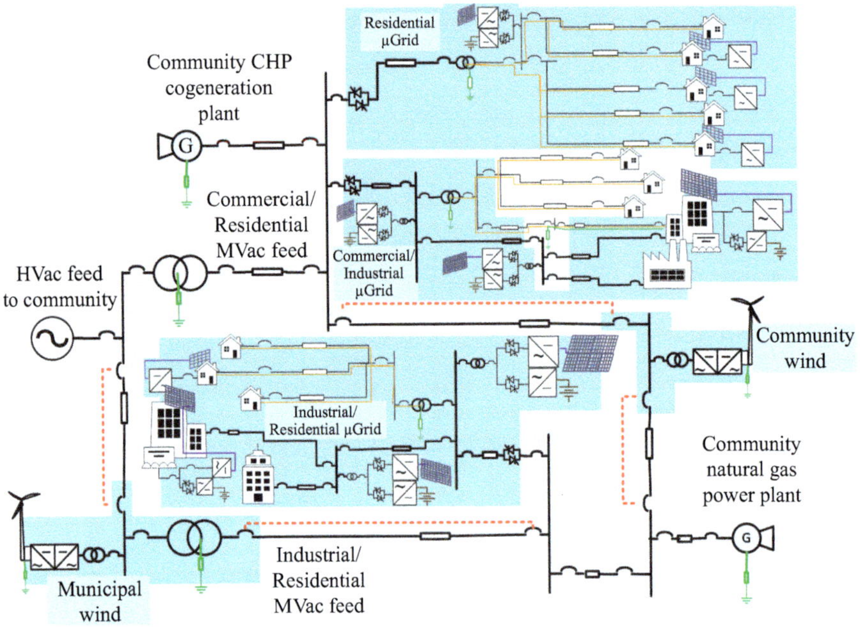

Figure 12.7 Microgrid installations in a conventional AC grid

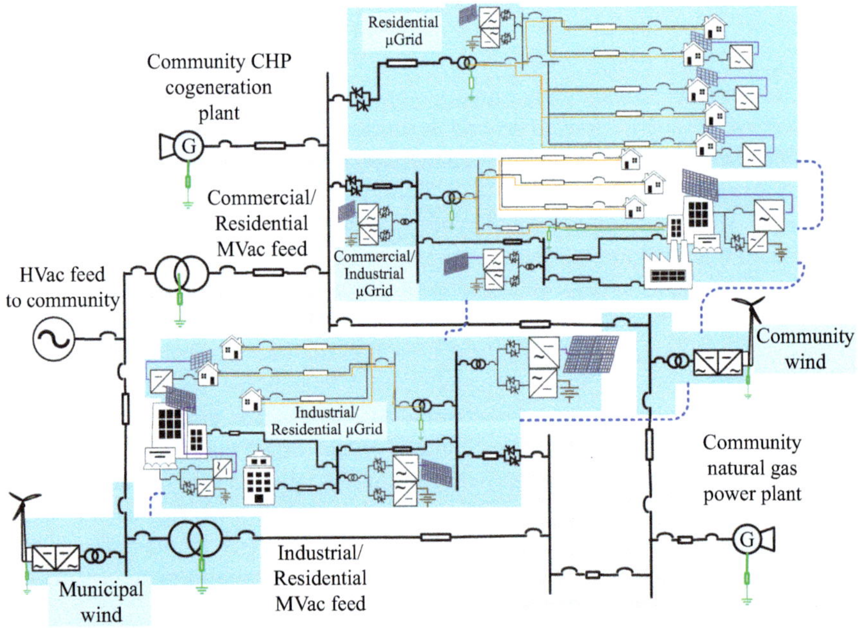

Figure 12.8 Conventional AC grid including networked local microgrid installations or AC community microgrid

implementation. The system of Figure 12.8 will be used as a reference for understanding the challenges that are introduced as DERs with power electronics interfaces being introduced and as the grid, under various configurations, becomes a dynamically changing meshed grid network. The requirements, features and structure of microgrid must necessarily take into account numerous protection challenges including DER impact on power flow, DER impact on fault current magnitude, various connected and islanded configurations of the microgrid and the impacts of cyberattack through the distributed control elements within the microgrid [30].

12.3.1 Impact of distributed energy resources on power flow

Typically, distribution systems are operated in a radial mode. Even though they have closing loops, the normally open switches, which are used for providing energy to un-energized loads after operation of CBs make the loops open [31]. Due to the distribution system structure, once the fault occurs, current flows from a single source to the fault location. As a result, the majority of distribution systems into which the microgrids are installed, without diligent attention to the protection issue, will be OCR protection-based. With the introduction of DERs into the microgrid structure, power flow will be bidirectional. As a result, the OCRs would be no longer effective in protecting of the microgrids. There will be a greater need for the use of DOCRs and differential relays, reducing the economics and simplicity of the protective system. Furthermore, the complexity of the protective system design process associated with coming up with an optimal protection system design comes into play.

12.3.2 Impact of distributed energy resources on fault current magnitude

Installation of DERs in the microgrid causes variations in fault current magnitudes, which could deteriorate the reliability of an existing conventional protection system underlying the microgrid structure. The degree of contribution of DERs to fault current depends upon the number, type and placement of DERs. There are generally four types of DERs: synchronous generators (SGs), asynchronous generators (AGs), doubly fed induction generators (DFIG) and power electronic converter interfaced units [32] (associated with wind, PV solar and the ESS). The fault current magnitude associated with SGs, AGs and DFIGs may reach up to ten times of the nominal current. This may not be a significant problem for distribution protection because the surge current provided to the fault may even aid protective tripping of OCRs, as long as the generators are aligned with the main grid supply. However, the excitation system design of the respective generator may not adequately take into account the fault event response. The duration of the injected current during the fault event may not be long enough to activate the OCR or, in islanded modes, the remaining connected generator(s) may lose regulation of the bus following fault recovery. The DFIG will include a crowbar, which is a series-resistance controlled device between the rotor and rotor-side converter to prevent high rotor currents and speed. During the activation of the crowbar, DFIGs may behave in similar to conventional AG with approximately the same current contribution to the grid [33]. However, the fault characteristic of the DFIG-fed grid is a function of the self-protective behaviors of the DFIG and not the system protective behaviors of an adequately designed protective system. For the power electronic converter-interfaced DER, the issue of fault current magnitude is even more serious. The power electronic converter-interfaced DER's own *unit-based protection* features will limit their fault current to two to three times its nominal current rating in order to prevent equipment damage [34].

Grid operators may impose a set of regulations (grid code) on the DERs to improve the stability of the grid [35]. In fact, this has been the trend for addressing the limitations to the protection system reliability. According to [35], DERs have to be connected to the grid during the fault event for a specific period and are required to actively contribute toward stability of the power system through the contribution of reactive power until fault clearance. Unfortunately, injection of the reactive power during fault events may lead to mis-coordination of the protective relays [36]. In addition, DERs have to control the output current in such a way to meet the requirements. Since there is no standard on control methods addressing this requirement, there is no certainty in fault current contributions of the DERs [33]. Therefore, the likelihood of ensuring reliability, speed and selectivity of protection within conventional distribution system by imposing requirements solely upon the DER units is very low.

12.3.3 Impact of microgrid connection modes and changing configurations

The above concerns of variability in power-flow direction during faults and fault magnitude contributions of the DERs are only aggravated by the fact that the microgrids must be able to operate in both grid-connected and islanded modes. Compared to grid-connected mode, the available fault current magnitude level is significantly reduced in islanded mode, especially when the majority of DERs are power electronics interfaced. As a consequence, when a microgrid transits from grid-connected to islanded mode, the existing OCRs, with parameters set based upon grid-connected mode, will have very poor selectivity, speed and selectivity. More definite attributes (falling out of dependability theory) are also a concern such as security and recoverability. All of this assumes that the microgrid configuration (i.e., radial) has remained in islanded mode. However, it is also very likely that the microgrid structure will change from, say, radial to meshed or meshed to ring, depending upon how the microgrid or, in the case of the networked community AC microgrids of Figure 12.8, microgrid(s) are islanded from the rest of the system. Not only will the settings of the protective equipment be inadequate but an entirely new protective scheme may also be required. As a result, a proper protection system has to be adaptive/independent to any change in microgrid configuration [37]. In general, the current measured at relay point are affected by many factors, including operation mode of microgrid (islanded and grid-connected modes), number of DERs connected to the microgrid, control mode, microgrid configuration and fault impedance. This could be formulated as follows:

$$I_{relay} = I_{fault-grid}(d, GC, R_f) \times Operation\ mode + \sum_{i=1}^{n}\left(k_i(d, GC) \times I_{faultDG_i}(CM, DG'stype, R_f) \times Status_{DG_i}\right)$$

Various islanding modes are illustrated in Figure 12.9.

To illustrate the above, two different scenarios that may occur within the hypothetical network of Figure 12.8 are discussed. One results when the system is islanded from the HVac and community natural gas power plant feeds and the commercial residential MVac feed is isolated. The resulting structure is a ring bus having only the municipal and community wind DERs. The industrial/residential microgrid is considered as a load feed off of the resultant ring bus. When an LL fault occurs on the ring bus, three possible maloperations of protective equipment scenarios may result as shown in Figure 12.10 [38]. In all cases, the correct protective action should be initiated by the tripping of the CB closest to the municipal wind feed, followed by tripping of the CB on the other side of the fault location. However, in maloperation scenario, (a) the DER(s) from the industrial/residential microgrid and community wind reduce the fault current, thereby inhibiting operation of the OCR associated with the municipal wind feeder CB. In (b), one or more

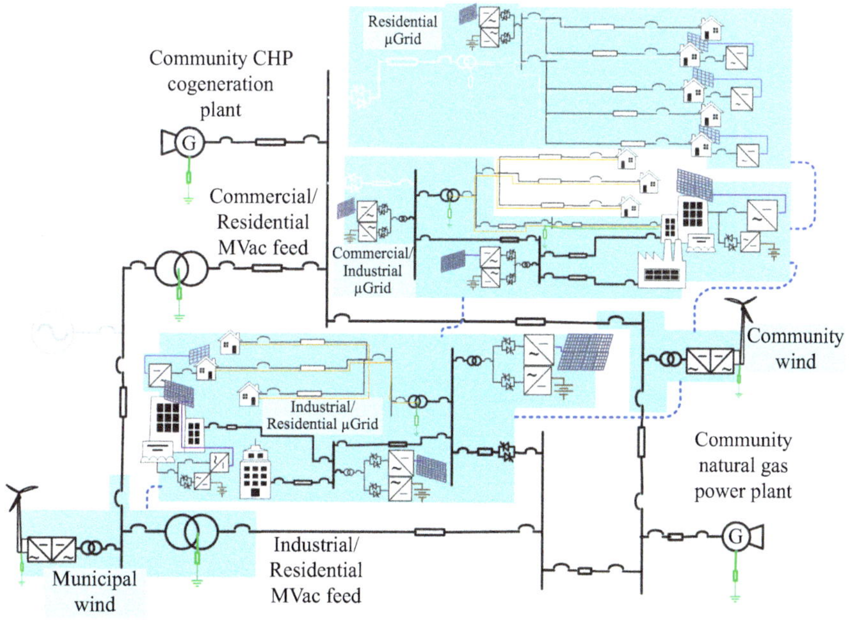

Figure 12.9 AC community microgrids islanded from the main utility grid and non-power electronic interfaced DERs

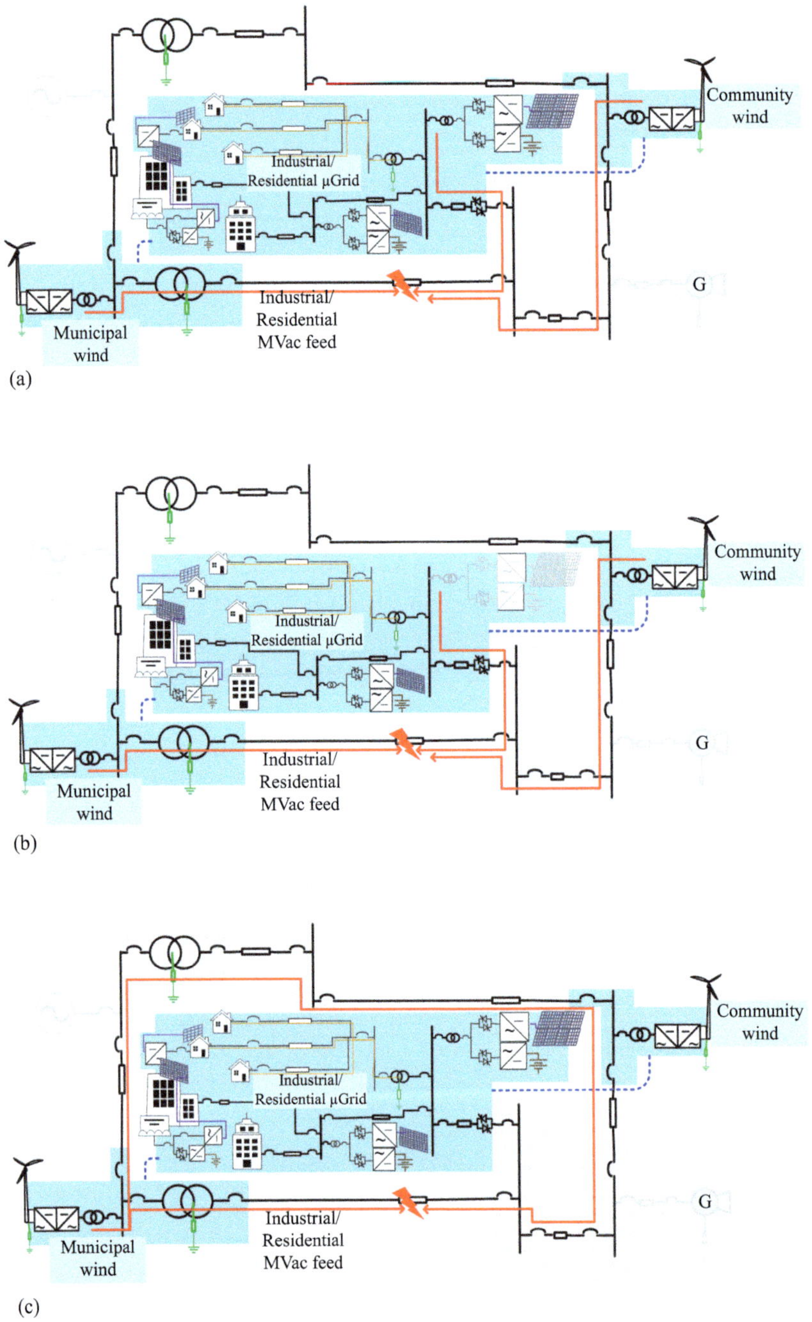

Figure 12.10 Protection failures in an islanded configuration: (a) blinding of protection, (b) sympathetic tripping and (c) weak-infeed loop fault

DER(s) within the industrial/residential microgrid fault(s) off on unit protection, reducing the availability of power to users within that microgrid. In (c), one or more CBs around the ring trip off instead of the required CBs on either side of the fault.

12.3.4 Earthing considerations

A critical consideration to microgrids is the earthing arrangement—particularly for LV microgrids. The power electronic and protection equipment making up the microgrid are contained within enclosures. The system safety is greatly affected by the grounding strategy, starting at the enclosure level where the enclosure chassis forms a ground reference or chassis ground. The earthing arrangement choices are related to the connections between the chassis ground, earth ground and the electrical reference points within the system circuit. The earth ground is usually made through electrodes at the above ground surface part of a ground rod (outside) or through designated electrodes within an installation that in some way make a contact connecting the actual earth beneath the installation (i.e., through rods, sewer pipes or various connections to metal rods distributed in concrete at the ground interfacing level of a building). The response of the system to LG and LLG faults depends upon the earthing arrangement. In AC systems, LG faults occurring within equipment are typically referred to as phase to chassis ground faults. For microgrid installations, the indoor part of the system will route cabling through trays which, in order to meet grounding and bonding safety requirements, are typically electrically connected to the chassis of interconnected equipment. Therefore, LG faults occurring at cable locations in cable trays are also, essentially, phase-to-chassis faults. Outside of the installation if cables are buried in the ground, the LG fault represents varying levels of phase-to-earth ground faults.

12.3.4.1 TN, TT and IT systems

The current level of phase-to-chassis ground faults is highly affected by earthing arrangement. There are two earthing wires in power systems. One of them is protective earthing (PE) which is the meshed equipotential bonding connecting devices and equipment metal frame and structures to the earth solidly in order to personal safety (i.e., chassis and cable trays). PE is one of the main concerns in LV distribution system designs for a wide class of consumers. The other earthing wire, which does not play a significant role in earth fault current level, is neutral earthing (NE). NE means the connection type of transformer and generator neutral point to ground. Three earthing arrangements are defined using two letters, TN, TT and IT. The first letter indicates the generators (or supply transformers) neutral point earthing connection:

- *T*, standing for Terra, means solid ground connection of neutral point.
- *I*, standing for Isolated, means no connection of neutral point to earth.

The TN system, as shown in Figure 12.11, may have three different configurations, TN-S, TN-C and TN-C-S, which do not affect earth fault current level significantly. In TN systems, the neutral point of the supply source (either transformer or generator) is solidly earthed with a conductor. The body of the supply source is earthed through the same conductor as well. Either one conductor for both PE and NE (TN-C system) or two conductors, for each NE and PE (TN-S system) are routed all the way from supply source to user along with the three phases, as shown in Figure 12.11. All metal frames of the consumers' devices must be connected to the PE conductor, and they do not require individual earthing system.

In TT systems, four conductors, including three phases and neutral, are routed between supply source and user. However, the PE for the supply source and user must be provided by an individual earth electrode at the point of supply or usage. There is no wire between separate PEs. The IT system is shown in Figure 12.12.

The principle of IT systems is that the neutral point of the supply source is not intentionally connected to the earth ground at all. The IT system is typically referred to as an *ungrounded* or *floating* system. Although there is no intentional direct connection, such systems will always make a connection to chassis (inside enclosures and trays) or earth ground (underground cables) through stray capacitance between current carrying cables, terminations and chassis or earth. By this same principle, the chassis (if not directly connected through safety ground) is always capacitively coupled to earth ground. For this reason, safety requirements within a facility always require a direct wire connection between chassis or trays and building ground in order to avoid the buildup of high touch-voltage on the outside metallic surfaces of chassis and trays. In short, in correctly installed

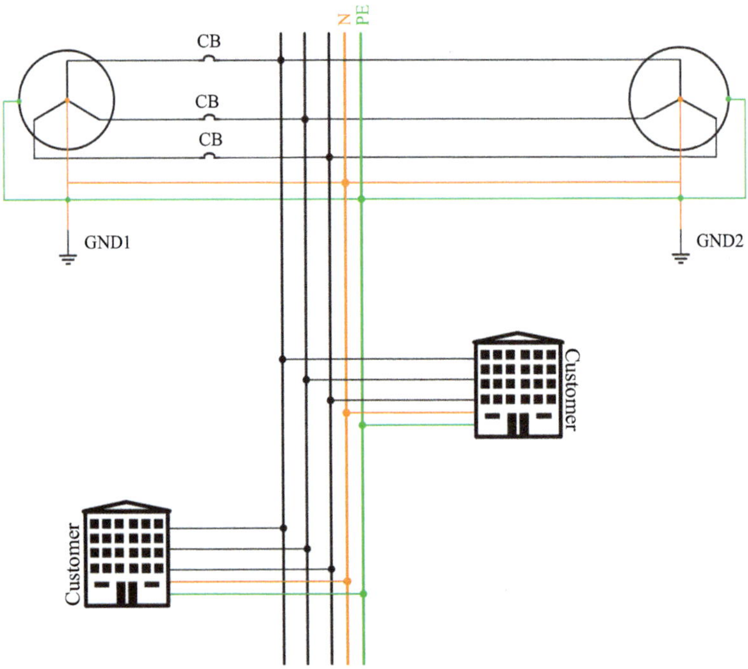

Figure 12.11 TN low voltage microgrid

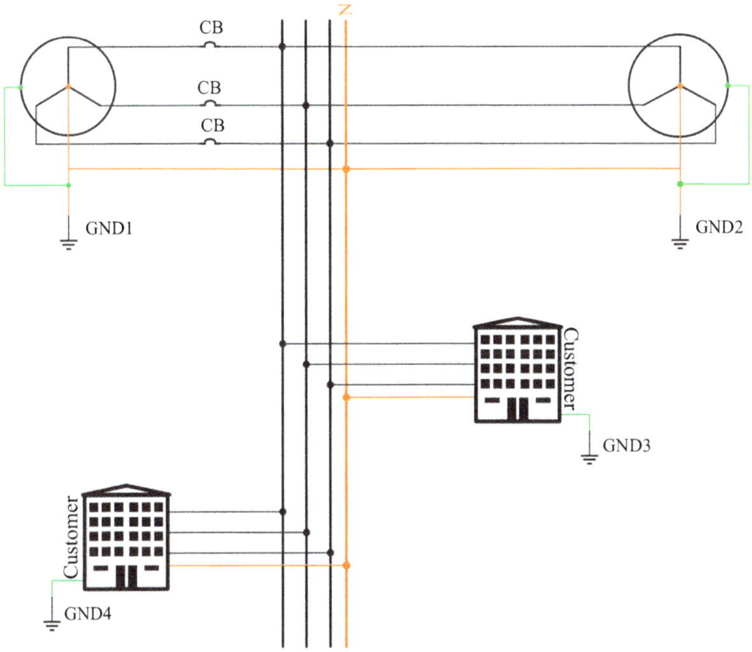

Figure 12.12 TT low voltage microgrid

systems, parasitic capacitive coupling will effectively occur between a current carrying cable or termination and earth ground. The IT system is shown in Figure 12.13.

The IT system has a very large neutral to ground impedance, $Z_{neutral}$. In such systems, common mode impedance, consisting of the leakage capacitance and resistance between each live conductor and the earth, must be taken into account. Fault current in three different earthing systems will be equal to

$$I_{fault-TN} = \frac{Vgeneration}{(Z_{cable} + Z_{fault} + Z_{cable})} \tag{12.3}$$

$$I_{fault-TT} = \frac{Vgeneration}{(Z_{cable} + Z_{fault} + Z_{earth-consumer} + Z_{neutral})} \tag{12.4}$$

$$I_{fault-IT} = \frac{Vgeneration}{(Z_{cable} + Z_{fault} + 3jC\omega Z_{fault}Z_{neutral} + Z_{neutral})} \tag{12.5}$$

where C is the phase-to-earth capacitors, assuming they are balanced.

12.3.4.2 Line-to-ground faults in radial LVac microgrid

Consider the distribution network shown in Figure 12.14. This distribution network has four community microgrids feeders connected to an upstream MV to LV transformer that interfaces to the utility. Assumptions are that the buildings have

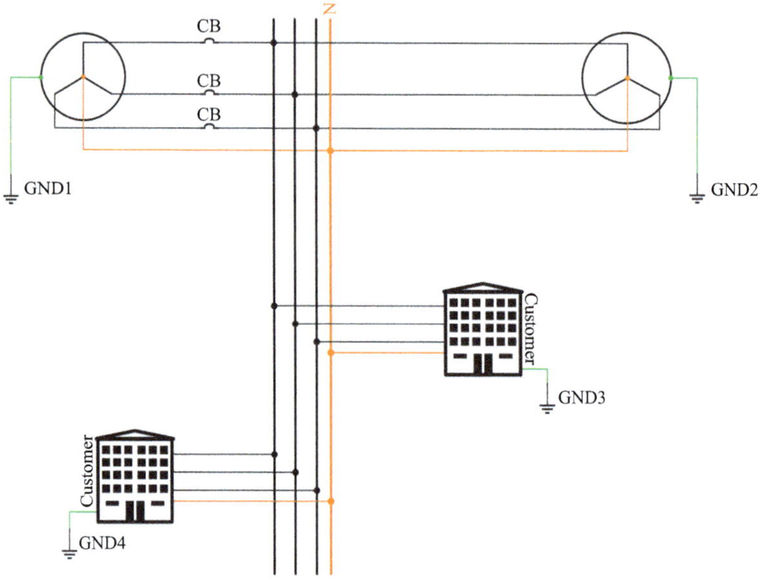

Figure 12.13 IT low voltage microgrid

their own solar PV installations and that each community microgrid is fed by two SGs in the islanded mode. The generators are droop controlled. The load assumption for each building is a heating ventilation and air conditioning (HVAC) system consisting of multiple three-phase induction motors and additional loading (i.e., lights, appliances) that are assumed to be resistive. Figure 12.15 shows the fault characteristics of a phase to chassis fault occurring at the first building on the bottom left of Figure 12.14. The difference between the fault current levels, dependent upon the earthing arrangements, indicates the need for different protection settings for each earthing arrangements in order to reliably detect and clear the fault.

In the normal condition, the current path is from the upstream feeder to the downstream loads. The red components, shown in Figure 12.14, represent OCRs at upstream locations and the blue components represent OCRs at downstream locations. These cascaded OCRs must be coordinated as explained in Section 12.2.3.1, based on the CTI that will provide a sufficient difference in time-trip setting to reliably discriminate the fault. Often descriptive statistics are necessary for coordination, where data feeding these statistics come from fault characterization applied to the system compared to the nominal load level calculations. The maximum and minimum values of the rated current levels in addition to the maximum and minimum value of the fault current levels relating to the points with the overcurrent device connection must be derived. The current settings must be greater than the maximum load current level while they must be lower than the minimum fault current level of the downstream relay. Eventually, an optimization method such as

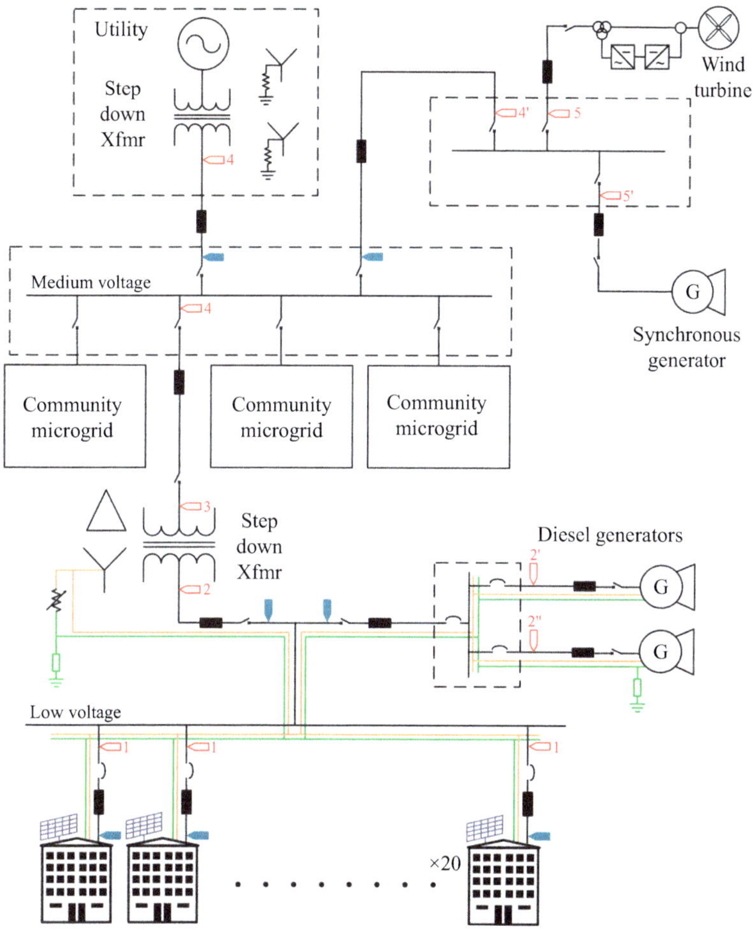

Figure 12.14 AC radial community microgrid

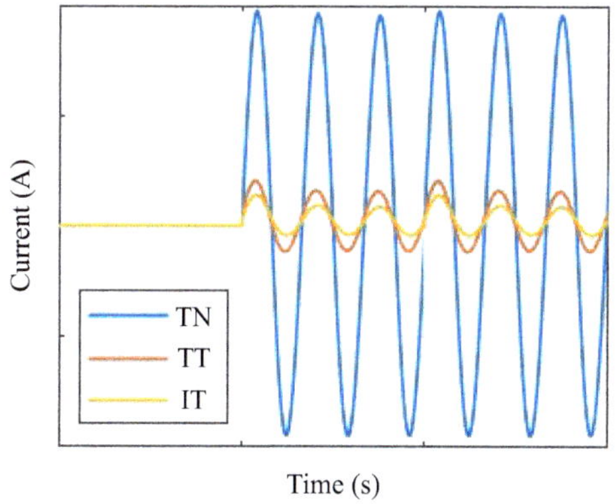

Figure 12.15 Characteristic LG fault currents in radial AC microgrid for various grounding schemes

particle swarm optimization and genetic algorithm may be required to find the optimized value for the manually set-able time-trip or "time-dial" settings.

As a result of varying islanded configurations that could occur within the microgrid of Figure 12.14, the optimal settings for the blue OCRs could vary significantly. For example, when a community microgrid is islanded from the utility, the OCRs associated with the local diesel generators must coordinate with downstream OCRs. In this case, the fault current sourcing capacity has changed and it may be necessary for the system to dynamically adjust CTIs in order to ensure fault discrimination. Even more extreme is the case where an individual building is islanded and operates off of its local solar PV generation. In this case, the grounding structure will change from TN or TT to IT, because inverter-based DERs will usually be isolated from ground to minimize the high-frequency circulating current that might otherwise circulate through the system (assuming that the DERs are not transformer isolated from the system). This change in configuration could cause an even more significant need to make configuration-based changes to the coordination scheme that requires real-time response.

For each microgrid, fault directional issues may occur as well. For example, the local diesel generators must be prohibited from causing OCR trips in the upstream grid. Also, the solar PV installed in one microgrid should not interact with the protection systems in the other microgrid. Therefore, DOCR is required at various levels to ensure proper protective coordination. Here again, the current settings will be configuration dependent.

12.3.5 Cyberattacks

Future microgrids will be equipped by communication networks for the purpose of data accumulation, control, monitoring, protection or energy management. Wide utilization of communication systems makes microgrids susceptible to malicious cyberattacks and may worsen the selectivity of the protection system [39–41]. Typically, cyberattacks are categorized into denial-of-service (DOS) attacks, eavesdropping attacks, account cracking, etc. [42]. These cyberattacks could be launched simultaneously, in a coordinate manner from multiple points in either space or in time, or as a combination of both simultaneous and coordinated attacks. Accordingly, they can influence on at least one of the security parameters including confidentiality, availability, integrity and non-repudiation/accountability of information [43].

12.4 Promising solutions for microgrid protection

At the core of any solution to microgrid protection is protective system design process. This protective system design process has to recognize that the protective equipment and scheme must be adaptive and independent to any change in the microgrid configuration [30]. Adaptive protective schemes require an awareness of the state of the system. Achieving this awareness will be aided by evolving communication standards. Data analytics and machine learning, applied during the design phase and in the field, will play a significant role. In general, the microgrid must be viewed as power/energy delivery service. The principles of dependability theory will apply, expanding the design objectives for protective system design, i.e., to include attributes such as security and availability, and re-formalizing the way in which protective systems are tested before deployment. Unit protection of the DERs and energy storage may become an integral part of the protective system design. And, finally, considering the objective of achieving an *energy secure* microgrid, i.e., maximizing *power availability*, *resiliency* and *efficiency*, the protective system design becomes the principle means by which *resiliency* of an energy secure system is achieved.

In order to mitigate DER impacts on resiliency during fault absorption and recovery, several microgrid protection methods have been proposed. These include limiting the maximum DER capacity; evolution of communication and interface standards as a means of imposing requirements on systems and components; utilization of fault current limiters (FCLs); ESSs-based fault discrimination and modifying the DER control. These protection approaches attempt to neutralize the DER effects on fault protection where the conventional AC distribution protection equipment is relied upon. A more comprehensive strategy for microgrid protection would involve revisiting altogether the protection system design process and coming up with a revised process that can be applied, leading to the right protection equipment and schemes for the microgrid installation—as opposed to adapting existing protection and equipment and schemes to the microgrid after it is installed. The end result of the protective design process may very well be an application of techniques for microgrid protection that are already in use, it will certainly keep pace with and, over time, influence evolving microgrid communication standards and illuminate new and unique approaches to microgrid protection, tailored to the objectives and constraints associated with the specific microgrid installation.

12.4.1 Limiting maximum DER capacity

As explained in previous sections, DERs will affect fault current levels. One of the simplest and primitive methods for mitigating this effect is to limit the maximum capacity of DERs in order to prevent mis-coordination of protective relays [44,45]. The main drawback of this method is that it also imposes a limitation on the penetration rate of DGs in the future power system. This approach is essentially "kicking the ball down the road" when it comes to solving the protection problem.

12.4.2 Evolving communication standards

With the introduction of power industry deregulation, a rising number of intelligent electronic devices (IEDs), such as relays, metering devices and digital disturbance recording devices, are being integrated into distribution grids [46]. According to this evolution, one of the upcoming challenges of the substation automation system (SAS) is the establishment of a comprehensive communication standard to support interoperability, interchangeability and scalability in the power system. It follows that as these standards incorporate protection consideration into the development of standards for emerging IEDs that, as a minimum, future IEDs and even protection equipment products could emerge having the capability addressing protection needs.

IEC 61850 is a widely used standard for the SAS [47]. In IEC 61850, three main classes of communications have been proposed including Generic Object Oriented Substation Event (GOOSE); sampled measured values (referred to as SVM but not to be confused with support vector machine (SVM) which uses the same acronym in this chapter) based on the publisher/describer mechanism and client–server communication between the electrical network monitoring and control system; and IEDs operated based on manufacturing message specification (MMS). GOOSE is an unsolicited and asynchronous message that transmits data from a single IED to single or multiple IEDs in peer-to-peer or one-to-many modes, respectively. Process bus technology transfers SMVs of currents and voltages to the IEDs [48]. MMS is used for no time requirement applications such as communication between controllers, between stations and controllers (see Figure 12.16).

These IEC 61850-based communication services utilizing for power utility automation are shown in Figure 1.9. Different classes of the IEC 61850 are defined as follows:

1. IEC 61850-7-4xx is for modelling of hydro power and DGs such as diesel generators, solar panels, fuel cells and CHP.
2. IEC 61850-5xx is for user guide.

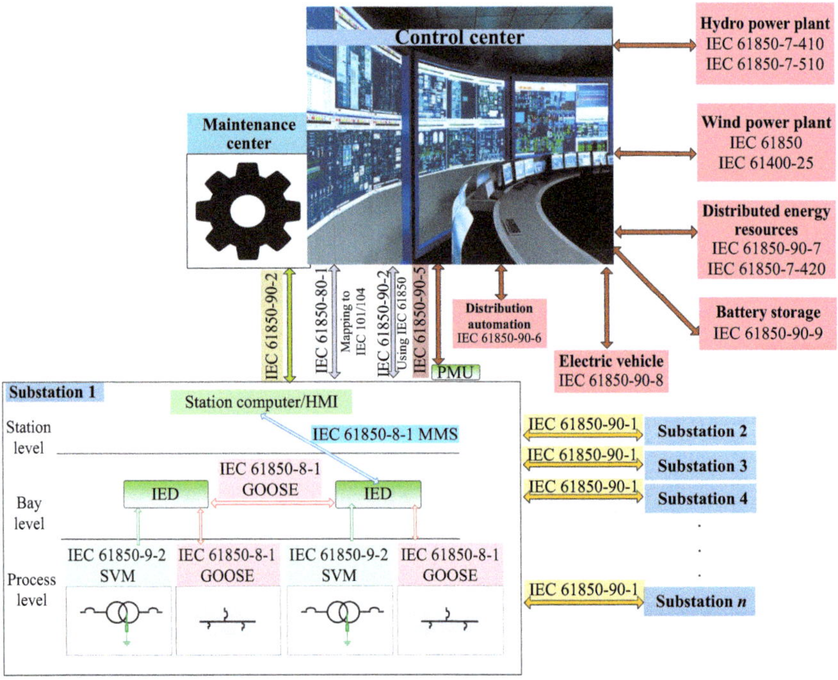

Figure 12.16 Communication networks for power utility automation

3. IEC 61850-80-x is for mapping to IEC 60870-5-101 and DNP3.
4. IEC 61850-90-x is for communication between substations, communication between substations and control centers, condition monitoring, transmission of synchrophasor information, network engineering guide for substations, distribution automation for PV, storage and electric vehicles schedules.

IEC 61850 is specifically designed for exchanging information between IEDs and modelling system's elements. However, one shortcoming in the IEC 61850 standard is lack of standardization of sequential, combinational, rule-based or any other forms of power system control and automation logic, such as interlocking logic for control operation [49].

In addition to IEC 61850, Common Information Model (CIM) standards including IEC 61970/61968/62325 are widely used to allow interoperability in smart grid domain [50]. CIM standards present data models, which are based on Unified Modelling Language bringing interoperability into the wide range of energy-management systems (EMS). Since there is massive data exchange between the EMS and substation automation, CIM and IEC 61850 must have high compatibility. Although these two class of standards are different, both in nature and evolution, many methods have been proposed to unify them [51–53].

The IEC 61499 standard, on the other hand, is presented to model-distributed industrial process measurement and control systems. The architecture of the IEC 61499 standard is based on event-driven function blocks (FBs) encapsulating functionalities, behaviors and their signal interconnections. These FBs can be combined to constitute a complex and hierarchical system description. Use of FBs facilitates implementation of the control system. Since a protection scheme of microgrids may consist of communication links, control system and intelligent management center, a promising standard must completely cover communications, modelling and distributed control. The integration of IEC 61850, IEC 61499 and CIM standards could meet the mentioned requirements. Recently, IEC 61850/61499 was successfully implemented to enhance flexibility and adaptability of automation systems.

12.4.3 Fault current limiters

FCL is another method to suppress DER fault current contribution and help the conventional protective relays to operate without changing their settings [54]. An ideal FCL is a device the impedance of which rapidly increases in the faulty condition from zero to a high value to limit fault current contribution of the DER [55,56]. Although FCLs allow a higher penetration of DGs, some main disadvantages such as the initial cost of FCL, the high recovery time of SFCL, and switching losses of SSFCL still exist.

12.4.4 Utilization of the ESS for fault discrimination

In the islanded mode, fault current level will be dropped considerably. One of the promising solutions is utilizing ESSs to provide a sufficient amount of fault current into the microgrid while in the islanded mode [57,58]. Similar to employing FCLs, this method does not provide a cost-effective solution.

12.4.5 Distributed generation control modifications

Another solution is to limit fault current by designing a proper control of DGs to control fault current contribution [36,59]. This method has many benefits such as being cost-effective, enabling higher integration of DGs, less need for adaptive settings of relays, and improving fault-ride through capability of DGs. However, one of the main disadvantages is that controlling fault current in indirect current control is difficult and overcurrent transients appear at the beginning of the limiting process [60].

12.4.6 Protective system design process for microgrids

Protective system design is undergoing an enormous paradigm shift away from the conventional process of Figure 12.2. The revised process for protective system design is shown in Figure 12.17. Here the microgrid is viewed as a power/energy delivery service with dynamic states of configuration subject to various fault type behaviors that are, in turn, affected by those configurations. The fault characteristic cannot be modeled with simple time-current expressions but require detailed control-hardware-in-the-loop (CHiL) simulations of the system under all possible configurations. Such conditions will include all possible fault types (that can be modeled) in every possible configuration. The CHiL simulations will represent sets of training cases that will be used to identify specific traits or characteristics unique to the fault condition. These traits or characteristics are identified through *features* that can be extracted by performing mathematical operations or data analytics on captured voltage and current *snapshots* associated with the fault.

In islanded configurations, the margin of difference between the most severe load transients and the limitations of fault sourcing current imposed by power electronics-interfacing DERs will be very small. This makes it difficult to discern load transients, such as start-up inrush currents and large load shifts, from various fault types. This challenge is illustrated in Figure 12.18, which shows the inverse time–current relationships between the grid feeder (in play while in grid-connected mode), a branch feed (which is also where the DER is connected) and a downstream load feed. The inverter-based DER imposes a limit on the branch feed that cannot be exceeded without tripping off the DER. Load transients are superimposed upon this curve in relation to the OCR time-trip settings. The close relationships between these current trajectories and protection equipment and protection scheme limits illustrate the need for additional local intelligence to be built into the PD to discern between a fault and a normal transient. *Features* such as change of energy, Shannon entropy and standard deviation from each phase of current signal can be extracted from the measured signals at the PD to aid in the fault discernment process. Furthermore, the distinctive shape of a fault or non-behavior comes into play, so signal-processing tools, such as the wavelet transform may be required to preprocess data before feature extraction. All of this requires a certain amount of high resolution field programmable gate array (FPGA)-based

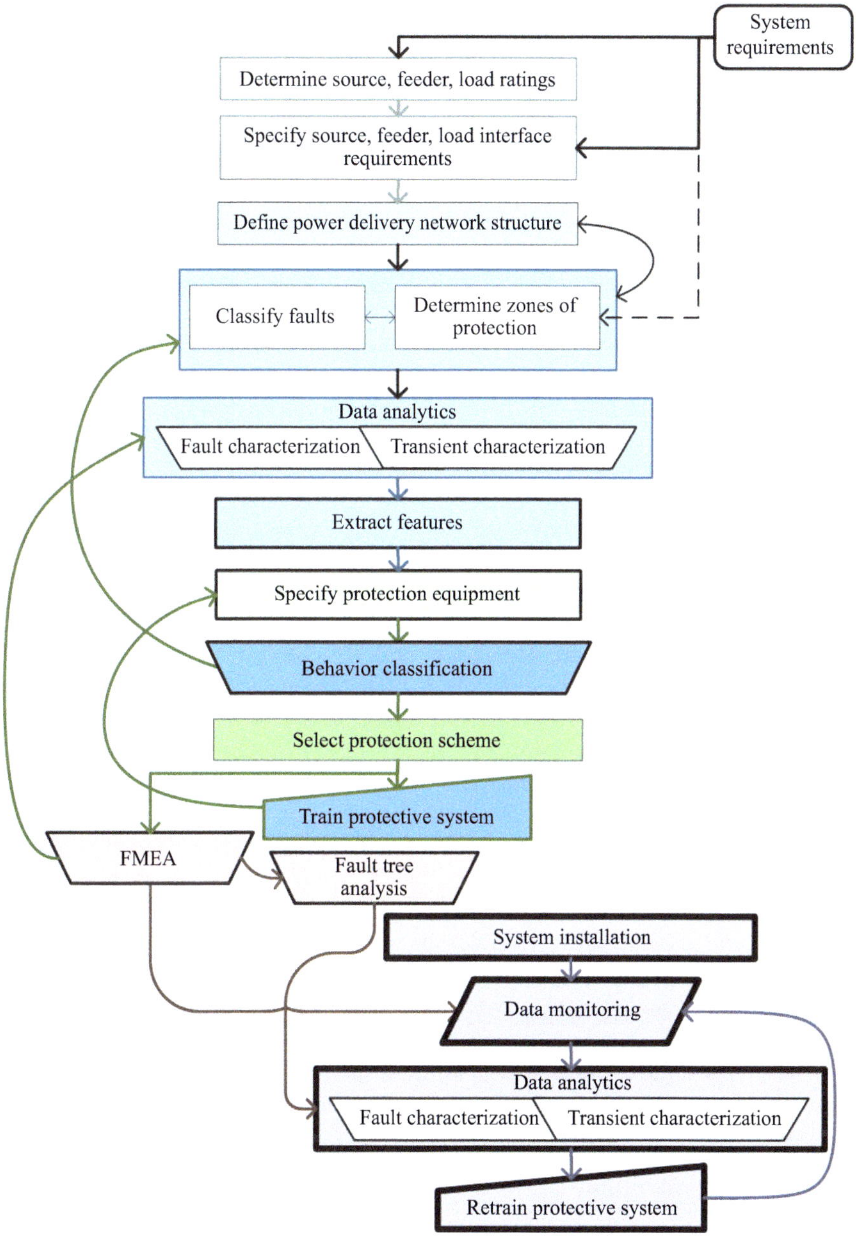

Figure 12.17 Microgrid protective system design process

local processing at the PD. Also, for each possible configuration, sets of transient events should be simulated.

The CHiL-based fault characterization will be used to identify features that can be extracted captured data, analyze those features and then classify them as faults

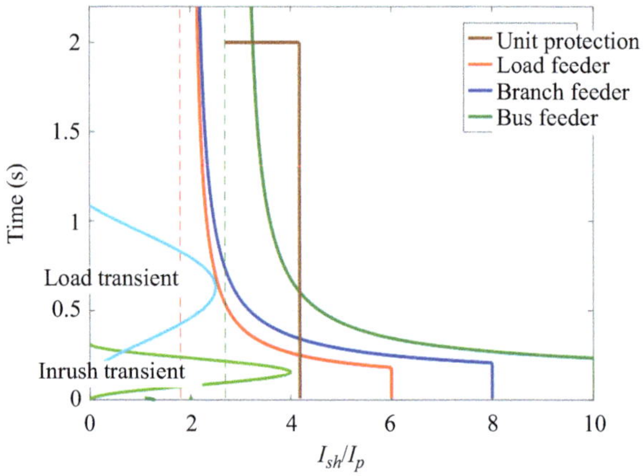

Figure 12.18 Inverse time-current curves for OCRs in a radial grid with load transients superimposed

or non-faults. Machine-learning techniques, such an SVM, can be utilized to perform this classification [61]. The CHiL simulations generate training cases for the SVM. The combination of feature extraction plus SVM can then form the basis of the protection scheme design. The proper protective equipment is selected to support stress mitigation, adaptive controls and data monitoring requirements derived during the protective equipment tailoring process. The utility of the CHiL in this process is the execution of real-time simulations that includes protective equipment and scheme time-delay responses. The requirements for data acquisition device interconnection sampling rates, latencies and resolution can also be derived and actual protective equipment selected can be tested before the microgrid is deployed.

This design process is then followed by a failure modes and effects analysis (FMEA) and fault tree analysis (FTA) that can be used to adapt the protective equipment and scheme during the design phase and that can inform the implementation of the deployed equipment. The final deployed protective functionality of the power/energy deliverable system must continue to be adaptive, learning from its environment and therefore maximizing resiliency.

The following subsections provide further insight into the steps applied in Figure 12.17.

12.4.6.1 Data analytics, feature extraction and behavior classification

As mentioned above, the fault characterization process of Figure 12.2 is replaced with data analytics, feature extraction and behavior classification steps. The data to be analyzed must be generated by detailed simulations of the microgrid, preferably through use of a CHiL platform that can simulate the right level of detail for fault characterization. Because of the sheer size of such simulation models, real-time or near real-time execution is needed to generate the required data in reasonable timeframes—hence the need for CHiL simulations. For systems where the majority of the energy and power is processed through power electronic converters, such as will be the case for the islanded microgrid, the distinguishing characteristics of fault behavior versus normal operational transients will be based upon signal shapes as opposed to signal levels or they may be a complex combination of shape and level. The protective scheme may influence the fault behavior, as will be the case for harmonic-injection-based techniques for fault location and discrimination of the parts of the system to be isolated. As a result, multiple stages of iteration may occur throughout the process (as indicated in Figure 12.17). Methods for data analysis for the purposes of fault characterization, fault location and fault discrimination may be categorized as pattern based, harmonic based, and traveling wave based.

In a pattern-based method, signal-processing tools such as wavelet transform and S transform may be utilized to extract desired components, features or signals from the captured data associated with a fault event. In [62], different features such as change of energy, Shannon entropy and standard deviation are extracted from each phase of the current signal. Then, a decision tree classifier uses these features to detect and classify the fault. It is shown that the integration of wavelet transform and decision tree enhance the reliability of the protective response. For transient disturbances, such as the transition from islanded mode to grid-connected mode, and connection to and starting of induction motors, frequency components are located in low-frequency half band with fixed locations and decaying magnitude. In faulted conditions, frequency components may be located in both low-frequency and high-frequency half bands with changing locations and magnitude. Subsequently, the wavelet transform extracts relocated frequency components of the signal to identify the type of disturbance [63]. There is a high degree of utility in using pattern recognition techniques to identify and locate high impedance faults and LG faults of varying impedance levels.

Harmonic-based methods are applied in combination with the injection of a high-frequency harmonic signal upon detection of a fault event. High-frequency injection requires a high-frequency injection source, which is provided by power electronic converters associated with DERs in the system. This approach adds system complexity and begins to transform the grid-protective approach to a more unit-protection centric approach (where the *units* are the power/energy source interfacing power electronic converters in the system). However, such approaches may make better sense as they embrace the changing nature of the power-distribution system to take advantage of inherent processing and sensing capabilities in the DERs as a means toward significantly simplifying the PDs in the system. For example, the closeness of a DER to a fault may be determined through simple communications between adjacent DERs in a meshed system after which the closest DERs on either potentially faulty line inject a specific frequencies used by very simple OCR devices (tuned to those frequencies) to faulted section of the faulty line [64].

Once the fault occurs, voltage and current traveling waves propagate along the distribution lines. Thanks to this phenomenon, the fault can be located by analyzing the traveling wave magnitudes, polarity and time intervals between the arriving waves. In [65], time and polarity of initial current traveling waves are extracted by a mathematical morphology method to identify a fault in a meshed microgrid. Although this method has a fast-time response, locating fault requires extremely high-sampling rate and very fast processor to be effective in the short line lengths associated with most microgrids.

12.4.6.2 Adaptive protection

As has been clearly stated, a defining capability of future microgrid protection schemes will be their ability to adapt to changing conditions. To enable such capability, protective relays must be able to sense or predict microgrid operation and configuration modes and DER status in order to adjust the relay settings to make itself compatible with the current conditions of the microgrid [66–68]. If any change occurs, new optimal relay settings must be sent to the protective relays. Generally, the algorithms for calculating the relay settings are classified into mathematical [69,70] and intelligent methods [71,72].

There are three essential attributes of an adaptive protection system. First of all, *it is critical that the protective scheme has the capability to recognize the conditions associated with the need for adaptation*. If the protective system design process of Figure 12.17 is followed, this capability will be a natural outcome of the data analytics, feature extraction and behavior classification process. Once the protection scheme is determined or designed, it will be necessary to train the protection scheme to respond correctly to fault events. Correct response will necessarily require readaptation of the system to stimulus according to detectable changes in configuration, connection or status (i.e., extracted from features from a decision-tree process) and a retraining of the system. The CHiL simulations are essential to this process because they provide the training cases.

Second, *the protection scheme must have the capability to adapt in the field and to learn from new behaviors encountered in the field, followed by an autonomous retraining process*. This requirement highlights a significant need for future research in order to combine machine learning and artificial intelligence capabilities that can be partitioned and deployed into the constituent parts of the microgrid power/energy delivery network. It is suggested in Figure 12.17 that once protection equipment and protection scheme are determined, FMEA and FTA should be performed as part of the system-design process. The results of FMEA and FTA can then inform the algorithms that will ultimately be deployed into the field and will form a critical basis upon which autonomous field adaptability is achieved.

Finally, it is recognized that the functionality of adaptive protection is enabled by the allocation of functionality to the components of the protection equipment and scheme which make up the constituent parts of the protective system. *In order for the system to adapt, all of these constituent parts of the protective system must have an awareness and a knowledge of the system configuration, connection and status*—this is the third essential attribute of an adaptive protective scheme. A brute force approach, and one that is still limited by the conventional approach of allocating all protective system functionality to distribution system components, is for the current, voltage and status of CBs in the system to be monitored via a centralized communication system [73]. At some central point in the system, the collected information is analyzed by an efficient method to identify the microgrid structure and determine new relay settings, and new settings are communicated to CBs. This approach is ultimately impractical, reduces the protective system economics and simplicity and introduces a *reliability paradox* wherein the mechanism introduced to increase reliability of the protection system (its ability to discern and isolate fault locations). The essence of the reliability dilemma is that the increased amount of hardware associated with the communication system will bring an increased probability of failure within the protection system. The centralized communication-based protective system relies upon accurate collection of data and correct action based upon that data, without which the correct protection decisions are not made and the likely result is poor-system recoverability.

The best means for achievement of the third adaptive protection system attribute of system awareness and knowledge is to recognize that the protection equipment making up the protective system must act as *quasi-autonomous agents* that must act quickly upon information acquired from the system using a minimum level of communication with other system members. Here again is an area of research need. One approach is for the components of the protection equipment and scheme within the system to communicate with its adjacent neighbor(s) via a high-speed serial input/output (HS-SIO) connection. Through information gained from its neighbors, each protective system component autonomously reconstructs its independent picture of the system configuration, connection and state using algorithms derived from complex network theory [74]. These components can act nearly autonomously, but there will be a need for restraining signals between adjacent elements containing real-time directionality information, provided over the HS-SIO link(s), to enhance the speed of protection—hence designation of protective system components acting as quasi-autonomous agents. Since the directional information can be readily provided via the HS-SIO links between components, it makes sense to constrain the actions of the protection equipment and scheme using the near instantaneous awareness provided by this information. The achievement of a resilient system based upon this approach will involve trade-offs between the speed at which the power/energy delivery service network can be reconstructed and the accuracy of the assumed network configuration, connection and status. Alternate approaches that utilize complex network theory may incorporate adjacent HS-SIO communication capability between protective system agents. This capability may be combined with a cloud-enabled means for centralized communication. In this way, a *digital twin* of the power/energy delivery network can be constantly updated at a centralized location in the cloud-utilizing information that is received through cloud communications with all of the protective system agents. This approach could enhance the accuracy of network reconstruction at the cost of speed. There will inevitably be instances in which the system using this approach will be subject to fault events before the *digital twin* in the cloud is fully updated. In these instances, the protective system reliability and selectivity will be degraded.

There is one final point regarding the achievement of adaptive systems for microgrids. The greatest success will be achieved from the standpoint of attaining the

Pareto-optimal intersection of reliability-speed-selectivity-simplicity-economics of protective system design by recognizing that the microgrid brings additional players into the role of protection equipment. These players include the DERs themselves, which typically have significant sensing, monitoring and processing capabilities. Much of what is proposed for future adaptive protective systems assume the utilization of all of the parts of the power/energy delivery system—both power conversion and distribution components—to ensure the dependability of the system.

12.4.7 *Addressing cybersecurity*

In order to have a secure and reliable cyber physical system, different areas including risk assessment, attack prevention, detection, mitigation and resilience, have to be designed carefully [75]. Regarding this upcoming issue in microgrid, ability to deal with cyberattack will be a critical part of the protective system design. In order to have a secure and reliable cyber-physical system, different areas, including risk attack prevention, detection, mitigation and resilience, have to be designed carefully (see Figure 12.19) [76]. Therefore, the ability to deal with cyberattack has to be involved in designing of the microgrid protection.

Since a cyberattack operates using a combination approaches, the protective system design should include measures such as risk assessment, attack resilience measurement design and moving target-inspiring algorithms in order to prevent attack [75]. Even with these measures, some types of attacks are not preventable. In such cases, detection systems have to be devised to identify the attack. For example, the *digital twin* of the adaptive system is one method that can be deployed to

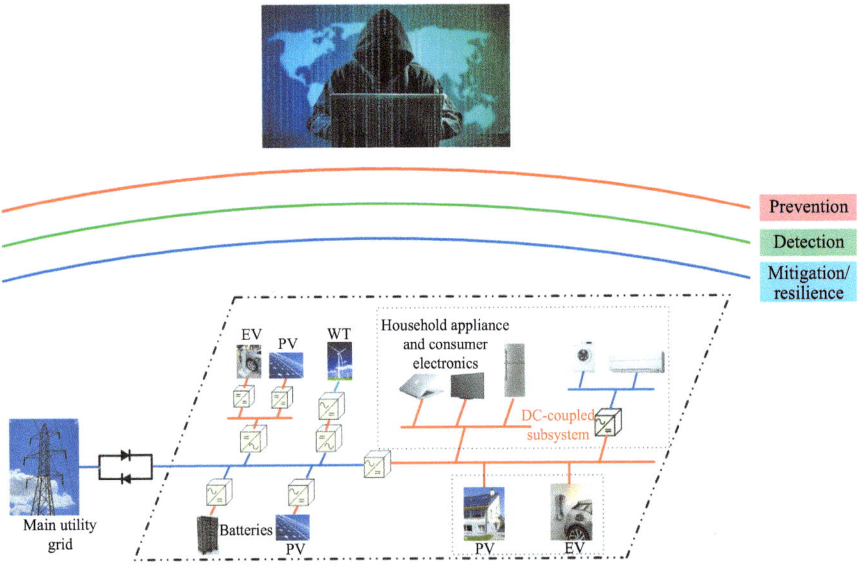

Figure 12.19 Defense lines for attack resilient microgrids

detect anomalous behavior. The digital twin can present a continuously update-able baseline against which measurements made within the physical system can be compared. Once the attack is detected, one of the intrusion-detection algorithms such as signature-based, anomaly-based counter-measured-based algorithms can be utilized to identify the cyberattack [77]. Finally, in the deepest line of defense, the system reacts by employing either of model-based mitigation or dynamic reconfiguration and resiliency algorithms to the minimize the negative cyber effects and guarantee the key functions of the protected systems [75].

The International Electrotechnical Commission (IEC) Technical Committee 57 (TC57) is responsible for development of international power system information exchanges through six accepted communication standards, including IEC 60870-5, IEC 60870-6, IEC 61850, IEC 61968, IEC 61970 and IEC 61334. By 1997, IEC TC57 recognized that security would be vital for these protocols. In 1999, IEC TC57 Working Group 15 (WG15) was established under the title of "power system control and associated communication data and communication security" to develop security standards that increase the overall information security assurance aspects of utility infrastructures. IEC TC57 WG15 published the first parts of IEC 62351 in 2007, more recent parts have been published in 2010, and some parts still being in progress to cover both information infrastructure and communication security [78]. The main goal of the IEC 62351 is to preserve confidentiality, availability, integrity and non-repudiation in a system through the presentation of authentication mechanisms. IEC 62351 standard defines security measures for the security of specific communication protocols, including IEC 60870-5 series, the IEC 60870-6 series, IEC 61850 series, IEC 61970 series and IEC 61968 series [9]. The interrelationships between IEC TC57 standards and IEC 62351 security standards are represented in Figure 12.16. The standard is split into eleven parts (see Table 12.2). IEC 62351-3 presents security for Transmission Control Protocol/ Internet Protocol (TCP/IP)-based protocols including IEC 60870-6 (also known as

Table 12.2 Excerpts from IEC 62351

Part	Title
IEC 62351-1	Communication network and system security introduction to security issues
IEC 62351-2	Glossary of terms
IEC 62351-3	Communication network and system security profiles including TCP/IP
IEC 62351-4	Profiles including MMS
IEC 62351-5	Security for IEC 60870-5 and derivatives
IEC 62351-6	Security for IEC 61850
IEC 62351-7	Network and system management (NSM) data object models
IEC 62351-8	Role-based access control
IEC 62351-9	Cyber security key management for power system equipment (unpublished)
IEC 62351-10	Security architecture guidelines
IEC 62351-11	Security for XML files (unpublished)

TASE.2), IEC 61850 ACSI over TCP/IP, and IEC 60870-5-104. This part of IEC 62351-3 utilizes Transport Layer Security encryption, message authentication and security certificated to provide protection against eavesdropping, man-in-the-middle and spoofing attacks, respectively. However, other measures have to be considered to protect against DOS. Transport and application profiles are provided by IEC 62351-4 to support the security of MMS-based protocols such as IEC 61850-8-1 and IEC 60870-6. IEC 62351-5 offers authentication mechanisms for IEC 60870-5 and derivatives to address spoofing, replay and DOS. Security of IEC 61850-81 (MMS) is provided by IEC 62351-3 and IEC 62351-4. As GOOSE is designed for protective relaying, the messages have to be transmitted in less than 4 ms, and implementation of encryption and other security measures are not applicable. IEC 62351-6 uses Rivest, Shamir and Adleman (RSA) public-key cryptosystem to digitally sign GOOSE and SV messages in order to provide security with minimal computing requirements. Although the RSA digital signature decreases computational burden, it does not meet the maximum transfer time criteria (within 4 ms). Elliptic Curve Digital Signature Algorithm, which uses elliptic curve cryptography, is an alternative to meet latency time requirement [79]. Despite parts 3–6 that is mainly for providing the security of communications and applications within a substation, parts 7–11 focus on the broader scopes. Part 7 focuses on management and control of communication and information infrastructures but has not provided details on mapping to an underlying protocol. Part 8 gives wide range of rule-based access control in power systems. Part 9, which has not been published, is assumed to cover digital certificate management. In part 10, a comprehensive guideline on power system security based on essential security controls is presented. Finally, IEC 62351-11, which has not been published, aims at defining the security for XML-based files by providing mechanism for authentication of the file source, tamper decision and through maintenance of maximum compatibility with different XML formats. Although IEC 62351 covers several security issues for communication standards in smart grids, following set of challenges or requirements have to be addressed in the next editions:

- Providing application layer end-to-end security when multiple transport layer connections are used [80].
- Requiring newer cryptographic algorithms to stay within the maximum transfer time of a GOOSE message [81].
- A common security framework, which should consider a hybrid solution with hop–hop group authentication and source authentication for end-to-end communications to protect all substation communications with different types of traffic and different timing requirements [82].

Widespread utilization of the IEDs raises concerns about communication standards as well as cyber security. Three main standards, including IEC 61850, CIM and IEC 61499, are introduced. These standards are designed for information exchange between IEDs, data modelling as well as modelling distributed industrial-process measurement and control systems, respectively. The comprehensive comparison of IEC 61850, IEC 61499 and CIM is presented in Table 12.2.

12.5 DC microgrid considerations

From a power/energy management standpoint, the DC microgrid presents many options that are more easily integrateable than would be the case with a purely AC microgrid. This is mainly because due to the fact that the need for phase synchronization of the power electronics-based DERs connected to the AC grid (and associated timing challenges) is limited only to AC to DC converter interface(s) between the main grid and the microgrid. DC connected DERs, such as solar, battery energy storage and wind (which are inherently DC), are placed where they can most naturally act as an energy buffer to address resiliency issues. It can be argued, therefore, that the DC microgrid has enhanced reliability and resiliency when compared to the AC microgrid.

There are a growing number of applications that benefit from LVdc and MVdc distribution as a means of distributing and integrating multiple DERs. In the LVdc realm, there are increasing examples of DC systems that are proving to be more efficient, less complex, have a high-power transfer ratio and are lower cost than competing AC systems [83–86]. Starting at the point of use, DC makes sense because the majority of loads require DC power [87–89]. For example, in India, homes, villages and hamlets are being electrified through local DC microgrids that incorporate PV and energy storage at faster rates than the government can install centralized electrical grid infrastructures [90,91]. This same model for electrification can be used to accelerate the adoption of renewables in urban residential neighborhoods and to use combined DER and energy storage to drive down the energy costs of low-income households [92,93]. Today, growth in the capacity of MVdc systems is driven principally by the need to increasing the hosting of DGs connected to the system [94,95]. For transportation applications, MVdc distribution will require a range of current capacities with the principle requirement being the capability of delivering MVdc through multiple paths [96].

The greatest impediment to wide-spread adoption of DC distribution and inherent benefits is the availability and knowledge of feasible and viable fault protection [97]. Many of the protection challenges are common in both AC and DC microgrids, i.e., changes in configuration, DER impacts on fault current capability, and cyberattack. However, protection of DC systems, when compared to AC systems, necessitates not only new approaches to protective system design and protection schemes but also a completely different protection equipment paradigm principally because of the following complicating factors:

No inherent current zero crossing: The lack of a current zero crossing changes the paradigm for CB design. The molded case CB (MCCB), the staple for AC power-distribution protection, is optimized to take advantage of this zero crossing to extinguish the arc that occurs during MCCB contact opening. This reduces the need for buildup of arc voltage to the level of the applied system level voltage in order to drive fault current to zero. MCCBs may be applied to LVdc systems; however, sustained arc voltage buildup to above the level of the system voltage is required [98]. If commercially available MCCBs are applied, two to three MCB poles that are equivalently rated to the LVac application must be connected in series in order to ensure sufficient arc voltage buildup. In MVdc systems, the vacuum CB (VCB) is the only viable conventional option [99], prompting the need for solid state CBs (SSCBs) for MVdc systems, which take a different approach to fault current limiting as will be explained.

Time compression of the fault time-current characteristic: Since the DERs and loads in a DC microgrid will have a power electronic interface, the sudden application fault characteristic is dominated by the capacitance of DERs and loads on the same bus as the fault and the cable inductance between the initial capacitor discharge source(s) of fault current and the fault location [97]. The initial capacitor discharge current into the fault has a very high di/dt and is limited only by the fault resistance and capacitor parasitic resistance(s). Once the initial capacitor discharge event is over, the DERs connected to the faulted bus will hard limit output current to ensure unit-protection for a short-time period, after which DER(s) will fault off to mitigate equipment damage. In LVdc systems, the sudden inception fault current discharge is over in 10s of microseconds, and in MVdc systems, the time frame for complete discharge is milliseconds. High peak currents and potential damage to the system can result in equipment damage unless the system (all connected DERs and loads) is over-designed for high peak current let-through, or designed to include fast-acting fault mitigation, such as SSCB, are applied.

Shifting ground reference potentials throughout the system: Unless every DER and load connected to the DC microgrid distribution bus is transformer isolated, conventional NT and TT-grounding approaches cannot be easily applied to a DC system. All DER and load-interfacing power electronic converters connected to the DC microgrid, i.e., non-isolated DC–DC and AC–DC converters, should only refer to a single ground point in the system to avoid the possibility of significant circulating common mode current throughout the system and nonzero voltage shifting with respect to ground. A proper grounding system has to be proposed with respect to safety, fault ride-through capability and ease of ground fault detection and location [100,101].

This section addresses the issues associated with DC microgrid protection, starting with the distinctive DC fault characteristics and then followed by discussions on protective system strategy, PDs and grounding practices.

12.5.1 DC fault characteristics

Most DC microgrids interface to the AC side of a DER or the AC utility connection through a voltage source converter (VSC). The minimal VSC AC interface is through an inductor (L_{ac}) and the VSC DC microgrid interface is through a capacitor (C) as shown in Figure 12.20(a). Because of VSC structure, when a short circuit fault is suddenly applied, first the DC side capacitor discharges into the fault through the DC microgrid network, then the contribution from the VSC interfaced sources (renewable or utility) forms the latter part of the response (see Figure 12.20(b)). Capacitor discharge will result in high current amplitude that could damage VSC components and other components in series with the fault. If the fault current let-through is a part of the protective strategy, then the excessive peak fault current must be taken into consideration as part of both system and component design processes. In order to understand and analyze the DC fault characteristics, the nonlinear system is solved by defining three different stages: capacitor discharge, AC-side current feeding and diode freewheeling. As shown in Figure 12.20(a), in each stage, the fault current circulates in different loops. The overall DC-side fault current associated with these stages is shown in Figure 12.20(b) for a grid-connected VSC. The response of this circuit will be analyzed in order to understand the DC fault characteristic response.

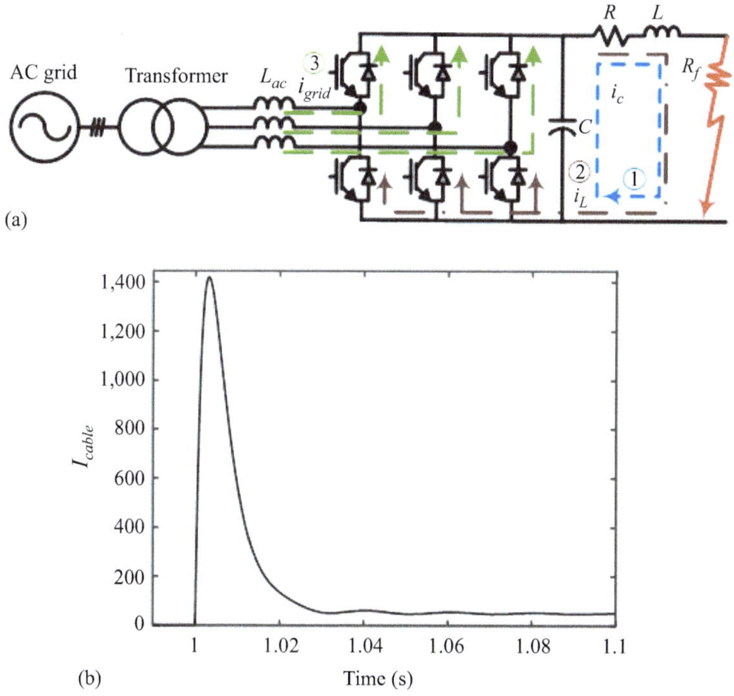

Figure 12.20 (a) Grid-connected VSC current flows into a DC-side fault showing fault current stages 1–3, (b) DC-side current during a sudden inception, DC-side short circuit fault

12.5.1.1 Stage 1: capacitor discharge stage (natural response of DC-side RLC circuit)

Once a sudden inception fault occurs in the DC microgrid, the capacitor starts discharging through the cable impedance as shown in Figure 12.21(a). In this stage, the peak value of fault current may go up to 100 times of the VSC rated current, depending on the internal resistance of the DC filter capacitor, capacitor value and cable inductance from the capacitor source to the fault location [see Figure 12.20(b)]. According to Figure 12.21(a), the RLC circuit response in Laplace domain can be written as [102]:

$$i(s) = \frac{V_C(0)/L + i_L(0)s}{s^2 + (R/L)s + (1/LC)} \tag{12.6}$$

where $i_L(0)$ and $V_C(0)$ are the initial current through the inductor and voltage across the capacitor, respectively. r and L are the resistance and inductance of the cable from the converter to the fault point. R_f is fault resistance and R is the sum of r and R_f. In the time domain, the fault current $i(t)$ can be expressed as

$$i(t) = \frac{v_C(0)}{L(s_2 - s_1)}\left[e^{(-s_1t)} - e^{(-s_2t)}\right] + \frac{i_L(0)}{s_2 - s_1}\left[-s_1e^{(-s_1t)} + s_2e^{(-s_2t)}\right] \tag{12.7}$$

where s_1 and s_2 are the roots of the characteristic (12.7) and are equal to

$$s_{1,2} = -\alpha \pm \left(\sqrt{\alpha^2 - \omega_0{}^2}\right) \tag{12.8}$$

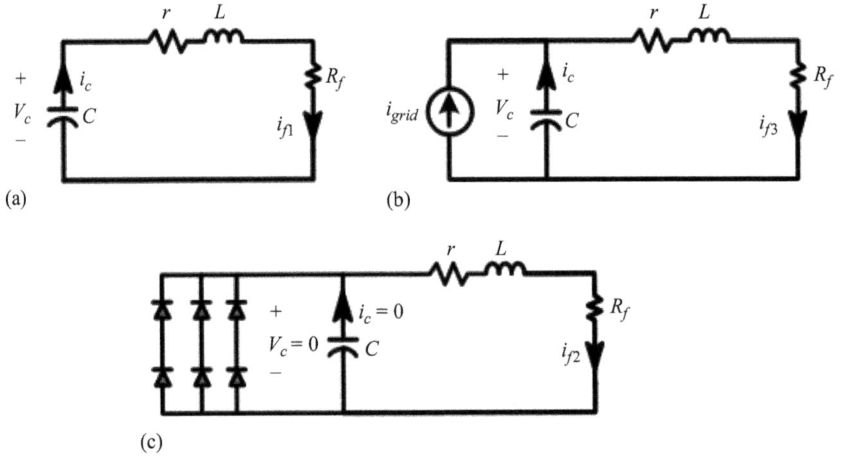

Figure 12.21 Grid-fed VSC sudden DC-side fault characteristic stages: (a) 1-capacitor discharge, (b) 2-AC-side current feeding and (c) 3-freewheeling diodes

In (12.8), α and ω_0 are, respectively, the damping factor and the resonance frequency defined as

$$\alpha = \frac{R}{2L} \tag{12.9}$$

$$\omega_0 = \frac{1}{\sqrt{LC_f}} \tag{12.10}$$

Based on the relationship between magnitudes of α^2 and ${\omega_0}^2$, the form of the current response is determined. For $\alpha^2 > \omega_0^2$, $\alpha^2 = \omega_0^2$ and $\alpha^2 < \omega_0^2$ the fault current response will be over-, critically- and under-damped, respectively. For example, the current response is obtained as follows for an under-damped system:

$$i(t) = \frac{V_C(0)}{L\omega_d} e^{(-\alpha t)} \sin(\omega_d t) + i_L(0) e^{(-\alpha t)} \left[\cos(\omega_d t) - \frac{\alpha}{\omega_d} \sin(\omega_d t)\right] \tag{12.11}$$

where $\omega_d = \sqrt{\omega_0^2 - \alpha^2}$.

12.5.1.2 Stage 2: AC-side current feeding stage (forced response)

In this mode, the VSC acts as an uncontrolled full-bridge rectifier and contributes to the fault current through the forward diode paths [see Figure 12.21(c)] [103]. The fault current in this stage is calculated as

$$i_{VSC} = i_{D1} + i_{D2} + i_{D3} = i_{ga,(>0)} + i_{gb,(>0)} + i_{gc,(>0)} \tag{12.12}$$

where $i_{ga,(>0)}$, $i_{gb,(>0)}$ and $i_{gc,(>0)}$ are, respectively, positive value of the phase-a, b, and c currents passing through the freewheeling diodes. For phase-a, the i_{ga} (>0) is calculated as

$$\begin{aligned}
i_{ga} &= I_g \sin(\omega_e t + \alpha - \varphi) + I_{gn} e^{(-t/\tau)} \\
\varphi &= \text{actan}\left[\frac{\omega_e(L_{ac} + L)}{R}\right] \\
\tau &= \frac{(L + L_{ac})}{R} \\
I_{gn} &= \left[I_{g|0|} \sin(\alpha - \varphi_0) - I_g \sin(\alpha - \varphi)\right]
\end{aligned} \tag{12.13}$$

where $I_{g|0|}$, φ_0 and L_{ac} denote the initial grid current amplitude and phase angle, and the grid-side inductance, respectively. The grid angular electrical frequency is ω_e.

12.5.1.3 Diode freewheeling stage (natural response of DC-side inductive circuit)

If the source of AC power is lost at any point during the fault response process, then the capacitor will be discharging through the cable until its voltage reaches zero. This will be the situation if, for example, AC-side CBs trip are upstream of the

VSC. In this case, the cable current commutates to the VSC freewheeling diodes [see Figure 12.21(b)]. Thus, the current response through the DC-side cable is

$$i(t) = I_0' \cdot e^{-R/L \cdot t} \tag{12.14}$$

where I_0' is the current in the cable at the instant that the capacitor current, $i_c = 0$.
The diode current is

$$i_{D1} = \frac{i(t)}{3} \tag{12.15}$$

The remaining inductive energy in the system may be significant, while the dissipative loss in the system may be very low. As a result, the freewheeling stage duration can be very long. This result has the following two negative effects: (1) if the freewheeling stage is included as a part of fault mitigation, the time to fault isolation will be significant and (2) the initial current I_0' can be on the order of ten times the rated current, resulting in thermal overload of the diodes during the long decay time. It is highly desirable to detect and isolate before the VSC enters this fault stage—unless the fault mitigation strategy is to inhibit fault current from the AC side and then restore AC-side power after the DC side fault is isolated. This latter approach is the only possible option for *unit-based* protection when VSCs are utilized at the AC to DC interface(s). This protective strategy is described in the following section.

12.5.2 DC protective system approaches

The DC microgrid protective system approaches generally fall into two categories: (1) *unit based* and (2) *non-unit based* or, more appropriately, *breaker based*. The principal focus of this chapter is on *breaker-based* approaches; however, *unit-based* protection is worthy of mentioning because this is the principal protective system approach currently utilized in shipboard systems [104], automotive electrification and DC powered homes [105]. The internal self-protection of the DERs is commonly referred to as *unit protection*. A *unit-based* protection strategy recognizes that this protection functionality exists within the DERs and utilizes it to either limit the DC-side fault current or drive it to zero.

An illustration of this approach is the shipboard DC Zonal Electric Distribution Systems (DC-ZEDS) shown in Figure 12.22. In the shipboard DC-ZEDS, the system is physically subdivided into zones of protection and DC buses span across all of these zones [106]. For example, Figure 12.22 shows buses extending through the zones on both sides of the ship that are directly fed by power generators through phase-controlled rectifiers (PCRs). Unlike the VSC, the PCR has the capability of limiting its output current to zero when there is a DC-side fault. A DC fault on either the port (PORT) or starboard (STBD) bus will cause the voltage on that bus to fold back to zero. Such a system will utilize the let-through current from the sudden inception fault current, sensed at distribution panels along the bus in each zone upon fault inception and then utilize a simplified DC version of DOCR protection to locate the bus fault [107,108]. Each PCR will communicate with all of

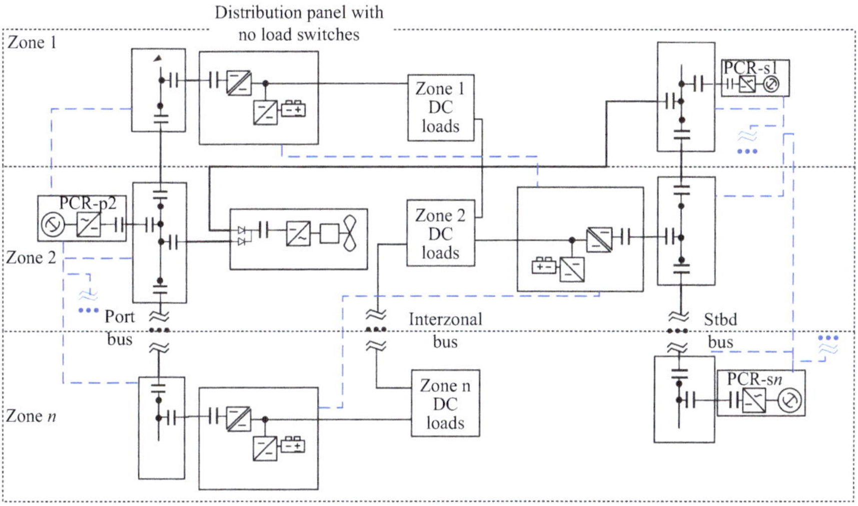

Figure 12.22 DC zonal electrical distribution system with unit-based protection

the distribution panels and all other PCRs on the same bus to which it is connected (as indicated by the blue dashed lines in Figure 12.22). Algorithms implemented collectively among the PCRs will utilize captured fault information to determine the fault location and a command will be sent collectively to the distribution panels on either side of the fault (if the fault is located on the interzonal bus) or to the affected distribution panel (if the fault is located to an intra-zone branch connected). No-load electromechanical switches (or DC no-load switches as they will be referred to hereafter) in the distribution panels will open to isolate the fault. Once it has been verified that the fault is isolated, the PCR(s) will reenergize the bus and restore power to non-faulted portions of the system. This entire sequence is referred to a *fault detection, isolation and recovery* or FDIR. During the entire FDIR sequence, downstream loads have lost their input power. The interzonal buses are deliberately isolated from each other at each end, as shown in Figure 12.22 so that a single bus does not bring down the power to the entire ship. The DC loads within each zone are fed by DC–DC converters and have short-term battery back-up. The DC loads are also interconnected between zones. This approach will enable the short-term power sourcing to DC loads connected from the non-faulted bus during FDIR sequence. The same approach is taken for loads connected to the main buses. For example, the propulsion motor drive in zone 2 is fed from both sides through auctioneering diodes.

For DC protection, the most effective method of fault isolation is to utilize some sort of power electronic mechanism to drive the fault current to zero and then, once it is verified that the zero current condition is present, the air gap isolates the fault from the system using an electromechanical device, such as a set of DC no-load isolating switches on positive and negative rails of the DC bus. Because many types of power electronic converters, such as PCRs, DC–DC buck converters and current source converters have inherent current limiting capability there are many advocates for unit-based protection of DC distribution. The ideal architecture for unit-based protection is shown in Figure 12.23. Such a system appears to be radially distributed from groups of power electronic converter interfacing DERs to point of load interfacing converters on a common bus, with unidirectional power flow from source(s) to loads. Such a system is based upon a narrow-minded definition of a fault as being caused a failure within installed equipment connected to the bus and geographically contained interconnecting bus. The mistaken assumption is made that the protection equipment is inserted between the power/energy source and the bus or between the bus and load, leading to an oversimplification of the scope of protective system design. Such an oversimplification may be valid for *DC nanogrids* that fan off of a larger DC microgrid in futuristic concepts for DC powered homes. In reality, the microgrid is made up of power electronics equipment contained within enclosures that are interconnected by cabling as shown in Figure 12.24. The most likely fault locations are on the interconnecting cables between DER, not within the DERs themselves.

If the protective strategy is unit based, then the DC microgrid designer must recognize that the system must be capable of withstanding the entire capacitor discharge stage, as described in Section 12.5.1.1. Also, it must be recognized that the FDIR sequence required to isolate the fault and restore the system could take several seconds, so critical loads must be capable of withstanding long interruptions. Localized energy storage will likely be required or alternate, redundant power/energy sourcing paths should be provided within the distribution architecture (similar to the DC-ZEDS approach). AC-side protection of VSCs is a form of unit-based protection. In this case, the system will be subjected to all three stages

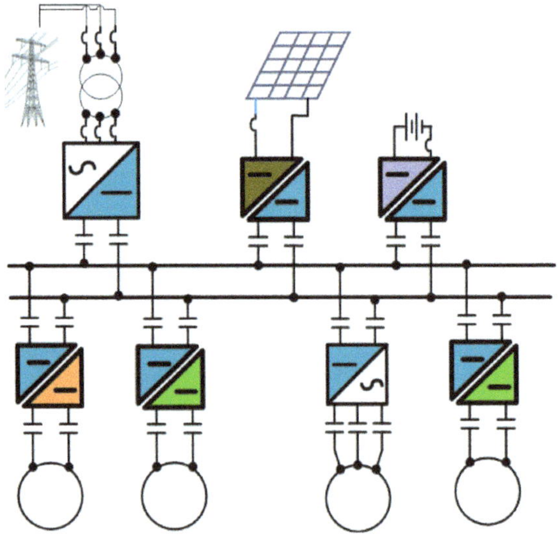

Figure 12.23 A simplistic radially distributed DC microgrid utilizing unit-protection within the DERs and DC no-load switches

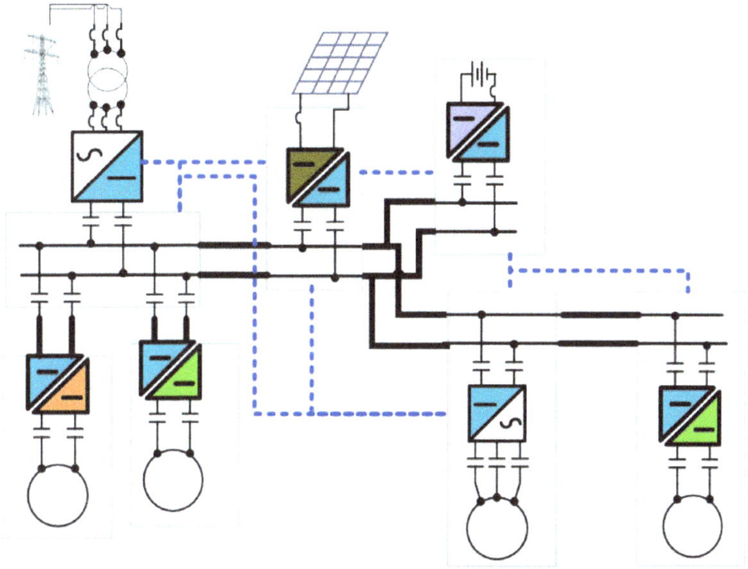

Figure 12.24 A realistic layout of DERs in a DC microgrid with interconnecting cable interconnects between DER enclosures utilizing unit-based protection

of the fault characteristic described in Section 12.5.1. The system will be required to withstand very high surge currents for a longer period of time than the capacitor discharge stage while being subjected to the AC-side feeding and diode freewheeling stages of the fault current characteristic. The unit-based protective strategy will also be highly dependent upon centralized communications between fault-feeding DERs and isolating DC no-load switches. Any loss of communication capability will degrade the reliability of the protective system.

A more desirable protective approach is one that will inhibit fault current before the current reaches high levels during the capacitor discharge stage and before the stored energy associated with the output filters of the DERs discharges completely into the fault. Such an approach will require fast responding SSCBs as the power electronic current limiting devices in the system. A DC system protective strategy that is based upon SSCBs is a *breaker-based* protective system. The system of Figure 12.24 utilizing breaker-based protection is shown in Figure 12.25. To provide context to Figure 12.25, Figure 12.26 shows a notional implementation of a two-pole DC protection unit utilizing SSCBs and how the SSCB-based DC protection unit would interact with other protection units and larger network communications to locate and isolate faults. In this case, the protection units could act as independent agents to localize the fault. The speed and reliability of the protective system are enhanced by HS FPGA-based communications between adjacent protective units. The network communications would only be necessary to adapt the system to changing conditions.

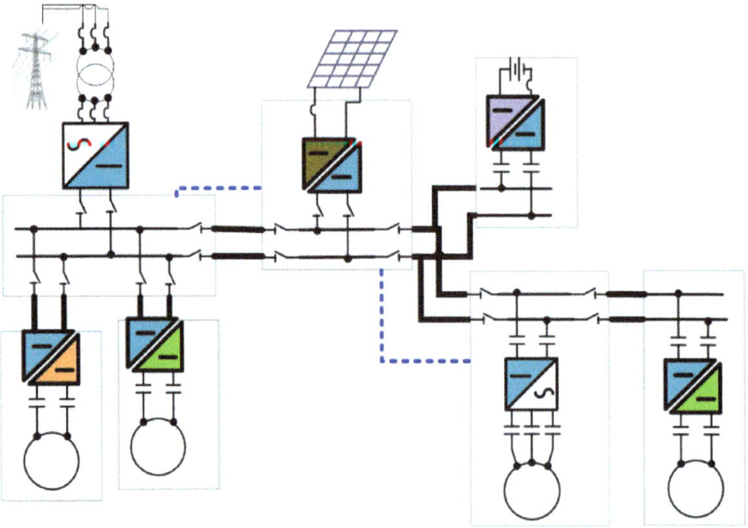

Figure 12.25 A realistic layout of DERs in a DC microgrid with interconnecting cable interconnects between DER enclosures utilizing breaker-based protection

So far, the protective scheme for DC microgrids has addressed the *protection equipment* aspect of the design by subdividing strategies into two approaches to achieve the necessary fault current limiting capability: unit-based protection, which relies upon the unit-protections within the DERs, and breaker-based, which relies upon external equipment (normally in enclosures that are external to and distinct from the DERs). The protective schemes that may be applied to DC systems are very similar to what has been covered so far for protective schemes—only time responses must be significantly time compressed. This latter characteristic of DC protective schemes must be enabled by protective equipment that is not readily available in the commercial market, is often highly developmental and is still the subject of much research. The speed of actuation of the DC no-load switches will also have a significant impact on the selectivity of the protective system. For LVdc systems, feasible solutions to the no-load switch are commercially available. However, for high-current LVdc systems (i.e., greater than 1,000 A) and MVdc systems, the no-load switch is also a developmental item and the subject of much research. There has been considerable attention to the utilization of Thompson-coil-based switches in the literature, but the simultaneous achievement of speed of isolation and power density is still an issue. Custom, highly developmental approaches to the no-load switch have been proposed in the literature achieving actuation times of less than 10 ms, but much work is needed to bring this capability into the realm of commercial viability.

The protective system design approach of Figure 12.17 applies to DC microgrid protective but with some caveats. First of all, the protection scheme developed

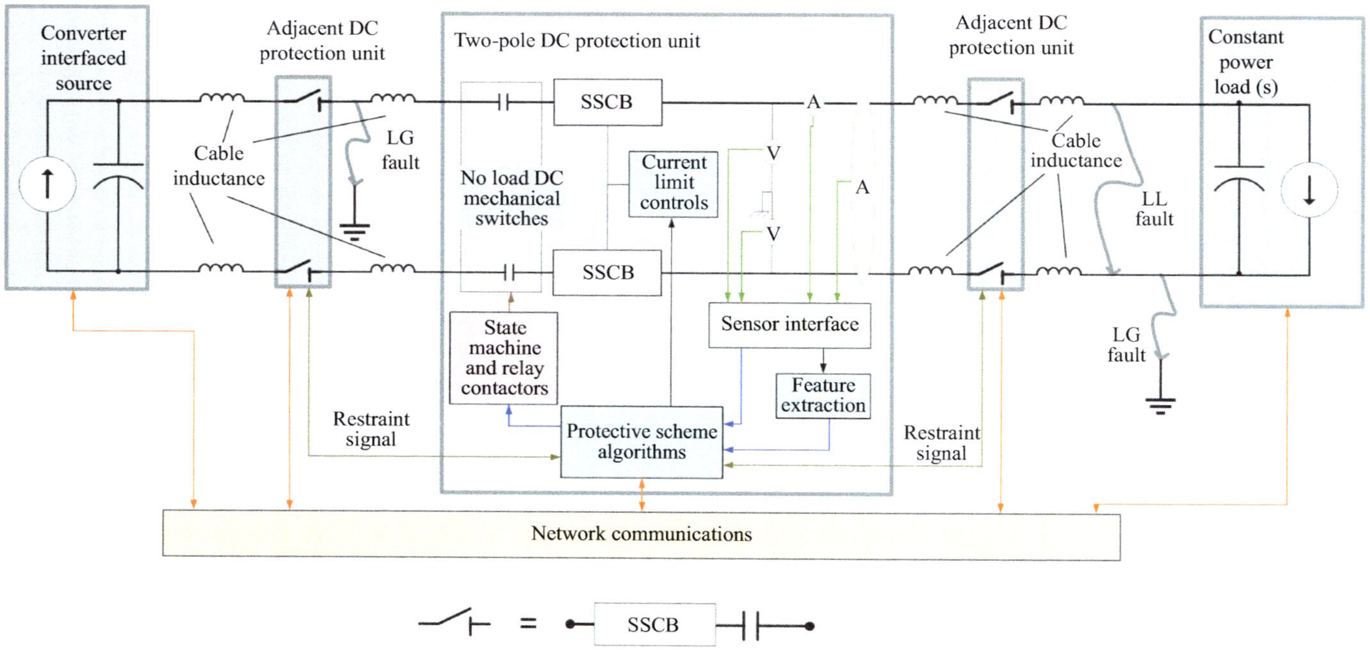

Figure 12.26 Breaker-based protective system utilizing SSCB-based DC protection units

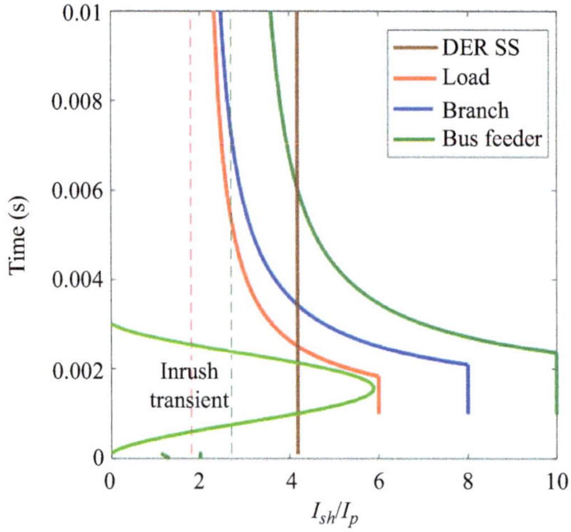

Figure 12.27 DC system inverse time-current curves

as a result of the process in Figure 12.17 must execute with latencies and rates on the order of microseconds or 10s of microseconds to properly discriminate fault location and mitigate loss of power to large parts of the system during the fault isolation process. The need for compressed time responses is illustrated by Figure 12.27 which shows notional inverse time-current curves that may be applied to the SSCBs configured as OCRs. Here the load feeder is within a radial distribution off of a branch feeder connected to a main bus and the bus feed interfaces a large capacity source to the bus, such as the AC–DC interface to the utility. These curves mimic the curves for MCCBs; only they are very time compressed. The DER SS limit would be imposed if the DC system is in, say, an islanded configuration which emphasizes the need to retune the curves in different system configuration states (just as is the case for AC microgrids in different configuration states). Any fault behavior that is downstream of the load feeder should cause an OCR trip if the load feeder line is exceeded.

Because of the nature of the DC-connected loads, it is very difficult to discriminate between normal transient behaviors, such as distinguishing between the inrush during connection and subsequent start-up transient of a downstream load from an LL fault. For the design stage of a DC protective system (the green blocks shown in Figure 12.17), the HiL system is required in order to generate the training cases and perform feature extraction from the fault and non-fault characteristics, just as is the case for AC protective system design. However, the time resolution of feature-extraction drives the need for much smaller HiL simulation time steps than for AC systems. Also, the tolerance of the protection equipment and scheme to timing latency is also a critical requirement that should be derived during the design phase. Therefore, an HS FPGA-based CHiL platform is even more critical for the DC system than for the AC system. The protective system deployment (the purple blocks shown in Figure 12.17 will require similar low latency, high sample-rate or analog time response). The utility of the FPGA-based CHiL platform is the ability to derive the requirements for hardware deployment. It will be both feasible and viable to distribute low-latency, high-sample rate processing capability among the protection units and DERs in a DC microgrid.

12.5.3 DC protective devices

PDs used in the DC system are broadly divided into ACCBs and DCCBs. ACCB is a simple and economic solution for VSC-based DC system (i.e., the *unit protection* allocated to the VSC). However, as has been discussed, the ACCB may not be fast enough to prevent damage to the VSC's freewheeling diodes. In addition, employing ACCB leads to disconnection of the whole network and long interruption times during FDIR sequences that can only be mitigated through the use of localized and distributed energy storage and multiple power/energy cross-connections to critical DC loads. A meshed DC microgrid with multiple connections to AC-based DGs and utility AC-feed connections will require unit-based protection that utilizes the coordinated combination of ACCB and DC no-load disconnect switches to isolated faults [109]. In this method, DC system has to be completely de-energized until the fault is removed. On the other hand, the increase in penetration of DGs combined with increased power/energy demand results in a rise in fault current levels that may exceed the rating of the existing ACCBs and DC no-load switches and loss of coordination of the overcurrent protection. Such a system cannot sustain growth over a long period of time—completely defeating the plug and play advantage of DC systems.

The DCCB generally refers to a range of options having the capability of both limiting fault current and isolating faults. Protection equipment in this category includes fuses, MCBs, MCCBs, SSCBs and Hybrid CBs (HCBs). For LVdc systems, the most feasible solutions are fuses, MCCBs and MCBs [110]. For MVdc systems VCBs, SSCBs and HCBs are feasible [111]. For the purposes of this discussion, LVdc is defined as in the range of 50 V DC–2 kV DC and MVdc is defined as from 2 to 60 kV DC.

12.5.3.1 Fuse

The fuse, which consists of a link and heat-absorbing material inside a ceramic cartridge, is used as a simplest PD in the protection of DC systems for voltages up to 4,200 V. Fuses are ideal to be applied in DC systems with a low inductance or (high deviation of current) because the time for the fuse to reach melting point would be minimum [112]. Although the fuse is a low-cost PD with a simple structure, it has disadvantages such as replacement after successful operation and inability to discriminate between transient and permanent fault.

12.5.3.2 Mechanical circuit breaker

MCB is a PD that uses a mechanical device or the so-called interrupter to stop the current. The MCCB and VCB fall under the general category of MCB. The interruption process of the current in the MCB is always accompanied with an electric arc between switching contacts. Suppression of the arc can occur through the passive process native to device design of arc voltage buildup above the system voltage (as is the case with the MCCB) or through an auxiliary circuit that generates a voltage opposing the system voltage to drive current to zero [98]. As has been discussed, in the former case, it may be necessary to connect multiple MCCBs in series to achieve sufficient voltage buildup because these devices have been optimized to complete arc extinction through the natural zero crossing of AC systems. In the latter case, when the interrupter opens, current flows through an LC circuit, with a capacitor that has not been pre-charged. The LC circuit oscillation creates current zero points. For actively commutated MCBs, the un-precharged capacitor is replaced with the precharged capacitor. In this type of MCB implementation, when the interrupter opens, the charged capacitor injects a negative current equal to fault current to make a zero-crossing current.

During the interruption process of MCBs with auxiliary or active commutation circuits, the magnetic energy stored in the system inductance must be dissipated. Metal oxide varistors (MOVs) are connected in parallel with interrupter to relieve overvoltage stress and absorb this inductive energy in the path of the fault. The main advantages of MCBs are low power loss and relatively low cost. However, slow response time and limited current interruption capability are their main disadvantages. The mechanical passive and active resonant CBs are shown in Figure 12.28(a) and (b), respectively.

12.5.3.3 Solid-state circuit breaker

In order to cope with the problem of low-time response and other limitations, semiconductor-based switches are used in SSCBs. A typical scheme of SSCB is shown in Figure 12.29. A cooling system must be employed to extract heat from the SSCB during the conducting condition.

Different semiconductor devices, such as gate-turn-off thyristors (GTOs), silicon insulated-gate bipolar transistor (IGBT), and integrated gate-commutated thyristor (IGCT) have been utilized in SSCBs, each one of them having their own advantages and disadvantages. It is reported that GTOs and IGCTs have much lower on-state losses than IGBTs. The IGCT offers highest reliability due to its

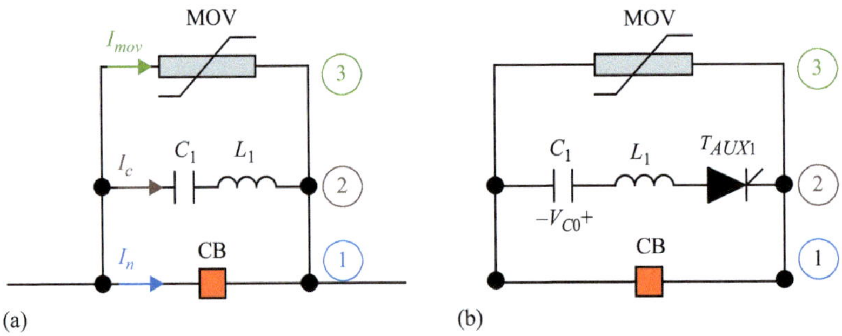

Figure 12.28 (a) MCB with a passive commutation circuit and (b) MCB with an active commutation circuit

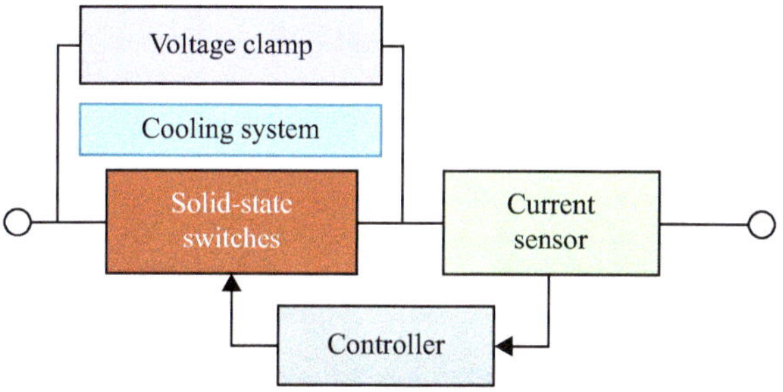

Figure 12.29 A typical SSCB

high surge current handling capability and robust mechanical design. Based on IGCT and GTO characteristics, Various SSCB topologies have been compared in the literature with respect to selectivity and economic considerations [111,113–115]. Utilizing IGBTs, GTOs and IGCTs increases both cost and conduction losses of SSCBs. Therefore, in industrial applications, high voltage and power of SSCBs are seriously hampered. One solution is to replace IGBTs, GTOs and IGCTs with thyristors. The main drawback of thyristors is not being able to actively turn-off current. However, this issue is less critical in SSCB applications, because there is no need to switch at high frequency. The on-state losses of thyristor switch are much less that IGCT switch. This would result in more reduction of overall life cycle costs of the SSCB and decrease in investment on the cooling system of thyristor-based SSCBs. SSCBs are viable for DC application when speed of fault isolation is a major objective. Depending upon the topology, voltage level and power semiconductor SSCBs can achieve fault commutation response in the range of 10–100 μs.

With the advancement in technology and introduction of new semiconductor devices such as wide bandgap (WBG) power semiconductors, the speed and efficiency of SSCBs is increasing [116,117]. Recently, WBG semiconductors such as silicon carbide (SiC) and gallium nitride (GaN) have been introduced as ideal materials for switching devices in high-voltage, low on-state losses, high-power, high-temperature and high-frequency applications [see Figure 12.30] [118]. WBGs semiconductors have the following unique characteristics including higher breakdown voltage thanks to higher field strength (Si:0.3, SiC:1.2–2.4, GaN:3.3), being thinner and having lower on-resistance and higher thermal conductivity. Although, high device cost and reliability concerns are two main obstacles that prevent applying these materials in SSCBs, research and development are pursued intensively in this area.

12.5.3.4 Hybrid circuit breaker

According to Table 12.3, the MCB and SSCB have their own drawbacks and benefits. The HCB combines MCB and SSCB to take advantages of both [119] but at the expense of increased response time when compared to the SSCB. The HCBs have advantages such as fast response when compared to MCBs, low power loss

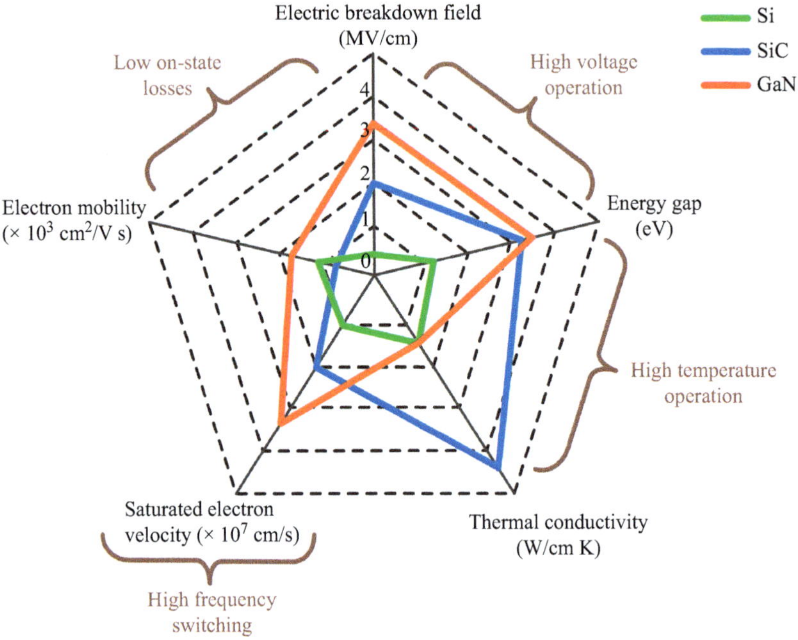

Figure 12.30 Summary of Si, SiC and GaN relevant material properties

Table 12.3 Protective devices used in DC microgrids

Protective device	Disadvantages	Advantages
Fuse	• Not able to distinguish between a transient and a permanent fault • Fuse needs to be replaced for successful operation	• Low cost • Simple structure
Mechanical CB	• Long operating times (30–100 ms) • Limited interruption current capability	• Relatively low cost • Very low power loss
Solid state CB	• Expensive • High power loss • Big due to heatsink needed	• Fastest response time (<100 μs) • Very long interruption lifetime
Hybrid CB	• Very expensive	• Low power loss • No arcing on mechanical contacts • Fast response time (few ms)

and minimal arcing at the mechanical contacts. However, the high cost of HCBs is still the main drawback. If the high selectivity and speed are major objectives, optimizing the SSCB to achieve a higher level of efficiency should be carefully considered before considering the HCB. Typically, as shown in Figure 12.31, the HCB consists of four main parts including a mechanical switch, lower voltage rated commutation switch, a high-voltage rated solid state switch, or main Breaker, and MOVs [120]. In the normal mode, the current flows through the main breaker to reduce power loss. At the instant of fault, the commutation switch opens, forcing the fault current into the main breaker branch. The commutation switch voltage stress is only limited by the main breaker branch conduction voltage and can therefore have a significantly lower voltage rating and resultant low loss during normal conduction. The main breaker usually consists of high voltage rated power semiconductors or a series connection of lower voltage rated devices in order to provide voltage withstand capability in excess of the system voltage. The fault current is driven to zero by a combination of the main breaker opening and voltage buildup across the MOV when the fault current commutates into that branch. The commutation process is over once the current through the mechanical switch is driven to zero and the inductive energy of the system inductance to the fault is fully dissipated by the MOV. The mechanical switch may be a DC no-load switches.

It should be noted that in the case of both the SSCB and HCB, an additional DC no-load switch is required to be in series in order to fully isolate the fault from the system. Without this switch, leakage current may still flow between the SSCB or HCB and the fault, resulting in unpredictable and potentially unsafe behavior. Depending upon the architecture of the DC distribution the speed of DC no-load opening may or may not be a factor that is critical to the selectivity of the protective system. If the SSCB or HCB is feeding a branch off of a main bus, ring bus or meshed bus, then the speed is probably not critical, and a slow acting switch (or in some cases a manually operated switch) may be applied. However, if the SSCB or

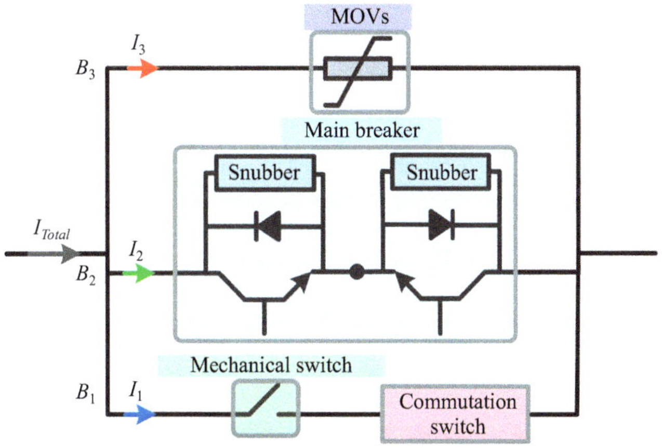

Figure 12.31 A typical HCB

HCB is on a ring bus, a meshed bus interconnect or DC-ZEDS main bus opening time of the DC no-load switch could have a significant impact on protective system selectivity. In systems where a significant amount of current must be handled by the SSCB or HCB, such as multi-MW-rated MVdc systems, the DC no-load switch can present a significant technical challenge.

12.5.4 DC system grounding

Three main goals of the grounding system are to ease detection of fault, to minimize DC stray current, and to increase personnel/equipment safety by reducing common mode voltage [120]. Stray current and common mode voltage are related to each other by impedance in the grounding path. In DC microgrids, this impedance may be highly capacitive resulting in high-frequency stray currents as power electronic converters interact with the system and cause switching frequency and switching frequency harmonic current to flow. Uncontrolled LG capacitance can lead to poor electromagnetic compatibility (EMC) (due to resonances), safety issues and difficulty in location of LG faults (in IT grounding systems) [101]. The high LG capacitance in such DC systems is usually a result of poorly designed electromagnetic interference (EMI) filters that are incorporated in the DERs in order shunt conducted emissions to ground at the DER interfaces to the DC bus. This approach is often taken by manufacturers of power electronic equipment because it is inexpensive and space efficient when compared to the smaller LG capacitors in combination with current blocking common mode inductors in the EMI filter. System integrators of DC systems and DC microgrids are extremely important to system inter-compatibility. Several trade-offs must be considered such as the very low stray current and potentially high common mode voltage in systems that are ungrounded or high-resistance grounded versus the low common mode voltage and high stray current of low resistance grounded systems [121].

Figure 12.32 shows NT, TT and IT grounding approaches that are suitable for DC systems. It is acceptable to create a ground reference point that is common to the shared DC bus of the microgrid. Grounding on the AC side feed(s) to the DC microgrid will result in common mode shifts, increase in stray ground current and increased system vulnerability to ground faults. For the TT grounding system, the DER power electronic converter output midpoint voltage (neutral) and chassis are separately connected to the ground points. The TT grounding system is straight forward to install and fault does not transfer to other parts of grid; however, circulation of current and potentially high voltage stress are the main drawbacks [122]. TN is the most commonly used DC grounding system. In this configuration, the neutral is connected to chassis of DER(s) and DC loads and a hard connection is made to both the neutral and PE. The TN topology has three subclasses including TN-S, TN-C and TN-C-S. In TN-S, separate PE and N conductors are used [see Figure 12.32(b)]; however, TN-C combines these two conductors to protective earth and neutral conductor in order to offer a cost-effective ground system [see Figure 12.32(c)]. TN-S system has highest electromagnetic EMC among different types of TN grounding systems. In addition, TN-S has higher safety than TN-C, because, if the conductor gets disconnected, the protective features still remain.

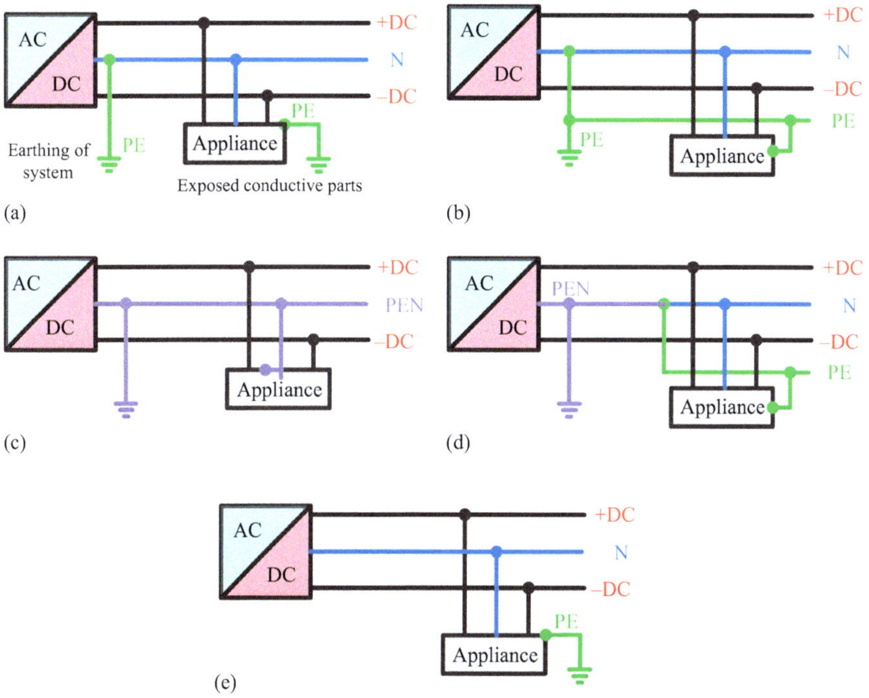

Figure 12.32 DC microgrid grounding systems: (a) TT, (b) TN-S, (c) TN-C, (d) TN-C-S and (e) IT

As a result, this grounding system is suitable for information technology and communication networks. TN-C-S grounding topology is the combination of TN-C and TN-S to have maximum benefit from two systems [see Figure 12.32(d)]. The advantages of TN grounding systems include having sufficient amount of fault current to be detected, requiring low-grounding impedance, and the limiting fault current by adjusting the ground resistance; however, for high voltage applications, the touch voltage is high. IT grounding system has no grounding point for the neural point and the appliance body is grounded separately [see Figure 12.32(e)]. This configuration has advantages such as small current under LG fault and ability to continue providing energy to the loads under the condition of a single LG fault; however, its disadvantages include hard-to-locate LG faults and unpredictable fault current paths through the DERs when a second line to ground fault occurs [101].

12.6 Conclusion: future of microgrid protection

This chapter has outlined the challenges associated with microgrid protection for both AC and DC systems. A clear message is that the approach to protection for microgrids has to be different than conventional approaches. Until recently, microgrid protection has received relatively little attention. Presently, it is a fertile area for research but, more importantly, the large-scale deployment of microgrids for both environmental and human benefit is being impeded in many cases by perceived and real risks associated with microgrid installation, integration and operation. To a large degree, these risks really come down to the protection issue. The behavior of microgrids under the various potential faulted conditions still presents a large number of unknowns and uncertainties. Although in the realms of communication standards and renewable energy interface standards protection is beginning to be addressed, there is still a great deal of work to do before standard engineering practice for microgrid protection is solidified. There has been a false sense of comfort regarding the integration of microgrids into systems using conventional distribution equipment and practices. This has resulted in a tacit acceptance of significantly lower levels of protective system selectivity and lack of inclusion of protection in the overall system design. This latter shortcoming is often overcome once the system is installed, resulting in add-on fixes that reduce protective system simplicity. As a result, the protection system complexity and maintenance increase. In general, it may seem to industry practitioners that the microgrid is a "one-off" system that must be uniquely designed to each installation—leading to a hesitancy in system deployment.

With DC microgrids, there is at least a general acceptance that protection is a challenge and a general predisposition to consider the DC microgrid installation as a system design. As a result, there will be a higher level of systems engineering practice applied (which is what is needed for all microgrid deployments) and acknowledgment that non-predicted behaviors may be manifested during integration and even operational phases. There will also be a greater level of hesitancy to embark on DC microgrid projects. When it comes down to it, the integration issues and risks associated with DC microgrids are related to behaviors during fault events. Otherwise, every other aspect of DC microgrid development is actually more in line with the mind-set of the power-conversion systems engineer: to deliver the most energy secure system—one that maximizes the time that reliable power can be provided to loads and does so with the highest level of efficiency possible. The protection problems of DC microgrids often fall into the general category of system survivability and the response of the system to forced events. There is presently an almost complete lack of standards for DC protection, but there are movements afoot to change that fact. On the other hand, the DC microgrid is often recognized from the beginning to be a power-delivery network structure with its own set of interface and performance requirements that must be validated and verified.

What is needed is a more holistic approach to the design of microgrids as a system. Standards and specifications must be developed keeping in mind the interrelationship between the system performance and the protective system design multiple and competing objectives of reliability, speed, selectivity, economics and simplicity. The microgrid implementation inherently has a tremendous capacity for real-time data capture and analysis. This capability must be exploited. Emerging technologies associated with data analysis, management and machine learning must be combined with advances in computational efficiency and network communications. Old paradigms regarding the hierarchy of data sampling and management may need to be set aside. The ever present threat of cyberattacks must be built into power system, communication system and protective system design processes, practices and standards. Microgrids and similar advanced distribution systems that manage the energy and power flow between multiple, re-configurable loads and sources must be viewed as power/energy delivery networks that must meet levels of dependability and security through verifiable practices and processes.

References

[1] Elmore WA. Protective Relaying: Theory and Applications. vol. 1. New York: CRC Press; 2003.

[2] Avizienis A, Laprie JC, Randell B, *et al.* Basic concepts and taxonomy of dependable and secure computing. IEEE Transactions on Dependable and Secure Computing. 2004;1(1):11–33.

[3] Alstom Grid. Network Protection and Automation Guide—Protective Relays, Measurement & Control; 2011.

[4] Zeineldin HH, Mohamed YARI, Khadkikar V, *et al.* A protection coordination index for evaluating distributed generation impacts on protection for meshed distribution systems. IEEE Transactions on Smart Grid. 2013;4(3):1523–1532.

[5] Benmouyal G, Meisinger M, Burnworth J, *et al.* IEEE standard inverse-time characteristic equations for overcurrent relays. IEEE Transactions on Power Delivery. 1999;14(3):868–872.

[6] Hussain M, Rahim S, and Musirin I. Optimal overcurrent relay coordination: A review. Procedia Engineering. 2013;53:332–336.

[7] Zeienldin H, El-Saadany E, and Salama M. A novel problem formulation for directional overcurrent relay coordination. In: Power Engineering, 2004. LESCOPE-04. 2004 Large Engineering systems Conference on. IEEE; 2004. p. 48–52.

[8] Noghabi AS, Sadeh J, and Mashhadi HR. Considering different network topologies in optimal overcurrent relay coordination using a hybrid GA. IEEE Transactions on Power Delivery. 2009;24(4):1857–1863.

[9] Bashir M, Taghizadeh M, Sadeh J, *et al.* A new hybrid particle swarm optimization for optimal coordination of over current relay. In: Power System Technology (POWERCON), 2010 International Conference on. IEEE; 2010. p. 1–6.

[10] Uthitsunthorn D, and Kulworawanichpong T. Application of artificial bees colony algorithm for optimal overcurrent relay coordination problems. ECTI Transactions on Electrical Engineering, Electronics, and Communications. 2014;10(1):98–107.

[11] Ziegler G. Numerical Distance Protection: Principles and Applications. John Wiley & Sons; 2011.

[12] Pradhan A, Routray A, and Gudipalli SM. Fault direction estimation in radial distribution system using phase change in sequence current. IEEE Transactions on Power Delivery. 2007;22(4):2065–2071.
[13] Jalilian A, Hagh MT, and Hashemi S. An innovative directional relaying scheme based on postfault current. IEEE Transactions on Power Delivery. 2014;29(6):2640–2647.
[14] Ukil A. Detection of direction change in prefault current in current-only directional overcurrent protection. In: Industrial Electronics Society, IECON 2016—42nd Annual Conference of the IEEE. IEEE; 2016. p. 3829–3833.
[15] Knable A. A standardized approach to relay coordination. In: IEEE Winter Power Meeting. vol. 69; 1969. p. 58.
[16] Rao VB, and Rao KS. Computer aided coordination of directional relays: Determination of break points. IEEE Transactions on Power Delivery. 1988; 3(2):545–548.
[17] Sharifian H, Abyaneh HA, Salman SK, *et al.* Determination of the minimum break point set using expert system and genetic algorithm. IEEE Transactions on Power Delivery. 2010;25(3):1284–1295.
[18] Albrecht R, Nisja M, Feero W, *et al.* Digital computer protective device co-ordination program I-general program description. IEEE Transactions on Power Apparatus and Systems. 1964;83(4):402–410.
[19] Zocholl S, Akamine J, Hughes A, *et al.* Computer representation of overcurrent relay characteristics: IEEE committee report. IEEE Transactions on POWER Delivery. 1989;4(3):1659–1667.
[20] Noghabi AS, Mashhadi HR, and Sadeh J. Optimal coordination of directional overcurrent relays considering different network topologies using interval linear programming. IEEE Transactions on Power Delivery. 2010;25(3):1348–1354.
[21] Urdaneta AJ, Pérez LG, and Restrepo H. Optimal coordination of directional overcurrent relays considering dynamic changes in the network topology. IEEE Transactions on Power Delivery. 1997;12(4):1458–1464.
[22] Abdelaziz AY, Talaat H, Nosseir A, *et al.* An adaptive protection scheme for optimal coordination of overcurrent relays. Electric Power Systems Research. 2002;61(1):1–9.
[23] Amraee T. Coordination of directional overcurrent relays using seeker algorithm. IEEE Transactions on Power Delivery. 2012;27(3):1415–1422.
[24] Birla D, Maheshwari RP, and Gupta H. A new nonlinear directional overcurrent relay coordination technique, and banes and boons of near-end faults based approach. IEEE Transactions on Power Delivery. 2006;21(3):1176–1182.
[25] Abyaneh HA, Al-Dabbagh M, Karegar HK, *et al.* A new optimal approach for coordination of overcurrent relays in interconnected power systems. IEEE Transactions on Power Delivery. 2003;18(2):430–435.
[26] Mansour MM, Mekhamer SF, and El-Kharbawe N. A modified particle swarm optimizer for the coordination of directional overcurrent relays. IEEE Transactions on Power Delivery. 2007;22(3):1400–1410.
[27] Singh M, Panigrahi B, and Abhyankar A. Optimal coordination of directional over-current relays using teaching learning-based optimization

(TLBO) algorithm. International Journal of Electrical Power & Energy Systems. 2013;50:33–41.

[28] Phadke AG, and Thorp JS. Computer Relaying for Power Systems. New York, NY, USA: John Wiley & Sons, Inc.; 1988.

[29] Wang Y, Zhang P, Li W, *et al.* Online overvoltage prevention control of photovoltaic generators in microgrids. IEEE Transactions on Smart Grid. 2012;3(4):2071–2078.

[30] Beheshtaein S, Savaghebi M, Vasquez JC, *et al.* Protection of AC and DC microgrids: Challenges, solutions and future trends. In: Industrial Electronics Society, IECON 2015—41st Annual Conference of the IEEE. IEEE; 2015. p. 005253–005260.

[31] Mirsaeidi S, Said DM, Mustafa MW, *et al.* Progress and problems in microgrid protection schemes. Renewable and Sustainable Energy Reviews. 2014;37:834–839. Available from: http://www.sciencedirect.com/science/article/pii/S1364032114003694.

[32] Manditereza PT, and Bansal R. Renewable distributed generation: The hidden challenges. A review from the protection perspective. Renewable and Sustainable Energy Reviews. 2016;58:1457–1465. Available from: http://www.sciencedirect.com/science/article/pii/S1364032115016597.

[33] Boutsika TN, and Papathanassiou SA. Short-circuit calculations in networks with distributed generation. Electric Power Systems Research. 2008;78(7):1181–1191. Available from: http://www.sciencedirect.com/science/article/pii/S0378779607002131.

[34] Haj-ahmed MA, and Illindala MS. The influence of inverter-based DGs and their controllers on distribution network protection. IEEE Transactions on Industry Applications. 2014;50(4):2928–2937.

[35] Tsili M, and Papathanassiou S. A review of grid code technical requirements for wind farms. IET Renewable Power Generation. 2009;3(3):308–332.

[36] Yazdanpanahi H, Li YW, and Xu W. A new control strategy to mitigate the impact of inverter-based DGs on protection system. IEEE Transactions on Smart Grid. 2012;3(3):1427–1436.

[37] Beheshtaein S, Savaghebi M, Vasquez JC, *et al.* Protection of AC and DC microgrids: Challenges, solutions and future trends. In: IECON 2015 – 41st Annual Conference of the IEEE Industrial Electronics Society; 2015. p. 005253–005260.

[38] Che L, Khodayar ME, and Shahidehpour M. Adaptive protection system for microgrids: Protection practices of a functional microgrid system. IEEE Electrification Magazine. 2014;2(1):66–80.

[39] Liu Y, Ning P, and Reiter MK. False data injection attacks against state estimation in electric power grids. ACM Transactions on Information and System Security (TISSEC). 2011;14(1):13.

[40] Hashimoto H, and Hayakawa T. Distributed cyber attack detection for power network systems. In: Decision and Control and European Control Conference (CDC-ECC), 2011 50th IEEE Conference on. IEEE; 2011. p. 5820–5824.

[41] Ross KJ, Hopkinson KM, and Pachter M. Using a distributed agent-based communication enabled special protection system to enhance smart grid security. IEEE Transactions on Smart Grid. 2013;4(2):1216–1224.

[42] Young SE, and Aitel DA. The Hacker's Handbook: The Strategy Behind Breaking Into and Defending Networks. CRC Press Company; 2004. Available from: https://books.google.dk/books?id=u2AOmAEACAAJ.

[43] Cardenas AA, Amin S, and Sastry S. Secure control: Towards survivable cyber-physical systems. In: 2008 the 28th International Conference on Distributed Computing Systems Workshops; 2008. p. 495–500.

[44] Zeineldin HH, Mohamed YAI, Khadkikar V, *et al.* A protection coordination index for evaluating distributed generation impacts on protection for meshed distribution systems. IEEE Transactions on Smart Grid. 2013;4(3):1523–1532.

[45] Chaitusaney S, and Yokoyama A. Prevention of reliability degradation from recloser fuse miscoordination due to distributed generation. IEEE Transactions on Power Delivery. 2008;23(4):2545–2554.

[46] McDonald JD. Substation automation. IED integration and availability of information. IEEE Power and Energy Magazine. 2003;99(2):22–31.

[47] Tilaro FM, Copy B, and Gonzalez-Berges M. IEC 61850 Industrial Communication Standards under Test. 2014 Mar. Available from: https://cds.cern.ch/record/1696980.

[48] Mazur DC, Kreiter JH, Rourke ME, *et al.* Developing protective relay faceplates: Taking advantage of the benefits of IEC 61850. IEEE Industry Applications Magazine. 2015;21(1):33–40.

[49] Higgins N, Vyatkin V, Nair NC, *et al.* Distributed power system automation With IEC 61850, IEC 61499, and intelligent control. IEEE Transactions on Systems, Man, and Cybernetics, Part C (Applications and Reviews). 2011; 41(1):81–92.

[50] McArthur SDJ, Davidson EM, Catterson VM, *et al.* Multi-agent systems for power engineering applications Part II: Technologies, standards, and tools for building multi-agent systems. IEEE Transactions on Power Systems. 2007;22(4):1753–1759.

[51] Santodomingo R, Rodriguez-Mondejar JA, and Sanz-Bobi MA. Using semantic web resources to translate existing files between CIM and IEC 61850. IEEE Transactions on Power Systems. 2012;27(4):2047–2054.

[52] Strasser T, Andrn F, Vyatkin V, *et al.* Towards an IEC 61499 compliance profile for smart grids review and analysis of possibilities. In: IECON 2012 – 38th Annual Conference on IEEE Industrial Electronics Society; 2012. p. 3750–3757.

[53] Santodomingo R, Rohjans S, Uslar M, *et al.* Facilitating the automatic mapping of IEC 61850 signals and CIM measurements. IEEE Transactions on Power Systems. 2013;28(4):4348–4355.

[54] Pan Y, Steurer M, Baldwin TL, *et al.* Impact of waveform distorting fault current limiters on previously installed overcurrent relays. IEEE Transactions on Power Delivery. 2008;23(3):1310–1318.

[55] Jo H, Joo S, and Lee K. Optimal placement of superconducting fault current limiters (SFCLs) for protection of an electric power system with distributed generations (DGs). IEEE Transactions on Applied Superconductivity. 2013;23(3):5600304–5600304.
[56] El-Khattam W, and Sidhu TS. Restoration of directional overcurrent relay coordination in distributed generation systems utilizing fault current limiter. IEEE Transactions on Power Delivery. 2008;23(2):576–585.
[57] Haj-ahmed MA, and Illindala MS. Investigation of protection schemes for flexible distribution of energy and storage resources in an industrial microgrid. IEEE Transactions on Industry Applications. 2015;51(3):2071–2080.
[58] Cruden A, and Dudgeon GJW. The impact of energy storage devices used in conjunction with renewable embedded generators, on the protection and control system. In: 2001 Seventh International Conference on Developments in Power System Protection (IEE); 2001. p. 230–233.
[59] Beheshtaein S, Savaghebi M, Guerrero JM, *et al.* A secondary-control based fault current limiter for four-wire three phase inverter-interfaced DGs. In: IECON 2017 – 43rd Annual Conference of the IEEE Industrial Electronics Society; 2017. p. 2363–2368.
[60] Piya P, Ebrahimi M, Karimi-Ghartemani M, *et al.* Fault ride-through capability of voltage-controlled inverters. IEEE Transactions on Industrial Electronics. 2018;65(10):7933–7943.
[61] Sarlak M, and Shahrtash S. High impedance fault detection in distribution networks using support vector machines based on wavelet transform. In: Electric Power Conference, 2008. EPEC 2008. IEEE Canada. IEEE; 2008. p. 1–6.
[62] Mishra DP, Samantaray SR, and Joos G. A combined wavelet and data-mining based intelligent protection scheme for microgrid. IEEE Transactions on Smart Grid. 2016;7(5):2295–2304.
[63] Saleh SA, Ahshan R, Rahman MA, *et al.* Implementing and testing d–q WPT-based digital protection for micro-grid systems. In: 2011 IEEE Industry Applications Society Annual Meeting; 2011. p. 1–8.
[64] Beheshtaein S, Savaghebi M, Vasquez JC, *et al.* A hybrid algorithm for fault locating in looped microgrids. In: 2016 IEEE Energy Conversion Congress and Exposition (ECCE); 2016. p. 1–6.
[65] Li X, Dyko A, and Burt GM. Traveling wave-based protection scheme for inverter-dominated microgrid using mathematical morphology. IEEE Transactions on Smart Grid. 2014;5(5):2211–2218.
[66] Horowitz SH, Phadke AG, and Thorpe JS. Adaptive transmission system relaying. IEEE Transactions on Power Delivery. 1988;3(4):1436–1445.
[67] Piesciorovsky EC, and Schulz NN. Fuse relay adaptive overcurrent protection scheme for microgrid with distributed generators. IET Generation, Transmission Distribution. 2017;11(2):540–549.
[68] Laaksonen H, Ishchenko D, and Oudalov A. Adaptive protection and microgrid control design for Hailuoto Island. IEEE Transactions on Smart Grid. 2014;5(3):1486–1493.

[69] Shih MY, Salazar CAC, and Enríquez AC. Adaptive directional overcurrent relay coordination using ant colony optimisation. IET Generation, Transmission Distribution. 2015;9(14):2040–2049.
[70] Papaspiliotopoulos VA, Korres GN, Kleftakis VA, *et al.* Hardware-in-the-loop design and optimal setting of adaptive protection schemes for distribution systems with distributed generation. IEEE Transactions on Power Delivery. 2017;32(1):393–400.
[71] Wan H, Li KK, and Wong KP. An adaptive multiagent approach to protection relay coordination with distributed generators in industrial power distribution system. IEEE Transactions on Industry Applications. 2010;46(5):2118–2124.
[72] Najy WKA, Zeineldin HH, and Woon WL. Optimal protection coordination for microgrids with grid-connected and islanded capability. IEEE Transactions on Industrial Electronics. 2013;60(4):1668–1677.
[73] Coffele F, Booth C, and Dyko A. An adaptive overcurrent protection scheme for distribution networks. IEEE Transactions on Power Delivery. 2015;30(2): 561–568.
[74] Milanović JV, and Zhu W. Modeling of interconnected critical infrastructure systems using complex network theory. IEEE Transactions on Smart Grid. 2018;9(5):4637–4648.
[75] Ashok A, Govindarasu M, and Wang J. Cyber-physical attack-resilient wide-area monitoring, protection, and control for the power grid. Proceedings of the IEEE. 2017;105(7):1389–1407.
[76] Cardenas AA, Amin S, and Sastry S. Secure control: Towards survivable cyber-physical systems. In: Distributed Computing Systems Workshops, 2008. ICDCS'08. 28th International Conference on. IEEE; 2008. p. 495–500.
[77] Ntalampiras S. Detection of integrity attacks in cyber-physical critical infrastructures using ensemble modeling. IEEE Transactions on Industrial Informatics. 2015;11(1):104–111.
[78] Cleveland F. IEC TC57 WG15: IEC 62351 security standards for the power system information infrastructure. White Paper; 2012.
[79] Fuloria S, Anderson R, McGrath K, *et al.* The protection of substation communications. In: Proceedings of SCADA Security Scientific Symposium; 2010.
[80] Fries S, Hof HJ, Dufaure T, and Seewald MG, Security for the smart grid – Enhancing IEC 62351 to improve security in energy automation control. International Journal on Advances in Security. 2010;3.
[81] Schlegel R, Obermeier S, and Schneider J. A security evaluation of IEC 62351. Journal of Information Security and Applications. 2017;34:197–204.
[82] Moreira N, Molina E, Lázaro J, *et al.* Cyber-security in substation automation systems. Renewable and Sustainable Energy Reviews. 2016;54: 1552–1562.
[83] Xu C, and Cheng K. A survey of distributed power system AC versus DC distributed power system. In: Power Electronics Systems and Applications (PESA), 2011 4th International Conference on. IEEE; 2011. p. 1–12.

[84] Tzimas A, Antoniou D, and Rowland S. Low voltage DC cable insulation challenges and opportunities. In: Electrical Insulation and Dielectric Phenomena (CEIDP), 2012 Annual Report Conference on. IEEE; 2012. p. 696–699.

[85] Antoniou D, Tzimas A, and Rowland SM. Transition from alternating current to direct current low voltage distribution networks. IET Generation, Transmission & Distribution. 2015;9(12):1391–1401.

[86] Lotfi H, and Khodaei A. AC versus DC microgrid planning. IEEE Transactions on Smart Grid. 2017;8(1):296–304.

[87] Ton M, Fortenbery B, and Tschudi W. DC Power for Improved Data Center Efficiency. Lawrence Berkeley National Laboratory; 2008.

[88] Jeon JY, Kim JS, Choe GY, *et al.* Design guideline of DC distribution systems for home appliances: Issues and solution. In: Electric Machines & Drives Conference (IEMDC), 2011 IEEE International. IEEE; 2011. p. 657–662.

[89] Wunder B, Ott L, Szpek M, *et al.* Energy efficient DC-grids for commercial buildings. In: Telecommunications Energy Conference (INTELEC), 2014 IEEE 36th International. IEEE; 2014. p. 1–8.

[90] Kaur P, Jain S, and Jhunjhunwala A. Solar-DC deployment experience in off-grid and near off-grid homes: Economics, technology and policy analysis. In: DC Microgrids (ICDCM), 2015 IEEE First International Conference on. IEEE; 2015. p. 26–31.

[91] Jhunjhunwala A, Lolla A, and Kaur P. Solar-dc microgrid for Indian homes: A transforming power scenario. IEEE Electrification Magazine. 2016;4(2):10–19.

[92] Cuzner RM, Palaniappan K, and Shen ZJ. System specification for a DC community microgrid and living laboratory embedded in an urban environment. In: Renewable Energy Research and Applications (ICRERA), 2015 International Conference on. IEEE; 2015. p. 1119–1125.

[93] Palaniappan K, Veerapeneni S, Cuzner R, *et al.* Assessment of the feasibility of interconnected smart DC homes in a DC microgrid to reduce utility costs of low income households. In: DC Microgrids (ICDCM), 2017 IEEE Second International Conference on. IEEE; 2017. p. 467–473.

[94] Bathurst G, Hwang G, and Tejwani L. MVDC – The New Technology for Distribution Networks; 2015.

[95] Long C, Wu J, Smith K, *et al.* MVDC link in a 33 kV distribution network. CIRED – Open Access Proceedings Journal. 2017;2017(1):1308–1312.

[96] Kankanala P, Srivastava SC, Srivastava AK, *et al.* Optimal control of voltage and power in a multi-zonal MVDC shipboard power system. IEEE Transactions on Power Systems. 2012;27(2):642–650.

[97] Cuzner RM, and Venkataramanan G. The status of DC micro-grid protection. In: Industry Applications Society Annual Meeting, 2008. IAS'08. IEEE. IEEE; 2008. p. 1–8.

[98] Meyer JM, and Rufer A. A DC hybrid circuit breaker with ultra-fast contact opening and integrated gate-commutated thyristors (IGCTs). IEEE Transactions on Power Delivery. 2006;21(2):646–651.

[99] Alferov D, Budovsky A, Evsin D, *et al.* DC vacuum circuit-breaker. In: Proceedings of the Xxiiird International Symposium on Discharges and Electrical Insulation in Vacuum. vol. 1; 2008. p. 173–176.

[100] Mobarrez M, Fregosi D, Bhattacharya S, *et al.* Grounding architectures for enabling ground fault ride-through capability in DC microgrids. In: DC Microgrids (ICDCM), 2017 IEEE Second International Conference on. IEEE; 2017. p. 81–87.
[101] Cuzner RM, Sielicki T, Archibald AE, *et al.* Management of ground faults in an ungrounded multi-terminal zonal DC distribution system with auctioneered loads. In: Electric Ship Technologies Symposium (ESTS), 2011 IEEE. IEEE; 2011. p. 300–305.
[102] Fletcher S, Norman P, Galloway S, *et al.* Determination of protection system requirements for DC unmanned aerial vehicle electrical power networks for enhanced capability and survivability. IET Electrical Systems in Transportation. 2011;1(4):137–147.
[103] Yang J, Fletcher JE, and O'Reilly J. Short-circuit and ground fault analyses and location in VSC-based DC network cables. IEEE Transactions on Industrial Electronics. 2012;59(10):3827–3837.
[104] Jin Z, Sulligoi G, Cuzner R, *et al.* Next-generation shipboard dc power system: Introduction smart grid and dc microgrid technologies into maritime electrical networks. IEEE Electrification Magazine. 2016;4(2):45–57.
[105] Fletcher SD, Norman PJ, Galloway SJ, *et al.* Optimizing the roles of unit and non-unit protection methods within DC microgrids. IEEE Transactions on Smart Grid. 2012;3(4):2079–2087.
[106] Ciezki JG, and Ashton RW. Selection and stability issues associated with a navy shipboard DC zonal electric distribution system. IEEE Transactions on Power Delivery. 2000;15(2):665–669.
[107] Cuzner R, and Jeutter A. DC zonal electrical system fault isolation and reconfiguration. In: Electric Ship Technologies Symposium, 2009. ESTS 2009. IEEE. IEEE; 2009. p. 227–234.
[108] Rose MW, and Cuzner RM. Fault isolation and reconfiguration in a three-zone system. In: Electric Ship Technologies Symposium (ESTS), 2015 IEEE. IEEE; 2015. p. 409–414.
[109] Tang L, and Ooi BT. Locating and isolating DC faults in multi-terminal DC systems. IEEE Transactions on Power Delivery. 2007;22(3):1877–1884.
[110] Cuzner R, MacFarlin D, Clinger D, *et al.* Circuit breaker protection considerations in power converter-fed DC systems. In: Electric Ship Technologies Symposium, 2009. ESTS 2009. IEEE. IEEE; 2009. p. 360–367.
[111] Li G, Liang J, Balasubramaniam S, *et al.* Frontiers of DC circuit breakers in HVDC and MVDC systems. In: Energy Internet and Energy System Integration (EI2), 2017 IEEE Conference on. IEEE; 2017. p. 1–6.
[112] Brozek JP. DC overcurrent protection-where we stand. IEEE Transactions on Industry Applications. 1993;29(5):1029–1032.
[113] Meyer C, Schroder S, and De Doncker RW. Solid-state circuit breakers and current limiters for medium-voltage systems having distributed power systems. IEEE Transactions on Power Electronics. 2004;19(5):1333–1340.

[114] Mu J, Wang L, and Hu J. Analysis and design of topological structure for DC solid-state circuit breaker. In: World Non-Grid-Connected Wind Power and Energy Conference, 2009. WNWEC 2009. IEEE; 2009. p. 1–5.
[115] Cuzner RM, and Singh V. Future shipboard MVdc system protection requirements and solid-state protective device topological tradeoffs. IEEE Journal of Emerging and Selected Topics in Power Electronics. 2017;5(1): 244–259.
[116] Juvekar S, Compton B, and Bhattacharya S. A fast acting DC solid state fault isolation device (FID) with Si and SiC devices for MVDC distribution system. In: Energy Conversion Congress and Exposition (ECCE), 2012 IEEE. IEEE; 2012. p. 2005–2010.
[117] Shen ZJ, Sabui G, Miao Z, *et al.* Wide-bandgap solid-state circuit breakers for DC power systems: Device and circuit considerations. IEEE Transactions on Electron Devices. 2015;62(2):294–300.
[118] Chow TP. Wide bandgap semiconductor power devices for energy efficient systems. In: Wide Bandgap Power Devices and Applications (WiPDA), 2015 IEEE 3rd Workshop on. IEEE; 2015. p. 402–405.
[119] Shukla A, Demetriades GD. A survey on hybrid circuit-breaker topologies. IEEE Transactions on Power Delivery. 2015;30(2):627–641.
[120] Peng C, Huang AQ, and Song X. Current commutation in a medium voltage hybrid DC circuit breaker using 15 kV vacuum switch and SiC devices. In: Applied Power Electronics Conference and Exposition (APEC), 2015 IEEE. IEEE; 2015. p. 2244–2250.
[121] Dragičević T, Lu X, Vasquez JC, *et al.* DC microgrids—Part II: A review of power architectures, applications, and standardization issues. IEEE Transactions on Power Electronics. 2016;31(5):3528–3549.
[122] Park JD, and Candelaria J. Fault detection and isolation in low-voltage DC-bus microgrid system. IEEE Transactions on Power Delivery. 2013;28(2): 779–787.

Chapter 13

Black start and islanding operations of microgrid

Clara Gouveia[1], Carlos Moreira[2], André G. Madureira[1], José Gouveia[1], Diego Issicaba[1], and João Abel Peças Lopes[2]

This chapter describes different control techniques applied in microgrids (MGs) with responsive distributed energy resources (DER) for intentional and automatic islanding, deployment of a local black start strategy and including the reconnection to the upstream distribution network. The coordination of MG local resources, achieved through an appropriated network of controllers and communication system, enables some cells of the distribution system, with sufficient autonomy, to operate interconnected to the upstream network or autonomously in some circumstances. In this case, the potentialities of DER can be truly realized if the islanded operation is allowed and bottom-up black start functionalities are implemented. In particular, black start will be addressed for medium voltage (MV) MGs, made of several low voltage (LV) MGs, leading to the concept of a multi-MG (MMG). The chapter will then describe the black start sequence that can be implemented in such a grid considering also the contribution from electric vehicle storage.

13.1 Microgrid operational modes

MGs are widely regarded as an effective mean to promote the integration of DER – namely using local renewable energy sources (RES) – in distribution grids, particularly at the lower voltage levels. Moreover, its specific characteristics make it an interesting option for remote and isolated areas where resorting to conventional centralized generation is very expensive or even near impossible.

MGs can be autonomous in specific circumstances or remain grid connected, providing services to the distribution system, either individually or grouped in clusters (MMGs).

[1]INESC Technology and Science (formerly INESC Porto), Porto, Portugal
[2]INESC Technology and Science (formerly INESC Porto), Faculty of Engineering of the University of Porto, Porto, Portugal

13.1.1 The microgrid

The definition of MG according to the FP5 EU project 'MICROGRIDS – Large Scale Integration of Microgeneration to Low Voltage Grids' is the following [1]:

Definition 13.1: *A MG can be defined as a low voltage distribution system to which small modular generation systems are to be connected. In some sense, a MG corresponds to an association of electrical loads and small generation systems through a low voltage distribution network. This means that loads and sources are physically close so that a MG can correspond for instance to the network of a small urban area, to an industry or to a large shopping centre.*

The key feature of an MG system is the coexistence of the distribution network infrastructure, together with microgeneration devices, responsive electric/thermal loads and storage devices as well as dedicated control and management systems.

Figure 13.1 shows the typical hierarchical structure of an MG system.

The MG is managed by an MG Central Controller (MGCC), installed at the LV side of the MV to LV secondary substation, which serves as an interface between the main distribution system and the MG components. Several control functionalities to support MG operation are embedded in the MGCC. At the lower control level, two different types of controllers are present: load controllers (LCs), responsible for managing locally each load/group of loads, and microsource controllers (MCs), which manage locally each microgeneration unit or storage device.

Communication between these two levels of control controllers is required since LCs serve as interfaces to control loads (exploiting load flexibility), and MCs manage the microgeneration active and reactive power production levels, following requests from the MGCC, that ensures the coordinated management of the whole MG by providing adequate set points to both LC and MC. Also, the MGCC requires communication upstream to the central distribution management system (DMS) in order to support the operation at the MV network level.

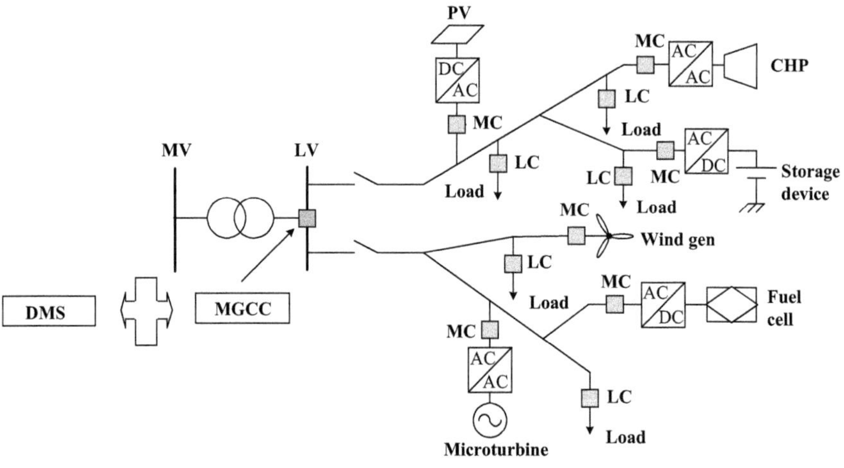

Figure 13.1 MG system architecture [2]

These characteristics enable an MG to operate in two distinct operating modes [2]:

- *Normal interconnected mode*, when the MG system is electrically connected to the main MV distribution network either being supplied by this network totally or partially or injecting power into the main grid.
- *Emergency mode*, when there is a failure in the main MV distribution network, and the MG system operates in an isolated and autonomous mode, similar to the electrical systems of geographic islands.

While in normal interconnected mode, the MGCC is responsible for collecting data from both microsources and loads in order to monitor and manage the operation of the MG, namely by performing forecasting studies, economic scheduling of microgeneration, security assessment evaluations, demand side management (DSM) functions and interface with the DMSs [3].

In emergency mode, the MGCC acts as a secondary control loop in order to request a change in the output power control of microgenerators that change from a dispatched power mode to a frequency control mode in the isolated grid. Also, the MGCC has an accurate knowledge of the type of loads in the grid (to eventually adopt interruption strategies) and to use support from storage devices. Finally, the MGCC can be responsible for local black start strategies, which is essential in order to ensure that the MG has an impact in terms of improving reliability and continuity of service, by reducing interruption times [3].

Finally, MCs and LCs aim at contributing to the economic scheduling activities, to local control of storage devices, to load tracking activities and to manage loads with interruption or peak shaving capabilities [3].

13.1.1.1 Distributed energy resources in a microgrid

The MG can integrate distinct types of DER, for instance microgeneration units, storage and responsive loads. The microgeneration units most commonly used in MGs are primarily DC generation sources based on RES and low-carbon technologies for micro-combined-heat-and-power (micro-CHP) applications. In some cases, small synchronous generators are also used. In this context, microsources are usually classified according to their degree of controllability:

- *Non-controllable microsources*, which are based on RES, namely PV systems and micro wind turbines. Given the variability of the primary energy source, the output power of the microgenerator is also highly variable, being controlled to extract the maximum power considering the availability of the energy resource.
- *Controllable microsources*, which include dispatchable sources such as microturbines and fuel cells that have been adopted in the development of the MG concept, due to their high efficiency and low-emission levels compared to other technologies.

Storage is also a key element of the MG for enabling its autonomous operation. These technologies are characterized by their power capacity, energy storage capacity, response time and cost. For MG autonomous operation high-power density and fast response (typically some milliseconds), storage technologies are preferred, in order to be able to follow load variations in the MG and compensate for transitory disturbances. However, medium-to-high energy density may also be required to compensate for low microgeneration capacity.

The loads connected to LV networks are mainly single-phase with the exception of larger loads that require three-phase connections. An MG can supply residential, commercial or small industrial customers. Residential and commercial loads are usually resistive loads such as electrical heaters and cooking appliances, lighting, electronic devices, thermostat-controlled loads such as air conditioners, refrigerators and space and water heaters and small induction motors or heaters. Some of these loads have some degree of controllability, due to their thermal storage capability or long-running cycles, thus enabling the implementation of DSM strategies.

Electric vehicles (EVs) are also expected to be connected to LV network for charging through a single-phase charger (e.g. home charging), representing a new load for the MG system. At the same time, the EV load flexibility and distributed storage capacity can be exploited through smart charging strategies and vehicle-to-grid (V2G) concepts in order to support the system operation. In this context, a new type of local controller is required in addition to MCs and LCs – the vehicle controller.

The flexibility of EV and loads can be exploited in order to support MG autonomous operation. When operating autonomously, the MG relies in local storage and generation capacity to supply the loads and maintain power balance. If the MG has insufficient generation reserve capacity to supply load, part of the MG loads will have to be disconnected until the MG is reconnected to the main MV distribution network.

13.1.1.2 Power electronic interfaces for distributed energy resources

The majority of MG components, namely the microsources, storage units and EV batteries require power electronic interfaces (either DC/AC or AC/DC/AC), in order to convert their power output into grid-compatible AC power. Contrarily to the conventional power system, in a power electronic-based MG, the system can be inertialess due to the absence of synchronous machines.

The grid-side inverter core functionalities include the control of the active and reactive power flow with the grid, DC-link voltage and synchronization between the microsource and the grid. However, additional functionalities can be added to provide grid support, namely local voltage regulation, frequency regulation and voltage harmonic and unbalance compensation. The grid-side converter can be classified in two groups, depending on the approach that is used to control its operation:

- *Grid-feeding inverter*, controlled through an active and reactive (*P–Q*) control strategy. The inverter is a current-controlled voltage source, designed to deliver a controlled amount of power (active and reactive) to the grid (these inverters are usually referred as grid-tied inverters). The *P–Q* controlled inverter does not have the ability of imposing voltage and frequency. The grid-side inverters of microsources (controllable and non-controllable) are usually controlled as grid-feeding inverters for both islanded and interconnected operation.

- *Grid-forming inverters* are controlled as voltage sources establishing the phase angle and magnitude of the output voltage. The inverter is usually fed by a DC power source with energy buffering capabilities, such as batteries. It is usually referred to as voltage source inverter (VSI) and can also incorporate grid supporting functionalities for voltage and frequency regulation, which is especially important during emergency operation.

Following the two possible strategies for power electronic inverters control, two operational strategies have been proposed for MG islanded operation:

- *Single master operation* where the MG operates with only one grid-forming inverter. The other inverters are controlled as grid-feeding inverters coupled to non-controllable or controllable microsources.
- *Multi-master operation* where the MG operates with more than one grid-forming inverter. This strategy can be applied when the MG integrates more than one fast-acting storage unit.

The MG is composed of several inverters coupling distinct DER and following different control strategies. The coordination between them is achieved through the MG management and control architecture, described previously.

13.1.1.3 Microgrid control and management

As shown in Figure 13.1, the MG is divided in two main layers constituted by the MGCC and the local controllers. The intelligence/autonomy level attributed to each layer may vary according to the group of functionalities to perform. Considering the technical and economical operation requirements, the MG control functions can be divided in different groups:

- *Long-term energy-management tools* which will be responsible for dispatching the controllable resources of the MG such as generation, distributed storage and loads and control the power flow between the MG and the main grid. The management objective can vary according to the MG business model and operation modes. In case the system can operate islanded, the energy management tools should be able to ensure that the MG system has enough reserve capacity to ensure a secure islanded operation.
- *Short-term balancing tools*, responsible for the ensuring frequency and voltage stability of the system in the case of disturbances, including islanding. Control functionalities include frequency and voltage regulation, load-following capability, power-quality compensation and load-shedding schemes.
- *Security assessment tools*, as proposed in [4], are responsible for identifying if the MG is operating in an insecure operating point, which in the case of an unplanned islanding could lead to the system collapse. In case the MG is considered insecure, preventive load shedding is implemented to reduce the power disturbance during the islanding.
- *MG restoration procedure* will be triggered by the MGCC when the MG islanding fails or a general blackout occurs. The algorithm was first presented in [5] and consists of a fully automatic procedure to restore service locally

based on the coordination of local resources. During the procedure, short-term balancing tools are required to ensure the stability of the system.

13.1.2 Microgrid hierarchical control for emergency operation

The MG emergency control will result in a combination of local and centralized control strategies, implemented at the local controllers and at the MGCC, respectively. Local control acts in short-term responding immediately to disturbances affecting the balance and voltage of the MG. Therefore, local control strategies do not depend on the MG communication infrastructure, instead relying in electric variables (i.e. voltages, power and frequency) measured locally. On the other hand, centralized control strategies will act in a longer term interacting with the local control layer in order to produce the required changes in the resources' behaviour. The centralized control is designed to support the action of local control strategies, monitoring the MG operating state and providing an adequate coordination between the MG resources during emergency conditions.

A multilevel hierarchical control architecture is proposed, which is based in European Network of Transmission System Operators (ENTSO-E) load-frequency control. This architecture is represented in Figure 13.2 and includes additional frequency, voltage and power quality functionalities at the different levels of control [6].

Level 0 and Level 1 control are implemented at the local controllers and at the MG resources grid-coupling inverters. The provision of primary voltage and frequency regulation (i.e. Level 1 in Figure 13.2) is vital for the stability of the MG operating in emergency mode, requiring a robust and fast response from the controllers. Therefore, these functions are implemented locally as external control

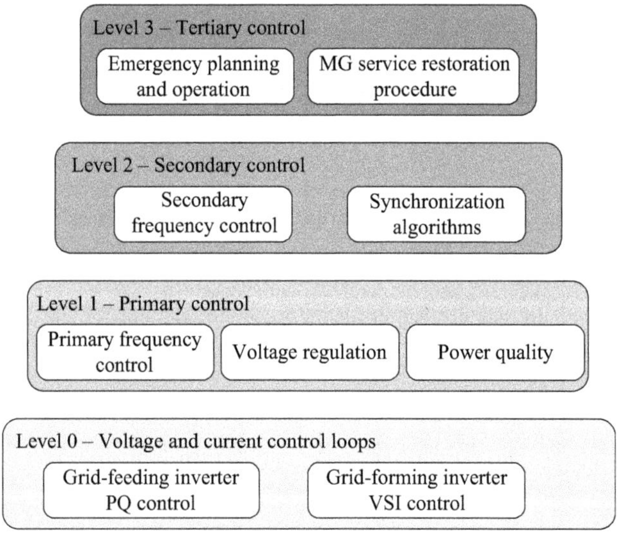

Figure 13.2 MG hierarchical control for emergency operation

loops of the VSIs coupled to energy storage devices. Considering the capability of VSI for imposing voltage and frequency, specific voltage-balancing strategies can also be adopted in order to improve MG power quality.

The controllability of dispatchable microsources can also be explored to perform secondary frequency regulation (Level 2), reducing the power requirements from storage units providing primary frequency regulation. This control can be implemented locally or centralized at the MGCC.

Considering the limited resources of the MG to maintain the stability of the system, specific strategies were adopted for EV complemented by active demand response schemes, specifically designed for the MG emergency operation. The proposed strategies take advantage of the flexibility of EV and responsive loads in order to provide grid support during MG autonomous operation and ultimately increase MG resilience.

The main objective of tertiary control proposed is to manage the power flow between the MG and the upstream network when operating interconnected. However, by taking advantage of the smart metering infrastructure, new functionalities can be derived in order to ensure a safe islanding and manage the resources during autonomous operation. For that, it is necessary to update the emergency operation strategies according to the actual operating state of the MG, allowing a better coordination of the available resources and consequently improving the MG resilience and success of the autonomous operation. If islanding fails or a general blackout occurs, the MGCC will also be responsible for triggering and coordinating the MG restoration procedure, taking advantage of flexible resources such as the EV.

13.1.3 Extending the concept – the multi-microgrid

In a foreseeable future scenario where there is a massive integration of microgeneration in LV distribution networks, a new operational paradigm must be considered based on the development and extension of the MG concept. The connection of other distributed generation (DG) sources directly to the MV distribution network, together with the deployment of LV MGs able to operate as active cells, requires that a coordinated hierarchical control system is implemented. This new type of system/entity is known as an MMG [7].

The definition of MMG according to the FP6 EU project 'Advanced Architectures and Control Concepts for MORE MICROGRIDS' is the following [3]:

Definition 13.2: *The massive integration of MGs and DG units connected to the MV level, and consequent need for coordinated management of these units, lead to the development of the MMG concept. This requires a higher-level structure, formed at the MV level, consisting of LV MGs and DG units connected on several adjacent MV feeders. For the purpose of control and management, MGs, DG units and MV-connected loads under DSM control can be considered as active cells in this new type of system.*

In Figure 13.3, the hierarchical structure of an MMG system is shown.

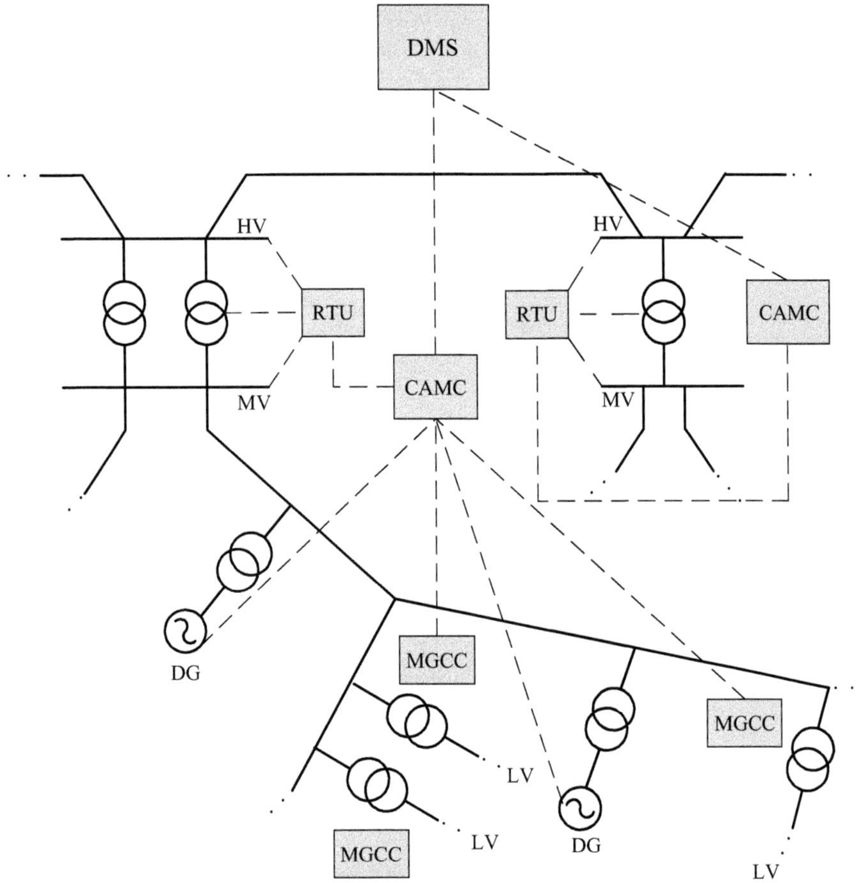

Figure 13.3 MMG system architecture [3]

The MMG concept assumes that some of the traditional functions assigned to a DMS are transposed to a new control layer located at the level of the MV distribution system. The controller responsible for managing the MV network as whole, including DG connected to the MV grid, High Voltage (HV)/MV transformers with On Load Tap Changers and MGs (through the corresponding MGCC), is the central autonomous management controller (CAMC). This controller is to be installed at the MV bus level of the HV/MV primary substation, under the responsibility of the DSO. In practice, the CAMC will operate as a 'mini-DMS' that coordinates all the resources at the MV system level, enabling operation under normal and emergency operating conditions. In a similar way to what happens for the MG system, the MMG may also be operated in normal mode when the MMG is operated interconnected to the main HV-distribution network and in emergency mode when the MMG is operated autonomously, i.e. isolated from the main HV distribution network, or following a blackout when the MMG system is contributing to service restoration by triggering a black start procedure [7].

13.2 Microgrid islanding and reconnection

A distinctive characteristic of MGs is the capability of islanding and therefore operating autonomously. As mentioned earlier, this mode of operation is enabled by the MG hierarchical management and control structure that can combine local and centralized control strategies for voltage and frequency regulation of the islanded system.

13.2.1 Microgrid primary frequency and voltage control

Since the MG is an inverter dominated system, the MG power electronic interfaces must be explored in order to provide grid support, ensuring frequency and voltage regulation. The adoption of droop control functionalities in grid-forming inverters for primary frequency and voltage regulation has been proposed in Consortium for Electric Reliability Technology Solutions (CERTS) MG and European MG concepts [2,6,8,9]. The main objective is to emulate the behaviour of synchronous generators by establishing the system frequency, according to the active power output (power–frequency (*P–f*) droop) and the voltage magnitude according to the reactive power output (*Q–V* droop) [2,6,8,9].

Enabling MG autonomous operation requires the use of (at least) a grid-forming inverter operating as a VSI usually coupled to an energy storage unit, where the reference frequency and voltage magnitude will be determined through droops as in the following equation:

$$\begin{aligned} \omega &= \omega_0 - k_P \times P \\ V &= V_0 - k_Q \times Q \end{aligned} \tag{13.1}$$

where ω, V is the reference angular frequency (rad/s) and voltage values (V); P, Q is the inverter active (W) and reactive power (var) outputs; k_P, k_Q is the droop slopes (rad/s W and V/var); ω_0, V_0 is the idle values of the angular frequency (rad/s) and voltage (V) at no load conditions.

When a disturbance occurs, the MG frequency will reflect the power unbalance between the local generation and load, and the grid-forming inverters will adjust its power output based on their droop characteristics.

The droop strategy ensures a good load sharing between the grid-forming inverters, without requiring fast communication infrastructures between the inverters [2,10–12]. However, some disadvantages have been identified regarding droop strategies [13], namely

- Active power–frequency droop (*P–f* droop) imposes a permanent frequency error which has to be corrected through additional control functions.
- Droop control assumes that the coupling impedance between the inverter and the MG is mostly inductive. However, MG LV feeders usually have high *R*/*X* ratio which means that the decoupling between active power and voltage is no longer verified.

The first issue can be solved by adopting a secondary frequency control strategy acting at the MG level and taking advantage of the reserve capacity of existing resources. For the second issue, adaptations to existing droop-based solutions as well as alternative droop methods have been proposed in the literature for dealing with the coupling between active power and node voltage magnitude. A detailed review of those methods is provided in [13]. The three more relevant strategies are identified below:

- Reverse power droops: *P–V* and *Q–f* droops have been demonstrated through numerical simulation in [14]. However, no proportional load sharing would be obtained [12] since the loads would be supplied by the nearest generator, without allowing for power-dispatch strategies.
- Compensation of line impedance effect. In [15], no decoupling between active and reactive power is considered. The MG frequency and voltage is determined based on the droop coefficients and considering the *R*/*X* ratio of the line impedance, which can be difficult to determine. Virtual impedance methods have been also proposed in [16,17] as an additional control concept that is integrated at the power converter control level. In [16], a virtual inductance is introduced in order to decouple active and reactive power and enable the adoption of conventional droop, even for high *R*/*X* ratio. In order to enable a proportional share of reactive power, a compensating term is added to the *Q–V* droop, reflecting the effect of the line voltage drop and local load effect on voltage. The implementation of this method requires determining droop coefficients as a function of the voltage at the MG Point of Common Coupling. However, it is not clear that the introduction of additional MS and loads changing voltages in the MG feeder would produce such an equal sharing of reactive power.
- Power-angle droop control was proposed in [18,19]. The VSI active power output is controlled through the voltage angle, without changing the frequency of the system as in conventional droop strategy. The implementation of this method requires a global positioning system signal for angle referencing.

Additional changes to conventional droop controllers have been proposed to deal with non-linear loads [15,20], improve small signal stability [19] and increase inertia of the system [21].

13.2.2 Electric vehicles contribution to primary frequency support

Considering the specific case of MG operation, EV contribution for the provision of primary reserve would reduce the central storage capacity required to assure MG system robustness, during islanded conditions. In order to provide primary reserve, the EV charger can be controlled through the frequency-active power droop (*f–P*) control strategy proposed in [22–24] which will change the EV charging power according to the MG frequency.

The EV droop characteristic is represented in Figure 13.4. For frequencies around the nominal value, the EV will charge the battery at a predefined charging rate (P_{ref}). If a disturbance occurs and the frequency drops below the dead-band minimum, the EV reduces its power consumption, reducing the load of the system. When the MG frequency overpasses the frequency dead-band maximum, the EV can also increase its power consumption. For large disturbances, causing the frequency to

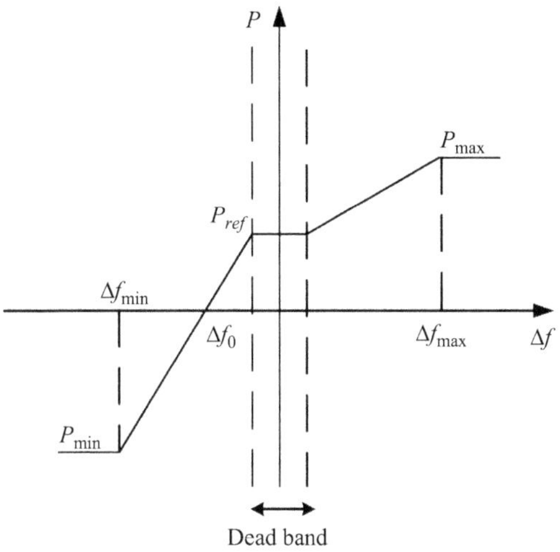

Figure 13.4 EV f–P droop characteristic

go below the zero-crossing frequency (Δf_0), the EV starts to inject power into the grid (V2G functionality). When the MG frequency becomes out of the predefined frequency range, the vehicle will inject/absorb a fixed power (P_{max} or P_{min}).

The definition of the EV control parameters such as P_{max} or P_{min} and P_{ref} will depend of the EV charger characteristics. Considering the charger technical limitations and the willingness of EV owners to participate in such services, the MGCC may remotely change the EV charging rate (P_{ref}) and the frequency parameters such as Δf_0, Δf_{min} and Δf_{max}, in order to promote adequate coordination with the MG frequency regulation mechanisms (load shedding schemes, availability of energy storage devices and their state of charge (SOC)).

13.2.3 Secondary control and emergency dispatch strategies

Secondary control is intended to provide the mechanism for power dispatch during MG autonomous operation while restoring the islanded MG frequency to the nominal value. The strategies for secondary control can be performed locally by the grid forming unit(s) or by each controllable microsource based on local frequency measurement or by a centralized algorithm housed at the MGCC.

Local secondary control reduces the communication requirements between the inverters [25]. However, it may impose stability concerns requiring the coordination of the local controller's parameters.

Centralized strategies such as the ones propose in [26] allow the dispatch of the microsource based on optimal power flow formulations and take advantage of the processing capabilities of the MGCC. In [27], the authors propose an optimal power dispatch for droop-controlled generators. The optimal power outputs are implemented by changing the droop parameters, ensuring that they are able to maintain the system stability and achieve the desired level of damping. The optimal power dispatch algorithm is complemented by two additional tools: the droop stability analysis which defines the stability region of the MG operation and the droop selection tool. An alternative centralized secondary control strategy based on active power unbalance is proposed in [28]. The real-time data provided by the smart meters and local controllers will be considered to characterize the microsources current power injection and its reserve capacity. The algorithm developed enables the coordination between microsources controlled as grid feeding (*P–Q*) and grid forming (VSI) as well as with other MG DER, such as flexible loads or EV.

The algorithms presented in [25–28] only consider the power limits of the microsources and storage devices. However, this may not lead to a feasible solution, since the participation of distributed storage unit will depend both on its power and energy-reserve capacity. In [29], an emergency balance algorithm is proposed to maximize the time the LV system can operate autonomously, by maximizing the MG power and energy-reserve capacity.

The MG emergency balance algorithm can be set to plan the control for the next time step or in a longer term (from minutes to a few hours), considering the time resolution of the real-time and forecasting data. The algorithm will evaluate the MG operating state and then dispatch the distributed storage units providing grid support. If the power and/or energy reserve capacity is insufficient, the algorithm will schedule additional control actions for mobilizing residential flexibility. The algorithm comprises three different steps:

1. Characterize the MG operating state: The tool will collect the data relative to the MG storage devices, loads and microgeneration state and determines the available power and energy reserves.
2. Emergency dispatch of MG controllable resources: The algorithm identifies the operation plan for the defined timeframe, based on the current state of the network and in the load and microgeneration forecasts.
3. Schedule emergency control actions: In case the MG does not have enough power and/or energy reserve capacity to guarantee the power balance, it is necessary to exploit emergency demand response strategies to ensure power balance and thus avoid the system collapse.

The storage unit's active power set points for the time horizon considered for the islanded operation will be calculated through an iterative approach as illustrated in Figure 13.5. The algorithm will return a plan for the time horizon considered, t_n, consisting on a set of active power set points for each time step, t.

In the first step, the algorithm will determine an initial solution for the power dispatch of the DER participating in the secondary control, including distributed storage units. Then, in the second stage, the algorithm will redistribute the amount of power dispatched for the distributed storage unit(s) according to the SOC estimated at the end of the considered time horizon t_n.

The power dispatch is determined in order to ensure the power balance of the MG for each time step t, as in the following equation:

$$\Delta P_t = P_{MS_t} - P_{L_t} - P_{EVt} + \sum_{i=1}^{m} P_{i_t} + \sum_{i=1}^{m} \Delta P_{Flex_t} \tag{13.2}$$

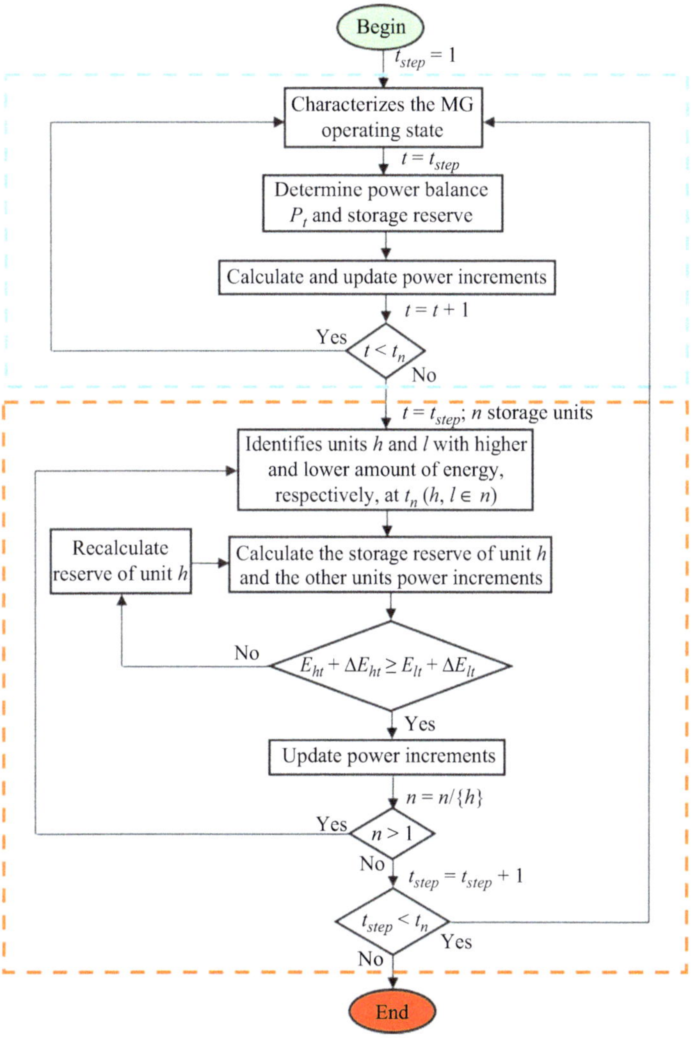

Figure 13.5 Emergency dispatch algorithm of MG distributed storage units providing grid support

where ΔP_t is the power unbalance that will need to be provided by the VSI, P_{MS_t} is the active power forecast of PV generation, P_{L_t} is the active power load forecast P_{EVt} is the active power load forecast resulting from EV charging in time step t, P_{i_t} is the active power set point of controllable resources including controllable generation and grid supporting storage, ΔP_{Flex_i} active power flexibility set point defined for grid supporting loads and EV.

The initial solution dispatches each controllable microgeneration unit and grid supporting storage unit based on its power reserve upward and downward.

Loads providing flexibility services as well as EV will also be considered in the dispatch strategy, modelled as smaller equivalent distributed storage units. The dispatch of this flexibility will result in increments/decrements to the actual load consumption (ΔP_{Flex}).

When the MG generation exceeds the load, the VSI will charge in order to balance the system. In this case, the algorithm will request a power consumption increase to the grid-supporting storage units, considering the reserve up available. On the contrary, when the VSI is discharging in order to supply the remaining load, the dispatch will define new set points either to reduce the power consumption from the storage units or even to reverse its power output and inject power in the LV system.

The contribution of each controllable resource (including flexibility) will be defined for unit i by the ratio between its reserve (R_i) and the total reserve of the system, as explained before, according to the following equation:

$$P_{it} = \frac{R_i}{\sum Ri} \times \Delta P_t \tag{13.3}$$

At the end of the first stage of the algorithm, the final SOC of the storage units are determined for the time horizon t_n.

Based on final SOC, the algorithm will then adjust the storage units' power set points by increasing the power provided by the unit with the highest SOC in order to reduce the power provided by units with lower SOC. The power change for the other units will be determined as shown in in the following equation:

$$\Delta P_{it} = \frac{(C_i - E_i)}{\sum (C_i - E_i)} \times \Delta R_{\max} \tag{13.4}$$

where ΔP_t is the active power change for unit i at time step t, $\Delta R_{\max}$ is the reserve capacity provided by the unit with the maximum SOC, E_i is the energy capacity available at the end of the time horizon.

The reserve is distributed by the other unit(s) considering the ratio between the energy capacity available at the end of the time horizon (E_i) and the total energy capacity available.

The balancing of the distributed storage dispatch is performed iteratively. A constraint was imposed in order to ensure that the SOC of the re-dispatched units does not get lower than the unit with the lowest SOC. This prevents the algorithm from getting a worst solution than the one determined initially.

For each time step, the algorithm will verify the VSI limits and the network technical limits (e.g. under and over voltages or feeder congestion). However, in case a violation is detected, the algorithm will schedule other consumers load flexibility in order to support the MG islanded operation.

13.2.4 Black start strategies in multi microgrids

In a black start process, it is important to identify a sequence of actions that minimizes the loss of power supplied. Therefore, it is crucial to shorten as much as possible the time spent on this process, while ensuring the safety of the utility crew and equipment, as well as trying to keep within acceptable limits the frequency and voltage values [30].

13.2.4.1 Typical strategies

Traditionally, in a black start process in a large system, two strategies can be applied, depending on the system characteristics: the build-down strategy and the build-up strategy [31,32].

The build-down strategy consists in energizing first the network as a whole, followed by a progressive synchronization of the remaining sources without black start capability and the connection of loads in an attempt to avoid wasting time for the posterior synchronization of the network portions.

In the build-up strategy, the system is divided into islands that are put back into service individually and then synchronized in order to reconstruct the system as a whole. For this accomplishment, each island must contain at least one unit with black start capability. Furthermore, blocks of load must be connected to each subsystem following small increments aiming at avoiding large frequency and voltage deviations.

13.2.4.2 Synchronization issues

Before closing a circuit breaker between two islanded systems, it is essential to match voltages on both sides of the circuit breaker. If this synchronizing process is not performed properly, a power system disturbance will take effect and equipment can be damaged. In order to synchronize the voltage waveform correctly, three different aspects must be closely monitored [33]:

- the voltage magnitudes
- the voltage frequency
- the phase difference between the voltages waveforms

Whether the voltage magnitudes are not closely matched, a sudden rise in reactive flow will appear across the circuit breaker as it is closed. The allowable voltage magnitude differences across the open circuit breaker are system specific. However, for general guidance, a difference of a few per cent is unlikely to cause any serious problem [33].

Whether the voltage frequencies on either side of an open circuit breaker are not matched, a sudden change in active power flow will appear across the circuit breaker as it is closed. The sudden active power flow change is in response to the initial frequency difference as the system seeks to establish a common frequency [33]. Figure 13.6 illustrates the voltage frequencies at two islands which are connected by closing a circuit breaker at $t =$ s.

At last, whether the phase difference between voltages on either side of the open circuit breaker is not reduced to a small value, a large active power flow increase will suddenly occur once the circuit breaker is closed. The voltage phase difference is the difference between the zero crossings of the voltage waveforms on either sides of the open circuit breaker. Ideally, the voltage phase should be as close to zero degrees as possible before closing the circuit breaker [33].

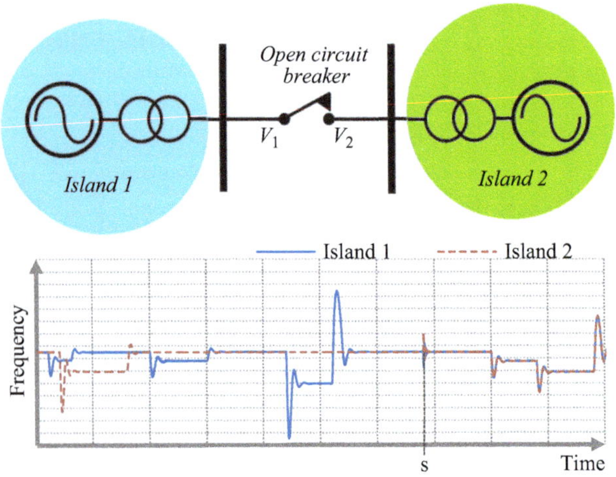

Figure 13.6 Closing a circuit breaker between two islands

Since the two islands are independent electrical systems, all three of the synchronizing variables (voltage magnitudes, frequency and phase difference) must be monitored to ensure they are within acceptable limits prior to closing the open circuit breaker.

13.2.5 Black start procedure

The service restoration of any power system is always a delicate process. Regarding an MMG, it is expected that this procedure can be done more expeditiously since the adequate exploitation of the hierarchical control, management and communication infrastructure will allow the automation of the whole process, reducing (or even eliminating) the manual intervention that still persists in conventional systems.

The CAMC, installed at the HV/VM substation level, will coordinate the process by a set of rules and conditions (previously defined) which should be verified during the black start process. To these rules and conditions, there will be a corresponding set of control actions that the CAMC will take on the elements under its control. In general, the procedure involves the MV network preparation (energization), the connection of the DG units with voltage and frequency control capacity, the interconnection with downstream MGs, the progressive connection of controllable loads (switching with other available production to balance the demand) and, finally, the synchronization with the upstream HV network (when it became available).

The black start process can be initiated by the CAMC in case of a general blackout (in all MMG) or local blackout (only in a portion of the MMG). In both cases, the black start is justified if it is not possible to feed the affected part of the grid using an alternative path towards the upstream network. Throughout coordination between the CAMC and DMS, the service restoration process in the MMG should not start (wasting time and resources) if, for example, the upstream network will be available in a short term or there is an alternative way to continue feeding all MMG or portion of the affected MMG via a reconfiguration of the MV network.

The diagram in Figure 13.7 shows the steps followed by the CAMC to detect and decide in proceeding with the black start process, adapted from [32].

In order to perform the first stages of the restoration in an MMG, the presence of generators with black start capability is required. The availability of the communication system in order to give updated information to the CAMC and to give orders to the devices under its control is another basic requirement.

13.2.5.1 Initial assumptions

First, it is assumed that faults are cleared and isolated, as well as the network is available for all energization procedures. This involves performing a sequence of actions that places the network in an acceptable starting point for the service restoration. The following actions are then implemented after the system black out, as soon as the decision of proceeding with the black start was taken [32].

- The MMG is separated from the upstream HV network;
- Each MG is isolated from the MV network and has passed to autonomous operation. In this work, it is assumed that this is done automatically and the

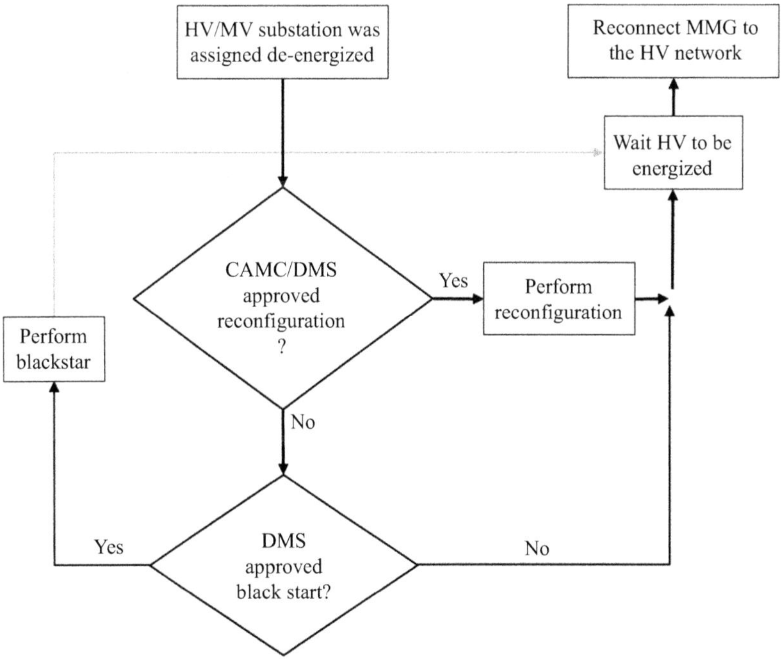

Figure 13.7 Diagram that allows the CAMC to decide about the starting of the black start process, adapted from [32]

MG would be prepared to synchronize with MT network when requested, as demonstrated in [34].

- All sources of reactive power (capacitor banks, for example) are turned off.

13.2.5.2 Sequence of actions

After the occurrence of a blackout, the CAMC proceeds with the service restoration in the MMG using the available information obtained through the communication infrastructure and its own database, following generically the next stages [35]:

1. Disconnect all loads, separate the corresponding MV/LV transformers from the network and turn off the reactive power sources. This allows limiting the unpredictability and magnitude of the frequency deviations during the service-restoration phase, as well as minimizing the voltage transients to energize the MV network.
2. Note that in actual networks, this is mostly done by opening remote-controlled sectionalizers across the network, rather than separating transformer by transformer. Particular attention must be taken in re-setting the protection devices (breakers, sectionalizers, auto-reclosers and so forth).
3. Segment the MMG around each MG and each generator with black start ability, creating islands.
4. Construction of the MV network. The initial network energization is performed gradually in order to avoid large deviations in frequency and voltage. In general, this process starts from one of the units with black start capability, energizing MV lines.
5. Synchronization of various islands. This allows stiffening the system and it is done as soon as the lines that connect the islands are energized and the synchronism conditions are verified.
6. Connection of priority loads. The loads considered of utmost importance are connected only if the local generation has capacity to serve them. The quantity and time lag in this procedure depend upon the permitted frequency and voltage deviations.
7. MG synchronization with the MV network. Once all transformers are energized, the MG operating autonomously and already black started can be synchronized with the MMG. In case storage devices in the MG are envisioned to smooth frequency deviations in the black start process, this step can be anticipated.
8. Gradual energization of the remaining MV/LV transformers. This will allow a progressive reintegration of the remaining load.
9. Load reposition. At this stage, with all network energized, load can be finally connected. This is done alternately with the increase of power from controllable generation sources or, alternatively, from non-controllable dispersed sources.
10. Connection of non-controllable dispersed sources to MV network. If the MMG is strong enough to absorb the power fluctuations from this kind of sources, they can be connected.

11. Load increase. The remaining load can be connected since all production existent is available.
12. Activation of the automatic frequency control, in order to assure an operation closer, as much as possible, to the nominal frequency, while the MMG operates in islanded mode.
13. Finally, connect the MMG to the upstream HV network, provided that the HV grid is prepared for it and the DMS hierarchical control level verifies all conditions to proceed with this synchronization.

13.3 Case study

13.3.1 Microgrid islanding case study

In order to illustrate the concepts previously presented for MG autonomous operation, it was considered a test system consisting in a typical Portuguese urban LV network (Figure 13.8), aiming at analysing the active participation of DER in the energy and the power-balance management following MG islanding.

The LV network is operated under unbalanced conditions with both single-phase and three-phase loads and microgeneration, having a peak power of 184 kW. The following resources were considered to be available in the grid for controlling purposes:

- A 100 kW/50 kW h storage unit connected to the LV bus of the LV/MV substation, controlled as grid forming unit.
- Two storage units controlled as grid feeding, being operated with fixed active and reactive power set points. A 30 kW/30 kW h storage system is connected to node 29 and a 15 kW/15 kW h storage unit installed at node 74.
- A 30 kW Single shaft microturbine (SSMT) connected to node 74.

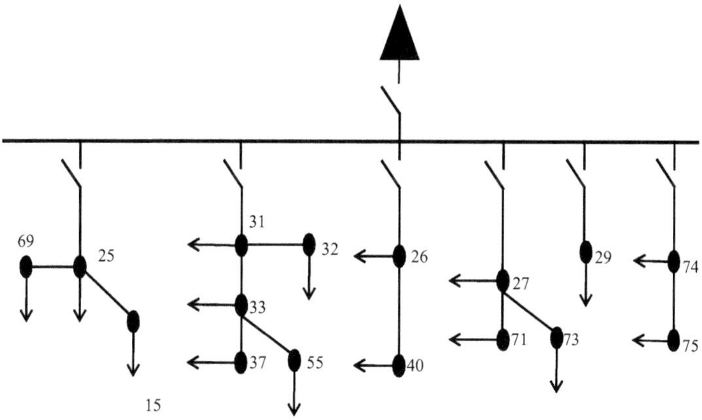

Figure 13.8 Single-line diagram of the test case network

- 63 kW from single-phase distributed EV and 45 kW from single-phase residential storage were distributed by the three phases of the system. Each individual EV and residential storage unit were considered to have a power capacity of 3 kW and an energy capacity of 3 kW h.

Table 13.1 presents the load and microgeneration forecasting for a 2-h period, considered as the maximum time the MG can operate in islanded mode. As observed, there exists an excess of load compared to PV generation.

In order to demonstrate the effectiveness of the emergency balancing tool, two distinct simulation scenarios were considered:

- Scenario A, where the primary and secondary control is only ensured by VSI.
- Scenario B, where the dispatch performed by the MG energy balance management tool is considered.

Before islanding, all the loads were being fed by MV grid. An unplanned islanding occurs in the beginning of the simulation time. The MG frequency can be observed in Figure 13.9 for both scenarios. At the moment of the islanding transient, a

Table 13.1 Operational scenario for test system

Time interval	**1**	**2**	**3**	**4**	**5**	**6**	**7**	**8**
Load (kW)	75	83	77	90	110	100	110	120
PV generation (kW)	−15	−13	−12	−5	0	0	0	0
Difference (kW)	60	70	65	85	110	100	110	120

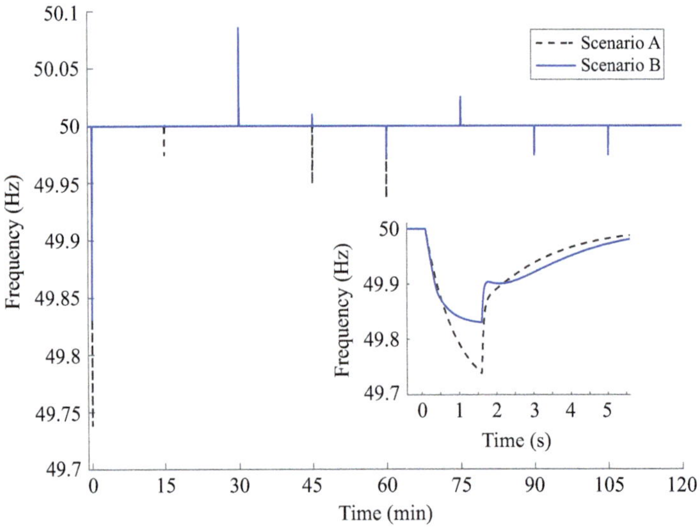

Figure 13.9 MG frequency during islanding transient and autonomous operation

frequency drop of 49.74 Hz was verified in Scenario A, while in Scenario B, it has attained a minimum value of 49.83 Hz. The reduction in the frequency deviation is obtained in scenario B due to the participation of the grid-supporting storage and EVs in primary control, through the *f*–*P* droop characteristic presented in Figure 13.4.

The VSI is equipped with a local secondary frequency control loop that allows to correct the frequency deviation. However, as shown in Figure 13.10, following the natural response of the converter in the islanded system, the battery bank will continue to charge/discharge in order to maintain power balance between generation and loads. Consequently, as shown in Figure 13.11, when no coordinated control is considered, the system would collapse after 60 min of islanded operation, since the VSI would be completely discharged.

In Scenario B, the controllable DER will have an active participation in MG power balance, which allows the VSI to inject a less amount of active power in the grid (see Figure 13.10), and consequently it will not present such an extreme discharge when compared to Scenario A (Figure 13.11). It is observed that in Scenario A, the VSI will fully discharge approximately at the end of the first hour, while in Scenario B it will present an SOC of 18%.

The active power injection of grid feeding units and of the SSMT is depicted in Figure 13.12. Since SSMT has no energy limitation, it operates in its maximum power set point during emergency operation (approximately 30 kW). On the other hand, due to the limited energy reserve, 30 and 15 kW energy storage systems active power injection in the grid is 0 from 45 min to the end of the islanding process, since both reached their minimum SOC of 20% (see Figure 13.12).

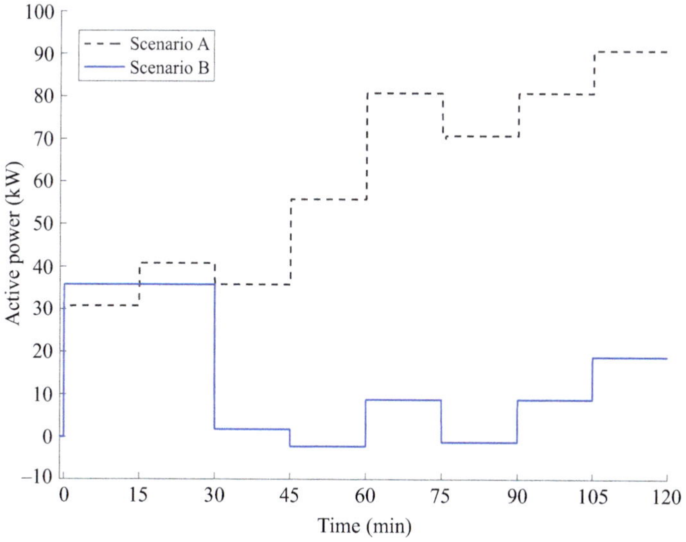

Figure 13.10 Active power injected by the VSI during islanded operation

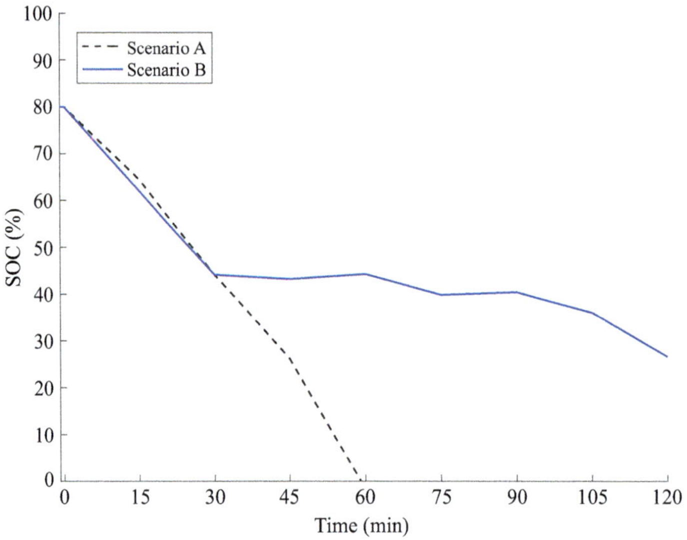

Figure 13.11 State of charge of VSI during islanded operation

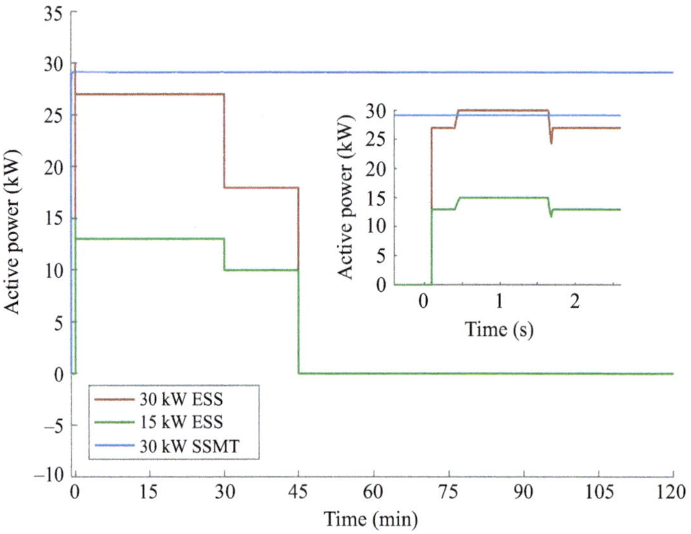

Figure 13.12 Active power injection provided by grid feeding units and by the SSMT in Scenario B

Relatively to EV and residential storage, it is possible to observe in Figure 13.13 their aggregated active power injection during islanded operation. The aggregated units will discharge in the first periods (system with lower load levels) in order to discharge in the last periods where the system feeds a larger amount of

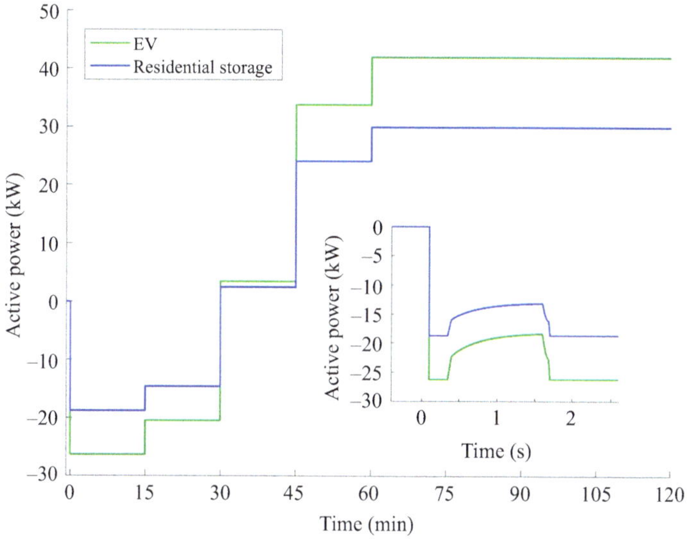

Figure 13.13 Active power injection provided by EV and residential storage units in Scenario B

load (achieving a maximum active power injection of approximately 40 kW in EV aggregate and 30 kW in residential storage aggregate).

13.3.2 Multi microgrid black start case study

This case study will demonstrate the feasibility of the black start process in a rural MV distribution network that incorporates two LV MG, several DG units and two EV charging stations, as presented in Figure 13.14.

For this accomplishment, one has assumed the worst case scenario of a general black out at the HV level, followed by an MMG unsuccessful islanding process that in turn caused a total MMG black out.

13.3.2.1 Initial remarks

The MG islanded operation and black start strategies were demonstrated in previous sections in this chapter. Hence, in this case study, the black start takes place fundamentally at the MV level, assuming the technical issues related MG black start and/or islanded operation were adequately solved. In addition, it is considered that fast transients are out of the scope of this work, since the implemented dynamic models were not prepared to tackle with this sort of phenomenon.

A specific dynamic simulation platform with the corresponding dynamic models was developed to simulate all the dynamics during the black start procedure. For that purpose, it was considered that initially the MMG is connected to an upstream HV network, modelled by an infinite bus. When the simulation starts, this connection is broken following the MMG blackout. Furthermore, all generation units are disconnected from the MMG after the black out. A system island is

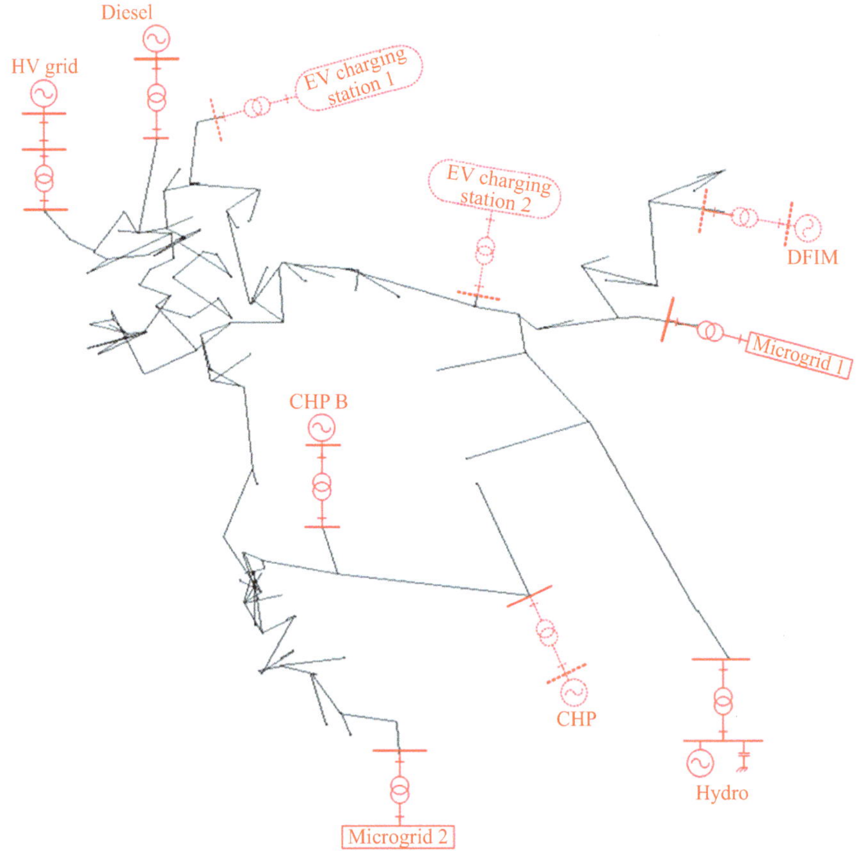

Figure 13.14 Single line diagram of the MMG network operated at 13.2 kV [32]

created for each MG injector by using an auxiliary synchronous generator, guaranteeing that these generators will not influence the dynamic simulation itself.

13.3.2.2 Simulation and results analysis

The black start sequence is also addressed from the perspective of an EV charging station, frequency and voltage values in the system (or portions of the system) for the same load level.

Network collapse and initial energization

When the simulation starts, all remotely controlled switches are opened and the DGs are separated from the MMG as well as MG and EV charging stations. Then, the MMG is disconnected from the HV network and DG units are shut down. The system voltage falls to zero, and at this point, the MMG is de-energized. Because of the absence of information of priority loads and following the principle of fostering the process automatism, we have chosen to manipulate only the remotely controlled switches in order to increment blocks of loadings to the system islands.

The CHP units are initially started and selected to energize two different portions of the MV network which are electrically isolated. These units are applied to start the network energization since they have black start capability. As expected, after the CHP units are connected to the portions of the network, at 10 and 20 s, the frequency stabilized at a value inferior to the 50 Hz nominal value.

After that, the productions of the CHP units are increased in order to bring the frequency closer to the nominal value. Before synchronizing the islands, at 50 s, one of the network buses is also energized by closing a remotely controlled switch. At 100 s, the diesel unit is connected and at 150 s the power production of the CHP unit is increased again to bring the frequency value around 50 Hz.

The impacts caused by these events on the temporal evolution of the frequency and voltage are shown in Figure 13.15. The buses 61 and 2 are the buses where the CHP and Diesel units are connected, respectively.

Synchronization

Before closing the switch between the two energized islands, it is essential to match voltages on both sides of the tie-switch. Although a similar value of frequency and voltage magnitude existed between the two islands, the phase angle difference between the voltage waveforms was still too high. Then, in order to perform a correct synchronization, the phase angle difference between the voltages is reduced by connecting blocks of loads and increasing the production from the diesel unit at 220 and 260 s, respectively. This allowed the synchronization between the two isolated systems at 340 s, stiffening the system (Figure 13.16).

After synchronization, the load blocks are connected followed by the increase in the diesel unit production at 490 s. This is performed gradually to minimize frequency and voltage deviations and to allow the energization of the remaining MV branches.

Electric vehicles integration

Next an EV control station with capability to participate in primary frequency control via a droop control approach is connected to the grid at 690 s. Load continues to be increased alternately with an increase in diesel and CHP production at 590 and 750 s, respectively. This is performed gradually to minimize frequency and voltage deviations and energize the remaining MV branches. The CHP unit also increased its output to allow charging some EV load charging at 750 s. Latter at 790 s, another load is connected to the grid leading to a clear frequency drop. The responses on grid frequency with and without the EV charging station participating in primary frequency control are depicted in Figure 13.17. The primary control provided by the EV charging station clearly allows smoothing the frequency and voltage deviations along load connection, as it can be seen from the shorter timescale detail of frequency at 790 s. As expected, with EV charging station droop control, the frequency behaviour is smoothed and the frequency deviations are minimized.

Frequency and voltage profile

Figure 13.18 represents the complete sequence of the black start events' impact on frequency profile. The frequency is always kept around 50 Hz, which demonstrates the feasibility of the restoration procedure.

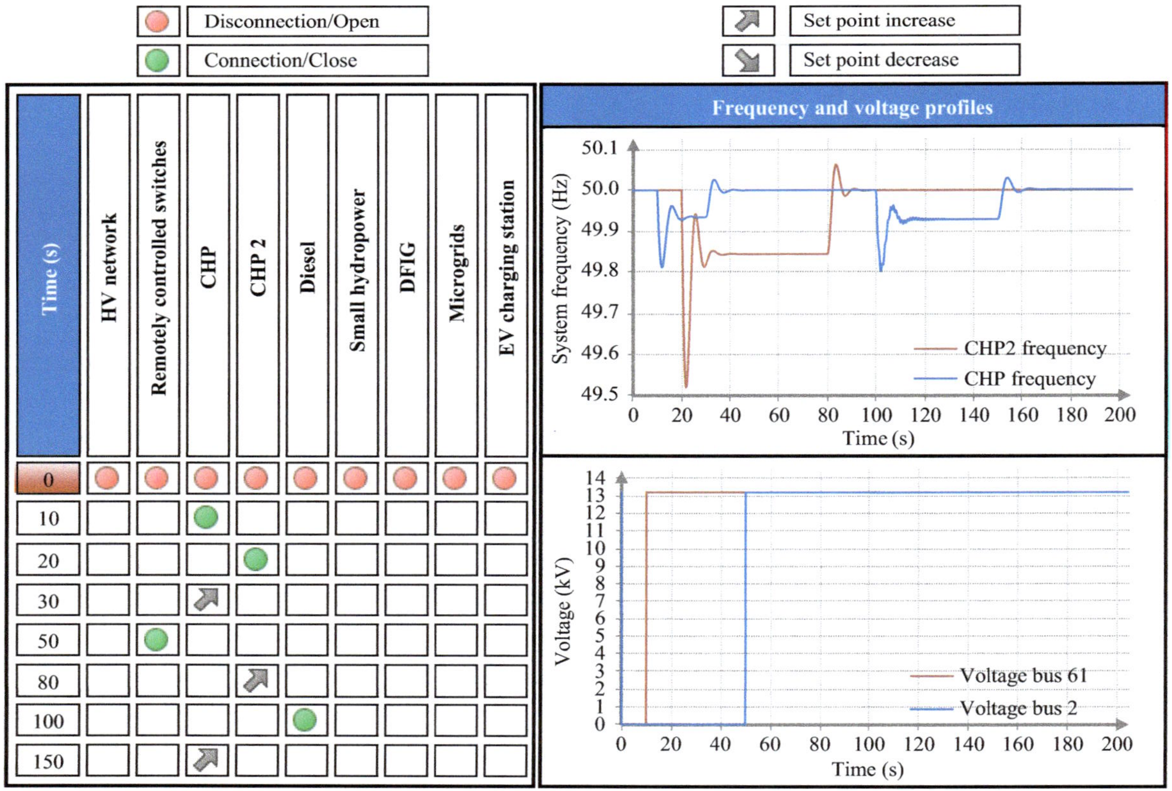

Figure 13.15 Frequency and voltage profiles during the black start process from 0 to 200 s

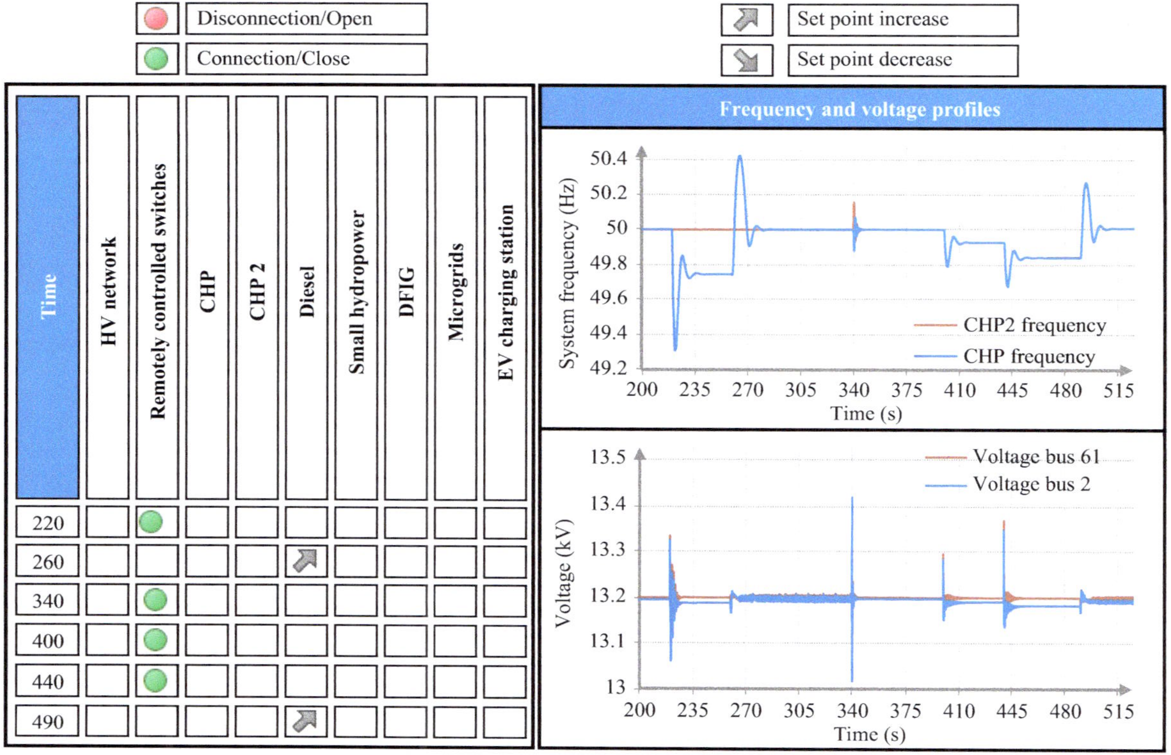

Figure 13.16 Frequency and voltage profiles during the black start process from 200 to 515 s

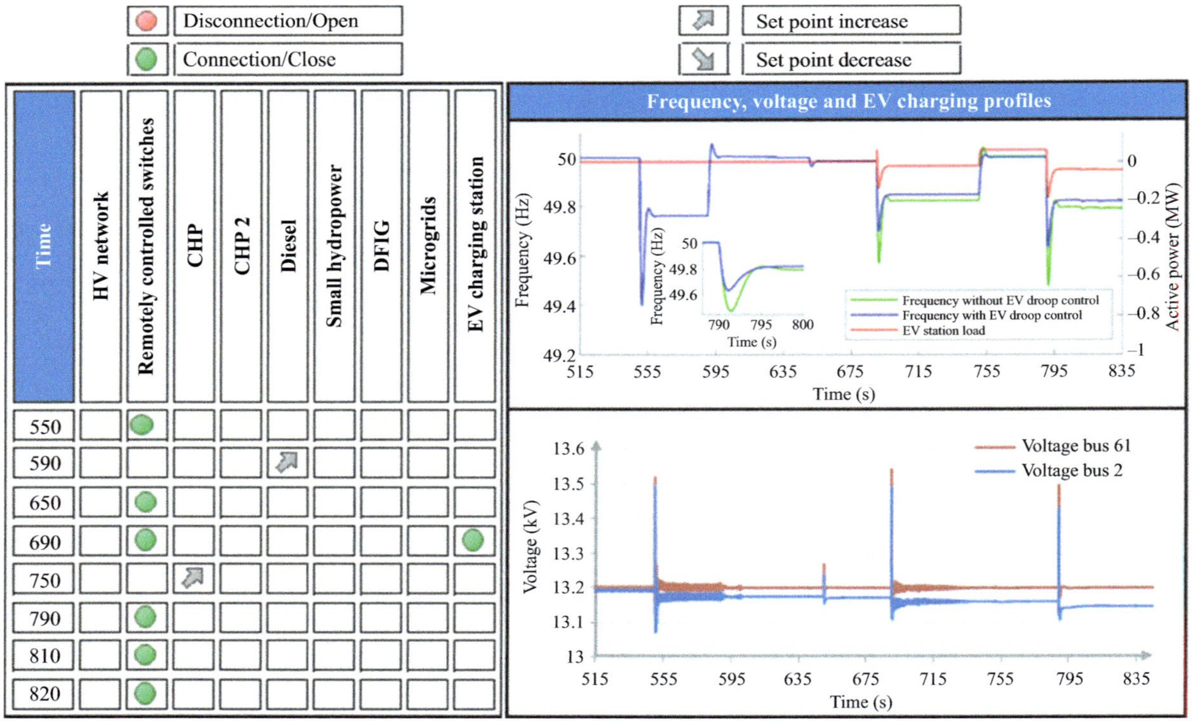

Figure 13.17 Frequency and voltage profiles during the black start process from 515 to 835 s

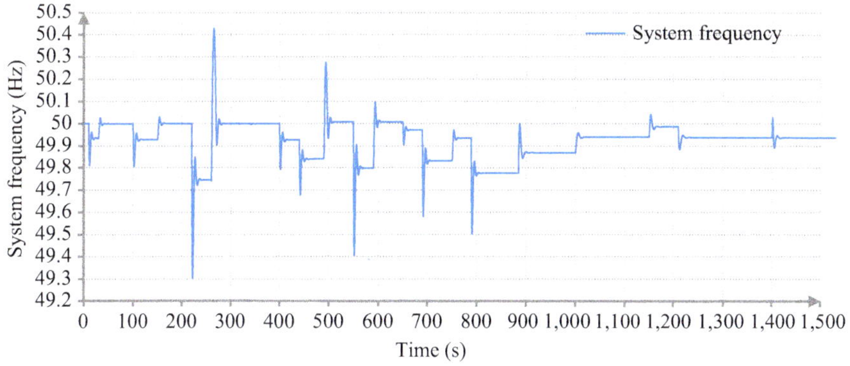

Figure 13.18 Frequency variations along black start

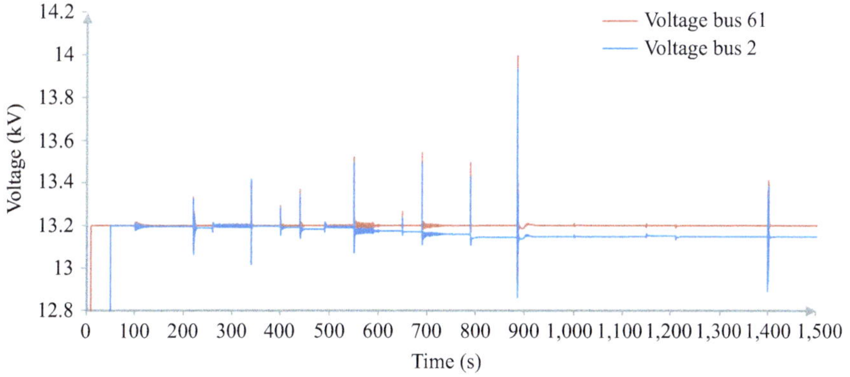

Figure 13.19 Voltage variations along black start

It is important to stress that all this sequence of actions can be performed much faster since, in this simulation, the interval between actions was enlarged to derive graphs where the individual contributions of different sources can be visualized in detail.

Figure 13.19 depicts the impact of the complete sequence of the black start actions in the voltage profile. After energization, the voltage never strays too far from 13.2 kV.

13.4 Concluding remarks

The MG control functionalities presented and discussed in this chapter take advantage of EV and load flexibility to complement the action of primary and secondary control within an MG. The exploitation of such flexibility constitutes key-resources-enabling MG autonomous operation. By promoting the coordination of MG resources and adapting, the system operation to real-time conditions improves MG resilience during autonomous operation, taking into account the MG limited energy storage capability and frequency response. The tools developed are intended to manage the system in short term. For larger time frames of operation in islanded conditions, complementary approaches need to be considered, involving forecasting of loads with different degrees of flexibility (including EV), as well as forecasts for renewable-based microgeneration.

By extending the MG concept, it is also presented in this chapter the possibility of exploiting MMG capabilities in terms of its contribution to black start. For this purpose, it exploited not only the EV charging control but also the possibility of taking advantage of the EV storage capability to inject power into the grid (V2G mode). When a large amount of EVs is integrated as a static load without the autonomous proportional frequency (droop) control, frequency deviations reached high values. Nevertheless, when the proportional frequency (droop) control is considered, the EV battery charging behaves as dynamic load reducing or even injecting active power in order to support frequency control, hence diminishing frequency excursions and smoothing the frequency behaviour during the restoration phase. More specifically, with the support of EV control, the probability of successful MMG islanding increases since the maximum/minimum frequency deviations are reduced and, consequently, relay tripping due to frequency deviations is reduced as well. At last, islanded operation (load-following) and black start are improved since EV control can smooth frequency variations caused by load ramps and (re-)connection of blocks of loads and generators.

References

[1] J. A. Peças Lopes, J. T. Saraiva, N. Hatziargyriou, and N. Jenkins, "Management of Microgrids," presented at the International Electric Equipment Conference, Bilbao, Spain, 2003.

[2] J. A. Peças Lopes, C. L. Moreira, and A. G. Madureira, "Defining Control Strategies for Microgrids Islanded Operation," IEEE Transactions on Power Systems, vol. 21, pp. 916–924, 2006.

[3] J. A. Peças Lopes, A. G. Madureira, and C. C. L. M. Moreira, "A View of Microgrids," Wiley Interdisciplinary Reviews: Energy and Environment, vol. 2, pp. 86–103, 2013.

[4] C. Moreira and J. A. Peças Lopes, "Microgrids Dynamic Security Assessment," in Clean Electrical Power, 2007. ICCEP'07. International Conference on, 2007, pp. 26–32.

[5] C. Gouveia, C. L. Moreira, J. A. Peças Lopes, D. Varajão, and R. E. Araújo, "Service Restoration in Low Voltage MicroGrids with Plugged-in Electric Vehicles," IEEE Industrial Electronics Magazine, vol. 7, no. 4, pp. 26–41, 2013.

[6] J. M. Guerrero, J. C. Vasquez, J. Matas, L. G. de Vicuna, and M. Castilla, "Hierarchical Control of Droop-Controlled AC and DC Microgrids—A General

Approach Toward Standardization," IEEE Transactions on Industrial Electronics, vol. 58, no. 1, pp. 158–172, 2011.
[7] J. A. Peças Lopes, A. Madureira, N. Gil, and F. Resende, "Operation of Multi-Microgrids," in *Microgrids: Architectures and Control*, N. Hatziargyriou, Ed., Chichester: John Wiley and Sons Ltd, 2014, pp. 165–205.
[8] B. Lasseter, "Microgrids [Distributed Power Generation]," IEEE Power Engineering Society Winter Meeting 2001, vol. 1, 28 January–February 2001, pp. 146–149.
[9] R. H. Lasseter and P. Piagi, "Control of small distributed energy resources," US Patent 7 116 010, October 3, 2006.
[10] M. C. Chandorkar, D. M. Divan and R. Adapa, "Control of Parallel Connected Inverters in Standalone AC Supply Systems," in IEEE Transactions on Industry Applications, vol. 29, no. 1, pp. 136–143, 1993.
[11] A. Arulampalam, M. Barnes, A. Engler, A. Goodwin, and N. Jenkins, "Control of Power Electronic Interfaces in Distributed Generation Microgrids," International Journal of Electronics, vol. 91, no. 9, pp. 503–523, 2004.
[12] A. Engler, "Applicability of Droops in Low Voltage Grids," International Journal of Distributed Energy Resources, vol. 1, no. 1, pp. 1–5, 2005.
[13] J. M. Guerrero, M. Chandorkar, T. Lee, and P. C. Loh, "Advanced Control Architectures for Intelligent Microgrids—Part I: Decentralized and Hierarchical Control," IEEE Transactions on Industrial Electronics, vol. 60, no. 4, pp. 1254–1262, 2013.
[14] H. Laaksonen, P. Saari, and R. Komulainen, "Voltage and Frequency Control of Inverter based Weak LV Network Microgrid," 2005 International Conference on Future Power Systems, 18 November 2005, pp. 6.
[15] K. De Brabandere, B. Bolsens, J. Van den Keybus, A. Woyte, J. Driesen, and R. Belmans, "A Voltage and Frequency Droop Control Method for Parallel Inverters," IEEE Transactions on Power Electronics, vol. 22, no. 4, pp. 1107–1115, 2007.
[16] Y. W. Li and C. Kao, "An Accurate Power Control Strategy for Power-Electronics-Interfaced Distributed Generation Units Operating in a Low-Voltage Multibus Microgrid," IEEE Transactions on Power Electronics, vol. 24, no. 12, pp. 2977–2988, 2009.
[17] J. M. Guerrero, L. Garcia de Vicuna, J. Matas, M. Castilla, and J. Miret, "Output Impedance Design of Parallel-Connected UPS Inverters With Wireless Load-Sharing Control," IEEE Transactions on Industrial Electronics, vol. 52, no. 4, pp. 1126–1135, 2005.
[18] R. Majumder, A. Ghosh, G. Ledwich, and F. Zare, "Angle Droop Versus Frequency Droop in a Voltage Source Converter based Autonomous Microgrid," 2009 IEEE Power & Energy Society General Meeting, 2009, pp. 1–8.
[19] R. Majumder, B. Chaudhuri, A. Ghosh, R. Majumder, G. Ledwich and F. Zare, "Improvement of Stability and Load Sharing in an Autonomous Microgrid Using Supplementary Droop Control Loop," in IEEE Transactions on Power Systems, vol. 25, no. 2, pp. 796–808, 2010.

[20] D. De and V. Ramanarayanan, "Decentralized Parallel Operation of Inverters Sharing Unbalanced and Nonlinear Loads," IEEE Transactions on Power Electronics, vol. 25, no. 12, pp. 3015–3025, 2010.
[21] N. Soni, S. Doolla, and M. C. Chandorkar, "Improvement of Transient Response in Microgrids Using Virtual Inertia," IEEE Transactions on Power Delivery, vol. 28, no. 3, pp. 1830–1838, 2013.
[22] J. A. Peças Lopes, F. J. Soares, and P. M. R. Almeida, "Integration of Electric Vehicles in the Electric Power System," Proceedings of the IEEE, vol. 99, no. 1, pp. 168–183, 2011.
[23] J. A. Peças Lopes, P. M. Rocha Almeida, F. J. Soares, and N. Hatziargyriou, "Electric Vehicles Grid Integration Under the MicroGrid Concept," Smart Grids and Mobility – Smart Grids and Mobility, Würzburg, Germany, June 2009.
[24] J. A. Peças Lopes, S. A. Polenz, C. L. Moreira, and R. Cherkaoui, "Identification of Control and Management Strategies for LV Unbalanced Microgrids with Plugged-In Electric Vehicles," Electric Power Systems Research, vol. 80, no. 8, pp. 898–906, 2010.
[25] A. Madureira, C. L. Moreira, and J. A. Peças Lopes, "Secondary Load-Frequency Control for MicroGrids in Islanded Operation," Proceedings of ICREPQ'05 – International Conference on Renewable Energies and Power Quality, Zaragoza, Spain, March 2005.
[26] S. Conti, R. Nicolosi, S. A. Rizzo, H. H. Zeineldin, and A. Member, "Optimal Dispatching of Distributed Generators and Storage Systems for MV Islanded Microgrids," IEEE Transactions on Power Delivery, vol. 27, no. 3, pp. 1243–1251, 2012.
[27] J.-Y. Kim, J.-H. Jeon, S.-K. Kim, *et al.*, "Cooperative Control Strategy of Energy Storage System and Microsources for Stabilizing the Microgrid during Islanded Operation," IEEE Transactions on Power Electronics, vol. 25, no. 12, pp. 3037–3048, 2010.
[28] C. Gouveia, J. Moreira, C. L. Moreira, and J. A. Peças Lopes, "Coordinating Storage and Demand Response for Microgrid Emergency Operation," IEEE Transactions on Smart Grids, vol. 4, no. 4, pp. 1898–1908, 2013.
[29] J. Gouveia, C. Gouveia, J. Rodrigues, *et al.*, "MicroGrid Energy Balance Management for Emergency Operation," 2017 IEEE Manchester PowerTech, Manchester, 2017, pp. 1–6.
[30] J. J. Ancona, "A Framework for Power System Restoration Following a Major Power Failure," IEEE Transactions on Power Systems, vol. 10, no. 3, pp. 1480–1485, 1995.
[31] M. M. Adibi and L. H. Fink, "Power System Restoration Planning," IEEE Transactions on Power Systems, vol. 9, no. 1, pp. 22–28, 1994.
[32] N. J. Gil, "Identificação de Condições de Operação em Modo Isolado e Procedimentos de Controlo de Emergência para Multi-Micro-Redes," PhD thesis (in Portuguese), Faculty of Engineering of the University of Porto, 2009.
[33] Operations-Training-Solutions. http://www.o-t-s.com/synchronizing.html (last accessed on 02/05/2019).

[34] J. A. Peças Lopes, C. L. Moreira, and F. O. Resende, "MicroGrids Black Start and Islanded Operation," 15th Power Systems Computation Conference (PSCC 2005), Liège, Belgium, 2005.
[35] J. A. Peças Lopes, N. Gil, F. Resende, E. Voumvoulakis, and N. Hatziargyriou, "More MicroGrids Project Deliverable DD3: Strategies for Emergency Functions (TD4. Emergency Functions – Islanding with Several Microgrids and Blackstart)," INESC Porto & ICCS/NTUA, 2008.

Chapter 14

Microgrid feasibility study and economics

Alessandra Parisio[1], Luigi Glielmo[2] and Evangelos Rikos[3]

The microgrid concept is a promising approach to enable smarter energy grids. Microgrids are capable of managing and coordinating distributed generation, storage devices and loads in a more decentralized way reducing the need for centralized coordination and management entities. The increased control over communication network and digitization makes microgrids increasingly active, i.e. they can generate, sense, compute, communicate and actuate. This will also enable a more proactive role of consumers. Hence, efficient optimization and control of microgrid operation is extremely important. This chapter discusses how advanced control and optimization techniques can be used to realize the technical, economical and environmental potential of microgrids and facilitate more active demand management. Several aspects related to the feasibility study with particular focus on the control and operational aspects are discussed. The principal sources of uncertainty (e.g. energy demand, renewable generation) are taken into account to render microgrid operation less sensitive to unforeseen events. Simulation and experimental results show the potential economic and environmental benefits of the illustrated operational strategies.

14.1 Overview

The microgrid concept could efficiently support high penetration of renewable energy sources (RES), storage systems and the integration of demand response policies in future energy grid [1,2]. The microgrid can be seen as a *cell* of the energy network, i.e. an integral element of a dynamically sizable energy system having the inherent ability to balance itself through the integrated and ICT-monitored and -controlled use of all components [2]. A self-adjusting and efficient whole-energy system can be designed and operated by interconnecting multiple of these cells,

[1]School of Electrical and Electronic Engineering, University of Manchester, Manchester, UK
[2]Department of Engineering, Università degli Studi del Sannio, Benevento, Italy
[3]Department of PVs and DER Systems, Center for Renewable Energy Sources and Saving, Athens, Greece

facilitating the management of distributed energy resources (DERs) and reducing the need for more expensive and less reliable centralized coordination. In 2050, some parts of the electricity grids are expected to be fully automated and local energy communities based on microgrids to be fully developed [2]. Among the several value streams identified in microgrids, the participation in demand response programs and local energy markets as well as the capability to export on-site generation to the electricity grid are the key economic ones [3]. Since a microgrid has the capability to use on-site generation and storage resources, it can reduce the amount of electricity exchanged with the utility grid as well as export power back to it. In addition, a microgrid can be highly effective at offering peak shaving, load levelling and other demand response services without service interruption [3]. Microgrid can benefit from less expensive local renewable generation and from generating power surplus in high market price hours and also at times of congestion in the utility grid. Reduced energy purchase from the utility grid, and then also reduced T&D charges, would be translated into lower energy costs for each individual consumer within the microgrid [4]. The ability to adjust loads (e.g. demand response) via building and/or microgrid controllers is beneficial also to consumer supply reliability [4].

Hence, an efficient energy management of microgrids can potentially benefit the entire system and enable a more effective economic dispatch of available energy resources in the utility grid [4]. The optimization of the microgrid operation is extremely important in order to cost-efficiently manage its energy resources. This requires new modeling capabilities, e.g. storage dynamics must be incorporated into the optimization and control design in order to coordinate storage use with RES generation and energy prices and address the complexity of the charging/discharging schedule [5]. Namely, microgrid modeling needs both continuous (such as storage charge/discharge rates) and discrete (such as on/off states of distributed generators (DGs) and demand response compatible loads) decision variables. The modeling capabilities and the computational advances of mixed integer problem (MIP) algorithms have led several independent system operators and regional transmission organizations to implement MIP-based solution methods in order to find a better solution to solve day-ahead and real-time market problems [6]; however, not solving unit commitment problems to complete optimality can cause several issues, such as incorrect generator payoff deviations and surplus issues [7]. Due to the problem complexity and because of the large economic benefits that could result from its improved solution, considerable attention has been and is being devoted to the development of suitable modeling frameworks and efficient optimization algorithms, which account for system uncertainty and are based on predicted future conditions and demand requirements, include demand response and storage systems and apply optimal instead of heuristic-based approaches.

In addition, demand for new sources of flexibility are also increasing interest in the interaction between energy sectors, like electricity, heating/cooling and gas, through microgeneration, for instance [8,9]. This interaction, in combination with demand response services and energy storage technologies, has been proven to offer numerous benefits, e.g. providing flexibility to counteract the intermittency of RES [10,11]. The growing need of reducing carbon emissions makes the concept of microgrid even more attractive from an environmental viewpoint. Thus, electrical energy and thermal energy must be jointly managed, along with demand response services and emissions.

The main objectives from an energy and control perspective are then to coordinate various DGs, RES, storages and loads in order to determine the optimal energy schedule depending on running costs, thermal and electrical energy requirements, demand response and consumers preferences. The main challenges are represented by controllable loads, which account for consumers preferences, and uncertainties due to RES and loads. Most of the management and control techniques proposed in the literature do not include uncertainty handling and experimental assessment. However, these are the two important milestones before a possible implementation in real systems [12,13].

In this chapter, we tackle the optimal economic/environmental operation management of microgrids under supply and load uncertainties and illustrate promising control-oriented approaches to microgrid modeling and operation optimization taking technical, economic and environmental aspects into account with particular focus on the grid-connected mode. We discuss also feasibility study aspects and experimental results.

Furthermore, there is a growing interest also in the interaction of microgrids (microgrid clusters) and in the economic and environmental benefits that can derive from their cooperation [4]. In fact, microgrids could exchange energy with neighbouring microgrids and cooperate to facilitate a more efficient operation of the whole grid. These local interactions can enable additional microgrid value streams [3]. The operation optimization of microgrid clusters is one of the most important future research directions. At the same time, technological development in monitoring, communication and actuation technology is increasing the viability of flexibility provision by DERs available within urban districts, and many markets are now becoming more accommodating to demand-side resources [14,15]. Residential consumers can be contracted by a third-party aggregator for their flexibility, in addition to its contract with the supplier. Aggregators may then have a contract with the distribution system operator (DSO) for providing it with peak shifting or demand-adjustment services [16]. In the last section of this chapter, we will then briefly illustrate an energy management system for urban districts comprising multiple microgrids, which have the opportunity to offer flexibility services to the DSO through an aggregator and be paid for it. Several types of flexibility and technologies available in urban districts will be considered (e.g. storage, demand-side management, heat pumps, combined heat and power (CHP)), as well as the interaction between electricity, heating and gas energy vectors. In addition, by adopting a model predictive control (MPC) technique, the proposed energy management system can take advantage of the opportunity for re-optimization based on real-time measurements and/or state estimation and compensate more effectively for uncertainty (e.g. in RES power output).

14.1.1 Outline of the chapter

In Section 14.2 adopted methods and techniques are outlined, while models of the main microgrid components and operational modes are described in Section 14.3. Advanced operational strategies are illustrated in Section 14.4, and relevant feasibility aspects are discussed in Section 14.5. Case studies and experimental results are analysed in Section 14.6, while some conclusions are drawn in Section 14.7.

14.2 Theoretical background

In this section we will outline the methods and techniques adopted to develop the microgrid energy management systems illustrated in this chapter: (i) model predictive control (MPC) and (ii) two-stage stochastic programming.

14.2.1 Model predictive control

MPC is an efficient and advanced approach to constrained control that has been successfully applied in many areas over the last decades [17,18]. During each sampling period, a finite horizon optimal control problem is formulated and solved over a finite prediction horizon, based on current measurements and updated information. In terms of microgrid control this means that, at each current point in time, an optimal plan is formulated for the next hours/day, based on current power measurements and updated predictions of the upcoming demand, production from renewable energy units and energy prices. After computing the optimal control sequence, only the first control step is implemented, and subsequently the horizon is shifted (see Figure 14.1). At the next sample the new state of the system is measured or estimated, and a new optimization problem is solved using this new information. A feedback mechanism is inherently introduced by this receding horizon approach, since the new optimal control problem solved at the beginning of the next time interval will be a function of the new state at that point in time and hence of any disturbance that has meanwhile acted on the microgrid.

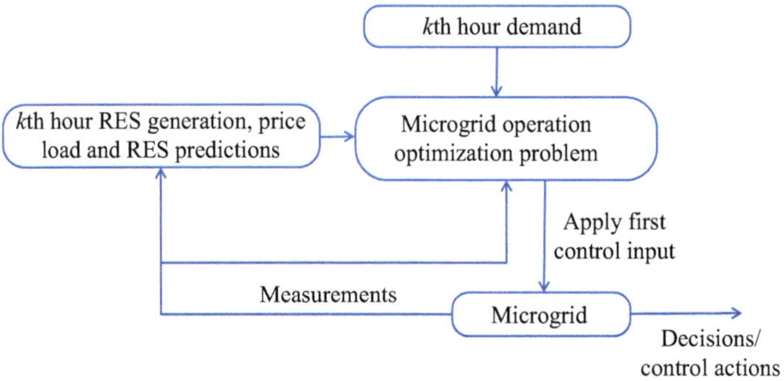

Figure 14.1 MPC scheme for microgrid control

An MPC algorithm for microgrid operation optimization consists of the following steps, to be repeated for each time step k, $k = 0, 1, \ldots, T$ (T is the prediction horizon):

1. at the current time k, the microgrid model is initialized to the measured/estimated current state of the microgrid components, i.e. the storage current energy level, on/off state and power levels generated by the controllable generation units;
2. the MPC problem is solved and the optimal input sequence is computed for the next T time steps, based on the demand and RES generation forecasts and on predictions of the upcoming microgrid behaviour;
3. the first sample of the input sequence is applied;
4. the new state of the system at time $k+1$ is measured/estimated, the forecasts of energy demand and RES generation are updated over a shifted horizon and the MPC problem is solved using this updated information. This *receding horizon* philosophy helps the control strategy to partially compensate for disturbances acting on the microgrid;
5. $k \leftarrow k + 1$, go to step (2).

14.2.2 Two-stage stochastic programming

This sections is extracted from [19] and provides the basic concepts of *stochastic programming with recourse*, also known as *two-stage stochastic programming*. For further details, the reader is referred to [19] and the references therein.

In many practical applications, as in the studied problem, there are several sources of uncertainty in the decision-making process, which affects the decision variables. In the two-stage stochastic programming approach the decision variables are partitioned into two sets. The first-stage variables are those that have to be decided before the actual realization of the uncertain parameters becomes available; once the random events occur, the values of the second stage or recourse variables can be decided. These recourse variables are also interpreted as *correction actions* as they are used to compensate any infeasibility from the first-stage decisions; thus, violations are accepted, but their costs affect the choice of the first-stage variables. The objective is to choose the first-stage variables in order to minimize the sum of first-stage costs and the expected value of the random second-stage or recourse costs. In the two-stage model illustrated in this chapter, the non-anticipativity constraints, stating that the first-stage decisions do not depend on the random variables realizations, are implicitly modeled by not having first-stage decision vectors depending on the random vector.

A two-stage linear stochastic program with simple continuous recourse can be formulated as follows:

$$\begin{aligned} &\min \boldsymbol{c}'\boldsymbol{u} + \mathcal{Q}(\boldsymbol{u}), \\ &\text{s.t.} \\ &\boldsymbol{A}\boldsymbol{u} = \boldsymbol{b}, \\ &\boldsymbol{u} \in \mathcal{U}, \end{aligned} \tag{14.1}$$

where $\boldsymbol{u} \in \mathbb{R}^n$ is the first-stage decision vector, $\mathcal{U}$ is the first stage constraints set, $\boldsymbol{A} \in \mathbb{R}^{s \times n}$, $\boldsymbol{b} \in \mathbb{R}^s$. The random vector is denoted by $\boldsymbol{\omega} \in \mathbb{R}^r$, with probability distribution P_ω (assumed being independent from $\boldsymbol{u}$), and $\mathcal{Q}(\boldsymbol{u}) = \boldsymbol{E}_\omega[\Phi(\boldsymbol{u}, \boldsymbol{\omega})]$ is the *expected recourse*, with $\Phi(\boldsymbol{u}, \boldsymbol{\omega})$ being the *second-stage function*:

$$\begin{aligned}
&\Phi(\boldsymbol{u}, \boldsymbol{\omega}) = \min(\boldsymbol{q}^+\boldsymbol{\zeta}^+ + \boldsymbol{q}^-\boldsymbol{\zeta}^-),\\
&\text{s.t.}\\
&\boldsymbol{\zeta}^+ \geq \boldsymbol{H}\boldsymbol{u} - \boldsymbol{\omega},\\
&\boldsymbol{\zeta}^- \geq -(\boldsymbol{H}\boldsymbol{u} - \boldsymbol{\omega}),\\
&\boldsymbol{\zeta}^+, \boldsymbol{\zeta}^- \geq 0,
\end{aligned}$$

where the inequalities are meant componentwise, $\boldsymbol{q}^+$ and $\boldsymbol{q}^-$ are penalty coefficient row vectors of appropriate size and $\boldsymbol{\zeta}^+$ and $\boldsymbol{\zeta}^-$ are the recourse vectors.

By introducing $\boldsymbol{y} := \boldsymbol{H}\boldsymbol{u}$ and considering P_ω as a finite discrete probability distribution with the marginal distribution of ω_i given by $p_{ij} = P_\omega(\omega_i = \hat{\omega}_{ij})$, $j = 1, \ldots, S_i$, S_i being the number of possible outcomes and $i = 1, \ldots, r$ the ith component, it is straightforward rewriting the problem (14.1) as the following equivalent deterministic linear problem:

$$\begin{aligned}
&\min \boldsymbol{c}'\boldsymbol{u} + \sum_{i=1}^{r}\sum_{j=1}^{S_i} p_{ij}\left(q_i^+\zeta_{ij}^+ + q_i^-\zeta_{ij}^-\right),\\
&\text{s.t.}\\
&\boldsymbol{A}\boldsymbol{u} = \boldsymbol{b},\\
&\zeta_{ij}^+ \geq y_i - \hat{\omega}_{ij},\\
&\zeta_{ij}^- \geq -(y_i - \hat{\omega}_{ij}),\\
&\zeta_{ij}^+, \zeta_{ij}^- \geq 0,\\
&\boldsymbol{u} \in \mathcal{U}.
\end{aligned}$$

Problem (14.1) is said to have *complete recourse* [19], i.e. the feasible set of the second-stage problem stays non-empty since it is always possible to respond to any possible disturbance realizations in a reasonable time (reasonable with respect to the sampling period).

14.3 Microgrid component modeling and constraints

In this section, we outline a control-oriented microgrid modeling framework, which comprises continuous-time dynamics of energy flows and storage units, on/off status of microgrid units, CHP capabilities, flexible loads. The modeling set-up is developed with the goal of maintaining the problem tractable.

The interested reader is referred to [20–22] for detailed modeling of DERs and demand, and to [24] for detailed modeling of thermal energy storage.

14.3.1 Nomenclature

The parameters, the disturbance forecast and the decision variables used in the proposed formulation are described, respectively, in Tables 14.1, 14.2, 14.3; in Table 14.1, for simplicity, the subscripts referring to the *i*th unit or load and the subscript denoting the time k are omitted. The fuel consumption cost for a DG unit is traditionally assumed to be a quadratic function of the generated power, i.e., $C^{DG}(P) = a_1 P^2 + a_2 P + a_3$. Power variables represent the average power over a given sampling period. In the following sections, vectors and matrices are denoted in bold.

14.3.2 Loads

In the proposed modeling framework, we include both fixed and flexible loads.

14.3.2.1 Fixed loads

Fixed loads are demand levels related to essential processes that must be always met. They can be both electrical and thermals.

14.3.2.2 Flexible loads

The Flexible loads are controllable loads, which can respond to price signals or incentives/penalization mechanisms. The power requirements of these types of

Table 14.1 Decision and logical variables

C^{grid}	Cost of power exchanged with the utility grid
P^{gas}	Gas power input to the CHP
P^{grid}	Power imported(positive)/exported(negative) from/to the utility grid
$P^{\text{CHP,el}}$	CHP electrical power
$P^{\text{CHP,heat}}$	CHP thermal power
$P^{\text{HP,el}}$	HP electrical power
$P^{\text{HP,heat}}$	HP thermal power
P^{ESS}	Power exchanged (positive for charging) with an ESS
P^{TSS}	Thermal power exchanged with a TSS
x^{ESS}	ESS energy level
x^{TSS}	TSS energy level
δ^{ESS}	Discharging (0)/charging (1) mode of an ESS
δ^{TSS}	Discharging (0)/charging (1) mode of a TSS
δ^{grid}	Exporting (0)/importing (1) mode to/from the utility grid
β	Curtailed power fraction
$D_i^{\text{f,heat}}$	Power level of a flexible thermal load
P	Electrical power level of a DG
δ	Off (0)/on (1) state of a DG
ξ^-, ξ^+	Recourse variables associated to electrical balance
ξ_q^-, ξ_q^+	Recourse variables associated to thermal balance

Table 14.2 Parameters and sets

T	Prediction horizon
DG, L	Number of DGs and fixed electrical loads
FL	Number of flexible electrical loads
TL, FTL	Number of fixed and flexible thermal loads
c^{gas}	Fuel (natural gas) cost for a CHP system
α^{sb}	Storage coefficient of degradation/ageing
η^{cE}, η^{dE}	Electric storage charging and discharging efficiencies
η^{cT}, η^{dT}	Thermal storage charging and discharging efficiencies
β_{min}, β_{max}	Minimum and maximum allowed curtailment
OM^{ESS}	Operating and maintenance cost of an ESS
OM	Operating and maintenance cost of a DG
$C^{DG}(P)$	Fuel consumption cost curve of a DG
C^{b}_{max}	Storage power limit
T^{g}	Maximum grid interconnection power limit
ρ_{f}	Penalty weight on curtailments
q^{+}, q^{-}	Costs on recourse actions
$\mathcal{X}^{CHP}, \mathcal{X}^{DG}, \mathcal{X}^{HP}$	Sets defined by the CHP, HP and DG related constraints
$\mathcal{X}^{grid}$	Set defined by the constraints related to the interaction with the grid
$\mathcal{X}^{ESS}, \mathcal{X}^{TSS}$	Sets defined by the ESS and TSS related constraints
$\mathcal{X}^{f,el}, \mathcal{X}^{f,heat}$	Sets defined by the constraints related to flexible electrical and thermal loads
$D^{f,el}, D^{f,heat}$	Power level of a flexible electrical load and of a flexible thermal load
U_{tank}	Heat transfer coefficient
A_{tot}	Total heat exchange area of the water storage
M	Mass of the water inside the thermal energy storage
c_{p}	Specific heat
$D^{heat,min}, D^{heat,max}$	minimum and maximum limits on the flexible thermal demand

Table 14.3 Disturbances

P^{RES}	Total power from RES
D^{el}, D^{heat}	Power level of a fixed electrical and thermal loads

loads can be shifted in time and optimized within given limits. We consider both electrical and thermal flexible loads:

- *electrical flexible loads*, i.e. electrical loads that can be reduced or shed during supply constraints or emergency situations. A continuous-valued variable, $0 \leq \beta_i(k) \leq 1$, associated to each controllable load i and to each sampling time k is defined. This variable represents the percentage of preferred power level to be curtailed at time k. Flexible demand related to smart appliances can be also included in the proposed modeling framework, taking user-specified time preferences into account. The reader is referred to [23] for detailed modeling of these types of flexible loads;

- *thermal flexible loads*, i.e. demand levels related to thermal indoor comfort. Limits on thermal power required to keep the internal temperature within a given comfort range for each microgrid can be computed through a dynamic building model based on forecasts of the weather conditions and forecast occupancy patterns [25].

14.3.3 Distributed generators

DGs are modeled by introducing one binary signal, $\delta(k)$, to represent its off(0)/on(1) state at each sampling time. The operating constraints on the minimum amount of time for which a controllable generation unit must be kept on/off (minimum up/down times) can be expressed by the mixed integer linear inequalities without resorting to any additional variable (see [20] for details). The DG unit start-up and shut-down behaviour are also modeled in order to account for the corresponding costs. For this reason, two auxiliary variables, $SU(k)$ and $SD(k)$, are introduced, representing respectively the start-up cost and the shut-down cost for each DG unit at time k. These auxiliary variables must satisfy a set of mixed integer linear constraints (see [20] for details).

14.3.4 Energy storage systems

Here we detail the modeling of both electrical and thermal energy storage systems.

14.3.4.1 Electrical energy storage

The following discrete time model of an electrical storage system (ESS) is considered:

$$x^{\mathrm{ESS}}(k+1) = \alpha^{sb} x^{\mathrm{ESS}}(k) + \eta P^{\mathrm{ESS}}(k), \tag{14.2}$$

where

$$\eta = \begin{cases} \eta^{\mathrm{c}}, & \text{if } P^{\mathrm{ESS}}(k) > 0 \text{ (charging mode)}, \\ \eta^{\mathrm{d}}, & \text{otherwise (discharging mode)} \end{cases} \tag{14.3}$$

Typically $\eta^{\mathrm{c}} < 1$ and $\eta^{\mathrm{d}} = 1/\eta^{\mathrm{c}}$ account for energy losses.

By using the MLD framework described in [26], a binary variable $\delta^{\mathrm{ESS}}(k)$ and an auxiliary variable $z^{\mathrm{ESS}}(k) = \delta^{\mathrm{ESS}}(k) P^{\mathrm{ESS}}(k)$ are introduced to model the logical conditions provided in (14.3); the storage dynamics and the corresponding constraints are rewritten in the following compact form:

$$x^{\mathrm{ESS}}(k+1) = \alpha^{\mathrm{sb}} x^{\mathrm{ESS}}(k) - (\eta^{\mathrm{dE}} - \eta^{\mathrm{cE}}) z^{\mathrm{ESS}}(k) + \eta^{\mathrm{dE}} P^{\mathrm{ESS}}(k), \tag{14.4a}$$

$$\text{subject to} \quad \mathbf{E_1}^{\mathrm{ESS}} \delta^{\mathrm{ESS}}(k) + \mathbf{E_2}^{\mathrm{ESS}} z^{\mathrm{ESS}}(k) \leq \mathbf{E_3}^{\mathrm{ESS}} P^{\mathrm{ESS}}(k) + \mathbf{E_4}^{\mathrm{ESS}}, \tag{14.4b}$$

where the column vectors $\mathbf{E_1}^{\mathrm{ESS}}, \mathbf{E_2}^{\mathrm{ESS}}, \mathbf{E_3}^{\mathrm{ESS}}, \mathbf{E_4}^{\mathrm{ESS}}$ are provided in the Appendix.

14.3.4.2 Thermal energy storage

A thermal storage system (TSS) can be described similarly as follows:

$$x^{\mathrm{TSS}}(k+1) = x^{\mathrm{TSS}}(k) + \eta_T P^{\mathrm{TSS}}(k) - U_{\mathrm{tank}} A_{\mathrm{tot}} [T_{\mathrm{mean}}(k) - T_{\mathrm{env}}(k)], \quad (14.5)$$

where

$$\eta_T = \begin{cases} \eta^{\mathrm{cT}}, & \text{if } P^{\mathrm{TSS}}(k) > 0 \text{ (charging mode)}, \\ 1/\eta^{\mathrm{dT}}, & \text{otherwise (discharging mode)}, \end{cases} \quad (14.6)$$

with $0 < \eta^{\mathrm{cT}}, \eta^{\mathrm{dT}} < 1$. Denoting by $T_{\mathrm{ref}}(k)$, the reference temperature at time k, the energy stored in the tank can be evaluated as

$$x^{\mathrm{TSS}}(k) = Mc_p[T_{\mathrm{mean}}(k) - T_{\mathrm{ref}}(k)], \quad (14.7)$$

and the thermal storage dynamics can be written as

$$x^{\mathrm{TSS}}(k+1) = \left(1 - \frac{U_{\mathrm{tank}} A_{\mathrm{tot}}}{Mc_p}\right) x^{\mathrm{TSS}}(k) + \eta_T P^{\mathrm{TSS}}(k) \\ - U_{\mathrm{tank}} A_{\mathrm{tot}} [T_{\mathrm{ref}}(k) - T_{\mathrm{env}}(k)]. \quad (14.8)$$

The reference temperature $T_{\mathrm{ref}}(k)$ can be chosen equal to the return water temperature from the network. As for the electric storage case, we model the charging and discharging behaviour through an auxiliary variable $z^{\mathrm{TSS}}(k) = \delta^{\mathrm{TSS}}(k) P^{\mathrm{TSS}}(k)$. Again, the logical conditions can be expressed in terms of mixed integer linear inequalities as in [26]. Consequently, the storage model can be reformulated as

$$x^{\mathrm{TSS}}(k+1) = x^{\mathrm{TSS}}(k) + (\eta^{\mathrm{cT}} - 1/\eta^{\mathrm{dT}}) z^{\mathrm{TSS}}(k) + (1/\eta^{\mathrm{dT}}) P^{\mathrm{TSS}}(k) \\ - U_{\mathrm{tank}} A_{\mathrm{tot}} \left(T_{\mathrm{ref}}(k) + \frac{x^{\mathrm{TSS}}(k)}{Mc_p} - T_{\mathrm{env}}(k) \right), \quad (14.9a)$$

$$\text{subject to} \quad \mathbf{E}_1{}^{\mathrm{TSS}} \delta^{\mathrm{TSS}}(k) + \mathbf{E}_2{}^{\mathrm{TSS}} z^{\mathrm{TSS}}(k) \le \mathbf{E}_3{}^{\mathrm{TSS}} P^{\mathrm{TSS}}(k) + \mathbf{E}_4{}^{\mathrm{TSS}} \quad (14.9b)$$

where the column vectors $\mathbf{E_1}^{TSS}, \mathbf{E_2}^{\mathrm{TSS}}, \mathbf{E_3}^{\mathrm{TSS}}, \mathbf{E_4}^{\mathrm{TSS}}$ are provided in the Appendix.

In addition to the storage dynamics and the logical conditions expressed in (14.9a) and (14.9b), the storage bounds reflecting the physical constraints of the device need to be modeled [24]. The next inequality models this constraint:

$$x_{\mathrm{min}}^{\mathrm{TSS}} \le x^{\mathrm{TSS}}(k) \le x_{\mathrm{max}}^{\mathrm{TSS}}. \quad (14.10)$$

To guarantee reliability of the power network, the storage should be able to handle the power demand for 1 h.

14.3.5 Multi-energy components

We here provide models of components where different energy vectors interact (e.g. heat, gas, electricity), which can be (but are not limited to) heat pumps (HPs) and CHP systems.

14.3.5.1 Heat pumps

The behaviour of a heat pump at each time k is modeled as $P^{\mathrm{HP,heat}}(k) = \mathrm{COP}(k) \cdot P^{\mathrm{HP,el}}(k)$, where $COP(k)$ is the forecast coefficient of performance (COP) of the heat pump at time k based on weather forecasts over the prediction horizon. At each time step, COP forecasts can be integrated in the proposed control framework in order to predict the future heat generation from each heat pump. A constant COP can also be considered, if such approximation is acceptable.

14.3.5.2 Combined heat and power systems

A CHP system can be modeled by defining the CHP system efficiency as *power-to-heat ratio* (PHR). The PHR indicates the proportion of power (electrical or mechanical energy) to heat energy (steam or hot water) produced in the CHP system [27]. A CHP system can be then represented by the equation $P^{\mathrm{CHP,el}}(k) = PHR \cdot P^{\mathrm{CHP,heat}}(k)$, where $P^{\mathrm{CHP,el}}(k)$ is the electrical and $P^{\mathrm{CHP,heat}}(k)$ is the thermal power levels of the CHP system are each time k.

A data-driven modeling approach can also be adopted, as shown in [22], where a validated micro-CHP model is described by the equations $P^{\mathrm{CHP,el}}(k) = \alpha_1 P^{\mathrm{gas}}(k) + \alpha_0$, and $P^{\mathrm{CHP,heat}}(k) = \beta_1 P^{\mathrm{gas}}(k) + \beta_0$, where $P^{\mathrm{gas}}(k)$ is the gas power level at time k.

14.3.6 Electrical and thermal balance

The balance between electrical energy production and consumption must be met at each time k. Without loss of generality, we consider one electrical storage unit and one CHP system as multi-energy DGs. The electrical balance can be modeled as

$$\begin{aligned} P^{\mathrm{grid}}(k) &= \sum_{i=1}^{L} D_i^{\mathrm{el}}(k) + \sum_{i=1}^{FL} [1 - \beta_i(k)] D_i^{\mathrm{f,el}}(k) + \sum_{i=1}^{APP} D_i(k)^{\mathrm{app}} \\ &\quad - \sum_{i=1}^{DG} P_i(k) + P^{\mathrm{ESS}}(k) - P^{\mathrm{RES}}(k) - P^{\mathrm{CHP,el}}(k) + P^{\mathrm{HP,el}}(k). \end{aligned} \tag{14.12}$$

The following linear constraints must be included into the MPC problem to consider thermal energy:

$$\begin{aligned} & P^{\mathrm{CHP,heat}}(k) + P^{\mathrm{HP,heat}}(k) = \sum_{i=1}^{TL} D_i^{\mathrm{heat}}(k) + \sum_{i=1}^{FTL} D_i^{\mathrm{f,heat}}(k) + P^{\mathrm{TSS}}(k), \\ & D_i(k)^{\mathrm{heat,\ min}} \le D_i^{\mathrm{f,heat}}(k) \le D_i(k)^{\mathrm{heat,\ max}}. \end{aligned} \tag{14.13}$$

We remark that gas and waste heat boilers can be easily included in the left-hand side of the thermal balance constraints and in the objective function to account for their running costs.

In order to clearly describe the two-stage stochastic approach, it is more convenient to rewrite the electrical and thermal power balance equations in a compact form, $\boldsymbol{F}(k)\boldsymbol{u}(k) + \boldsymbol{f}\boldsymbol{w}(k) = 0$, where $\boldsymbol{u}(k)$ is the column vector collecting all decision variables (see Table 14.1), $\boldsymbol{w}(k)$ is the column vector containing all disturbances (see Table 14.3), $\boldsymbol{F}(k)$ and $\boldsymbol{f}$ are two vectors of appropriate sizes, the former being time-varying since it contains the time-varying preferred energy level for the flexible loads, which is assumed to be known.

14.3.7 Interaction with the utility grid

By employing the MLD framework, we introduce a binary variable $\delta^{grid}(k)$ and an auxiliary variable $C^{grid}(k)$ to model the possibility either to purchase or to sell energy from/to the utility grid. Then, the purchasing/selling microgrid behaviour can be compactly expressed by the following mixed integer linear inequalities:

$$\mathbf{E_1}^{\text{grid}}\delta^{\text{grid}}(k) + \mathbf{E_2}^{\text{grid}}C^{\text{grid}}(k) \leq \mathbf{E_3}^{\text{grid}}(k)P^{\text{grid}}(k) + \mathbf{E_4}^{\text{grid}}. \tag{14.14}$$

The column vectors $\mathbf{E_1}^{\text{grid}}, \mathbf{E_2}^{\text{grid}}, \mathbf{E_3}^{\text{grid}}(k), \mathbf{E_4}^{\text{grid}}$ are provided in the Appendix. The matrix $\mathbf{E_3}^{\text{grid}}(k)$ is generally time-varying due to the time-varying energy prices.

14.4 Microgrid operational strategies

14.4.1 MPC-based energy-management system for operational optimization

The energy management for optimizing microgrid operation is a complex problem, which necessitates a systematic control approach, integrating dynamic models, forecasts, constraints and objectives, to make optimized decisions. Both modeling and optimization pose challenges. Various research groups have focused on different aspects of the overall problem [13]. Research contributions to demand response modeling and optimization for microgrids have made significant progress in addressing these challenges but much remains to be done, in particular, with regard to uncertainty management and effective solving methods for mixed integer, stochastic formulations needed to capture all relevant dynamics and technical/operational constraints. Several algorithmic approaches have been used, such as particle swarm and genetic algorithms, mixed-integer programming, Lagrangian relaxation, *Q*-learning and Lyapunov optimization, dynamic programming, multi-objective optimization (MOO), stochastic programming (see [13,28–33] and references therein).

Typically, the available energy management frameworks either are computationally intensive and not suitable for real-time applications or can produce suboptimal solutions and focus on some specific aspects, e.g. economic dispatch or demand response.

MPC [18] has drawn the attention of the power system community because of its capability to handle the future behaviour of the system, demand and renewable energy generation forecasts, systems constraints, as well as the feedback mechanism it provides, making the controlled system more robust against uncertainty [12].

In this section, we illustrate a stochastic MPC controller for economical/environmental microgrid operation optimization. Since microgrid control requirements involve different control approaches and time scales, the typical control hierarchy of a microgrid requires a centralized, high-level controller on the top of the hierarchy and local controllers associated with each DG, storage and load units [34]. The high-level controller deals with longer time scales and aims at generating optimal set points for DERs and controllable load units, while local controllers should ensure that the power reference values are tracked, keeping voltage stability and power quality (which has higher priority). The high-level controller is very weakly dependent on the transient behaviour of the fast dynamics. Then a steady state assumption for microgrid components can be safely made without much loss of accuracy.

We focus on the high-level control design and propose the use of MPC in combination with mixed integer linear programming (MILP) [35] and two-stage stochastic programming [19]. By taking the environmental concern into account, the microgrid operation optimization problem can be formulated as a MOO problem with conflicting objectives (operative costs and emissions), for which a set of optimal solutions (Pareto optimal set) and the corresponding objective function values (Pareto curve) are computed. In this paper, the *weighted min–max method* is applied to compute Pareto optimal points; this approach provides both necessary and sufficient conditions for Pareto optimality [36]. The desired number of Pareto-optimal (non-dominated) solutions can be obtained by varying the weights on the objective functions.

14.4.1.1 Uncertainty modeling

In order to make the proposed control action effective, the stochastic nature of RES and demand is incorporated in the MPC problem through scenarios. Denote by ω_k the uncertainty during the sampling period k, which can represent either the energy demand or the RES generation. The random variable ω_k is decomposed as $\omega_k = \bar{\omega}_k + \tilde{\omega}_k$ where $\bar{\omega}_k$ is the forecast and $\tilde{\omega}_k$ is the forecast error at time k.

14.4.1.2 Stochastic MPC formulation

In this section, the two-stage stochastic problem for microgrid economic and environmental operation management is formulated. The conflicting objectives considered in this problem are running costs and emissions.

Microgrid running costs

The microgrid running costs at time step k is denoted by $V_r(\boldsymbol{u}(k))$ and is given by

$$\begin{aligned} V_r(\boldsymbol{u}(k)) := \sum_{i=1}^{DG}\left[C_i^{\mathrm{DG}}(P_i(k)) + OM_i P_i(k) + SU_i(k) + SD_i(k)\right] + C^{\mathrm{grid}}(k) \\ + OM^{\mathrm{ESS}}\left|P^{\mathrm{ESS}}(k)\right| + c^{\mathrm{gas}}(k)P(k)^{\mathrm{gas}} + \rho_{\mathrm{f}}\sum_{i=1}^{FL}\beta_i(k)D_i^{\mathrm{f,el}}(k). \end{aligned} \tag{14.15}$$

We recall that $C^{\mathrm{grid}}(k)$ can be negative, i.e. energy is sold to the utility grid, representing a profit for the microgrid system. Depending on the DG, maintenance costs can be also expressed by the term $OM_i\delta_i(k)$. The absolute value $|P^{\text{ESS}}(k)|$ can be easily expressed as $(2z^{\text{ESS}}(k) - P^{\text{ESS}}(k))$. Similar cost terms as the ones related to the ESS can be included for the TSS.

Note that $V_r(\boldsymbol{u}(k))$ is a quadratic cost due to the presence of the quadratic terms $C_i^{\text{DG}}(P_i(k))$. We therefore approximate every function $C_i^{\text{DG}}(P_i)$ with a convex piecewise affine function, which provides very similar results, but can be included into a mixed integer linear program [35].

Microgrid emissions

The emission rate (kg/h) of atmospheric pollutants such as sulphur oxides SO_2, carbon oxides CO_2, and nitrogen oxides NO and NO_2 can be modeled separately; the total emission of the ith unit can be expressed as [37]:

$$E_i(P) = \nu_i + \varsigma_i P + \gamma_i P^2 + \eta_i e^{\lambda_i P} \tag{14.16}$$

where ν_i, ς_i, γ_i, η_i and λ_i are nonnegative coefficients of the emission characteristics. We emphasize that, when multi-energy components are considered, the total emission of the unit can still be expressed as function of the electrical power generated. The microgrid emission at time step k is denoted by $V_e(\boldsymbol{u}(k))$ and is given by

$$V_e(\boldsymbol{u}(k)) := \sum_{i=1}^{N_g} E_i(P_i(k)).$$

Each function $E_i(P)$ can also be approximated with a piecewise affine function in order to keep the problem linear.

We point out that we do not account explicitly for the pollutants content in the electricity purchased from the grid, which is hard to estimate and out of the scope of this study. However, the proposed optimization model can easily include emissions from utility grid (for instance, utility grid emissions can be modeled by a linear function of the purchased power whose coefficients represent time-varying emission rates).

14.4.1.3 A stochastic programming approach with simple recourse

In the microgrid scenario, recourse actions are needed when the actual energy demand is different from the actual energy generation in the microgrid and system performance naturally degrades: higher demand than generation leads to energy *shortage*, while lower demand than generation results in unexpected energy *surplus*. In the case of power imbalances, technically, the utility grid can cover it and charge the microgrid operator according to special agreement between the microgrid operator and the DSO. Power imbalances are represented through recourse variables in the stochastic optimization problem. For instance, in the case of energy deficit during a certain time, a recourse action can be taken, i.e. the needed amount of energy is purchased from the utility grid at real-time energy prices: the corresponding recourse variable represents the needed amount of energy to be purchased in order to keep the power balance at that point in time.

The power balance constraints, defined above in compact form as $\boldsymbol{F}(k)\boldsymbol{u}(k) + \boldsymbol{fw}(k) = 0$, are the random constraints or second-stage constraints, whilst all the other constraints can be considered as first-stage constraints.

Given an initial time step κ and a time duration T and denoting the initial storage levels at time κ as $x^b(\kappa)$, we define the vectors $\boldsymbol{\xi}_\kappa^{+,T}, \boldsymbol{\xi}_\kappa^{-,T}, \boldsymbol{u}_\kappa^T, \boldsymbol{w}_\kappa^T$, whose elements are, respectively, the recourse variables associated to electrical and thermal loads, the decision variables and the disturbances at each time step over the prediction horizon, from κ to $\kappa + T - 1$.

First-stage function

We remind that, in the two-stage stochastic framework, the first-stage cost function and constraints do not depend on random variables. The first-stage constraints are the ones defining the sets in Table 14.2. In those sets, unit capacity limits, ramp limits and all technical and operational bounds on the decision variables are also included.

Since we account for both economical and environmental costs, to compute the first-stage plan of microgrid operations, we apply the weighted min–max method [36]. Define $\hat{V}_i(\boldsymbol{u}_\kappa^T) = \sum_{k=0}^{T-1} V_i(\boldsymbol{u}(\kappa + k))$, $i \in \{r, e\}$.

A normalization of objectives is required to get a Pareto optimal solution consistent with the assigned weights since different objective functions can have different magnitude. Consider then the following cost function:

$$J_{fs}(x^b(\kappa), \boldsymbol{u}_\kappa^T) = \max_{i \in \{r,e\}} \{\theta_i \tilde{V}_i(\boldsymbol{u}_\kappa^T)\}, \tag{14.17}$$

where $\tilde{V}_i(\boldsymbol{u}_\kappa^T)$ is the normalized objective function (we refer the reader to [21] for details). All the objective functions after normalization will take values between zero and one.

Second-stage function

At the second stage, we aim at minimizing the expected value of the costs due to power imbalances. In the case of discrete distribution of random variables with a finite number S of scenarios, $\hat{\boldsymbol{w}}_{1,\kappa}^T, \ldots, \hat{\boldsymbol{w}}_{S,\kappa}^T$, with corresponding probability $p_1, \ldots, p_S$, the second-stage function can be written as

$$J_{ss}(\boldsymbol{u}_\kappa^T, \boldsymbol{w}_\kappa^T) = \min \sum_{i=1}^{S} p_i \left(\boldsymbol{q}^+ \boldsymbol{\xi}_{i,\kappa}^{+,T} + \boldsymbol{q}^- \boldsymbol{\xi}_{i,\kappa}^{-,T} \right)$$

s.t.

$$\boldsymbol{\xi}_{i,\kappa}^{+,T} \geq \tilde{\boldsymbol{F}}_\kappa \boldsymbol{u}_\kappa^T + \tilde{\boldsymbol{f}} \hat{\boldsymbol{w}}_{i,\kappa}^T$$

$$\boldsymbol{\xi}_{i,\kappa}^{-,T} \geq -(\tilde{\boldsymbol{F}}_\kappa \boldsymbol{u}_\kappa^T + \tilde{\boldsymbol{f}} \hat{\boldsymbol{w}}_{i,\kappa}^T)$$

$$\boldsymbol{\xi}_{i,\kappa}^{+,T}, \boldsymbol{\xi}_{i,\kappa}^{-,T} \in \mathbb{R}_+,$$

$$\forall i = 1, \ldots, S,$$

where $\boldsymbol{q}^+$ and $\boldsymbol{q}^-$ are costs related to energy surplus and shortage respectively, $\tilde{\boldsymbol{F}}_\kappa = \mathrm{diag}(\boldsymbol{F}(\kappa), \ldots, \boldsymbol{F}(\kappa + T - 1))$ and $\tilde{\boldsymbol{f}} = \boldsymbol{f} \otimes I_{T \times T}$, with $\otimes$ denoting the *Kronecker* product.

Problem formulation
The two-stage stochastic problem for microgrid economic/environmental operation management can be formulated as follows:

$$\begin{aligned}&\min_{\boldsymbol{u}_{\kappa}^{T},\boldsymbol{\xi}_{i,\kappa}^{+,T},\boldsymbol{\xi}_{i,\kappa}^{-,T}} J_{fs}(x^b(\kappa),\boldsymbol{u}_{\kappa}^{T}) + J_{ss}(\boldsymbol{u}_{\kappa}^{T},\boldsymbol{w}_{\kappa}^{T})\\ &\text{s.t.}\\ &\text{first-stage constraints.}\end{aligned} \tag{14.18}$$

We apply the weighted min–max method [36], hence the optimization problem above can be reformulated as the following mixed integer linear program:

$$\begin{aligned}&\min_{\boldsymbol{u}_{\kappa}^{T},\boldsymbol{\xi}_{i,\kappa}^{+,T},\boldsymbol{\xi}_{i,\kappa}^{-,T}} \mu + J_{ss}(\boldsymbol{u}_{\kappa}^{T},\boldsymbol{w}_{\kappa}^{T})\\ &\text{s.t.}\\ &\theta_i\tilde{V}_i(\boldsymbol{u}_{\kappa}^{T}) \le \mu \quad \forall i \in \{r,e\}\\ &\text{first-stage constraints,}\end{aligned} \tag{14.19}$$

where μ is a scalar auxiliary variable introduced to model the max in (14.17).

Reactive power
The first-stage constraints can be extended in order to model the reactive power management and solve a combination of active and reactive power-management problem for a microgrid. In order to guarantee the power quality and the grid stability, once the optimal active power set points have been computed, the reactive power has to be adjusted for all the microgrid components to maintain voltages within limits. In order to define suitable set points of the reactive power, either local droop controllers can be used, which is the case considered in this work, or optimal values can be computed through optimization, by including in Problem (14.19) the reactive power balance and the relationship between reactive and active power for each microgrid component, which can be modeled as a quadratic constraint given by the capability curve of the component (see [21] for details).

14.4.1.4 MPC-based energy management for microgrid clusters

In this section, we just outline an MPC-based cooperative energy management system for microgrid clusters. The reader is referred to [22] for details and a comprehensive illustration of the proposed methodology. We focus on local energy systems at distribution network level, which can represent, for instance, urban districts or university campuses. Through an incentivization/penalization mechanism embedded in the proposed framework, the microgrids comprised in the local energy system shape their electrical and thermal energy profile to cooperate and achieve a common goal. The common control objective is to minimize the running costs of the entire local energy system, thus reducing as much as possible the energy exchange with the distribution grid, while keeping an acceptable level of comfort. A fair allocation of profits and benefits of the available shared resource is guaranteed; in addition, microgrids adopting the proposed energy management scheme are ensured to achieve a minimum cost saving. By doing so, each microgrid will optimally manage its flexibility resources not only for reducing its energy costs but also for supporting the overall energy system and being rewarded for that.

A hierarchical architecture is proposed: each microgrid is equipped with a local MPC-based energy management system, which computes set points to the local controllers of the DERs and controllable loads available in that energy subsystem. The shared resource is equipped with a local controller to track the power set point calculated by the proposed framework. The interconnected microgrids cooperate through an aggregator, which manages also the shared resource. The information exchange is based only on the power profiles computed by the local decision systems. Thus, detailed information at microgrid level, e.g. user preferences, number of appliances, local generation capabilities, can be kept private.

The overall energy management problem is solved in a distributed fashion, based on the use of MPC in combination with a MILP [38]. The feedback mechanism introduced through the MPC scheme will partially compensate for the uncertainty associated with RES power outputs, time-varying load and energy prices. The proposed algorithm consists of three phases: (i) an initialization phase, where local optimization problems are solved for each microgrid in parallel; (ii) a power distribution phase, where first an optimization problem is solved to coordinate the microgrids and then solutions of the local problems at microgrid level are computed in parallel based on the energy management plan obtained by the aggregator; (iii) a power redistribution phase, where problems at microgrid level are sequentially solved in order to redistribute, whereas feasible, power deviations from the power profile computed at aggregator level.

Simulation results show that the potential cost saving increases as the number of microgrids and the simulation horizon increase. It can be seen that the energy exchange with the distribution grid resulting from the cooperation of the microgrids is beneficial not only in terms of cost saving but also for the substantial reduction in the amount of energy exchanged with the grid.

The reduction of the energy exchange with the utility grid through a more efficient energy management and coordination of microgrids and their flexibility sources has been proved to relieve power distribution line congestions and potentially improve the reliability of the main grid [39]. This aspect is under study.

14.4.2 MPC-based multi-objective AC optimal power flow

The MPC-based AC optimal power flow (OPF) problem is the optimal operation schedule for the electrical network or microgrid that takes decision on the quantity of power to be produced from various local generators and exchanged with storage devices each hour to minimize the operating cost and line losses, while respecting the load balance, physical and operating constraints. The problem is somewhat located at a lower hierarchical level with respect to the problem seen above.

The objective function can be constructed using two kinds of instantaneous costs, i.e:

$$f^E(k) = C_0(k) + \sum_{i=1}^{N} \left(C_i^g(k) + C_i^b(k) \right), \tag{14.20a}$$

$$f^L(k) = \sum_{(i,j)\in\mathcal{E}} P_{(i,j)}^{\text{loss}}(k). \tag{14.20b}$$

In the above equations, $f^E(k)$ and $f^L(k)$ are the network-operating cost and the line losses, respectively, at a generic time k; the summation in (14.20a) is taken over all nodes/buses and counts the costs related to local generators possibly located at node i, say C_i^g, or to storage systems, say C_i^b; the summation in (14.20b) is taken over all lines, instead. Additionally, the cost C_0 models the cost due to the utility and is given by

$$C_0 = c_0(k)P_0^g, \tag{14.21}$$

where c_0 is actually time-varying and depends on the day-ahead energy prices provided by the grid based on market-clearing prices. To account for the bidirectional power flows, a negative lower bound $\underline{P}_0^g \leq 0$ on the injected active power is employed. A negative value of P_0^g indicates selling to utility leading to negative contribution to the OPF objective function and vice versa. Notice that P_0^g is essentially the quantity previously denoted by P^{grid}; we use the different notation to emphasize the different context. The subscript $_0$ reminds the special role played by the point of common coupling as a reference node in the power equations described below.

The two instantaneous costs may be combined into one multi-objective function $J\hat{k}$ computed at time $\hat{k}$, observing a horizon of duration T and consisting of a convex combination for some $\phi \in [0, 1]$:

$$J\hat{k} := \sum_{k=\hat{k}}^{\hat{k}+T} \phi f^E(k) + (1-\phi)f^L(k). \tag{14.22}$$

At each time step $\hat{k}$ we are given the initial state of charge (SoC) of the storages, and a forecast of loads and renewable production. The controller's task is to compute, over the horizon, the optimal sequence of bus variables characterizing the network working, i.e. active and reactive powers, voltages, phases and powers exchanged with the storage units. The optimization problem is equipped with a number of constraints describing the power flows over the transmission lines (basically, Kirchhoff's laws in power terms), the interaction with the external utility grid, the storage dynamics. Details can be found in [40].

14.5 Feasibility study aspects

In order to determine the potential technical/economic/environmental benefits of microgrids and, in particular, of its optimized operation, several technical and economic considerations have to be addressed, e.g. characterization of microgrid requirements, costing of microgrid components, data collection and processing, control strategies. We need to analyse the current power system infrastructure, microgrid topology and existing DERs such as solar, wind, CHP, fuel cells and energy storage, operating conditions, short-term reliability improvement, data collection and processing.

We would like to mention that other specific technical topics may be part of the microgrid feasibility study, such as short circuit and dynamic simulations to address transients, noise issues, microgrid protection, ancillary services.

In this chapter we focus on the potential advantages coming from optimized microgrid operation and thus on topics more relevant to that.

14.5.1 Design and operation

Microgrids are to be designed and operated in order to fulfil specific requirements. They are operated based on the concepts of distributed generation and DERs. Thus, the basic assumption behind a microgrid operation is its self-reliance on its own resources in order to minimize the utilization and power transfer from the upstream network. In order to meet this requirement, a microgrid should be able to operate either in grid-connected or in islanded mode. In both cases, it is essential that the microgrid ensures the short and long-term power balancing, so as to guarantee stable and safe operation. Technically speaking, the balancing is achieved by various means such as good design and sizing of the microgrid units as well as controllability of sources and loads. This way, a microgrid can perform as an uninterruptible power system for its end users, yet it can also support the upstream network with the provision of ancillary services. It is worth noting that aside from the benefits towards the electrical network, a microgrid can also contribute in optimization of energy usage when it comes to CHPs. The latter allow the exploitation of waste heat by end users as opposed to generation by thermal power plants in which waste heat is dissipated in the environment.

14.5.2 Components and topology

Microgrids are usually but not exclusively defined as parts of LV distribution networks with specific geographical boundaries that act as single control entities and, thusly, incorporate various types of DERs under the same control scheme. From the energy resources portfolio viewpoint, the most characteristic types are the following:

1. intermittent RES, which include mainly small-scale photovoltaics (PVs) and wind generators (WGs);
2. dispatchable sources which can be controlled in terms of the output power. This category of sources includes CHPs, fuel cells, diesel generators;
3. storage units which are essential for meeting both the short- and the long-term balance requirements of a microgrid. Among different storage technologies various types of batteries (e.g. lithium-ion, lead–acid (LA)), ultracapacitors and flywheels can be used in a microgrid;

4. flexible loads, which can vary their consumption profile based on different control strategies including demand-side management and demand response techniques;
5. critical loads, which are non-controllable and require a high quality of power supply (i.e. uninterruptible power supply).

In terms of grid configuration, microgrids are typically radial grids with one point of common coupling with the upstream MV distribution network. The interconnection to the MV grid is done via LV/MV transformers and, in order to ensure that optimal and safe operation, microgrids are usually equipped with a controllable switch (preferably static), which allows the rapid disconnection of the microgrid in the case of disturbances in the upstream network.

14.5.3 Active and reactive control strategies

The controllability of a microgrid largely depends on the controllability of its individual resources. Among the different types of resources, we can distinguish the three major categories that follow:

1. inverter-based units, which are used by microgrid components such as PVs, WGs, FCs, battery storage, and which allow for a variety of operating actions in terms of active and reactive power;
2. rotating generators, which follow well-known operating rules of classic electric machines;
3. switchable loads, which can vary their active/reactive power on an on–off basis.

The first two types of resources provide significant capability in terms of active and reactive power control. The basic control principle that can be used in such systems is the droop control, which can modify the active(P)/reactive(Q) power yield of a unit in proportion to the frequency (f)/voltage (V) at the unit connection point. This results in a number of benefits in both grid-connected and islanded modes. In the grid-connected mode deviations of frequency from the nominal value lead to change in the output power of the DERs in a way that they support the primary frequency stability of the system. Likewise, the Q–V control of the DER units can support the local voltage stability of the grid. In the islanded mode, the P–f droop can be used for the power sharing among units without the need for external communication/control signals. Essentially, in this case, over/under frequency is used as an activation signal for increase of curtailment of either the generation or the consumption. Thus, faulty situations such as excessive battery charging can be avoided. In addition to the controllability of these types of DER, switchable loads can be used as extreme measures to ensure power balancing. The activation/shedding of these loads can also be done based on frequency/voltage criteria. It should be pointed out that the inverter-based units mentioned above could be one of the following types: grid-forming, grid-feeding and grid-supporting inverters. Each of these plays a specific role in the operation of the microgrid. For example, grid-forming devices are useful when islanded operation is desired. In this case, at least one unit of the system (usually a battery storage system) acts as a grid-forming unit while all the others are tied to this unit. All the above control features are related to the local operation of units. Therefore, the above controllers are located at device level. These controllers are also known as local controllers and are divided into load controllers and microsource controllers. However, the optimal operation of a microgrid requires the existence of an even higher level controller, also known as microgrid central controller (MGCC), which is responsible for the optimal operation management of the microgrid as a whole. The MGCC can be used for secondary frequency/voltage control strategies, economic and CO_2 optimization strategies, as the ones described in Section 14.4, participation of the microgrid in the energy market and so on.

14.5.4 Data collection and processing

Data collection in a microgrid can be implemented at device and at microgrid central level. Electrical and other environmental data can be collected locally by the DER controller, smart meters and other local devices. This data is processed locally if it regards local control or communicated to the MGCC in order to be used for specific centralized processes (e.g. forecasting) or algorithms (e.g. optimization problem). The infrastructure and requirements for data collection in a microgrid do not differ from a more generic smart grid application.

Data can be employed to model uncertainty sources, such as renewable generation and demand, and incorporate it into the control design. By doing so, the control performance can be improved, in particular at supervisory, whole system level. Uncertainty can be modeled by deriving forecasts and estimating forecast errors, as illustrated in Section 14.4.1.1, from data. Integrating into the control design predictive capabilities can help achieve more efficient and reliable microgrid operation. In order to do that, renewable power and demand forecasts can be computed based on historical measurements. The renewable power and demand data series generally exhibit high-frequency fluctuations and peak shifting, also influenced by meteorological factors, such as outside temperature and irradiance. Broadly exploited methodologies for nonlinear forecasting are neural networks [41] and support vector machines, which have been successfully applied to demand forecasting and for predicting renewable generation [42,43].

Historical and current measurements can be also utilized to derive models of forecast errors, incorporate them into a stochastic optimization problem, which can be solved to find the best operational strategy under uncertainty. The main issue for finding a solution of a stochastic program is the necessity to model uncertainty such that the corresponding optimization problem is numerically tractable. One possible solution is to approximate the actual probability distribution of the uncertainty as a discrete distribution with a tractable number of representative outcomes, which occur with a certain probability. This means that a trade-off is needed between the accuracy of uncertainty model and the tractability of the corresponding optimization problem. To build the discrete distribution for the uncertainty, it is quite natural to use historical data, if available.

14.5.5 Costing of microgrid components

It is essential to estimate the system parameters, such as investment and operating costs and efficiencies. In this section it is outlined the estimation of the parameters for the microgrid components used within the experimental studies.

The various microgrid units present a characteristic diversity in terms of cost due to the nature of the units themselves as well as the market trends at the time of implementation. Nowadays, these costs may differ yet only from a quantitative rather than a qualitative point of view. The investment cost for a PV system based on polycrystalline silicon technology is estimated to be between 3 and 4 €/W for small-scale systems (< 5 kW). The maintenance cost concerning PVs is rather negligible and does not overcome 1% of the initial investment cost. This cost may also include replacement of one power inverter during the system's lifetime (i.e. 25 years). Another important component in terms of costs is the lead acid battery storage of the microgrid. This type of battery involves an investment cost estimated at 150 €/kWh plus the cost for a four-quadrant inverter estimated at 1 €/W. The AC battery storage system adds an additional cost of approximately 0.05–0.10 €/kWh to each kW h being stored and returned to consumption, mainly because the battery needs to be replaced periodically (every 5–10 years), and the related cost needs to be recovered over the limited lifetime of the battery. The lifetime of the battery bank is sensitive to the depth of discharge (DOD, which should not be more than 60% of nominal capacity) as well as the operating temperature. The best maintenance practice requires the electrolyte level to be always kept over the plates. Depending on the DOD, the manufacturer provides the number of cycles that is expected from the battery, e.g. for OPzS batteries at 50% DOD, 2,500 full cycles can be achieved. Beside the replacement of cells, the remaining maintenance cost is rather negligible. The diesel generator, on the other hand, presents a rather low investment cost (approximately 0.4 €/W for generators < 10 kVA). The operating cost is instead significant mainly due to the fuel cost (at the time of the experiments it was 1.5 €/L). The diesel genset presents also a maintenance cost proportional to the investment cost. The lifetime of this type of DER is rather small and amounts to less than 5,000 h of operation. The cost of a natural-gas micro-CHP is 1.5 €/W, while the operational and maintenance costs are 48 €/kW h (gas consumption) and 0.01 €/kW h, respectively. The fuel cell unit used in the case study has a cost of 5 €/W (for a 5 kW unit) and 40% of the investment cost is maintenance cost after 20,000 h of operation. It is worth noting that currently some costs have significantly changed. Thus, according to [44], the average investment cost for a utility scale PV system over the past 8 years (2010–17) was approximately 1.6 $/W. According to [45], lithium iron phosphate ($LiFePO_4$) as well as other types of lithium-based batteries show a high potential in terms of distributed storage, gradually replacing other technologies such as lead acid ones. Although the investment cost for these batteries is still high as opposed to lead acid batteries (i.e. 500–600 $/kW h), this cost is expected to drop to 40% by the year 2030.

The microgrid parameter estimation conducted in the Greek experimental case study described in this chapter is detailed in [46]. All data used in these experiments are realistic; they are based on measurements, datasheets and market prices.

14.6 Case studies

In this section, we discuss the experimental evaluation of the MPC-based energy-management system illustrated in Sections 14.4.1 and 14.4.2.

14.6.1 Experimental evaluation in Athens, Greece

This case study focuses on the application of the MPC-based energy management system illustrated in Section 14.4.1. The experimental set-up is first described, then experimental results are discussed.

14.6.1.1 Description of the experimental set-up

The experimental evaluation of the control algorithm was performed at the Center for Renewable Energy Sources and Saving (CRES), Pikermi-Athens, Greece. The microgrid runs on a low-voltage three-phase electric network with all DER components connected, an *Interbus* communication and control network, a *Modbus*-based monitoring of active and reactive power at the mains, a graphical interface for supervision, monitoring and control. The experimental microgrid of CRES comprises the following units (see Figure 14.2):

- two PV generators 1.1 and 4.4 kWp (the second on a single-axis tracking system);
- two battery-storage systems with total capacity 80 kW h, each consisting of 2 V flooded LA battery cells and interconnected to the mains through battery

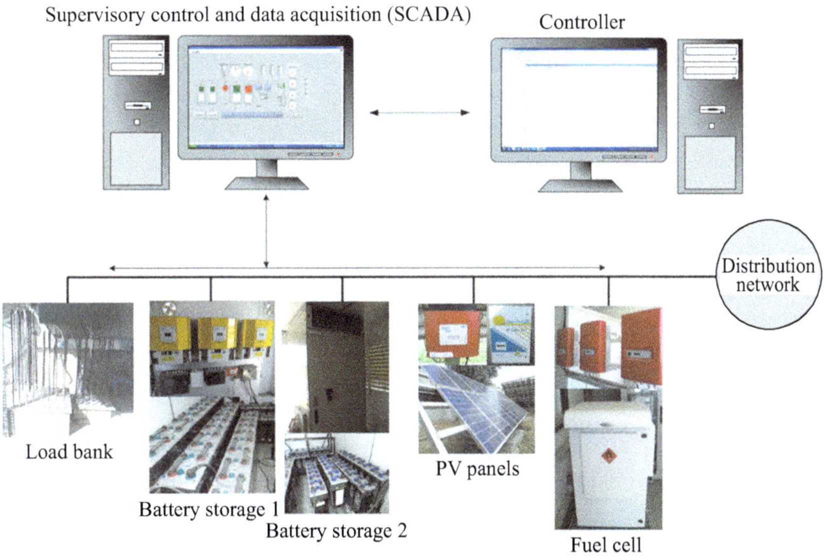

Figure 14.2 Experimental microgrid set-up. Reprinted, with permission, from Reference [20]

inverters which offer numerous control capabilities, such as grid forming, grid tied and droop mode; it is worth mentioning that during the tests one of the two battery systems was used to simulate the operation of a micro-CHP;
- a load bank of resistors of maximum active power 13 kW. This load is equally distributed into the three phases. The loads are grouped into flexible and critical loads and programmed by using data files of consumption so as to produce the desired profiles;
- one 5 kW proton exchange membrane (PEM) fuel cell equipped with a DC/AC three-phase system. Hydrogen consumption is 40 NL/min at 3 kW and 75 NL/min at 5 kW. The fuel cell is supplied with a compressed hydrogen storage tank at a maximum pressure of 16 bar, a physical volume of 3,000 L and a nominal hydrogen storage capacity of 50 N m^3;
- one PEM electrolyser producing hydrogen at 0.5 Nm^3/h at a pressure of up to 14 bar.

Low-level controllers for microgrid components are available:

- a local PI controller is employed to regulate the battery inverter frequency according to the active power set point;
- a local PI controller is used for the μCHP inverter;
- the fuel cell unit is regulated by using a local controller and a comparator with the mean power reference value;
- as to the load bank, a local controller selects the most appropriate combination of resistors which provides the closest consumption value according to the set point.

The control algorithm is implemented on a PC interfaced with the microgrid SCADA system through the local area network.

In this study, a least squares support vector regressor [42] is applied to compute forecasts $\bar{\omega}$, while the forecast errors, $\tilde{\omega}$, are modeled by using a hidden Markov model (HMM). Historical data are employed to train a HMM for generating the finite number of paths, i.e. scenarios, and the corresponding probability of occurrence [47]. During the experiments, the regressor and the HMM are updated at each time step (15 min), with the most recent available measurement, while the oldest data are discarded. To properly decrease the scenario number and make the stochastic problem tractable, scenario reduction algorithms can be applied. The *backward reduction algorithm* has been proved to provide very good performances in the two-stage mixed integer stochastic programming framework [48].

The accuracy of the scenario generation approach will affect the quality of decision-making. We will not argue about the statistical property of the computed approximation, but we evaluate its effectiveness through the improvement in the solution quality. In order to evaluate the suitability of the scenario-generation method for the proposed stochastic programming model, stability tests as described in [48] have been performed. It is found that a fan of 80 scenarios can guarantee good performance.

However, we remark that a more effective scenario-generation mechanism can be easily integrated in the proposed framework.

14.6.1.2 Experimental evaluation

We denote the stochastic MPC controller design in Section 14.4.1 as MG-SMPC. The deterministic MPC control scheme is obtained by neglecting the forecast errors (hence, the second-stage objective in the MPC problem formulation) and just considering the disturbance forecasts. The results of four experiments (run over a 2-month period) are discussed as follows:

- Experiment $MG - DMPC_1$: the microgrid is operated by the deterministic MPC control scheme on day 1. The controller is set with total weight on operation costs.
- Experiment $MG - SMPC_1$: the microgrid operation is managed by the MG-SMPC control scheme on day 2. The controller is set with total weight on operation costs.
- Experiment $MG - SMPC_2$: the microgrid operation is managed by the MG-SMPC control scheme on day 3. The controller is set with $\theta_r = 0.7$ (weight on operating costs) and $\theta_e = 0.3$ (weight on emissions).
- Experiment $MG - SMPC_3$: the microgrid operation is managed by the MG-SMPC control scheme on day 4. The controller is set with $\theta_r = 0.3$ (weight on operating costs) and $\theta_e = 0.7$ (weight on emissions).

The daily spot prices are from the European energy exchange (EEX). Loads follow a realistic profile derived from real-consumption measurements from CRES; in particular, thermal loads determine the energy demand profile of a heat pump employed in the research centre. We consider both curtailable and critical electrical loads. During all the four experiments, the system is operated with the same demand profiles.

The MPC-based controllers are run with a planning horizon of 48 time steps. We consider a 15-min sampling period and the experiments are run over 6 h. The experiments started after dawn so the PV units generate some power at all time steps.

Table 14.4 reports the experimental results in terms of operating costs and emissions. Costs reported in Table 14.4 include the costs of corrective actions that have to be taken to meet the power balance at each time step, i.e. costs to compensate for actual power imbalances. In the case of power deficit or surplus, several options are available: the deficit of power can be purchased from the utility grid or discharged from the battery, or controllable loads can be further curtailed, while the excess of power can be either sold to the grid, or used to charged battery, or

Table 14.4 Results evaluation. Reprinted, with permission, from Reference [21]

	PV yield (kW h)	**Costs (€/kW h)**	**Emissions (kg/kW h)**
$MG\text{-}DMPC_1$	3.43	0.2705	0.1083
$MG\text{-}SMPC_1$	9.93	0.1747	0.0778
$MG\text{-}SMPC_2$	13.76	0.2204	0.0777
$MG\text{-}SMPC_3$	14.37	0.47	0.0438

possible curtailments can be reduced. The feasibility of the aforementioned options must be verified, since constraints on curtailments and power exchange with battery and grid must be satisfied. Nearly all the imbalances during the experiments are due to power deficit, which is then purchased from the grid. However, we associate a cost to any of the options described above, in order to account for possible costs and penalties due to real-time imbalances, as well as for the non-optimality of the resulting power dispatch.

In order to assess the performance of the proposed control framework independently of disturbance profiles (i.e. PV power output in this case study), costs reported in Table 14.4 are average costs per kWh of supplied energy, which includes the energy sold to the grid but not the energy generated by the PV plant.

We can see that the stochastic strategy focusing on operating costs, $MG - SMPC_1$, yields a 30.5% savings over the deterministic one, $MG - DMPC_1$.

During all experiments with stochastic controllers, imbalances occur mainly in the morning, at the beginning of the experiments, when less data is available to train the regressor and the HMM. During the experiments with the MG-SMPC controller the imbalances are reduced by 54% compared to the experiments with the MG-DMPC.

The cost of power shortage or surplus in the stochastic controllers is set to 10, which is at least one order of magnitude higher than the other costs. We set a not so high weight on curtailment, $\rho_f = 0.1$: this way the optimization algorithm uses curtailments when more economically convenient.

We observe the following during the experiments:

1. the battery is employed mainly to gain some profit by selling power to the utility grid. The battery is always discharged at the same time steps when curtailments are required; this allows to sell power to the utility grid while the PV plant and the DGs are employed to support the loads;
2. the fuel cell is heavily used on day 4, when the preferred objective is to reduce emissions, while it is much less utilized on day 3, when the preferred objective is to reduce costs, due to its high maintenance costs (see Figures 14.5);
3. because of the weight on the running costs, the controller aims at compensating the higher costs due to the use of the fuel cell by increasing the curtailments and using the local generated power to sell power. Hence, on day 4, some power is sold during the price peak at 11 a.m., which requires to curtail the flexible load during the price peak at 1 p.m. This happens because the ultimate purpose of this experiment is to minimize the emissions, but with an eye on the costs (see Figure 14.4);
4. curtailments are usually penalized since they lead to user discomfort, so they are not performed unless strictly convenient or necessary. Curtailments usually occur when there are power peaks in the curtailable loads and the prices are higher (see Figure 14.3).

Further experiments are run, with the same demand profiles, to compare the deterministic MPC control scheme with the current practice, whose ultimate purpose is the power balance between production and consumption and to minimize the power/energy flow to and from the public grid. The deterministic MPC

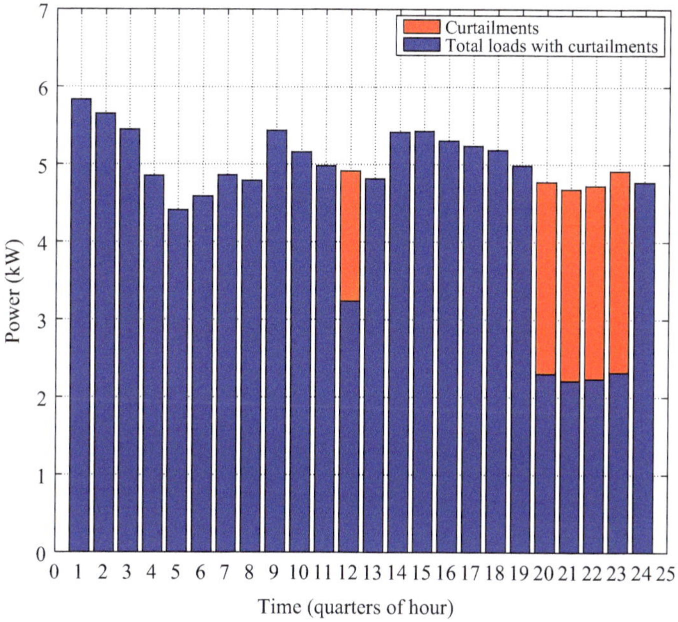

Figure 14.3 Curtailments over on day 4, 6 h. Reprinted, with permission, from Reference [21]

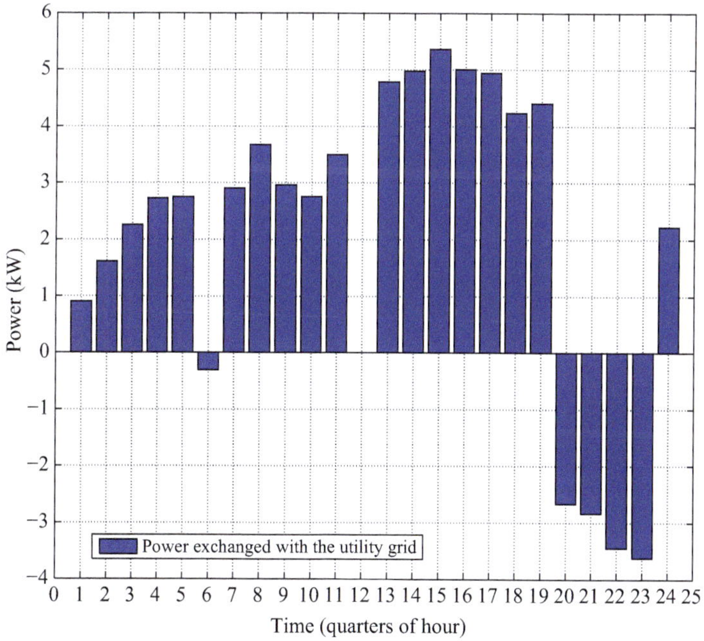

Figure 14.4 Power exchanged with the utility grid on day 4, 6 h. Reprinted, with permission, from Reference [21]

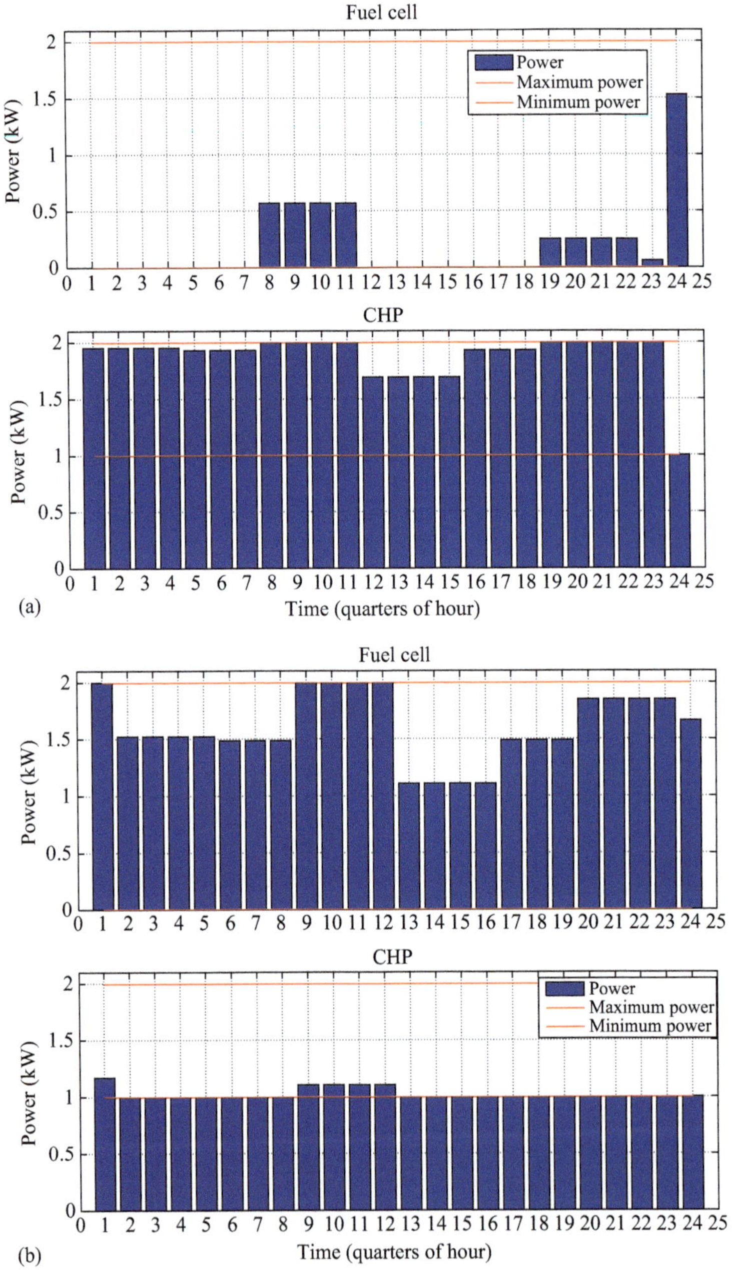

Figure 14.5 *CHP and fuel cell power generation over 24 time steps, 6 h: (a) MG-SMPC on day 3 and (b) MG-SMPC on day 4. Reprinted, with permission, from Reference [21]*

controller yielded 28.5% saving with $T = 24\%$ and 34.7% saving with $T = 72$ over the current practice in microgrid operation.

Computational time for each iteration is less than 1.5 s on average, which would make the proposed control algorithm suitable for faster online operation and much shorter sampling periods.

14.6.2 Steinkjer microgrid

This case study focuses on the application of the MPC-based energy management system illustrated in Section 14.4.2. The middleware architecture is first described, then the experimental evaluation is presented.

14.6.2.1 Semantic middleware architecture

In [40] a layered middleware architecture called the *I3RES Middleware Architecture* was proposed (see Figure 14.6). It is a software that alleviates the complexity associated with the heterogeneity and interoperability of lower level devices. Thus, it transforms the data collected from various devices into homogeneous information for the application layer. The middleware provides the applications with the services and functionalities they need for accomplishing their tasks. A good

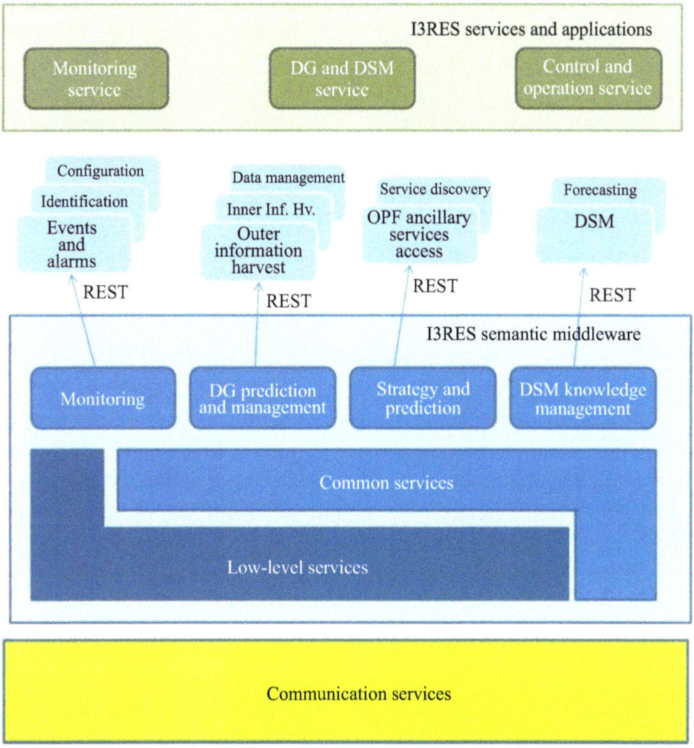

Figure 14.6 High level, low level and common services of the middleware. Reprinted, with permission, from Reference [40]

approach in the backdrop is to define a set of 'high-level services' with a common API that is used by different applications. This provides significant simplicity as the changes and updates in the lower layer become transparent to the application.

Thus, the proposed middleware has the features of the hardware abstraction solution and solves the heterogeneity and interoperability issues. It is organized into three layers: high-level, low-level and communication services (see, Figure 14.6). The 'high-level services layer' provides the necessary services and applications required for implementing the functional and non-functional requirements of the smart grid control algorithms. The low-level services are connected to the hardware and communications layer. These services are used by the high-level services that provide information needed by the application layer. Middleware communication services include ancillary services that are used by both high-level and low-level services to provide control functionalities required by the applications layer. Finally, the lower layer consists of the physical components such as devices from various vendors and proprietary data formats.

The proposed middleware architecture provides smart grids desirable features such as flexibility, interoperability and scalability. In addition, by providing semantic features, it provides the capability to apprehend the meaning of the information being transmitted, thus increasing the 'intelligence' as well as the performance of the entire system. The semantic middleware, as a basic part of the architecture, will embed the common information model (CIM) [49,50], as well as provide the mechanisms (set of common services) for monitoring the context and integrating and supporting the services. Details of the CIM profiles for the microgrid model used in RH ACOPF is reported in [51].

14.6.2.2 Experimental evaluation: a Demo Network in Steinkjer, Norway

This section presents the deployment results of the middleware and the multi-objective RH OPF algorithm in a microgrid located at Steinkjer, Norway. It is a radial network that consists of 3 hydro-generators, 32 aggregating load stations, 50 links and 84 nodes. The RES generation is largely due to solar panels and has grid-level storage units. Table 14.5 shows the network parameters in per unit (p.u.) with a base of 100 MVA and absolute values (a.v.) of storage parameters. The hydro-generation supplies around 10% of the total energy depending on the season and weather condition. To achieve power balance in the microgrid, the network operator buys the energy from a utility at the day-ahead market.

To illustrate the advantages of the ACOPF, its performance is compared with that of DCOPF, its linearized version, only coping with active power and small-phase

Table 14.5 Network Parameters

Parameter	Value
$\underline{V}$	0.95 p.u.
$\overline{V}$	1.05 p.u.
$\underline{B}$	0.025 MWh a.v.
$\overline{B}$	0.5 MWh a.v.
r^{rated}	1 MW a.v.

shifts. The ACOPF has multiple objectives of reducing the line losses and operating cost; the DCOPF aims at dispatching the power generation so as to reduce the operating cost. In our analysis, we set $\phi = 0.5$, i.e. equal weights are assigned to both economic and line-loss objectives.

Figure 14.7 compares the RHC-based ACOPF (in thick lines) and DCOPF (thin and dashed lines). The variations in normalized energy prices provided by the utility in the day-ahead electricity market during the test period is shown in Figure 14.7(a). The energy consumption from the utility, RES generation during the test period, and the total demand on the grid with the ACOPF is shown in Figure 14.7(b). It is interesting to note that there is a difference between the total power generated and demand due to the power supplied by the ESS. Furthermore, Figure 14.7(c) shows that the power is bought/sold from/to the utility grid during

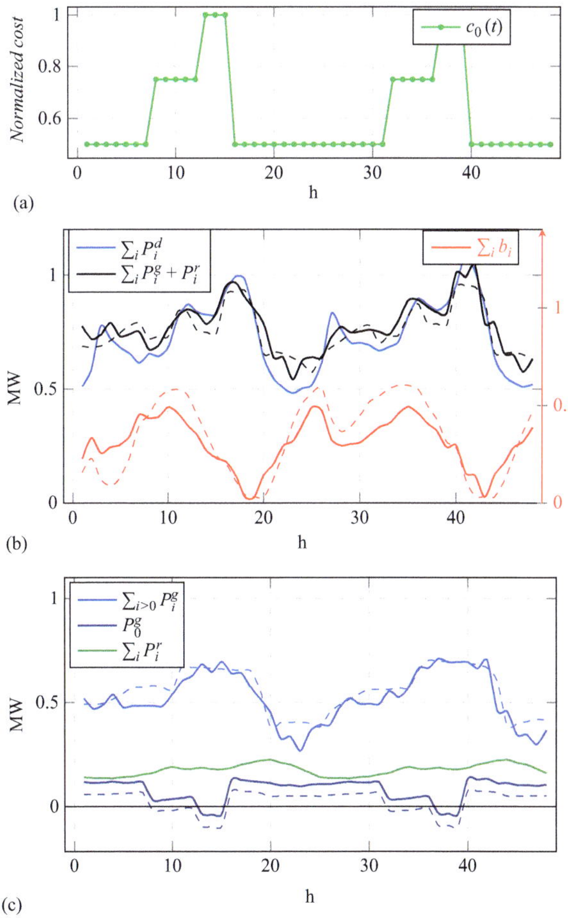

Figure 14.7 Simulation with time-varying generation cost using MPC-based ACOPF controller: (a) prediction of power cost function in the interaction with the utility grid, (b) power balance in the distribution grid and (c) contribution of different generation to the power balance. Reprinted, with permission, from Reference [40]

the periods of decreasing/increasing energy prices. The behaviour is observed both in ACOPF and DCOPF due to the receding horizon approach. Results revealed that the ACOPF reduces the cost and line losses by 6.54% and 15.6%, respectively. We remind that DCOPF cannot include the line-losses. Furthermore, the voltage profiles and power quality maintained using the ACOPF, whereas the DCOPF model does not provide any guarantee. The results also proved that judicious integration of RES and ESS provided significant cost and operational benefits.

14.7 Conclusions

In the chapter, we describe how advanced control and optimization techniques can be used to realize the technical, economical and environmental potential of microgrids. The feasibility and potential benefits are discussed from different perspectives, such as microgrid design and operation, data collection and analysis, cost saving. Experimental results show that the advanced control schemes are able to economically and/or environmentally optimize the microgrid operations and save money and/or reduce emissions, as well as power losses and imbalances, compared to the current practice. The results also evidence that a trade-off between demand peak reduction and user comfort can be achieved through allowable curtailments.

Currently, stability and feasibility properties are under studies though extensive numerical computations have given reassuring results. In the specific context, stability and constraints satisfaction should be guaranteed under uncertainty; intuitively speaking, it should be ensured that enough energy is stored in storage elements to counteract an unpredicted change in demand and RES generation. This can be ensured by a terminal set for the storage level which is a robust control invariant set, or by constraints tightening.

Appendix A

A.1 Matrices

$$\begin{aligned}
\mathbf{E_1}^{\mathrm{ESS}} &= \left[C^{\mathrm{ESS}} \quad -(C^{\mathrm{ESS}}+\varepsilon) \quad C^{\mathrm{ESS}} \quad C^{\mathrm{ESS}} \quad -C^{\mathrm{ESS}} \quad -C^{\mathrm{ESS}}\right]' \\
\mathbf{E_2}^{\mathrm{ESS}} &= \mathbf{E}_2^{\mathrm{TSS}} = \mathbf{E}_2{}^{\mathrm{grid}} = [0 \quad 0 \quad 1 \quad -1 \quad 1 \quad -1]' \\
\mathbf{E_3}^{\mathrm{ESS}} &= \mathbf{E}_3^{\mathrm{TSS}} = [1 \quad -1 \quad 1 \quad -1 \quad 0 \quad 0]' \\
\mathbf{E_4}^{\mathrm{ESS}} &= \left[C^{\mathrm{ESS}} \quad -\varepsilon \quad C^{\mathrm{ESS}} \quad C^{\mathrm{ESS}} \quad 0 \quad 0\right]' \\
\mathbf{E}_1^{\mathrm{TSS}} &= [M^s \quad -(M^s+\varepsilon) \quad M^s \quad M^s \quad -M^s \quad -M^s]' \\
\mathbf{E}_4^{\mathrm{TSS}} &= [M^s - \varepsilon \quad M^s \quad M^s \quad 0 \quad 0]' \\
\mathbf{E_1}^{\mathrm{grid}} &= \left[T^{\mathrm{grid}} \quad -(T^{\mathrm{grid}}+\varepsilon) \quad M^{\mathrm{grid}} \quad M^{\mathrm{grid}} \quad -M^g \quad -M^g\right]' \\
\mathbf{E_3}^{\mathrm{grid}}(k) &= \left[1 \quad -1 \quad c^P(k) \quad -c^P(k) \quad c^S(k) - c^S(k)\right]' \\
\mathbf{E_4}^{\mathrm{grid}} &= \left[T^{\mathrm{grid}} - \varepsilon \quad M^{\mathrm{grid}} \quad M^{\mathrm{grid}} \quad 0 \quad 0\right]'
\end{aligned}$$

where $M^{\mathrm{grid}} := T^{\mathrm{grid}} \max_k(c^P(k), c^S(k))$, ε is a small tolerance (typically the machine precision).

References

[1] European Commission. 2030 Framework for Climate and Energy Policies. European Commission; 2013. Available from: http://ec.europa.eu/energy/en/topics/energy-strategy/2030-energy-strategy.

[2] ETIP SNET Vision 2050. European Technology & Innovation Platform Smart Networks for Energy Transition (ETIP SNET); 2018. Available from: https://mailchi.mp/etip-snet/vision2050download.

[3] Stadler M, Cardoso G, Mashayekh S, *et al.* Value Streams in Microgrids: A Literature Review. Applied Energy. 2016;162:980–989.

[4] Parhizi S, Lotfi H, Khodaei A, *et al.* State of the Art in Research on Microgrids: A Review. IEEE Access. 2015;3:890–925.

[5] Ilic J, Prica M, Rabiei S, *et al.* Technical and Economic Analysis of Various Power Generation Resources Coupled with CAES Systems. National Energy Technology Laboratory, DOE/NETL-2011/1472; 2011.

[6] O'Neill RP, Dautel T, and Krall E. Recent ISO Software Enhancements and Future Software and Modeling Plans. Staff Report. Federal Energy Regulatory Commission; 2011.

[7] Sioshansi R, O'Neill R, and Oren SS. Economic Consequences of Alternative Solution Methods for Centralized Unit Commitment in Day-Ahead Electricity Markets. IEEE Transactions on Power Systems. 2008;23(2):344–352.

[8] Strbac G, Pudjianto D, Djapic P, *et al.* Understanding the Balancing Challenge. Imperial College London; 2012.

[9] Capuder T, and Mancarella P. Techno-economic and Environmental Modelling and Optimization of Flexible Distributed Multi-generation Options. Energy. 2014;71:516–533.

[10] Neves D, Pina A, and Silva CA. Demand Response Modeling: A Comparison between Tools. Applied Energy. 2015;146:288–5297.

[11] Strunz K, Abbasi E, and Huu DN. DC Microgrid for Wind and Solar Power Integration. IEEE Journal of Emerging and Selected Topics in Power Electronics. 2014;2(1):115–126.

[12] Olivares DE, Mehrizi-Sani A, Etemadi AH, *et al.* Trends in Microgrid Control. IEEE Transactions on Smart Grid. 2014;5(4):1905–1919.

[13] Samad T, and Annaswamy AM. Controls for Smart Grids: Architectures and Applications. Proceedings of the IEEE. 2017;105(11):2244–2261.

[14] Good N, Ellis KA, and Mancarella P. Review and Classification of Barriers and Enablers of Demand Response in the Smart Grid. Renewable and Sustainable Energy Reviews. 2017;72:57–72.

[15] Nolan S, and O'Malley M. Challenges and Barriers to Demand Response Deployment and Evaluation. Applied Energy. 2015;152:1–10.

[16] Koraki D, and Strunz K. Wind and Solar Power Integration in Electricity Markets and Distribution Networks Through Service-centric Virtual Power Plants. IEEE Transactions on Power Systems. 2018;33(1):473–485. Online.

[17] Mayne DQ, Rawlings JB, Rao CV, *et al.* Constrained Model Predictive Control: Stability and Optimality. Automatica. 2000;36:789–814.

[18] García CE, Prett DM, and Morari M. Model predictive control: Theory and practice—A survey. Automatica.1989;25(3):335–348. Available from http://www.sciencedirect.com/science/article/pii/0005109889900022
[19] Shapiro A, Dentcheva D, and Ruszczyński A. Lectures on Stochastic Programming: Modeling and Theory. Mathematical Programming Society-SIAM Series on Optimization, Philadelphia: Society for Industrial & Applied Mathematics; 2009.
[20] Parisio A, Rikos E, and Glielmo L. A Model Predictive Control Approach to Microgrid Operation Optimization. IEEE Transactions on Control Systems Technology. 2014;22(5):1813–1827.
[21] Parisio A, Rikos E, and Glielmo L. Stochastic Model Predictive Control for Economic/Environmental Operation Management of Microgrids: An Experimental Case Study. Journal of Process Control. 2016;43:24–37.
[22] Parisio A, Wiezorek C, Kyntäjä T, *et al.* Cooperative MPC-Based Energy Management for Networked Microgrids. IEEE Transactions on Smart Grid. 2017;8(6):3066–3074.
[23] Paridari K, Parisio A, Sandberg H, *et al.* Robust Scheduling of Smart Appliances in Active Apartments With User Behavior Uncertainty. IEEE Transactions on Automation Science and Engineering. 2016;13(1):247–259.
[24] Verrilli F, Srinivasan S, Gambino G, *et al.* Model Predictive Control-based Optimal Operations of District Heating System with Thermal Energy Storage and Flexible Loads. IEEE Transactions on Automation Science and Engineering. 2017;14(2):547–557.
[25] Parisio A, Wiezorek C, Kyntäjä T, *et al.* An MPC-based Energy Management System for Multiple Residential Microgrids. In: 2015 IEEE International Conference on Automation Science and Engineering (CASE); 2015. p. 7–14.
[26] Bemporad A, and Morari M. Control of Systems Integrating Logic, Dynamics, and Constraints. Automatica. 1999;35(3):407–427.
[27] Catalog of CHP Technologies. U.S. Environmental Protection Agency Combined Heat and Power Partnership; 2008.
[28] Samad T, Koch E, and Stluka P. Automated Demand Response for Smart Buildings and Microgrids: The State of the Practice and Research Challenges. Proceedings of the IEEE. 2016;104(4):726–744.
[29] Takeuchi A, Hayashi T, Nozaki Y, *et al.* Optimal Scheduling Using Metaheuristics for Energy Networks. IEEE Transactions on Smart Grid. 2012; 3(2):968–974.
[30] Hytowitz RB, and Hedman KW. Managing Solar Uncertainty in Microgrid Systems with Stochastic Unit Commitment. Electric Power Systems Research. 2015;119:111–118.
[31] Kuznetsova E, Ruiz C, Li YF, *et al.* Analysis of Robust Optimization for Decentralized Microgrid Energy Management Under Uncertainty. International Journal of Electrical Power & Energy Systems. 2015;64:815–832.
[32] Prodan I, and Zio E. A Model Predictive Control Framework for Reliable Microgrid Energy Management. International Journal of Electrical Power & Energy Systems. 2014;61(0):399–409. Available from: http://www.sciencedirect.com/science/article/pii/S0142061514001197.

[33] Palma-Behnke R, Benavides C, Lanas F, *et al.* A Microgrid Energy Management System Based on the Rolling Horizon Strategy. IEEE Transactions on Smart Grid. 2013;4(2):996–1006.

[34] Bidram A, and Davoudi A. Hierarchical Structure of Microgrids Control System. IEEE Transactions on Smart Grid. 2012;3(4):1963–1976.

[35] Richard A, and How J. Mixed-Integer Programming for Control. In: American Control Conference. vol. 4. Portland, Oregon (USA); 2005. p. 2676–2683.

[36] Marler RT, and Arora JS. Survey of Multi-objective Optimization Methods for Engineering. Structural and Multidisciplinary Optimization. 2004;26(6): 369–395.

[37] Abido M. Environmental/Economic Power Dispatch Using Multiobjective Evolutionary Algorithms. IEEE Transactions on Power Systems. 2003;18(4): 1529–1537.

[38] Floudas CA. Nonlinear and Mixed-Integer Programming and Fundamentals and Applications. Oxford, UK: Oxford University Press; 1995.

[39] Liang H, and Zhuang W. Stochastic Modeling and Optimization in a Microgrid: A Survey. Energies. 2014;7:2027–2050.

[40] Maffei A, Srinivasan S, Castillejo P, *et al.* A Semantic-Middleware-Supported Receding Horizon Optimal Power Flow in Energy Grids. IEEE Transactions on Industrial Informatics. 2018;14(1):35–46.

[41] Yafeng Y, Yue L, Junjun G, *et al.* A New Fuzzy Neural Networks Model for Demand Forecasting. In: Automation and Logistics, 2008. ICAL 2008. IEEE International Conference on; 2008. p. 372–376.

[42] Osuna E, Freund R, and Girosi F. Support Vector Machines: Training and Applications. Massachusetts Institute of Technology: Computer Science and Artificial Intelligence Lab (CSAIL), AIM-1602, CBCL-144; 1997. http://hdl.handle.net/1721.1/7290.

[43] Sharma N, Sharma P, Irwin D, *et al.* Predicting Solar Generation from Weather Forecasts Using Machine Learning. In: Proceedings of the Second IEEE International Conference on Smart Grid Communications (SmartGridComm). Brussels, Belgium; 2011.

[44] IRENA. Renewable Power Generation Costs in 2017. Abu Dhabi: International Renewable Energy Agency; 2018.

[45] IRENA. Electricity Storage and Renewables: Costs and Markets to 2030. Abu Dhabi: International Renewable Energy Agency; 2018.

[46] Parisio A, Rikos E, Tzamalis G, *et al.* Use of Model Predictive Control for Experimental Microgrid Optimization. Applied Energy. 2014;115:37–46.

[47] Zucchini W, MacDonald IL, and Langrock R, Hidden Markov Models for Time Series: An Introduction Using R, Second Edition, Chapman & Hall/CRC Monographs on Statistics and Applied Probability, FL: CRC Press; 2017.

[48] Kaut M, and Wallace SW. Evaluation of Scenario-Generation Methods for Stochastic Programming. Stochastic Programming E-Print Series. 2003; 14(1). Available from: http://edoc.huberlin.de/series/speps/2003-14/PDF/14.pdf.

[49] Gungor VC, Sahin D, Kocak T, *et al.* Smart Grid Technologies: Communication Technologies and Standards. IEEE Transactions on Industrial Informatics. 2011;7(4):529–539.

[50] Naumann A, Bielchev I, Voropai N, *et al.* Smart Grid Automation Using IEC 61850 and CIM Standards. Control Engineering Practice. 2014;25:102–111.

[51] i3RES Consortium. Deliverable D6.1: Smart Grid Model; 2015. Available from: http://http://www.i3res.eu/v1/.

Chapter 15
Power electronics—microgrid interfacing

Saeed Peyghami[1], Mohammed Alhasheem[1], and Frede Blaabjerg[1]

Power electronics is the key enabling technology for modern power systems. Power converters are increasingly used in a wide range of applications from generation to consumption levels. Due to the significant importance of power electronics in modern power systems, this chapter presents the possible structure of the future power systems employing the microgrid (MG) technologies. Different MG topologies and converter structures are introduced. Moreover, the control and operation principles of power electronic converters in MGs are discussed. Two case studies are provided in order to show the importance of power electronics in operation and control of modern power systems. Finally, the chapter is summarized with highlighting challenges encountering modern power electronics-based power systems.

15.1 Importance of power electronics in a microgrid

The electricity demand is growing every year even faster than overall energy demand worldwide. This requires developing energy conversion and delivery systems to be able to supply the demand reliably. Developing conventional power systems with central generation systems has not been cost effective due to the highest investment on transmission lines, expanding distribution networks and installing large thermal power plants. Furthermore, a large amount of power loss in the traditional power plants and transmission lines together with the type of prime energy sources oppose the global warming issues. Hence, decarbonizing and economizing the power systems have represented a paradigm shift in the phase of evolution and development of energy conversion and delivery systems. First, the small-scale energy resources especially renewable energies have locally been installed nearby the loads. Distributed generations (DGs) introduce high efficiency and reliability, low costs and emissions. However, the DG technology exposes the power systems to the operation, control and stability challenges. In order to add more flexibility and controllability to the DGs, the concept of MGs as distributed networks has been used in the distribution systems. Conceptually, the MG

[1]Department of Energy Technology, Aalborg University, Aalborg East 9220 Denmark

technology has been employed in aircraft, spacecraft, space stations, ships, etc. for many years. However, it has gained increasing interest in recent years for developing power systems.

Technically, MG is a technology to operate a group of energy resources and storages to supply the demand with different loads and especially the critical demand in the presence or absence of utility. The MGs will be different segments of future power grids from generation, transmission down to distribution levels. Optimal and reliable operation of MGs together with DGs connected to the conventional power system requires communication infrastructures to collect the information of loads, generations, climate condition, grid topology, etc. to economically control the power systems. Hence, as the last phase of power system development, the smart grid technology has been presented to form the modern power systems.

It appears that the MGs will be the essential segments of modern power systems. Generally, it can be classified as ac and dc MGs. Future power system structure might be a multiterminal hybrid ac/dc grid. AC and dc MGs can be implemented at different power conversion levels. For example, in wind generation systems, both ac and dc grid structures can be employed to collect the power of wind turbines and inject it into a transmission system. Power electronic converters interface the wind turbines to the local ac or dc grids, and local loads.

Meanwhile, power electronics is the key enabling technology for the development of MGs. It has a broad range of applications in MGs at different power conversion processes. Most of the DGs require an interface converter to connect into the grid. An interface converter in a photovoltaic (PV) array or wind turbine is responsible for operating it at maximum power point tracking (MPPT) mode, as well as synchronizing it with the output voltage—voltage amplitude in dc grids and phase angle and voltage amplitude in ac grids—with the grid at the point of common coupling (PCC). Other types of energy sources such as fuel cell stacks and micro-turbines must be connected into the grid through a power electronic converter to meet the voltage/frequency requirements. Battery storage systems can improve the availability of the energy delivery systems especially in the presence of renewable resources. Power electronic converters are in charge of energy management and power flow control in, e.g., battery storage systems in order to control its state-of-charge level.

From the power system perspective, increasing the presence of renewable energies compared to the conventional thermal power plants enforces them to participate in grid voltage and frequency support. Power electronic interfaces, thanks to their flexibility and controllability, provide such an applicability for these units in order to maintain power system stability and availability.

Interconnecting different MGs including ac and dc as well as ac grids with different operating frequencies is only possible by employing power electronic converters known as interlinking converters. Proper converter topologies can be utilized to interconnect the dc grids with the same or different voltage levels, ac with the same or different frequencies and other possible structures. For instance, dc homes can be considered a dc MG connected to a neighboring ac or dc grids.

Interlinking converters control the power flow between interconnected grids according to the requested demand by the corresponding system operators. They can also support the interconnected grids' voltage and/or frequency.

Taking into account the significant role of power electronics in MGs, this chapter represents the MG architectures and different ac–dc and dc–dc converter topologies. Afterward, a number of switching schemes for converters and hierarchical control strategies for converters operating in ac and dc MGs are explained. Also active and passive filters for power converters are introduced. Two case studies for MG operation and control are provided. Finally, the chapter is summarized with addressing the challenges posed to the modern power electronics-based power systems.

15.2 Classifications of microgrids

Generally, MGs can be classified as ac, dc and hybrid MGs regardless of the voltage or power ratings. This section presents the different MG architectures and their applications.

15.2.1 AC microgrids

A typical structure of an ac MG is shown in Figure 15.1, where ac sources such as wind turbines are connected through an ac–ac converter (typically ac–dc–dc–ac) to the local grid. Furthermore, the dc sources and storages are connected through a dc–ac converter or an inverter to the local grid. The MG supplies its critical and noncritical loads, and it can be connected to the utility grid though a transformer. It can also be connected to the medium-voltage (MV) dc grid by an ac–dc converter. AC grids can be implemented in either low voltage (LV) or MV depending on the power level. For instance, power distribution systems with DGs can be operated as an LV MG. An example of MV applications can be a power system in a ship.

15.2.2 DC microgrids

DC grids are the second type of MGs, which have been gaining significant interest in recent years. A typical structure of a dc MG is shown in Figure 15.2. DC sources and storages are connected to the local grid by a dc–dc converter. AC sources should supply the dc grid through an ac–dc converter or rectifier. DC MGs can be supplied by the utility grid, or it can be interconnected to the neighboring ac or dc MGs. Space stations, aircrafts, dc ships, data centers, dc homes, dc offshore oil drilling systems, high voltage direct current (HVDC) systems, MV distribution systems and dc onshore wind farms are some applications of dc MGs. For instance, an medium voltage direct current (MVDC) ship-board structure is shown in Figure 15.3, and an HVDC system is shown in Figure 15.4.

15.2.2.1 Hybrid microgrids

Hybrid MGs, or generally hybrid grids, are clusters of ac and dc grids with different voltage, frequency and power levels. It seems that the future power systems will be formed as an interconnection of different MGs. Different structures of hybrid MGs

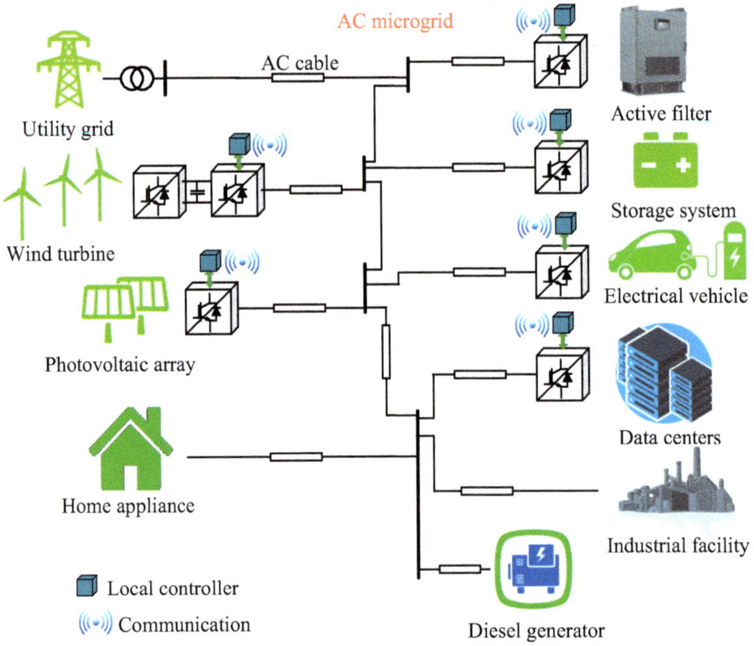

Figure 15.1 Typical structure of an ac microgrid

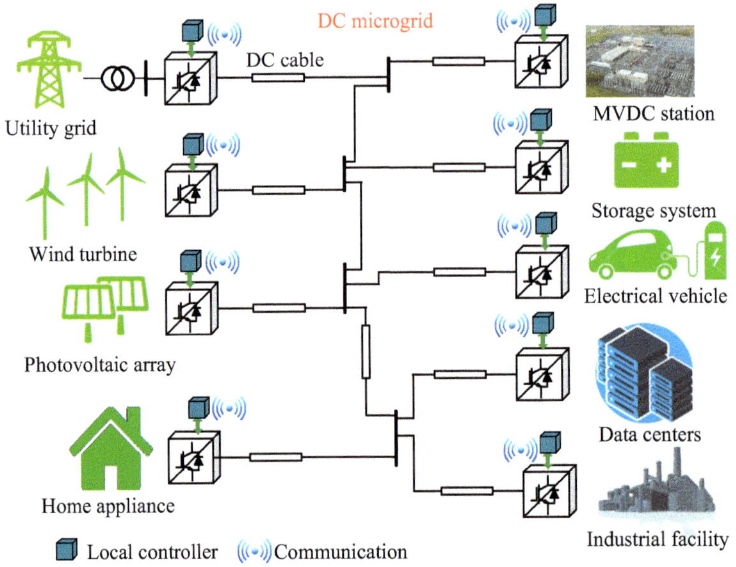

Figure 15.2 Typical structure of a dc microgrid

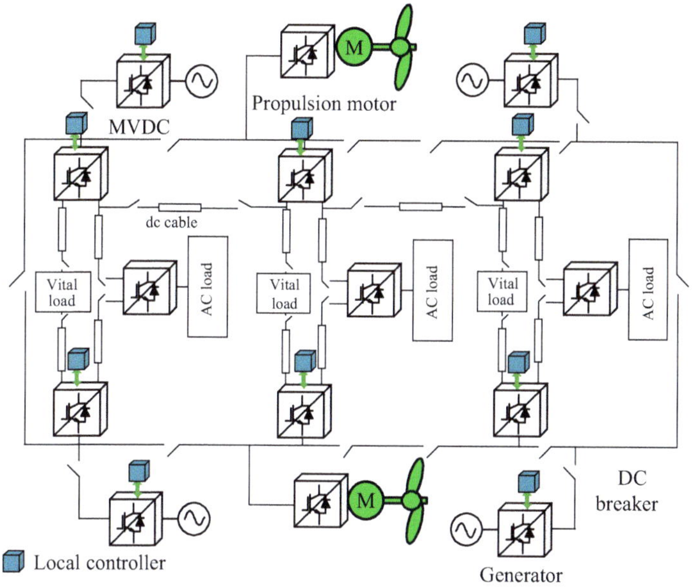

Figure 15.3 MVDC structure of an electric ship-board

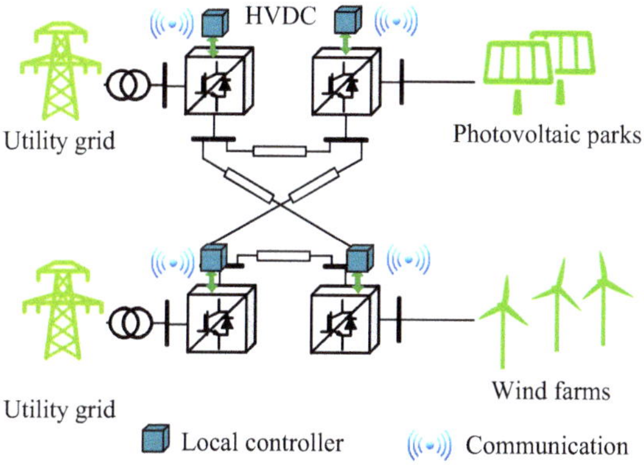

Figure 15.4 HVDC transmission system

are illustrated next. Figure 15.5 shows a hybrid ac–ac MG, which is applicable for interconnecting two ac grids with different frequencies. The structure of future distribution systems can be a hybrid ac–dc MG as shown in Figure 15.6. Another hybrid MG topology is interconnecting two dc grids such as supplying an LV dc grid from an MV dc grid as shown in Figure 15.7. The interconnection of different types of MGs will form the modern power systems as shown in Figure 15.8, which is called inter-grid [1]. The MGs are connected to the neighboring grid with the same or different voltage/frequency levels through power electronic converters, and hence, they are dynamically decoupled from each other.

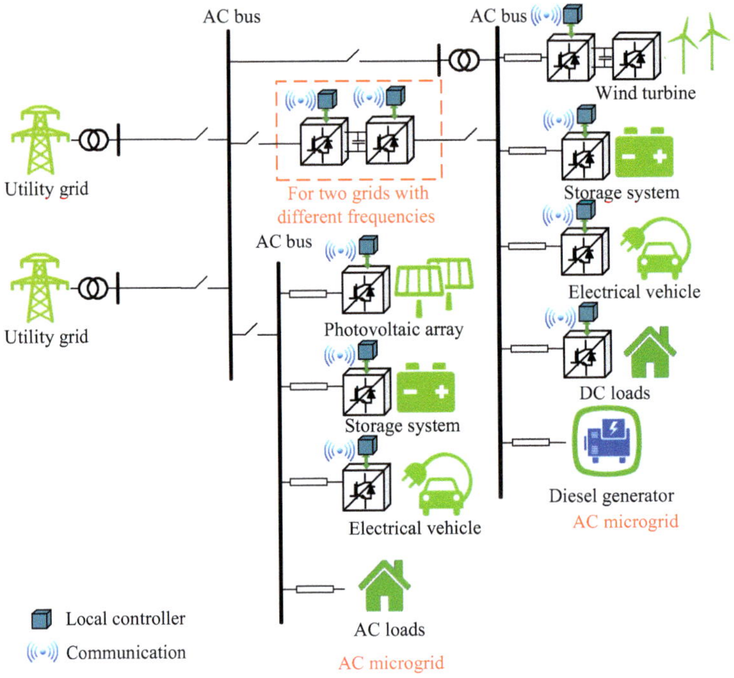

Figure 15.5 Hybrid ac–ac microgrid structure

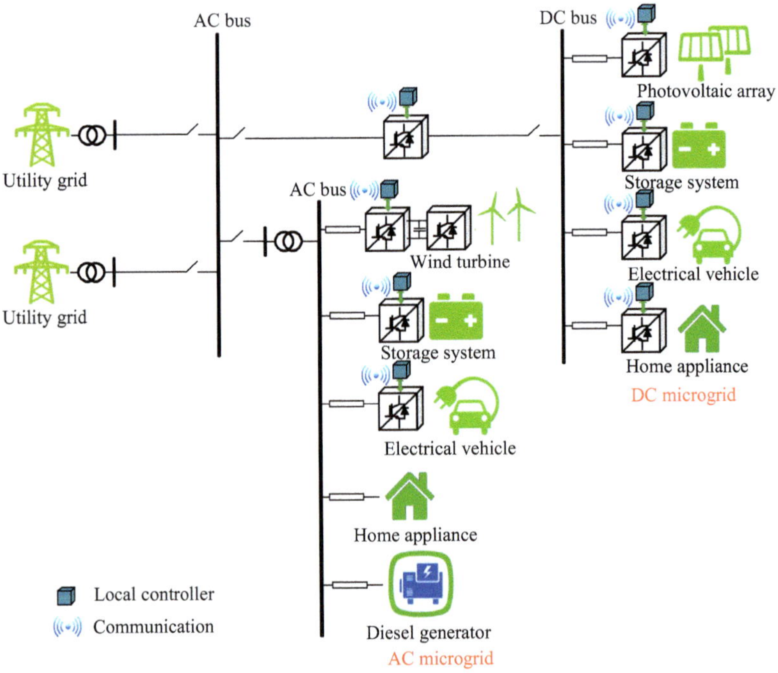

Figure 15.6 Hybrid ac–dc microgrid structure

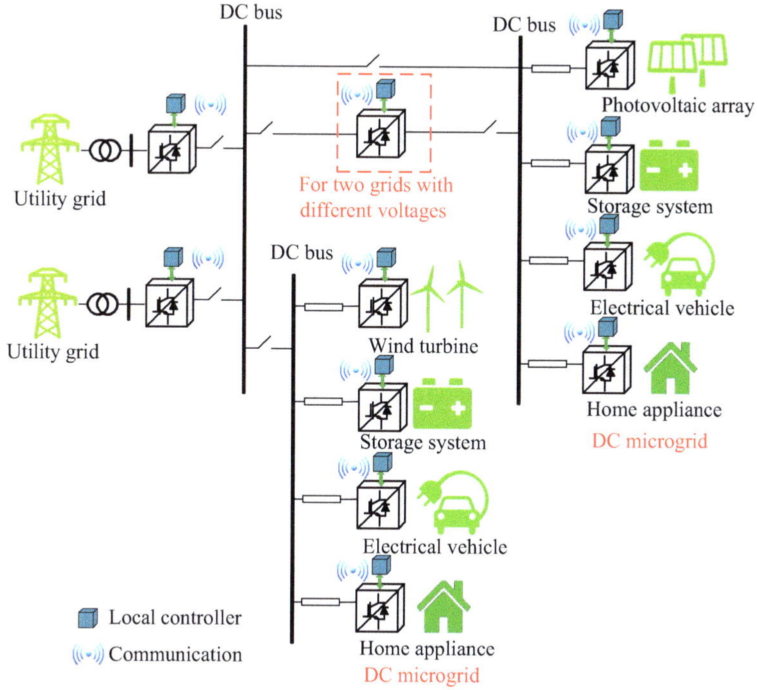

Figure 15.7 Hybrid dc–dc microgrid structure

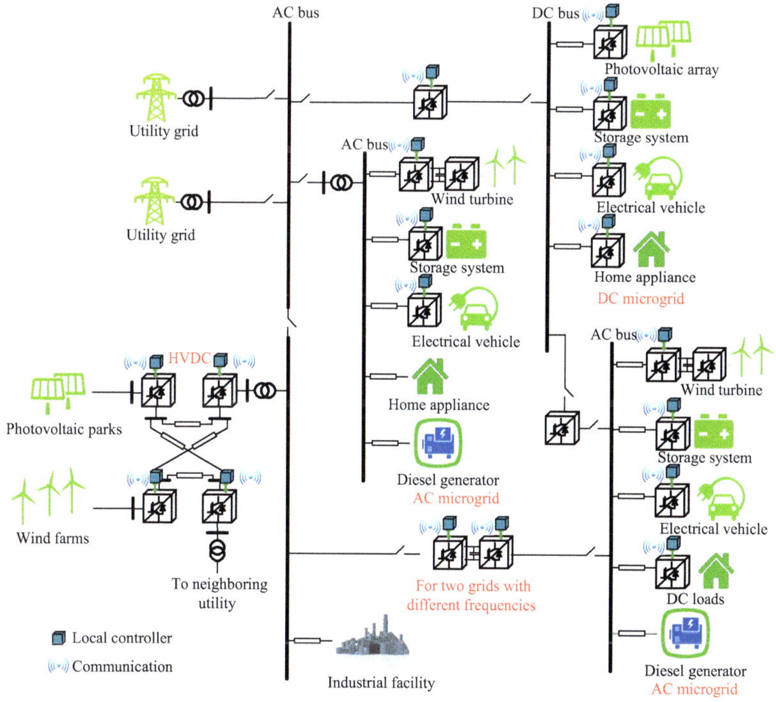

Figure 15.8 Inter-grid: the future power systems

15.3 Power electronic converters

Numerous converter structures have been presented for power conversion in different applications. According to the input/output voltage levels and power ratings, power electronic converters can be classified into four classes as dc–dc, dc–ac, ac–ac and ac–dc converters. The dc–dc and dc–ac converters are the most common structures used in MGs. Hence, this section presents the general concept of power conversion and some basic structures of dc–dc and dc–ac converters employed in MGs.

15.3.1 General power conversation concept

A power electronic converter has at least one semiconductor switch, which is controlled to be turned on or off. This switch ideally has two states of operation including short-circuit mode during on time and open-circuit during off time. Figure 15.9(a) shows the basic structure and operation of a converter with an ideal switch (Q). When Q is in state "1," the output voltage is equal to the input voltage, and when it is in position "2," the output voltage is zero. The output voltage waveform $v_o(t)$ for an ideal input voltage is shown in Figure 15.9(b). If the time of being in position "1" is DT, the average output voltage can be obtained as

$$V_o = \frac{1}{T}(D \cdot T \cdot V_i + (1 - D) \cdot T \cdot 0) = D \cdot V_i \tag{15.1}$$

in which, T is the switching period, and D is the duty cycle. Since $D < 1$, the input voltage will be stepped down, while employing the converter in reverse direction will step up the voltage level. This basic operation principle can be used to figure out the operation mechanisms of different converters.

For instance, a converter with two ideal switches is shown in Figure 15.10(a). Whenever Q_1 is ON, the output voltage is equal to $V_i/2$, and when Q_2 is ON, the output voltage is $-V_i/2$. Depending on the switching duty cycle, the average output voltage can be positive (if $D > 0.5$) and negative (if $D < 0.5$). If D is a constant value, the output voltage will be dc. Moreover, if D has a sinusoidal form, the output voltage will be sinusoidal. Different modulation scheme can be used to generate sinusoidal waveforms as discussed in [2] and will be presented in the next section. In the following sections, the converter structures for dc–dc and dc–ac applications are introduced.

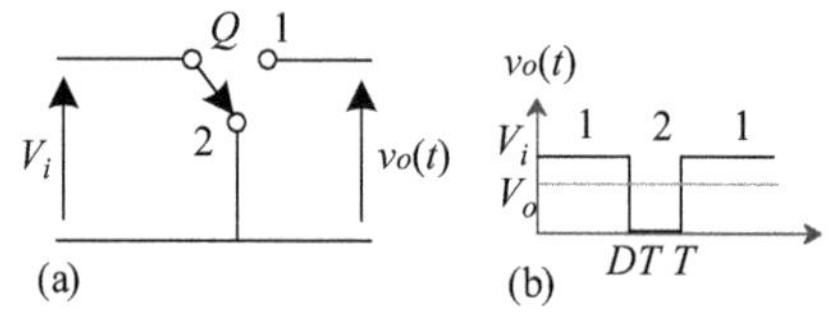

Figure 15.9 Basic concept of voltage conversion with an ideal switch: (a) structure, (b) output voltage waveform

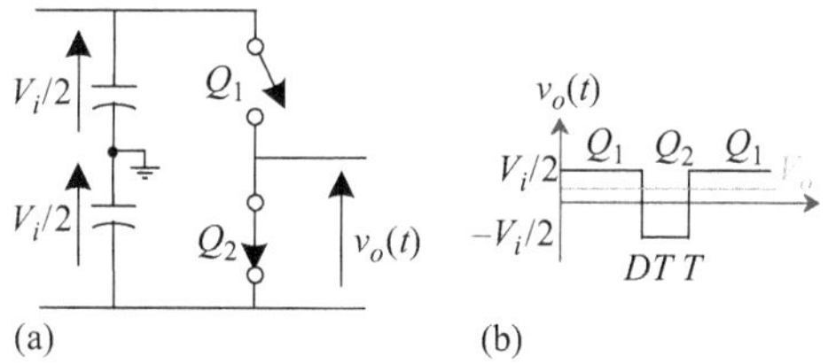

Figure 15.10 Basic concept of voltage conversion with two ideal switch series connected in one leg: (a) structure, (b) output voltage waveform

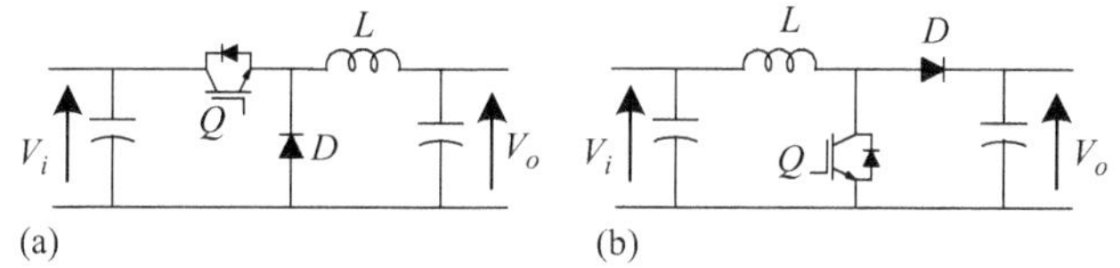

Figure 15.11 Step-up and step-down dc–dc converters: (a) buck converter, (b) boost converter

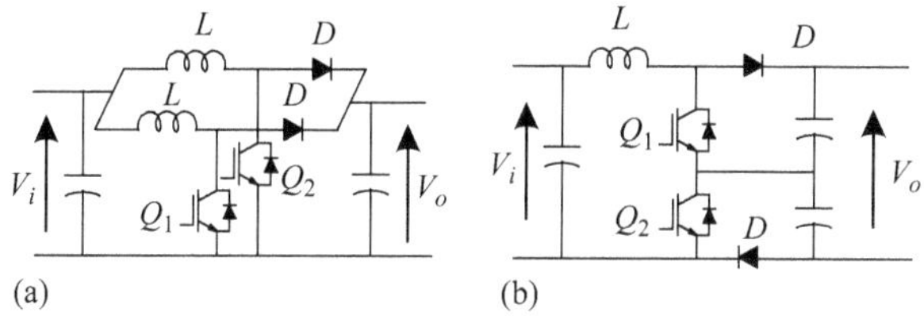

Figure 15.12 Interleaved dc–dc boost converters: (a) parallel interleaved boost, (b) series interleaved boost

15.3.2 DC–DC converters

Buck and boost converters, as shown in Figure 15.11, are the popular dc–dc converters for a wide range of applications, where the first one is used for step down of the input voltage, and the second one steps up the input voltage. These converters can be used in dc MGs for interfacing PV and fuel cell units to the grid. Some interleaved topologies have also been presented for PV or fuel cell stacks in which a high-voltage conversion gain is required. For instance, a parallel interleaved boost is shown in Figure 15.12(a), and series interleaved boost is shown in Figure 15.12(b). Moreover, Figure 15.13 shows a bidirectional boost converter for battery storage applications in dc MGs.

A boost converter is also widely used in multistage grid-connected PV converters as an MPPT controller. For example, the structure of a commercial 5 kW transformer-less inverter is shown in Figure 15.14.

The structure of a buck–boost converter is shown in Figure 15.15. The output voltage of this converter can be less or greater than the input voltage according to the switching duty cycle.

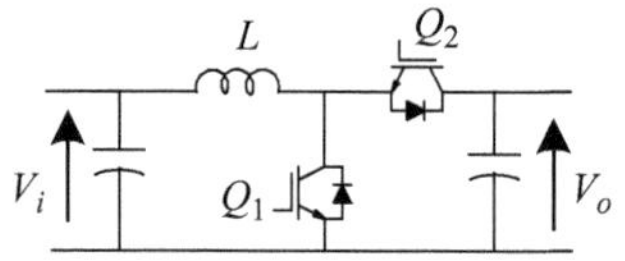

Figure 15.13 Bidirectional dc/dc boost converter

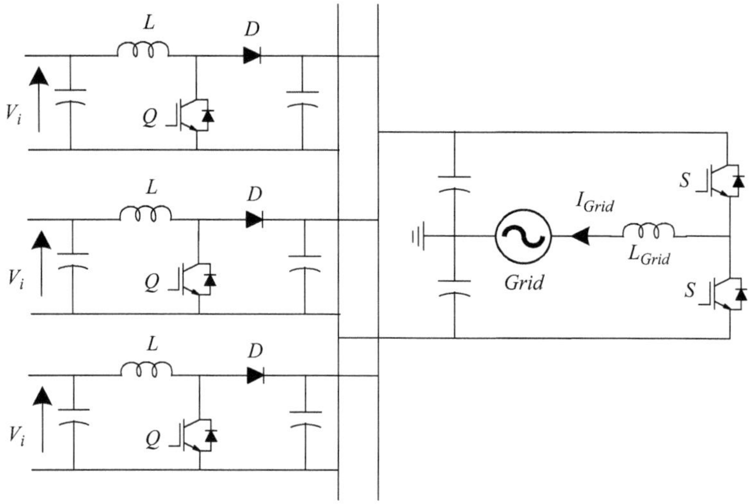

Figure 15.14 SMA Sunny Boy 5000 TL dc–dc converters

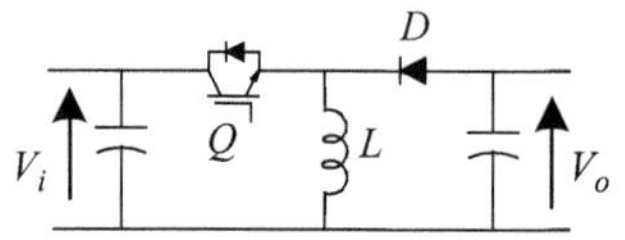

Figure 15.15 Buck–boost dc–dc converter

In some applications, galvanic isolation between the input and output stage is necessary. Therefore, some dc–dc converter structures include high-frequency transformers in the conversion stage. The output voltage is stepped up or stepped down by the transformer winding turns ratio (N_2/N_1) and the switching duty cycle. These converters can be categorized as unidirectional transformer core excitation, such as flyback and forward converters, and bidirectional transformer core excitation such as push–pull, half-bridge and full-bridge topologies. The corresponding converter circuit schematics are shown in Figures 15.16–15.20. More details regarding circuit analysis and design of these converters can be found in [2].

In order to increase the transformer winding utilization, a double-stage dc–ac–dc converter with high-frequency transformer can be used as shown in Figure 15.21, where the first stage converts the dc voltage to a high-frequency ac

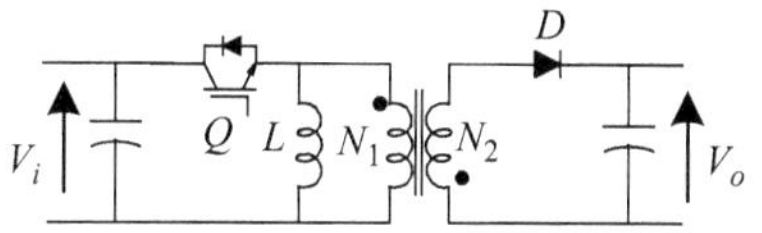

Figure 15.16 Flyback dc–dc converter

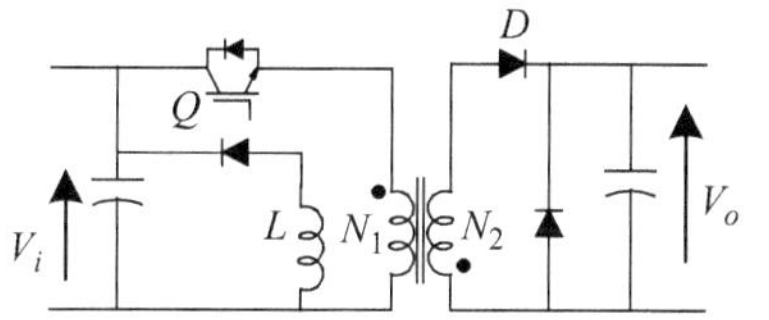

Figure 15.17 Forward dc–dc converter

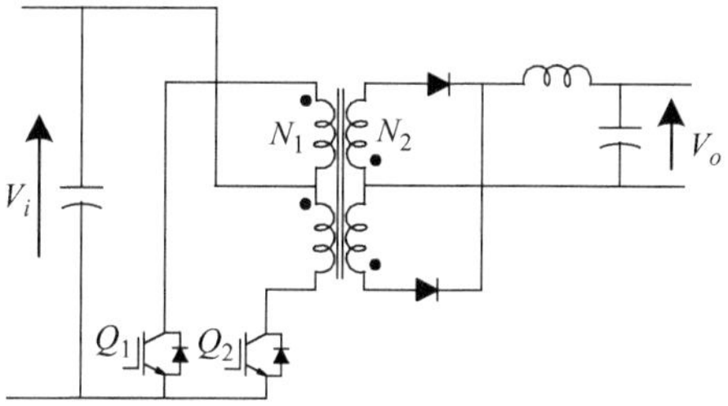

Figure 15.18 Push–pull dc–dc converter

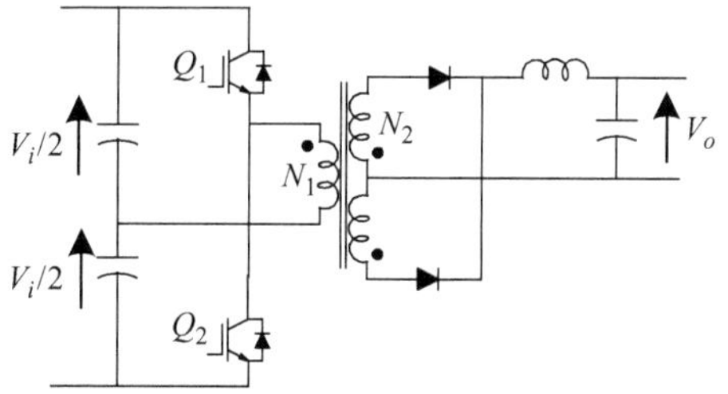

Figure 15.19 Half-bridge dc–dc converter

voltage. The ac voltage level is changed in the transformer, and the output voltage of the transformer is rectified by a diode bridge. Compared to the full-bridge converter shown in Figure 15.20, the transformer windings are always under load, while in the center tap transformer, each of the secondary windings is loaded during one-half cycle. In other words, the transformer size in the diode bridge is less than the one with the center tap transformer.

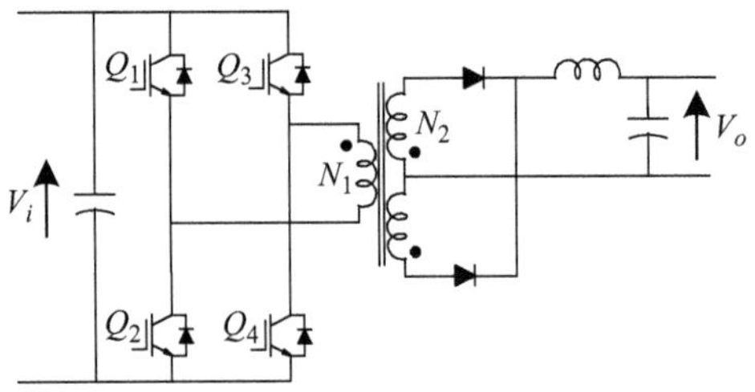

Figure 15.20 Full-bridge dc–dc converter

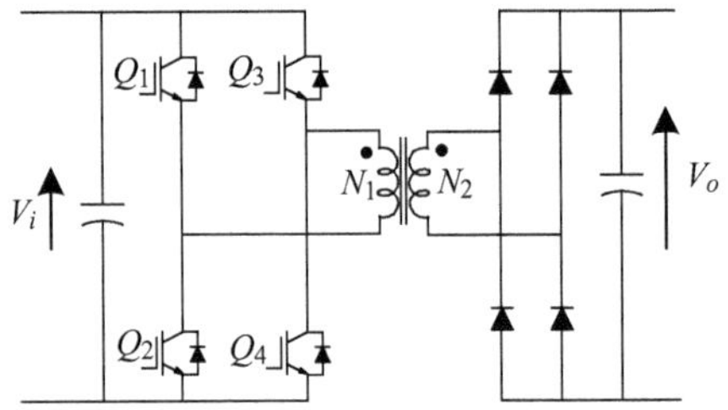

Figure 15.21 A double-stage dc–ac–dc converter with high-frequency transformer

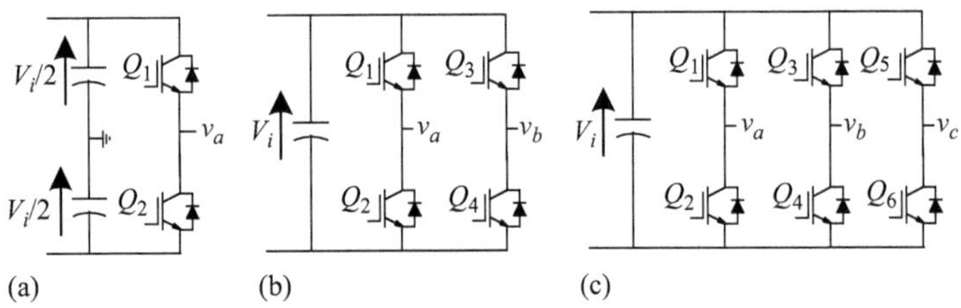

Figure 15.22 Two-level inverters: (a) single-phase half-bridge inverter, (b) single-phase full-bridge inverter, (c) three-phase inverter

15.3.3 DC–AC converters

The dc–ac converters are known as inverters and used to convert the dc voltage to ac voltage. A basic structure of a single-phase half-bridge inverter is shown in Figure 15.10(a) and redrawn in Figure 15.22(a). The load is connected to the neutral point of the input side. Two large capacitors are required to keep the voltage of the midpoint constant with respect to the positive or negative pole of input source. However, in most cases, neutral point is not available, and using such large capacitors is not applicable. Hence, a single-phase full-bridge inverter as shown in Figure 15.10(b) can be employed. Furthermore, for high power applications and three-phase loads, three-phase inverters as shown in Figure 15.10(c) is utilized.

In two-level converters, the output voltage is equal to $\pm V_i/2$. However, for high-voltage/high power applications, multilevel inverters have gained much attention due to their advantages over the two-level inverters. They have features like less common-mode voltage, less current harmonics and less voltage stress on switches. The most common structure of multilevel converters can be categorized as diode clamped, flying capacitor and cascaded H-bridge multilevel inverters. Three-level inverter structures are shown in Figure 15.23. The first two types of multilevel inverters require dc capacitors to properly form the output voltage, while for the cascaded H-bridge, the operation does not depend on dc capacitors.

According to the power and voltage ratings, multilevel inverters can be implemented with different voltage levels. For instance, a five-level cascaded H-bridge inverter is shown in Figure 15.24. In this structure, the input dc sources are isolated from each other.

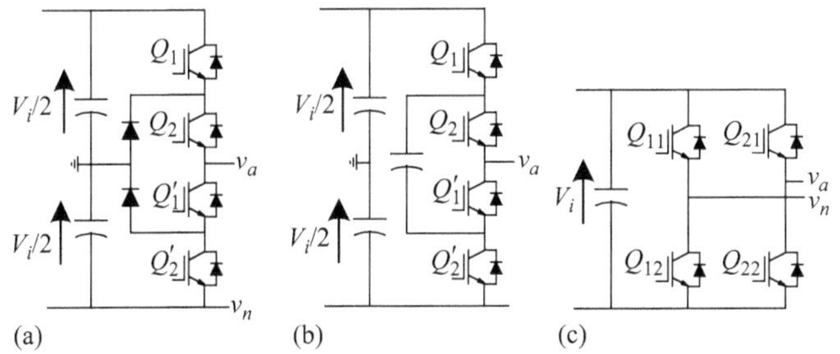

Figure 15.23 Multilevel inverter topologies (one leg is shown): (a) three-level neutral point clamped, (b) three-level flying capacitor, (c) cascaded H-bridge

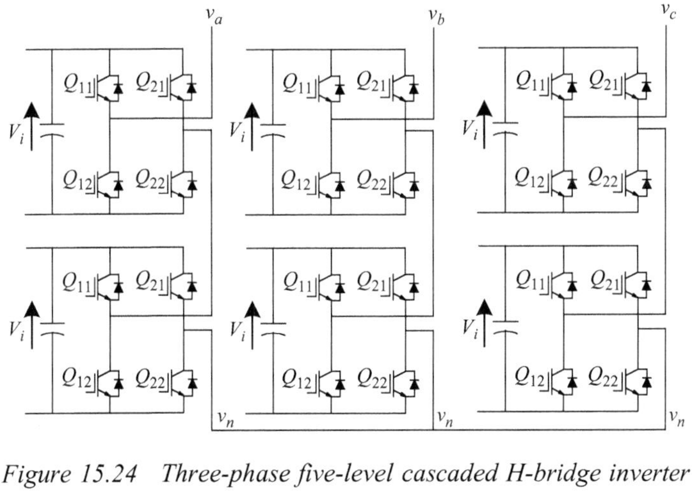

Figure 15.24 Three-phase five-level cascaded H-bridge inverter

Moreover, for very high-power systems, modular multilevel inverters are introduced. Modular structure, easy scalability, fault-tolerant capability, low device ratings are the key advantages of these generation of converters for high power applications. A schematic of this kind of inverters is shown in Figure 15.25, where the sub-module blocks can have different topologies such as half bridge, full bridge. Representing different topologies and operation principles of modular multilevel converter can be found in [3–5].

Different converter structures have been presented for MG applications. Besides voltage and power levels, reliability is another key factor should be taken into account in decision-making for the planning of MGs. Recent achievements on design for reliability approaches make an opportunity to select a suitable converter structure for a specific application in MGs. For instance, the applicability of different type of dc–dc converters for PV arrays considering reliability constraints has been addressed in [6]. This approach can also be employed for planning and operation of converters, i.e., design and control of multi-converter systems, which may have a significant impact on the system reliability [6–9].

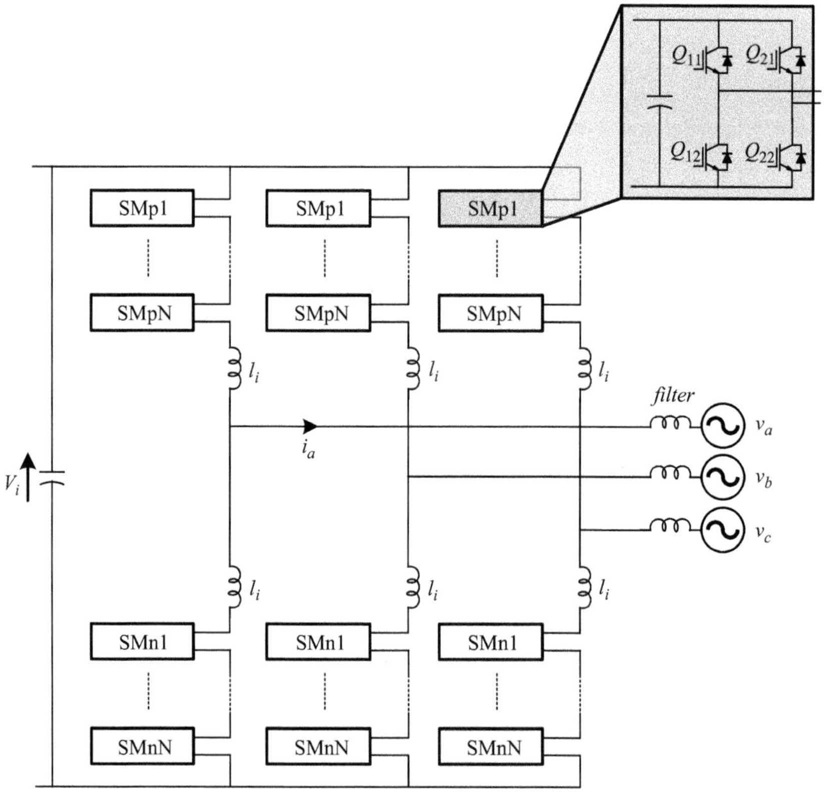

Figure 15.25 General circuit structure of a modular multilevel inverter

15.4 Power converter switching schemes

Two-level converters consist of one–three legs, each one is made up of two switches and two antiparallel diodes connected to them as shown in Figure 15.22. The switches are normally Insulated gate bipolar transistors (IGBTs) or Metal-Oxide-Semiconductor Field-Effect Transistors (MOSFETs). The legs are connected to a dc link that includes a capacitor and provides almost constant dc voltage in inversion mode. When this structure is used in rectifying mode, the capacitor voltage is controlled to be constant. An L or inductor (L)- capacitor (C)- inductor (L) (LCL) filter is typically used in the output to filter the voltage harmonics and allow connection to the grid in the case of grid-connected operation.

The converter is controlled by an appropriate switching on and off of its six switches. The switching signals are enabled by a driver that gets the gating signals from a controller, which turns the switch on/off. The dc-link capacitor will be short circuited if two switches of a leg are switched on simultaneously. Therefore, the gating signals of the two switches of a leg are complementary. The output voltage of a single leg will vary on the basis of its switching state. The output voltage of the legs will be $+V_i/2$ when the upper switch is on and $-V_i/2$ when the lower switch is on. The output current can pass through the conducting switch or its antiparallel diode in each case, depending on the current direction. This also shows the importance of antiparallel diodes, considering that they provide a path for currents to pass, which is necessary because of the series filter inductances.

By modulation of the two output values of each leg, it is possible to control the effective output voltage, and hence, the output voltage of the converter. This is called pulse width modulation (PWM), and this can be done in different ways, as explained in the following subsections.

15.4.1 Pulse width modulation

PWM is used to control the average value of a waveform over a switching period by controlling the pulse width. The PWM generation methods can be classified into carrier-based and space vector modulation (SVM) methods. Both of these methods provide high-quality output voltages and currents as well as good transient response.

15.4.2 Carrier-based pulse width modulation

This is the most widely used PWM method because of its ease of implementation. Many modern controller devices have dedicated hardware for this type of PWM methods. In this method, a set of carrier and reference waveforms are compared to generate the PWM signals. Figure 15.26 shows an example where the gating signals, which are controlling the semiconductor devices, are generated by comparing a sinusoidal reference signal with a triangular carrier wave of frequency, f_c. f_r is the frequency of the reference signal, which is determining the converter output frequency, i.e., f_o. The amplitude of the reference signal, controlled by a modulation index m_a, gives the output voltage to the converter. However, the output voltage can be varied by varying m_a. Bearing in mind that the number of

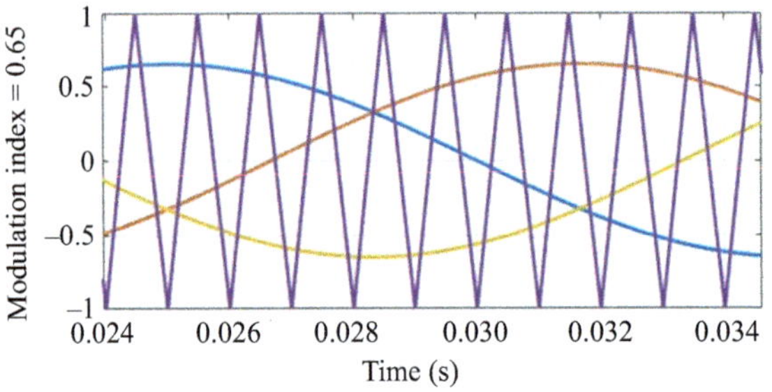

Figure 15.26 Triangular signals compared with the modulation index to generate the appropriate switching signals for the power converter

pulses per half-cycle depends on the carrier frequency. Note that a constraint of two transistors of the same leg exists (e.g., Q_1 and Q_2) as they cannot conduct at the same time.

As shown in Figure 15.26, the idea is when the amplitude of the reference sinusoidal signal is higher than the triangular signal, the switching signal is conducted. Contrarily, when the amplitude of the reference signal is lower than the triangular signal, the switches are disconnected. Bearing in mind, when one of the switches in each leg is conducting, the other switch in the same leg (complementary) should be off. If both of them are conducting at the same time, that means a short circuit will occur on the dc supply. The three references are given in the following equation:

$$\begin{aligned} V_{aref} &= M\cos\theta \\ V_{bref} &= M\cos\left(\theta - \frac{2\pi}{3}\right) \\ V_{cref} &= M\cos\left(\theta + \frac{2\pi}{3}\right) \end{aligned} \tag{15.2}$$

where M is the modulation index and affects the magnitude of the output phase voltages, and θ is the phase angle offset. High-order harmonics are centered on multiples of the switching frequency, and this can be filtered to have the desired sinusoidal output voltage. The use of higher frequencies will result in higher frequency harmonics, which are easier to filter, but this comes at the expense of higher switching losses and lower efficiency.

15.4.3 Zero-sequence injection

The use of three sinusoidal voltage references in sinusoidal pulse width modulation (SPWM) will result in sinusoidal phase and line voltages (after the filtering of higher order harmonics). It is possible to add a zero-sequence (ZS) signal (ZSS) to these reference values to form new modulation signals. The addition of the same ZSS to all three reference voltages does not change the output line-to-line and phase voltages. Therefore, it is used as a degree of freedom to reduce the current harmonics to improve the dc bus utilization.

For SPWM with three phase-shifted voltage references, third harmonic injection of the following form is a classic example of ZS injection:

$$V_{zss,3h} = -\frac{1}{6}\cos(3\theta) \quad (15.3)$$

In a more general case, the most widely used ZSS for a three-phase system with any type of voltage references is calculated as follows:

$$V_{zss} = -\frac{1}{2}\left(\max\left(V_{aref} + V_{bref} + V_{cref}\right) + \min\left(V_{aref} + V_{bref} + V_{cref}\right)\right) \quad (15.4)$$

The modulation index may be increased further without overmodulation resulting. It can be shown that the maximum modulation index can be increased in this way to

$$M_{\max} = \frac{2}{\sqrt{3}} \quad (15.5)$$

Therefore, with the same dc-link voltage, larger sinusoidal output voltages can be constructed, and therefore, dc-link utilization can be improved.

15.4.4 Space vector modulation

For the conventional voltage source converter (VSC) i.e., three-leg two-level converters, each leg has two possible switching states; therefore, the converter has eight possible states. Using the space vector theory, the three-phase voltages of each of these states can be represented by the following equation:

$$V = 2\sqrt{3}\left(V_a + V_b e^{j2\pi/3} + V_c e^{-j2\pi/3}\right) \quad (15.6)$$

Table 15.1 shows all the eight possible states, as well as their corresponding output voltages and space vectors. In Figure 15.27, the space vectors of a two-level converter are shown. The space vector is first calculated to realize a specific set of three-phase voltages using the previous equation. Then, a number of calculated

Table 15.1 Line voltages and space vectors of a two-level converter

State	Line voltages	Space vector
0 0 0	0, 0, 0	0
1 0 0	V_{dc}, 0, $-V_{dc}$	$(2/\sqrt{3})V_{dc}$
1 1 0	0, V_{dc}, $-V_{dc}$	$(2/\sqrt{3})e^{j\pi/3}V_{dc}$
0 1 0	0, $-V_{dc}$, V_{dc}	$(2/\sqrt{3})e^{2j\pi/3}V_{dc}$
0 1 1	$-V_{dc}$, 0, V_{dc}	$(2/\sqrt{3})e^{j\pi/1}V_{dc}$
0 0 1	0,$-V_{dc}$, V_{dc}	$(2/\sqrt{3})e^{4j\pi/3}V_{dc}$
1 0 1	V_{dc}, $-V_{dc}$, 0	$(2/\sqrt{3})e^{5j\pi/3}V_{dc}$
1 1 1	0, 0, 0	0

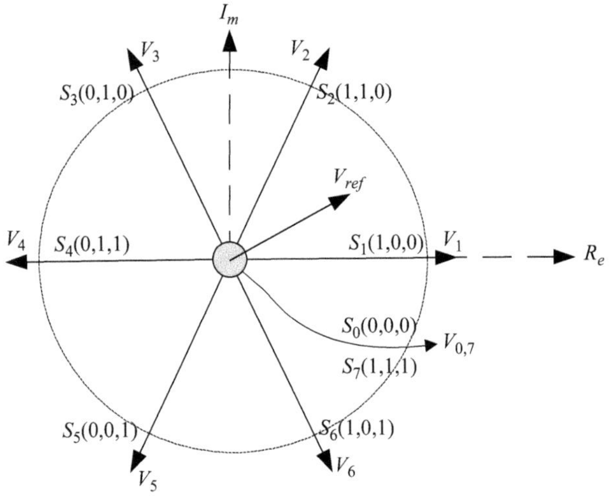

Figure 15.27 Space vectors of a two-level converter

vectors are used with duty cycles according to the following equation so that their average resulting vector is equal to V:

$$V = (d_1 V_1 + d_2 V_2 + \ldots + d_n V_n) \tag{15.7}$$

where $d_1 + d_2 + \ldots + d_n = 1$.

15.5 Power converter basic control schemes

Energy resources in MGs are classified into two major categories: (1) dispatchable and (2) non-dispatchable units. The dispatchable sources can supply any requested active/reactive power under the corresponding ratings, while the output power of non-dispatchable units cannot be controlled. The non-dispatchable units cannot supply any reference power due to the limitation on the input power, such as solar irradiance for PV arrays. Nevertheless, they can be controlled to operate under the available maximum power. This classification is summarized in Table 15.2 with some examples for each category.

15.5.1 Electrical model of converters

From control perspective, energy resources are modeled as a voltage source, current source, controlled voltage source and controlled current source as shown in Figure 15.28. Technically, the voltage source units are called grid-forming units, and they are employed to form the grid voltage, where grid voltage has an amplitude and a frequency of ac systems as well as a voltage magnitude of dc systems. A very famous example of grid-forming units is a synchronous generator connected to the *slack bus* in conventional power systems. Furthermore, the current sources

Table 15.2 Classification of different distributed sources and storages

Source type	Specification	Example
Distributed generators	Dispatchable	Fuel cell, micro-turbine
	Non-dispatchable	Photovoltaic, wind turbine
Distributed storages	Dispatchable/Fast	Battery
	Dispatchable/Ultrafast	Supercapacitors
	Dispatchable/Slow	Flywheel, regenerative fuel cell

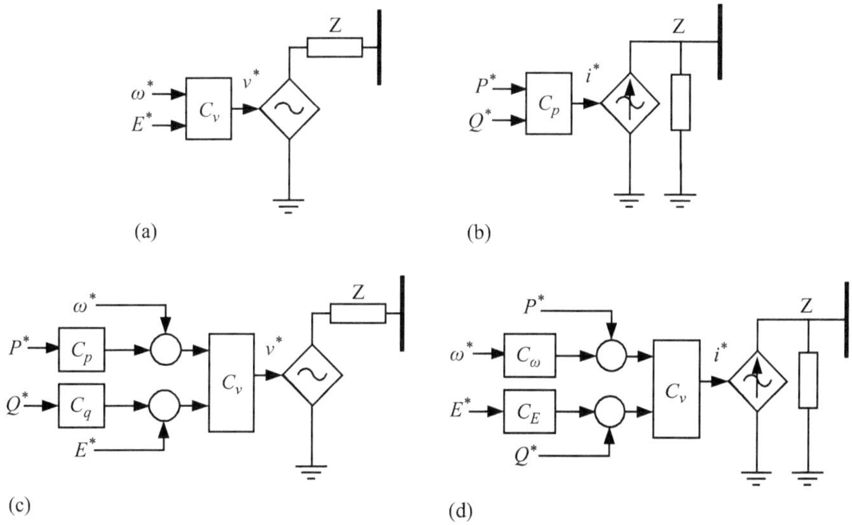

Figure 15.28 Electrical model of (a) grid-forming converter, (b) grid-feeding converter, (c) grid-supporting voltage source converter and (d) grid-supporting current source converter in ac microgrids

are called grid-feeding units, where they inject the reference active/reactive power into the grid in which its voltage has been formed by a grid-forming unit. The generators in conventional power systems connecting to the *P–V* (power controlled–voltage controlled) busses are the well-known grid-feeding units.

Two other types of energy sources are introduced by the development of MGs, which are grid-supporting voltage source and grid-supporting current source. In the MGs with low power ratings, the existence of a slack bus is almost impossible. Hence, in order to prevent overstressing the parallel converters, the droop control is introduced. The idea of droop control can be implemented in two ways; forward and reverse droop. The forward droop is utilized for the dispatchable units where they can supply any requested power. Hence, they can form the grid voltage, and their output power can be managed. These types of units behave as a master source in the MGs, and they are called grid-supporting voltage source units. Meanwhile, the non-dispatchable units cannot supply any reference power, and they cannot form the grid voltage in general. However, they can support the grid and participate in power sharing under the corresponding maximum available power. For instance, large PV and wind units are operated, e.g., 10% under MPPT in order to participate in frequency regulation, which means it is voltage forming. However, they cannot supply the load beyond the MPPT. These units are controlled in reverse droop mode and behave as a grid-supporting current source. They are indeed slave units where they use the voltage and frequency, which are formed by a grid-forming converter, or grid-supporting voltage sources.

Basically, the energy resources are controlled to regulate the output voltage and/or current by inner voltage and/or current control loops. The reference voltage and/or current for the inner loops are provided by the power-sharing control loop, which can have different strategies according to the type of unit. In the hierarchy of the power/energy management system, power-sharing control is known as primary controller. The secondary controller is utilized in order to restore the voltage and/or frequency of the grid after load change, generation variation and grid connection/ disconnection. Furthermore, the tertiary controller is also implemented in MGs to fulfill the market demands such as economic dispatch and reliability enhancement. The hierarchical control strategy is conceptually illustrated in Figure 15.29, where the controllers at level III is in charge of tertiary control and defines the reference for the secondary control. The secondary control in level II determines the voltage/ frequency restoration terms feedforwarded to the primary controller in level I. These controllers can be implemented in centralized, decentralized and distributed manners. More detailed discussions in this area are provided in [10,11], while it is out the scope in this chapter.

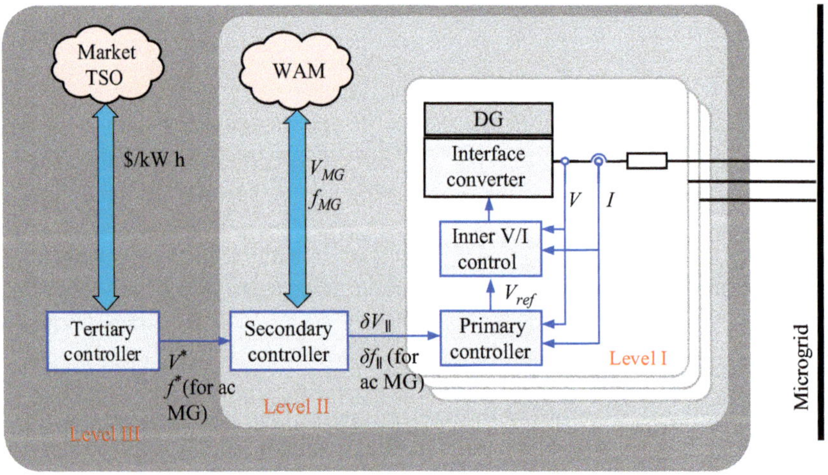

Figure 15.29 Hierarchical control (three levels) structure of microgrids

15.5.2 Control of converters in ac grids

The basic control structure of the grid-forming converters for a three-phase source is shown in Figure 15.30. This control system forms an ac voltage of $V(t) = E^* \sin(\omega^* t)$ at PCC. The control system contains two inner voltage and current regulators. In a power system, only one grid-forming converter must be considered. As this converter forms the grid voltage, there is no need for a phase locked loop (PLL) in the control structure, and the rotating angle can be obtained by integrating the reference frequency ω^*.

The grid-feeding control structure in a three-phase synchronous reference frame is shown in Figure 15.31. This converter injects the reference active and reactive powers P^* and Q^* into the grid at PCC. Since it cannot control the grid voltage, a PLL is required to generate the rotating angle for the converter, and the reference currents in the synchronous reference frame can be obtained by dividing the power references by the grid voltage. Furthermore, it can be controlled in the stationary reference frame as shown in Figure 15.32. Hence, the rotating angle of the grid voltage is directly translated into the reference current angles by Park transformation (*abc* to $\alpha\beta$). The reference currents can be achieved by the instantaneous power theory as shown in Figure 15.32.

As an example of the grid-feeding converter, the control structure of a single-phase inverter for a PV application is shown in Figure 15.33. In this case, the reference current amplitude is dictated by the MPPT unit, and its angle is obtained by a PLL from the grid voltage at PCC.

Moreover, two types of converters are used in the ac MG including grid-supporting voltage source and current source. The applications of these converters have already been presented. The control structure of grid-supporting voltage source converter is shown in Figure 15.34. For the sake of reliability, multiple converters are controlled to form the grid voltage in ac MGs. The operation of

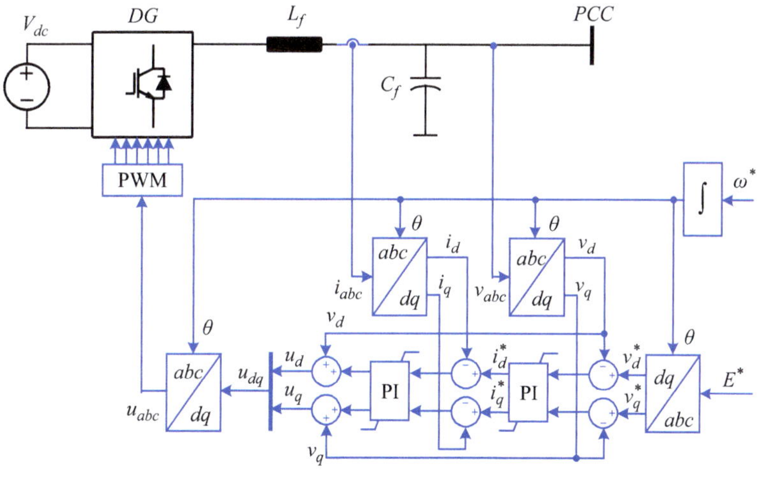

Figure 15.30 Basic control structure of a grid-forming converter: E^* is the nominal voltage amplitude and ω^* is nominal frequency

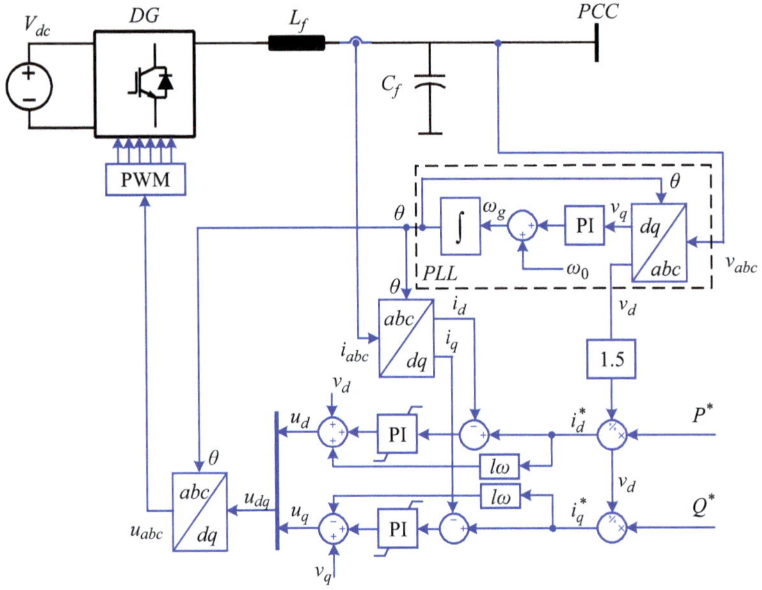

Figure 15.31 Basic control structure of a grid-feeding converter in synchronous reference frame: P^* is the reference active power and Q^* is the reference reactive power

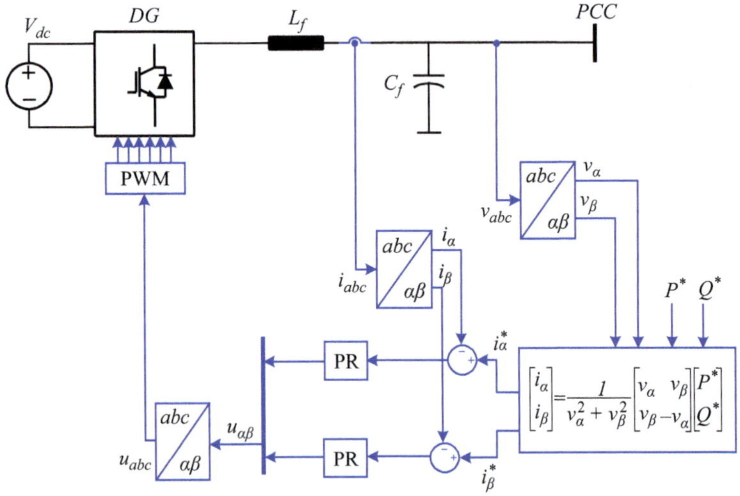

$$\begin{bmatrix} i_\alpha \\ i_\beta \end{bmatrix} = \frac{1}{v_\alpha^2 + v_\beta^2} \begin{bmatrix} v_\alpha & v_\beta \\ v_\beta & -v_\alpha \end{bmatrix} \begin{bmatrix} P^* \\ Q^* \end{bmatrix}$$

Figure 15.32 Basic control structure of a grid-feeding converter in stationary reference frame: P^* is the reference active power and Q^* is the reference reactive power

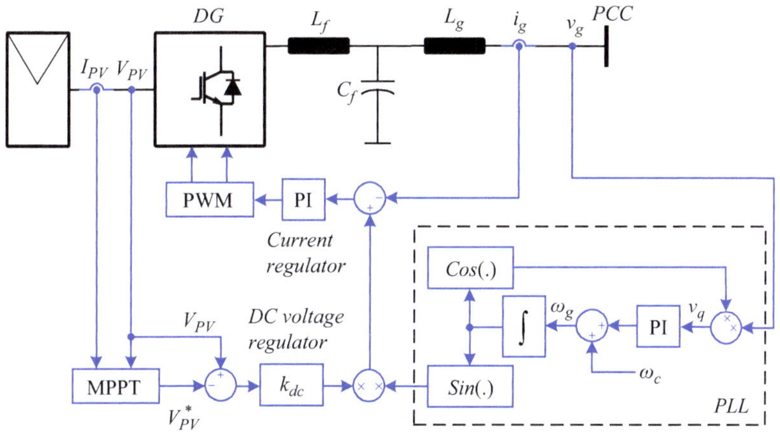

Figure 15.33 Basic control structure of a single-phase grid-feeding converter for a PV application

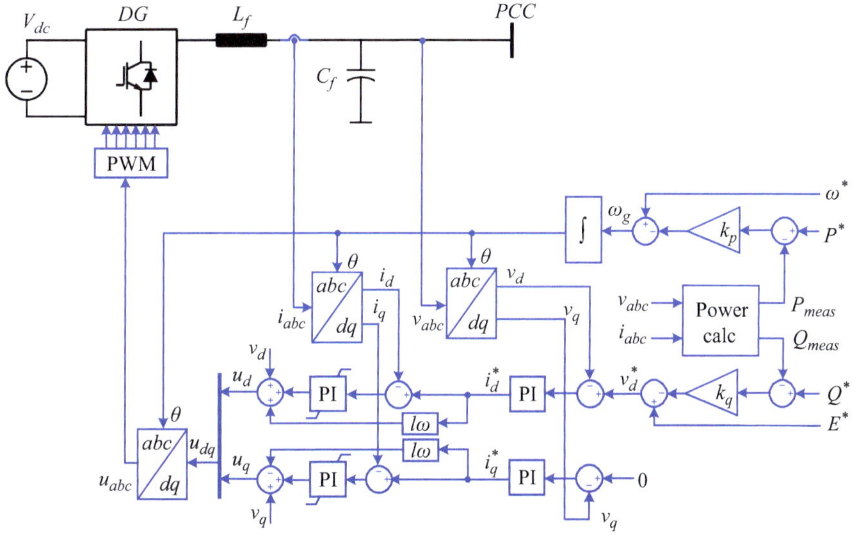

Figure 15.34 Basic control structure of a three-phase grid-supporting voltage source converter with forward droop control

multiple parallel-connected converters in grid-forming mode is possible only if forward droop method is employed as shown in Figure 15.34. This droop control determines the reference frequency and voltage for the converter considering the load sharing among the different converters. The load sharing can be carried out considering the converter ratings, operation costs [12] and reliability [7]. Indeed, this control structure is similar to the grid-forming strategy while it is applied for

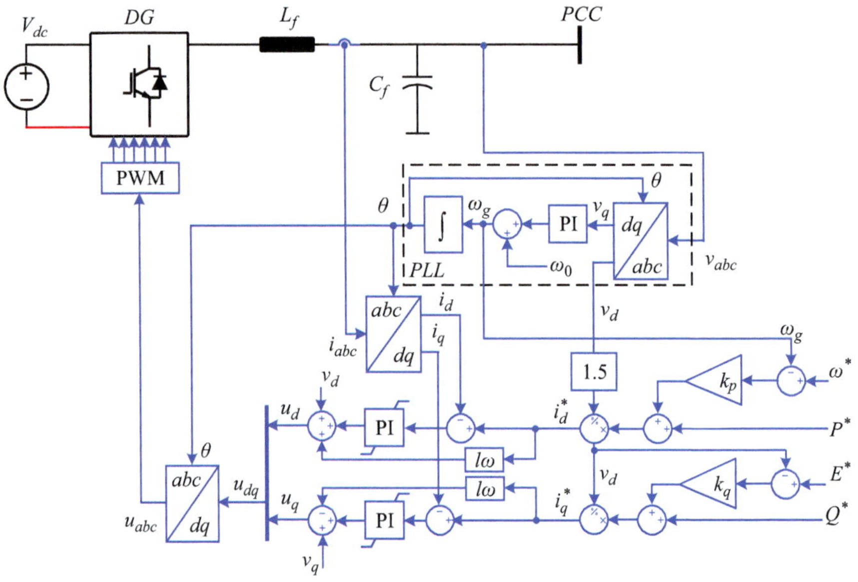

Figure 15.35 Basic control structure of a three-phase grid-supporting current source converter with reverse droop control

more than one converter. This cluster of converters behaves as a master unit in the MG for the grid-feeding and grid-supporting current source units.

The control approach of grid-supporting current source converter is shown in Figure 15.35. The reference powers, P^* and Q^*, are determined by the tertiary control or by the MPPT unit depending on the conveyor application. For frequency-supporting purpose, for instance, PV and wind converters are operated under MPPT. Therefore, this type of converter can participate in frequency and voltage regulation. It utilizes the grid frequency (ω_g) and voltage (E^*) to change the reference active and reactive power by using the reverse droop method as shown in Figure 15.35. However, as they do not have enough capacity, they cannot control in forward droop mode unlike the grid-supporting voltage source units.

15.5.3 Control of converters in dc grids

Similar control strategies can be applied for dc converters operating in dc MGs. However, in dc grids, the only control variable is the dc-link voltage, which can be controlled by active power. The general schematics and control system of a dc source is shown in Figure 15.36. The small-signal model of the converter operating in a voltage-controlled mode is shown in Figure 15.37. The inner voltage and current controllers can be designed according to the dynamic model of the converter. More details regarding the dynamic modeling of converters are discussed in [13].

The grid-forming converter structure is shown in Figure 15.38(a) where it behaves as an ideal dc voltage source as shown in Figure 15.38(b). The PCC bus

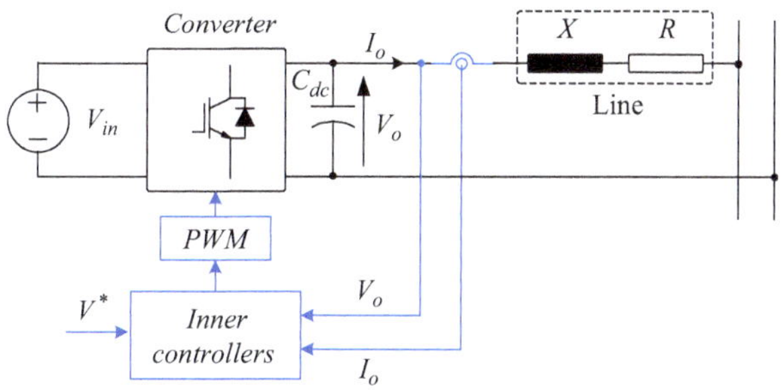

Figure 15.36 Schematics and control block diagram of a primary and secondary controller for the kth converter in a dc microgrid

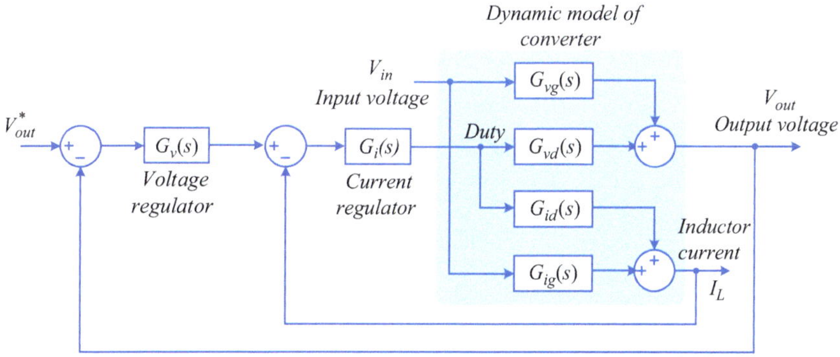

Figure 15.37 Dynamic model and control block diagram of a dc/dc converter with voltage and current regulators

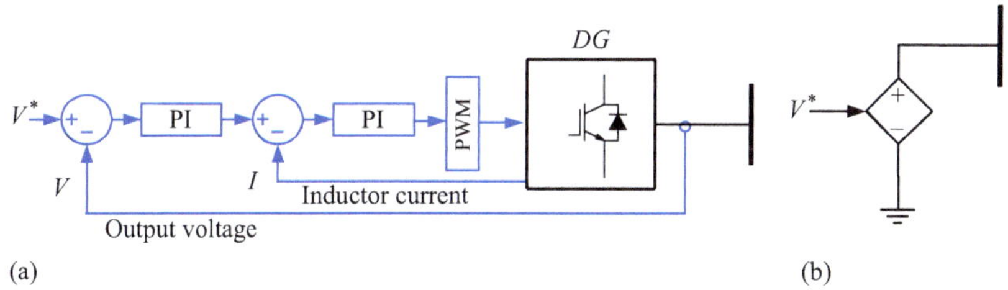

Figure 15.38 Grid-forming converter: (a) control block diagram, (b) simplified electrical model

voltage is regulated at V^* by a voltage regulator. The grid-feeding converter injects the dc current (or power) into the grid, which is determined by the MPPT unit or tertiary controller. The corresponding control structure is shown in Figure 15.39(a) and (b). This converter is electrically modeled by an ideal current source as shown in Figure 15.39(c).

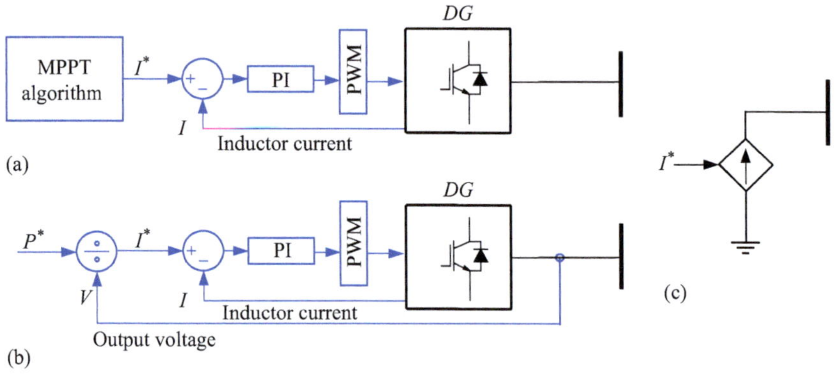

Figure 15.39 Grid-feeding converter: (a) control block diagram for MPPT-based units, (b) control block diagram for dispatchable units at constant power mode and (c) simplified electrical model

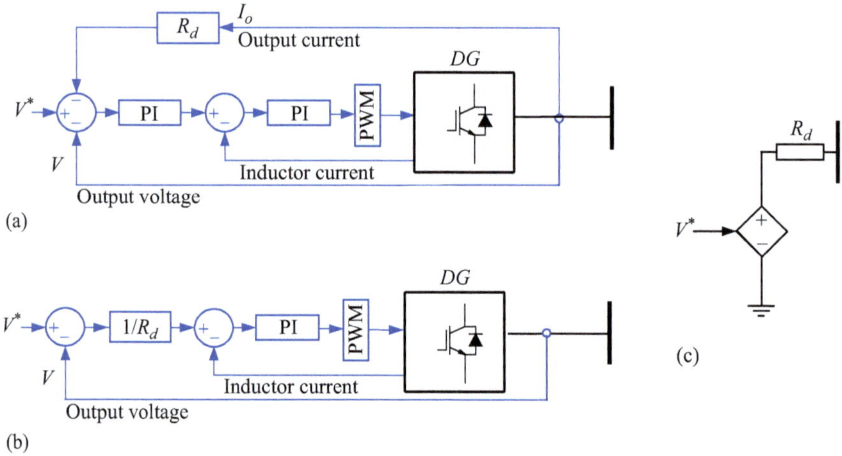

Figure 15.40 Grid-supporting converter: (a) control block diagram of the reverse droop, (b) control block diagram of the forward droop, (c) simplified electrical model

Moreover, the grid-supporting control structures employing the reverse and forward droop methods, respectively, are shown in Figure 15.40. This converter can be modeled as an ideal voltage source which is series connected by a resistor. This resistor is the corresponding droop gain, which controls the load sharing among the grid-supporting converters.

15.6 Filters for power converters—active and passive

This section provides a general discussion regarding passive and active filters in power converters.

15.6.1 Passive filters

Generally, a passive filter is a kind of electronic filter that is made from passive elements such as resistor, capacitor, inductor and transformer. More complex passive filters may involve nonlinear elements, or more complex linear elements, such as transmission lines. The pros of the passive filters are that they guarantee the stability, good scaling capability for large signals, and they are relatively cheap. Here, the focus will be on the grid filter as this part of the book spotting on the grid-connected application. The grid filter allows the control of active and reactive power by the inverter which is done by adding an impedance between the inverter and the grid. By doing so, the inverter output voltage and the grid voltage can have different amplitude and phase angles. In that sense, the active and reactive power flow are possible due the difference in amplitudes and angles. Another feature of including the impedance is that current spikes are limited, since the impedance between the grid and the inverter is increased [14,15].

The filtering stage should provide sinusoidal output current from the converter to the grid with a desired frequency. Due to the high-frequency PWM switching, the grid-connected converter introduces harmonics to the system. Harmonics in the system can cause disturbance in the loads or equipment and increase the losses [14]. Thus, harmonics in the grid-injected current should be suppressed, to a level stated in the respective grid codes or standards.

The degree of harmonic distortion in a signal can be measured and called total harmonic distortion (THD). THD compares the r.m.s. value of the fundamental harmonic to all other harmonic components. It is defined as

$$THD = \sqrt{\sum_{h=1}^{\infty} \left(\frac{yh}{yi}\right)^2} \tag{15.8}$$

where y_i is the r.m.s. value of the fundamental harmonic component, and y_h is the r.m.s. value of the *h*th harmonic component. In order to reduce the harmonics in an electrical signal supplied to the grid, a filter must be introduced to the system. Usually, the most commonly used is an LCL filter to mitigate the harmonics. A single-phase LCL low-pass filter can be seen in Figure 15.41. The focus will be on the LCL due to the following reasons:

- Compared to an L filter some LCL filters provide better harmonic attenuation, which is necessary to comply with strict grid codes.
- The LCL filter can be designed with small values of inductance and capacitance compared to the L filter.
- The LCL filter has the capability to cut off the high-frequency components, without attenuating the fundamental component, due to the low-pass characteristics.
- Since it is a third-order filter, the LCL filter attenuates with 60 dB/decade after the resonance frequency.

There are several considerations in the filter process designs. This includes current ripple, resonance problems, harmonic limitations and practical things such as size,

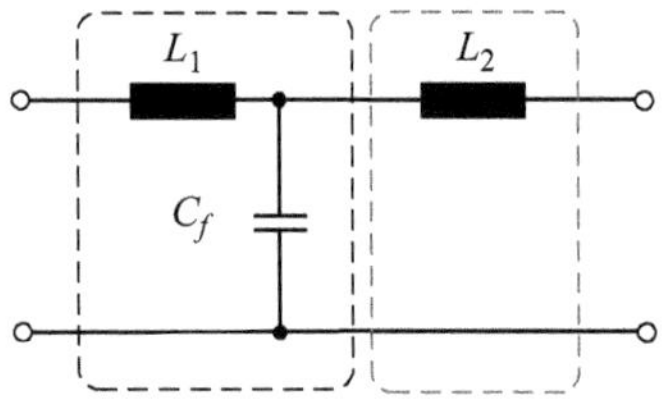

Figure 15.41 LCL filter used in voltage source converter

material and cost. The design procedure can be split into three parts. First, the converter-side inductor, L_1, is chosen based on the current ripple limit for the converter. The inductance L_1 can be determined from the following equation:

$$\Delta I_{\max} = \frac{1}{n}\frac{V_{dc}}{L_1 \cdot f_s} \tag{15.9}$$

where $\Delta I_{\max}$ is the maximum allowed current ripple, V_{dc} is the dc-link capacitor voltage, f_s is the switching frequency and n is dependent on the modulation [15].

Second, the resonance frequency of the LCL filter is chosen, and thereby the characteristics of the harmonic attenuation of the filter. On the grid side of the filter, the limits to comply with are based on harmonics rather than ripple amplitude; hence, the harmonic attenuation is evaluated by frequency analysis. The harmonic attenuation can be evaluated using the transfer function of grid inductor L_2 and the capacitor C_f. The relation is given by

$$\frac{i_g(w)}{i_i(w)} = \frac{Z_{LC}^2}{\left|Z_{LC}^2 - w^2\right|} \tag{15.10}$$

where i_g is the grid current, i_i is the inverter current and $Z_{LC}^2 = \left[\left(L_g + L_2\right)\right]$. The resonance frequency for

$$w_{res}^{LC} = \sqrt{\frac{1}{C\left(L_g + L_2\right)}} \tag{15.11}$$

Based on that, this method can be used for attenuation checking from the filter input current to the filter output current at each harmonic individually. Therefore, this method of using both equations can be used to design the filter for specific harmonic limits from a standard or a grid code.

The transfer function for the inverter output voltage and the filter output current can be seen in the following equation:

$$\frac{i_g}{v_i} = \frac{1}{C_f \cdot L_1 \cdot \left(L_2 + L_g\right) \cdot s^3 + \left(L_1 + L_2 + L_g\right) \cdot s} \tag{15.12}$$

And the resonance frequency is given by

$$w_{res}^{LCL} = \sqrt{\frac{Z_{LC}^2 (L_g + L_2 + L_1)}{L_1}} \tag{15.13}$$

The resonance frequency should be chosen so that the resonance problems at lower harmonics around the fundamental frequency and upper harmonics around the switching frequency are limited. Notably, this can be realized by using the suggested criteria in as follows:

$$f_g \cdot 10 < f_{res} < \frac{1}{2} \cdot f_s \tag{15.14}$$

where f_g is the grid frequency, which in this case is 50 Hz, and f_s is the switching frequency of the inverter.

The third part of the design steps is the optimization process of the filter components and used damping methods. Since the filter consist of reactive components, therefore, the power factor decreases. To ensure that the capacitor is not influencing the power factor, the installed capacitive reactive power should be less than 5% of the rated power. Therefore, the capacitor can be designed from

$$Q_{C_f} = \frac{V_{C_f}^2}{X_{C_f}} \tag{15.15}$$

As an approximation of the capacitor voltage, V_{C_f}, can be assumed to be equal to the grid voltage, V_g, by assuming that the voltage across L_2 and L_g is neglectable. In that case, the capacitance can be estimated by

$$5\% \cdot S_N = \frac{V_g^2}{(1/2 \cdot \pi f_g C_f)} \tag{15.16}$$

where S_N is the rated power, and f_g is the frequency of the grid. The resonance frequency depends on the grid inductance. Therefore, it is possible to examine the change in resonance frequency when the grid inductance changes. This relation is described as

$$\Delta w_{res} = \frac{1}{2 w_{res} C_f} \left(\frac{1}{L_2 + L_{g1}} - \frac{1}{L_2 + L_{g2}} \right) \tag{15.17}$$

where L_{g1} is the initial grid inductance, and L_{g2} is the new grid inductance. From (15.17), it is seen that an increase in filter capacitance decreases the influence of grid inductance changes on the resonance frequency. From evaluating the installed reactive power and the change in resonance frequency when the grid inductance changes, the filter capacitor C_f can be chosen. Finally, L_2 can be determined when L_1 and C_f are known to fulfill the desired resonance frequency. However, in this chapter, L_2 is known since this is the leakage inductance of the transformer.

Even though an LCL filter can reduce unwanted harmonics, it can also amplify some harmonics due to resonance between the inductors and the capacitor. A resistor can be added to dampen the resonance. The damping resistor can be added in series with the filter capacitor or in parallel with the grid-side inductor. An initial design value of the damping resistor can be estimated by the following equation [15]:

$$R_{damp} = X_{C_f}(w_{res}) \tag{15.18}$$

where the damping resistor is chosen to be in the same order of magnitude as the impedance of the capacitor at the resonance frequency. However, adding a resistor will also increase the losses, and it will reduce the attenuating performance. Therefore, the value of the damping resistance has to be evaluated by looking at the resonance peak and attenuation after the resonance frequency. The transfer function of the LCL filter with a damping resistor, R_d, in series with the filter capacitor is

$$\frac{i_g}{v_i} = \frac{C_f R_d \cdot s + 1}{C_f{\cdot}L_1 \cdot (L_2 + L_g){\cdot}s^3 + C_f{\cdot}R_d \cdot ((L_1 + L)_2 + L_g){\cdot}s^2 + (L_1 + L_2 + L_g) \cdot s} \tag{15.19}$$

The resonance of the resistor damped LCL filter is the same as given in (15.13) since the resistor is not frequency dependent. A possible solution to the power loss is to use active damping, by adjusting resonance from the inverter controller.

15.6.2 Active filters

Current harmonics are one of the most significant power quality issues, which have attracted tremendous research interest. Shunt active power filter (APF) is the best solution to minimize the harmonic contamination, but its effectiveness is strictly dependent on how quickly and accurately its control algorithms can be performed. The focus here is to review various types of existing control algorithms, which have been employed for controlling the operation of SAPF. Figure 15.42 shows the overall circuit configuration of a typical voltage-source-inverter-based SAPF and illustrates its four main control algorithms. The focus here is on the SAPF using the two-level voltage source converter as the case study.

In Figure 15.42 the SAPF consists of two main elements namely a two-level VSC, which can be any other power converter, and a controller. Meanwhile, the controller involves four control algorithms as follows:

Harmonic extraction algorithm: The harmonic-polluted power system has the distortion signals, and the key idea is to extract them and separate the harmonics from the fundamental ones. After achieving the harmonic isolation, a new current reference is generated to be fed to the power converter controller. The generated reference current signal is used to govern the operation of the SAPF in reducing the harmonic distortion. This is known as a reference current generation algorithm.

DC-link capacitor voltage regulator: The working principle of this algorithm is to take the instantaneous dc-link capacitor voltage u_{dc} and compare it with a desired reference value. The resulting error is used to estimate the suitable magnitude I_{dc} of the instantaneous dc-link charging current i_{dc}. The estimated I_{dc} is the amount of i_{dc} needed to be drawn by the APF to regulate its switching operation in order to constantly maintain the dc-link capacitor voltage u_{dc}

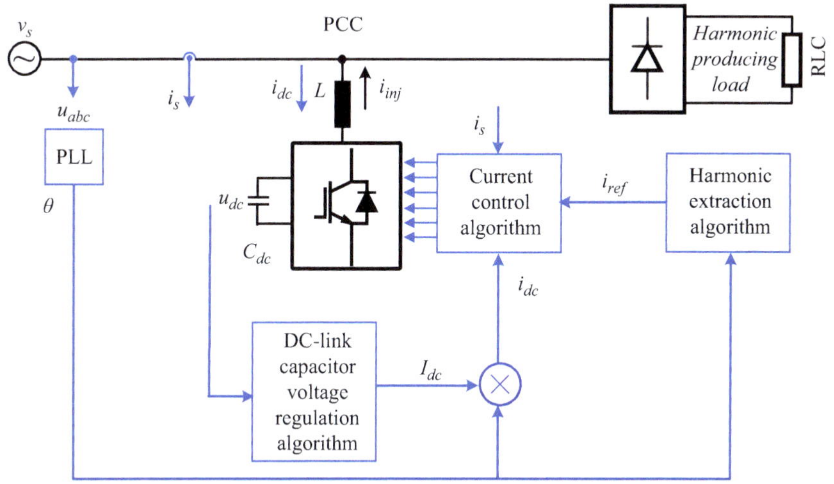

Figure 15.42 Active filter example using a controlled VSC

at the desired reference value. The voltage regulation process is said to have achieved when the real power drawn by the APF is made equal to its switching losses.

Current control algorithm: The harmonic output is taken by the current control algorithm, which is separated by both harmonic extraction and dc-link capacitor voltage regulation algorithms. The proper switching pulses are provided for controlling the operation of the inverter so the power converter functioning as an APF [16].

On the other hand, the current flow in a harmonic-polluted power system can mathematically be expressed as

$$i_s = i_L = i_{fund} + i_H \tag{15.20}$$

where i_{fund} is the fundamental current, and i_H is the harmonic current generated by the harmonic-producing loads. Note that the source current i_S is distorted due to the addition of i_H and has moved away from the source voltage v_S. However, after connecting the SAPF at the PCC (as shown in Figure 15.43), there are two additional current flows in the harmonic-polluted power system: (1) injection current I_{inj}, which is the opposition mitigation current injected by the SAPF into any harmonic-polluted power system for i_H cancelation; and (2) dc-link charging current i_{dc}, which is a small amount of current drawn by the SAPF to regulate its switching losses and to maintain V_{dc} at constant level. Hence, with the installation of an SAPF, (15.20) can be rewritten according to Kirchhoff's current law as

$$i_s = \left[i_{fund} + i_H\right] - i_{inj} + i_{dc} \tag{15.21}$$

To put it simply, the function of SAPF is to effectively recover the sinusoidal shape of i_S, by injecting i_{inj} into the harmonic-polluted power system to cancel out i_H contained in i_L. This can simply be achieved by making i_{inj} equal to i_H. In this way,

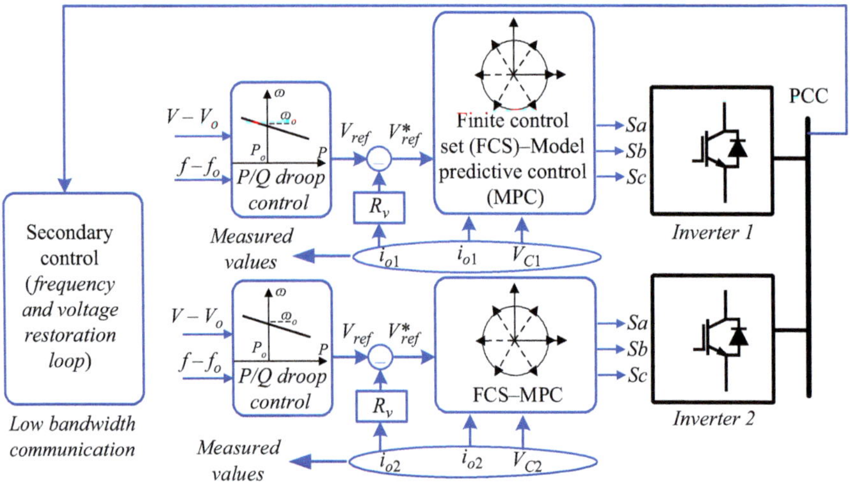

Figure 15.43 Primary, secondary control, circuit structure for an ac microgrid

i_S will eventually regain its sinusoidal characteristic and work in-phase with v_S. Hence, (15.21) can now be simplified as

$$i_s = i_{fund} + i_{dc} \tag{15.22}$$

15.7 Case studies

In this section, two case studies of microgrids are provided. The first illustrates the concept of MPC in ac MGs, and the second one is associated with the concept pf power sharing in dc MGs.

15.7.1 Case I: MPC-controlled converters in ac microgrids

Following the general concept illustrated in Figure 15.43, here a microgrid system comprising two VSCs are considered. Voltage amplitudes and frequencies of the control structure need to be adapted on line in order to ensure a proper power sharing among the different converter units by using the conventional droop control methodology. This case is called a primary control since it tracks the local variables and implements the droop and virtual impedance (VI) strategies. The concept of the VI is to subtract a small portion from the original voltage reference, and the generated reference is sent to the FCS–MPC controller. For the VI designation, the precision of power sharing depends on how much the R_v is larger than the output impedance. In the same context, larger R_v means extra voltage drop and distortion. As a result, a small R_v value is a compromise between improving the power sharing and keep the voltage drop limited. On the other hand, droop controller receives the active and reactive power measurements and provides the reference voltage and frequency for the inverters based on the tuning of the K_p and K_q parameters to ensure the stability of the overall system.

This design is very critical especially during the nonlinear load and parameter mismatch operations. Additionally, droop control can be designed as explained in order to ensure the system stability [17,18]. Hence, the conservative designation of the droop control coefficients can have a higher level of freedom by using the FCS–MPC. As an explanation, the fast dynamic response of the FCS–MPC leads to an order of magnitude smaller bandwidth compared to the cascaded linear loop structure. Additionally, in the classical strategy, the synchronized switching among the parallel inverters is usually an assumption which leads to a conflict with the decentralized control systems desires. This case study shows also a new effect on the circulating power which is shared using the improved FCS–MPC, which is explained previously in this chapter. Circulating current between two inverters exists due to the small differences between the output voltages of the inverters. The currents not only depend on the voltage differences but also on the impedance of the system and the output frequency, which is determined by the droop controller. Hence, a primary control structure with a proper tuning is used in order to ensure an improvement of the transient capabilities and lower circulating power of the FCS–MPC regulated VSCs in the ac MG. The study aims to prepare this system to be a grid-connected MG in the future works. Notably, this structure needs to be controlled by a secondary control strategy, which restores the voltage and frequency deviations of the controlled variables. And this can also be significantly fast due to the fast dynamic response of the FCS–MPC. The droop control is correspondingly designed by

$$V_{ref} = V'_{ref} - R_v i_o \tag{15.23}$$

$$E_i = E^* - K_p P_{cal} \tag{15.24}$$

$$w_i = w^* - K_q Q_{Cal} \tag{15.25}$$

where E_i and ω_i are the reference voltages amplitude and frequencies for both converters used to synthesize E^*, which is fed by a VI loop, R_v is the resistive VI, V'_{ref} is the filter capacitor voltage reference, while V_{ref} is the voltage reference provided by the outer droop control. E^* and ω^* are the nominal voltage amplitude and frequency, respectively. The droop coefficients, K_p and K_q, determine the slope of the droop curves. Finally, P_{cal} and Q_{cal} are the calculated instantaneous active and reactive powers. Accordingly, Figure 15.44 shows that the transient time is less than 10 ms during supplying the *RL*; and as a result, the time will be less in the case of supplying only *R* load. The figures also show the accuracy in the power sharing using the droop control and VI approaches. It can be seen from Figure 15.45 that the circulating current is reduced by using the improved FCS–MPC scheme. As pointed out before, the differences in the voltage are the main reason for this incident. Therefore, the improved FCS–MPC scheme minimizes the difference by improving in the THD of each converter although the output impedance is different. Additionally, the reference frequency for each converter is generated by the

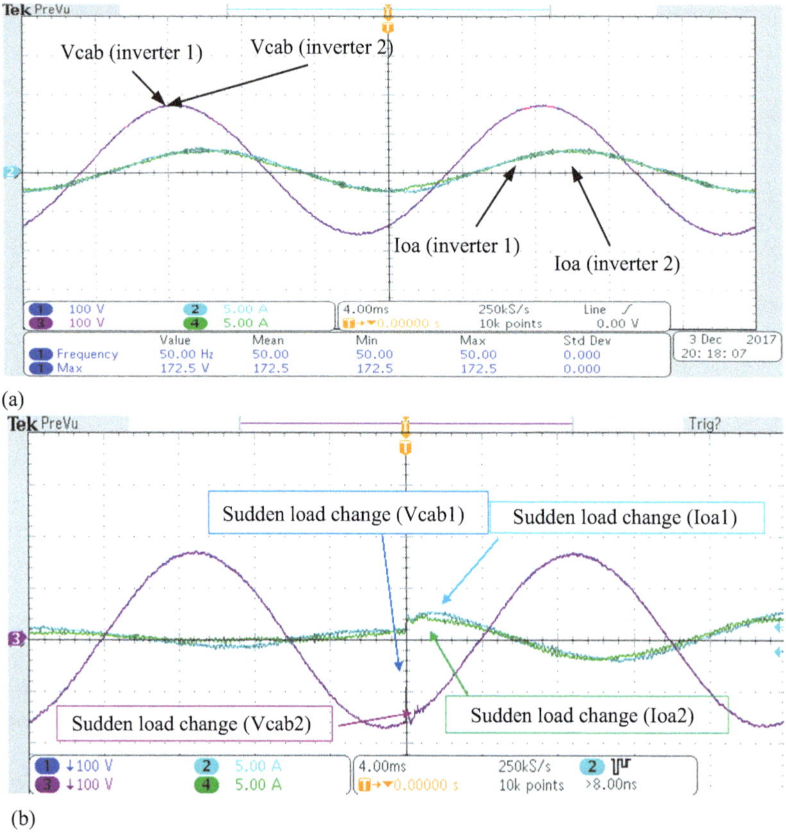

Figure 15.44 Experimental results validating the improved MPC scheme: (a) steady-state operation for two VSCs regulated by MPC, (b) step response test for the parallel converters

droop controller and fed to the FCS–MPC controller units. Finally, it is very important to mention that the weighting factor values for each cost function (CF) should be designed to have good results in terms of THD, losses and circulating power in parallel operation conditions [19].

15.7.2 Case II: Power-sharing control in a dc grid

Power converters can provide some supplementary functions for control, operation, protection and identification in power systems, thanks to their flexibility and controllability. For instance, modulating an extra signal for islanding detection and coding operating condition to communicate with other converters for economic dispatch are some opportunities that can be provided by the converters. In this case study, the possibility of frequency droop control in dc MGs is illustrated

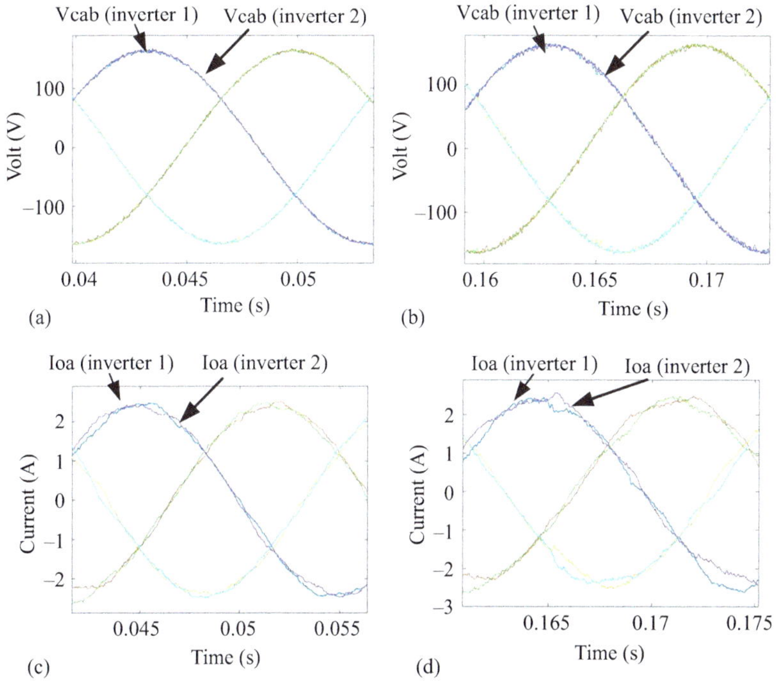

Figure 15.45 (a) Output voltage using conventional MPC scheme, (b) output voltage using improved MPC scheme, (c) output current using conventional MPC scheme, (d) output current using improved scheme

[20–22]. The dc grid structure and control system are shown in Figure 15.46, where for simplicity, two dc converts are connected to a load through the lines. The idea is to inject a small ac voltage by the dc converters over the dc voltage, where the frequency of the injected voltage is proportional to the output power/current of the converter. Then, the injected reactive power due to the ac voltage is employed to adjust the dc voltage level. The reactive power is proportional to the injected frequency, and the frequency is changed according to the output current. Therefore, if the output current of the converter is not proportional to its rated value, the frequency will be changed, and then the reactive power adjusts the dc voltage. By adjusting the dc voltage, the dc current will be controlled. Therefore, the output current of the converters can be regulated according to the corresponding rated currents. More details are discussed in [20–22]. The frequency droop control and inner voltage and current controllers are shown in Figure 15.46(b). As a consequence, the main advantage of this approach is to carry out the power sharing among the converters without utilizing any communication infrastructure. Indeed, the converters are communicating with each other through the injected ac voltage.

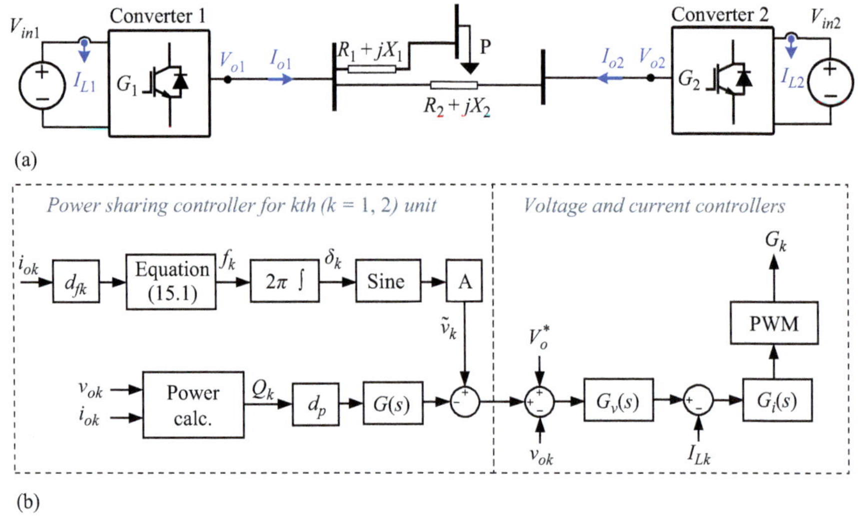

Figure 15.46 Block diagram of (a) a dc microgrid, (b) frequency droop control structure—$G_v(s)$ and $G_i(s)$ are PI-based inner voltage and current controllers (PI, Proportional–integrator)

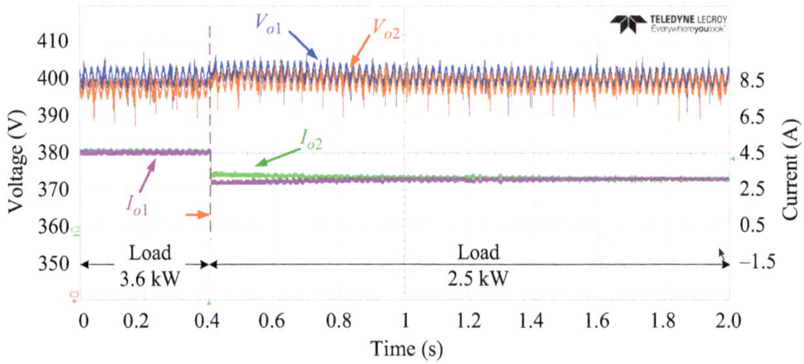

Figure 15.47 Output current and voltage of converters with equal converter ratings: load variation from 3.6 to 2.5 kW

The power-sharing performance of the frequency droop control is shown in Figures 15.47 and 15.48 for equal, and unequal load sharing between the two converters is shown in Figure 15.46(a). As can be seen, the converter voltages have small ac voltage over the dc voltage with the amplitude of 2.5 V and a frequency of almost 50 Hz. In both cases, the load is changed from 3.6 to 2.5 kW. The output voltages are regulated close to the rated voltage, i.e., 400 V before and after load change. Furthermore, the output currents of converters are equally shared as shown in Figure 15.47 as they have the same ratings. In the simulation case given in

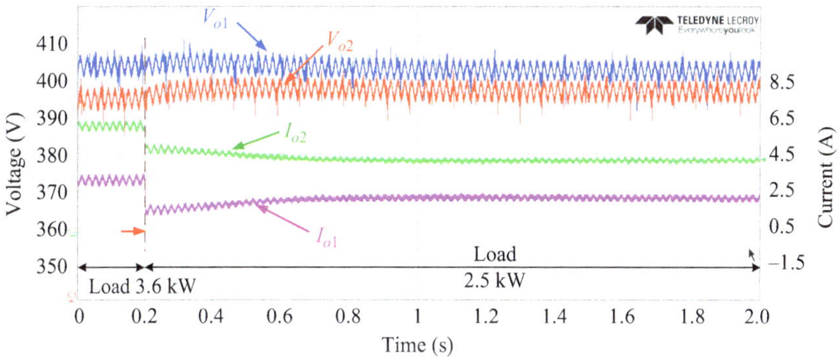

Figure 15.48 Output current and voltage of converters with unequal converter ratings: load variation from 3.6 to 2.5 kW

Figure 15.48, the rated current of the second converter is two times the first one, and hence, the first converter supplies the load two times more than the first one.

15.8 Conclusions

Power electronic converters are employed in MGs for interfacing energy resources and storages to the grid/local grid, and also interconnecting different MGs, as well as connecting the loads to the grid, e.g., electrical vehicle chargers. The converters not only serve as a physical interface between sources/loads and a grid but also support the grid stability by an appropriate control strategy. In this chapter, the importance of the power electronic converters in modern power systems, known as MGs, has been discussed. Different MG architectures have been introduced, which include ac MGs, dc, hybrid ac–ac, hybrid ac–dc and hybrid dc–dc MGs. Furthermore, some basic converter structures have been presented for dc–dc and dc–ac conversion applications, which are the most common used in MGs.

Afterward, the switching schemes and control strategies for power converters operating in MGs have been presented. The hierarchical control structure, type and role of different converters in MGs from the operation point of view have been explained. Furthermore, different types of active and passive filters for converters have been discussed. Finally, two application case studies for control and operation of power converters in MGs are provided.

Due to the key of the MGs in developing the power systems, optimal design, control, and operation of ac, dc and hybrid MGs taking into consideration the reliability constraints are further required. Suitable control strategies in converter level and power system level can improve the system reliability. Furthermore, proper converter structures as one of the fragile components of the MGs should be developed in order to fulfill a certain level of reliability in planning and operation of MGs.

So far, converter structures, control strategies and grid architectures have been presented for MGs in order to appropriately interconnect sources/loads and adequately balancing the generations and demands. However, increasing the use of power electronics in the development of power systems poses new challenges to reliable planning and operation of modern power systems. The future research on MGs could be reliability, control and operation of power converters in MGs [6,8,9]. System level design for reliability of power converters for a specific application can also enhance the overall system reliability. A suitable selection of converter topologies and control strategies considering the climate and operating conditions may have a significant impact on the overall system reliability.

References

[1] Boroyevich, D., Cvetkovic, I., Burgos, R., and Dong, D.: 'Intergrid: A Future Electronic Energy Network?' *IEEE J. Emerg. Sel. Top. Power Electron.*, 2013, **1**, (3), pp. 127–138.
[2] Mohan, N., and Undeland, T.M.: 'Power Electronics: Converters, Applications and Design' (John Wiley & Sons, Delhi, 2007).
[3] Malinowski, M., Gopakumar, K., Rodriguez, J., and Pérez, M.A.: 'A Survey on Cascaded Multilevel Inverters' *IEEE Trans. Ind. Electron.*, 2010, **57**, (7), pp. 2197–2206.
[4] Rodriguez, J., Lai, J.-S., and Peng, F.Z.: 'Multilevel Inverters: A Survey of Topologies, Controls, and Applications' *IEEE Trans. Ind. Electron.*, 2002, **49**, (4), pp. 724–738.
[5] Zhao, Z., Li, K., Jiang, Y., Lu, S., and Yuan, L.: 'Overview on Reliability of Modular Multilevel Cascade Converters' *Chin. J. Electr. Eng.*, 2015, **1**, (1), pp. 37–49.
[6] Peyghami, S., Davari, P., Wang, H., and Blaabjerg, F.: 'The Impact of Topology and Mission Profile on the Reliability of Boost-type Converters in PV Applications', in 'Proc. IEEE COMPEL' (2018), pp. 1–8.
[7] Peyghami, S., Davari, P., and Blaabjerg, F.: 'System-Level Lifetime-Oriented Power Sharing Control of Paralleled DC/DC Converters', in 'Proc. IEEE APEC' (2018), pp. 1890–1895.
[8] Peyghami, S., Wang, H., Davari, P., and Blaabjerg, F.: 'Mission Profile based Power Converter Reliability Analysis in a DC Power Electronic based Power System', in 'Proc. IEEE ECCE' (2018), pp. 1–7.
[9] Peyghami, S., Davari, P., Wang, H., and Blaabjerg, F.: 'System-Level Reliability Enhancement of DC/DC Stage in a Single-Phase PV Inverter' *Microelectron. Reliab.*, 2018, **88–90**, pp. 1030–1035.
[10] Rocabert, J., Luna, A., Blaabjerg, F., and Rodriguez, P.: 'Control of Power Converters in AC Microgrids' *IEEE Trans. Power Electron.*, 2012, **27**, (11), pp. 4734–4749.
[11] Peyghami, S., Mokhtari, H., and Blaabjerg, F.: 'Distributed and decentralized control of dc microgrids', in Dragičević, T., Wheeler, P., Blaabjerg, F. (Eds.): 'DC Distribution Systems and Microgrids' (IET, London, 2018), pp. 23–42.

[12] Nutkani, I.U., Loh, P.C., and Blaabjerg, F.: 'Droop Scheme with Consideration of Operating Costs' *IEEE Trans. Power Electron.*, 2014, **29**, (3), pp. 1047–1052.

[13] Erickson, R., and Maksimovic, D.: 'Fundamentals of Power Electronics' (Kluwer, New York, 2001, second).

[14] Teodorescu, R., Liserre, M., and Rodriguez, P.: 'Grid Converters for Photovoltaic and Wind Power Systems' (John Wiley & Sons, Hoboken, 2011).

[15] Liserre, M., Blaabjerg, F., and Hansen, S.: 'Design and Control of an LCL-Filter-Based Three-Phase Active Rectifier' *IEEE Trans. Ind. Appl.*, 2005, **41**, (5), pp. 1281–1291.

[16] Ferreira, S.C., Gonzatti, R.B., Pereira, R.R., da Silva, C.H., da Silva, L.B., and Lambert-Torres, G.: 'Finite Control Set Model Predictive Control for Dynamic Reactive Power Compensation with Hybrid Active Power Filters' *IEEE Trans. Ind. Electron.*, 2018, **65**, (3), pp. 2608–2617.

[17] Yu, X., Khambadkone, A.M., Wang, H., and Terence, S.T.S.: 'Control of Parallel-Connected Power Converters for Low-Voltage Microgrid—Part I: A Hybrid Control Architecture' *IEEE Trans. Power Electron.*, 2010, **25**, (12), pp. 2962–2970.

[18] Peng, F.Z., Li, Y.W., and Tolbert, L.M.: 'Control and Protection of Power Electronics Interfaced Distributed Generation Systems in a Customer-Driven Microgrid', in '2009 IEEE Power & Energy Society General Meeting' (IEEE, 2009), pp. 1–8.

[19] Alhasheem, M., Dragicevic, T., Blaabjerg, F., and Davari, P.: 'Parallel Operation of Dual VSCs Regulated by FCS-MPC Using Droop Control Approach', in 'Proc. ECCE EUROPE EPE' (2018).

[20] Peyghami, S., Mokhtari, H., and Blaabjerg, F.: 'Decentralized Load Sharing in a Low-Voltage Direct Current Microgrid with an Adaptive Droop Approach Based on a Superimposed Frequency' *IEEE J. Emerg. Sel. Top. Power Electron.*, 2017, **5**, (3), pp. 1205–1215.

[21] Peyghami, S., Mokhtari, H., Loh, P.C., Davari, P., and Blaabjerg, F.: 'Distributed Primary and Secondary Power Sharing in a Droop-Controlled LVDC Microgrid with Merged AC and DC Characteristics' *IEEE Trans. Smart Grid*, 2018, **9**, (3), pp. 2284–2294.

[22] Peyghami, S., Davari, P., Mokhtari, H., Loh, P.C., and Blaabjerg, F.: 'Synchronverter-Enabled DC Power Sharing Approach for LVDC Microgrids' *IEEE Trans. Power Electron.*, 2017, **32**, (10), pp. 8089–8099.

Index

Absorbent Glass Mat (AGM) 10
AC-coupling microgrid 270, 272
ac grids, control of converters in 553–6
AC microgrids 535
  typical structure of 536
AC nanogrid
  decentralized control in 184
    real power/frequency droop control for battery 187–9
    real power/frequency droop control for fuel cell 191–2
    real power/frequency droop control for PV 189–91
    simulation case study 192–4
  parameter design for centralized hierarchical control for 204–11
active power filter (APF) 562
active power sharing 28
adaptive hill climbing (AHC) control 36
adaptive protective schemes 420
advanced energy management system (AEMS) 189
asynchronous generators (AGs) 411
Attributed Goal-Oriented Requirements Analysis Model 79
autonomous control of microgrids 10

backtracking search optimization (BSO) algorithm 226, 237–40, 251
backup VPP (BVPP) 116–17
backward reduction algorithm 520
basic open-source nonlinear mixed integer programming (BONMIN) solver 278
batteries advantages, disadvantages and applications 267
battery energy management system (BEMS) 313
battery energy storage system (BESS) 5, 145–8, 153, 221, 251, 265, 378
  lead–acid 267
  lithium-ion 266–7
  sodium–sulphur 267
  structure of 265
  vanadium redox batteries 268
battery-mastered control of simple photovoltaic/battery system 181–2
battery storage VPP, in Australia 110–11
beacon redundancy protocol 121
black start procedure 275, 478
  initial assumptions 479–80
  sequence of actions 480–1
black start strategies in multi microgrids 476
  synchronization issues 477–8
  typical strategies 477
break point set (BPS) 406
building area network (BAN) 296, 298
building automation network 298–9

CANopen 121
carbon-foam-based Absorbent Glass Mat battery 10

carrier-based pulse width modulation 547–8
CB4 108
cellular communication 304–5
central autonomous management controller (CAMC) 470, 478–9
central controller (CC) 109
central direct load control 34
centralised control layer (CCL) 314
centralised secondary control 23–4, 42
  corrective controller 43–8
  dynamic adequacy objective function 52–4
  operation objective function 49–51
  preventive controller 57–8
  problem formulation 48, 58–9
  simulation results 54
    over-frequency event 57
    over-voltage event 57
    performance under prediction errors 61–3
    time-series study 59–61
    under-frequency event 54–7
    under-voltage event 57
  sustainability objective function 51
  technical objective function 48–9
centralized controlled VPP (CVPP) 113
centralized VPP 8–9
circuit breakers (CBs) 401, 415, 428–9
clustered microgrid (CMG) 143–7
  case studies 160
    Study case I (an overloaded MG with primary and secondary actions only) 165
    Study case II (an overloaded MG with all actions) 162–5
    Study case III (an overloaded MG with primary and tertiary actions only) 165
    Study case IV (an overgenerating MG with primary and secondary actions only) 165–6
    Study case V (an overgenerating MG with all actions) 166
    Study case VI (an overgenerating MG with primary and tertiary actions only) 166–7
    Study case VII (multiple PMGs and HMGs with all actions) 167–8
  control and operation of 152
    droop-regulated strategy 152–4
    modeling of non-dispatchable DERs 156–7
    optimization solver 155–6
  modeling of 147–52
  optimization problem formulation and technical constraints 157–60
cogeneration system, nanogrid based on (case study) 217
combined heat and power (CHP) systems 507
common data classes (CDCs) 310
Common Information Model (CIM) standards 423, 526
communication delay 42
communication infrastructure 301
  wired communication 301
    optical fibre 302–3
    power line 301–2
    twisted pair 302
  wireless communication 303
    cellular communication 304–7
    satellite communication 304
    wireless local area network (WLAN) 303
    wireless mesh 303–4
    worldwide interoperability for Microwave Access 305
    ZigBee 303
    Z-wave 304
communication protocols in microgrid environment 291
  case study 315–20
  communication layer 295
    building automation network 298
    field area network (FAN) 300
    home area network (HAN) 298

local area network (LAN) 299–300
neighbourhood area network (NAN) 299
wide area network (WAN) 300–1
communication objectives and requirements 294–5
Distributed Network Protocol version 3.3 309–10
IEC 61850 310–11
importance of communication technology in microgrid control 311–15
Internet communications protocol suite 306–8
Modbus protocol 308–9
communication standards 120, 421–3
components of microgrid 221–2
case studies 241–59
distributed energy resources, models of 226
dynamic battery energy storage power output model 232–3
micro-turbine power output model 233–4
probabilistic photovoltaic power output model 228–32
probabilistic wind power output model 226–8
microgrid sizing and profit maximization 222–6
optimal sizing of 234
backtracking search optimization (BSO) algorithm 237–40
mathematical formulation 235–7
solution approach 240–2
comprehensive control 35
comprehensive controller 42–3
control-hardware-in-the-loop (CHiL) simulations 424–6
controllable DGs 18
controllable microsources 465
control of MG 15
application of grid-tied MGs 18–20
case studies 25
centralised secondary control 42–63
demand-side primary frequency control 34–42
droop-based power control 25–34
primary control 20–2
secondary control 22–5
tertiary control 25
uncertainty of the generation and demand 17–18
control techniques 7
converters used in nanogrid 179–81
coordinate time interval (CTI) 402
corrective controller 43–8
cost-effective operation of MG 16, 23
cost minimization of MGs 145
cross-network redundancy protocol 121
current control algorithm 563
current control mode (CCM) 181
current transformers (CTs) 401
cyberattacks 420, 430
cybersecurity, addressing 430–2

Danish EDISON VPP project 110
DC–AC converters 544–6
DC-coupling microgrid 270, 272
DC–DC converters 541–4
dc grids, control of converters in 556–8
DC-link capacitor voltage regulator 562
DC microgrid considerations 433–4, 452
DC fault characteristics 434–5
AC-side current feeding stage 437
capacitor discharge stage 436–7
diode freewheeling stage 437–8
DC protective devices 445
fuse 445
hybrid circuit breaker (HCB) 447–50
mechanical circuit breaker (MCB) 445–6
solid-state circuit breaker (SSCB) 446–7

DC protective system approaches 438–45
DC system grounding 450–1
DC microgrids 535
hybrid microgrids 535–9
typical structure of 536
DC nanogrid 178
decentralized control in 183–4
decentralized control
in AC nanogrid 184
real power/frequency droop control for battery 187–9
real power/frequency droop control for fuel cell 191–2
real power/frequency droop control for PV 189–91
simulation case study 192–4
in DC nanogrid 183–4
for multiple battery-based nanogrid 182–3
for multiple energy storage units based nanogrid 195
from conventional droop control to improved droop control (case study) 199–200
load step changes in active and reactive power (case study) 200–1
local load (case study) 201–4
demand response (DR) resources 15
demand-side control 22
demand-side primary control schemes 22
demand-side primary frequency control 34
adaptive hill climbing (AHC) control 36
simulation results 37–42
step-by-step (SBS) control 37
dependability 396
depth of discharge (DOD) 518
Dgens 144, 146, 167–8
diesel generator 327
directional overcurrent relays (DOCR) 405–7
direct load control (DLC) 118
dispatchable DGs (DDGs) 18, 44–6
dispatchable distributed energy resources (DDERs) 146
Distributed Energy Resource Management System (DEMS) 110
distributed energy resources (DERs) 1, 3, 8, 74, 99, 103, 144–5, 153, 175, 179, 221, 223, 291, 395, 463
distributed energy resources, models of 226
dynamic battery energy storage power output model 232–3
micro-turbine power output model 233–4
probabilistic photovoltaic power output model 228–32
probabilistic wind power output model 226–8
distributed energy resources in microgrid 465–6
distributed energy storage units (DESUs) 175, 179, 195
distributed generation control modifications 424
distributed generations 176, 533
distributed generators (DGs) 17, 327, 395, 505
Distributed Network Protocol version 3.3 309–10
distribution management system (DMS) 464
distribution system operator (DSO) 499
DNP protocol 120
Doliwo–Dobrowolski three-phase generator 100
double attached node for PRP (DANP) 122–3
doubly fed induction generator (DFIG) 327, 354, 360, 363, 367–8, 411–12
droop-based approach 385–6
droop-based power control 25
active power sharing 28
formulation of control algorithms 26–8

reactive power sharing 28
selection of the control gains 28–30
simulation results 30–4
droop coefficient 60, 188–9, 207–9, 565
droop-controlled MGs 21–2
droop controller 312
droop-regulated system (DRS) 146–8
droop stability analysis 473–4
DR program (DRP) 44
dynamic adequacy objective function 52–4
dynamic battery energy storage power output model 232–3

economic control center (ECC) 178
electrical and thermal balance 507–8
electrical flexible loads 504
electricity networks 106
electricity power provider infrastructure 131
electric vehicles (EVs) 466
contribution to primary frequency support 472–3
electromagnetic compatibility (EMC) 450
electromagnetic interference (EMI) 450
El Toqui 223
Emergency mode 465
emergency operation, microgrid hierarchical control for 468–9
energy capacitor system (ECS) 5–6
energy management in nanogrid 181
AC nanogrid, decentralized control in 184
real power/frequency droop control for battery 187–9
real power/frequency droop control for fuel cell 191–2
real power/frequency droop control for PV 189–91
simulation case study 192–4
battery-mastered control of a simple photovoltaic/battery system 181–2
DC nanogrid, decentralized control in 183–4
multiple battery-based nanogrid, decentralized control for 182–3
multiple energy storage units based nanogrid, decentralized control for 195
from conventional droop control to improved droop control (case study) 199–200
load step changes in active and reactive power (case study) 200–1
local load (case study) 201–4
parameter design for a centralized hierarchical control for AC nanogrid 204–11
energy management system (EMS) 16–17, 23–4, 125–7, 291, 423
energy markets, liberalization of 102–6
short-circuit protection issue 105–6
voltage control issues 103–5
energy storage system (ESS), optimal sizing of 263
battery energy storage systems 265
lead–acid 267
lithium-ion 266–7
sodium–sulphur 267
vanadium redox batteries 268
flywheel 268
fuel cell 268–9
necessity of energy storage in microgrids 273
black start 275
demand shifting and peak shaving 274
frequency regulation 274
non-spinning reserve 275
power smoothing 274–5
reliability enhancement 274
storage trades/arbitrage 275
voltage support 274

numerical results 282
comparison 284–6
microgrid with non-optimised ESS 282–4
microgrid with optimised ESS 284
microgrid without ESS 282
problem formulation (case study) 279–82
supercapacitor 269–70
superconducting magnetic energy storage 269
system description and input data (case study) 277–8
technology comparison 270–2
uncertainty modelling (case study) 278
combined wind and load uncertainty 279
demand uncertainty 279
wind power generation uncertainty 279
energy storage systems (ESSs) 3–6, 395, 505
electrical energy storage 505
thermal storage system (TSS) 506
types of 5
utilization of, for fault discrimination 423–4
energy storage technology, future trends of 10
energy technologies 101
European energy exchange (EEX) 521
European Network of Transmission System Operators (ENTSO-E) 468

failure modes and effects analysis (FMEA) 426
fault current limiters (FCLs) 421, 423
fault event 398
fault tree analysis (FTA) 426
feasibility study analysis 497
case studies
experimental evaluation in Athens, Greece 519–25
Steinkjer microgrid 525–8
feasibility study aspects 514–15
active and reactive control strategies 516–17
components and topology 515–16
costing of microgrid components 518
data collection and processing 517
design and operation 515
microgrid component modeling and constraints 502
distributed generators 505
electrical and thermal balance 507–8
energy storage systems 505–6
interaction with utility grid 508
loads 503–5
multi-energy components 507
nomenclature 503
microgrid operational strategies 508
MPC-based energy-management system for operational optimization 508–9
MPC-based multi-objective AC optimal power flow 513–14
theoretical background 500
model predictive control (MPC) 500–1
two-stage stochastic programming 501–2
Fenix project 110
Fibre Check 121
field area network (FAN) 295, 300
firewall communication and security servers 131
fixed loads 503
flexible loads 503–5
floating system 416
flywheel energy storage 268
flywheel ESSs (FESSs) 5–6
forward droop control (FDC) 21

frequency–power-response-based topologies 384–5
frequency stability 7, 13
case study 386–91
effect of low inertia on 379–80
enhancement 380
droop-based approach 385–6
frequency–power-response-based topologies 384–5
swing equation based topology 383–4
synchronous generator (SG) model-based topologies 381–3
issues 377–9
fuel cell 268–9
fuse 445
future form of microgrid communication 10–11
future microgrids 9–10

gallium nitride (GaN) 447
gate-turn-off thyristors (GTOs) 446
generation and demand, uncertainty of 17–18
generation-side primary control schemes 21–2
Generic Object Oriented Substation Event (GOOSE) 310–11, 422
genetic algorithm (GA) 155
Global Position System 7
global reference frame (GRF) 337–8
goal-oriented RE (GORE) 79
Goal-Oriented Requirement Language [31] Model 79
grid-connected MGs 16
grid-connected microgrid architecture 2
grid-feeding inverter 466
grid-forming inverters 467
grid-tied microgrids 16–17, 377–8
application of 18–20
GridWise Architecture Council 80

hardgoals 82
harmonic-polluted power system 562
healthy MG (HMG) 144
heating ventilation and air conditioning (HVAC) system 418
heat pumps 507
hidden Markov model (HMM) 520
high-seamless redundancy (HSR) grid mechanism 122
high-speed serial input/output (HS-SIO) connection 429
high voltage direct current (HVDC) systems 535, 537
home area network (HAN) 3, 6, 293, 296, 298
HTTPS server web application server 131
HV/MV transformer 103–5
hybrid CBs (HCBs) 445
hybrid circuit breaker (HCB) 447–50

*i** goal model 79, 81–3
formalisation of multi-objective optimization functions of 86–96
I3RES Middleware Architecture 525
IBM CPLEX optimisation tool 84, 93
IEC 60870-5 121
IEC 61158 quality of service (QoS) 121
IEC 61850 121, 310–11
IEC 61850-based communication services 422–3
IEPE's topology 382–3
industrial area network (IAN) 296
information communication technologies (ICT) 6–7, 10, 111, 120
control system operation and states 127–8
database management process 129–30
databases 128–9
distribution and dispatching centre 130–2
energy management system 125–7
HSR grid mechanism 122
microgrid/VPP cybersecurity 123–5

PRP grid mechanism 122–3
RSTP grid mechanism 121–2
SHP grid mechanism 122
supervision control and data acquisition 127
information data flow in energy power networks 131
insulated gate bipolar transistors (IGBTs) 547
integrated gate-commutated thyristor (IGCT) 446
intelligent electronic devices (IEDs) 11, 110, 293, 421
interconnection devices (ICDs) 298
inter-grid 537, 539
internal control 20
Internet communications protocol suite 306–8
Internet Protocol (IP) 307
interoperability 11
inverter-based micro-sources 336
ISE lab's topology 384
islanded MGs 16–18
islanding operations of microgrid 3, 471
black start procedure 478–9
initial assumptions 479–80
sequence of actions 480–1
black start strategies in multi microgrids 476
synchronization issues 477–8
typical strategies 477
electric vehicles contribution to primary frequency support 472–3
microgrid islanding case study 481–5
microgrid primary frequency and voltage control 471–2
secondary control and emergency dispatch strategies 473–6
isolated MGs 144

Kawasaki Heavy Industries (KHI) topology 382–3
Knowledge Acquisition in Automated Space Model 79

large-scaled intelligent nanogrid (case study) 211–13
large-signal voltage stability 328
latency 42
lead–acid (LA) battery 267
limiting maximum DER capacity 421
line impedance effect 472
lithium-ion batteries 6, 266–7
load controllers (LCs) 464
load-following strategy 21
loads 503
fixed loads 503
flexible loads 503–5
local area network (LAN) 3, 122–3, 293, 299–300
local control 20
local VPP (LVPP) 111–13
low-voltage (LV) network 101, 106, 466
low-voltage AC (LVac) 396

Manufacturing Messaging Specification (MMS) standard 310, 422
master controller (MC) 312
MATLAB 160, 369
maximum power point tracking (MPPT) 179, 183–4, 378, 534, 552
mechanical circuit breaker (MCB) 445–6
media redundancy protocol (MRP) 121
medium-voltage (MV)-distribution system 221
medium-voltage AC (MVac) 396
medium voltage direct current (MVDC) 535, 537
Metal-Oxide-Semiconductor Field-Effect Transistors (MOSFETs) 547
metal oxide varistors (MOVs) 446
metropolitan VPP (MVPP) 113
micro-CHP (μCHP) 112, 118
micro-combined-heat-and-power (micro-CHP) applications 465

microgrid, defined 464
microgrid central controller (MGCC) 517
microgrid control issues 2–3
microgrid-embedded distribution system 222
microgrid protection 395
challenges 408
cyberattacks 420
earthing considerations 415–20
impact of distributed energy resources on fault current magnitude 411–12
impact of distributed energy resources on power flow 411
impact of microgrid connection modes and changing configurations 412–15
conventional protective system design practice 398–400
fault characterization 400–1
fault coordination analysis and protective relaying 402–8
protective equipment and scheme components 401–2
DC microgrid considerations 433–4
DC fault characteristics 434–5
DC protective devices 445
DC protective system approaches 438–45
DC system grounding 450–1
future 451–3
promising solutions for 420–1
addressing cybersecurity 430–2
distributed generation control modifications 424
evolving communication standards 421–3
fault current limiters 423
limiting maximum DER capacity 421
protective system design process for microgrids 424–6
utilization of ESS for fault discrimination 423–4
protective system design objectives 396–7
microgrids, classifications of 535
AC microgrids 535
DC microgrids 535
hybrid microgrids 535–9
microgrid/VPP cybersecurity 123–5
micro hydropower plants (MHPP) 386
microsource controllers (MCs) 107, 109–11, 464
micro-turbine power output model 233–4
micro-turbines (MTs) 225
mixed integer linear programming (MILP) 509
mixed integer non-linear programming (MINLP) 225, 277, 404, 406
mixed integer problem (MIP) algorithms 498
Modbus 121, 308–9
model predictive control (MPC) 500–1
model predictive control (MPC)-based energy-management system for operational optimization 508–9
MPC-based energy management for microgrid clusters 512–13
stochastic MPC formulation 509–10
stochastic programming approach with simple recourse 510–12
uncertainty modeling 509
model predictive control (MPC)-based multi-objective AC optimal power flow 513–14
modified particle swarm optimisation (MPSO) method 47, 53–4, 58
moment of inertia 378
multi-agent systems (MAS) 23, 314
multi-energy components 507
combined heat and power (CHP) systems 507
heat pumps 507
multi-master operation 467

multi-microgrid (MMG) 469–70
  black start strategies in 476
    synchronization issues 477–8
    typical strategies 477
multi microgrid black start case study 485
  initial remarks 485–6
  simulation and results analysis 486
    electric vehicles integration 487
    frequency and voltage profile 487–91
    network collapse and initial energization 486
    synchronization 487
multiple battery-based nanogrid, decentralized control for 182–3
multiple energy storage units based nanogrid, decentralized control for 195
  from conventional droop control to improved droop control (case study) 199–200
  load step changes in active and reactive power (case study) 200–1
  local load (case study) 201–4
multiport converter (MPC) 178
MV/LV transformer 103–5

nanogrid 1, 12, 175
  architecture of 176–9
  case study 211
    large-scaled intelligent nanogrid 211–13
    nanogrid based on cogeneration system 217
    nanogrid installed in remote villages 213–17
    small-scaled intelligent nanogrid 213
  concept of 175–6
  converters used in 179–81
  energy management in 181
    battery-mastered control of a simple photovoltaic/battery system 181–2
    multiple battery-based nanogrid, decentralized control for 182–3
    multiple DG units based nanogrid, decentralized control for 183–94
    multiple energy storage units based nanogrid, decentralized control for 195–204
    parameter design for a centralized hierarchical control for AC nanogrid 204–11
  typical structure of 177
National Electricity Market 2, 143
NegaWatt cost 114
neighbourhood area network (NAN) 6, 295–6, 299
neutral earthing (NE) 415
Nickel–metal hydride 6
non-controllable microsources 465
non-dispatchable DGs (NDDGs) 18, 44–6
non-dispatchable distributed energy resources (NDDs) 156–7
non-droop-based control methods 21
nonfunctional goals 82
nonfunctional requirements (NFRs) 74, 79–80
normal interconnected mode 465

objective function (OF) 17
off-grid modes 327
Oman, microgrids in 350
  electricity sector restructure 350–1
  Masandum Power System
    dynamic analysis 358–60
    static analysis 354–7
  Musandam power network 354

rural area electricity company 351–3
Sahim PV-Rooftop Project 353–4
OPC 121
Open Systems Interconnection (OSI) reference model 306
operational modes, microgrid 463
distributed energy resources in microgrid 465–6
microgrid control and management 467–8
microgrid hierarchical control for emergency operation 468–9
multi-microgrid 469–70
power electronic interfaces for distributed energy resources 466–7
operation objective function 49–51
optical fibre 302–3
optimal operation algorithms 16–17
optimal power flow (OPF) 117
output active power 185
overcurrent relays (OCRs) 403–4, 407, 411
over frequency relay (OFR) 408
overvoltage relay (OVR) 408

parallel redundancy protocol (PRP) grid mechanism 122
passive optical network (PON) 302
peak intervention VPP (PVPP) 116
penalty 158
phase locked loop (PLL) 553
phasor measurement unit (PMU) 7
photovoltaic (PV) cells 312
plug-in hybrid electric vehicles (PHEVs) 175
point of common coupling (PCC) 292, 534
point-to-point topology 309
Polish electricity sector 100–2
power-angle droop control 472
power control algorithms (PCAs) 17–18
power converters 1
basic control schemes 550
ac grids, control of converters in 553–6
dc grids, control of converters in 556–8
electrical model of converters 550–2
filters for 558
active filters 562–4
passive filters 559–62
power converter switching schemes 547
carrier-based pulse width modulation 547–8
pulse width modulation 547
space vector modulation 549–50
zero-sequence injection 548–9
power converter system (PCS) control 195
power electronic converters 5, 292, 540
DC–AC converters 544–6
DC–DC converters 541–4
general power conversation concept 540–1
power electronic interfaces for distributed energy resources 466–7
power electronics 533
case studies 564
MPC-controlled converters in ac microgrids 564–6
power-sharing control in a dc grid 566–8
classifications of microgrids 535
AC microgrids 535
DC microgrids 535–9
filters for power converters 558
active filters 562–4
passive filters 559–62
importance of power electronics in a microgrid 533–5

power converter basic control schemes 550
control of converters in ac grids 553–6
control of converters in dc grids 556–8
electrical model of converters 550–2
power converter switching schemes 547
carrier-based pulse width modulation 547–8
pulse width modulation 547
space vector modulation 549–50
zero-sequence injection 548–9
power electronic converters 540
DC–AC converters 544–6
DC–DC converters 541–4
general power conversation concept 540–1
power electronics in microgrid 3–4
reliability 4–5
power factor correction (PFC) circuit 176
power-flow analysis (PFA) 153
power–frequency swing equation 384
power line 301–2
power-system protection encompasses 395
power-to-heat ratio (PHR) 507
powertrain system analysis toolkit (PSAT) program 369
power transaction operator (PTO) 146–7, 150, 168
preventive controller 57–8, 61
primary control 20–2
primary frequency and voltage regulation, of microgrid 471
probabilistic photovoltaic power output model 228–32
probabilistic wind power output model 226–8
probability density function (PDF) 226
problem MG (PMG) 144, 148, 150
production planning 117–18
Profibus 121
prohibited operation zones (POZs) 225
promising solutions for microgrid protection 420–1
proportional power sharing 21–2
protection scheme 7
protective devices (PDs) 401, 403, 448
protective earthing (PE) 415
protective system design objectives 396–7
pulse width modulation (PWM) 547
pumped storage 5

rate of change of frequency (ROCOF) 378–80, 386
RCI 53
reactive power sharing 28, 45
Real Virtual Network Computing 124
're-electrification of rural areas' 101
remote villages, nanogrid installed in (case study) 213–17
renewable-based DGs 17
renewable energy-based sources 3
renewable energy sources (RESs) 100, 377, 395
global penetrations of 377, 381
requirements analysis and modelling of the TEM system 84
formalisation of multi-objective optimization functions of the *i** goal model 86–96
goal modelling of the TEM system 84–5
methodology 86
requirements engineering (RE)
in transactive energy management 78–83
residential customers 76
residential nanogrids 3
resiliency 399–400
resilient microgrids 430
retail customers 76
reverse droop control (RDC) 21

RMS (Royal Mail Steamer) 75
rooftop photovoltaic 1
RSTP (rapid spanning tree protocol) 121–2

sampled value (SV) messages 310
satellite communication 304
scalability and variability, addressing 8
secondary control 22–5
secondary control and emergency dispatch strategies 473–6
security, defined, 396
self-healing protocol (SHP) grid mechanism 122
signal-processing tools 427
silicon carbide (SiC) 447
silicon insulated-gate bipolar transistor (IGBT) 446
single master operation 467
single shaft microturbine (SSMT) 481, 483
sinusoidal pulse width modulation (SPWM) 548
small-scaled intelligent nanogrid (case study) 213
small-signal model of a microgrid for voltage stability analysis 335
small-signal stability analysis 332
small-signal voltage stability 328
smart grid (SG) 127, 292
smart power grid (SPG) 120
Smart Power Hamburg 111
smart sensors 6
sodium sulfide (NaS) battery 6
sodium–sulphur battery storage system 267
softgoals 82
software development life cycle (SDLC) 74
solar power conversion 4
solid-state circuit breaker (SSCB) 434, 446–7
space vector modulation 549–50
spinning reserve 52, 263
stability and protection issues 7
standalone MGs 16
Standish Group Report 75
states of charge (SoC) 221
steady state (SS) 402
step-by-step (SBS) control 37
stochastic MPC formulation 509–10
stochastic programming approach with simple recourse 510–12
strategic dependency (SD) model 82
strategic rationale (SR) model 82
substation automation system (SAS) 421
sudden inception fault 398, 401
supercapacitor (SC) energy storages 269–70
superconducting magnetic energy storage (SMES) 5–6, 269
supervisory control 23
supervisory control and data acquisition (SCADA) 7, 127
  operator's system 123
support vector machine (SVM) 422, 426, 547
survivability, 399
  objective function 51
swing-equation-based topology 383–4
synchronous generators (SGs) 3, 378, 411
  SG model-based topologies 381–3
synchronous optical networking (SONET)/synchronous digital hierarchy (SDH) 302
synchronverters 381–2
synthetic inertia 114–15, 380

technical objective function 48–9
tertiary control 25
thermal flexible loads 505
thermal storage system (TSS) 506
TN-C-S grounding topology 451
total harmonic distortion (THD) 559
traditional grid versus smart grid 76
transactive energy (TE) 73

transactive energy management (TEM) 74–8
goal modelling of 84–5
methodology 86
transactive energy systems (TESs) 74
transformation of microgrid to virtual power plant 8–9, 99
evolution of electricity 100–2
information communication technologies 120
control system operation and states 127–8
database management process 129–30
databases 128–9
distribution and dispatching centre 130–2
energy management system 125–7
HSR grid mechanism 122
microgrid/VPP cybersecurity 123–5
PRP grid mechanism 122–3
RSTP grid mechanism 121–2
SHP grid mechanism 122
supervision control and data acquisition 127
liberalization of the energy markets 102
future problem identification 102–6
microgrid and VPP, difference between 118–19
microgrid configuration 107–9
microgrid turns to virtual power plant 106
MGs structure and application 106–7
microsource controller 109
virtual power plant general concept 109–11
regulation of VPP and MGs 132–8
virtual power plants, types of 111
area-based approach to virtual power plants 111–13
grid support and ancillary services 113–17
VPP model and algorithms 117–18
Transmission Control Protocol (TCP) 307–8, 313
TroposModel 79
twisted pair 302
two-stage stochastic programming 501–2

uncontrolled DGs: *see* non-dispatchable DGs (NDDGs)
under frequency relay (UFR) 408
under voltage relay (UVR) 408
ungrounded system 416
User Datagram Protocol (UDP) 307–8
utility grid 146, 175–7, 181, 213, 216–17, 401, 498, 508–10, 513–14, 535

vanadium redox batteries 268
violation, defined 158
virtual power plant (VPP) 1–3, 11, 99–100, 109–11
area-based approach to 111–13
general concept 109–11
grid support and ancillary services 113–17
load management in 118
microgrid and VPP, difference between 118–19
regulation of 132–8
transformation of microgrid to 8–9
VPP model and algorithms 117–18
virtual power plant for regulation (VPPR) 113
load-frequency control and balancing consumption and generation 114
reactive power and voltage control 115–17
synthetic inertia 114–15
virtual synchronous generators (VSGs) 385

virtual synchronous machine (VISMA) topology 382–3
voltage control mode (VCM) 181
voltage source converter (VSC) 434–5, 549
voltage source inverter (VSI) 176, 333, 467, 483
voltage stability of microgrids 327
assessment 329
dynamic analysis 331–2
proximity to voltage instability 332–4
P–V and *Q–V* curves 329–30
static voltage stability analysis 329
V–Q sensitivity analysis 330–1
case studies 336
inverter control system 346–7
inverter model 345–6
microgrids, in Oman 350–60
model of the whole system 347
models of the induction generator and the whole system 349
network model 347
small signal model of inverter 338–9
small signal model of the load 341–2
small signal model of the power network 339–40
small signal model of the whole system 342
stability analysis 342–3, 349–50
synchronous generator model 343–5
voltage stability analysis 347–8
concept 328
issues 328–9
small-signal model of microgrid 335
voltage stability enhancement 335
VSYNC's topology 384–5

wavelength-division multiplexing 302
Weibull distribution 226–8
wide-area control (WAC) 7, 11
wide-area networks (WANs) 3, 6, 131, 300–1
WiMAX: *see* worldwide interoperability for Microwave Access (WiMAX)
wind generators (WGs) 515
wired communication 301
optical fibre 302–3
power line 301–2
twisted pair 302
wireless communication 303
cellular communication 304–7
satellite communication 304
wireless local area network (WLAN) 303
wireless mesh 303–4
worldwide interoperability for Microwave Access 305
ZigBee 303
Z-wave 304
wireless local area network (WLAN) 303
wireless mesh 303–4
worldwide interoperability for Microwave Access (WiMAX) 295, 305

zero-sequence injection 548–9
zero-sequence signal (ZSS) 548
zero voltage switching (ZVS) 180
ZigBee 296, 298, 303, 319
Z-wave 6, 304